SUSTAINABLE MARITIME TRANSPORTATION AND EXPLOITATION OF SEA RESOURCES

PROCEEDINGS OF THE 14TH INTERNATIONAL CONGRESS OF THE INTERNATIONAL MARTIME ASSOCIATION OF THE MEDITERRANEAN (IMAM), GENOVA, ITALY, 13–16 SEPTEMBER, 2011

Sustainable Maritime Transportation and Exploitation of Sea Resources

Editors

Enrico Rizzuto
University of Genoa – DICAT, Genoa, Italy

Carlos Guedes Soares
Centre for Marine Technology and Engineering (CENTEC)
Technical University of Lisbon, Lisboa, Portugal

VOLUME 2

CRC Press is an imprint of the
Taylor & Francis Group, an **informa** business
A BALKEMA BOOK

Cover photo credit (front and back cover): 'Views of the port of Genoa', courtesy of the Genoa Port Authority

CRC Press/Balkema is an imprint of the Taylor & Francis Group, an informa business

Typeset by Vikatan Publishing Solutions (P) Ltd., Chennai, India
Printed and bound by CPI Group (UK) Ltd, Croydon, CRO 4YY

Published by: CRC Press/Balkema
P.O. Box 447, 2300 AK Leiden, The Netherlands
e-mail: Pub.NL@taylorandfrancis.com
www.crcpress.com – www.taylorandfrancis.co.uk – www.balkema.nl

ISBN: 978-0-415-62081-9 (set of 2 volumes + CD-ROM)
ISBN: 978-0-415-62082-6 (Vol 1)
ISBN: 978-0-415-68393-7 (Vol 2)
ISBN: 978-0-203-13033-9 (eBook)

Sustainable Maritime Transportation and Exploitation of Sea Resources – Rizzuto & Guedes Soares (eds)
© 2012 Taylor & Francis Group, London, ISBN 978-0-415-62081-9

Table of contents

3 *Structures & materials*

3.1 *Analysis of local structures*

3.2 *Analysis of the global structures*

3.3 *Static and dynamic structural assessment*

3.4 *Materials*

3.5 *Degradation/Defects in structures*

VOLUME II

5 *Propulsion plants*

5.1 *Prime movers*

5.2 *Machinery*

6 *Safety & operation*

6.1 *Safety*

© 2012 Taylor & Francis Group, London, ISBN 978-0-415-62081-9

Preface

This book presents the proceedings of the XIV Conference of the International Maritime Association of the Mediterranean (IMAM), held in Genoa on September 13th to 16th 2011 under the theme 'Sustainable Maritime Transportation and Exploitation of Sea Resources'.

The book covers the most updated aspects of maritime transports and of coastal and sea resources exploitation, with focus on the Mediterranean area. Vessels for transportation are analysed from the viewpoint of ship design in terms of hydrodynamic, structural and plant optimisation, as well as from the perspective of construction, maintenance, operation and logistics. The exploitation of marine and coastal resources is covered in terms of fishing, aquaculture and renewable energy production as well as of subsea resources extraction.

The characterisation of the marine environment is seen under the twofold perspective of providing reference loads and conditions for the design of means for the resources exploitation, but also of setting limits to the design in order to preserve the natural ambient and minimise the impact of anthropogenic activities related to both transportation and exploitation.

Efficiency, reliability, safety and sustainability of sea- and Mediterranean-related human activities are the focus throughout the book. The text is mainly devoted to technical operators of the various field involved, coming from shipbuilding and ship-owner companies, research organisations, Universities, certifying bodies, but it represents an updated reference text of interest also for government Agencies and other institutional and educational bodies.

ABOUT THE INTERNATIONAL MARITIME ASSOCIATION OF THE MEDITERRANEAN (IMAM)

The Association was established in 1974 with the acronym of IMAEM (International Maritime Association of East Mediterranean), initially including institutions from six countries (Bulgaria, Egypt, Greece, Italy, Turkey and Yugoslavia). The membership was later progressively enlarged to most of the Mediterranean countries and neighbouring areas and, since the 1990 conference, the acronym was changed to IMAM, dropping the reference to the eastern part of the basin.

The International Maritime Association of the Mediterranean (IMAM) is proud of its more than thirty-year-long history committed to the enhancement and dissemination of technical knowledge related to study, design, construction and lifetime operation of ships and other marine structures. The focus of the Association is on the development of marine transports and on the exploitation of sea resources in the Mediterranean area, in line with the principles of a sustainable growth.

The IMAM Congresses are privileged forums for the maritime technical community of the Mediterranean. They have been hosted in the following key locations:

Istanbul (1978)	Trieste (1981)	Athens (1984)	Varna (1987)
Athens (1990)	Varna (1993)	Dubrovnik (1995)	Istanbul (1997)
Ischia (2000)	Crete (2002)	Lisbon (2005)	Varna (2007)
Istanbul (2009)	Genoa (2011)		

The IMAM conferences are traditionally attended by qualified representatives from the Academia and from Professional and Technical Associations of the various fields involved. The conferences represent therefore a window for areas of high potential growth such as the Balkans, Eastern Europe, the Middle and the Near East as well as North Africa.

Sustainable Maritime Transportation and Exploitation of Sea Resources – Rizzuto & Guedes Soares (eds)
© 2012 Taylor & Francis Group, London, ISBN 978-0-415-62081-9

IMAM Organisation

IMAM EXECUTIVE COMMITTEE

IMAM TECHNICAL COMMITTEES

Hydrodynamics

Marine structures

Sustainable Maritime Transportation and Exploitation of Sea Resources – Rizzuto & Guedes Soares (eds)
© 2012 Taylor & Francis Group, London, ISBN 978-0-415-62081-9

XIV IMAM conference organisation

The conference was organised by the University of Genoa in the seat of the Faculty of Engineering, Villa Cambiaso Pallavicini in Genoa.

CONFERENCE COMMITTEE

Enrico Rizzuto (Chairman) – *University of Genoa*
Carlo Podenzana Bonvino – *University of Genoa*
Bruno Della Loggia – *Atena*

CONFERENCE SECRETARIAT

Isa Traverso – *University of Genoa*
Michela Tizzani – *Promoest*
Elisabetta Gembillo – *University of Genoa*

5 *Propulsion plants*

5.1 *Prime movers*

Sustainable Maritime Transportation and Exploitation of Sea Resources – Rizzuto & Guedes Soares (eds)
 ISBN 978-0-415-62081-9

Combustion chamber of the Diesel engine—theory and numerical simulation

L.C. Stan & N. Buzbuchi
Constanta Maritime University, Constanta, Romania

ABSTRACT: The simulation of the processes inside the marine engines was a permanent subject of study for the specialists. The complexity of all the associated phenomena and their strong correlation is making out of this a very difficult task. This paper tried to simulate the combustion inside the marine diesel engine using the newest computer methods and technologies with the result of a diverse and rich palette of solutions, extremely useful for the study and prediction of complex phenomena of the fuel combustion. The paperwork is tridimensional modeling of the geometry of the combustion chamber and the CFD (Computational Fluid Dynamics) net/grid of the finite elements involved and tridimensional simulation of the thermo-chemical behavior of the combustion process and calculating the interesting parameters and the distribution of the mass fraction for the burning by-products. The CFD model was issued for the combustion area and a rich palette of results interesting for any researcher of the process were deduced.

1 INTRODUCTION

The combustion process is an important factor for the estimation of the level of pollution and emissions. The efficiency of such an engine is closely linked of the quality of the combustion process. The combustion is the process in any engine functioning when it is delivered the energy via the chemical combustion processes; hence the importance of this phase is the most important in the cycle of the engine, all the technical and economical parameters heavily depending on it (Heywood 1988). The combustion process is important but, equally, complex. There are a lot of parameters involved in it, highly dependent on the conditions of combustion, and, as a consequence, the combustion process is not easy to be mastered and investigated. The unknown aspects related to the combustion chemistry and conditions in which ignition and propagation of flames occurs are still some problems that need to cope with any modeling attempt of the combustion process (Benson & Whitehouse 1983).

Via numerical models/simulations, the behavior of the marine engines and propulsion systems is easy to be done, more economical, allowing various design solutions and optimization studies (Ramos 1989).

In the present paper, Finite Element Analysis (FEA) was applied to a complex and interdisciplinary theme, pertaining on the simulation of the combustion in the combustion chamber of a diesel engine, the most important parameters of this process being deduced.

2 GETTING STARTED

The inlet process described using the first law of thermodynamics states for an open system, the continuity equation and the equations of the pressure of steady flow from orifices (Buzbuchi & Stan 2008).

The below equations were solved using different approaches in the equation flow and thermodynamics constants. After the calculation of the stoichiometric combustion, the pressure p_{in} and temperature T_{in} are obtained at theoretical end of the inlet process:

$$p_{in} = \frac{p_s}{\sigma}\left(1 - \frac{n^2}{1800\frac{n_a}{n_a-1}\frac{RT_s}{M_{aer}}}\frac{\left(\frac{\varepsilon-\mu\rho}{\varepsilon-1}\right)^2}{\phi^2(fm\_Vs\cdot n)^2}\right)^{\frac{n_a}{n_a-1}} \quad (1)$$

$$T_{in} = T_s\frac{p_{in}}{p_s}\frac{\varepsilon}{\varepsilon-1}\left[1+\frac{\rho p_{out}T_s v p_{in}}{p_s^2 T_{out}(\varepsilon-1)^2\theta n_a}\times\left(\varepsilon+(n_a-1)(\varepsilon-1)\sigma-\frac{p_{out}}{p_{in}\rho}\right)\right] \quad (2)$$

T_s; p_s—inlet air parameters,
T_{out}; p_{out}—exhaust gas parameters,
ε—compression ratio,
θ—ratio of temperature increase,
ρ—scavenge coefficient,
ϕ—debit and velocity coefficient flow in inlet valve,
fm_V_s—ratio of section time and piston displacement,

$$\mu = \frac{p_{out}}{p_{in}} \frac{T_{in}}{T_{out}},$$

ν —coefficient of inertial inlet,
n_a—isentropic index,
n—speed engine,

The compression process may be calculated using the first law of thermodynamics for a closed system. In order to calculate the state changes p and T it is subdivided the stroke volume into a number of intervals of small volume change and equate the correlation at the beginning and end of each interval using the first law of thermodynamics. The smaller the volume changes in the interval the more accurate the calculation (Buzbuchi et al., 1997).

It is considered a small change in volume $dV = V_{j+1} - V_j$. It can be applied the first law in the form:

$$dQ = \Delta E + dW \tag{3}$$

where dQ is the heat transfer to the cylinder walls. If the cylinder wall temperature is Tw and the exposed surface area is A then the heat transfer is:

$$\frac{dQ_t}{dt} = \alpha_c A(T_{mj} - T_w) + \alpha_r A(T_{mj}^4 - T_w^4) \qquad \left[\frac{j}{s}\right]$$
$$dQ_t = \left[\alpha_c A(T_{mj} - T_w) + \alpha_r A(T_{mj}^4 - T_w^4)\right]\frac{1}{6n}d\alpha \quad [j] \tag{4}$$

α_c — convective heat transfer coefficient,
α_r — radiative heat transfer coefficient.

3 NUMERICAL SIMULATION

The naval engines have specific characteristics and their functioning is quite different in comparison with the normal diesel engines. The numerical models developed until now use as theoretical basis, the zero-dimensional thermodynamic models capable to describe globally the phenomena within the marine diesel engines, the result being a lack of the significant detail.

Figure 1. The diesel engine Caterpillar, type 3606.

The simulation of the combustion started with the definition of the parameters used in the numerical simulation with a short introduction on the processes forerunning the combustion as the admission, the compression, targeting a general frame for the combustion process.

The attention is shifted on defining the FEA concepts used in modeling the fluid mechanics, along with the general equations of flow and heat exchange for ideal and real fluids, the types of finite elements used in the paperwork, the possible models of materials fully mathematically described, and the treatment of the boundary conditions (Stan 2010).

The numerical simulation has as a departure point a real engine, Caterpillar, type 3606, given in the following figure.

4 WORK RESULTS AND VALIDATION

The simulation is conducted in 2D (Bidimensional) approach, the FEA net (grid) is presented and then all the results beginning with the pressure in the combustion chamber, velocities, temperatures distribution, diffusion coefficients, heat conductivity, Prandtl number, entropy, enthalpy, internal energy, mass fractions (Stan 2010). Some of the results are shown in the following figures.

The maximum static pressures have the value of 1.40e7 Pa and the dynamic pressures range from 3.18e-2 Pa up to a maximum of 9.7e2 Pa in the areas of fuel injection and at the piston surface.

The velocities of the jet fuel have the highest values in the middle area (max. 21.8 m/s) and the lowest ones in the upper wall area of the combustion chamber.

The highest temperatures are obtained in the middle area of the combustion chamber and the piston area, with the value of 2380^0 K.

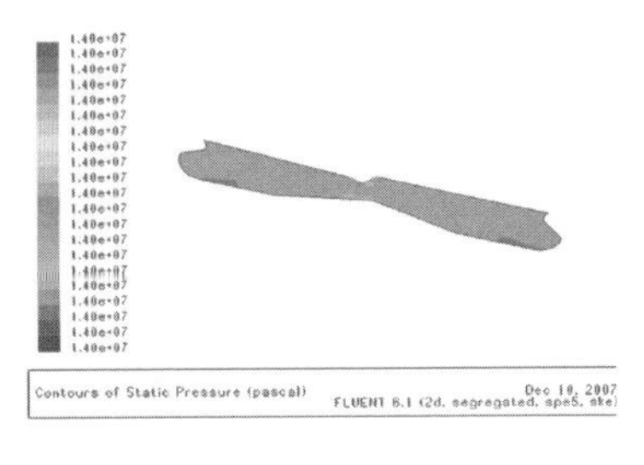

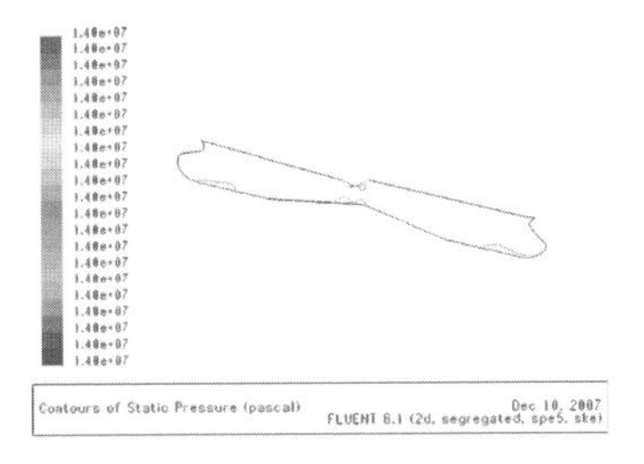

Figure 2. Static pressures in the combustion chamber.

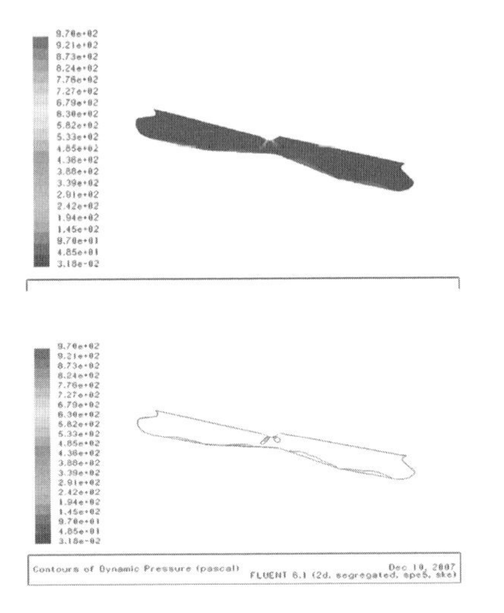

Figure 3. Dynamic pressures in the combustion chamber.

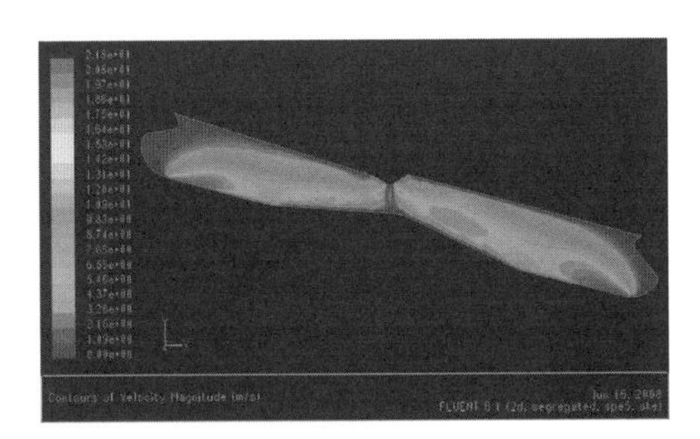

Figure 4. Velocity from the combustion chamber.

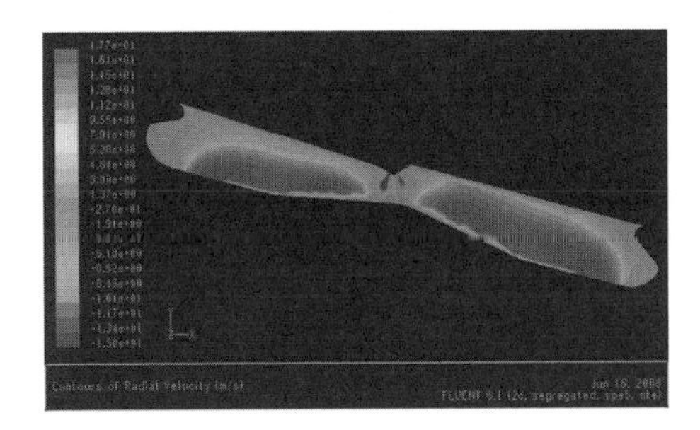

Figure 5. Radial velocity from the combustion chamber.

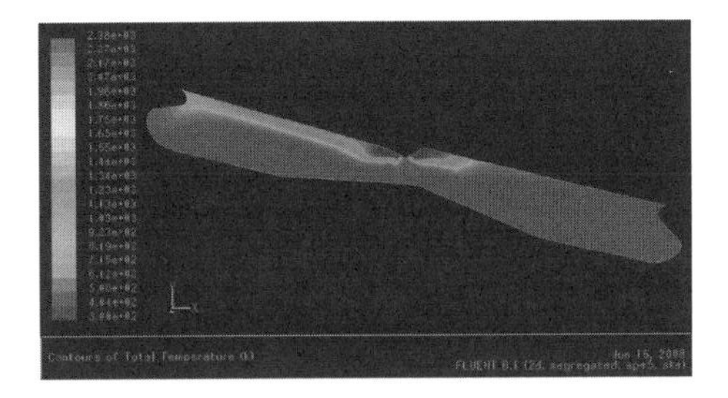

Figure 6. Temperatures distribution in the combustion chamber.

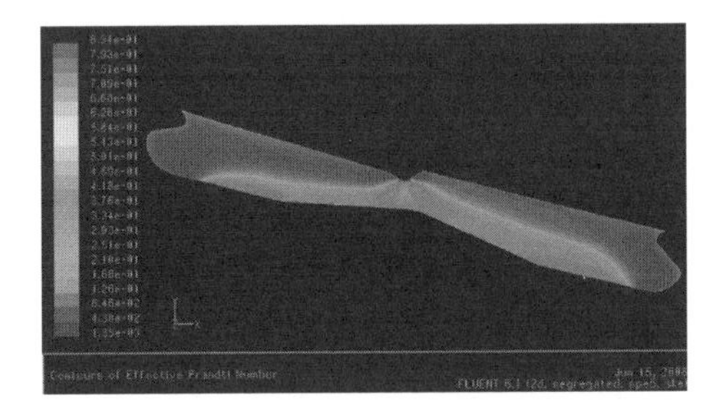

Figure 7. Prandtl number distribution in the combustion chamber.

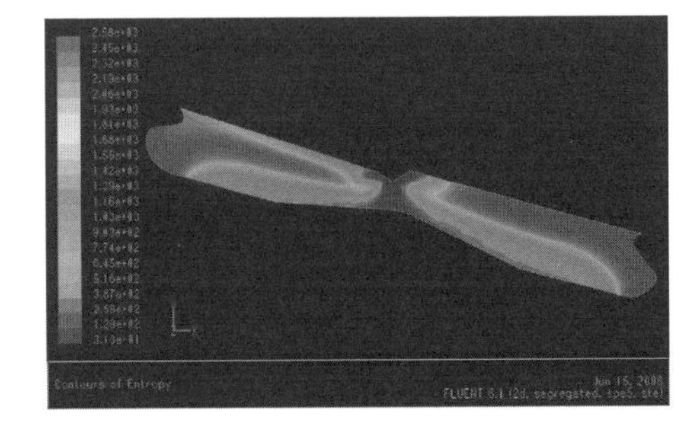

Figure 8. Entropy distribution in the combustion chamber.

At the cylinder area, the temperatures vary between 1130–2380⁰ K, there is a slight unbalance due to the tilted position of the injector.

Combined with the velocities distribution, it can be concluded that the tilted injector causes the unbalanced conditions in the combustion chamber, but that does not affect too much the pressure distribution, which is nearly symmetrical and uniform.

The maximum values of the Prandtl number are reached in the upper area of the combustion chamber (max. $5.43e^{-1}$) and the lowest values are reached in the proximity of the piston surface (min. $1.54e^{-4}$).

The entropy reaches the maximum values in the external—median area of the combustion chamber where the combustion processes are most intense and where the temperatures reach the higher values.

The stream functions show the influence of the tangential velocities; their distribution is almost similar in the two halves of the combustion chamber.

The effective viscosity is higher in regions of injection and the extreme areas of the combustion chamber (max. 9.18 e-4 kg/ms); in other regions, the values have an average of 2.64 e–5 ... 2.49 e-4 kg/ms.

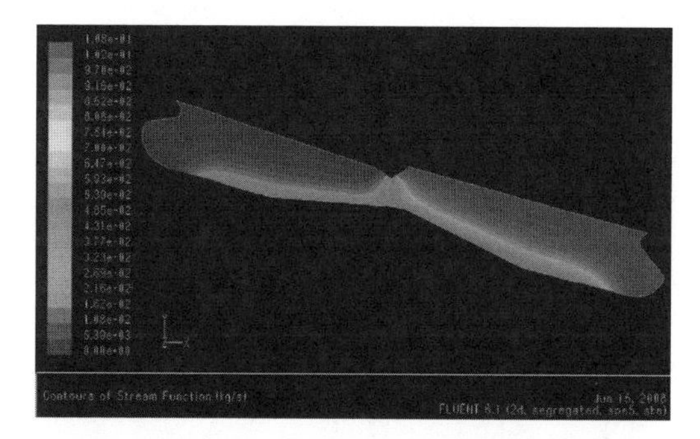

Figure 9. Stream function in the combustion chamber.

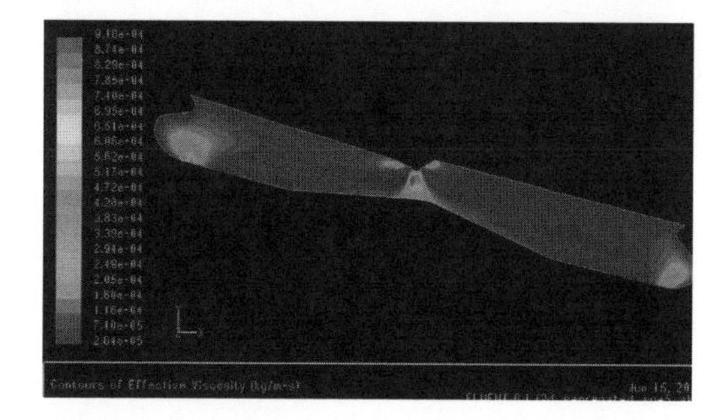

Figure 10. Effective viscosity in the combustion chamber.

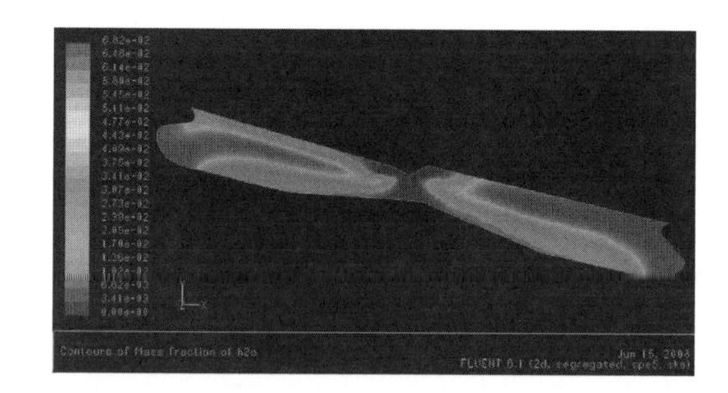

Figure 11. H_2O mass fraction distribution in the combustion chamber.

The diffusion coefficients are of great theoretical importance, their knowledge allowing various quantitative and qualitative analysis of the combustion process. From the distribution point of view, the diffusion coefficients of all species have the same shape, with the maximum values in the injection area and the ends of the combustion chamber.

The results were validated by comparing them with the existing databases and work of other authors concerning the same area of interest.

For the future works, there is recommended an extensive approach in which the processes preceding and afterwards of combustion in the engine cycle, may be treated in the FEA-CFD approach, an intensive direction where the combustion chamber may be modeled in the 3D (tridimensional approach) with the explicit simulation of mixed phases as fuel droplets which are evolving in the combustion chamber.

5 CONCLUSIONS

The original code is an useful tool in the early design stage necessary to the prediction of the running parameters and performance indicators of direct injection compression ignition engine. The precision of calculus depends on initial data and can be improved if we set up properly these data for the specified engine. Even if the numerical results have a satisfactory precision it is not possible to select immediately the final constructive solution. We can make only global evaluations and indicate the optimization directions that must be followed because the code contain many simplified hypothesis, empirical constants in the mathematical expressions of the model and minimum initial parameters (good for design stage but insufficiently for a final optimum solution). The present theoretical approaches offer good solutions for nominal regime and only satisfactory solutions for other regimes. This observation underlines that some initial parameters are not constant and are not optimum for the partial engine rates.

The Wiebe functions combustion model is more flexible than the fuel rate combustion model and offer a higher precision. This flexibility and the precision of this model are based on larger number of constants that must be set up as initial data. The constants calibration is a very difficult task instead. The fuel rate combustion model has a smaller number of constants and it offer a quickly solution but the precision is lower.

REFERENCES

Benson, R.S. & Whitehouse, N.D. 1983. *Internal combustion engine*, Pergamon Press, London.

Buzbuchi, N., Manea, L. & Dinescu, C. 1997. *Motoare navale. Procese si caracteristici*, Editura Didactică si Pedagogică, Bucuresti.

Buzbuchi, N & Stan, L. 2008. *Procese si caracteristici ale motoarelor navale*, 211 pg A4, Editura Nautica, Colectia Masini Navale, ISBN 978-973-7872-78-4, Constanta.

Heywood, J.B. 1988. *Internal Combustion Engine Fundamantals,* McGraw-Hill Book Company.

Ramos, J.I. 1989. *Internal Combustion Engines Modeling* Hemisphere Publishing Corporation, New York.

Stan, L.C. 2010. Marine propulsion systems and the influence of the operational factors, pp. 346–351 2010. *The 2nd International Conference on Computer and Automation Engineering (ICCAE 2010), Session 71 Fluid Dynamics and Thermodynamics Technologies*, Vol. 5, ISBN 978-1-4244-5585-0, IEEE Catalog Number: CFP1096F-PRT, February 26–28, Singapore.

Sustainable Maritime Transportation and Exploitation of Sea Resources – Rizzuto & Guedes Soares (eds)
© 2012 Taylor & Francis Group, London, ISBN 978-0-415-62081-9

Mean value modeling and model predictive control of a turbocharged diesel engine airpath

Safak C. Karakas & Oguz S. Sogut
Faculty of Naval Architecture and Ocean Engineering, Istanbul Technical University, Istanbul, Turkey

Can Ozsoy
Faculty of Mechanical Engineering, Istanbul Technical University, Istanbul, Turkey

ABSTRACT: This paper reports the results of a study on mean value modeling and model predictive control of a turbocharged Diesel engine airpath. The objective of this study, and also in general the control of Diesel engines, is to increase the usefulness of the electronic engine control units by using effective control strategies, such as model predictive control. The system used in this study consists of a four stroke, four cylinder high speed Diesel engine, a Variable Geometry Turbocharger (VGT) and an Exhaust Gas Recirculation (EGR) system. A controller is designed to control the Mass Air Flow (MAF) and Manifold Absolute Pressure (MAP) by changing the positions of exhaust gas recirculation valve and varying the geometry of turbocharger. Simulations are developed to enable the system work at different setpoints and also to include the effects of disturbances and wrong model order.

1 INTRODUCTION

Diesel engines are among the most efficient units of internal combustion engines. Thermal efficiencies of two stroke low speed Diesel engines can reach 50% and 40% efficiency is very common for four stroke Diesel engines. Therefore they have been extensively employed in power production since their discovery in 1892 and research studies have continued to increase to develop more efficient, cleaner and more useful engines. Ongoing research especially focuses on reducing emissions and it is now clearly observed that usage of control systems is compulsory to reach that aim. Control systems respond very quickly to changes in ambient conditions and disturbances acting on the engine and prevent the system deviations from the nominal working conditions.

Three-term (PID) controllers are usually used on Diesel engine control systems. However, in recent years, a number of new control approaches have been initiated to get satisfactory results for controlling and identifying the engine systems. These methods may be listed as:

- Multivariable techniques,
- Model-based control systems technology,
- Time delay compensation control,
- Adaptive control,
- Model predictive control,
- Optimal control,
- Robust control,
- Neural and fuzzy control,
- Variable structure control,
- Extremum control.

There have been many studies on modeling and control of Diesel engines. The most prominent ones are Heywood's and Assanis' (1986), Hendricks' (1989), Guzzella's (1998) and Jung's and Glover's (2003) studies. Heywood and Assanis set up a complete thermodynamic model of Diesel engines, Hendricks, Guzzella and Glover have constructed Diesel engine models by using Mean Value Modeling (MVM) approach and have used different control strategies, such as H-∞ control.

Model Predictive Control (MPC) is more advantageous than classical control approach about the transient-response of the controller. Classical control approach is a calibration problem. In that method, engine maps are derived by using the specifications of the engine. Calibration problem is tried to be solved for the minimum-fuel-criteria, taking the constraints into consideration. Then, optimum variables are uploaded to the Engine Control Unit (ECU). This method is an empirical approach so it is not a dynamic method; for this reason, it does not have the ability of controlling the transient-response.

MPC is also more advantageous than standard PI control approach because of the interaction between the variables of the model's being taken into consideration. A Multiple Inputs Multiple Outputs (MIMO) system is divided into Single

Input Single Output (SISO) systems and every system is controlled by a different controller. So, the interaction between the variables of the model is not taken into consideration in this method. But in MPC approach, the optimum operating point of the whole system is calculated by the controller. Moreover, the chance of changing the setpoint easily in the operating region is an advantage of MPC, too.

2 MATHEMATICAL MODEL OF THE TURBOCHARGED DIESEL ENGINE AIRPATH

The configuration diagram of the turbocharged Diesel engine with model variables is shown in Figure 1.

Air goes through the compressor and gets into the engine passing through the intercooler and intake manifold. After combustion process, some of the exhaust gas goes out of the system passing through the Variable Geometry Turbine (VGT) of the turbocharger and some amount of the exhaust gas goes back into the engine passing through the EGR system (EGR-cooler and EGR valve). VGT is used instead of a waste-gate system since VGT systems' having the expectation of good acceleration performance and high turbocharger efficiency over the whole range of operation (Guzzella et al., 2004). EGR system is used for NO_x reduction.

2.1 *Mean value modelling*

Mean value models are continuous Control Oriented Models (COM) which neglect the discrete cycles of the engine and assume all processes and effects are spread out over the engine cycle. They contain empirical formulas. The method of MVM is a mixture of *quasi-steady* and *filling-and-emptying* engine simulation models.

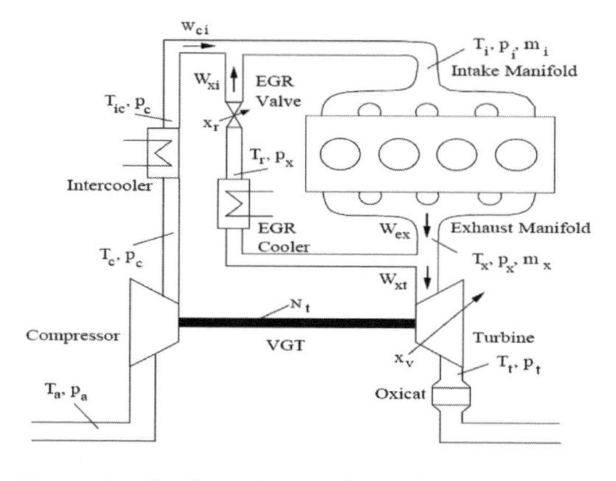

Figure 1. Configuration of the turbocharged Diesel engine (Jung 2003).

2.1.1 *Turbocharger*

The rotational speed of the turbocharger shaft N_t is derived as:

$$\frac{dN_t}{dt} = \left(\frac{60}{2\pi}\right)^2 \frac{P_t - P_c}{I_t N_t} \tag{1}$$

where P_t and P_c are the power of turbine and compressor respectively and I_t is the inertia of the turbocharger.

$$T_c = T_a + \frac{1}{\eta_c} T_a \left(\left(\frac{p_c}{p_a}\right)^{\frac{k-1}{k}} - 1 \right) \tag{2}$$

$$P_c = W_{ci}(h_c - h_a) = W_{ci} c_p (T_c - T_a) \tag{3}$$

$$T_t = T_x + \frac{1}{\eta_t} T_x \left(\left(\frac{p_t}{p_x}\right)^{\frac{k-1}{k}} - 1 \right) \tag{4}$$

$$P_t = W_{xt} c_p T_x \eta_t \left(1 - \left(\frac{p_t}{p_x}\right)^{\frac{k-1}{k}} \right) \tag{5}$$

T_t and T_c are the temperatures downstream of the turbine and compressor. k is the specific heat ratio (c_p/c_v), η_c and η_t are the isentropic efficiencies of the compressor and turbine, W_{ab} is the time based mass flow rate from a into b and p is the pressure. Subscripts of T and p denote the location where they are measured. (See Figure 1 for the locations.)

W_{ci}, η_c, p_t/p_x and η_t are determined from compressor and turbine maps.

2.1.2 *Other subsystems*

The crankshaft dynamics is expressed as

$$\frac{dN}{dt} = \frac{60}{2\pi} \frac{\tau_B - \tau_l}{I_B} \tag{6}$$

where N is the rotational speed of the engine shaft, τ_e and τ_l are the engine and load torques and I_e is the inertia of the engine. Engine torque can be approximately determined from:

$$\tau_e = \frac{V_d p_{mB}}{v 2\pi} \tag{7}$$

$$p_{me} = e p_{mf} - p_{mr} \tag{8}$$

$$p_{mf} = \frac{H_u m_f}{V_d} \tag{9}$$

where V_d is the displacement volume of the engine, p_{me} is the engine's mean effective pressure, p_{mf} is the mean fuel pressure, p_{mr} is the mean friction pressure, m_f is the mass of fuel injected in the cylinder in one cycle and H_u is the fuel's lower heating value. $v = 1$ for two-stroke and $v = 2$ for four-stroke engines ($e \in [0.38, 0.58]$, $p_{mr} \in [10^{2.4} \times 10^2]$) (Guzzella et al., 1998).

Mass flow rate of the air passing through the intake manifold to the cylinders is determined from:

$$W_{ie} = \eta_v \frac{m_i}{V_i} \frac{N}{60} \frac{V_d}{2} \tag{10}$$

where m_i is the mass of the air in the intake manifold and V_i is the volume of the intake manifold. η_v is the volumetric efficiency and it must be fitted as a polynomial in engine speed and intake manifold pressure (Wanscheidt 1998, Outbib et al., 2002).

Manifolds are modeled as open thermodynamic systems.

$$\frac{dp_i}{dt} = \frac{kR}{V_i}(T_{ic}W_{ic} + T_r W_{xi} - T_i W_{ie}) \tag{11}$$

$$\frac{dp_x}{dt} = \frac{kR}{V_x}(T_e W_{ex} - T_x(W_{xi} + W_{xt})) \tag{12}$$

$$\frac{dm_i}{dt} = W_{ci} + W_{xi} - W_{ie} \tag{13}$$

$$\frac{dm_x}{dt} = W_{ex} - W_{xi} - W_{xt} \tag{14}$$

where R is gas constant $(c_p - c_v)$ and V_x is the volume of exhaust manifold. Subscript i stands for the intake manifold and x stands for the exhaust manifold. Temperatures of manifolds are determined from the ideal gas law as:

$$T_i = \frac{V_i}{Rm_i} p_i \tag{15}$$

$$T_x = \frac{V_x}{Rm_x} p_x \tag{16}$$

Exhaust gas temperature is approximately calculated from:

$$T_e = T_i + a_1 \lambda^{a_2} + a_3, \qquad \lambda > \lambda_{\min} \leq 1.8 \tag{17}$$

$a_1 \approx 2500$ K
$a_2 \approx -2$
$a_3 \approx 100$ K

where λ is the air-fuel ratio ($\lambda_{\min} \leq 1.8$) (Guzzella et al., 1998).

Mass rate flow of the air passing through the EGR valve is calculated from the orifice flow equation:

$$W_{xi} = \frac{A_r(x_r) p_x}{\sqrt{RT_x}} \sqrt{\frac{2k}{k-1}\left(pr^{2/k} - p_r^{(k+1)/k}\right)} \tag{18}$$

$$p_r = \max\left(\frac{p_i}{p_x}, \left(\frac{2}{k+1}\right)^{\frac{k}{k+1}}\right) \tag{19}$$

$A_r(x_r)$ is the effective area of the EGR valve and is a quadratic polynomial of the valve opening ratio x_r.

The downstream temperatures of the coolers are determined from:

$$T_{\text{down}} = \eta_s T_{cool} + (1 - \eta_s) T_{up} \tag{20}$$

where η_s is the efficiency of the cooler and T_{cool} is the temperature of the coolant.

3 SIMULATION OF THE SYSTEM

Model inputs and outputs are given in Figure 2.

The nonlinear model consisting of Equations (1), (6), (11), (12), (13) and (14) was attempted to be solved by the *Runge-Kutta method* but the results were unstable (Karakas 2009). So, a linearised model given in (Jung 2003) was used for simulations and also for the controller design. The nominal operating point was chosen as 1500 rpm for engine speed and 85 Nm for load torque. Conditions at the nominal operating point are given in Table 1. The linearised model is given as:

$$\dot{x} = Ax + Bu \tag{21}$$

$$y = Cx + Du \tag{22}$$

$$A = \begin{bmatrix} -0.4125 & -0.0248 & 0.0741 & 0.0089 & 0.0000 & 0.0000 \\ 101.5873 & -7.2651 & 2.7608 & 2.8068 & 0.0000 & 0.0000 \\ 0.0704 & 0.0085 & -0.0741 & -0.0089 & 0.0000 & 0.0200 \\ 0.0878 & 0.2672 & 0.0000 & -0.3674 & 0.0044 & 0.3692 \\ -1.8414 & 0.0990 & 0.0000 & 0.0000 & -0.0343 & -0.0330 \\ 0.0000 & 0.0000 & 0.0000 & -359.0000 & 187.5364 & -87.0316 \end{bmatrix}$$

$$B = \begin{bmatrix} -0.0042 & 0.0064 \\ -1.0360 & 1.5849 \\ 0.0042 & 0 \\ 0.1261 & 0 \\ 0 & -0.0168 \\ 0 & 0 \end{bmatrix}$$

$$C = \begin{bmatrix} 0 & 0 & 0 & 0 & 0 & 3.6 \\ 0 & 0 & 0 & 1 & 0 & 0 \end{bmatrix}$$

$$D = \begin{bmatrix} 0 & 0 \\ 0 & 0 \end{bmatrix}$$

Results of the first simulation are shown in Figure 3. VGT actuator position is kept constant (60%) during the simulation. In the second simulation (Fig. 4), EGR actuator position is kept constant at 20%.

In the third simulation (Fig. 5), EGR and VGT actuator positions are given as PRBS (Pseudorandom binary sequence) signals. MAF and MAP responses to the PRBS signals take values around their nominal working conditions.

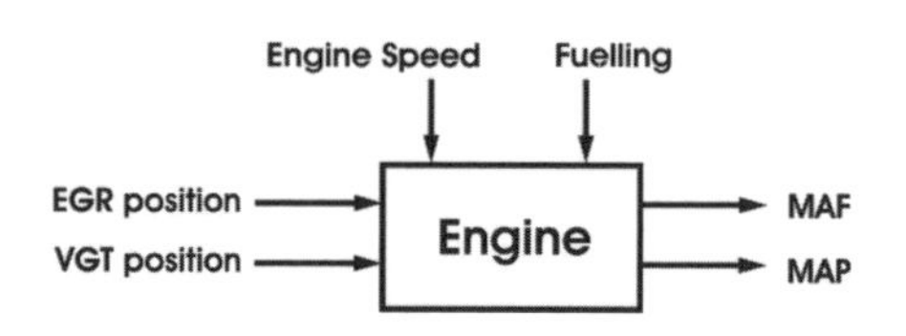

Figure 2. Model inputs and outputs.

Table 1. Conditions at the operating point.

	Parameter	Unit	Nominal value	State
	Inputs			
	Engine speed	$[N] = \text{rpm}$	1500	
	Fuelling	$[W_f] = \text{mg/st}$	16.7	
u_1	EGR actuator position	$[x_r] = \%$	20	
u_2	VGT actuator position	$[x_v] = \%$	60	
	Outputs			
y_1	MAF	$[W_{ci}] = \text{kg/h}$	78.5	x_6
y_2	MAP	$[p_i] = \text{kPa}$	111.3	x_4
	Internal variables			
	Intake manifold mass	$[m_i] = \text{g}$	7.15	x_3
	Exhaust manifold pressure	$[p_x] = \text{kPa}$	118.0	x_2
	Exhaust manifold mass	$[m_x] = \text{g}$	0.66	x_1
	Turbocharger speed	$[N_t] = \text{rpm}$	58400	x_5

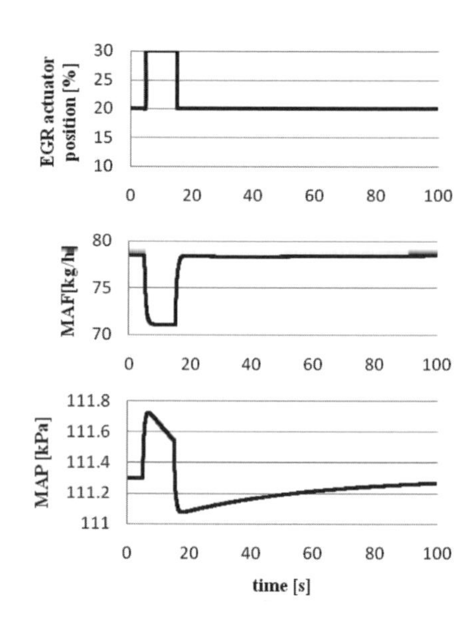

Figure 3. Results of the first simulation.

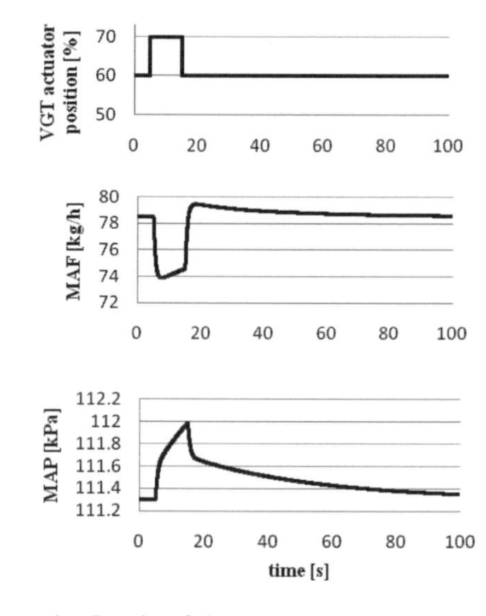

Figure 4. Results of the second simulation.

4 MODEL PREDICTIVE CONTROL OF THE SYSTEM

The cost function used in the Model Predictive Control (MPC) algorithm is:

$$V = \mathbf{Q}\Sigma_{i=n_w}^{n_y} \| \mathbf{r}_{k+i} - \mathbf{y}_{k+i} \|_2^2 + \mathbf{R}\Sigma_{i=0}^{n_u-1} \| \Delta\mathbf{u}_{k+i} \|_2^2 \quad (23)$$

$$u_{\min} \le u \le u_{\max} \quad (24)$$

where $\mathbf{Q}$ and $\mathbf{R}$ are weight matrices, n_w, n_y and n_u are the parameters of the receding horizons and

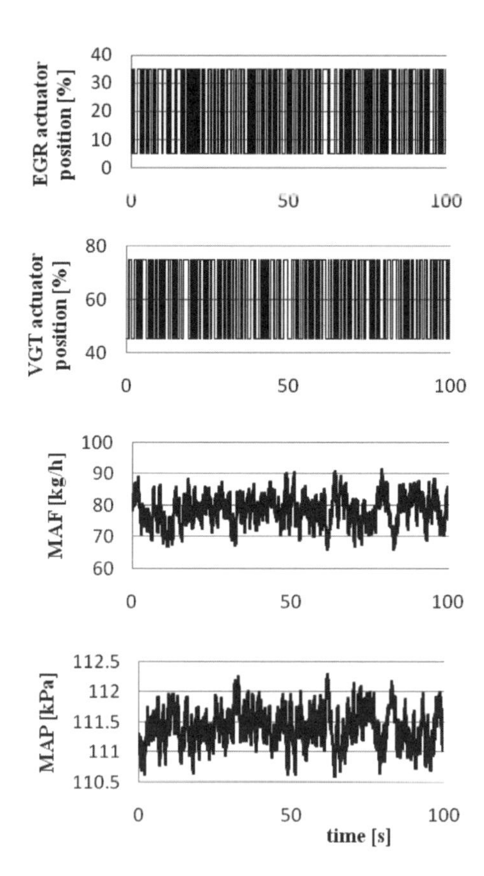

Figure 5. Results of the third simulation.

r is the setpoint. A quadratic programming problem, that is about finding the minimum value of V while taking the constraints into consideration, is solved at each sampling instant. Sampling period of the simulations is 0.1 s. Constraints of the EGR actuator are [−20; 80] and VGT actuator are [−60; 40]. These constraints are the closed or open conditions of the valves.

Results of the first MPC simulation are shown in Figure 6. In this simulation, MAF is increased by 3 kg/h and MAP is increased by 0.3 kPa while the system is working at the nominal setpoint.

Tuning parameters for this simulation are:

$$n_w = 1$$
$$n_y = 200 \times 0.1s = 20s$$
$$n_u = 100 \times 0.1s = 10s$$

$$\mathbf{Q} = [2\ 1]$$
$$\mathbf{R} = \begin{bmatrix} 0.1 \\ 0.5 \end{bmatrix}$$

The results indicate that, in the beginning, controller decreases the EGR actuator position to

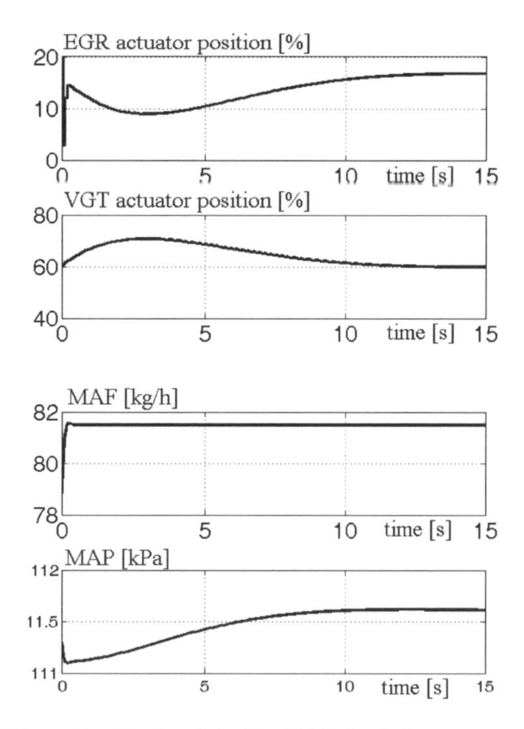

Figure 6. Results of the first MPC simulation.

increase the amount of exhaust gas passing through the turbine and to increase the speed of the turbocharger. Then MAF increases because of the compressor's starting to rotate faster but MAP decreases since the velocity of the air is increasing. In the end MAF increases by 3 kg/h so the objective is achieved.

After that moment, EGR valve position decreases and VGT actuator position increases. So that turbocharger speed is kept constant and MAP increases because of the backpressure of the exhaust gas in the exhaust manifold.

Results of the second MPC simulation are shown in Figure 7. In this simulation, a disturbance of 10 kPa is acting on the exhaust manifold gas pressure and the controller tries to make the system work at its nominal setpoint. Tuning parameters for this simulation are:

$$n_w = 1$$
$$n_y = 600 \times 0.1s = 60s$$
$$n_u = 30 \times 0.1s = 3s$$

$$\mathbf{Q} = [1\ 4]$$
$$\mathbf{R} = \begin{bmatrix} 1 \\ 4 \end{bmatrix}$$

The results indicate that, the 10 kPa increase of pressure in the exhaust manifold increases MAP

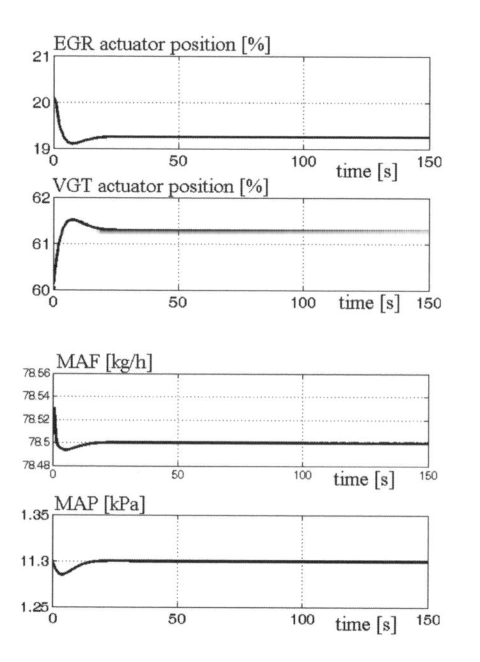

Figure 7. Results of the second MPC simulation.

because of the backpressure. Hovewer, the MPC controller can put the system into the nominal conditions again in about 30 seconds.

Results of the third MPC simulation are shown in Figure 8. In this simulation, a noise of −1 kg/h for MAF and −1 kPa for MAP are acting and the controller tries to make the system work at its nominal setpoint. These noises are thought as the measurement errors at the MAF and MAP sensors.

Tuning parameters for this simulation are:

$$n_w = 1$$
$$n_y = 40 \times 0.1s = 4s$$
$$n_u = 3 \times 0.1s = 0.3s$$

$$\mathbf{Q} = [1 \ \ 4]$$

$$\mathbf{R} = \begin{bmatrix} 1 \\ 1 \end{bmatrix}$$

The results indicate that, the MPC controller can put the system into the nominal conditions again in about 15 seconds. MAF is 1 kg/h and MAP is 1 kPa greater than their nominal values at the end of the simulation because the sensors measure these values wrong, so the system is again working at its nominal condition at the end of the simulation.

Results of the fourth MPC simulation are shown in Figure 9. In this simulation, the controller is designed by using a wrong order model.

The state-space model of the system is sixth order. For constructing a wrong order model,

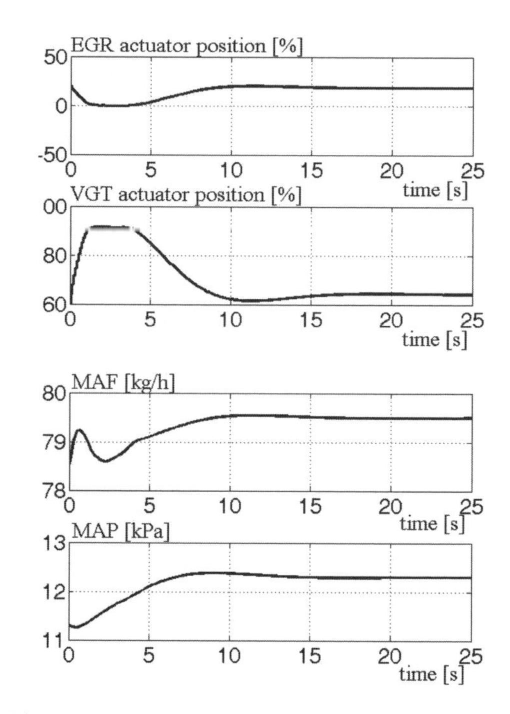

Figure 8. Results of the third MPC simulation.

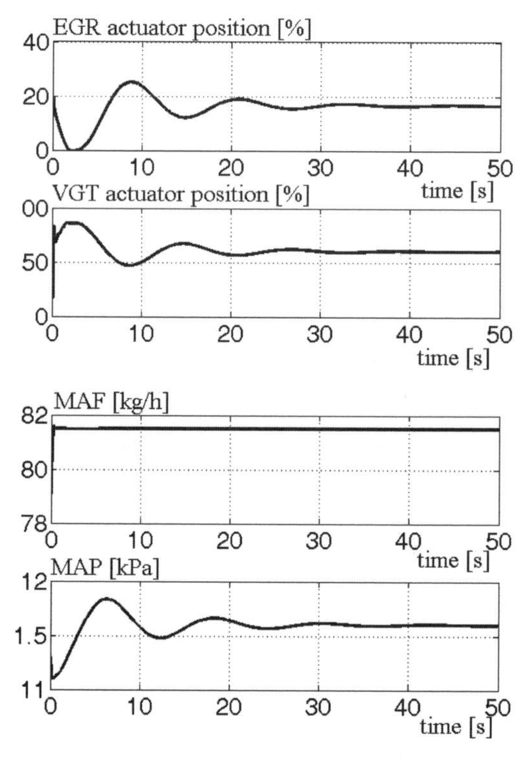

Figure 9. Results of the fourth MPC simulation.

one state of the model is ignored and the MPC controller is designed by using this new fifth order model, but it tries to control the original (sixth order) model. Tuning parameters for this simulation are:

$$n_w = 1$$
$$n_y = 10 \times 0.1s = 1s$$
$$n_u = 3 \times 0.1s = 0.3s$$

$$\mathbf{Q} = [1\ 1]$$
$$\mathbf{R} = \begin{bmatrix} 0.1 & \\ & 0.01 \end{bmatrix}$$

The results indicate that, the MPC controller can put the system into the nominal conditions again in nearly 35 seconds.

5 CONCLUSION

A model predictive control algorithm has been developed for a turbocharged Diesel engine airpath in this study. The results of the simulations indicate that MPC controller has the ability to control the system while a disturbance is acting on the variables. It can control the system even if the controller is designed by using a wrong order model and this situation shows that the controller is insensitive for incorrect models. Considering the results of this study, the method of model predictive control is a right choice for controlling the airpath of turbocharged Diesel engines.

For future work, inclusion of combustion process and injection timing will be considered to the turbocharged Diesel engine airpath model and a new MPC controller will be designed. Moreover, experimental results from a Diesel engine test bed, together with engine simulation, will be used to investigate whether MPC is the correct choice to be used in turbocharged Diesel engine systems.

REFERENCES

Assanis, D.N. 1986. *A Computer Simulation of the Turbocharged Turbocompounded Diesel Engine System for Studies of Low Heat Rejection Engine Performance*, Phd Thesis, Massachusetts Institute of Technology. Dept. of Ocean Engineering.

Assanis, D.N. & Heywood, J.B. 1986. Development and Use of a Computer Simulation of the Turbocompounded Diesel System of Engine Performance and Component Heat Transfer Studies, *SAE Paper 860329*.

Challen, B. & Baranescu, R. 1999. *Diesel Engine Reference Book Second Edition*, Delhi: Elsevier.

Christen, U., Vantine, K. & Collings, N. 2001. *Event-based mean-value modelling of DI Diesel engines for controller design*, Detroit: SAE International congress.

Guzzella, L. & Amstutz, A. 1998. Control of Diesel Engines, IEEE, Vol. 18,5.

Guzzella, L. & Onder, C.H. 2004. *Introduction to Modeling and Control of Internal Combustion Engine Systems*, Zurich: Springer.

Hendricks, E. 1989. Mean Value Modelling of Large Turbocharged Two-Stroke Diesel Engines, *SAE Paper 890564*.

Horlock, J.H. & Winterbone, D.E. 1986. *The Thermodynamics and Gas Dynamics of Internal-Combustion Engines Vol. 2*, Oxford: Clarendon Press.

Jung, M. 2003. *Mean-Value Modelling and Robust Control of the Airpath of a Turbocharged Diesel Engine*, Phd Thesis, Cambridge University.

Jung, M. & Glover, K. 2003. Control-Oriented Linear Parameter Varying Modelling of a Turbocharged Diesel Engine *Proceedings of the IEEE Conference on Control Applications*.

Karakas, S.C. 2009. *Turbosarjli Diesel Motor Hava Akis Sisteminin Ortalama Deger Yontemi Temel Alinarak Model Tabanli Ongorulu Kontrolu*, Master's Thesis, Istanbul Technical University.

Maciejowski, J.M. 2002. *Predictive control with constraints*, Harlow: Prentice-Hall, Pearson Education Limited.

Ogata, K. 1978. *Modern Control Engineering*, New Delhi: Prentice-Hall.

Ortner, P. & del Re, L. 2007. Predictive Control of a Diesel Engine Air Path, *IEEE Transactions on Control Systems Technology* Vol. 15, 3.

Outbib, B., Dovifaaz, X., Rachid, A. & Ouladsine, M. 2002. Speed Control of a Diesel Engine: A Nonlineer Approach, *Proceedings of the American Control Conference, Anchorage*, AK, 8–10 May.

Oz, I.H., Borat, O. & Surmen, A. 2003. *Icten Yanmali Motorlar*, Istanbul: Birsen Yayinevi.

Pulkrabek, W.W. 2003. *Engineering Fundamentals of the Internal Combustion Engine Second Edition*, New Jersey: Prentice Hall.

Rossiter, J.A. 2003. *Model-Based Predictive Control A Practical Approach*, Boca Raton: CRC Press.

Sánchez-Peña, R.S., Casín, J.Q. & Cayuela, V.P. 2007. *Identification and Control: The Gap between Theory and Practice*, London: Springer-Verlag.

Wanscheidt, W.A. 1968. *Theorie der Dieselmotoren*, Berlin: Veb Verlag Technik Berlin.

Sustainable Maritime Transportation and Exploitation of Sea Resources – Rizzuto & Guedes Soares (eds)
© 2012 Taylor & Francis Group, London, ISBN 978-0-415-62081-9

Evaluation of ignition and combustion quality of different formulations of heavy fuel marine oil by laboratory and engine test

C.R.P. Belchior & J.P.V.M. Bueno
COPPE-EP-Federal University of Rio De Janeiro, Rio de Janeiro, Brazil

ABSTRACT: Federal University of Rio de Janeiro installed a laboratory regarding the possibilities of obtaining performance tests of heavy fuel oil in a marine medium speed engine with the aim of improving its ignition quality and combustion properties. The paper describes the facilities, equipments installed, test methodology and obtained results. The fuel is also analyzed in a FIA test instrument and the results compared to those obtained in the engine laboratory. Special concerns are the ignition delay, estimated cetane number, maximum combustion pressure and others variables that defines the fuel ignition quality. A diagnostic system monitors the combustion and operation cycle of the engine using information such as cylinder pressure versus crank angle, ignition delay, maximum combustion pressure, mean indicated pressure and indicated work. A thermodynamic simulation model of the operation cycle was also developed to compare with the experimental results. The model allows engine performance estimation of different heavy fuel oil compositions.

1 INTRODUCTION

Current commercial marine fuel specifications may on occasion be insufficient to plot fuel purchasers from obtaining fuel oils with unfavorable ignition and combustion properties to meet specific ship engine operating requirements.

The development of fuel quality in recent years shows the following general trends:

- Fuels are becoming increasingly aromatic which is influencing ignition and combustion properties especially Heavy Fuel Oil (HFO).
- Average density of fuels world-wide is increasing.
- During the past years we have seen a large number of so called epidemic fuel problems.

Each year millions of tons of bunker fuels are sold into the marine market globally supplied in accordance with the ISO 8217 Standard. Nevertheless this fuel presents poor ignition performance in marine engines (Steernberg et al., 2007).

The objective of the study is to present a methodology consisting of the determination of ignition and combustion quality of the heavy fuel in the FIA equipment and also obtaining ignition delay of the fuel in marine engine performance test.

This paper describes de ignition quality determination of heavy fuel oil and the installation necessary for obtaining the acceptable results.

2 IGNITION QUALITY AND COMBUSTION OF HFO

In the absence of an indicator for ignition quality of HFO fuels, like the Cetane Number for distillate fuels, extensive research by SHELL resulted in the concept of the Calculated Carbon Aromaticity Index (CCAI). The research developed, had the objective of gaining an understanding of the factors controlling the ignition performance of residual fuel oils and to identify means of quantifying ignition quality (Schenk et al., 1983).

Both the physical and chemical properties of residual oil were found to have an influence on ignition performance.

Physical properties are viscosity and temperature since atomization is affected by them.

Chemical properties are aromaticity. The resulting CCAI index, can be calculated by a following formula:

$$CCAI = d - 81 - 141 Log[Log Vk + 0.85] - 483 Log\left[\frac{T+273}{323}\right] \quad (1)$$

where d = density at 15°C (Kg/m²); Vk = kinematic viscosity (mm²/s) at temperature T (°C).

It must be stressed that CCAI was adopted for allowing ranking the ignition qualities of different residual fuels oils. But, we should note that CCAI does not give an absolute measure since it

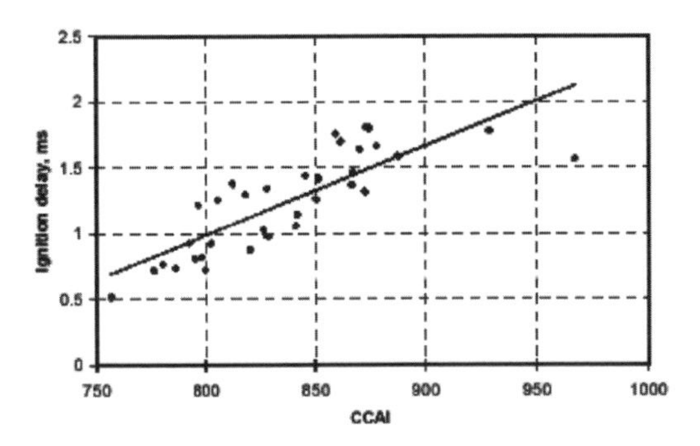

Figure 1. Correlation Ignition delay-CCAI.

is dependent upon other variables as engine design and operating conditions.

For instance, modern medium speed engines will tolerate quite satisfactory values by to 870 to 875 of CCAI. Low speed engines may be more tolerant of higher CCAI values.

The ignition quality of a fuel for diesel engine is normally based in the ignition delay period defined as the period counted from the start of the fuel injection to the point of start of combustion.

Typical results obtained in a research engine (Schenk et al., 1983) of ignition delay dos not presents a proper correlation with CCAI values.

Figure 1 presents these results (Valencia & Armas 2005):

The Figure 1 presents a scatter of data points around regression line rather large, increasing at lower engine output.

The use of CCAI has been evaluated and it has been concluded that this parameter gives a rough estimate of the ignition performance of heavy residuals fuels.

Other index was presented considering the aromaticity of the fuel (Calculated Vapour Aromaticity Index-CVAI) (Schenk et al., 1983) which also presented not acceptable correlation with ignition delay. The British Petroleum presented the Calculated Ignition Index (CII) (Espadafor, 2009). These Index have often been found to be inaccurate and inadequate to detect problem fuels.

3 THE LABORATORY TEST APPARATUS

In order to improve the ignition and combustion of the HFO, a Cooperative Research System has established between a Petroleum Research Center and the Thermal Engines Laboratory of the Federal University of Rio de Janeiro. In the Research Center the physical-chemical analysis of the fuel is obtained, defining its viscosity, density, metals, etc.

The CCAI index is also calculated. Ignition and combustion test of a fuel sample in a combustion

Figure 2. MAN 4L 16/24 diesel engine and control room with engine diagnosis system.

chamber FIA 100/3 is followed. Detailed explanation of the equipment is described in the next item. Complementary test are developed in the MAN 4L16/24 Diesel Engine installed in the Thermal Engine Laboratory.

The engine fuel supply system consists of the following main equipments in accordance to CIMAC Recommendations:

- Sediment Tank-valve the fuel and its water content is separated.
- Centrifuging to remove solid as well as liquid contaminants.
- Pre-heating the HFO to a temperature corresponding to the injection viscosity of 13 cSt.

The Diesel generator is supplying and linked to an electrical charge board, that can vary electric load of the system, according to the standard rules according to ISO 8178 test cycles. The engine monitoring system presents the cylinder pressure versus crank angle curve (Figure 2).

4 FUEL IGNITION ANALYZER

In order to improve the ignition and combustion quality of heavy fuel oil, an equipment is been used, called FIA-100/3 Fuel Ignition Analyzer.

The ignition and combustion conditions in FIA-100/3 combustion chamber are simulated by automatic preconditioning of the charge condition using external electrical power and compressed air. A sample of the fuel to be investigated is injected into this combustion chamber using mechanically activated conventional high pressure fuel injection pump and one orifice fuel nozzle. During the injection, the fuel spray self-ignites and burns in the constant volume combustion chamber. The test condition and combustion process are carefully monitored, and measured data are collected by a separate electronic unit containing an advanced microprocessor.

The test data is immediately displayed on the computer screen as well as analyzed, stored and field automatically in the PC. All combustion conditions can be altered at a touch of keyboards controls. The optimum conditions for best combustion can be established using this instrument.

The test procedure, consists of a investigation performed at a change air pressure of 45 bars and air temperature of 500°C. Depending on the viscosity of the sample the fuel can be preheated in order to obtain 15 cSt fuel injection viscosity.

The Ignition Delay in the FIA Test is defined as a time delay (in milliseconds) from start of injection until an increase in pressure of 0.2 bar above initial chamber pressure has been detected or a increase of 1% of maximum combustion pressure.

Start of Main Combustion phase is determined as the time (in milliseconds) when an increase in pressure of 3 bars above initial chamber pressure has been detected or an increase to 10% of maximum pressure. The start of Main Combustion is used in order to establish the ignition quality of the fuel tested as a FIA CN (Cetane Number).

The basis for FIA CN is a reference curve for the instrument in question, showing the ignition properties for mixtures between the reference fuels with cetane number 18.7 and 74.8 in variable percentage, in according with the fuel to be measured.

In general FIA CN (Cetane Number) should be above 28 to be considered presenting acceptable ignition properties.

Figures 3, 4 presents the characteristics curves obtained in a regular FIA 100/3, during the ignition and combustion of the fuel sample (CIMAC, 2010).

The curves indicate several variables that are defined in the symbology of the Figures 3 and 4.

The Estimate Cetane Number (ECN) can be calculated by the following equation (CIMAC, 2010):

$$ECN = 153.15e^{-0.2861MCD} \quad (2)$$

The results of the tests performed for chemical analysis and in the FIA 100/3 for the sample tested is described in the Table 2 and the Table 3.

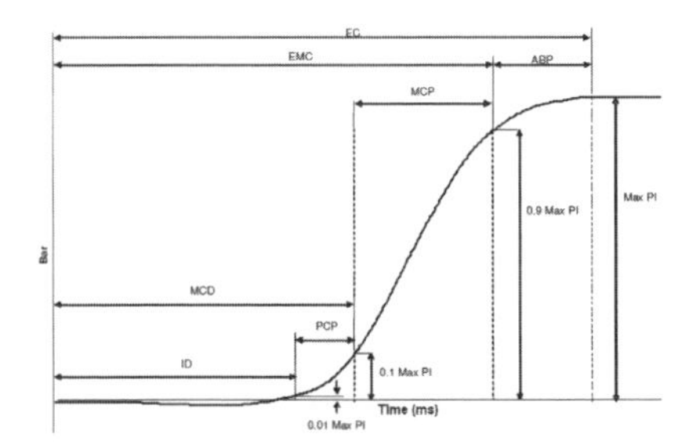

Figure 3. Pressure curve in function of time.

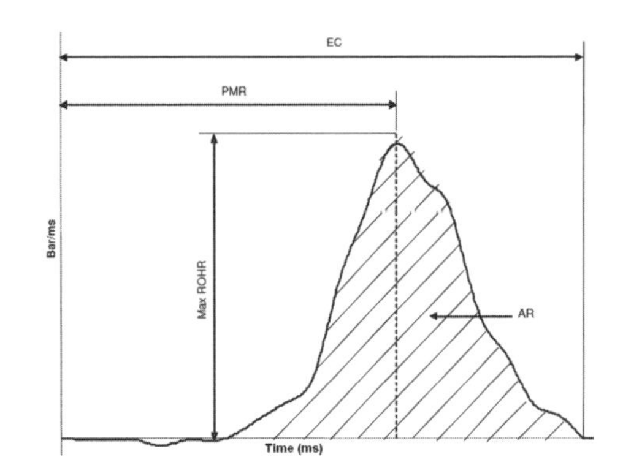

Figure 4. Rate of heat release as function of time.

	Figures 3 and 4 Symbology
ID	Ignition Delay
Max PI	Max Pressure Increase
MCD	Main Combustion Delay
PCP	Pre Combustion Period
MCP	Main Combustion Period
EMC	End of Main Combustion
EC	End of combustion
ABP	After Burning Period
ROHR	Rate of Heat Release
AR	Accumulated ROHR (Surface area)
PMR	Position of Maximum of Heat Release
CP	Combustion Period (positive Area of ROHR)
SMC	Start of Main Combustion

Table 1. Gives the test results of two typical heavy fuel oils.

Sample	HFO 1	HFO 2
FIA CN	50.7	<18.7
ID (ms)	4.25	12.55
SMC (ms)	6	19.35
CP (ms)	13.2	30.1
PMR (ms)	5.5	19.5
Quality comments	Good fuel	Bad fuel

Table 2. Presents the results of the chemical compositions of the HFO sample.

CCAI	852.7
LHV	41.06 MJ/Kg
Specific energy (ISO 8217)	40.96 KJ/Kg
O	0.6% m/m
N	0.7% m/m
C	88.8% m/m
H	11.1% m/m
S	0.8% m/m

Table 3. Presents the results of the FIA test for the fuel defined in table 2.

ID (ms)	6.38
MCD (ms)	8.49
EMC (ms)	14.42
EC (ms)	22.13
PCP (ms)	2.11
MCP (ms)	5.93
ABP (ms)	7.71
Maxi (bar)	6.98
ECN	13.5

Note: Acceptable ECN Values > 14 (CIMAC, 2010).

5 ENGINE TEST

The HFO test in the diesel engine is performed in the accordance to ISO 8178 test cycles and presented in the monitoring system of the engine.

The Figure 5 presents the curves of cylinder pressure versus crank angle for 50% of engine power for the diesel and the HFO, obtained by diesel engine monitoring system (Figure 2).

We can estimate the different ignition quality of the fuel by measurement the ignition delay (period between the start of the injection and start of ignition or pressure rise) on the Figures 5 and 6.

We can notes the different in those values for HFO and diesel oil.

We must note the ignition delay of the diesel oil is lower than the ignition delay of the bunker, denoting better ignition quality. By this method we can rank fuels samples presenting different ignitions qualities.

Figure 6 shows that the ignition delay of the fuel decreases with the increasing of the engine load.

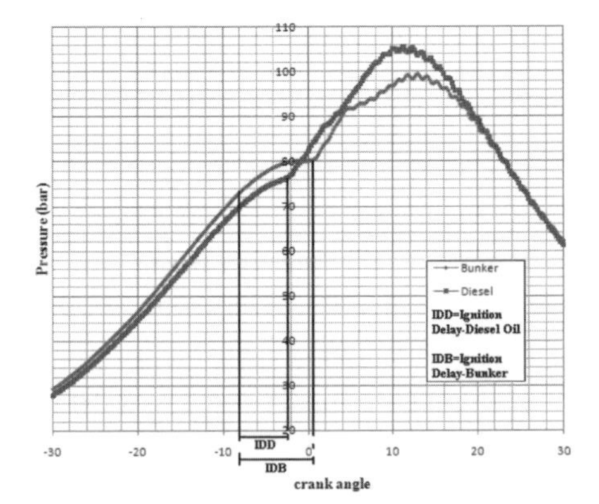

Figure 5. Experimental curves of pressure versus crank angle for diesel oil and bunker oil.

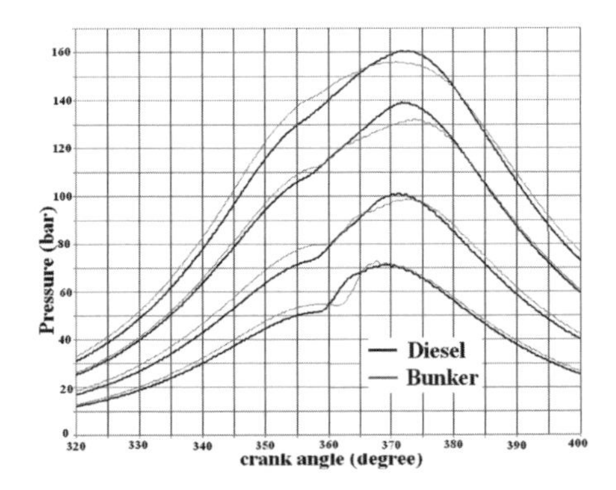

Figure 6. Experimental curves of pressure versus crank angle for diesel oil and bunker oil for different loads.

6 SUMMARY OF THE THERMODYNAMIC MODEL

A zero dimensional simulation model for the engine operation cycle is based in the first low of the thermodynamics, complemented by the equations of the different phases of the cycle.

The model variables are depending only on the time and the crank angle.

The model includes compression, combustion and expansion; begin in the time after the closing of the admission valves to the opening of the exhaust valves, as presented in the Figure 7.

In the Figure 7, δQ_{comb} is the rate of combustion heat release (KJ/degree of crank angle), δQ_p is the thermal energy (heat) transferred through the cylinder wall to external environmental (heat losses) (KJ/degree of crank angle), dU is the internal energy variation and δW is the net indicating work.

The compression process is analyzed as a polytropic transformation and the combustion and the expansion process is considered to behave as ideal gases.

According to Heywood (1988) and Lanzafame et al. (2003), the energy equation applied to the systems gives us an equation system as followed:

$$\frac{dW}{d\theta} = p\frac{dV}{d\theta} \tag{3}$$

$$\frac{\delta Q_{total}}{\delta\theta} = \frac{\delta Q_{comb} - \delta Q_p}{\delta\theta} \tag{4}$$

$$\frac{1}{T}\frac{dT}{d\theta} = \frac{1}{p}\frac{dp}{d\theta} + \frac{1}{V}\frac{dV}{d\theta} \tag{5}$$

where T = temperature; p = pressure; W = work; Q_{total} = total heat transferred to the system; θ = crank angle.

To solve the equations system we have to obtain complementary equation in function of the crank angle.

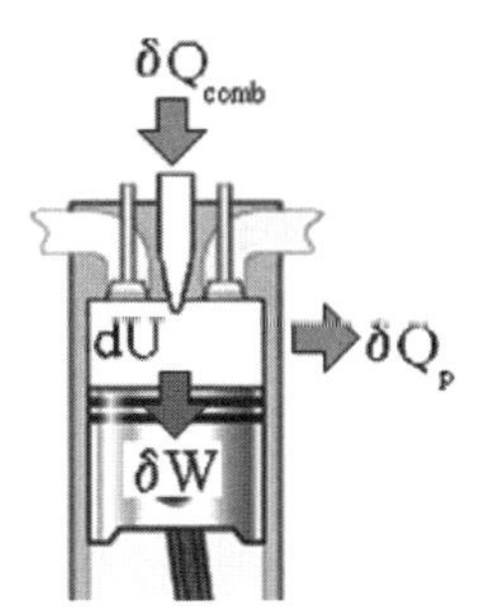

Figure 7. Cylinder energy balance.

Those equations will be presented in the next items.

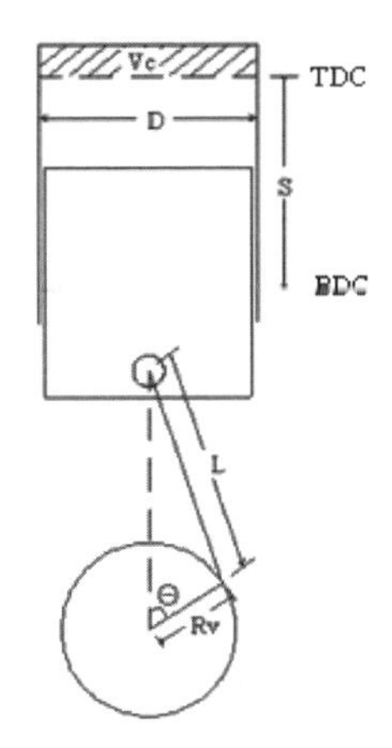

Figure 8. Engine geometry.

7 ENGINE GEOMETRY

Figure 8 presents the engine geometry:

From the figure 8 we can obtain the following equations. (Heywood, 1988; Brambila, 2006)

$$V(\theta) = V_C \left\{ \begin{array}{l} 1 + \frac{1}{2}(r-1)[BM + 1 - \cos\left(\frac{\theta\pi}{180}\right) \\ -\sqrt{BM^2 - sen^2\left(\frac{\theta\pi}{180}\right)}] \end{array} \right\} \quad (6)$$

$$A(\theta) = \frac{2\pi D^2}{4} + \frac{\pi DS}{2} \left[\begin{array}{l} BM + 1 - \cos\left(\frac{\theta\pi}{180}\right) \\ -\sqrt{BM^2 - sen^2\left(\frac{\theta\pi}{180}\right)} \end{array} \right] \quad (7)$$

where r = compression ratio of the engine; V = combustion chamber volume; D = cylinder diameter; A = piston area; S = stroke; Rv = crankshaft radius; L = connecting rod length.

8 IGNITION DELAY

The ignition delay is calculated by the following equation (Assanis et al., 1999):

$$\tau_{AI}(ms) = 2.4 * \phi^{-0.2} * p_c^{-1.02} * e^{\left(\frac{Ea}{\bar{R}*Tc}\right)} \quad (8)$$

where τ_{AI} = ignition delay; ϕ = equivalent ratio.

The milliseconds (ms) can be transformed in angle by the equation:

$$\tau_{AI}(\Delta\theta) = 0.006 * Rot * \tau_{AI}(ms) \quad (9)$$

where Rot = RPM of the engine; R = universal gas constant; T_c = temperature inside the combustion chamber obtained by:

$$Tc = T_1 * \varepsilon_{ef}^{\ c-1} \quad (10)$$

where ε_{ef} = effective compression ratio; T_1 = admission temperature; c = polytropic constant.

$$\varepsilon_{ef} = \frac{V_d}{V_{SOI}} \quad (11)$$

where V_d = displacement volume; V_{SOI} = volume in start of the ignition.

$$c = 1.4 - \frac{0.4}{1.1 * S_P + 1} \quad (12)$$

$$Ea = \frac{618840}{NC + 25} \quad (13)$$

where S_P = piston speed; NC = cetane number of the fuel determined by ECN in the FIA.

9 COMBUSTION RELEASE RATE

The total energy introduced in the system by the fuel (Q_{comb}) can be expressed:

$$Q_{comb} = m_f * LHV \quad (14)$$

where m_f = fuel mass admitted in the combustion chamber in an operation cycle; LHV = lower heat value of the fuel.

The heat release rate:

$$\frac{dQ_{comb}}{d\theta} = Q_{comb} \frac{dx}{d\theta} \quad (15)$$

where $\frac{dx}{d\theta}$ = rate of fuel burned; x = fraction of burned fuel.

In order to calculate the fraction of the fuel burned $x(\theta)$, it is used in the model the equation obtained by Miyamoto et al. (1985) modified by the correlation presented by Watson et al. (1980).

$$x(\theta)=1-\left[\beta_c.e^{-a\left(\frac{\theta-\theta_c}{\Delta\theta_p}\right)^{m_p+1}}+(1-\beta_c)e^{-a\left(\frac{\theta-\theta_c}{\Delta\theta_d}\right)^{m_d+1}}\right] \quad (16)$$

where $\Delta\theta_p$ = premixed combustion duration; $\Delta\theta_d$ = diffusive combustion duration; θ_c = start of combustion; m_p = form factor of the combustion chamber in the premixed combustion; m_d = form factor of the correlation chamber in the diffusive combustion.

Miyamoto (1985) found the following values for compression ignition engines: $m_p = 3$; $m_d = 1$; $\Delta\theta_p = 10°$; $\Delta\theta_d = 90°$; $a = 4.605$.

$$\beta_c = 1-\frac{a\phi^b}{\tau_{AI}^c} \quad (17)$$

where $0.8 < a < 0.95$; $0.25 < b < 0.45$; $0.025 < c < 0.5$ (Kannan et al., 2009).

10 HEAT TRANSFER

For the calculation of the heat transfer through the cylinder wall, we must have that heat transfer coefficient (h).

The Eichelberg (Shudo et al., 2002) correlation is the following:

$$h(\theta)=2.48*S_P{}^{0.23}*(pT)^{0.5} \quad (18)$$

11 SPECIFIC HEAT CALCULATION

The calculation adopted for the thermodynamic simulation value of c_p (function of T) is the following (Lanzafame et al., 2003):

$$c_p(T)=a_0+a_1(LnT)+a_2(LnT)^2+a_3(LnT)^3 + a_4(LnT)^4+a_5(LnT)^5 \quad (19)$$

The coefficients a_i have different values for CO_2, H_2O, O_2, N_2, CO, NO and SO_2 as stated by Lanzanfame et al. (2003).

12 THE HEAVY FUEL OIL

The equivalent fuel definition (Heywood, 1988) was used in the research, denoted by $C_jH_kO_lN_mS_n$, where C, H, O, N and S are the atoms of carbon, hydrogen, oxygen, nitrogen and sulfur as follows: j = %C/12; k = %H/1; l = %O/16; m = %N/14; n = %S/32.

The incomplete combustion of this fuel can be expressed as:

$$C_jH_kO_lN_mS_n+a(O_2+3.76N_2)\rightarrow n_{CO_2}CO_2+n_{H_2O}H_2O + n_{N_2}N_2+n_{SO_2}SO_2+n_{O_2}O_2 + n_{CO}CO+n_{NO}NO+n_{HC}HC \quad (20)$$

Heywood (1988) presents the solution for the equation balance of this equation and obtains the number of mol for the products and they can be used for γ equivalent of the reagents and the products, taking into account that:

$$\gamma=\frac{c_p}{c_p-R} \quad (21)$$

Then the γ of the mixture will be calculated by:

$$\gamma=\frac{1}{n_t}\sum_i n_i\gamma_i \quad (22)$$

where n_t = total numbers of moles of the mixture; n_i = number of moles of each component; γ_i = relation of specific heat each components.

In the simulation model was adopted the equivalent fuel (γeq), as:

$$\gamma_{eq}=(1-x)\gamma_r+x\gamma_p \quad (23)$$

where x = fraction of the reagents burned in a linear function.

13 SIMULATOR INPUT DATA

- Information of the geometry of engine as D, Rv/L, S, r, etc.
- Operational data of the engine as P_1, T_1, RPM, θ_{inj}, T_p, m_f.
- Combustion data as $\Delta\theta_p$, $\Delta\theta_d$, τ_{AI}, constants a, b and c.
- $c_p(T)$ in chemical reaction.
- Chemical analysis of the fuel as %C, %H, %O, %N, %S, viscosity and density.

14 THERMODYNAMIC MODEL CALCULATION

The equations system is solved by the WOLFRAM MATHEMATICA 7.

The output of the simulation model will be the curve of the cylinder pressure versus crank angle.

Using the curve obtained, W_i will calculated (indicated work) (Heywood, 1988):

$$W_i = \int_{\theta_{fv}}^{\theta_{av}} P(\theta)dV \tag{24}$$

where θ_{av} = crank angle where the exhaust valve opens; θ_{fv} = crank angle where the admission valve closes.

The mean indicated pressure IMP (N/m²):

$$IMP = \frac{W_i}{V_d} \tag{25}$$

where V_d = engine displacement.

The indicated power Pot_i (KW):

$$Pot_i = \frac{IMP * V_d}{120} * Rot \tag{26}$$

Table 4 presents the experimental and simulated values comparison.

For the 50% of engine power the ignition delay of the simulated model presents a value of 5.4°, near the experimental test value of 8° presented in figure 5 for the sample of the bunker fuel.

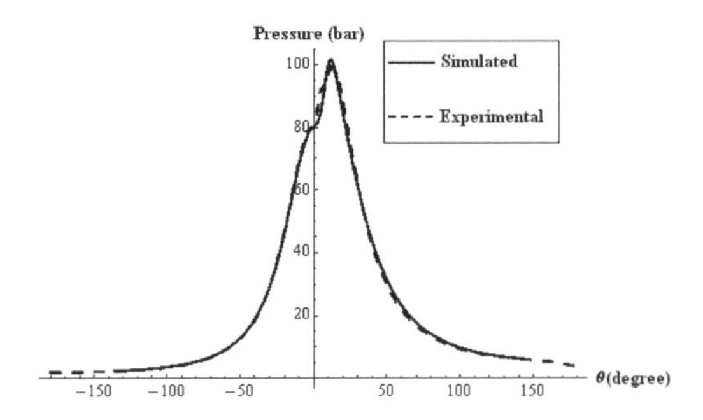

Figure 9. Cylinder pressure versus crank angle for experimental date and simulation model for the HFO at 50% of the maximum power.

Table 4. Performance parameters.

Power		50%	75%
Pmax (bar)	Experimental	99.41	132.3
	Simulated	101.71	136.19
	Difference %	−2.32	−2.94
IMP (bar)	Experimental	11.47	15.6
	Simulated	11.46	15.33
	Difference %	0.09	1.73
W_i (KW)	Experimental	276.74	376.39
	Simulated	276.58	369.85
	Difference %	0.06	1.73

15 CONCLUSIONS

The ignition quality for a Heavy Fuel Oil cannot be determined in the present days estimated formulas (CCAI, CII, etc) for the residual fuels.

It is required to establish a methodology in order to obtain the physical—chemical characteristics of the fuel.

It should be followed by the combustion chamber tests (FIA and engine test for ignition delay determination).

The two last tests are not equivalent, but they useful for rating different fuel samples.

A thermodynamics simulation model can predict the performance of different fuels in a specific engine, as a preliminary performance evaluation.

The FIA Test can be obtained several index and a estimated cetane number that must be in the range of acceptable results (CIMAC, 2010).

There is not a direct correlation between FIA/ECN and ignition delay obtained by engine performance. But, a ranking can be obtained when testing a standard heavy fuel marine oil and comparing the ignition delay obtained with different heavy fuel composition.

The lower is the ignition delay the better is the ignition quality.

16 FUTURE WORKS

The use of more precise models like the one dimensional thermodynamic model—BOOSTER, can predict performance and also emissions of the engine using different fuels. Its results should be compared to experimental data.

The installation in the engine devices like the injection pressure measuring and the lift of the injection valves can provide the point of the injection start. The calculation of the ROHR curve from the cylinder pressure versus crank angle curve will be also useful. Then, the ignition delay can be defined as the time from the start of injection (injection needle star to open and a drop in injection pressure) to start of combustion (rate of heat release has reached a given level).

We can develop a correlation for different fuels samples of FIA-ECN and ignition delay in engines.

The laboratory is also installing a variable speed Wartsila engine connected to a dynamometer in order to perform complementary tests of the heavy fuel marine oil for propulsion engines units for obtaining the complete combustion performance analysis of the fuel.

17 SYMBOLOGY

d	Density
D	Cylinder Diameter
Vk	Kinematic viscosity
ID	Ignition Delay
Max PI	Max Pressure Increase
MCD	Main Combustion Delay
PCP	Pre Combustion Period
MCP	Main Combustion Period
EMC	End of Main Combustion
EC	End of combustion
ABP	After Burning Period
ROHR	Rate of Heat Release
AR	Accumulated ROHR (Surface area)
PMR	Position of Maximum of Heat Release
CP	Combustion Period (positive Area of ROHR)
SMC	Start of Main Combustion
O	Oxygen
N	Nitrogen
C	Carbon
H	Hydrogen
S	Sulfur
IDD	Approximate Diesel Oil Ignition Delay
IDB	Approximate Bunker Ignition Delay
Q	Heat
W	Work
p	Pressure
T	Temperature
θ	Crank Angle
r	Compression Ratio of the engine
V	Combustion Chamber Volume
A	Piston Area
S	Stroke
Rv	Crankshaft Radius
L	Connecting Rod Length
τ_{AI}	Ignition Delay
ϕ	Equivalent Ratio
Rot	Rotation per Minute of the engine
R	Universal Gas Constant
ε_{ef}	Effective Compression Ratio
c	Polytropic Constant
V_d	Displacement Volume
V_{SOI}	Volume in Start
S_p	Piston Speed
NC	Cetane Number of fuel
ECN	Estimated Cetane Number (FIA)
m_f	Fuel Mass
θ_{av}	Crank Angle where the exhaust valve opens
θ_{fv}	Crank Angle where the admission valve closes
IMP	Mean Indicated Pressure
W_i	Indicated Work per cycle

REFERENCES

Assanis, D.N. et al.1999. A predictive Ignition Delay correlation under Steady-State and transient operation of a Direct Injection Diesel Engine. *Journal of Engineering for Gas Turbines and Power*: 450–457.

Brambila, J.A. 2006. Estudo experimental e simulação termodinâmica de desempenho em um motor de Combustão Interna operando com óleo diesel e Etanol. Master's Thesis. University of São Paulo. Brazil.

Bueno, J.P.V.M. 2011. Análise do desempenho de motores diesel utilizando óleo combustível pesado e combustível destilado marítimo. Federal University of Rio de Janeiro. Brazil.

Cimac 2010. Fuel Quality Recommendations at the Engine Inlet-Ignition and Combustion Properties. *CIMAC Working Group Heavy Fuel-WG7.*

Espadafor, F.J. 2009. The viability of pure vegetable oil as an alternative fuel for large ships. *Transportation Research*: 461–469.

Heywood, J.B. 1988. Internal Combustion Engine Fundamentals. New York:McGraw-Hill.

Kannan, K. et al. 2009. Modeling of Nitric Oxide Formation in Single Cylinder Direct Injection Diesel Engine Using Diesel-Water Emulsion. American Journal of Applied Sciences: 1313–1320.

Lanzafame, R. & Messina, M. 2003. ICE gross heat release strongly influenced by specific heat ratio values. *Internal Journal of Automotive Technology*. Vol. 3: 125–133.

Miyamoto, N. et al. 1985. Description and Analysis of Diesel Engine Rate of Combustion and Performance Using Wiebe's Functions. *SAE International Congress and Exposition*. Detriot. Michigan.

Schenk, C. et al. 1983. Ignition Quality of Residual fuel Oils. *Shell International Oil Products.*

Shudo, T. & Suzuki, H. 2002. Applicability of heat transfer equations to hydrogen combustion. *Society of Automotive Engineers of Japan*. Vol. 23: 303–308.

Steernberg, K. et al. 2007. The effects of a changing oil industry on marine fuel quality and how new and old analytical techniques can be used to ensure predictable performance in marine diesel engines. *CIMAC Congress 2007, Vienna.*

Valencia, F.A. & Armas, I.P. 2005. Ignition quality of Residual Fuel Oils. *Journal of Maritime Research.* Vol. II No. 3: 77–96.

Watson, N. et al. 1980. A combustion correlation for Diesel Engine Simulation. *SAE Paper.*

Sustainable Maritime Transportation and Exploitation of Sea Resources – Rizzuto & Guedes Soares (eds)
© 2012 Taylor & Francis Group, London, ISBN 978-0-415-62081-9

COGAS plant as possible future alternative to the diesel engine for the propulsion of large ships

G. Benvenuto, D. Bertetta & F. Carollo
Dipartimento di Ingegneria Navale ed Elettrica, University of Genova, Italy

U. Campora
Dipartimento di Macchine, Sistemi Energetici e Trasporti, University of Genova, Italy

ABSTRACT: Strong restrictions on emissions of engines for marine propulsion (particularly SOx, NOx) will probably be adopted in the near future. In this paper, by using specifically developed simulation techniques, a COGAS plant solution is proposed in substitution of the originally adopted prime mover (a two stroke low speed diesel engine), for the propulsion of a large container ship, whose characteristics and performance are known. Starting from the performance and cycle characteristics of an existing gas turbine, the COGAS steam cycle characteristics and related components are defined through an original design procedure. The combined cycle is compared with a Diesel engine, not only with regard to the plant performance and efficiency at design and high load off-design conditions, in both steady state and dynamic situations, but also for what concerns: engines dimensions and weights, engine room layout, overall propulsion plant weight, ship hull loading capacity, economic considerations.

1 INTRODUCTION

The need to reduce fuel consumption, and therefore the cost of fuel used and, on the other hand, the need to comply with increasingly stringent regulations in terms of limitations on polluting emissions, suggest to check the possibility to transfer, also to the shipping and marine sector, solutions already experimented and proved successful in terrestrial plants for the production of energy, based on the use of gas and steam turbine combined cycles as prime movers.

Indeed, this idea is not new. Combined gas turbine and steam turbine cycles have been industrialized since the 1970s for land-based power plants. Since then, an use of this plant typology has been continuously studied and taken in consideration also for ship propulsion (Merz et al., 1972, Abbott et al., 1974, Abbott et al., 1977, Mills 1977, Brady 1981). However, in recent times, to the authors' knowledge (see also Haglind 2008 and Wiggins 2008), only few vessels have been built adopting this type of propulsion system: two Russian Ro-Ro vessels, powered with a combined cycle plant with direct power transmission (COGAS), and some cruise vessels such as the Celebrity Cruise's Millennium, Infinity and others as well as the Royal Caribbean's Radiance of the Seas, powered by gas and steam turbine combined plants with turbo electric transmission (COGES).

The reason of the low diffusion of combined gas and steam turbine plants for the ship propulsion is due, in authors opinion, mainly to costs considerations (Haglind 2008, Bertetta et al., 2010). The energy conversion efficiency of COGAS plants is in fact comparable to or slightly higher than that of two stroke slow speed diesel engines (the type of engines more widely adopted for the propulsion of large ships). For the COGES plants, efficiencies are expected to be a little lower because of the greater energy losses due to the double energy conversion; for instance considering the "Millennium" ship power plant, the energy utilization varies from 45% to 50%, and so it is a bit lower than that of an equivalent medium-speed Diesel engine (Haglind 2008). On the other hand the higher costs (in terms of equipment and fuel) of the COGAS (or COGES) solution, as compared to the diesel engines solution, counterbalance the many potential advantages of the first solution (mainly: reduced weight and dimensions, maintenance costs, pollutant emissions, etc.).

However, it has to be noted that the ever more attention regarding the environmental impact of the marine engines will cause in the next future a probable severe increase of the emission regulations in the shipping sector (MEPC 2008). This trend will produce, but already is beginning to produce, an increase in equipment costs of diesel engines (because of the apparatus to reduce pollutant emissions, if adopted), but most probably a significant increase in marine fuels costs due to more stringent specifications related to their chemical composition. These considerations, in addition to the intrinsic low emissions of the gas turbines if compared to the slow speed diesel engines (AVIO 1992), can lead to a reflection about the opportunity to reconsider the combined gas and steam turbine plant as an alternative to diesel engines, at least for the propulsion of large ships, where high power levels could allow high levels of thermal efficiency.

To this aim, in this paper the authors propose a quantitative comparison between a COGAS plant specifically designed and the originally adopted prime mover (a two stroke low speed diesel engine) for the propulsion of a large container ship, whose data and performance characteristics are known.

The developed propulsion system simulator allows to compare, for the considered ship, the above mentioned alternatives, in terms of plant performance and efficiency, not only at design but also at high load off-design conditions, in both steady state and dynamic situations. The presented analysis, however, takes into consideration many other not less important aspects, such as: machinery dimensions and weights, engine room layout, maintenance requirements, ship hull loading capacity, preliminary economic evaluation.

2 SHIP EMISSIONS AND REGULATORY FRAMEWORK

A detailed dissertation regarding the typology, origin and consequences of ship engines emissions is made in Haglind 2008, to which the reader is referred. Here it is worth mentioning that, with regard to international legislation on the limitation of ship emissions, promoted by International Maritime Organization (IMO), the most recent Marine Environment Protection Committee meeting (MEPC 2008) approved a proposal of amendment to the MARPOL Annex VI regulations concerning current and projected restrictions on fuel sulphur content and nitrogen oxides engine emissions. The proposed fuel sulphur limits and implementation dates are shown in Fig. 1, where 'Globally' indicates the world-wide average while 'SECAS' is referred to the special SOx Emission Control Areas of the world.

The NOx emissions limits versus the engine maximum operating speed are reported in Fig. 2, where 'Tier I' represents the global limits in force, 'Tier II' the global limits applicable to engines constructed after January 1, 2011 and 'Tier III' the emission limits to be applied to engines constructed after January 1, 2016 for ships operating in NOx Emission Control Areas. In addition, it is to be noted that in some world regions, as for instance: Baltic Sea, European North Sea and English Channel, more restrictive regulations on fuel sulphur content are adopted since 2006.

In this context it is interesting the comparison shown in Tab. 1 (Haglind 2008) between two-stroke, slow-speed diesel engines (prime movers widely adopted for large ships) and gas turbines, in terms of typical emission indices, corresponding

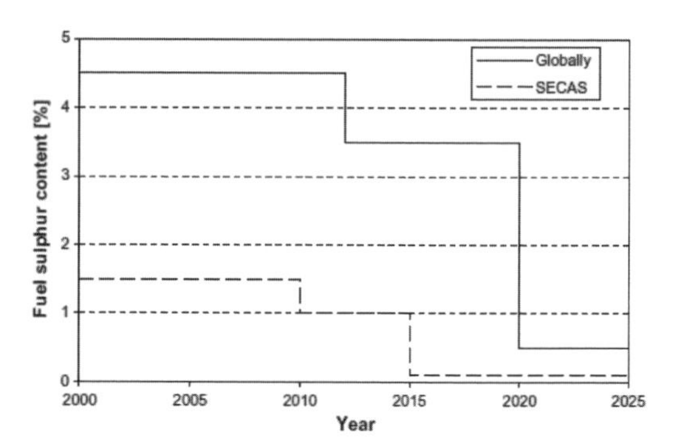

Figure 1. IMO current and projected restrictions on fuel sulphur content.

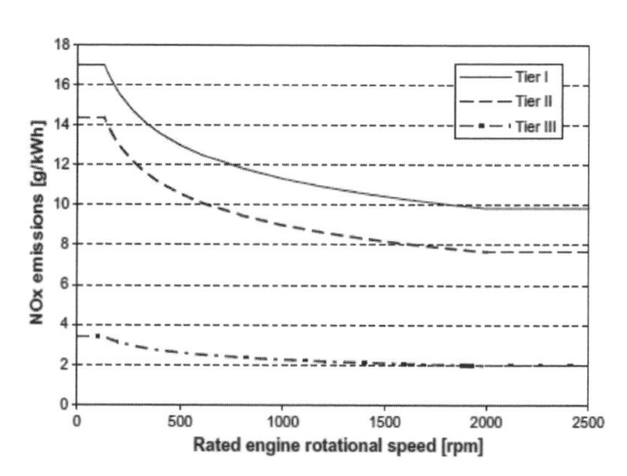

Figure 2. IMO current and projected NO_x emissions limits.

Table 1. Typical emission data for two stroke slow speed diesel engines and gas turbines.

Pollutant emission component	Diesel [g/kg fuel]	Gas turbine [g/kg fuel]	Differnce [%]
Sulphur oxides	54	7.6	–86
Nitrogen oxides	87	28.6	–67
Particulate matter	7.6	1.1	–85
Carbon monoxide	7.4	.14	–98
Hydrocarbons	2.7	0.05	–98

to high power conditions, for different emission components.

The gas turbines data refer to LM 2500+ gas turbine (without emission abatement), prime mover widely used for the propulsion of naval vessels. The presented data show that emissions from gas turbines are very much smaller than those of diesel engines, not only as regards sulphur oxides and nitrogen oxides, but also for all the pollutant emissions considered.

The major difference in emissions of sulphur oxides depends essentially on the percentage of sulphur in the fuels used by the engines in question: Heavy Fuel Oil (HFO) for the two stroke diesel engines with no more than 4.5% mass sulphur content and Marine Gas Oil (MGO) for the gas turbine, with a fuel sulphur percentage no greater than 1% mass, as General Electric (the LM2500 gas turbine manufacturer) requires (AVIO 1992). Obviously diesel engines sulphur oxides emissions can be reduced adopting fuels with lower sulphur content.

As regards the nitrogen oxides emissions of diesel engines, about triple with respect to gas turbines (see tab. 1), they depend on the different fuel combustion conditions of the respective combustion chambers. The nitrogen oxides emissions from the diesel engines can be reduced by adopting specific devices (i.e.: inlet air humidification, water or steam injection in the cylinders, post combustion gas treatment). However it has to be observed that also the gas turbine nitrogen oxides emissions can be further reduced, for instance by steam flow injection in the combustion chamber, as shown in (AVIO 1992).

The greater emissions of particulate matter from the diesel engines, as compared to the gas turbines (tab. 1), are mainly due to the intermittent nature of the diesel engines combustion, that is more inclined to produce this pollutant emission than the steady combustion in the gas turbines (Wiggins 2008). Similar considerations may be done as regards carbon monoxide and hydrocarbons emissions.

3 APPLICATION OF THE COGAS SOLUTION TO A CONTAINER SHIP

As known, today the great majority of prime movers of ocean-going large ships are two stroke slow speed diesel engines, because of their high efficiency (especially under load conditions near to the maximum engine power) and their ability to burn heavy fuel oil (HFO), fuel of low quality and hence of low cost. This choice is the best one, for economic reasons, for ships such as tankers and bulk carriers, used for the transportation of raw materials or goods of low added value. Different is the case of container vessels which, carrying goods with higher added value, semi-finished or finished products, can afford higher costs.

Table 2. Main ship data.

Length overall ($L_{o.A}$)	335 m
Length between perpendiculars (L_{BP})	319,92 m
Breadth (B)	45,32 m
Depth (D)	24,6 m
Draught (T)	14,5 m
Light Ship Weight (LSW)	34435,5 t
Full load displacement	144439,4 t
Engine power	79,55 MW
Ship speed (full load condition)	25,7 knots

As said before, for the propulsion of large container ships, a possible engine alternative to the two stroke diesel engines is the COGAS plant, which has efficiency comparable to or slightly higher, but has a more complex configuration and uses fuel of better quality, thus being penalized because of the higher plant and fuel costs.

However, in view of the emission regulations discussed above, it is reasonable to infer that there will be a necessary increase of the fuel quality for the diesel engines and this situation could favor the choice of the COGAS solution for large container ships also as regards the fuel costs (Haglind 2008).

In this perspective, the application of a COGAS plant specifically designed to a large container ship is proposed and discussed in the next sessions. A quantitative technical comparison with the original adopted prime mover (a two stroke low speed diesel engine) is carried out by using simulation techniques.

The main data of the considered 9000 TEU container ship (Bonfiglio 2009) are reported in the following Tab. 2.

The propeller is driven by a mechanical power transmission without reduction gear. The prime mover is a fourteen cylinders 14 K98MC-7 MAN two stroke low speed diesel engine whose maximum delivered power (MCR load condition) is 79,55 MW at 91 rpm.

4 COGAS PLANT BASIC CHOICES

Nowadays the efficiency of combined gas and steam turbine plants for electricity production is slightly below 60%. This performance is achieved at the expense of some complexity of the plant (mainly steam cycle with more pressure levels), justified by the high power levels and the great space availability. Different is the case of a marine propulsion system, characterized by relatively low power levels (ranging approximately, also in case of very large ships, between 50 and 100 MW) and the need to limit the size, weight and even the costs of the system components. Because of these requirements, COGAS plants with only one pressure level for the steam cycle have been proposed (Haglind 2008, Wiggins 2008), with a reduction of the energy conversion efficiency to values slightly above 50%, but still competitive with those of the two stroke diesel engines. In the paper the design procedure will be limited to the case of the single pressure level, but further development could also include the case of two pressure levels.

As far as the gas turbine is concerned, the choice has been oriented on an engine available in trade, the LM 2500 gas turbine, whose reliability is demonstrated by its large diffusion for the propulsion of naval ships.

Starting from the gas turbine cycle characteristics, the steam cycle features and related plant components are here defined by means of an original procedure described in the next sessions.

In the considered container ship the engine is located near the stern, connected to the propeller by a shaft line without reduction gear. The alternative propulsion system foresees the substitution of the original two stroke low speed diesel engine with a COGAS plant connected to the propeller through a reduction gear and a shaft line as shown in the scheme of Fig. 3.

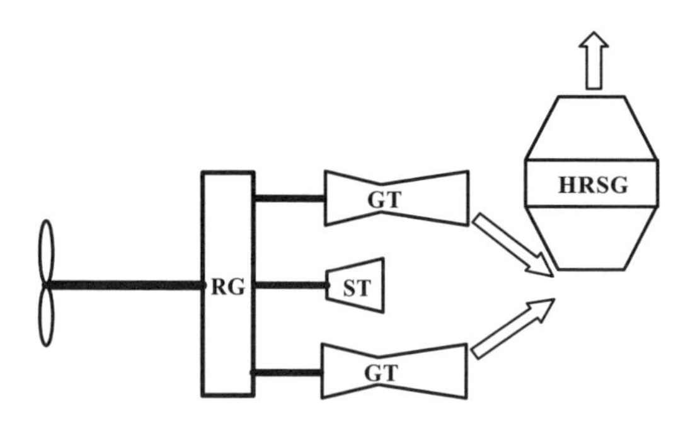

Figure 3. Scheme of a COGAS propulsion system with a mechanically driven propeller.

The choice of a mechanical transmission (COGAS) instead of an electric transmission (COGES) is preferable for the considered container ship because it reduces the complexity of the system and has a not negligible higher energy conversion efficiency.

Coming back now to the aforementioned comparison between the two alternative solutions (COGAS plant and diesel engine) for the propulsion of the selected ship, in what follows a brief description is given of the propulsion plant simulator, developed in MATLAB®-Simulink® environment, used to carry out the necessary quantitative evaluations not only at design conditions but also at off-design and dynamic situations.

5 GAS TURBINE SIMULATOR

The General Electric LM 2500 marine gas turbine is a single cycle, two shafts, high performance engine, of the aero-derivative type, consisting of a gas generator and a power turbine, whose main data, in the selected version, are as follows: 26.6 MW output power at 3600 rpm with a specific fuel consumption of 247 g/kWh.

By summing the powers delivered by two LM 2500 gas turbines with that obtainable by the steam turbine, according to the scheme of Fig. 3, it is possible to reach a power level very similar to that of the original two stroke diesel engine of the ship.

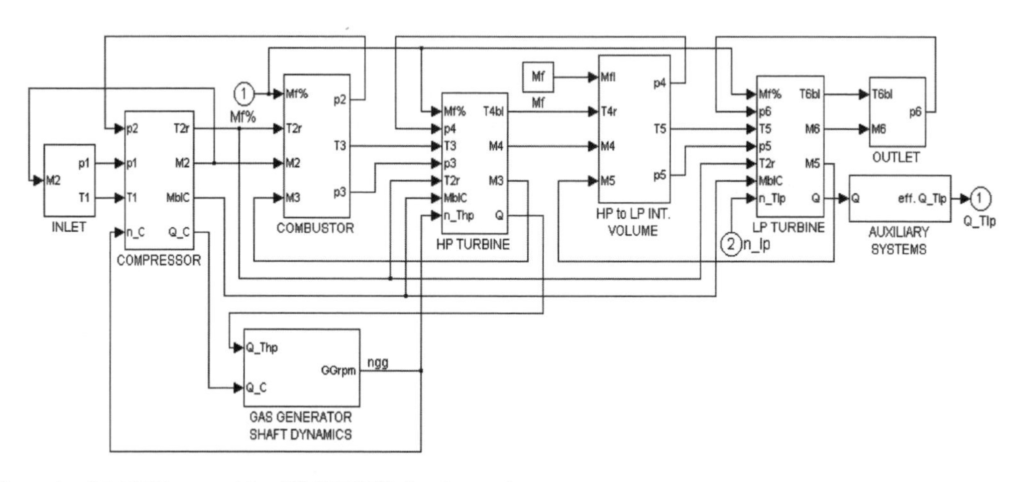

Figure 4. LM 2500 gas turbine SIMULINK simulator scheme.

The gas turbine simulation model, already presented in Benvenuto et al., 2005, is structured in a modular arrangement, each module being assigned to a gas turbine component (i.e.: compressor, turbine, combustor and so on) as shown in the SIMULINK scheme of Fig. 4. A short description of the gas turbine model is reported here, while for a more detailed explanation the reader is referred to the cited reference. In the compressor and turbines modules the steady state performance maps (Cohen 1987) are used. The combustor module is modeled as an adiabatic capacity, taking into account the time-dependent accumulation of mass and energy. In the shaft dynamics module the time variation of the shaft angular velocity is determined by the classic shaft dynamic balance equation.

A detailed validation of the gas turbine simulation model was reported in Benvenuto et al., 2005. Good agreement between experimental values and simulation results, in case of both steady state and transient situations, was observed (differences between simulated and reference data were within 0.4 to 3.5 per cent in the full gas turbine working range).

6 STEAM PLANT

Two specific computer codes have been developed in MATLAB®-Simulink®, the first is to define the steam cycle parameters, in order to obtain the maximum COGAS plant efficiency, and consequently to design the steam plant components (i.e.: Heat Recovery Steam Generator (HRSG), turbine, condenser). The second code, starting from the design data of the steam plant components (i.e.: turbine characteristics, pipes section and wall thickness, heat exchange coefficients), determines the steam plant performance at both design and off design load conditions. This second code works also in conjunction with the gas turbine simulator (described above) in order to simulate the COGAS plant behaviour in every steady-state and dynamic condition.

As already said, in this phase the analysis is limited to a steam generator with only one pressure level without reheater. In Fig. 5 a scheme of the considered steam plant is shown, where only the main components of the system are visualized.

Leaving the gas turbine (GT), the exhaust gas passes through a HRSG of drum type, consisting of three distinct heat exchangers: the economizer (ECO), the boiler (evaporator (EVA) connected to an external steam drum (SD)) and the superheater (SH). Steam leaving the superheater expands through a turbine (ST) and is exhausted to a condenser.

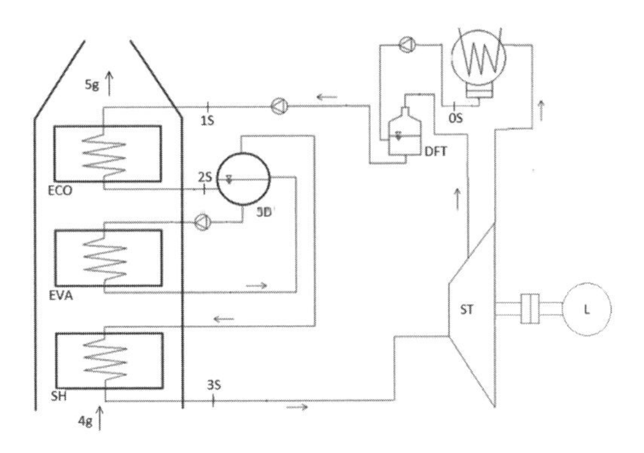

Figure 5. Steam plant scheme with single pressure HRSG.

The condensate pump moves the water to the Deareating Feed Tank (DFT) and the feed pump moves it to the economizer. This scheme is similar to that of a conventional steam plant, with the difference that in this case no fuel is burned in the steam generator.

6.1 *Steam cycle*

A short description of the calculation logic used to define the steam cycle parameters is given here, while a more detailed explanation of the developed computer code is reported in Bertetta et al., 2010.

Starting from the condenser, as known, a low steam condensing pressure contributes to raise the cycle efficiency, but this pressure depends on the cooling water temperature. In case of steam cycles for marine propulsion, taking into account a possible higher temperature of the cooling water and the advisability to reduce the condenser size and weight, a value of 0.08 bar could be a reasonable choice. As concerns the HRSG design variables, the chosen superheater outlet pressure (50 bar) is a compromise between the maximum attainable combined cycle efficiency and the costs of the steam generator pressurized components (Abbott 1974, Wiggins 2008).

With regard to temperature differences between gas and steam in the superheater and the economizer (ΔT approach point and ΔT pinch point, ΔTap and ΔTpp respectively in Fig. 7), the following values, typical of single pressure level steam generators, have been assumed: ΔTap = 49°C and ΔTpp = 9°C.

Starting from the above mentioned data and knowing the water temperature at the economizer inlet ($T_{1\,s}$ in Fig. 5), the gas turbine mass flow rate and exhaust temperature ($T_{4\,g}$ in Fig. 5), it is a simple matter to apply continuity and energy equations to economizer, boiler and superheater in order to obtain the steam mass flow rate and the gas temperature ($T_{5\,g}$) at the HRSG outlet.

In the Deareating Feed Tank (DFT in Fig. 5) the steam drawn from the turbine expansion warms the water introduced in the economizer. The pressure of the steam bled from the turbine is chosen in order to guarantee that its value remain greater than atmospheric pressure also at part load conditions.

The steam turbine power is then determined from the isentropic enthalpy drop and turbine efficiency.

In Tab. 3 the main COGAS plant data are reported.

6.2 *Steam plant components design*

A HRSG with vertical gas flow, mounted directly over the gas turbine has been selected, allowing to minimize ground floor and space requirements. The drum-type adopted solution foresees a circulation pump for the boiler and a disposition of the heat exchangers as shown in Fig. 5. The three heat exchangers (economizer, evaporator and superheater) are equipped with horizontal finned tubes and present a cross-counter flow configuration.

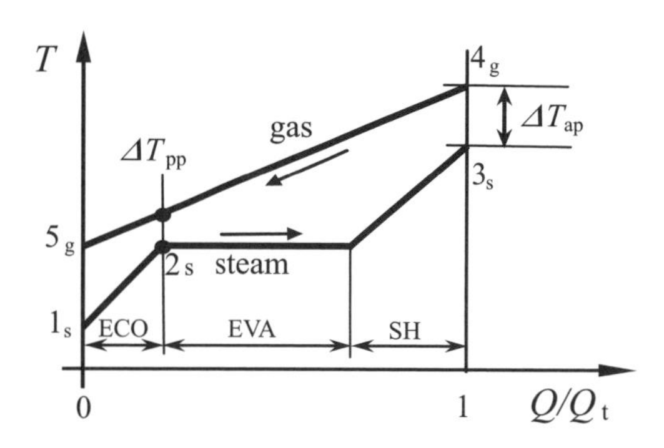

Figure 6. Typical temperature diagram for a single pressure HRSG.

Table 3. Main COGAS plant data.

Superheater pressure	50 bar
Condenser pressure	0.08 bar
DFT pressure	3 bar
Superheater temperature	517°C
Superheater mass flow	23.08 kg/s
Steam turbine power	24.9 MW
Steam Turbine efficiency	0.88
GT exhaust gas mass flow rate	167.86 kg/s
GT exhaust gas temperature	567.3°C
Stack exhaust gas temperature	211.3°C
Lower heating value of fuel	42800 kJ/kg

The heat exchange between the hot fluid and the cold fluid is determined, for each heat exchanger of the HRSG, with the procedure here summarized, described more in details in Bertetta et al., 2010.

The global heat exchange coefficient may be calculated as follows:

This coefficient may be calculated as follows:

$$\frac{1}{k_e} = \frac{1}{h_e} + R_e + \frac{s}{k}\frac{A_e}{A_{ml}} + R_i\frac{A_e}{A_i} + \frac{1}{h_i}\frac{A_e}{A_i} \tag{1}$$

where the various terms, which represent thermal resistances, are defined in nomenclature. Convective heat exchange coefficients are determined using correlations between non-dimensional numbers and k_e is evaluated taking into account the efficiency of finned tubes.

Once defined steam and gas temperatures in the various sections of the HRSG, for each heat exchanger, the total heat exchange area (A), necessary for the pipes length evaluation, is determined by means of the following heat exchange balance equation:

$$Q' = kFA(\Delta T)_{g-s} \tag{2}$$

where $(\Delta T)_{g-s}$ is the logarithmic mean temperature difference between gas and steam.

Knowing the thermal exchange area A it is possible to evaluate the total length of the pipes and so the dimension of the HRSG and the steam pressure losses. The design of the steam plant condenser is carried out with a procedure very similar to that adopted for the HRSG heat exchangers.

The procedure developed so far allows to obtain also the steam mass flow rate and the turbine enthalpy drop, from which, by considering the turbine efficiency, it is possible to determine the power delivered.

6.3 *Steam plant simulator*

As said before, a second computer code has been developed to determine the steam plant performance at both design and off design load conditions. Starting from the design data of the steam plant components (i.e., turbine characteristics, pipes section and wall thickness, heat exchange coefficients), this simulation code determines the steady state and dynamic performance of the steam plant by using a calculation procedure re-arranged and, in some way, inverted if compared with the design procedure.

More precisely the dynamic behaviour of the steam plant has been modelled by applying to the heat exchangers continuity and energy equations and taking into account the thermal inertia of the pipes walls. This second code may be used in conjunction with the gas turbine simulator allowing to evaluate the overall COGAS plant performance at any load condition.

In order to validate the developed simulation code, a comparison has been done between the results of the authors' code, on a specific single pressure HRSG of known characteristics, and those obtained by Bracco et al., 2007 with a different code, based on a similar but non coincident approach and already validated by experimental results.

A very good agreement between the results obtained respectively in correspondence of two examined load conditions of the turbine is observed (Bertetta et al., 2010).

7 SHIP PROPULSION SYSTEM SIMULATOR

In order to verify the applicability of the proposed COGAS solution to the considered container ship not only in correspondence of steady state design conditions but, obviously, also when the ship has to face different off-design and dynamic situations (typically during ship maneuvers), a simulation procedure of the propeller-hull system has to be developed and used in combination with the procedure describing the behavior of the prime mover. The authors have a long experience in the development of ship propulsion systems simulators (Benvenuto et al., 2000, Campora et al., 2003, Altosole et al., 2004, Altosole et al., 2009), therefore the interested reader is referred to the cited papers for a detailed description of the developed procedures and of the relative MATLAB®-Simulink® codes. Here only a brief synthesis is given of the simulator.

A generic ship propulsion plant may be considered made up by three main components: the engine, the transmission gear, the propulsor. The propulsor behavior (generally a screw propeller or a waterjet) is strongly influenced by the ship dynamics, primarily by the ship speed. Propulsion plant dynamics and ship dynamics can be represented by two equations of motions, derived from Newton's laws.

The propulsion plant is controlled by a control system that normally provides a set point for the rotational speed of the engine; the engine governor compares the desired and actual shaft speeds and adjusts the fuel flow rate in order to maintain the desired shaft speed. The behavior of the control system may be represented by a first order differential equation of the fuel flow m_f, as function of the shaft speed and its derivative. The dynamic performance of the ship may be obtained from the simultaneous solution of the three, previously described, non-linear differential equations in the time domain.

8 COGAS OR DIESEL: COMPARING TWO ALTERNATIVE SOLUTIONS

The main data of the two solutions considered for the propulsion of a 9000 TEU container ship (see Tab. 2) are summarized here.

Diesel solution:

A fourteen cylinders 14K98MC-7 MAN two stroke low speed diesel engine developing a maximum continuous power of 79.55 MW at 91 rpm.

COGAS Solution:

Two LM 2500 General Electric gas turbines developing 26.2 MW at 3600 rpm combined with a steam turbine developing 24.9 MW at 6300 rpm (see Tab. 4 for the steam plant data) for a total maximum power of 77.3 MW.

It is to be observed that the HRSG and chimney stack pressure losses increase the gas turbine exhaust pressure, thus reducing the LM 2500 power to 26.2 MW instead of the original 26.6 MW.

8.1 *Engines sizes and weights*

The data presented in Tab. 4 show clearly that the two stroke low speed diesel engine is characterized by sizes and weights much higher than those of the turbines. This is true also considering that the COGAS plant uses two gas turbines in order to obtain a total power of 77.3 MW, very similar to that delivered by the diesel engine (79.55 MW).

A further comparison of the considered alternative solutions in terms of weights and sizes of the overall propulsion system will be presented later.

8.2 *Engine room layout*

From the comparison above we can see that, in addition to the total weight, also the sizes relative to the COGAS plant elements are smaller and, anyway, the subdivision in different components allows a more flexible arrangement and a substantial reduction of the engine room dimensions, as shown in Fig. 7.

Table 4. Data on the considered prime movers.

Engine data	14 K98MC-7 diesel engine	LM 2500 gas turbine	Steam turbine
Length [m]	29.05	8.23	4.25
Width [m]	10.11	2.74	3.6
Height [m]	15	3.05	2.24
Weight [t]	2600	25	35
Power [MW]	79.55	26.2	24.9

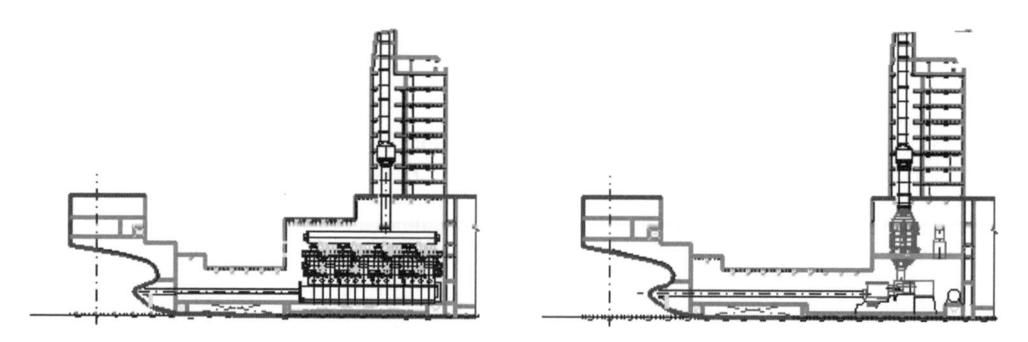

Figure 7. Ship's stern section with diesel engine (left) and COGAS plant (right).

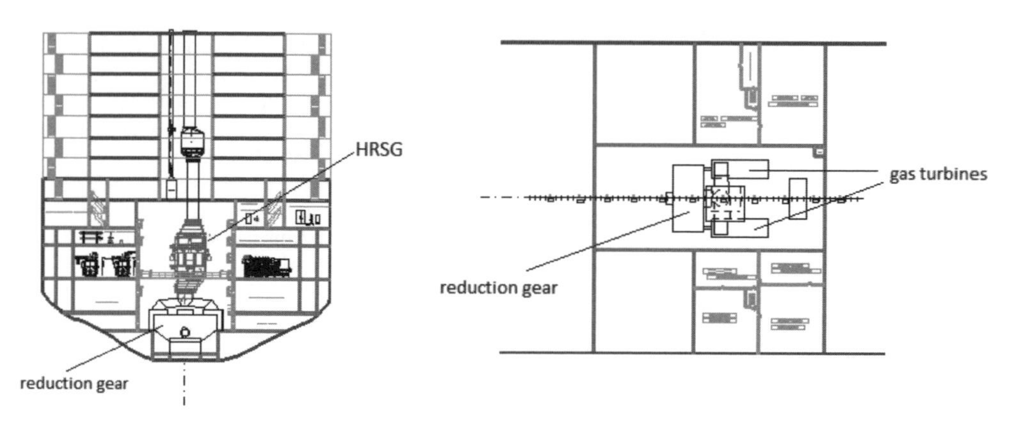

Figure 8. Engine room layout of the COGAS plant.

It has to be noted that the COGAS plant layout is conditioned by the choice of using the same chimney stack passage of the diesel solution, only adjusting the pipe section. For better clarity, the engine room layout of the COGAS plant is shown in Fig. 8 with a view from the propeller (left) and a plan view (right).

It is evident from Fig. 7 the ship's bulk increase allowed by the COGAS plant solution, as a consequence of the engine room size reduction and the lower weights of the plant components.

8.3 *Overall propulsion plant weight*

With reference to the considered alternative solutions, in Tab. 5 the weights of the propulsion plant components are shown, together with the total weights of the propulsion plants. From the table it is clear that, despite the greater number of components of the COGAS plant (i.e.: two gas turbines and one steam turbine, HRSG, condenser, reduction gear) and despite a longer and heavier shaft line, imposed by the choice of using the already existing chimney stack passage, the COGAS propulsion system is about 1900 tons lighter than the original diesel engine propulsion system.

Table 5. Comparison of the weights.

Propulsion plant component	Diesel engine	COGAS
14K98MC-7 diesel engine [t]	2600	
LM 2500 gas turbine (x2) [t]		50
Steam turbine [t]		35
Auxiliaryes [t]	290	290
Propeller [t]	142	142
Shaft line [t]	312	359
HRSG [t]		197.7
Condenser [t]		62.7
DFT [t]		13.8
Reduction gear [t]		278
Total weight [t]	3344	1428

8.4 *Ship's load capacity*

As shown in Fig. 7 and in Tab. 5, the reduced dimensions and weights of the COGAS plant solution allow a ship's load capacity increase of 1.25% in weight as reported in Tab. 6, where some parameters are included (LCG, TCG and VCG) showing

Table 6. Influence of the propulsion system choice on the ship's load capacity.

Ship data	Diesel engine	COGAS
LSW	34435.5 t	32520 t
Containers	90598,9 t	91729,3 t
Total tanks	19405 t	20190,1 t
Displacement	144439 t	144439,4 t
Full load LCG	147,12 m	147,12 m
Full load TCG	0 m	0 m
Full load VCG	19,484 m	19,468 m

that the proposed propulsion system substitution does not affect the vessel stability.

With regard to the ship trim variation, due the remarkable weight difference of the two engine solutions here compared (Tab. 5), particularly in case of unloaded ship conditions (see LSW in Tab. 6), this problem can be easily solved by ballast tanks.

8.5 *Propulsion plant performance*

Thanks to the developed ship propulsion simulator, the COGAS plant performance can be determined also at part load conditions. For the MAN two stroke diesel engine the performance in off design conditions is easily evaluated by using the procedures proposed by the manufacturer. In order to compare the performance of the considered propulsion systems, two load conditions are investigated, the Maximum Continuous Rating (MCR) and the Continuous Service Rating (CSR). For the COGAS plant, these two conditions are shown in the engine load diagram of Fig. 9, where also the curve of the required propeller power (including also the sea margin) is visualized. For the diesel engine the MCR and CSR load conditions are similar, as shown in Tab. 7. The data reported in the same table show that the specific fuel consumptions of the COGAS plant, referred to the engine power (ESFC), are always lower than those of the diesel engine, and that this COGAS plant advantage is reduced considering the specific fuel consumption referred to the delivered power (DSFC).

This is due to the greater COGAS plant friction losses caused by the reduction gear.

Table 7 shows also that the COGAS plant and the diesel engine produce equivalent effects in terms of ship speed and propeller speed.

As regards operational issues, Haglind (2008) points out that the COGAS plant is characterized by less vibration and noise level if compared to the diesel engine, while with regard to the start-up ability, the same author observes that in both engine types, this operation is very much dependent on engine parts temperature before the start-up.

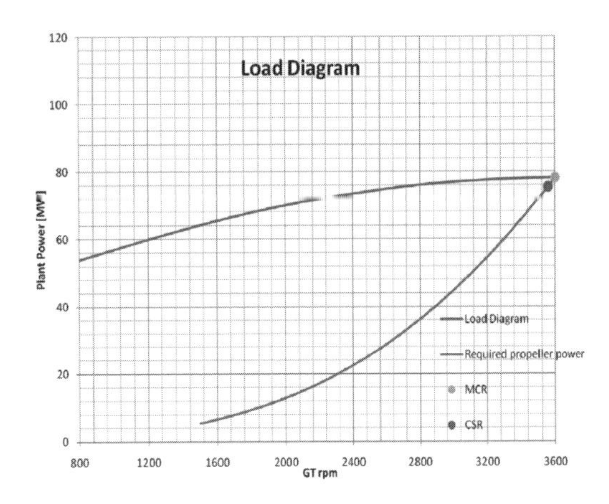

Figure 9. COGAS plant maximum load diagram and required propeller power curves.

Table 7. Diesel and COGAS performance comparison.

	Diesel engine		COGAS plant	
Ship data	CSR	MCR	CSR	MCR
Engine power [kW]	73660	79550	75339	77830
Delivered power [kW]	72926	78754	73079	75495
ESFC [g/kWh]	165.8	168.3	161.1	159.8
DSFC [g/kWh]	167.5	170	166	164.7
Propeller speed [RPM]	89.8	91.7	89.8	90.8
Ship speed [knots]	25.7	26.3	25.7	26

8.6 *Engine maintenance*

The required maintenance by crew personnel for the COGAS plant, according to Haglind (2008), is significantly less than for a typical marine diesel engine. The maintenance of the gas turbine (the most delicate plant component) can be summarized in a periodic (every 800 ÷ 1000 hours) compressor washing, necessary to maintain at high levels the gas turbine efficiency. With regard to the LM 2500 gas turbine Wiggins (2008) indicates, as further periodic maintenance intervention, a service of internal components visual inspection to be done approximately every 5000 hours.

It is at the authors knowledge that in an existing COGAS plant for electric energy and cogeneration production, using a LM 2500 gas turbine, a hot section repair interval of 25000 hours is scheduled. Normally this type of maintenance is carried out in a workshop, substituting the gas turbine under maintenance with an equal one, in order to maintain the plant operativeness. In the same time interval a light maintenance of the steam plant components can be performed, while a more important maintenance of these components can be required at time intervals also greater than 70000 hours.

On the other hand, according to information available to the authors, the normal maintenance service of a typical two stroke diesel engine, can be summarized as follows: an injection inspection every 3000 operative hours, a turbocharger and piston rings inspection every 6000 ÷ 8000 hours and, with regard to the piston rings, the substitution after 18000 hours. An important engine maintenance that requires the substitution of main engine components (i.e.: exhaust valves, cylinder cover, piston crown, main bearings, injection pump, turbocharger) is normally foreseen after 72000 ÷ 90000 hours.

8.7 *Preliminary economic analysis*

It has to be noted that it is very difficult to present reliable cost figures of marine COGAS plants, because costs of components are not always known and fuel costs are subjected to wide variations difficult to foresee. On the other hand, a detailed costs analysis is beyond the scope of this study. Presently the cost of a 14K98MC-7 MAN two stroke low speed diesel engine is about 21.3 MUSD, while the cost of a COGAS plant (including the reduction gear), calculated by summing the components costs, indicated by various manufacturer contacted by the authors, is comprised between 1.7 and 1.9 times that of the 14K98MC-7 MAN diesel engine.

With regard to operating costs, even if the specific fuel consumption of the two alternative propulsion plants is very similar in their typical operative conditions (see Tab. 7), the COGAS plant choice is presently not advantageous because of the price difference between the diesel HFO and the gas turbine distillate fuel. However, apart from the fuel cost, the lubricating oil consumption of a gas turbine is typically 1% of that of a two stroke diesel and, with regard to maintenance costs, the COGAS solution is convenient (Haglind 2008). Furthermore, as shown earlier, also the increased ship's load capacity allowed by the COGAS plant should be taken into account for an overall economic evaluation.

9 CONCLUSIONS

The conclusions that can be drawn from the presented analysis can be summarized as follows:

- With regard to ship emissions, as shown in Tab. 1, the quantities of pollutant emissions from gas turbines (essential components of the COGAS plant) are substantially lower than those of two stroke low speed diesel engines.
- The COGAS plant results more advantageous in terms of weights and dimensions if compared to a two stroke Diesel engine of equal power, allowing a more compact engine room to be adopted, with a consequent increase of the ship capacity (see Figs. 7–8 and Tab. 6).
- As concerns the propulsion plant efficiency, a slight COGAS plant advantage can be observed in terms of specific fuel consumption.
- The levels of vibration and noise of the COGAS plant are much lower than those of the diesel engine.
- With regard to engine maintenance, it appears that the required maintenance by crew personnel for the COGAS plant is significantly less than for a typical marine diesel engine. As concerns more deepened maintenance, the COGAS plants needs typically interventions after about the same running hours of the two stroke diesel engine.
- The costs of installation and operation are still in favour of the diesel engine. This is due to the greater complexity of the COGAS plant and to the greater gas turbine fuel cost. However, the adoption of stringent emission regulations in the shipping sector will produce, but already is beginning to produce, an increase in equipment costs of diesel engines (because of the apparatus to reduce pollutant emissions, if adopted), but most probably a significant increase in marine fuels costs due to more stringent specifications related to their chemical composition. This situation could significantly reduce the difference in the costs of installation and operation of the two examined propulsion plant solutions.

The points highlighted above allow to conclude that the COGAS plant could represent in a near future an advantageous, innovative alternative to the traditional two stroke diesel engine for the propulsion of large ships.

ACKNOWLEDGMENTS

The authors wish to gratefully acknowledge Eng. Stefano Bracco for his valuable support in the HRSG simulator development.

NOMENCLATURE

A_e pipe wall external area
A_i pipe wall internal area
A_{ml} pipe wall logarithmic mean area
h_e external pipe convective heat transfer coefficient

h_i internal pipe convective heat transfer coefficient
k wall thermal conductivity
LCG Longitudinal Center of Gravity
LSW Light Ship Weight
Q' heat flow
F exchanger geometry correction factor
R_e fouling external pipe thermal resistance
R_i fouling internal pipe thermal resistance
S HRSG pipe wall thickness
T temperature
TCG transversal Center of gravity
VCG vertical Center of Gravity

REFERENCES

Abbott W.J. & Baham G.J. 1974. COGAS-A New Look for Naval Propulsion, Naval Engineers Journal, Oct., pp. 41–55.

Abott J.W., McIntire J.G. & Rubis C.J. 1977. A Dynamic Analysis of a COGAS Propulsion Plant, Naval Engineers Journal, 89, pp. 19–34.

Altosole M., Benvenuto G., Campora U. & Figari M. 2004. Performance Prediction of a Waterjet Propelled Craft by Dynamic Numerical Simulation, Waterjet Propulsion 4 International Conference, London, UK, 26–27 May, 2004.

Altosole M., Benvenuto G., Campora U. & Figari M. 2009. Real-Time Simulation of a COGAG Naval Ship Propulsion System, Proc. Instn Mech Engrs Vol 223 Part M: Journal of Engineering for the Maritime Environment, ISSN: 1475-0902.

AVIO 1992. S.p.A., Report LME088ed, Torino, Italy, December, 1992.

Benvenuto G. & Campora U. 2005. A Gas Turbine Modular Model for Ship Propulsion Studies, HSMV, 7th Symposium on High Speed Marine Vehicles, Naples, Italy, 21–23 September, 2005.

Benvenuto G., Campora U., Carrera G. & Figari M. 2000. Simulation of Ship Propulsion Plant Dynamics in Rough Sea, 8 International Conference on Marine Engineering Systems (ICMES/SNAME 2000), New York, USA, May 22–23.

Bertetta D. & Carollo F. 2010. Dynamic Analysis of a COGAS Plant for Ship Propulsion, Naval Engineer Graduate Thesis, Genoa University, Engineering Faculty, Dipartimento di Ingegneria Navale ed Elettrica (DINAEL) Italy, July 16, 2010, in Italian.

Bonfiglio L. 2009. Progetto di massima di una nave portacontenitori da 9000 TEU, Naval Engineer Graduate Thesis, Genoa University, Engineering Faculty, Dipartimento di Ingegneria Navale ed Elettrica (DINAEL) Italy, Dec. 18 2009, in Italian.

Bracco S., Crosa G. & Trucco A. 2007. Dynamic Simulator of a Combined Cycle Power Plant. Focus on the Heat Recovery Steam Generator, ECOS 2007, 20th International Conference on Efficiency, Cost, Optimization, Simulation and Environmental Impact of Energy Systems, Padova, June 25/28. Vol. 1, pp. 189–196.

Brady E.F. 1981. Energy Conversion for Propulsion of Naval Vessels, Naval Engineers Journal, April, pp. 131–144.

Campora U. & Figari M. 2003. Numerical Simulation of Ship Propulsion Transients and Full Scale Validation, Proc. Instn Mech Engrs Vol. 217 Part M: Journal of Engineering for the Maritime Environment, ISSN: 1475–0902.

Cohen H., Rogers G.F.C. & Saravanamuttoo H.I.H. 1987. Gas Turbine Theory (Third Edition), Longman Scientific & Technical, Harlow, Essex, England.

Haglind F. 2008. A Review on the Use of Gas and Steam Turbine Combined Cycles as Prime Movers for Large Ships. Part I-II-III, *Energy Conversion and Management*, vol. 49, pp. 3458–3482.

Marx J. 2006. Analysis of combined gas turbine and steam turbine (COGAS) system for marine propulsion by computer simulation, Doctor of Philosophy in Marine Engineering Degree Thesis, School of Marine Science and Technology, University of Newcastle upon Tyne, September 2006.

MEPC. 2008. Report of the Marine Environment Protection Committee on its Fifty-Seventh Session. International Maritime Organization (IMO), Marine Environment Protection Committee (MEPC), 57th session, April 7, 2008.

Merz C.A. & Pakula T.J. 1972. The design and Operational Characteristics of a Combined Cycle Marine Power Plant, ASME Paper 72-GT90.

Mills R.J. 1977. Greater Ship Capability with Combined-Cycle Machinery, Naval Engineers Journal, Oct., pp. 17–25.

Nuovo Colombo, 'Manuale dell'Ingegnere' 1994. Ulrico Hoepli Editore, Milano, Italy, 82a edition.

Wiggins E.G. 2008. COGAS Propulsion for LNG Ships, SMTC-004-2008.

Sustainable Maritime Transportation and Exploitation of Sea Resources – Rizzuto & Guedes Soares (eds)

One solution for all emission challenges? An overview about the reduction of exhaust emissions from shipping

Dirk Thum
Head of Emission & Gas, Marine Medium Speed, MAN Diesel & Turbo SE, Denmark

Robert Brendel
Marine Medium Speed, MAN Diesel & Turbo SE, Denmark

ABSTRACT: In the light of the new IMO emission regulations, several options can be chosen to reach the requested limits. These options contain low-sulphur fuels or after treatment systems for desulphurization and SCR or exhaust gas recirulation to cope with the NO_x limits. The right technology of choice largely depends on the vessels properties and its operation profile. 4-stroke vessels operating only a limited time in an ECA, the right technology will be chosen based on the lowest investment costs. Then, the use of an SCR for NO_x-reduction while burning MGO is recommended. Increasing the time spent in an ECA will move the focus more to operating costs. Therefore the combination of cheap fuels like HFO with exhaust gas after treatment becomes reasonable. Additionally, the dual fuel option will be more of interest and due to its low fuel costs, it can be recommended for a lot of different applications in the foreseeable future.

1 INTRODUCTION

Shipping is by far the most efficient and environmental friendly form of transportation per transported ton of cargo. Due to the tremendous amount of cargo shipped, the overall emission generated by shipping does have a significant impact of the overall manmade emissions. In the recent years, shipping came more and more under pressure to reduce its emissions, especially focused on NO_x and SO_x. In line with land-based efforts to reduce air emissions from various sources, the International Maritime Organization (IMO) is setting limitations within MARPOL Annex VI for more than ten years now.

In 2008, the regulations were tightened to reduce sulphur oxides (SO_x) and nitrogen oxides (NO_x) emitted by internal combustion engines further. In addition, a time schedule was established to implement the different emission levels for NO_x and SO_x (based on the fuel sulphur content) in steps until 2020, Figure 1.

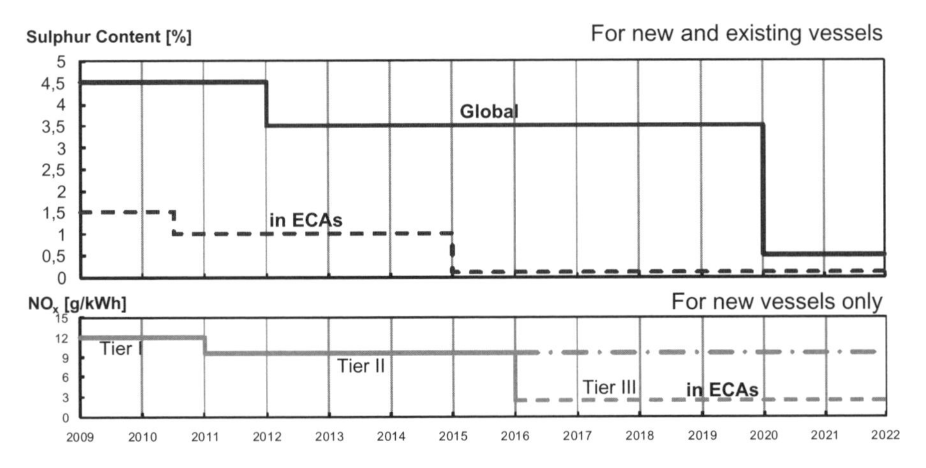

Figure 1. IMO emission limit implementation schedule.

In a first step, these new-founded limitations require a NO_x reduction by 16%–22% depending on the engine's speed. The so-called TIER II stage is already in force for vessels with a keel-laying date on and after 01.01.2011. In 2016, the next step will lead to a severe additional reduction for vessels build on and after 1st January 2016. In total, the NO_x reduction stage TIER III will lead to an emission reduction of about 80% compared to the emission TIER I.

In contrary to former regulations, TIER III is only limited to dedicated areas, the so-called Emission Controlled Areas, Figure 2. These areas are namely the North and Baltic seas and since August this year, the 200 nm continental coastal areas of the USA and Canada shown in the grey areas in Figure 2. All vessels entering an ECA have to comply with the IMO emission limits irrespective of their flag.

SO_x limitations are also differentiated in- and outside of emission control areas. Especially the 0.1% S-limit in an ECA and the 0.5% S limit globally from 2020 on cannot be reached with residual fuels without additional measures. These additional measures like scrubbers are able to fulfill the regulations in case the emissions at the funnel end are equivalent to those of suitable fuels. In contrary to the NO_x-limitations, the SO_x-limits are valid for all vessels irrespective the date of its keel-laying. Therefore, meeting this requirement is valid for all ship owner and charterers operating vessels in such areas.

Table 1. IMO NO_x-limits [g/kWh].

Rated engine speed	n < 130	130 ≤ n ≤ 2000	n > 2000
Tier I	17.0	$45n^{-0.2}$	9.84
Tier II	14.36	$44n^{-0.23}$	7.66
Tier III	3.4	$9n^{-0.2}$	1.97

2 EMISSION REDUCTION TECHNOLOGIES

To comply with the above mentioned regulations, several technologies are on the table.

2.1 *SO_x reduction techniques*

SO_x emissions cannot be influenced by the combustion process of internal combustion engines. Therefore, reducing of SO_x can only be realized on two different ways—using low sulphur fuels or using residual fuels in combination with an after treatment technology.

MGO as the liquid fuel of choice for fuel sulphur limits of about 0.5%S respectively 0.1%S has quite.

some advantages compared to residuals. This clean fuel will lead to a cleaner combustion resulting in fewer deposits in the engine. The fuel treatment is much less complicated and the installed equipment is minimized compared to any other measure including the today's use of HFO. Unfortunately, its significant higher prize compared to HFO or LNG raises the question about alternatives.

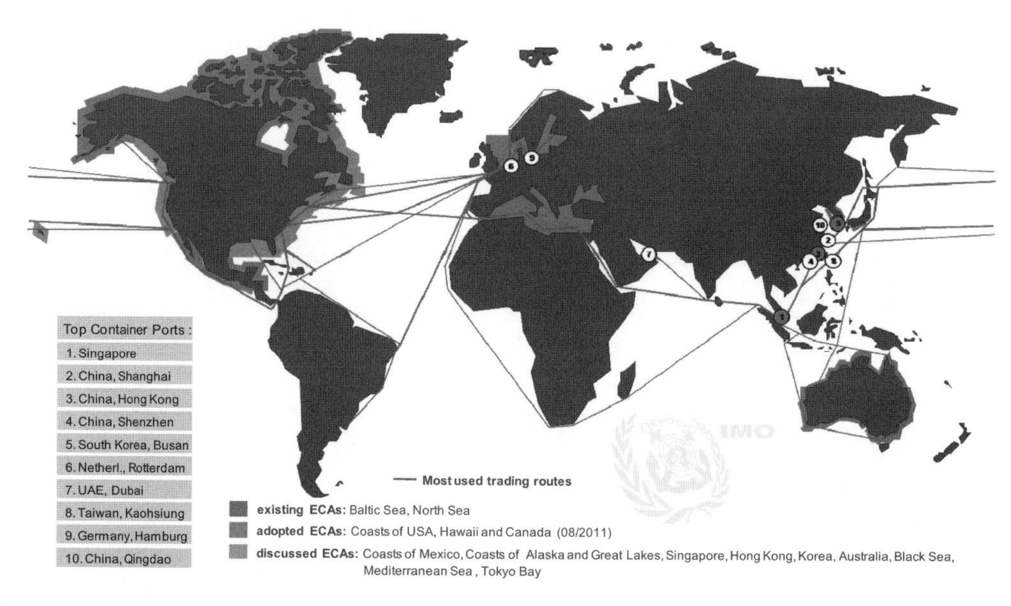

Figure 2. Emission controlled areas.

Especially for vessels already operating on HFO, after treatment techniques can be an option. These techniques are basically devided into two groups, wet scrubbing systems using sea water or fresh water with caustic soda and dry systems using calcium hydroxide.

Both types of systems are suitable to reduce sulphur emissions sufficiently and are technically on the same level of maturity. Nevertheless, some differences can be found.

Wet scrubbing systems wash the sulphur oxides out of the exhaust gas by using different types of wash waters. The so-called open loop system uses the natural salinity of sea water. The sea water is pumped from the sea chests to the scrubber, which is usually installed close to the funnel end, sprayed into the exhaust gas, re-collected and drained back into the sea. To meet the required IMO wash water criteria, the wash water is diluted by sea water before discharge over board. The closed-loop system uses fresh water enriched with caustic soda and the wash water circulates within the system. Only a small portion of wash water is drained out, cleaned by a water treatment system and discharged over board. The resulting sludge will be collected and disposed at reception facilities onshore. Due to the limited amount of discharged water (abt. 0.1 m^3/MWh), a temporary storage onboard can be realized when the vessel is operating in sensitive areas like harbors and a discharge of the drained wash water is not allowed.

Besides the use of HFO, several other advantages can be found. Open-loop scrubbers use only sea water, therefore no additional consumables have to be stored onboard. Even so scrubbers are bulky, due to their position close to the funnel, retrofits seems to be possible even on vessels with limited free space like RoPax and cruise vessels.

On the debit side, this high position of the scrubber has to be taken into account of the ship design. Additionally, some other aspects have to be considered. The large amount of water pumped through the vessel requires a significant amount of energy. In addition, some heat energy is required for a slight reheating of the exhaust gas to avoid white smoke. This white smoke can be formed due to the saturation of the exhaust gas in the scrubber. The continuous discharge of wash water in open loop scrubbers is coming into focus especially in sensitive areas like harbors and some coast lines. It can be expected that the discharge will be forbidden in such areas and therefore open-loop scrubbers cannot be operated there. Operating all the time on closed-loop requires a significant amount of caustic soda. To merge the advantages of both systems and solve the special disadvantages, hybrid systems which can operate both on open and closed loop are introduced.

Dry scrubbing systems like DryEGCS use a different principle of operation. Such systems absorb the sulphur in calcium hydroxide, a processed version of limestone, and remove it permanently out of the biosphere. This calcium hydroxide is provided in form of granulate in a packed bed absorber. In contrary to wet scrubbers, which cool down the exhaust gas below 100°C and even further, the temperature drop over such a dry system is insignificant. Therefore, the absorber can be installed close to the engine and measures for exhaust heat recovery and NO_x-reduction can be installed afterwards in the funnel. In this case, these components see a cleaned exhaust gas with low sulphur contents, which increases both recoverable heat amount and reliability. The granulate has a size of about 2–8 mm and will be slowly but continuously transported through the absorber. The used granulate will be stored onboard and discharged at the next bunkering stop free of charge. The residues will then be used for other desulphurization purposes in power plants and in steel production.

Additionally, the use of granulate avoids the extensive use of complex water treatment systems and operation limitation due to unaccepted discharge or white plume are not present.

On the debit side, DryEGCS requires a severe space onboard, even in comparison with wet scrubbing systems. Therefore, the place for installation has to be chosen carefully. Nevertheless, on most vessels, such a system can be installed.

Usually, all scrubbers have to be installed for each engine separately. An option to avoid overwhelming size penalties is to merge the exhaust of different funnels into one scrubbing system. This so-called multi streaming arrangement allows the installation of less scrubber capacity and limits the size of the whole system to the maximum required power onboard instead of the total installed power. Merging all the exhaust gas funnels into one scrubber requires the meaningful installation of reliable flaps with sealing air to avoid excessive corrosion in the shut down section.

2.2 *NO_x reduction technologies*

In contrary to SO_x emissions, the formation of nitrogen oxides can be influenced by proper adaptation of the combustion process. These measures like retarded injection or the use of miller cycle are suitable for a reduction in the range to fulfill TIER II. For further reduction levels, these techniques will show unacceptable fuel consumption. Consequently, additional measures have to be taken into account.

The introduction of exhaust gas into the combustion process can lead to a NO_x reduction suitable reaching TIER III. The reintroduced exhaust gas dilutes the combustion air in the combustion chamber and results in a slower combustion with less formation of NO_x. Tinschmann et al. showed, that the main challenges by using Exhaust Gas Recirculation (EGR) are increased particulate emissions and corrosion caused by sulphur in the recirculated gas. For the latter, the use of scrubbers or low sulphur fuels in combination with suitable materials can be a way out. Despite very promising results of inhouse and field tests, further development work especially for four stroke engines has to be executed before this technology reaches a sufficient maturity.

Another way to cope with NO_x emissions is the use of Selective Catalytic Reduction (SCR). This technique uses ammonia to reduce NO_x to nitrogen in the presence of a catalyst. The ammonia is generated in the exhaust gas by the decomposition of urea solution, which is sprayed into the funnel before entering the catalyst. SCR is the technology of choice in land-based applications for decades now and even in marine application; more than 100 installations were executed. Using ammonia in a sulphuric exhaust gas may lead to the formation of ammonia hydrogen sulphates. This sticky material can clog the catalyst; increases backpressure and can consequently take the SCR out of service. The generation of these sulphates depends on the sulphur content in fuel and the temperature of the exhaust gas. Figure 3 shows the relevant temperature curve.

To avoid the formation, the temperature should always be above the red curve. A close look to the exhaust gas temperature curve of a typical medium speed engine shows, that at several loads, the temperatures will not be sufficient. Switching off the urea injection is not an option at all. IMO will not accept NO_x emissions at load points at and above 25% MCR which are higher than 1.5 times of the cycle value. As a result, the engine will be tuned to exhaust gas temperatures right above the required

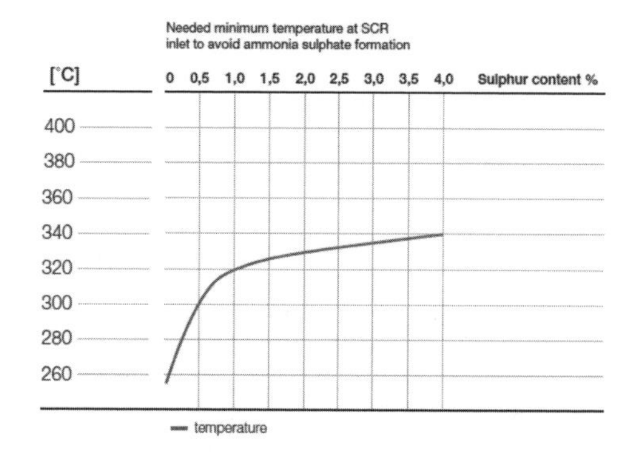

Figure 3. Ammonia hydrogen sulphate formation temperature curve.

temperature by using charge air blow off, waste gate or other measures like variable turbine area (VTA). Unfortunately, this temperature control affects efficiency and causes slightly increased fuel consumption. Besides general optimization measures, this temperature control will be of main focus for development in the near future.

2.3 *LNG as an option*

Another way reducing emissions can be the use of natural gas, stored liquid onboard (LNG).

LNG is free of sulphur and the smooth and clean combustion leads to tremendous low emission levels. NO_x is reduced far below TIER III and in addition, the PM emissions are also reduced significantly. Even a reduction of about 20% of CO_2 emissions can be seen. This very clean exhaust gas will also allow an increased recovery of heat out of the exhaust gas. In addition, the fuel prize seems to be of interest. The LNG-prize at several import terminals is far below the equivalent prize of HFO. It has to be recognized, that in addition to the prize levels at the terminals the distribution costs have to be added. Even so these costs cannot be determined finally. Nevertheless, general expectations are made that the prize level will not rise above the corresponding HFO level.

Besides the only basic infrastructure available, the storage of the LNG can be a challenge for several ship designs. LNG is stored in pressurized and isolated cryogenic tanks at abt. -162°C. Due to the pressure, only cylindrical, conical or bilobe tanks can be used, which has to be fitted into the ship design. In addition, LNG has roughly half the density of conventional fuels. In total, storing the same capacity of fuel for LNG compared to HFO requires roughly 3 to 4 times higher volume.

The engines using LNG are often dual fuel engines, able to burn both HFO/MDO and LNG. LNG is ignited by a small portion of pilot fuel. This engine type is able to switch seamless between the fuels. In special cases, the switch over from gas to liquid fuel can be realized within one crankshaft turn. This leads to a high redundancy, even higher than with conventional engines. The available dual-fuel engines base upon established diesel engine design. Mainly the cylinder head, liner and pistons differ from the conventional engines. Therefore, a retrofit from conventional to dual fuel engines is possible. Nevertheless, a slight power reduction has to be acknowledged.

3 ONE SOLUTION FOR ALL?

In the light of such a bunch of possible solutions to reach future limitations, the question may raise whether this bucket can be limited to a few options or even one solution for all applications can be a possibility. To answer this question, interaction between the potential solutions and commercial aspects have to be considered.

3.1 *Technical interactions*

For new building vessels, measures to reduce NO_x and SO_x cannot be seen as independent from each other. In contrary, the chosen SO_x reduction measure directly influences the method of choice for NO_x reduction. Figure 4 shows such an arrangement in principle.

Running SCR with distillate or DryEGCS will lead to lower operation costs on the SCR due to minimized soot blowing and minimized temperature control due to the low ammonium hydrogen sulphate generation temperature. In case of combining wet scrubbers and SCRs, the low exhaust gas temperature after the scrubber will lead to the upstream installation of the SCR. This will require a higher exhaust gas temperature and therefore a slight SFOC penalty of about 0.09 g/kWhK must be accepted. This has to be added to the overall energy requirements. In case of multi streaming arrangements, it has to be considered, that IMO requires in general one SCR per engine.

The combination of EGR with DeSOx technologies mainly depends on the used EGR system. In case of scrubber-free arrangements,

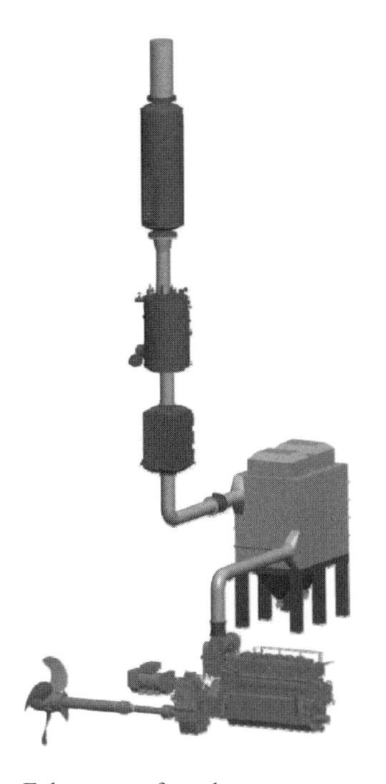

Figure 4. Exhaust gas funnel arrangement containing DryEGCS, SCR, Boiler and Silencer (from bottom to top).

MGO is mandatory to secure a safe and reliable operation. So, a full-flow scrubber after the engine is not required to meet the sulphur limits. In case of EGR operating with a scrubber to clean the recirculated exhaust gas, the full flow scrubber should also use the same principle in order to minimize the installation and operation efforts.

Regarding LNG, a dual-fuel 4-stroke engine operating on LNG will inherently fulfill both NO_x and SO_x requirements. In that case, no additional measures have to be considered as long as LNG operation is the normal operation mode for vessels operating in ECAs. The fast switch-over to liquid fuels in case of emergency is not affected at all. Keeping the full fuel flexibility respectively allow the liquid fuel operation even in normal operation in ECAs, all the other above mentioned measures have to be considered and installed. In the light of the excessive efforts of such a solution, this option is far from reasonable.

3.2 *Commercial aspects*

As a rule of thumb, solutions with low investment costs will lead to high operational costs and vice versa.

Using MGO is by far the DeSOx solution with lowest investment costs, even lower than today's HFO operation. Nevertheless, the high fuel costs affect the business case significantly. On the other hand, LNG has the highest investment costs, but for operation, the costs are by far the lowest of all possible options.

The investment costs of DryEGCS and LNG can be influenced by a careful choice of the required operational range. Typically, vessels in ECA operation can easily bunker more often without severe impact on vessel operation.

The biggest question mark is related to the costs of consumables. Basically, the fuel price difference between HFO/LNG and MGO defines the payback time of such a technology. Looking to the dynamic development of liquid fuel prizes in the recent years, it is hard to predict future development. Figure 5 shows the payback time of a DeSOx installation over the prize difference between HFO & MGO for a 10MW engine in full ECA operation.

This example indicates the basic principle of such a comparison. Below a certain fuel prize difference, the payback time increases exponentially towards a longer period. Nevertheless, above a prize level between 100 and 150 $/t the payback time moves into quite reasonable ranges and will become even better with additional fuel prize gap. In the light of today's fuel prize gap between HFO and MGO, this option could be of high interest. In case of part time ECA operation, the payback time will increase accordingly.

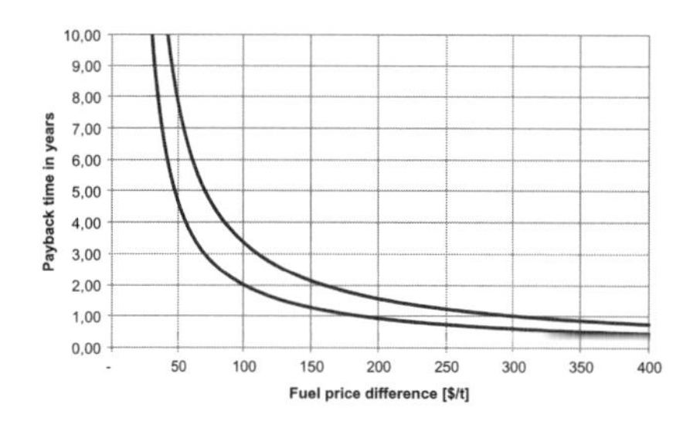

Figure 5. DeSOx payback time over fuel difference (lower curve: invest w/o installation; upper curve: invest including retrofit installation costs).

3.3 *What to do?*

In the light of the above mentioned aspects, it is obvious, that the choice of technology has to be made carefully.

At first, the operation profile of the vessel is crucial for a proper decision. Knowing the load profiles of the engines, minimum reasonable time between bunkering and the infrastructure to bunker the required consumables is essential to define the impact on ship design, cargo and the related investment costs.

Doing so today for desulphurization technologies will unveil two major problems, how to get payback in the time until the relevant limits will come into force and how to decide regarding consumable costs based on such assumptions. In addition, vessels are frequently changing the operation area depending on the charter or get sold after some years. Therefore the vessel should be prepared in a way that several technologies can be retrofitted. This preparation should at least include free space for the larger parts of these technologies like tanks or reactors to avoid large conversion costs of the vessel. Also, parts of the piping, pipe connections (e.g., to the sea chest) and any other installation efforts should be made, which can be realized at low cost during ship building and may cause a big conversion effort in case of later retrofit.

For new buildings on and after the 1st January 2016, mainly the ECA operation time will influence the choice of technology. Only a small ECA time share will require low invest options. Therefore SCR and the use of MGO are favorable. With an increased ECA share, SCR with desulphurization, EGR or LNG will come more and more into focus. The exact ECA share for such a shift towards the above mentioned technologies can only be done on a project basis.

4 CONCLUSION

The main focus on emission reduction in the next years is NO_x and SO_x, limited by the IMO's MARPOL Annex VI. For new building vessels, measures to reduce NO_x and SO_x cannot be seen as independent from each other. In contrary, the chosen SO_x reduction measure directly influences the method of choice for NO_x reduction. Reducing SO_x can be done on two different ways—low sulphur fuels or residual fuels in combination with exhaust gas after treatment. MGO is quite a clean fuel which leads to low emissions and deposits and reduces the fuel treatment to a minimum. Unfortunately, the prize level is by far higher than for HFO and therefore, after treatment techniques come into focus. The so-called DryEGCS uses calcium-hydroxide granules fitted in a packed bed reactor to absorb the SO_x out of the exhaust gas by converting the granules to gypsum. This technique removes the sulphur permanently and allows also its combination with SCR.

The latter is the favorite option to reduce NO_x for a 4-stroke medium speed engine. Established for years in shipping, some work is carried out to secure the reliable operation of SCRs under IMO Tier III. Another considerable option will be the use of liquefied natural gas (LNG) in dual-fuel engines. LNG leads to a very low emission level combined with a fuel prize in the range of HFO. Unfortunately, the equipment is bulky and requires significant investment.

At the end of the day, one option for all will be an illusion for future shipping. The right technology of choice largely depends on the vessels properties and its operation profile. 4-stroke vessels operating only a limited time in an ECA, the right technology will be chosen on the lowest investment costs. Then, the use of an SCR for NO_x-reduction while burning MGO is recommended. Increasing the time spent in an ECA the focus will move to operating costs. Therefore the combination of cheap fuels like HFO with after treatment becomes reasonable. Additionally, the dual fuel option will be of more interest and due to its low fuel costs, it can be recommended for a lot of different applications in the foreseeable future.

REFERENCES

Revised MARPOL Annex VI—Regulations for the prevention of air pollution from ships. 2008 London. IMO.

Thum, D. 2011. *The sulphur challenge—Technology for Ecology*. Paper No. 3, MariTech 2011.

Tischmann, G., Thum, D., Schlüter S., Pelemis, P. & Stiesch, G. 2010. *Sailing towards IMO Tier III - Exhaust Aftertreatment versus Engine-Internal Technologies for Medium Speed Engines*. Paper No. 274. Bergen, CIMAC Congress.

Sustainable Maritime Transportation and Exploitation of Sea Resources – Rizzuto & Guedes Soares (eds)
© 2012 Taylor & Francis Group, London, ISBN 978-0-415-62081-9

Evaluation of ship efficiency indexes

M. Figari
Naval Architecture, Marine Engineering, Electric Engineering Department, University of Genoa, Italy

M. D'Amico
D'Amico Shipping Company, Rome, Italy

P. Gaggero
Registro Italiano Navale, Genoa, Italy

ABSTRACT: IMO is introducing two different emission indexes for a vessel: the Energy Efficiency Design Index (EEDI) and the Energy Efficiency Operational Indicator (EEOI). The former will be used to assess the design of the vessel, the latter would be used to evaluate the vessel in operation. Both indexes represent the ratio between emissions, in mass of CO2, and the transported cargo quantity per sailed distance. At moment an important debate is focusing on the definition of the 'baseline' values for different ship categories. The collaboration between d'Amico Shipping Company, Registro Italiano Navale and Department of Naval Architecture, Marine Engineering, Electric Engineering of Genoa University provided the framework for a study aimed to evaluate the carbon footprint of the vessels of the D'Amico fleet, the analysis of various aspects of factors effecting the carbon dioxide emissions caused by ships and improvements of fleet energy management techniques. The results of the study can be divided into two main aspects: it is a picture of the actual carbon dioxide emission status of a cargo fleet and it gives the technical instruments and measure tools to start an emission control policy with reference to a ship energy efficiency management plan.

1 INTRODUCTION

All ships interfere with the environment, releasing pollutant substances into air, water and soil during their entire life cycle, from extraction of raw materials, to the ships' construction, operation and disposal. The efforts to reduce marine emissions, especially in coastal areas, are a constant commitment, accounted for in a broad range of measures, from political regulation to their practical implementation, e.g., through improved vessel design solutions.

Despite sea shipping being relatively environmental friendly when compared with other means of transport, it is still an important source of air pollutants, mainly due to exhaust emissions of sulphur dioxide (SO_2), carbon dioxide (CO_2) and nitrogen oxides (NO_x). The impacts on the environment represent an increasing concern in coastal areas and harbors with heavy traffic. So, the reduction of impacts from ship emissions to air is a constant commitment from all interested parties and increasingly a dominant policy driver.

IMO has introduced greenhouse gas (GHG) emission reduction in its agenda since 1995. In recent years the effort to reduce emissions from the shipping sector has strengthened. Many studies about CO2 shipping emission demonstrated that for many goods, shipping is the most efficient way of transport. Despite this, high potential of improvement certainly exists and IMO is thrusting the maritime sector toward a more rational approach to ship energy management and GHG emissions. IMO is developing some technical and economical instruments in order to introduce an emission regulation for the global fleet. As far as Technical Instruments are concerned, two different emission indexes for a vessel are proposed: the Energy Efficiency Design Index (EEDI) and the Energy Efficiency Operational Indicator (EEOI). The former used to assess the design of the vessel, the latter used to evaluate the vessel in operation. Both indexes are measured by the ratio between emissions, in mass of CO2, and the transported cargo quantity per sailed distance. At moment an important debate is focusing on the definition of the 'baseline' values for different ship categories.

The collaboration between d'Amico Shipping Company, Registro Italiano Navale and Department of Naval Architecture, Marine Engineering, Electric

Engineering (DINAEL) of Genoa University provided the framework for a study aimed to evaluate the carbon footprint of the vessels of the d'Amico fleet, the analysis of various aspects of factors effecting the carbon dioxide emissions caused by ships and improvements of fleet energy management techniques.

The paper contains the methodological approach that was adopted for the calculation and a critical analysis of the IMO Guidelines [IMO]. Design aspect influencing EEDI and all the operational aspects influencing EEOI are addressed. Port time, ballast legs and laden legs were treated together and then separately in order to give a clear explanation of the operational activities.

Output of the calculation are the EEDI values for the entire fleet and comparisons with literature results, the EEOI values for two ships evaluated over 1 year period.

The results of the study can be divided into two main aspects: it is a picture of the actual carbon dioxide emission status of a cargo fleet and, it gives the technical instruments and measure tools to start an emission control policy with reference to a ship energy efficiency management plan.

2 IMO TECHNICAL INSTRUMENTS *EEDI* AND *EEOI*

The ratio between the emissions and the benefit produced by the transport is an indicator of the transport activity impact, it can be expressed in ton of CO2 divided by cargo per distance.

IMO is promoting two different emission indexes for a vessel: the first is the Energy Efficiency Design Index (EEDI), the second is the Energy Efficiency Operational Indicator (EEOI). A brief introduction to the indexes is reported in the following.

2.1 *Energy Efficiency Design Index, EEDI*

The design index is be evaluated during the vessel's design process and then it has to be verified during the official sea trials. The GHG emissions are evaluated through the total fuel consumption (main and auxiliary engines), the capacity is represented by the deadweight (DWT), the sailed distance is represented by the ship speed. The fuel consumption is evaluated by the power and the specific fuel oil consumption of both main and auxiliary engines. The result is given in grams CO2/(miles*ton cargo).

The EEDI can be considered as a design characteristic of the vessel describing her emission efficiency. IMO is pursuing to set an upper limit of the EEDI value to thrust the design and build of more energy saving ships.

$$EEDI = \frac{\left(\sum_{i=1}^{nME} P_{ME(i)} \cdot C_{FME(i)} \cdot SFC_{ME(i)}\right) + \left(P_{AE} \cdot C_{FAE} \cdot SFC_{AE}\right)}{f_i \cdot Capacity \cdot V_{ref}} \quad (1)$$

where:

C_F is a non-dimensional conversion factor between fuel consumption and emission. It depends on the carbon content of the fuel.

V_{ref} is the ship speed, measured in knots, in the maximum load condition (Capacity) at the 75% of the Maximum Continuous Rating of the main engine The maximum load condition shall be defined by the deepest draught at which the ship is allowed to operate.

P is the power of the main and auxiliary engines, measured in kW.

The subscripts ME and AE refer to the main and auxiliary engine respectively.

2.2 *Energy Efficiency Design Indicator EEOI*

In this case the index is defined as an "indicator", in fact it is not calculated by the technical characteristics of the vessel. In EEOI formulation the actual fuel consumption, the distance sailed and the cargo transported during a real vessel's journey are taken into account. It is expressed, as for the EEDI, in grams CO2*mile-1*ton-1cargo.

EEOI is therefore a real indicator of CO2 emission considering all the fuel consumption on board coming from engines, boilers and diesel-engine driven cargo pumps.

$$EEOI = \frac{\sum_i \sum_j (FC)_{ij} \cdot (CF)_j}{\sum_i \left(M_{cargo} \cdot D\right)_i} \quad (2)$$

where:

- j is for fuel type
- i is for the voyage number
- FC is the mass of consumed fuel
- CF is the fuel mass to CO2 mass conversion factor

Mcargo is the cargo mass carried or the work done (number of TEU or passenger) or gross tonnage (Gt) for passenger ships.

D is the actual distance sailed expressed in nautical miles

2.3 *Baseline*

The calculation of the two indexes is not yet mandatory. At present EEDI and EEOI are well defined into MEPC guidelines but no previous experience exists about their values. IMO is promoting the

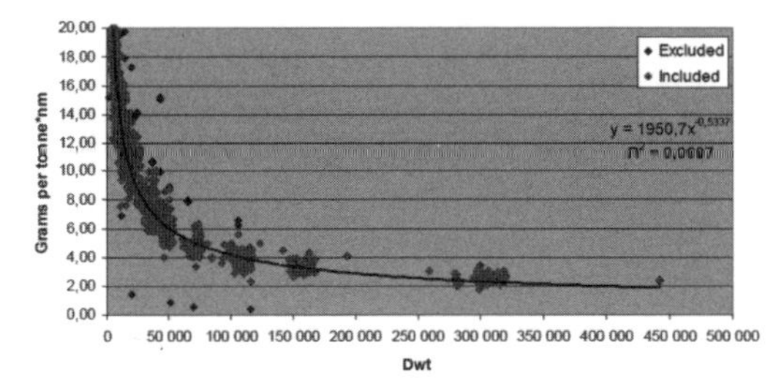

Figure 1. Tanker EEDI baseline source GHG_WG 2/2/7 submitted by Denmark.

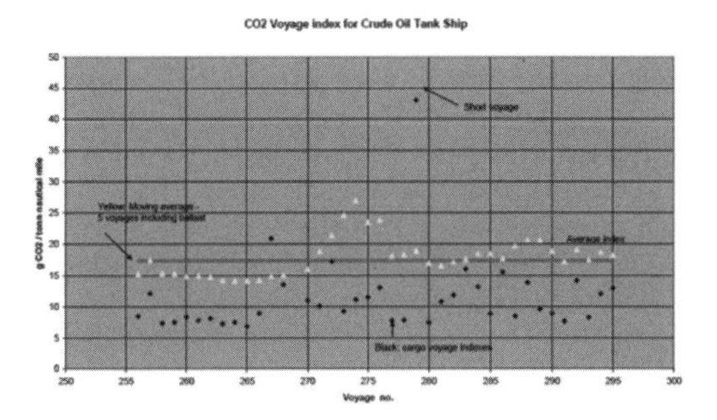

Figure 2. Sample EEOI values source MEPC 55/4/3 submitted by Norway and Germany.

voluntary application of the indexes to the existing fleets, in order to receive back useful results describing the actual situation about ship's energy efficiency. The enforcement of a limit on the EEDI or EEOI values should be done only after a good definition of the actual emission status of the world's fleet.

The circular MEPC.1circ.682, giving the "Voluntary guidelines for EEDI verification", has been published in order to push ship-owners and ship-builders to calculate the index and give back reliable results for the existing vessels. The result of a large statistical collection of actual EEDI values can be used for the definition of a baseline expressing the EEDI limit value versus ship's Deadweight.

Some GHG Working Groups have already conducted some studies and they have already suggested some baseline to the attention of IMO. Just to give an idea, the baseline calculated by Denmark, concerning Oil Chemical Tankers, is shown below.

Since EEDI would represent the efficiency of a ship 'as designed', the *EEOI* can be considered an indicator of the real vessel's efficiency, concerning the real CO_2 emitted during the voyages due to the real fuel consumption.

The *EEOI* can be calculated trip by trip during the vessel's life creating a monitoring panel which shows the efficiency of both commercial and technical fleet management.

3 EEDI CALCULATION

The goal of the presented study is the *EEDI* calculation for the cargo vessel fleet owned and managed by the *d'Amico Shipping Company S.p.A.* The considered fleet includes three different ships categories: *Oil/Chemical Tankers, Bulk Carriers* and *General Cargo Vessels*. The EEDI calculation is developed according to the IMO MEPC.1/Circ.681 "Interim guidelines on the method of calculation of the Energy Efficiency Design Index" published on 17th August 2009. As stated in the guidelines the EEDI has to be evaluated during the ship's design process.

The guidelines for the verification, MEPC.1/Circ.682, applies to new ships for which an application for the EEDI verification has been submitted to a verifier, at present, on a voluntary basis.

In the present work the calculation of the design index is not referred to a new building, but to an existing vessel. The index evaluation for vessels built in the last ten years can bring back a picture of the present situation of the shipping industry.

The data collected for this study were contained in the ships' drawings that the owner received at the delivery of the vessel. Thus these data are referred to the "*ship as built*" and not all the technical specifications managed during the design process were available.

Furthermore sea trials data were contained in the ship file (they are contractual documents) but they usually do not refer to the full load condition, which is the condition required by the *guidelines*.

In order to perform the EEDI verification for the d'Amico fleet, both the MEPC.1/Circ681 and MEPC.1/Circ682 were followed. The procedure can be resumed by three main steps:

1. Data collection through the ship's drawings
   a. Vessel's name and main particulars (dimension, dead weight …)
   b. Main and auxiliary engines technical files
   c. Sea trials speed test
2. Application of the EEDI formula according to IMO guideline
3. Comparison with other results already submitted to IMO (i.e., GHG WG 2/2/7)

The vessels considered in this study are cargo vessels equipped with two strokes diesel engine. They all are equipped with standard propulsion lay-out arrangement, without any shaft generator or any particular energy recovery system.

The general EEDI formulation which is contained into the guideline is composed by several terms in the numerator, but some of them are not considered in the calculation due to the types of vessels considered in this study (see Figure 3).

3.1 *The definition of V_{ref}*

The speed definition of the guideline usually does not match with the design speed of the ship. IMO requires a ship speed evaluated at maximum draft and at 75% of engine power.

The value can be calculated during the ship design, but may be difficult or impossible to evaluate for existing vessels, due to the lack of data. For existing ships, normally, sea trials data are available, but for some ships sea trial condition is far from the maximum draft, so, it becomes difficult to guess the IMO speed.

Different procedures have been considered during this study:

a. speed change due to the change of displacement using to the admiralty coefficient

$$\frac{\Delta_{design}^{2/3} \cdot V_{design}^{3}}{P_{75\%MCR}} = \frac{\Delta_{full\_load}^{2/3} \cdot V_{ref}^{3}}{P_{75\%MCR}} \tag{3}$$

This kind of correction was even proposed on the *MEPC 59/WP.8 par 6.32.*

b. determination of V_{ref} by the comparison of sea trials tests with full load speed prediction coming out from towing tank tests, or from full load power curve prediction (possible only for a few cases), as proposed in the guideline for the verification. The calibration of speed would be achieved as showed in Figure 4.

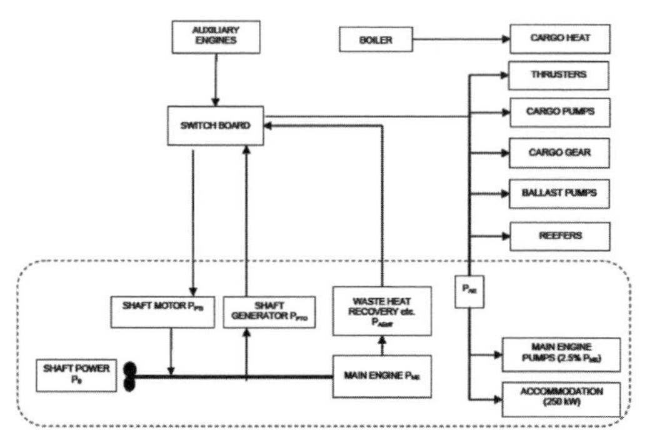

Figure 3. Energy flow scheme.

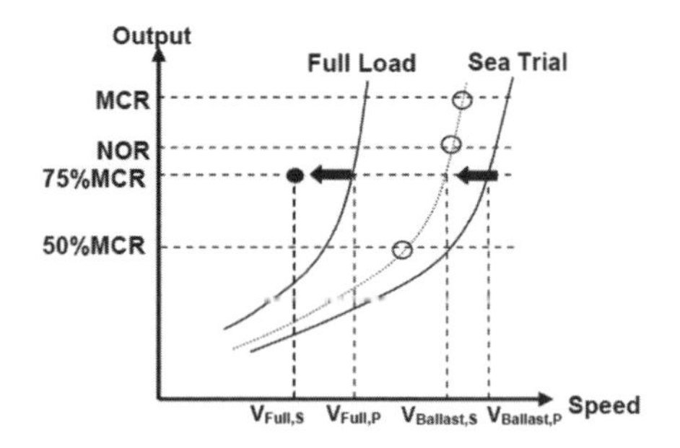

Figure 4. Ship speed definition according to IMO verification guideline Circ.682.

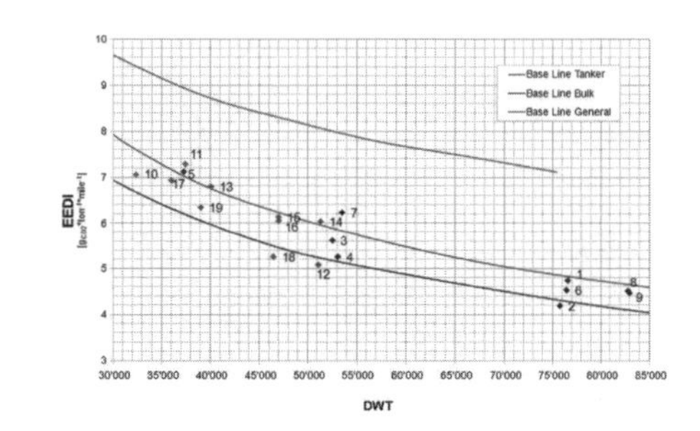

Figure 5. EEDI values vs. deadweight.

c. using the matching code developed by DINAEL (this method requires propeller and hull design characteristics, not always available for existing ships).

3.2 *EEDI results*

The considered ships were: 18 tankers with a DWT between 35000 and 52000 ton, 9 bulk carriers with a DWT between 35000 and 82000 ton, 3 general cargo vessels with a DWT of 37000 ton. In Figure 5 the EEDI values for each ship are reported: blue dots are bulk carriers, green dots are tankers and red dots refer to general cargo vessels. In the same figures the results presented by Denmark at IMO are reported: the blue line is the regression for bulk carriers results, the green line refers to the tankers and red line is related to the general cargo vessels.

4 EEOI CALCULATION

The EEOI, in its definition, contains the total fuel oil really burned on board, thus it is the actual measure of the emissions caused by the vessel operation. The principle at the basis of the EEOI definition is the ratio between the carbon dioxide mass emitted versus the benefit produced by the transport. The systematic EEOI calculation can become the vessel's efficiency monitoring tool during her life.

The calculation of the actual *EEOI* can be performed as shown by the formulas defined in the guidelines and reported in para 2.2. A main issue is the definition of the voyage, a definition that can heavily influence the results. Different possibilities exist that will be discussed in the following.

4.1 *Voyage definition*

The ship trade can be split in different voyages according to different definitions.

a. A voyage is every leg (in Figure 6 a period made by five legs is shown)
b. A voyage is the sum of the navigation time plus the time spent in the arrival port (in Figure 6 leg 2 and leg 3 are one voyage)
c. A voyage is defined by a time period with reference to all the cargo transported, all the distance sailed and all the fuel consumed in that reference period.

If definition "a." is assumed, the possible ship activities will be defined one by one as three main possible category: laden voyage, ballast voyage, port time. This characterization guarantees the maximum level of detail since the fuel consumed is registered separately for each working condition of the vessel.

If definition "b." is assumed (this is the IMO proposal) no trace of the port time will be ever recognized as ship activity. The consumption made at berth will be included in the travel leg, and this does not seem suitable for deeper investigations on vessel's fuel efficiency.

If definition "c." is assumed non-negligible differences could be found on the EEOI calculation. Since it is difficult to divide total fuel consumption to cargo units, this definition is not preferable. (Anyway it was already used as assumption at the basis of studies submitted to the IMO).

4.2 *Data collection*

The period considered for the EEOI calculation spans about one year and half for each vessel considered in the analysis. The data referring to ships travels were collected thanks to the collaboration of the Operation Office of the d'Amico Shipping Company based in Montecarlo.

The information about fuel consumption, cargo mass transported and actual distance sailed during ships' travels are normally recorded in the deck log-book by the ship's master. Contemporary, the master is asked to communicate these data to the Operation Office, which administrates these information for commercial and managing activities.

The format of the data was not suitable at the purpose of EEOI calculation. To overcome the problem it was necessary the support of the ICT department that worked hard to extract the required data and put them in a suitable format for the calculation. The analysis was not extended to the total fleet because of the difficulty found in the data handling. Thus, the first lesson learned is that: a well established procedure of data collection must be developed by the companies which wish to incorporate the EEOI calculation in their energy efficiency monitoring system.

4.3 *EEOI results*

The calculation addressed 1 general cargo vessel, 8 oil/chemical tankers and 4 bulk carriers. Data were collected using "a." definition of voyage, consequently considerations on the influence of each ship's activity on the EEOI will be reported.

Bulk carriers, indicated by red dots, are distributed from 35'000 to 82'000 DWT. Tankers are concentrated between 35'000 and 52'000 DWT.

Talking about vessels of same dimensions, in the area between 35'000 and 50'000 DWT, bulk carrier vessels seem to be more efficient than oil/chemical tankers. In this region, bulk carriers have the maximum EEOI which is about 11 gCO2*tonn-1*mile-1 against the minimum measured for the tankers which is about 14 gCO2*tonn-1*mile-1.

Different ships categories are characterized by different cargo handling operations, by different port operations, by different operational

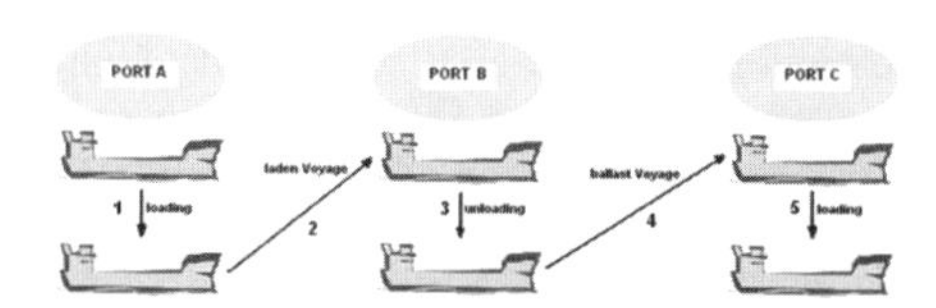

Figure 6. Ship trade and voyage definition.

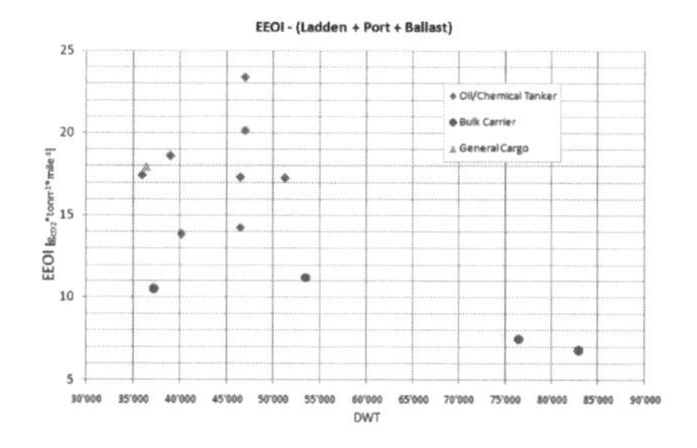

Figure 7. EEOI results.

conditions and, thus, by different actual fuel oil consumption which determines higher or lower CO2 actual emissions.

4.4 *Influence of ballast leg*

The EEOI calculation just for ballast voyages cannot be performed because the absence of cargo mass would make the denominator of the formula equal to zero.

However, the fuel consumption in ballast condition during the considered period has a big influence on the EEOI value since it contributes to add CO2 emissions at the numerator without the compensation of the mass transported at the denominator.

The influence of ballast legs on the EEOI ranges from a minimum of 15% of the total to a maximum of 30%, considering both tankers and bulk carriers.

The ballast voyages incidence has no correlation with the ship dimension, ballast trips do not depend on vessel technical characteristics but only on operational management and market.

4.5 *Influence of port time*

As well as for the ballast legs, the EEOI cannot be calculated just for periods spent in port: the distance travelled in this case is equal to zero, so the denominator of the formula is annulled.

Anyway, the fuel consumed in port contributes to the CO2 emissions and, thus, it has an influence on the EEOI value. The incidence of port time on the EEOI value can be the monitoring tool explaining the different impact that port activities have on different type of ships. The influence of the port time concerning the bulk carriers vessels is very low: it ranges from 2% to 7% of the total EEOI.

The port activities affect heavily the EEOI of tankers: the influence ranges from 12% to 26% of the total value.

The difference can be easily explained by the different cargo handling operations. Bulk carriers are normally loaded/unloaded by shore facilities. Just in few cases they uses their own cranes for the loading/unloading operations.

Tankers cargo handling operations involve cargo pumps and inert gas system thus requiring fuel consumptions.

5 EEDI VERSUS EEOI

This paragraph contains the comparison between the EEDI and the EEOI for the ships which are considered for both the calculations (Table 1).

Table 1. EEDI vs EEOI.

	EEDI	EEOI	EEOI laden
Tanker 1	6.34	18.59	11.08
Tanker 2	6.93	17.41	9.88
Tanker 3	6.80	13.86	9.30
Tanker 4	5.26	14.28	9.08
Tanker 5	6.14	23.34	11.04
Tanker 6	6.05	20.14	9.31
Tanker 7	6.03	17.26	8.81
Tanker 8	5.26	17.30	9.23
General	7.29	17.95	13.51
Bulk 1	4.53	7.51	4.63
Bulk 2	6.24	11.22	6.67
Bulk 3	4.47	6.80	4.57
Bulk 4	7.12	10.49	8.53

The comparison between the two indexes can give an important contribution in understanding the use of the indexes for an energy management policy.

The EEOI in Table 1 includes port time, ballast legs and laden voyages.

Bulk carriers show, in general, the lowest differences. Tankers are characterized by higher spread between the two indexes, a difference by 50% to 280% is found.

Variables such as the cargo quantity carried in laden voyages, the fuel consumption during ballast legs and port time are influencing the actual energy efficiency of vessels in a non-negligible measure. To account for this, EEOI evaluated only for trips in laden condition is also reported in Table 1.

The results, as expected, are characterized by a minor difference between the design estimate and the operational measures.

Bulk carriers are still the ships with the lowest difference: in two cases EEOI and EEDI seems to be comparable showing differences only by 2%.

Tankers and general cargo ships show higher differences, from 50% to 80%. Even if the spread is more reduced than in case of comparison with the total EEOI, the differences between the energy efficiency predicted with the design index and the energy efficiency evaluated by the operational indicator is non-negligible.

6 CONCLUSION

The emission indexes EEDI and EEOI (Technical Instruments) defined by IMO are discussed and evaluated for several ships of the D'Amico Shipping Company.

The presented results form the basis for a realistic picture of the actual carbon dioxide emission status of a cargo fleet.

The actual growing attention in emission reduction, in order to preserve our environment, and the Technical Instruments proposed by IMO can contribute to introduce a new politic of energy efficiency management in the maritime sector that will produce concurrent benefits for owners, users and the society.

ACKNOWLEDGEMENT

A particular thanks to Nicola D'Alesio and Stefano Gallo, graduated at Genoa University in 2010, for their contribution to this work during their Master Thesis.

REFERENCES

Altosole, M., Borlenghi, M., Capasso, M. & Figari, M. 2007. Computer-based design tool for a fuel efficient–low emissions marine propulsion plant. Proceedings of 2nd International Conference on Marine Research and Transportation ICMRT 07, 28–30 June, Italy.

Figari, M. & Guedes Soares, C. 2009. Fuel consumption and exhaust emissions reduction by dynamic propeller pitch control, MARSTRUCT 2009, Lisbon, Portugal, 16–18 March 2009.

IMO, MEPC 1/Circ. 681, Interim guidelines on the method of calculation of the energy efficiency design index for new ships.

IMO, MEPC 1/Circ. 682, Interim guidelines for voluntary verification of the energy efficiency design index.

5.2 *Machinery*

Sustainable Maritime Transportation and Exploitation of Sea Resources – Rizzuto & Guedes Soares (eds)
© 2012 Taylor & Francis Group, London, ISBN 978-0-415-62081-9

A new approach in engine-propeller matching

A. Coraddu, S. Gaggero, D. Villa & M. Figari
Department of Naval Architecture, Marine Engineering and Electrical Engineering (DINAEL), Genoa University, Genoa, Italy

ABSTRACT: Traditional engine-propeller matching techniques are mainly based on nondimensional parameters analysis: thrust/advance coefficient or torque/advance coefficient ratio are the most used variables to assess the ship propulsion point or to match, at each selected ship speed, the selected propeller with the ship engine. The advantages of this robust and well established procedure (standard propeller open water measures only at some pitch ratios around the design configuration, availability of large databases and extrapolation laws) however, turn into the drawbacks for the inclusion of different constraints and objectives, further than the minimum fuel consumption, in the matching algorithm. On the other hand modern numerical tools and available hardware resources let to partially substitute, in the design stage, experimental campaigns and to collect large amount of information on propeller performances, including cavitation. In this sense numerical computations make out of date approaches just developed to overcome the deficiency of experimental measures. On the basis of these numerical data new algorithms for the engine-propeller matching can be developed capable of investigate different objectives and the influence of different constraints on the traditional optimum points.

1 INTRODUCTION

Fuel saving, in a world of increasing oil cost and more strict regulations on environmental pollution, is one of the most important aspect for the design of the ship propulsion plant. Two sides of the problem worth to be investigated: the design of efficient propellers and the efficient matching of the designed propeller with the selected engine.

The former aspect has been investigated extensively in the last century. Lerbs' work (Lerbs, 1952) on optimum loaded propellers is a milestone for the numerical design of highly efficient propellers. More recently the work by Coney (1989) set the basis for a theoretically more accurate design, based on variation approaches and more accurate solvers, like lifting surface methods, able to take into account advanced propeller configurations, like contra rotating and ducted propellers. At the same time optimization strategies have been employed (Gaggero, 2009, 2010) to adopt panel methods (developed, primarily, as analysis tools) in an inverse design procedure, in order to more accurately consider the influence of the propeller geometry (blade thickness and the presence of the hub, mainly) and include more strict (and directly calculated by the numerical approach) constraints on cavitation and unsteady propeller performances.

The latter aspect, instead, has been traditionally solved in a more standard/simplified way, by means of nondimensional parameters, systematical series and extrapolation from experimental measures. The classical approach to match the open water propeller curves with the engine load diagram is based on the K_T/J^2 parameter: the propeller optimal rate of revolution is found as the equilibrium point between the provided (by the propeller) and the required (by the hull) nondimensional thrust value. If the propeller is of controllable pitch type, once the design point has been defined, with the same procedure it is possible to identify the working points for each ship speed and each pitch value and, consequently, select the best combination of propeller rate of revolution and pitch that, at a given speed, minimize the fuel consumption (Figari et al., 2007 & 2009).

The procedure highlighted above, although being a robust approach to the engine-propeller matching, is affected by some important limitations. Usually, in fact, in the preliminary design stage, all the data regarding the propeller performances as functions of the blade pitch are not fully available: extrapolation from few experimental measures or databases (well established but, however, far from the real propeller performances) are the only way to overcome this lack of information. Moreover, if other constraints further than minimum fuel consumption have to be addressed in the matching procedure, and occurrence of cavitation is one of the most influencing and dangerous phenomena that should be monitored at each pitch/speed, this traditional procedure fails.

Occurrence of cavitation may cause reduction of delivered propeller thrust, it increases propeller noise and structural vibrations and, definitely, alters the propeller working points, changing the best pitch/rpm combination of minimum fuel consumption. Only empirical formulae and criteria (Burril diagrams, for instance) can be adopted within a nondimensional analysis. Any other information (numerical as much as experimental) regarding the risk of cavitation cannot be included in the parameter KT/J^2: propeller performances (thrust and torque coefficients), in cavitating regime, depends from the cavitation index that, in turn, is a function of the propeller rate of revolutions. The nondimensional parameter, which is used to identify the propeller equilibrium point apart from the knowledge of the propeller revolutions, is consequently inherently inadequate to take into account the effects of cavitation on propeller performances. Ship speed and engine revolutions have to be considered as independent variables in the matching procedure, thus requiring a systematic knowledge of the propeller performances at each pitch value and for each possible combination of ship speed and propeller rate of revolution that realizes each considered advance coefficient. A so deep analysis of the propeller performances is experimentally prohibitive: it involves thousands and thousands of possible working conditions simply not affordable in any design stage (especially in the preliminary design phase), for time and costs reasons. Furthermore databases and extrapolations are far from being reliable tools when cavitation is the issue.

Nowadays, numerical tools represent an accurate and reliable way to predict propeller performances and, in the preliminary stage, they can effectively substitute model test measures. Ideally, thrust, torque and cavity extension can be computed for each pitch and each combination of rate of revolution and ship speed, giving the complete amount of data needed for an engine-propeller matching, based on dimensional rate of revolution (and thus cavity) dependent values of delivered thrust and required torque.

On the light of these considerations, in the present work a new algorithm for the engine-propeller matching is presented. The potential panel method developed at the University of Genova (Gaggero, 2009, 2010) has been adopted for the systematical analysis of steady cavitating controllable pitch propeller characteristics. A fully numerical approach, as a natural improvement of the classical nondimensional analysis, has been developed in order to identify, on the basis of the computed propeller performances, the optimal propulsive points (pitch and rpm for each selected ship speed), being the optimal a more general objective as a weighted combination of minimum fuel consumption and minimum cavity extension.

First, a preliminary non cavitating analysis will be presented. The numerical code will be applied in a classical way, having in mind only the minimization of fuel consumption. Results will be compared with the traditional nondimensional approach, in order to validate the new developed algorithm with respect to well established matching procedure.

Finally also propeller cavitation will be taken into account: new engine-propeller matching points will be calculated and compared with the noncavitating computations, in order to highlight the effect of cavitation on propeller performances and to point out the possibility to identify a new set of rpm/pitch combinations capable of grant working points free of cavitation.

2 THE COMPUTATIONAL TOOL

2.1 *Panel method for propeller analysis*

Panel/boundary elements methods model the flowfield around a solid body by means of a scalar function, the perturbation potential $\phi(x)$, whose spatial derivatives represent the component of the perturbation velocity vector. Irrotationality, incompressibility an absence of viscosity are the hypothesis needed in order to write the more general continuity and momentum equations as a Laplace equation for the perturbation potential itself:

$$\nabla^2\phi(x) = 0 \tag{1}$$

For the more general problem of cavitating flow, Green's third identity allows to solve the three dimensional differential problem as a simpler integral problem written for the surfaces that bound the domain. The solution is found as the intensity of a series of mathematical singularities (sources and dipoles) whose superposition models the inviscid cavitating flow on and around the body.

$$\begin{aligned} 2\pi\phi(x_p) = & \int_{S_B+S_{CB}} \phi(x_q)\frac{\partial}{\partial \boldsymbol{n}_q}\frac{1}{\boldsymbol{r}_{pq}}ds \\ & - \int_{S_B+S_{CB}} \frac{\partial\phi(x_q)}{\partial \boldsymbol{n}_q}\frac{1}{\boldsymbol{r}_{pq}}ds \\ & + \int_{S_W} \Delta\phi(x_q)\frac{\partial}{\partial \boldsymbol{n}_q}\frac{1}{\boldsymbol{r}_{pq}}ds \end{aligned} \tag{2}$$

Neglecting the supercavitating case (computation is stopped when the cavity bubble reaches the blade trailing edge) and assuming that the cavity bubble thickness is small with respect to the profile chord, (Gaggero, 2009) singularities that model cavity bubble can be placed on the blade surface instead than on the real cavity surface, leading to an integral equation in which the subscript q corresponds to the variable point in the integration, $\boldsymbol{n}$ is the unit normal to the boundary surfaces and $\boldsymbol{r}_{pq}$ is the distance between points p and q, S_B is the fully wetted surface, S_W is the wake surface and S_{CB} is the projected cavitating surface on the solid boundaries. This approach can be addressed as a partial nonlinear approach that takes into account the weakly nonlinearity of the boundary conditions (the dynamic boundary condition on the cavitating part of the blade and the closure condition at its trailing edge) without the need to collocate the singularities on the effective cavity surface. A set of boundary conditions is required to solve, iteratively, nonlinear problem of Eq. 2: kinematic b.c. on wetted blade areas, kinematic and dynamic b.c. on cavitating area, Kutta condition at blade trailing edge and cavity closure condition at cavity trailing edge. A typical discretized surface mesh for both cavitating and non cavitating computations is that of Figure 1: 1200 hyperboloidal elements panels each bladde plus its portion of hub, while the trailing vortical wake is discretized with an angular step of 6°.

2.2 *Matching algorithm*

The core of the developed matching algorithm is represented by a systematic knowledge of the propeller characteristics, including face and back cavity extension in addition to thrust, torque and efficiency coefficients.

Once established that the propeller rate of revolution is an independent variable in the matching algorithm, it is necessary to characterize the propeller performances at each pitch angle in terms of any possible pair of propeller *rpm* and ship speed. Panel method could help in collecting all these data that can be organized in a *n-dimensional* matrix array as shown in Figures 3–6.

A vector of ship speed and a vector of propeller revolutions define the parameter space of the propeller working conditions: for each working condition, at each pitch setting, all the propeller characteristics (thrust, torque, efficiency and cavitation extension) are stored in a structured form sketched as in Figure 2. As in the traditional engine-propeller matching the equilibrium points are computed one by one for all the ship speed:

1. As the velocity and the correspondent ship resistance have been fixed (in turn also the required propeller thrust is known through the thrust reduction factor), it is possible to identify all the pairs of propeller pitch and rate of revolutions that deliver the required thrust (Figure 7).
2. From the above structured collection of data, the one to one correspondence between propeller rpm and pitch with all the other propeller characteristics allows to assign to each equilibrium point all the information regarding cavitation, efficiency and required torque (Figure 8):

$$\left\{N, \frac{P}{D}\right\}_{T=\bar{T}} \Leftrightarrow \left\{A_{cav}, \eta_o, K_Q\right\}_{T=\bar{T}} \tag{3}$$

3. The efficiency chain lets to map all the propeller working points onto the engine layout and to assign at each point (one for each ships speed and pitch setting) the corresponding specific consumption value using the engine specific fuel consumption map (Figure 9).
4. The best combination of propeller rate of revolutions and pitch can be identified through a minimization procedure of a weighted function of absolute fuel consumption and cavitation area of the equilibrium points.

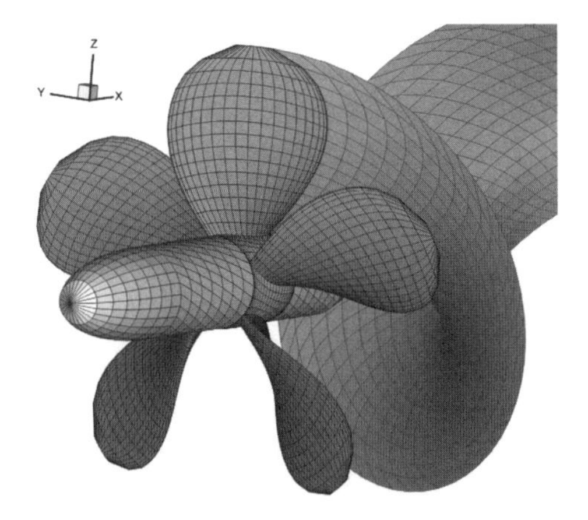

Figure 1. Typical propeller panel mesh.

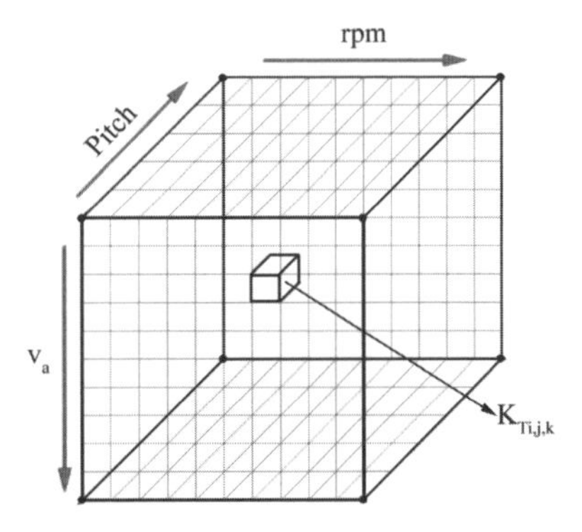

Figure 2. Arrangement of the propeller computed characteristics.

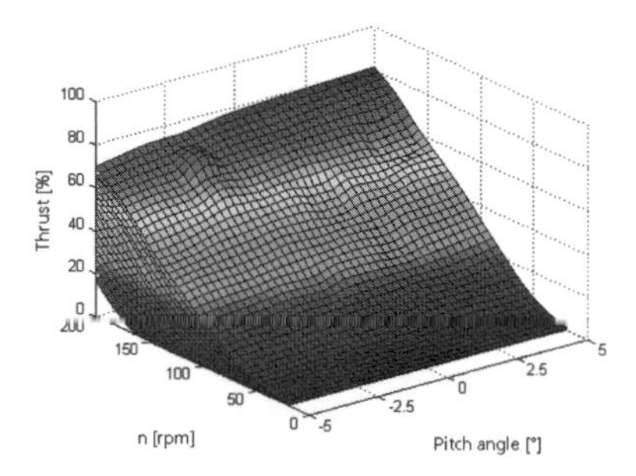

Figure 3. Thrust surface generate by propeller at different speed.

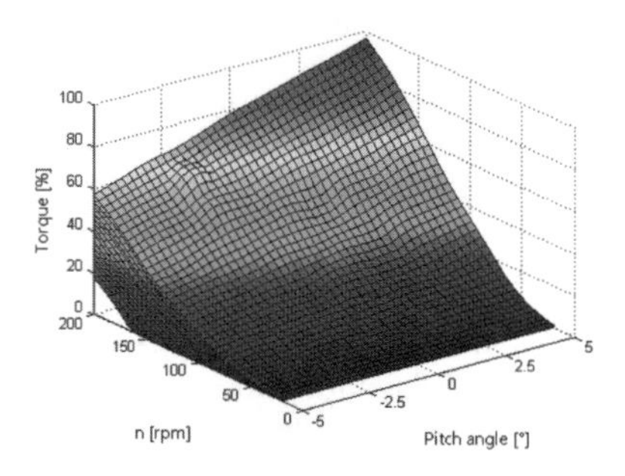

Figure 4. Torque surface generate by propeller at different speed.

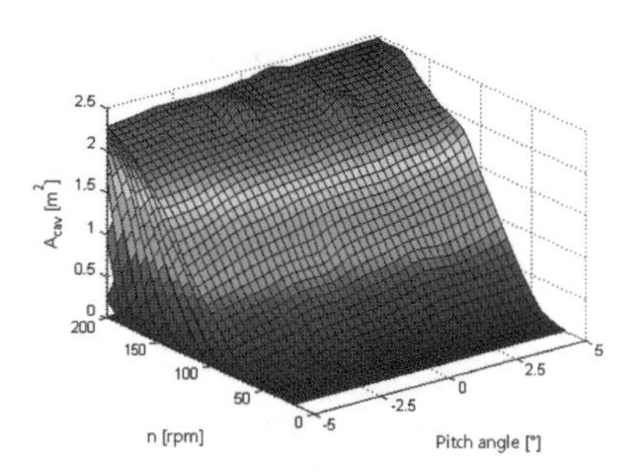

Figure 5. Back cavitating area generate by propeller at different speed.

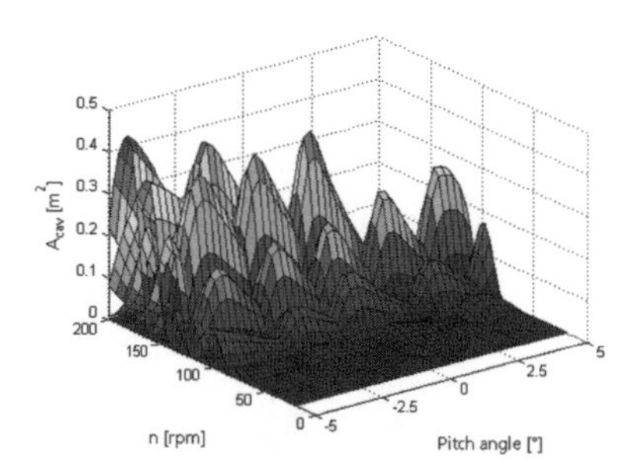

Figure 6. Face cavitating area generate by propeller at different speed.

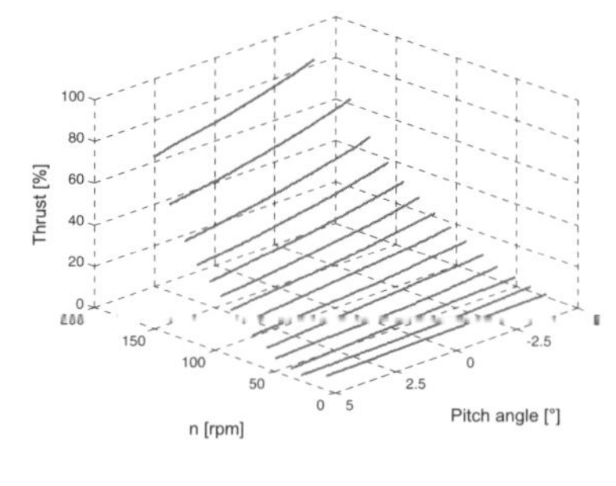

Figure 7. Propeller isothrust curves at different speeds.

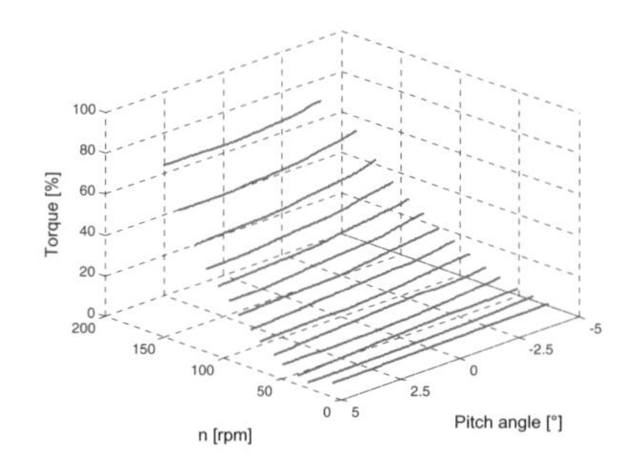

Figure 8. Propeller torque curves at different speeds.

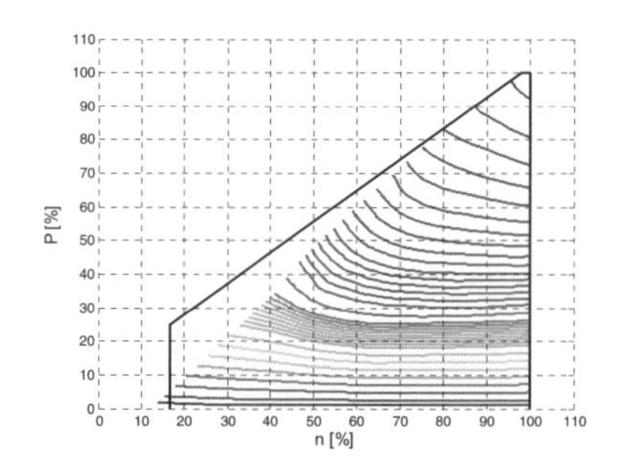

Figure 9. Engine consumption map.

Figure 10. Main dimension of FREMM frigate.

3 CASE STUDY

The FREMM frigate has been used as a test case in order to develop the matching algorithm.

The FREMM frigate is a multi-role ship developed in collaboration between Italian Navy and French Navy. The Italian version has the following main characteristics (Table 1):

The energy and propulsion system of the vessel consists of the following three main subsystems:

1. Main propulsion system, featured in a CODLAG arrangement, able to propel and manoeuvre the ship;
2. Auxiliary propulsion system, i.e. an azimuthal retractable thruster, located in the fore part of the hull in order to improve low speed manoeuvrability and to ensure mobility in emergency conditions (failure of the main propulsion system);
3. High Voltage (HV) Power plant, which consists of Diesel Generators (DG), an HV network with switchboards, the shore connection.

The general architecture of the whole energy and propulsion system is shown in Figure 11, where the three main subsystems are evident. The two Controllable Pitch Propellers (CPP) are driven by one Gas Turbine (GT) via a cross connected gearbox. Two Electrical Propulsion Motors (EPM), directly mounted on the two shaftlines can be used for the low speeds of the vessel during the silent running or together with the gas turbine for the full power. The use of the two different kinds of prime movers (EPM and/or GT) is ensured by two clutches between the gearbox and the two EPM and by another clutch between the GT and the gearbox.

The electric power is supplied by Isotta Fraschini DG located in two separated zones. Further electric power can be produced in GT mode by the two EPM, working as Shaft Generators (SG).

Table 1. FREMM main data.

FREMM		
L_{PP}	128.9	[m]
L_{OA}	140.4	[m]
B	19.70	[m]
D	11.00	[m]
T	5	[m]
Δ	6000	[t]
Speed	27	[kn]
Main Engine	GT GE/Avio LM 2500	

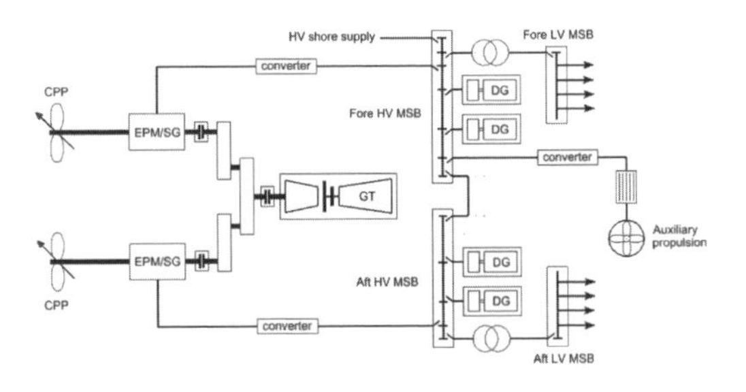

Figure 11. Propulsion plant.

4 APPLICATION AND RESULTS

The new developed matching procedure has been tested for the GT propulsion mode of the FREMM Frigate. Results are than compared with the matching procedure assessed by the traditional approach. In the following paragraph the numerical results have been presented for both the procedures starting from some Kt and Kq curved evaluated by the previous presented panel code. The matching has been carried out with an imposed blade propeller geometry with different pitch angle.

4.1 *Non cavitating standard matching procedure*

In the first stage, a traditional non-dimensional approach has been used in order to calculate total ship consumption for any pitch angle, as shown in the Figure 12. With this graph, as the traditional approach propose, the optimum working condition can be evaluated, taking in to account only the fuel flow rate. Should be noted that this procedure cannot take into account the real propeller condition as the Reynolds number of the propeller, because a nondimensional approach has been used.

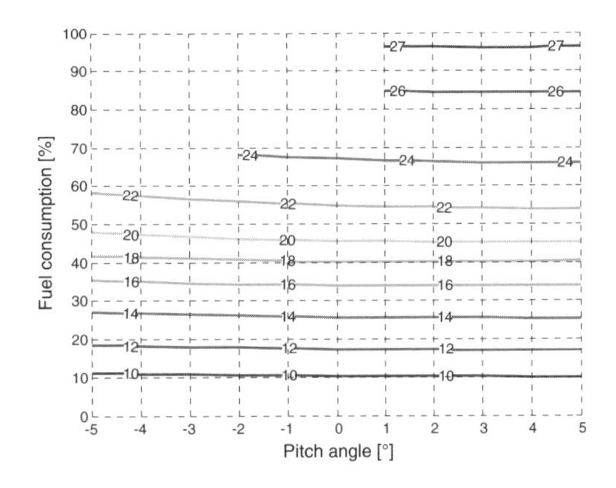

Figure 12. Fuel Consumption at constant speed.

4.2 *Non cavitating developed matching algorithm*

In the second stage, with the previous nondimensional curves, at each point of the structured matrix the trust and the torque has been re-arranged to be computed in the new code. This stage is useful to evaluate the capability of the methodology to predict the same equilibrium point with the same input data. The Figure 13 shown the fuel consumption for each pitch angle at the imposed velocity.

Can be noted that the results is in good agreement with the previous ones, the little gap is due to the approximation given by the numerical procedure and the interpolation used in the code.

Moreover, in theory, this procedure can take into account the different value of the working Reynolds number. In fact for the same advance coefficient the trust coefficient is affected with a small variation do to the different propeller working Reynolds number. The panel code, indeed, can take into account this effect throw a different viscous correction of the blade forces.

4.3 *Cavitating matching algorithm*

The last stage consist on the application of the new methodology proposed with the data evaluated from the cavitating panel code.

The Figure 14 shown the fuel consumptions in function of the pith angle, for the real working condition of the propeller taking in to account the cavitating phenomenon.

Could be interested present the results taking in to account the zones in which the cavitation occurs, as shown in the Figure 15. From that figure is simple choose the pitch angle in which the cavitation isn't present and in the same time the fuel consumption is the lower possible.

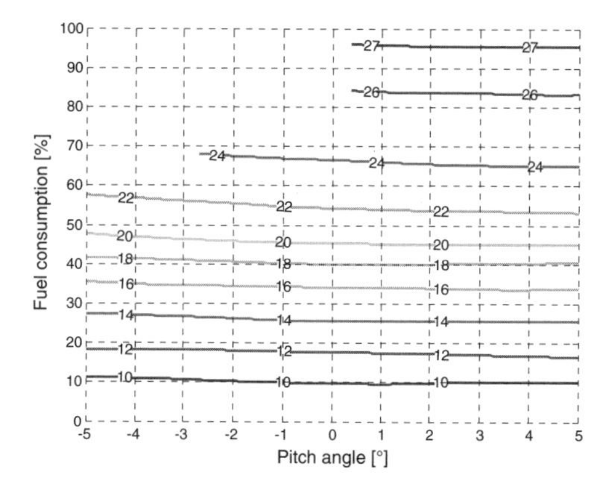

Figure 13. Fuel consumption at constant speed in non-cavitating condition.

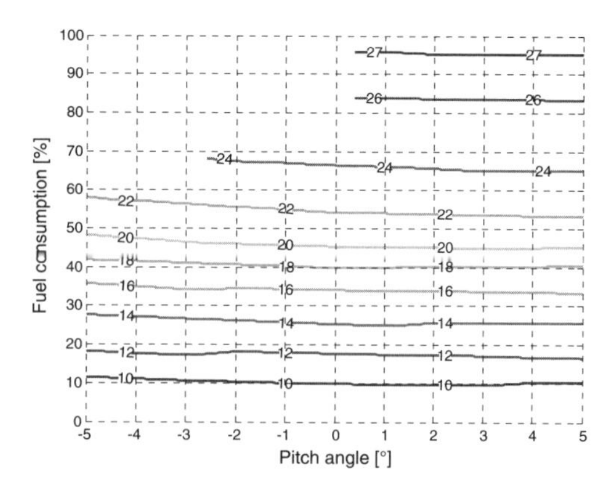

Figure 14. Fuel consumption at constant speed in cavitating condition.

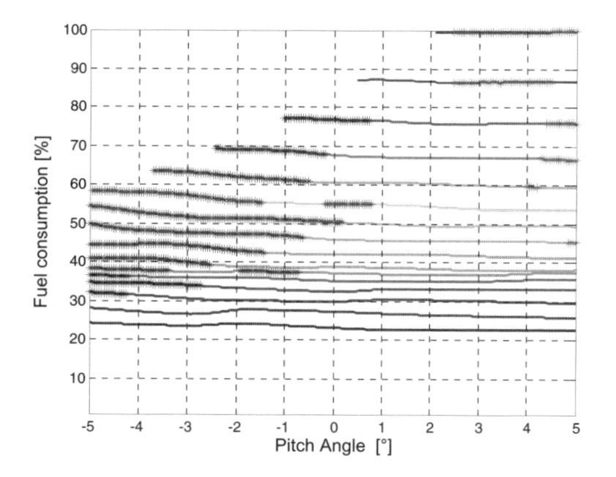

Figure 15. Fuel consumption at constant speed with evidence of cavitation zone.

5 CONCLUSIONS

The optimum performance of the propulsion system is a goal for any ship designer and ship owner. Fuel saving, low emissions and quiet running are key objectives for naval vessels as well as for cruise vessels and pax ferries. The authors established a minimum fuel consumption design methodology firstly applied to the Italian Navy Aircraft Carrier Cavour. Presently, with the proposed approach, the authors wish to extend the notion of 'optimum performance' not only to fuel consumption and exhaust emissions but also to propeller cavitation and noise.

The new methodology leave apart the nondimensional K_T/J^2 approach and uses a full computational approach for the propeller performance and the matching optimisation. The procedure give the possibility to select the 'weigth' of the optimization variables (consumption/cavitation) in order to identify the best speed/rpm/pitch combination for the different mission profiles of the naval vessels.

REFERENCES

Coney, W.B. 1989. A Method for the Design of a class of Optimum Marine Propellers, *Ph.D. Thesis, Massachusetts Institute of Technology, USA, 1989*.

Figari, M. and Altosole, M. 2007. Dynamic behavior and stability of marine propulsion systems, *Journal of the Engineering for the Maritime Environment, Proc. IMechE Vol. 221 Part M*, 2007.

Figari, M. and Guedes Soares, C. 2009. Fuel consumption and exhaust emissions reduction by dynamic propeller pitch control, *Proc. MARSTRUCT 2009, Lisbon, 16–18 March 2009*.

Gaggero, S. 2010. Development of a Potential Panel Method for the Analysis of Marine Cavitating and Supercavitating Propellers. *Ph.D. Thesis, University of Genova, Italy, 2010*.

Gaggero, S. and Brizzolara, S. 2009. A Panel Method for Trans-Cavitating Marine Propellers, *7th Int. Symposium on Cavitation, Ann Arbor, Michigan, USA, 2009*.

Gaggero, S. and Brizzolara, S. 2009. Parametric CFD Optimization of Fast Marine Propellers, *Proc. 10th International Conference on Fast Sea Transportation, Athens, Greece, October 2009*.

Lerbs, H.W. 1952. Moderately loaded propellers with a finite number of blades and an arbitrary distribution of circulation, *Transaction of the Society of Naval Architects and Marine Engineers, 60, 1952*.

Morgan, W.B. 1960. The design of counter-rotating propellers using Lerbs. *Transaction of the Society of Naval Architects and Marine Engineers*, 68, 1960.

Sustainable Maritime Transportation and Exploitation of Sea Resources – Rizzuto & Guedes Soares (eds)
 ISBN 978-0-415-62081-9

A new logic for controllable pitch propeller management

F. Balsamo, F. De Luca & C. Pensa
Università degli Studi di Napoli "Federico II", Naples, Italy

ABSTRACT: In this study we want to propose an active logic that, continuously, optimizes the configuration of the propeller and motor speed taking into account changes in resistance and wake.

The working principle of the control system is based on the measurement of the torque absorbed by the propeller and the engine speed, to obtain the actual thrust and advance speed coefficients.

Based on these data, the controller identifies the configuration of the propeller for the best performance of the entire propulsion chain, from engine to propeller. Moreover, in addition to torque and speed limits of the engine, the control system chooses pitch angle taking into consideration the propeller's cavitation.

1 INTRODUCTION

1.1 *The informing principle*

Controllable pitch propellers allow a greater flexibility in propulsion; at a given operating point the required thrust can be obtained with different pairs of propeller pitch and speed values. The same ship motion can be achieved with several pairs of speed and pitch but with different efficiencies.

This paper presents an active logic that, continuously getting measurements of propeller shaft torque, estimates the optimum CPP pitch and rotational speed to minimize fuel consumption.

The knowledge of propeller torque allows to identify the actual hydrodynamic working point, from open water characteristics of propeller, according to a procedure that is quite similar to a towing tank self propulsion test.

The optimization is based not only looking at the best propeller efficiency (η_0) but at engine efficiency (η_m) also, evaluating the combined efficiency ($\eta_0 \cdot \eta_m$).

1.2 *Services and working point suitable for the control system proposed*

For the estimation of high propulsive efficiency, the knowledge of the actual hydrodynamic working point of the propeller is more useful more are the diversified services and the sailing conditions of the ship.

The application of this control is quite effective because of unpredictable sailing point: as a case in point, when high thrust and low advance speed of the propeller is high over an extended period of time.

In these conditions the crossing velocity of the water through the propeller disk is strongly dependent on the suction of the propeller and on small variations of ship's speed. In other words, small variations of the working point causes significant variations of the propeller diagram, taking in to account the engine behaviour.

Sharing the above-mentioned considerations, is well-founded to suppose that same types of ship (Trawlers and Tugs but also Patrol ships and Cutters) are particularly suitable to resort to the logic here exposed.

Finally, the use of propeller as gauge to measure the V_A entails three further advantages:

- the evaluation of wake variations in time due to the increase of momentum given to the viscous field (typically for growth of fouling),
- the overcoming of the scale effects and inaccuracies due to the limitations of experimental and numerical methodologies and
- the online evaluation of η_0 (and its partial derivative) allows the system to consider values near the envelope of maximum efficiency. That values are avoided because of they are close to a zone that involves the greatest gradient of η_0.

2 PROPELLER AND ENGINE CHARACTERISTICS FORMULATION

2.1 *Controllable pitch propeller open water characteristic*

To achieve our purpose we need a flexible tool to describe the propeller; an unusual way to represent the characteristics of controllable pitch propellers is proposed.

The propeller open water characteristic is given by the following parameters:

Advance Coefficient:	J	$= V_A / (n\, D)$
Trust Coefficient:	K_T	$= T / (\rho\, n^2 D^4)$
Torque Coefficient:	K_Q	$= Q / (\rho\, n^2 D^5)$
Prop. efficiency:	η	$= V_A\, T/2\,\pi\, n\, Q$
		$= J/2\pi\, (K_T / K_Q)$

where:

D = propeller diameter (m)
ρ = water density (kg/m^3)
n = rotational speed of propeller (s^{-1})
T = thrust (N)
Q = torque (N m)
V_A = speed of advance (m/s)

For a fixed blades propeller the adimensional coefficients K_T and K_Q are functions of the advance coefficient J only. For a CPP these coefficients are also functions of the blade orientation, so they are expressed as follow:

$$K_Q = K_Q(J;\, p)$$
$$K_T = K_T(J;\, p)$$

where p is the blade orientation angle starting from a reference pitch P_0.

The orientation angle p has the same role of the *Pitch/Diameter Ratio* of a classical propeller systematic series.

The characteristics of fixed blade propellers are usually described with a polynomial form of J. Typically a polynomial of degree four is enough to describe a single quadrant propeller characteristic; a more complex form is necessary to describe a four quadrant characteristic.

Regarding the single quadrant the characteristics are expressed as shown:

$$K_T(J) = A_4\, J^4 + A_3\, J^3 + A_2\, J^2 + A_1\, J + A_0$$
$$K_Q(J) = B_4\, J^4 + B_3\, J^3 + B_2\, J^2 + B_1\, J + B_0$$

where A_i and B_i are constants.

To describe a CPP, coefficient A_i and B_i must be functions of blade orientation:

$$K_T(J;\, p) = A_m\,(p)J^m + \ldots + A_4\,(p)J^4 + A_3\,(p)J^3 + A_2\,(p)J^2 + A_1\,(p)J + A_0\,(p)$$
$$K_Q(J;\, p) = B_m\,(p)J^m + \ldots + B_4\,(p)J^4 + B_3\,(p)J^3 + B_2\,(p)J^2 + B_1\,(p)J + B_0\,(p)$$

It is possible to describe A_i and B_i as a polynomial form of blade angle

$$A_i\,(p) = a_{in}p^n + a_{in-1}p^{n-1} + \ldots + a_{i2}p^2 + a_{i1}p + a_{i0}$$
$$B_i\,(p) = b_{in}p^n + b_{in-1}p^{n-1} + \ldots + b_{i2}p^2 + b_{i1}p + b_{i0}$$

In this way the open water CPP characteristics are completely described by $2 \times n \times m$ constants.

To simplify the discussion we will use the following vector notation:

$$\mathbf{p}^T = \{1, p, p^2, \ldots, p^n\}$$
$$\mathbf{p_P}^T = \{0, 1, 2p, 3p^2, \ldots, np^{n-1}\}$$
$$\mathbf{J}^T = \{1, J, J^2, \ldots, J^m\}$$
$$\mathbf{J_J}^T = \{0, 1, 2\,J, 3\,J^2, \ldots, mJ^{m-1}\}$$

The coefficients are organized in the following matrices:

$$\mathbf{A} = \begin{pmatrix} a_{00} & \ldots & a_{0n} \\ \ldots & \ldots & \ldots \\ a_{m0} & \ldots & a_{mn} \end{pmatrix} \qquad \mathbf{B} = \begin{pmatrix} b_{00} & \ldots & b_{0n} \\ \ldots & \ldots & \ldots \\ b_{m0} & \ldots & b_{mn} \end{pmatrix}$$

We could describe K_Q, K_T and η_0 functions in a more simple way:

$$K_T(J;p) = (\mathbf{J^T A p})\quad K_Q(J;p) = (\mathbf{J^T B p})$$
$$\eta_0\,(J;p) = J/\,2\,\pi\,(\mathbf{J^T A p})\,(\mathbf{J^T B p})^{-1}$$

Also for the partial derivative:

$$\frac{\partial \eta_0}{\partial J} = 1/2\pi(\mathbf{J^T B P})^{-2}\{[(\mathbf{J^T A p}) + \mathbf{J}(\mathbf{J_J}^T\mathbf{A p})]\,(\mathbf{J^T B p}) - [\mathbf{J}(\mathbf{J^T A p})(\mathbf{J_J}^T\mathbf{B p})]\}$$

$$\frac{\partial \eta_0}{\partial p} = J/2\pi(\mathbf{J^T B P})^{-2}\left[(\mathbf{J^T A p_p})(\mathbf{J^T B p})] - (\mathbf{J^T B p_p})(\mathbf{J^T A p})\right]$$

$$\frac{\partial K_T}{\partial J} = (\mathbf{J_J}^T\mathbf{A p});\ \frac{\partial K_T}{\partial p} = (\mathbf{J^T A p}_{\mathrm{p}})$$

$$\frac{\partial K_Q}{\partial J} = (\mathbf{J_J}^T\mathbf{B p});\ \frac{\partial K_Q}{\partial p} = (\mathbf{J^T B p}_{\mathrm{p}})$$

Given propeller diameter, advance speed and trust, to find the propeller operating point, the adimensional coefficient K_T/J^2 is used. This coefficient does not depend on rotational speed of propeller:

$$K_T/J^2 = T\,/\,(\rho D^2 Va^2) = J^{-2}\,(\mathbf{J^T A p})$$

Whose partial derivatives are:

$$\frac{\partial \frac{K_T}{J^2}}{\partial p} = J^{-2}\{J^{-2}(\mathbf{J_J}^T\mathbf{A p}) - 2J(\mathbf{J^T A p})\}$$

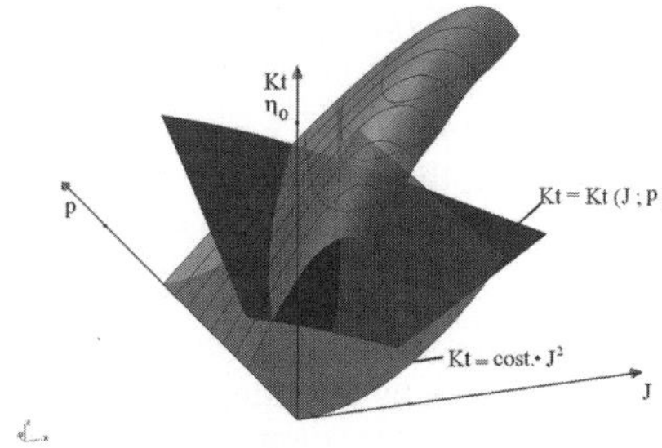

Figure 1. Shapes of open water characteristic of controllable pitch propeller.

$$\frac{\partial \frac{K_T}{J^2}}{\partial J} = J^{-2}\{J^2(\mathbf{J}^T\mathbf{B}\mathbf{p}_p) - (\mathbf{J}^T\mathbf{A}\mathbf{p}_p)\}$$

To set the coefficients of the matrices **A** and **B** the two unconstrained optimization problems

$$\min_{\mathbf{A}} \|\mathbf{y} - \hat{\mathbf{y}}\|^2$$
$$\min_{\mathbf{B}} \|\mathbf{w} - \hat{\mathbf{w}}\|^2$$

have to be solved.

In the above formulas **w** and **y** are the sets of experimental K_Q and K_T values while $\hat{\mathbf{w}}$ and $\hat{\mathbf{y}}$ are the corresponding sets plotted by the proposed formula.

Applying this procedure on many CPP open water experimental data, the order of magnitude of $\|\mathbf{y} - \hat{\mathbf{y}}\|^2$ and $\|\mathbf{w} - \hat{\mathbf{w}}\|^2$ are 10^{-4} and 10^{-5} respectively; moreover no value of $|y_i - \hat{y}_i|$ and $|w_i - \hat{w}_i|$ are greater than 0.4%.

The high effectiveness of the solution used to find matrices **A** and **B** is determined by the smoothness of the functions that describe the phenomenon.

2.2 *Engine characteristic*

To work simultaneously on engine and propeller effectiveness is necessary to obtain an agile formulation of engine fuel consumption. These data are usually expressed in engine maps, as shown in Figure 2.

With this purpose we follow the same steps of paragraph 2.1, finding the polynomial formulation of fuel consumption in terms of power and rotational speed

$$\mathbf{N}^T = \{1; N; N^2\}; \mathbf{n}^T = \{1; n_m; n_m^2; n_m^3; n_m^4; n_m^5\}$$
$$C_s = \mathbf{N}^T \mathbf{C}\, \mathbf{n} \in [n_{min}; n_{max}] \times [N_{min}; N_{max}]$$

Where:

- $\mathbf{C}$ = coefficients matrix
- n_m = rotational speed of engine (s^{-1})
- H_i = Net heating value (MJ/kg)
- Cs = specific fuel consumption (g/kWh)
- N = engine power (kW)

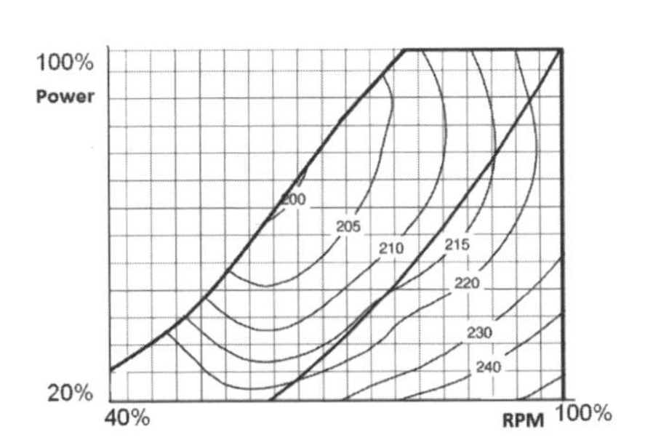

Figure 2. Typical engine performance diagram.

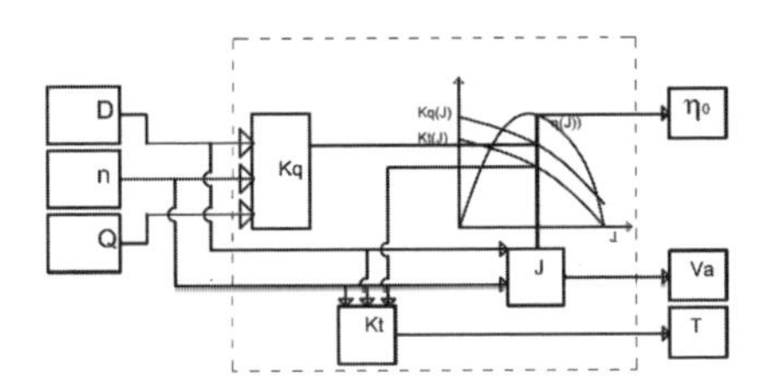

Figure 3. Thrust and advance velocity estimation, at a given pitch.

Finally the engine efficiency is given by:

$$\eta_m = 3600/H_i\, C_s$$

2.3 *Estimation of self propulsion coefficients*

To obtain a continuous estimation of the operating point, torque and rotational speed have to be measured on the shaft line. In this way it is possible to have an indirect estimation of wake using the propeller as a measurement instrument: basically like a self-propulsion towing tank test.

Compared to towing tank procedure, here the thrust will be indirectly estimated by the propeller characteristic and not measured.

Obviously in this way it is not possible to evaluate the wake distribution on disk propeller; usually this effect is taken into account by introducing the relative rotative efficiency η_R, that can be estimated through a direct thrust measurement. Thrust measurements are commonly done in towing tank tests but are not effective onboard.

3 OPTIMIZATION PROBLEM

3.1 *Objective function*

To minimize fuel consumption the only conscious manageable losses, varying the pitch, are those related to the engine performances and to isolated propeller efficiency. Therefore the objective function to be maximized is $\eta_0 \times \eta_m$.

3.2 *Equality constrain*

To operate an optimization without varying the operating point of the ship it is necessary to introduce an equality constrain that does not depend on propeller rotational speed, represented by the K_T^*/J^2 term, that represents the thrust coefficient required, function of trust, diameter and advance velocity.

$$K_T^*/J^2 = T / (\rho D^2 Va^2)$$

The equality constrain could be written in this form:

$$\frac{Kt(p;J)}{J^2} - \frac{Kt^*}{J^2} = 0$$

The first term is the thrust coefficient obtainable by propeller at different pitch and advance coefficient.

In an explicit form:

$$J^{-2}(\mathbf{J^T A p}) - T/(\rho D^2 Va^2) = 0$$

3.3 *Inequality constrains and parameters bounds*

The inequality constrains presented are substantially:

- the operational limit of the propeller and the engine;
- great variation of propeller efficiency;
- cavitation.

The propeller limits represent the boundary of experimental data expressed by:

$$\begin{cases} J - J_{\min} \geq 0 \\ J_{\max} - J \geq 0 \\ p - p_{\min} \geq 0 \\ p_{\max} - p \geq 0 \end{cases}$$

The engine limits, that are presented in paragraph 2.2, are:

$$\begin{cases} n - n_{\min} \geq 0 \\ n_{\max} - n \geq 0 \\ N_{\max}(n) - N \geq 0 \end{cases}$$

where $N_{max}(n)$ is a function that describes the upper power limit varying the rotational propeller speed.

In order to avoid working points subjected to great efficiency variations a gradient constrain on propeller efficiency can be introduced.

$$\frac{\partial \eta_0}{\partial J} \geq 0$$

This means that the zone with negative derivative, that involves the greatest variation, is neglected in whole pitch angle range. In Figure 4 is shown the cross out zone for a fixed pitch position.

The constrain could be written in the following explicit form:

$$1/2\pi\,(\mathbf{J^T B p})^{-2}\,\{[(\mathbf{J^T A p}) + J(\mathbf{J_J{}^T A p})]\} \geq 0$$

To implement a cavitation limit, if there are no experimental or numerical data for the considered propeller, an equivalent Burril cavitation curves can be considered:

$$\tau - \tau_{cr}(\sigma) \geq 0$$

3.4 *Constrained optimization*

The whole problem could be expressed in the following form

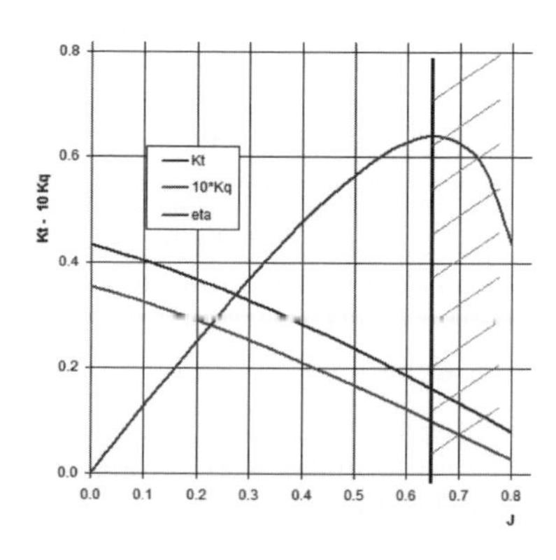

Figure 4. Cross out zone for a fixed pitch position.

Figure 5. optimal control block (identification logic).

$$\begin{cases} \max_p \eta_0(p;J)\cdot \eta_m(N(p,J);n(p,J)) \\ \begin{cases} J - J\min \geq 0 \\ J\max - J \geq 0 \\ p - p\min \geq 0 \\ p\max - p \geq 0 \end{cases} \\ \begin{cases} n - n\min \geq 0 \\ n\max - n \geq 0 \\ N - N\min \geq 0 \\ N\max(n) - N \geq 0 \end{cases} \\ \eta_{0J}(p;J) \geq 0 \\ \dfrac{Kt(p;J)}{J^2} - \dfrac{Kt^*}{J^2} = 0 \\ \tau - \tau_{cr}(\sigma) \geq 0 \end{cases}$$

Solving this problem is possible to obtain optimal surfaces that return the optimal propeller pitch and speed as functions of trust and advance speed. In Figure 5 the scheme of the optimal control system is shown.

4 CONTROL STRATEGIES

Onboard ships fitted with CP propeller the thrust is normally achieved by setting propeller pitch and speed according to a curve called propeller combinator curve.

This curve, for each command lever position assigns univocally a pairs of values for speed and pitch, controlling in this way the set point of main engine governor and of propeller pitch actuator, generally hydraulic.

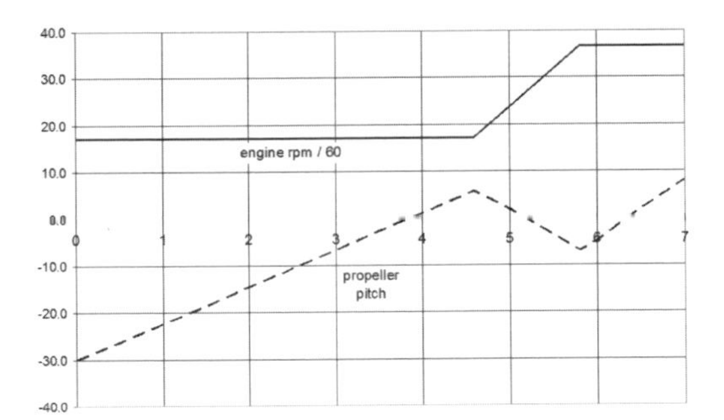

Figure 6. Classical combinator curve.

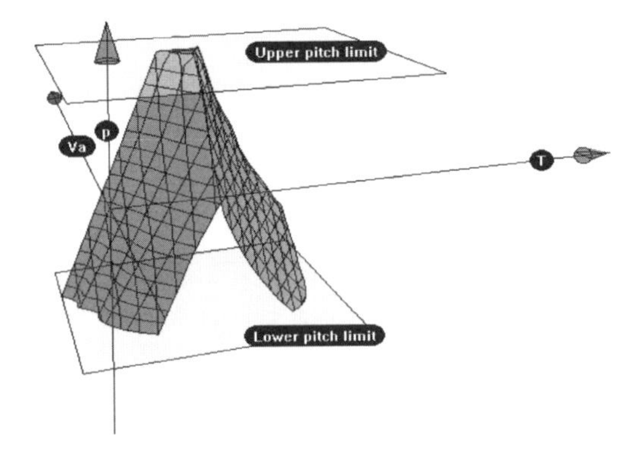

Figure 7. Combinator surface (Optimal pitch surface).

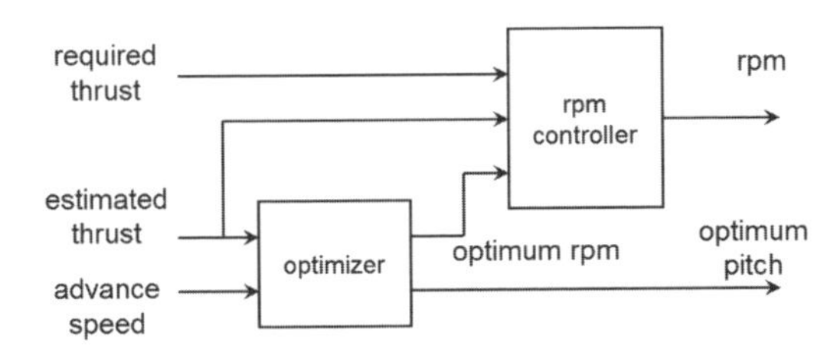

Figure 8. Control logic.

The combinator curve is determined at a design stage or at propulsion system commissioning stage.

Some times one or more combinator curve can be provided, but the master has to switch manually from one curve to another. In any case, according to ship operating conditions and to master sensitivity, the propeller runs closer or not to its optimum.

If getting the best propulsion system performances at any operating condition is desired, a propeller combinator surface must be considered instead.

An optimal CPP control system chooses the working condition not only on the basis of command lever settings but of ship running condition also. The command lever does not determine univocally the values of propeller pitch and speed, due to the degree of freedom that the system allows.

From the optimization procedure the optimal values of propeller speed and pitch are expressed as function of thrust coefficient and speed of advance, but none of these two parameters is suitable to be used directly as a command input.

To determine the ship running condition it seems reasonable to consider the thrust in a dimensional form.

The command lever acts by setting a desired value of thrust, so that the ship will reach a speed depending on loading and environmental condition; it is concern of ship master to select a proper thrust value. In this way the effectiveness of propulsion chain doesn't affect the ship speed.

To get the required thrust a controller is needed; it takes into account the actual thrust estimate and sets the propeller rotational speed, n. If the estimated thrust is different from the required one, the controller varies engine rpm setting in order to cancel any difference.

The corresponding pitch value is determined by the optimizer as function of actual thrust and advance speed estimate. The optimizer gives also the optimal propeller speed n*; the thrust controller must achieve the required thrust by minimizing the difference n*-n.

In this way, during the transient time the pairs of speed and pitch can be considered laying close to a Pareto frontier.

After transient time, the propulsion system will find a steady state working condition with an optimal pair of pitch and rotational speed.

Because the optimizer consider also a cavitation limit (in the sense that a limited amount of cavitation could be tolerated, i.e., 5%), the system will preserve the optimal working point by excessive cavitation.

The Figures 9a and 9b show the schemes of the whole control system.

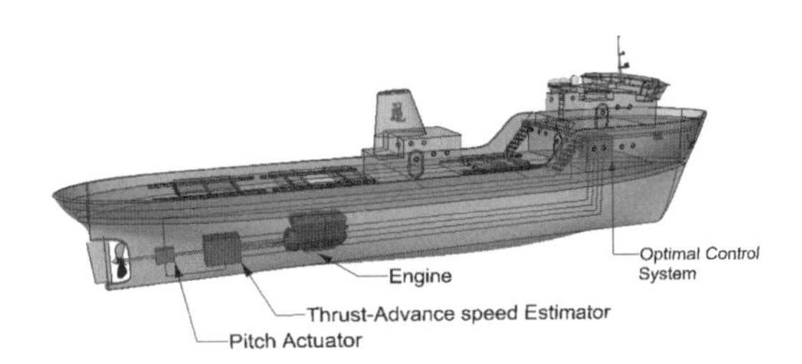

Figure 9a. Control system.

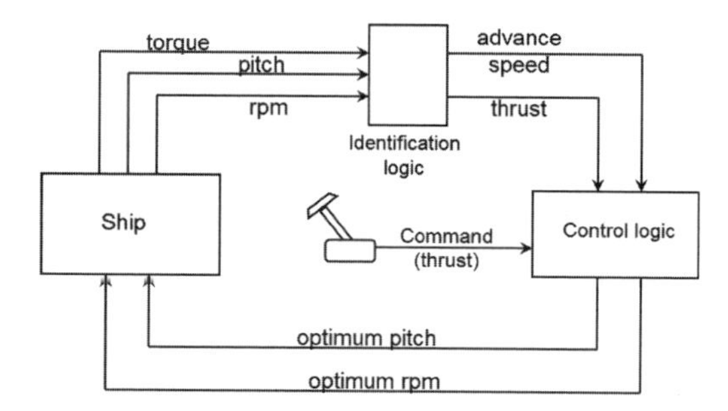

Figure 9b. Control system: working diagram.

5 TEST CASE

As a starting step to assess the feasibility of the proposed optimal control system a series of test on a simulation model has been conducted.

The ship modeled is a 24 meter long passenger ferry, having a maximum speed of 13 kn, powered by a 100 kW@2200 rpm diesel engine, subjected to payload variations (about 40% of displacement) and to wave resistance variations induced by sailing in shallow water (up to 15%). The propeller considered is the propeller E.028 tested in Towing Tank of *Dipartimento di Ingegneria Navale* of Naples.

The model allows to evaluate difference in terms of specific fuel consumption between the ship propelled with a typical combinator CPP curve and the control system proposed.

The simulations start from design conditions so that the output of optimal control system matches the combinatory output. After a certain time a resistance variation occurs and, unlike the classical combinator control, the optimal control system searches new values of pitch and speed, as shown in the following diagram.

Table 1 shows the results of simulation carried out with a traditional combinatory curve control. Starting from an initial sailing condition, because of an increasing in resistance imposed, the ship find a new steady condition, obviously a lesser speed occurs.

Table 2 shows the same conditions but with the optimal control working. The comparison shows, in that particular case, a improvement of efficiency with respect to the case of traditional control.

Because the optimal control maintains the thrust to a required value, when an increase of hull resistance occurs the ship will experience a speed reduction greater than in the case of a traditional control. To compare date at same speed an increasing of required thrust is considered when the resistance rises, so to obtain the same ship speed as in traditional command; the results are shown in

Table 1. Results of simulation with traditional combinator curve control.

	Engine Speed [rpm]	Pitch [deg]	Speed [m/s]	SFOC [g/kWh]
Initial condition 1	1398	−3.3	5.57	294.1
with disturbance	1398	−3.3	5.38	304.2
Variation			**−0.19**	**10.1**
Initial condition 2	1545	−4.9	5.78	299.1
with disturbance	1545	−4.9	5.58	311.5
Variation			**−0.20**	**12.4**
Initial condition 3	1704	−6.4	5.98	306.4
with disturbance	1704	−6.4	5.78	321.3
Variation			**−0.20**	**15.0**
Initial condition 4	1875	−7.9	6.19	315.6
with disturbance	1875	−7.9	5.98	333.2
Variation			**−0.21**	**17.6**
Initial condition 5	2060	−9.3	6.40	323.9
with disturbance	2060	−9.3	6.19	343.5
Variation			**−0.21**	**19.6**

Table 2. Results of optimal control without thrust correction.

	Engine Speed [rpm]	Pitch [deg]	Speed [m/s]	SFOC [g/kWh]
Initial condition 1	1398	−3.3	5.57	294.1
with disturbance	1379	−3.4	5.31	291.2
Variation			**−0.26**	**−2.9**
Initial condition 2	1545	−4.9	5.78	299.2
with disturbance	1519	−4.9	5.49	297.0
Variation			**−0.28**	**−2.2**
Initial condition 3	1704	−6.4	5.98	306.6
with disturbance	1664	−6.3	5.68	305.3
Variation			**−0.31**	**−1.3**
Initial condition 4	1875	−7.9	6.19	315.8
with disturbance	1817	−7.6	5.86	315.5
Variation			**−0.33**	**−0.3**
Initial condition 5	2060	−9.2	6.40	324.3
with disturbance	1980	−8.9	6.04	325.1
Variation			**−0.35**	**0.9**

Table 3, confirming that the system improves any case the whole system efficiency as the reduction of specific fuel consumption shows.

As the specific fuel consumption values shows, the initial operating condition for the ship tested are not the best, but this is the case where this approach has sense.

It has also to be pointed out that the simulation model is based on hull resistance, propeller

Table 3. Results of optimal control with thrust correction.

	Engine Speed [rpm]	Pitch [deg]	Speed [m/s]	SFOC [g/kWh]
Initial condition 1	1398	–3.3	5.57	294.1
with disturbance	1433	–4.0	5.38	293.2
Variation			**–0.19**	**–0.9**
Initial condition 2	1545	–4.9	5.78	299.2
with disturbance	1589	–5.6	5.58	300.5
Variation			**–0.19**	**1.3**
Initial condition 3	1704	–6.4	5.98	306.6
with disturbance	1752	–7.1	5.78	311.1
Variation			**–0.20**	**4.5**
Initial condition 4	1875	–7.9	6.19	315.8
with disturbance	1927	–8.5	5.98	322.3
Variation			**–0.20**	**6.5**
Initial condition 5	2060	–9.2	6.40	324.3
with disturbance	2117	–9.9	6.19	330.4
Variation			**–0.21**	**6.2**

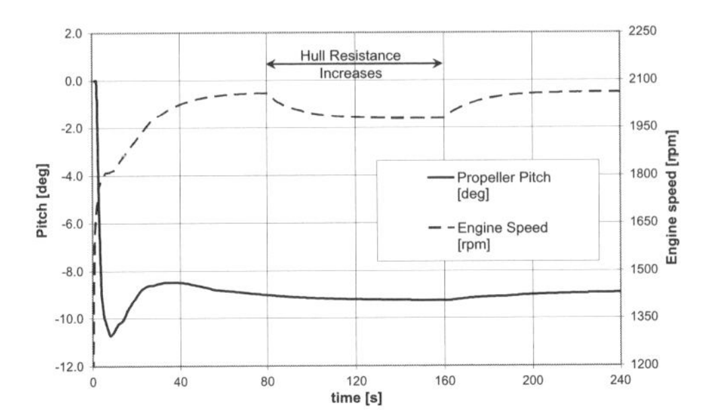

Figure 10. Test case simulation results.

and engine data curves that are not affected by scale effects.

In a real case there will be more uncertainties than in a simulation model: it is to be noted that the η_R coefficient, that represents in a certain manner an uncertainty coefficient of the propeller open water curves, has been considered in the model of ship but obviously the optimization routine doesn't consider it. This aspect leads to an error in the estimate of the effective propeller thrust.

Nevertheless the system finds a working condition that allows the best efficiency.

6 FURTHER COMMENTS ON CONTROL

More considerations must be done in order to improve the comprehension of the control system and to make it appropriate for a practical application onboard.

First of all, the control system proposed estimates propeller advance speed and thrust by torque measurement; the quality of acquired data is a fundamental aspect of the matter. The measured torque data can be affected by disturbances due to sensor noise and to environmental noise.

Many authors have dealt with this topic in depth, mainly in works concerning dynamic positioning plants.

Moreover another important aspect is the quality of ship, engine and propeller data used to carry out the optimization procedure.

In particular, the error propagation of thrust estimate must be considered.

To assesses the effectiveness of this proposed control its sensitivity to measurements errors, model uncertainty and external disturbance must be evaluated.

The controller used in the simulation has a simple PID structure and gives good results in a simplified model; more advanced controller may improve performance and stability robustness also considering more realistic disturbances and uncertainties.

7 OPTIMIZZATION TECNIQUES

Optimization techniques are used to find a set of design parameters, $\mathbf{x} = \{x_1, x_1, ..., x_n\}$, that can in some way be defined as optimal. In a simple case this might be the minimization or maximization of some system characteristic that is dependent on x. In a more advanced formulation the objective function, $f(x)$, to be minimized or maximized, might be subject to constraints in the form of equality constraints, $g_i(\mathbf{x}) = 0$ (i = 1, ..., n); inequality constraints, $h_i(x) = 0$ (i = 1, ..., m); and/or parameter bounds, x_l, x_u.

General Problem description is stated as

$$\begin{cases} \min_x f(x) \\ h_i(x) = 0 \\ g_i(x) \le 0 \end{cases}$$

methods could be focused on the solution of the Karush-Kuhn-Tucker (KKT) equations:

$$\nabla f(x_0) + \sum_{i=1}^{m} \mu_i \nabla g_i(x_0) + \sum_{i=1}^{l} \nu_i \nabla g_i(x_0) = 0$$
$$\mu_i g_i(x_0) = 0, \mu_i \ge 0, i = 1, ..., m$$

The KKT equations are necessary conditions for optimality of a constrained optimization problem. If the problem is a so-called convex programming problem, that is, $f(x)$ and $g_i(x)$, i = 1, ..., m and $h_i(x)$, i = 1, ..., n are convex functions, then the

KKT equations are both necessary and sufficient for a global solution point.

First a sequential quadratic programming is solved to obtain modification of **d**, where **d** is the descent direction and $\mathbf{x}^{(k+1)} = \mathbf{x}^{(k)}\ \alpha\mathbf{d}$ is the line search.

The SQP (sequential quadratic programming) is presented as follow:

$$P(d) = F(x^{(k)}) + \nabla F^T \left[x^{(k)}\right] \cdot d + \frac{1}{2} d^T \cdot H^{(k)} \cdot d$$
$$h_i\left[x^{(k)}\right] + \nabla h_i^T \left[x^{(k)}\right] \cdot d = 0 \qquad i = 1, N$$
$$g_i\left[x^{(k)}\right] + \nabla g_i^T \left[x^{(k)}\right] \cdot d \leq 0 \qquad i = 1, M$$

That represent a quadratic form of the objective function and a linearization of constrains.

Where **H** is the Hessian matrix of KKT equation:

$$H = \nabla^2 F + \sum_{i=1}^{M} \mu_i \nabla^2 g_i(x_0) + \sum_{i=1}^{N} \nu_i \nabla^2 g_i(x_0)$$

To find the solution the Quasi Newton method. Descent direction is defined as follow:

$$\mathbf{d} = -\mathbf{Q}\nabla\mathrm{F}$$

were Q is an approximation of the Hessian matrix. To find Q, BFGS (Broyden-Fletcher-Goldfarb,-Shanno) method was used.

8 CONCLUSIONS

Direct measurements of thrust and torque gives a knowledge of propeller hydrodynamic parameters, so that, in the case of CP propeller, the best pair of revolution speed and pitch values can be set. Torque measurements are quite easier and more practicable, than thrust.

Thrust estimation based on torque measures are used for positioning system control fitted with fixed pitch propellers (L. Pivano et al., 2009).

On the other hand, optimization of CPP has already been considered by other authors (R.A. Morvillo 1996), based only on an advance speed estimate, statistically or experimentally (in model scale) predetermined and referring to a limited number of propeller working points.

In the present work a procedure based on thrust estimate has been proposed, with aim to choose objectively the optimum propeller speed and pitch.

The problem has been dealt by creating a combinator surface that take into account the combined efficiency $\eta_0 \times \eta_m$, where optimum propeller pitch and rotational speed are expressed in terms of thrust and advance speed.

The procedure has been tested on a quite simple model making use of an unsophisticated controller.

On the test model the results shows that in the case the sailing condition varies, the optimum controller achieves better efficiencies than a traditional combinatory curve control, maintaining a stable behavior.

In many types of vessels that solution could be particularly suitable to offer large reduction of consumption because of unpredictable sailing condition.

Finally it is possible to synthesize that the procedure showed:

1. has highlighted the reliability of the informing principle;
2. has confirmed the validity of the interpolation technique and of the characteristic surfaces-computation procedures;
3. has demonstrated that a quite simple and unsophisticated controller is able to reach the appointed objective.

In the same time, herein it has been highlighted some critical point that should bee studied in depth. In particular the research will be developed towards three directions:

1. the individuation of ship types and services suited to take advantage of the potential of the logic presented;
2. the realization of robust procedures of measurement, signal treatment and computation of hydrodynamic coordinates indispensable to identify the working point of the propeller;
3. the implementation of new control logics directed towards different objective functions, for example the realization of the maximum trust.

REFERENCES

Doi, M., Nagamoto, K., Takehira, T. & Mori, Y. 2009. *Deay time of Propelling Force with Ship's Controllable Pitch Propeller (CPP)*. IGROS-SICE International Conference 2009 Fukuoka, Japan.

Fossen, T.I. & Blanke, M. 2005. Nonlinear output feedback control of underwater vehicle propellers using feedback from estimated axial flow velocity. IEEE Journal of Oceanic Engineering, vol. 25, no. 2.

Guibert, G., Foulon, E. & Ait-Ahmed, N. 2005. *Thrust control of electric marine thrusters*. Industrial Electronics Society, 2005. IECON 2005. 31st Annual Conference of IEEE.

Morvillo, R.A. 1996. Application of Modern Digital Controls to Improve the Operational Efficiency of Controllable Pitch Propellers In SNAME Transaction, Vol. 104, pp. 115–136.

Nordström, H.F. 1945. *Propellers With Adjustable Blades Results of Model Experiments*. Meddelande Från Statens Skeppsprovningsantalt (pubblication of the swedish state shipbuilding experimental tank), Göteborg.

Pivano, L., Joansen, T. & Smongelli, O.N. 2009. *A Four-Quadrant Thrust Controller for Marine propellers with Loss Estimation*. In Marine Technology, Vol. 46, No. 4, Otober, pp. 229–242.

Smogeli, Ø.N., Ruth, E. & Sorensen, A.J. 2005. *Experimental validation of power and torque thruster control*. Proc. IEEE 13th Mediterranean Conference on Control and Automation (MED), Cyprus.

Tsuchida, K. 1962. *Design Diagrams of Three-bladed Controllable Pitch Propellers*. Fourth Symposium on Naval Hydrodynamics- Ship Propulsion and Hydroelasticity, Washinton D,C. – USA.

1982, *"Curve caratteristiche di elica isolatea pale orientabili su Quattro quadranti. Elica E28"* Internal report—Dipartimento di Ingegneria Navale—Università degli Studi di Napoli "Federico II", Naples, Italy.

1994, *"Rapporto delle prove in mare del traghetto MB 54 operante nel Canal Grande a Venezia"* Internal report—Dipartimento di Ingegneria Navale—Università degli Studi di Napoli "Federico II", Naples, Italy.

Sustainable Maritime Transportation and Exploitation of Sea Resources – Rizzuto & Guedes Soares (eds)
© 2012 Taylor & Francis Group, London, ISBN 978-0-415-62081-9

A mathematical model of the propeller pitch change mechanism for the marine propulsion control design

M. Altosole & M. Martelli
Department of Naval Architecture and Electrical Engineering, Genoa, Italy

S. Vignolo
Department of Production Engineering, Thermal Energetic and Mathematical Models, Genoa, Italy

ABSTRACT: The paper is mainly focused on the mathematical model of the control pitch mechanism for a marine controllable pitch propeller (CPP), able to perform the propeller blade position change and to give a proper information about the oil pressures, produced inside the CPP hub. In fact, too high pressures can be responsible for the mechanism failure, then they should be always under examination by the ship automation. With regard to the traditional representation of the few spindle torque data reported in literature, in the proposed mathematical model the transportation inertial forces and the Coriolis inertial forces acting on the propeller blade are evaluated taking into account the yaw motion of the ship, the propeller speed (including shaft accelerations and decelerations) and the blade turning during the pitch change. On the basis of the introduced procedure, it is developed the CPP model which is part of an overall propulsion simulator, representing the dynamic behaviour of a twin-screw fast vessel. The aim of the work is to represent the ship propulsion dynamics by time domain simulation, on the ground of which the automation designers can develop and test several propulsion control options. A brief description of the simulation approach adopted for the vessel crash stop is illustrated at the end of this paper. In particular, the propulsion control action is studied taking into account machinery performance and constraints, including also the control pitch mechanism feedback in terms of allowable forces and pressures.

1 INTRODUCTION

Since several years University of Genoa is involved in the development of propulsion control systems for naval vessels, mostly propelled by powerful gas turbines driving controllable pitch propellers (CPPs). The increasing complexity of these recent propulsion systems requires control system functions able to manage high power in several propulsion configurations and during critical manoeuvres of the ship (slam start, crash stop, severe turning circles, etc.).

Simulation techniques may be a very useful tool to represent the marine propulsion dynamics, on the ground of which the automation designers can develop and test several propulsion control options (Altosole et al., 2008). From this point of view, a good design and optimization of the whole control system, based on simulation, entails the need to represent in details the dynamics of the control pitch mechanism of the CPP, by means of a reliable time domain numerical model. Although the CPPs are already in use for many years, a well known mathematical procedure, able to consider all the several involved phenomena, does not yet exist. The main problems are due to an approximate knowledge of the loads acting on the propeller blade and on the moving part of this mechanism. For this reason, a possible solution for this kind of problem is described hereinafter and consequently the corresponding simulation model of the control pitch mechanism has been developed. The model is able to perform the propeller blade position change and to give a proper information about the oil pressures, produced inside the CPP hub. In fact, too high pressures can be responsible for the mechanism failure, then they should be always under examination by the ship automation. In fact, optimizing the whole propulsion control system of the vessel, especially in critical conditions, means to find the proper compromise between performance and safety. In particular, each propulsive component, including the pitch change mechanism, has to be safeguarded from possible dangerous overloads; at the same time a prompt answer of the propulsion system, fuel saving and reduced cavitation phenomena should be pursued by the automation designer.

The proposed differential and algebraic equations, representing the CPP behaviour over time, form one of the several submodels of the overall propulsion simulator of the ship. By this simulation approach, it is possible to describe the dynamic performance of the CPP mechanism during soft or critical manoeuvres of the ship.

2 SIMULATION APPROACH

The ship performance is simulated by means of an overall mathematical model that is able to predict the interactions among the dynamics of the propulsion plant, the control system and the ship motions.

The model is depending on time and it consists of a set of differential equations, algebraic equations and numerical tables that represent the various elements of the propulsion system. All the main elements of the propulsion plant such as engines and their governors, hull, propeller, rudder, shaft line and telegraph are modelled as separated subsystems, linked each other. The whole simulation model is illustrated in Fig. 1, where it is possible to see the propulsion simulator block, managed by the overall controller through its main inputs.

The control functions regard both the propulsion and the electric power management (PMS), while the simulator is able to represent machinery dynamics and ship manoeuvrability (Altosole et al., 2010). Every numerical submodel is developed by Matlab-Simulink® software, a wide used platform for the dynamic systems simulation.

In particular, propeller thrust T and torque Q_o, calculated in the simulator block, are given by:

$$T = \rho K_t \dot{\theta}^2 D^4 \quad (1)$$

$$Q_0 = \rho K_q \dot{\theta}^2 D^5 \quad (2)$$

where ρ = sea water density; K_t = propeller thrust coefficient; K_q = propeller torque coefficient; $\dot{\theta}$ = propeller speed; and D = propeller diameter.

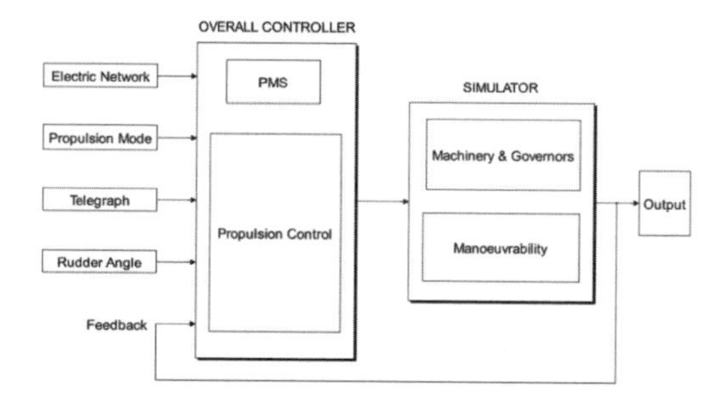

Figure 1. Simulation scheme.

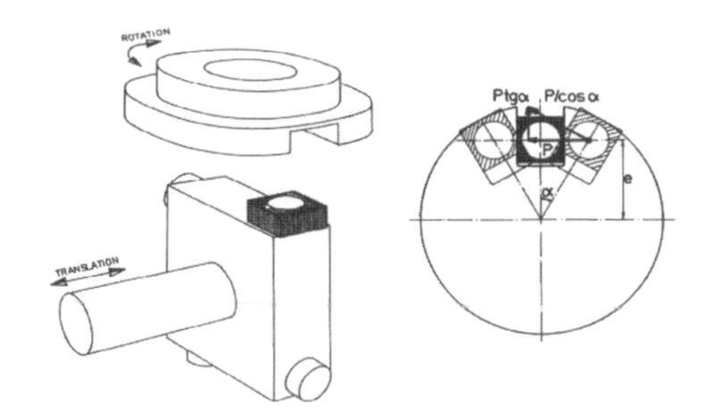

Figure 2. Pitch change mechanism.

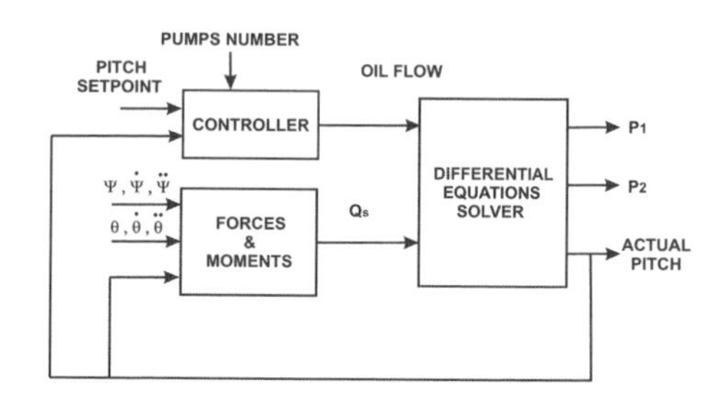

Figure 3. Calculation process of the actual pitch.

Coefficients K_t and K_q are derived from the open water propeller tests, as a function of the propeller advance coefficient and pitch. The actual value of the pitch depends on loads acting on a propeller blade and it is calculated by the simulation model of the pitch change mechanism.

The kind of the considered mechanism is illustrated in Fig. 2, where on the left it is possible to see the rotation of the blade carrier, driven by the piston inside the CPP hub, while on the right it is described the pin slot working (Wind 1978).

The calculation process for the propeller pitch dynamics is illustrated in Fig. 3. The pitch setpoint, depending on the bridge lever position, is transformed, by the mechanism controller, into a proper oil flow, acting on the piston; finally, by solving the pressure and motion equations, described in the following, it is possible to achieve the actual value of the propeller pitch.

3 MATHEMATICAL MODEL

3.1 *Main differential equations*

The proposed mathematical model relies on two main differential equations.

The first one is the motion equation of a blade around its $\mathbf{f}_3$ axis:

$$\ddot{\phi} = \frac{1}{I_{33}}(Q_h + Q_s + Q_{-\Phi}) \quad (3)$$

where $Q_s = Q_{SI} + Q_{SH} + Q_{SFr}$ = the total spindle torque acting on the blade: Q_{SI} = inertial forces torque, Q_{SH} = hydrodynamic forces torque, Q_{SFr} = frictional forces torque; $Q_{-\Phi}$ = torque due to the interaction forces between propeller blade and blade bearing; Q_h = hydraulic torque; and I_{33} = moment of inertia of the blade about the spindle axis $\mathbf{\underline{f}}_3$.

The second differential equation describes the motion of the cylinder:

$$m_{eq} \cdot \ddot{x} = A_1 p_1 - A_2 p_2 - B_p \dot{x} + \sum_{i=1}^{Z} \Phi_i \tag{4}$$

where x = cylinder position; A_1 and A_2 = yoke areas of the astern chamber and of the ahead chamber respectively; p_1 and p_2 = pressures inside the two chambers; B_p = damping coefficient; and $\sum_{i=1}^{Z} \Phi_i$ is the resultant of the reaction forces due to the interaction between each blade and the piston (Z denoting the number of blades).

To properly implement the differential Equations (3) and (4), we have to evaluate all the forces and moments acting on a single blade. This will be shown hereinafter.

3.2 *Reference frames*

The reference frames used in this paper are the following ones:

- The Inertial **n**-frame $(O_n, \underline{n}_1, \underline{n}_2, \underline{n}_3)$. It is a local geographical frame fixed to the Earth. The positive unit vector $\mathbf{\underline{n}}_1$ points towards the North, $\mathbf{\underline{n}}_2$ points towards the East, and $\mathbf{\underline{n}}_3$ points towards the centre of the Earth. The origin O_n is located on mean water free-surface at an appropriate location.
- The ship fixed **b**-frame $(O_b, \underline{b}_1, \underline{b}_2, \underline{b}_3)$. This frame is fixed to the hull. The positive unit vector $\mathbf{\underline{b}}_1$ points towards the bow, $\mathbf{\underline{b}}_2$ points towards starboard and $\mathbf{\underline{b}}_3$ points downwards. Often, for marine vehicles, the axes of this frame are chosen to coincide with the principal axes of inertia; this determines the position of the origin of the frame.
- The hub fixed **e**-frame $(O_e, \underline{e}_1, \underline{e}_2, \underline{e}_3)$. This frame is fixed to the hub. The positive unit vector $\mathbf{\underline{e}}_1$ points towards the bow, $\mathbf{\underline{e}}_3$ coincides with the spindle axis of a given blade and $\underline{e}_2 = \underline{e}_3 \wedge \underline{e}_1$. The origin O_e of the frame is in the center of the shaft line.

The blade fixed f-frame $(O, \underline{f}_1, \underline{f}_2, \underline{f}_3)$. This frame is fixed to any single blade. The unit vector $\mathbf{\underline{f}}_3$ coincides with the spindle axis of the blade, while the unit vectors $\mathbf{\underline{f}}_1$ and $\mathbf{\underline{f}}_2$ describe the rotation of the blade around its spindle axis as indicated in Fig. 7. The origin O coincides with O_e.

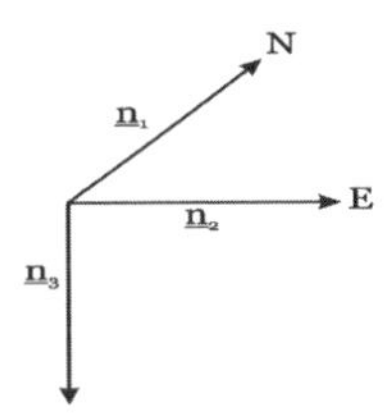

Figure 4. Inertial frame.

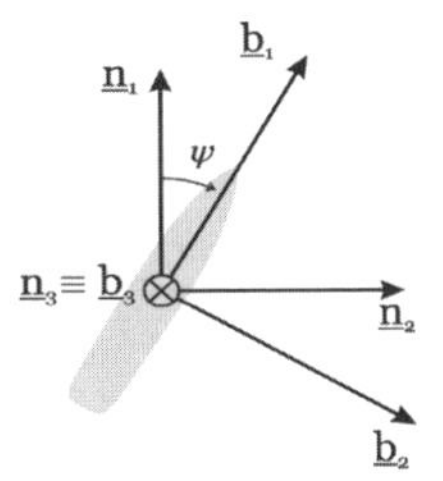

Figure 5. Ship fixed frame.

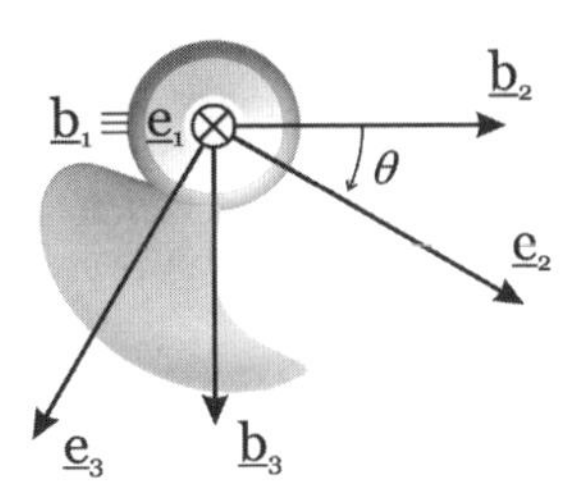

Figure 6. Hub fixed frame.

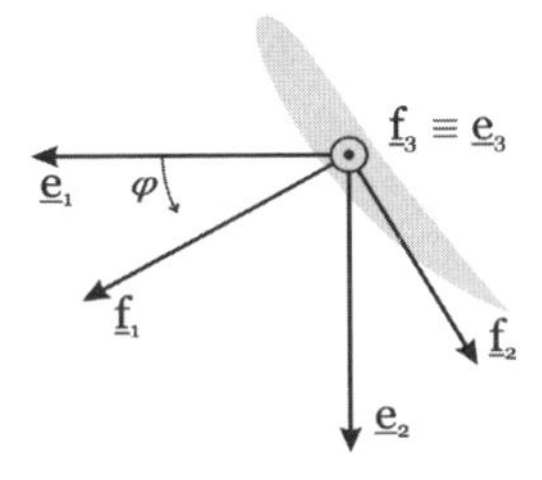

Figure 7. Blade fixed frame.

Our aim is to study the propeller blade motion in the blade fixed frame. Therefore, we need to describe any involved angular velocity. To this end, by using the angular velocity composition theorem, it is possible to write the angular velocities of a single blade and the hub respectively as:

$$\underline{\omega}_B = \underline{\omega}_\psi + \underline{\omega}_\vartheta + \underline{\omega}_\varphi = \dot{\psi}\underline{b}_3 + \dot{\vartheta}\underline{e}_1 + \dot{\varphi}\underline{f}_3 \tag{5}$$

$$\underline{\omega}_H = \underline{\omega}_\psi + \underline{\omega}_\vartheta = \dot{\psi}\underline{b}_3 + \dot{\vartheta}\underline{e}_1 \tag{6}$$

3.3 *Inertial and weight forces*

The yaw motion of the ship, the rotation of the propeller and the turning of the blade give rise to corresponding Coriolis and transportation inertial forces acting on each blade in the **e**-frame. Moreover, gravity yields a sinusoidal varying force.

More in detail, the Coriolis force is defined by:

$$\underline{F}^C = \int_\beta -2\rho_b \underline{\omega}_H \wedge \underline{v}_P^r d\tau \tag{7}$$

the transportation force is expressed as:

$$\underline{F}^S = \int_\beta -\rho_b \left[\begin{array}{c} \underline{a}_O + \underline{\omega}_H \wedge (\underline{\omega}_H \wedge (P-O)) \\ + \dot{\underline{\omega}}_H \wedge (P-O) \end{array} \right] d\tau \tag{8}$$

and the weight force is given by:

$$\underline{F}^W = \int_\beta \rho_b \underline{g} d\tau = M\underline{g} \tag{9}$$

where $\underline{\boldsymbol{a}}_o$ = acceleration of the origin O with respect to the inertial frame; ρ_b = mass density of the propeller blade; $(P–O)$ = position vector of a generic point P of the blade with respect to the origin O; and β = whole set of the points making up the blade.

$$\underline{v}_P^r = \underline{v}_0^r + \underline{\omega}_\varphi \wedge (P-O) = \dot{\varphi} \underline{f}_3 \wedge (P-O) \tag{10}$$

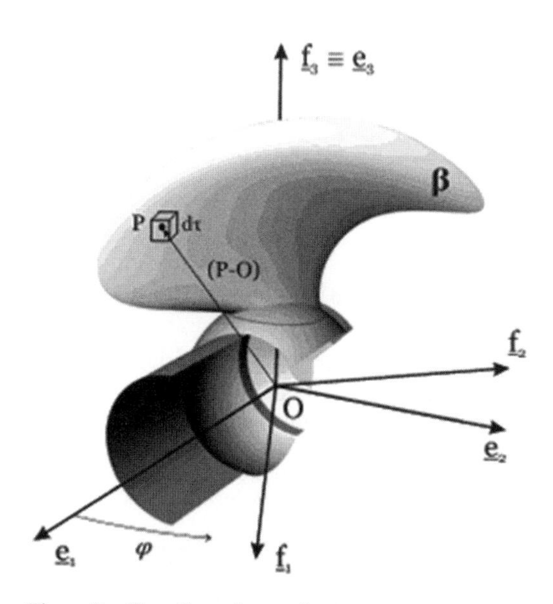

Figure 8. Propeller reference frames.

is the velocity of a generic point P of the blade evaluated in the **e**-frame.

Recalling the definition of centre of gravity for a blade:

$$(G-O) = \frac{1}{M} \int_\beta \rho_b (P-O) d\tau \tag{11}$$

where M = total mass of a blade.

We can express the inertial forces (7) and (8) in the simpler form:

$$\underline{F}^C = -2M\underline{\omega}_H \wedge \underline{v}_G^r \tag{12}$$

$$\begin{aligned} \underline{F}^S = & -M\underline{a}_o - M\underline{\omega}_H \wedge [\underline{\omega}_H \wedge (G-O)] \\ & - M\dot{\underline{\omega}}_H \wedge (G-O) \end{aligned} \tag{13}$$

The total contribution of inertial and weight forces acting on a blade is then given by the sum:

$$\underline{F}_I = \underline{F}^W + \underline{F}^S + \underline{F}^C \tag{14}$$

In addition to this, we need also to explicitly represent the moments with respect to the origin O of the above forces, namely:

$$\underline{M}_O = \int_\beta (P-O) \wedge \underline{F} d\tau \tag{15}$$

Making use of the definition of inertia tensor:

$$I_O(\underline{v}) = \int_\beta \rho (P-O) \wedge [\underline{v} \wedge (P-O)] d\tau \tag{16}$$

after some calculations, we get the final expressions for the moments of the inertial forces:

$$\begin{aligned} \underline{M}_O^C = & -\underline{\omega}_\varphi \wedge I_O(\underline{\omega}_H) - \underline{\omega}_H \wedge I_O(\underline{\omega}_\varphi) \\ & + I_O(\underline{\omega}_\varphi \wedge \underline{\omega}_H) \end{aligned} \tag{17}$$

$$\begin{aligned} \underline{M}_O^S = & -M(G-O) \wedge \underline{a}_O - \underline{\omega}_H \wedge I_O(\underline{\omega}_H) \\ & - I_O(\dot{\underline{\omega}}_H) \end{aligned} \tag{18}$$

where $\underline{\omega}_\varphi = \dot{\varphi} \underline{f}_3$ denotes the angular velocity of the blade evaluated in the **e**-frame. The moment due to the weight force is:

$$\underline{M}_O^W = M(G-O) \wedge \underline{g} \tag{19}$$

The total moment due to inertial and weight forces acting on a blade is then the sum:

$$\underline{M}_I = \underline{M}^W + \underline{M}^S + \underline{M}^C \tag{20}$$

3.4 *Hydrodynamic forces*

To predict hydrodynamic loads, different theoretical methods are proposed in literature but they are difficult to apply because the computation time cost is enormously high compared with the standard simulation time of the propulsive system. Therefore, the hydrodynamic forces are evaluated with a quasi-steady methodology that takes into account the change of force acting on a blade using the open water characteristic diagrams and the steady state position of the center of pressure. We assume that the sum of all forces in the $\underline{\mathbf{f}}_1$ direction results in a total thrust and the sum of all forces in $\underline{\mathbf{f}}_2$ direction results in a tangential force (Q/r). The forces in $\underline{\mathbf{f}}_3$ direction are neglected. The hydrodynamic spindle torque $M_{HD,f3}$ and the bending moments $M_{HD,f1}$, $M_{HD,f2}$ are defined as the components along the unit vectors $\underline{\mathbf{f}}_3$, $\underline{\mathbf{f}}_1$ and $\underline{\mathbf{f}}_2$ respectively of the resultant moment of hydrodynamic forces with respect to the origin O.

The resultant forces are assumed to act in the hydrodynamic centre (x_{CH}, y_{CH}, z_{CH}), so:

$$F_{HD,\underline{b}_1} = T^* \tag{21}$$

$$F_{HD,\underline{b}_2} = \frac{Q^*}{z_{CH}} \tag{22}$$

where T^* = single blade thrust; and Q^* = required torque by a single blade.

$$T^* = \frac{T}{Z} \tag{23}$$

$$Q^* = \frac{Q}{Z} \tag{24}$$

T and Q are evaluated using expression (1) and (2).

If we assume that the system of forces forms a parallel system of vectors, we obtain:

$$M_{HD,\underline{b}_1} = -F_{HD,\underline{b}_2} \cdot z_{CH} = -\frac{Q^*}{z_{CH}} \cdot z_{CH} \tag{25}$$

$$M_{HD,\underline{b}_2} = F_{HD,\underline{b}_1} \cdot z_{CH} = T^* \cdot z_{CH} \tag{26}$$

$$\begin{aligned} M_{HD,\underline{b}_3} &= -F_{HD,\underline{b}_1} \cdot y_{CH} + F_{HD,\underline{b}_2} \cdot x_{CH} \\ &= T^* \cdot y_{CH} + \frac{Q^*}{z_{CH}} \cdot x_{CH} \end{aligned} \tag{27}$$

3.5 *Frictional forces*

The blade bearing supports the propeller blade in the axial and radial direction. Here the friction forces will be derived for each direction separately, following the procedure proposed by Godjevac and coauthors (Godjevac et al., 2009).

In particular, from the statics forces and moments equations we derive the expressions of the radial component of the reaction force $F_{R,f3}$ along $\underline{\mathbf{f}}_3$ and the reaction moment components $M_{R,f1}$, and $M_{R,f2}$:

$$\begin{aligned} \sum F_{f3} = 0 \Rightarrow F_{I,f3} + F_{HD,f3} \\ + F_{R,f3} = 0 \Rightarrow F_{R.f3} = -F_{I,f3} \end{aligned} \tag{28}$$

$$\begin{aligned} \sum M_{f1} = 0 \Rightarrow M_{HD,f1} + M_{R,f1} + M_{I,f1} = 0 \Rightarrow \\ M_{R,f1} = -M_{HD,f1} - M_{I,f1} \end{aligned} \tag{29}$$

$$\begin{aligned} \sum M_{f2} = 0 \Rightarrow M_{I,f2} + M_{HD,f2} \\ + M_{R,f2} = 0 \Rightarrow M_{R,f2} \\ = -M_{I,f2} - M_{HD,f2} \end{aligned} \tag{30}$$

We define the vector $\underline{\boldsymbol{M}}_{RAD}$ as the projection of the reaction moment on the plane generated by $\underline{\mathbf{f}}_1$ and $\underline{\mathbf{f}}_2$. Its modulus is given by:

$$M_{RAD} = \sqrt{M^2_{R,f1} + M^2_{R,f2}} \tag{31}$$

The radial component $F_{R,f3}$ is supposed to be uniformly distributed (see q_F in Fig. 10), while the force distribution generating $\underline{\boldsymbol{M}}_{RAD}$ is considered not uniform (see q_M in Fig. 10). The sum of these two loads distributions will result in a total load distribution (see q in Fig. 10).

The radial part of a blade bearing can be considered as split in two portions: one in the fore direction and the other in aft one. If the load due to the radial force is higher than the load caused by $\underline{\boldsymbol{M}}_{RAD}$, only the fore portion will be loaded. On the contrary, the total load results to be the sum of the loads acting on both parts. We can define a coefficient that describes the ratio between the radial force and the maximum force generated by the $\underline{\boldsymbol{M}}_{RAD}$:

$$u = \frac{F_{RAD} \cdot r}{2 M_{RAD}} \tag{32}$$

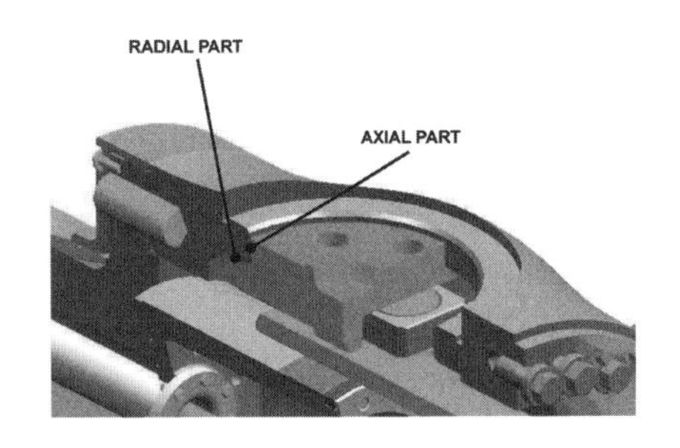

Figure 9. Scheme of blade bearing.

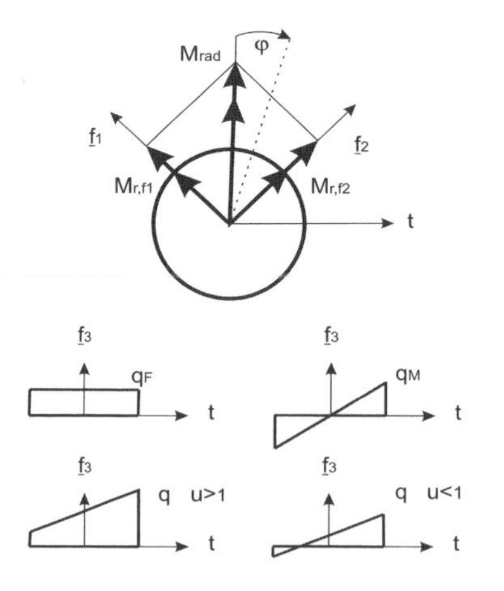

Figure 10. Distribution of bending moment and radial force over radial part of the blade bearing.

The total load distribution, shown in Fig. 10, is then:

$$q = q_F + q_M = \frac{F_{RAD}}{2\pi r} + \frac{2M_{RAD} \cdot \sin(\varphi)}{r \cdot 2\pi r} \tag{33}$$

where r = radius of the blade bearing.

The tangential component of the friction force density f_{FR} is defined as:

$$f_{FR} = \mu q \tag{34}$$

where μ = static friction coefficient.

The torque due to the radial friction is:

$$Q_{FR,RAD} = \int_0^{2\pi} f_{FR} r^2 d\varphi \tag{35}$$

The final expression of the radial friction torque depends on the value of the coefficient u; if u > 1 from (35) we obtain:

$$Q_{FR,RAD} = \mu r^2 \int_0^{2\pi} \left(\frac{F_{RAD}}{2\pi r} + \frac{2M_{RAD} \sin(\varphi)}{2\pi r^2} \right) \times d\varphi = \mu F_{RAD}\, r \tag{36}$$

On the contrary, if u < 1 we can proceed in the following way .

First find the angle φ_o where the load (q) equals zero is:

$$\frac{F_{RAD}}{2\pi r} + \frac{2M_{RAD} \cdot \sin(\varphi_0)}{r \cdot 2\pi r} = 0 \Rightarrow$$
$$\varphi_0 = -\arcsin\left(\frac{F_{RAD} \cdot r}{2M_{RAD}} \right) = -\arcsin(u) \tag{37}$$

The friction torque can be found by integrating:

$$Q_{FR,RAD} = -2 \cdot \mu \cdot r^2 \int_{-\pi/2}^{\varphi_0} q d\varphi + 2 \cdot \mu \cdot r^2 \int_{\varphi_0}^{\pi/2} q d\varphi \tag{38}$$

Substituting (33) into (38), we obtain:

$$Q_{FR,RAD} = -2 \cdot \mu \cdot r^2 \times \int_{-\pi/2}^{\varphi_0} \frac{F_{RAD}}{2\pi r} d\varphi - 2 \cdot \mu \cdot r^2 \int_{-\pi/2}^{\varphi_0} \frac{2M_{RAD} \cdot \sin(\varphi)}{r \cdot 2\pi r} + 2 \cdot \mu \cdot r^2 \int_{\varphi_0}^{\pi/2} \frac{F_{RAD}}{2\pi r} d\varphi + 2 \cdot \mu \cdot r^2 \int_{\varphi_0}^{\pi/2} \frac{2M_{RAD} \cdot \sin(\varphi)}{r \cdot 2\pi r} \tag{39}$$

The integration gives:

$$Q_{FR,RAD} = \frac{4}{\pi} \cdot \mu \cdot M_{RAD} \left[\sqrt{1-u^2} + u \cdot \arcsin(u) \right] \tag{40}$$

The friction in the axial part of the blade bearing depends on the sum of all forces in the axial direction as is show in Fig. 11:

Also in this case we impose the statics equations for forces and moments:

$$\sum F_{AX,f1} = 0 \Rightarrow F_{HD,f1} + F_{I,f1} + F_{AX,f1} = 0 \Rightarrow F_{AX,f1} = -F_{HD,f1} - F_{I,f1} \tag{41}$$

$$\sum F_{AX,f2} = 0 \Rightarrow F_{HD,f2} + F_{I,f2} + F_{AX,f2} = 0 \Rightarrow F_{AX,f2} = -F_{HD,f2} - F_{I,f2} \tag{42}$$

$$F_{AX} = \sqrt{F_{AX,f1}^2 + F_{AX,f2}^2} \tag{43}$$

We consider the interaction between the blade carrier and the hub as a pointwise contact.

So the torque yielded by the sum of the axial forces is given by:

$$Q_{FR,AX} = \mu \cdot F_{AX} \cdot d_a \tag{44}$$

where d_a = distance between the point where the axial force is supposed to be applied and the friction seat hub of the propeller root.

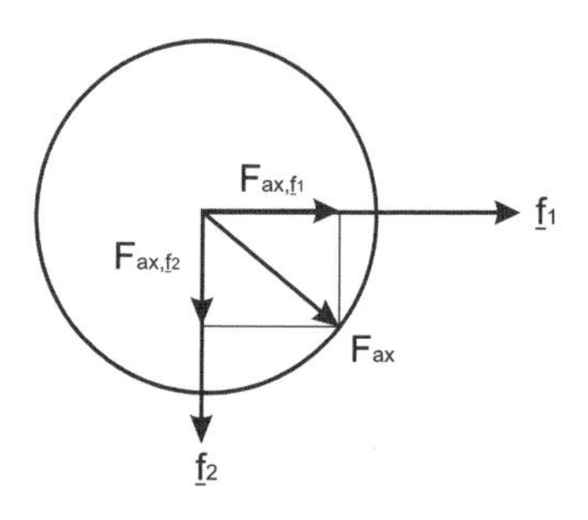

Figure 11. Axial contact situation.

3.6 *Hydraulic forces*

The oil pressure needed to turn the blade or, alternatively, to hold it in the right position is supplied by one or two pumps. The considered actuating system consists of a double effect actuator with single rod and circular section. The developed force is proportional to the yoke area and to the oil pressure, but there is a difference between the thrust action and the traction ones, with the following relationship:

$$F_{Thrust} = \frac{\pi D_p^2 p}{4} \tag{45}$$

$$F_{Traction} = \frac{\pi (D_p^2 - d_r^2) p}{4} \tag{46}$$

where D_p = piston diameter; d_r = rod diameter; and p = pressure produced inside the hub chamber.

To evaluate the dynamic change of pressure, it has been used:

- state equation:

$$\rho_{oil} = \rho_{i,oil} + \frac{\rho_{i,oil}}{B} p \tag{47}$$

- continuity equation:

$$\sum m_i - \sum m_o = g \frac{d(\rho_{oil} V)}{dt} = g\left(\frac{d\rho_{oil}}{dt} V + \rho_{oil} \frac{dV}{dt} \right) \tag{48}$$

where V = compartment volume; ρ_{oil} = oil density; and B = oil Bulk modulus.

We ignore the dependence of the density on the temperature and combining the two equations we obtain:

$$q_i - q_o = \frac{\dot{p}}{B} \cdot V + \dot{V} \tag{49}$$

where q_i = flow going in; and q_u = flow going out.

We define the leakage of the hydraulic actuator between the two chambers through the following coefficient:

$$C_{ip} = \pi \cdot \frac{D_p}{2} \cdot \frac{e^3}{L_p} \cdot \frac{1}{6\mu} \tag{50}$$

where e = orifice thickness; L_P = thickness of the piston head; and μ = oil dynamic viscosity.

Finally the differential pressure equation becomes:

$$\dot{p}_i = (q_i - C_{ip} \cdot p_i - A \cdot \dot{x}) \frac{B}{A \cdot x + V_0} \tag{51}$$

4 SIMULATION RESULTS

As an application of the model previously described, some simulation results are reported in this paragraph. In particular, they regard the time histories of some important characteristics during the crash stop of a twinscrew ship. The whole performance of the vessel is given by the propulsion simulator, including the overall control system, illustrated in Fig. 1. The oil flow q_i is properly provided by the model of the local governor of the CPP mechanism, where it is simply represented by the proportional action of the valve, acting on the pitch error. In fact, the bridge lever position sets the shaft speed and pitch setpoints, which are compared with the corresponding actual values; then, proportional and integral regulations act on the two errors, in order to calculate respectively the proper fuel flow of the engine and the oil flow acting on the CPP piston. The fuel flow is necessary for the engine torque calculation, on the ground of which it is possible to achieve the shaft speed by solving the shaft line dynamic equation (Altosole et al., 2009).

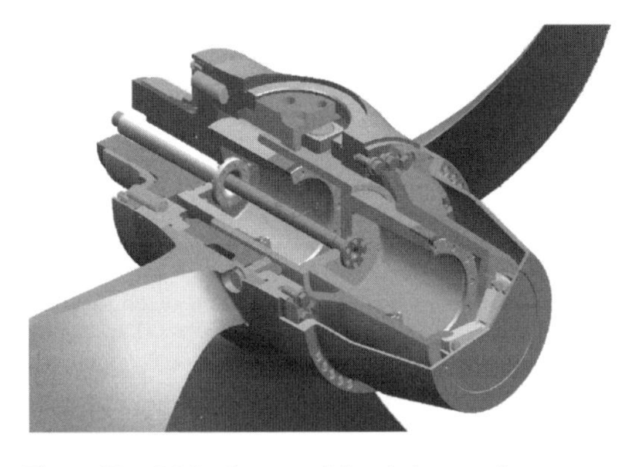

Figure 12. Main element of the pitch actuation system.

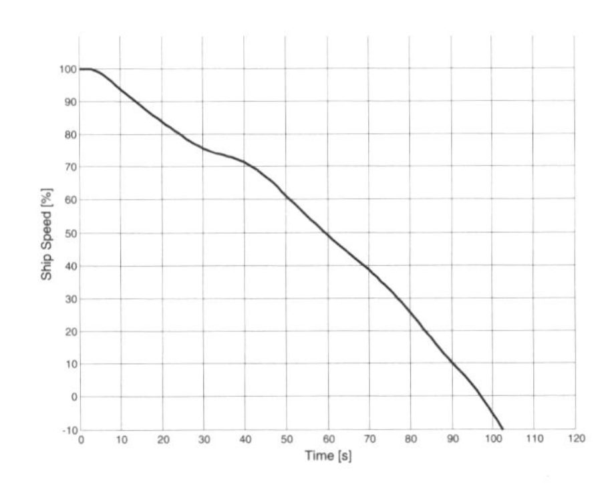

Figure 13. Ship speed vs. time.

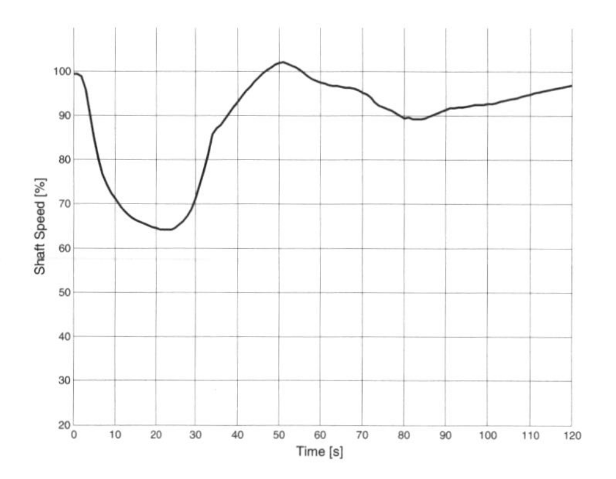

Figure 14. Shaft speed vs. time.

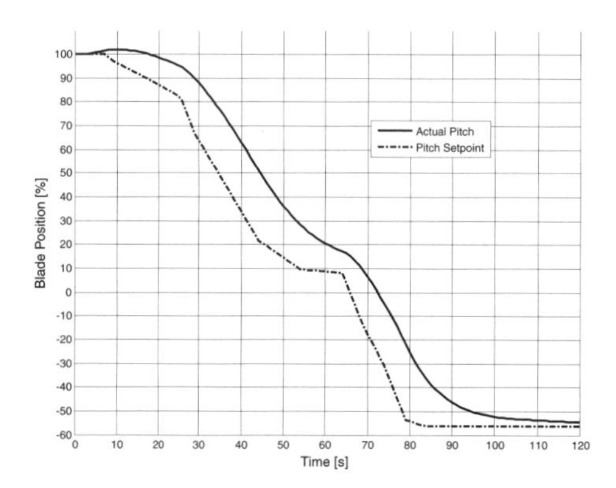

Figure 15. Blade position vs. time.

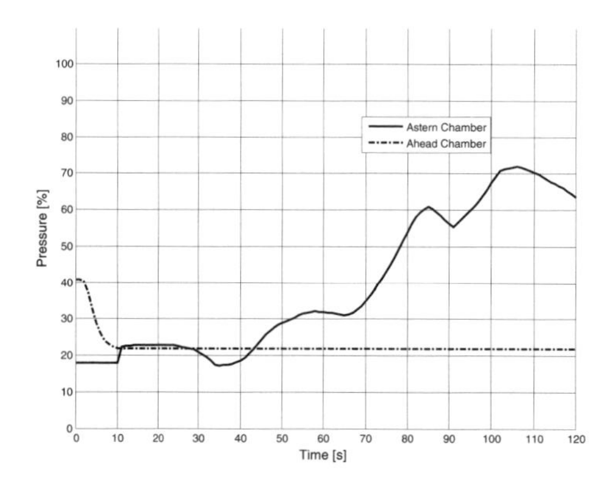

Figure 16. Oil pressures vs. time.

5 CONCLUSIONS

In this paper, a mathematical model able to represent the performance of the CPP mechanism has been illustrated. This brief description is mainly focused on the evaluation of the several loads acting on the piston and on the blades of the propeller, taking into account the influence of the ship and CPP blade motions (in terms of velocities, accelerations and decelerations). To this end, different reference frames have been introduced, in order to suitably consider each contribution of forces and moments.

The main aim of the proposed procedure is the assessment of the velocity of the pitch change and of the oil pressures produced inside the two hub chambers. This target is mainly important from the point of view of the propulsion control design of the ship: in fact, the overall controller has to provide the proper pitch setpoint in accordance with particular control parameters, such as possible ramps (fit to modulate some input signals), regulation gains, etc ... This aspect becomes crucial especially when the automation designer has to set the proper logic to manage some critical or emergency manoeuvres, as severe turning circles, slam start or crash stop of the vessel. In these conditions, the whole propulsion plant model is able to assess the numerical value of some important variables such as the oil pressure, the propeller torque, thrust, etc; the latter can be compared with the design limits or also to define such limits during design process.

ACKNOWLEDGEMENT

The authors wish to thanks Fincantieri for the support received during the research activity.

REFERENCES

Altosole, M., Benvenuto, G., Campora, U. & Figari, M. 2009. Real-time simulation of a COGAG naval ship propulsion system. *Journal of Engineering for the Maritime Environment*. Volume 223, Number 1/2009: 47–62.

Altosole, M., Figari, M., Michetti, S., Millerani Trapani, A. & Viviani, M. 2010. Simulation of the dynamic behaviour of a CODLAG propulsion plant. *Warship 2010 Conference; Proc. intern. symp., London, 9–10 June 2010*. London: The Royal Institution of Naval Architect.

Godjevac, M., Van Beek, T., Grimmelius, H.T., Tinga, T. & Stapersma, D. 2009. Prediction of fretting motion in a controllable pitch propeller during service. *Journal of Engineering for the Maritime Environment*. Volume 223.

Martelli, M. 2009. Dynamic simulation of controllable pitch propeller mechanism. *MSc Thesis, Department of naval architecture and marine engineering*. Genoa.

Wind, J. 1978. Principles of mechanism used in controllable pitch propellers. *3th Lips Propeller Symposium*. Drunen.

Sustainable Maritime Transportation and Exploitation of Sea Resources – Rizzuto & Guedes Soares (eds)
 ISBN 978-0-415-62081-9

Comparative analysis of basic procedures for shafting alignment calculation

I. Ančić, A. Šestan & N. Vladimir
Faculty of Mechanical Engineering and Naval Architecture, University of Zagreb, Croatia

ABSTRACT: Shafting alignment procedure is very important for safe and reliable operation of ship propulsion system and consequently for ship safety. It includes calculations, installation and testing, as well as possible readjusting to ensure appropriate loading of all bearings and no excessive bending in shaft sections. Three Moment Equation Method (TMEM), Finite Element Method (FEM) as well as Transfer Matrix Method have been compared within the numerical example which includes shaft line of a new generation oil tanker. Firstly, some general remarks on shafting alignment procedures and their importance in proper operation of a ship are given. Further on, ship main particulars and basic characteristics of shaft line are described, and their description is followed by outline of TMEM and FEM, adjusted for this particular case. The results obtained by both methods are compared to those obtained in the shipyard, which are based on application of Transfer Matrix Method. Finally, some improvements of shafting alignment procedure have been proposed.

1 INTRODUCTION

The purpose of the propulsion machinery (main engine, gearbox, propulsion shaft line, propeller and associated auxiliary systems) is to propel the ship and to control maneuvering, thus enabling the ship master to be in control of ship's speed and course (Vulić et al., 2008). Shafting alignment procedure is very important for safe and reliable work of ship propulsion system and consequently for ship safety. It includes calculations, installation and testing, as well as possible readjustment to ensure appropriate loading of all bearings. However, its primary goal is to ensure an acceptable static load distribution among the shaft-supporting bearings, which (as generally accepted) will be a prerequisite for satisfactory whirling and transverse vibration of the propulsion system (Šverko, 2006). When conducting shafting alignment calculation, the shafting is usually assumed to be continuous beam isolated from ship hull which is supported by bearings. Bearing reaction as well as bending and shear stresses at each section should be calculated to check their compliance with predetermined acceptance criteria.

Although there are numerous papers related to shafting alignment procedures, and the problem (and its solution) seems to be well-known, the frequency of shaft alignment related bearing damage has increased significantly in recent years (Šverko, 2006). According to Šverko (2006) this damage can be mostly attributed to inappropriate analysis tools, changes in the design of vessels, inadequate shipyard practices as well as lack of well-defined criteria (ABS, 2006, NKK, 2006).

In recent years the trend of building larger (and more flexible) ships with higher installed power is evident. In that sense discrepancy between hull and shaft line flexibility/rigidity is highly pronounced, and consequently the alignment of the propulsion system becomes more sensitive to hull girder deflections. Influence of hull girder flexibility on shafting alignment is already commented in the literature (Batalha et al., 2009, Murawski, 2006, Šverko, 2005, 2006), but it is evidently that the problem is still too far from its final solution. There are also some additional problems as for example those related to possible coupling of torsional and lateral vibration, which has been investigated by full-scale measurements by Šestan et al. (2010).

These facts inspired the authors to examine the possibilities of basic procedures as a starting point for development of a more advanced numerical model which should take into account some more complex issues as for example hull girder flexibility, proper modelling of bearing foundation, influence of bearing material on dynamic characteristics of propulsion system etc. In that sense the calculation presumptions, shafting parts modelling as well as material properties and loading are described and analysed in the paper. Outlines of Three Moment Equation Method (TMEM) and Finite Element Method (FEM), adjusted for this particular case, are given. TMEM and FEM results are compared to those obtained by the shipyard, which have been calculated using DNV Nauticus Machinery Shaft Alignment (NMSA) software (DNV Software, 2004), based on transfer matrix method. Beside analysis of the obtained results concluding remarks include comments on possible improvements of the methods related to proper modelling of bearing foundation stiffness and influence of a ship hull deformation on a shaft line alignment.

2 SHIP PARTICULARS

In this paper a shafting line of the oil product tanker Stena P-Max is under consideration, Figures 1 and 2. Few years ago a series of P-Max tankers has been under construction in the Brodosplit shipyard in Croatia. Main particulars of the ship are the following:

Length between perpendiculars	$Lpp = 175.5$ m
Moulded breadth:	$B = 40$ m
Moulded depth:	$H = 17.9$ m
Scantling draught:	$T = 13.5$ m
Deadweight:	$D = 65200$ DWT
Trial speed:	$v_{tr} = 17.3$ kn
Service speed:	$v_s = 14.5$ kn

The ship is equipped with two independent propulsion systems placed in two engine rooms, and consequently two shaft lines (one at portside and another at starboard side), Figure 3.

The ship is powered by two MAN B&W 6S46MC-C engines, 7860 kW each running at 129 rpm. The shaft line consists of intermediate shaft, OD-box servo shaft and propeller shaft. In service they are supported by 3 bearings i.e. aft and forward stern tube bearing and intermediate shaft bearing.

Physical model that describes shaft line in working condition includes 6 additional bearings inside engine. According MAN B&W (1995), the results obtained using this kind of model are more reliable. Since the obtaining of the exact solution for working condition is time consuming, it seems to be more rational to perform the comparative

Figure 1. Stena P-Max tanker.

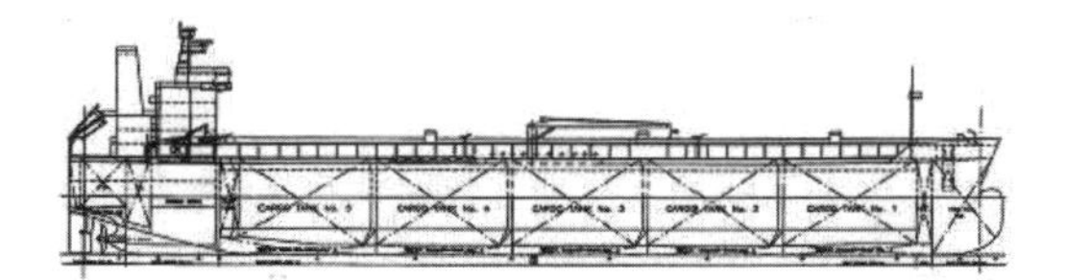

Figure 2. General arrangement of the considered ship.

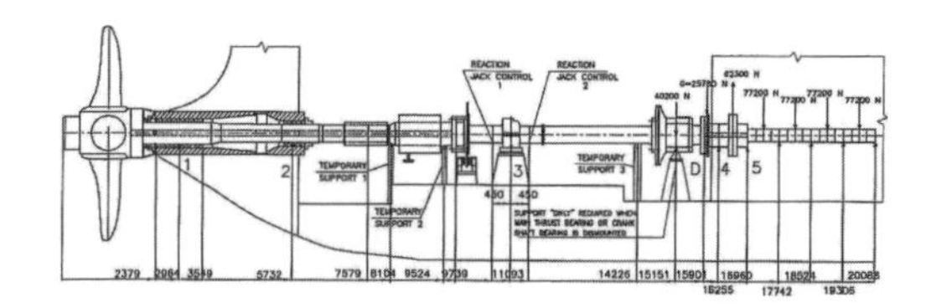

Figure 3. Shaft line.

analysis taking into account the shaft line during the assembly. That analysis is sufficient to validate FEM model. For the condition called Montage 1 shaft line consists of propeller shaft and OD-box servo shaft. These two shafts are coupled with SKF clutch and supported by 4 bearings i.e. aft and forward stern tube bearing and two temporary supports.

3 OUTLINE OF THREE MOMENT EQUATION METHOD

Three Moment Equation Method is described in details in (Chang & Juang, 2006), and here only the basics are given. Continuous beam, supported by three fixed supporting points is shown in Figure 4.

Supporting points are denoted by N, while symbols w_N and w_{N+1} are related to uniform load acting on length of left and right segments denoted by L_N and L_{N+1}, respectively. Concentrated loads are denoted by P_N and P_{N+1}, while symbols M_N and M_{N+1} denote moments acting on left side and right segment.

Supporting point N divides the beam into two segments, Figure 5, and the deflection angles and reactive forces can be expressed by following formulae:

$$\theta_N = -\frac{w_N L_N^3}{24EI_N} - \frac{M_N L_N}{3EI_N} - \frac{M_{N-1} L_N}{6EI_N} \quad (1)$$

$$R_N = \frac{w_N L_N}{2} + \frac{M_{N-1} - M_N}{L_N} \quad (2)$$

$$\theta_{N'} = \frac{w_{N+1} L_{N+1}^3}{24EI_{N+1}} + \frac{M_N L_{N+1}}{3EI_{N+1}} + \frac{M_{N+1} L_{N+1}}{6EI_{N+1}} \quad (3)$$

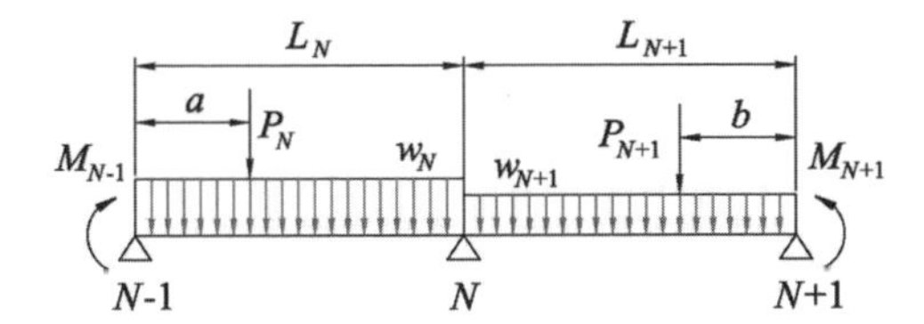

Figure 4. Continuous beam on three supports and related loads.

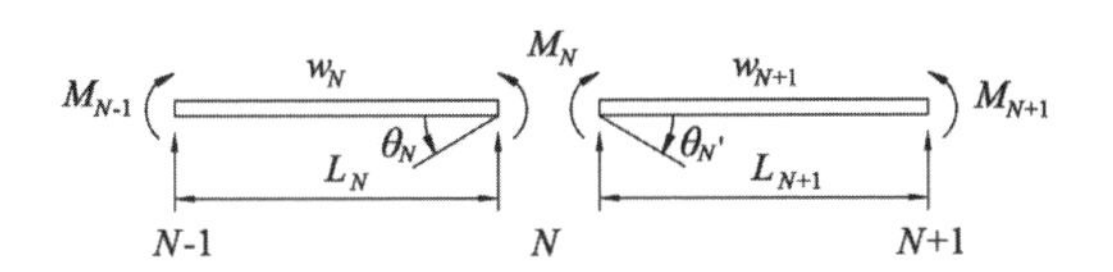

Figure 5. Deflection angles caused by continuous load.

$$R_{N'} = \frac{w_{N+1}L_{N+1}}{2} + \frac{M_{N+1} - M_N}{L_{N+1}}. \tag{4}$$

In the previous formulae the terms EI_N and EI_{N+1} represent flexural stiffness of left and right segment, respectively.

If concentrated loads P_N and P_{N+1} are applied either to left and right segment, Figure 6, the relevant relations for deflection angles and reaction forces are Eqs. (5), (6), (7) and (8).

$$\theta_N = -\frac{P_N a}{6I_N L_N}\left(L_N^2 - a^2\right) - \frac{w_N L_N^3}{24EI_N} - \frac{M_N L_N}{3EI_N} - \frac{M_{N-1}L_N}{6EI_N} \tag{5}$$

$$R_N = \frac{P_N a}{L_N} + \frac{w_N L_N}{2} + \frac{M_{N-1} - M_N}{L_N} \tag{6}$$

$$\theta_{N'} = \frac{P_{N+1}b}{6I_{N+1}L_{N+1}}\left(L_{N+1}^2 - b^2\right) + \frac{w_{N+1}L_{N+1}^3}{24EI_{N+1}} + \frac{M_N L_{N+1}}{3EI_{N+1}} + \frac{M_{N+1}L_N}{6EI_{N+1}} \tag{7}$$

$$R_{N'} = \frac{P_N a}{L_N} + \frac{w_N L_N}{2} + \frac{M_{N-1} - M_N}{L_N}. \tag{8}$$

If the beam is assumed to be continuous, one can write for the deflection angles $\theta_N = \theta_{N'}$ at the support N, and the reaction force R at the support N, can be simply obtained by superposition of R_N and $R_{N'}$.

If the continuous beam is not lied on a horizontal straight line, the offset influence angles β_N and β_{N+1}, Figure 7, should be taken into account.

According to Chang & Juang (2006) the formulae for angles β_N and β_{N+1} yield:

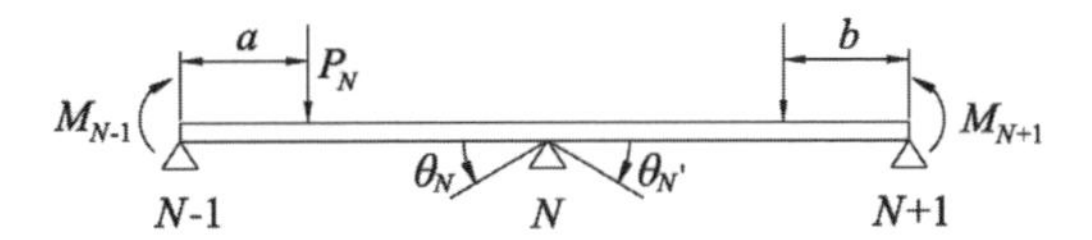

Figure 6. Deflection angles caused by concentrated loads.

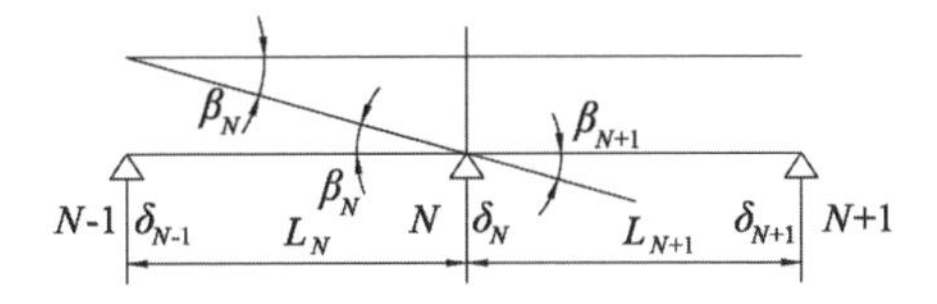

Figure 7. Schematic view of offset influence angles.

$$\beta_N = \frac{\delta_{N-1} - \delta_N}{L_N} \tag{9}$$

$$\beta_{N+1} = -\frac{\delta_{N+1} - \delta_N}{L_{N+1}}, \tag{10}$$

and the final expressions:

$$M_{N-1}\left(\frac{L_N}{I_N}\right) + 2M_N\left(\frac{L_N}{I_N} + \frac{L_{N+1}}{I_{N+1}}\right) + M_{N+1}\left(\frac{L_{N+1}}{I_{N+1}}\right) = -\frac{w_N L_N^3}{4I_N} - \frac{w_{N+1}L_{N+1}^3}{4I_{N+1}} - \frac{P_N a}{L_N I_N}\left(L_N^2 - a^2\right) - \frac{P_{N+1}b}{I_{N+1}L_{N+1}}\left(L_{N+1}^2 - b^2\right) - 6E\left(\frac{\delta_{N-1} - \delta_N}{L_N} + \frac{\delta_{N+1} - \delta_N}{L_{N+1}}\right) \tag{11}$$

$$R_N = \frac{M_{N-1} - M_N}{L_N} + \frac{M_{N+1} - M_N}{L_{N+1}} + \frac{P_N a}{L_N} + \frac{P_{N+1}b}{L_{N+1}} + \frac{w_N L_N}{2} + \frac{w_{N+1}L_{N+1}}{2}. \tag{12}$$

Based on the above described theoretical consideration, the Matlab code for calculation execution has been developed.

4 FEM ANALYSIS

FEM model of shaft line for the stage called Montage 1 is shown in Figure 8. It consists of 19 beam elements with 4 DOF (2 translations and 2 rotations). The finite element stiffness matrix yields (Senjanović, 1990):

$$K = \frac{2EI}{l^3}\begin{bmatrix} 6 & 3l & -6 & 3l \\ & 2l^2 & -3l & l^2 \\ & & 6 & -3l \\ Sym. & & & 2l^2 \end{bmatrix} \tag{13}$$

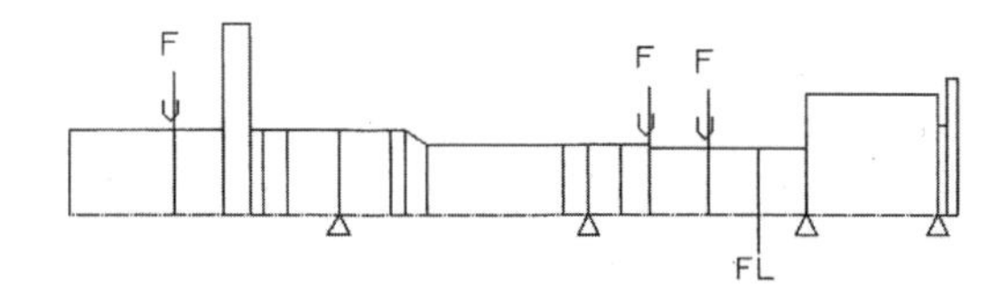

Figure 8. FEM model for Montage 1.

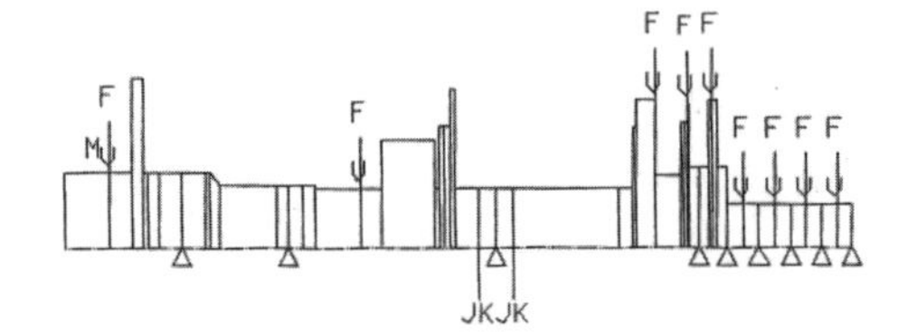

Figure 9. FEM model for working condition.

FEM model of shaft line in working condition is shown in Figure 9. It consists of 44 beam elements. The same topology of finite elements is used as that in NMSA software. External loads are also the same as those in calculation conducted by the shipyard. Mass of the crankshaft is neglected as recommended in MAN B&W (1995). Eccentricity bending moment is taken into account. Calculation is performed using Matlab for bearing offsets recommended by NMSA software.

5 COMPARATIVE ANALYSIS

Analysis is carried out for following external loads:

Table 1. External loads for Montage 1.

Node no.	Distance from node No. 1, m	Point load, kN	Bending moment, kNm
2	1.128	145.8	0
14	6.395	65.0	0
15	7.045	7.6	0
17	8.104	0	0

Table 2. External loads for working condition.

Node no.	Distance from node No. 1, m	Point load, kN	Bending moment, kNm
2	1.128	145.8	–91.0
15	7.579	7.6	0
28	15.151	40.2	0
31	15.966	25.8	0
35	16.595	–62.3	0
38	17.351	77.2	0
40	18.133	77.2	0
42	18.915	77.2	0
44	19.697	77.2	0

Reaction values for each bearing for zero bearing offset during Montage 1 by different methods are given in table:

Table 3. Bearing reactions for Montage 1.

	ST aft	ST forward	Temp 1	Temp 2
TMEM, kN	381.53	–128.78	147.45	–6.48
FEM, kN	384.63	–138.05	159.66	–12.53
Diff, %	0.8	7.2	8.3	93.2
NMSA, kN	381.89	–129.68	149.96	–8.46
Diff, %	0.1	0.7	1.7	30.5

Table 4. Bearing reactions for particular bearing offsets, Montage 1.

	ST aft	ST forward	Temp 1	Temp 2
Bearing offset, mm	0	+2.000	+1.246	+0.569
TMEM, kN	284.05	109.16	–37.62	38.12
FEM, kN	295.76	76.60	–2.96	24.32
Diff, %	4.1	29.8	92.1	36.2
NMSA, kN	295.73	75.24	0.80	21.95
Diff, %	4.1	31.1	102.1	42.4

Analysis showed that NMSA calculation matched remarkably with TMEM. But when the bearing offsets were included, reactions changed significantly.

That showed us, as it was expected, just how sensitive shaft line is to bearing offset. Very important in shaft line calculation is to determine reaction force influence coefficients i.e. reaction forces due to a unit offset of 0.1 mm. They were calculated by each method and are shown in following tables:

Table 5. Influence coefficients, TMEM.

TMEM	1	2	3	4
1	3.65	–9.91	9.61	–3.35
2	–9.91	34.0	–45.1	21.0
3	9.61	–45.1	76.1	–40.6
4	–3.35	21.0	–40.6	22.9

Table 6. Influence coefficients, FEM.

FEM	1	2	3	4
1	3.32	–8.95	8.58	–2.95
2	–8.95	30.8	–41.0	19.1
3	8.58	–41.0	69.9	–37.5
4	–2.95	19.1	–37.5	21.3

Table 7. Influence coefficients, NMSA.

NMSA	1	2	3	4
1	3.20	–8.52	7.97	–2.65
2	–8.52	29.2	–38.6	17.9
3	7.97	–38.6	66.3	–35.6
4	–2.65	17.9	–35.6	20.4

Discrepancies of the results as the percentage values are compared in Tables 8 and 9.

Additional analysis is carried out for working condition with recommended bearings offsets, but only using Nauticus Machinery Shaft Alignment software and FEM. Reaction values for recommended bearing offsets match pretty well and are given in Table 10.

Table 8. Discrepancies between TMEM and FEM,%.

TMEM vs. FEM	1	2	3	4
1	–9.0	9.7	–10.7	11.9
2	9.7	–9.4	9.1	–9.0
3	–10.7	9.1	–8.1	7.6
4	11.9	–9.0	7.6	–7.0

Table 9. Discrepancies between TMEM and NMSA,%.

TMEM vs. NMSA	1	2	3	4
1	–12.3	14.0	–17.1	20.9
2	14.0	–14.1	14.4	–14.8
3	–17.1	14.4	–12.9	12.3
4	20.9	–14.8	12.3	–10.9

Table 10. Bearing reactions in working condition.

Bearing No*.	NMSA	FEM
1	254.76	254.92
2	70.63	71.08
3	56.00	55.57
4	62.72	56.26
5	24.35	29.96

*Bearing numbers are defined in Figure 2.

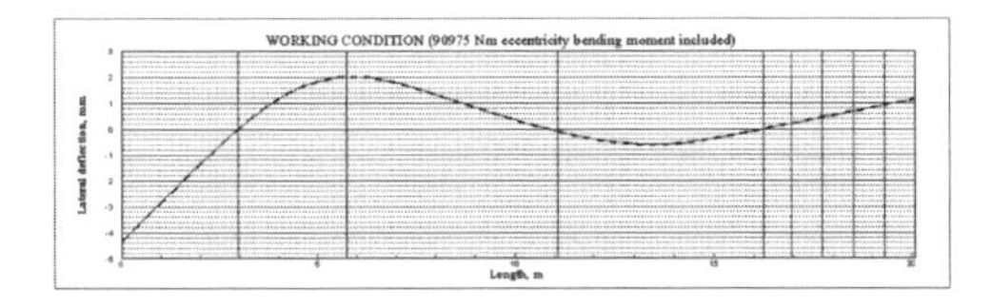

Figure 10. Elastic curve of shaft line obtained using FEM (solid line) and NMSA (dashed line).

Very good agreement of the results is obvious, Table 10. That leads to remarkably good agreement of elastic curve as shown in Figure 10.

6 CONCLUSION AND FUTURE WORK

Basic procedures for shafting alignment calculation have been compared within the numerical example which considers shaft line of a new generation oil tanker. Original design of shaft line (shipyard design), which has been assessed using Nauticus Shaft Alignment software (based on transfer matrix method) is verified against Three Moment Equation Method as well as classical FEM.

Analysis showed quite good agreement of the results. However, the Three Moment Equation method is not suitable when the shaft line is supported by many bearings and the shaft cross-sections are changing significantly along the shaft line. In these cases the calculation should be based on numerical methods which are proven to be reliable within the numerical example presented in this paper.

Future work should consider more realistic loading (exploitation condition) i.e. a reliable numerical tool which couples hull girder flexibility and shaft line behavior should be the next step in the development of shafting alignment calculation procedures.

ACKNOWLEDGMENT

The authors would like to express their gratitude to Mr. J. Vladislavić, Brodosplit Shipyard, Split, Croatia, for conducting shafting alignment calculations using DNV Nauticus Machinery Shaft Alignment software.

REFERENCES

ABS, 2006. *Guidance notes on propulsion shafting alignment*. American Bureau of Shipping.

Batalha, F.A., Silva Neto, S.F. & Belchior, C.R.P. 2009. *Shaft alignment Analysis: a study on shaft alignment and investigation of hull deflection influence*. 13th International Congress IMAM, Istanbul, Turkey, 359–362.

Chang, M. & Juang, S. 2006. *Study on the application of Three Moment Equation for design of shafting alignment of high speed craft*. 17th Symposium on Theory and Practice of Shipbuilding in memoriam prof. Leopold Sorta, Opatija, Croatia, 301–311.

DNV Software, 2004. *Nauticus Shaft Alignment User Guide*. Rev.2-2004-04-27, Det Norske Veritas.

MAN B&W, 1995. Shaft Alignment for Direct Coupled Low-speed Diesel Propulsion Plants, Copenhagen, 312-95.09.

Murawski, L., 2005. Shaft line alignment taking ship construction flexibility and deformations into consideration. *Marine Structures*. 18 (1): 62–84.

NKK, 2006. *Guidelines on shafting alignment*. Nippon Kaiji Kyokai.

Senjanović, I., 1990. *Ship Vibrations*. Part II, Textbook, Zagreb, University of Zagreb.

Šestan, A., Ljubenkov, B. & Vladimir, N. 2010. *Investigation of propulsion system torsional vibration resonance of a catamaran vessel in service*. Advanced Ship Design for Pollution Prevention, C. Guedes Soares & J. Parunov (eds.), Taylor & Francis, 269–274.

Šverko, D. 2005. *Investigation on hull deflection and influence on propulsion shaft alignment*. Houston, American Bureau of Shipping.

Šverko, D. 2006. *Shaft alignment optimization*. 17th Symposium on Theory and Practice of Shipbuilding in memoriam prof. Leopold Sorta, Opatija, Croatia, 399–424.

Vulić, N., Šestan, A. & Cvitanić, V. 2008. Modelling of propulsion shaft line and basic procedure of shafting alignment calculation. *Brodogradnja/Shipbuilding*. (3): 223–227.

Sustainable Maritime Transportation and Exploitation of Sea Resources – Rizzuto & Guedes Soares (eds)
© 2012 Taylor & Francis Group, London, ISBN 978-0-415-62081-9

Exhaust gas waste heat recovery in marine propulsion plants

G. Theotokatos & G. Livanos
Department of Naval Architecture, TEI of Athens, Egaleo, Greece

ABSTRACT: In the present paper, the waste heat recovery (WHR) installations used for the production of saturated steam and electric power for the cases of a two-stroke and a four-stroke engine ship propulsion plant are investigated. The examined waste heat recovery system is considered to be of the single steam pressure type with an external heat exchanger for the heating of feed water entering into the boiler drum. The option of using the engine air cooler for heating the feed water was also examined. The waste heat recovery installation was modeled under steady state conditions and the derived WHR installation parameters for various engine loads are presented an analyzed. Furthermore, using the simulation results, the improvement of energy efficiency design index (EEDI) of a typical merchant ship is calculated and the impact of the WHR on the ship EEDI is discussed.

1 INTRODUCTION

The raising of fuel oil prices throughout the last years, as well as the increased concern for the reduction of the CO_2 gaseous emissions owing to environmental issues, have resulted in the proposals of using mechanical layouts for recovering part of the main engine(s) exhaust gas energy and producing thermal and electric power in marine powerplants. Taking into account the efficiency of the marine Diesel engines used onboard ships for propulsion and electricity generation (about 50% for 2-s engines and slightly lower for 4-s engines), a considerable amount of energy is wasted to the ambient. A portion of that energy could be recovered to cover part of the ship thermal power and/or electric power requirements. Example of wasted energy is the ship main and auxiliary engines exhaust gas energy, which accounts for about 25% of the energy delivered to the engine with the fuel or equivalently about 40–60% of the engine produced power. In addition, the exhaust gas temperature is high enough, so that the exhaust gas wasted energy can be recovered using techno-economically efficient installations.

Usually, two typical options are used for exhaust gas waste heat recovery (WHR) onboard ships (SNAME 1990, MAN Diesel 2009a): a) the installation for the production of saturated steam indended for covering the thermal power requirements of the ship (heating services) and b) the installation for production of both saturated steam and superheated steam. The superheated steam is expanded in a steam turbine producing mechanical energy to drive an electric generator. For the WHR installations of case (b), simple steam pressure systems or more complicated double steam pressure systems can be used (Ioannidis 1984, SNAME 1990, Dzida 2009a). In the latter case, the installation is of higher cost but produces greater amount of electric power (recovering greater amount of the wasted heat), and therefore the installation payback period can be kept in a rational time. The double steam pressure systems often require the recovery of an additional amount of energy from the other ship waste heat sources, e.g., from the engine air cooler or the engine jacket water cooler. In such cases, the recovered heat is used for preheating the low and high pressure water (thus replacing the boiler preheaters). For further increasing the overall efficiency of ship propulsion plant (and thus further reducing the CO_2 gaseous emissions), more complicated installations have been designed comprising an exhaust gas turbine in addition to the steam turbine and exhaust gas boiler (Schmid 2004, MAN Diesel 2005, Rupp 2007, MAN Diesel 2008, Dzida 2009b).

Previous studies concerning the design, optimization and modeling of WHR installations have been presented in Ioannidis (1984), Valdes & Rapun (2001), Schmid (2004), Tien et al. (2007), Dizda (2009a,b), Aklilu & Gilani (2010), Dimopoulos & Kakalis (2010) and Grimmelius et al. (2010).

The last years, the International Maritime Organization (IMO) introduced the Energy Efficiency Design Index (EEDI) to measure the CO_2 efficiency of merchant ships (IMO 2009a,b). The EEDI is a simple formula that estimates CO_2 output per tonmile. The numerator of this formula represents CO2 emissions after accounting for innovative machinery and electrical energy efficiency technologies that are incorporated to the design, whereas the denominator is a function of the speed, capacity and ship specific factors.

The present work is focused on the performance investigation of the WHR system utilizing the exhaust gas of the propulsion engine of a typical handymax class bulk carrier. Two cases for the ship propulsion installation are considered. In the first case, a two-stroke marine Diesel engine, which is directly connected to the ship propeller via the ship shafting system, is used as the main engine of the ship. In the second case, a four-stroke marine Diesel engine connected to the ship propeller via a gear box/clutch unit and shaft is used as the ship propulsion engine. The engine efficiency is higher in the case of two-stroke engine, resulting in lower exhaust gas temperature (and thus lower available energy content in the exhaust gas). In the case of four-stroke engine, lower values of the propulsion plant efficiency is obtained taking into account the lower efficiency of the four stroke marine diesel engine and the efficiency of the shafting system, which includes a gearbox/clutch unit.

The WHR system is considered to be of the single steam pressure type with external heat exchanger for heating the feed water entering into the boiler drum. The option of heating the feed water tank using the engine air cooler is also investigated. The WHR system is used for the production of saturated steam for covering the requirement of the ship in thermal power for the ship heating services, and electric power for covering partly or totally the ship electric power demands.

The WHR system was modeled under steady state conditions for various engine loads in the range from 50% to 100% of the engine MCR point. The derived results are analyzed and the effect of the WHR system operating parameters on the produced electric power and the ship powerplant efficiency improvement is discussed. In addition, the EEDI of the considered ship is calculated and the influence of WHR system on the variation of ship EEDI is commented.

2 DESCRIPTION OF THE WHR INSTALLATION

A typical WHR installation of single pressure for saturated and superheated steam production is shown in Figure 1. The exhaust gas boiler consists of three stages; the economizer (preheater), the evaporator and the superheater.

The feed water is pumped by the feed water pump into the water/steam drum. In the installation shown in Figure 1, an external heat exchanger is used for preheating the feed water. The feed water having temperature in the range from 50 to 120°C enters into the heat exchanger, where it is heated reaching

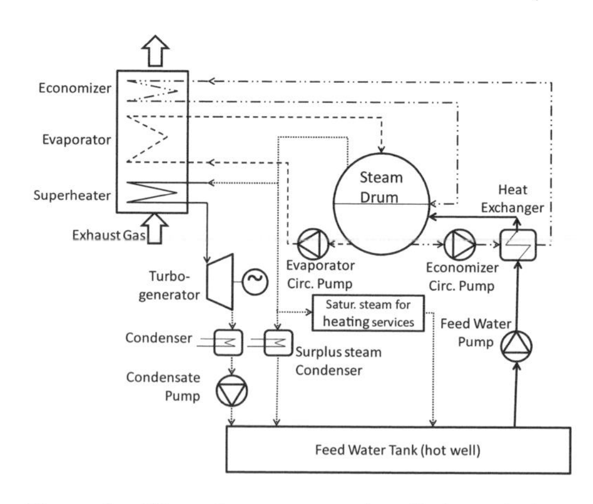

Figure 1. Waste heat recovery installation of single steam pressure for production of saturated steam and electric power.

a temperature value slightly lower than the water saturation temperature (165°C for pressure 7 bar), and subsequently enters into the water/steam drum. The heating medium is the saturated water contained in the drum, which is pumped by the economizer circulating water pump, enters the heat exchanger heating the feed water and leaves the heat exchanger with lower temperature. In order for the temperature of the circulating water exiting the external heat exchanger and entering into the economizer to be kept in the range of 130°C, the flow rate of the circulating pumps must be three or four times the flow rate of feed water. The circulating water exiting the economizer returns to the drum having temperature of approximately equal to the saturation temperature corresponding to the drum operating pressure.

An additional circulating pump is used to circulate the water through the evaporation section of the boiler. The pump flow rate should be two to four times the flow rate of feed water, so that the integrity of the evaporator section is not jeopardized due to overheating in cases where the evaporator tubes run out of water. Inside the tubes of the evaporation section, a portion of saturated water is evaporated and saturated steam is produced. The saturated water/steam mixture exiting the boiler returns into the drum, where the saturated steam is separated from the water and is accumulated in the upper part of the drum. A portion of the saturated steam is used to cover the ship heating services, whereas the rest is advanced into the superheater section of the boiler.

The temperature of the superheated steam exiting the exhaust gas boiler depends on the temperature of the exhaust gas entering the boiler. The temperature difference between the two fluid flows (i.e., the exhaust gas entering the boiler superheater section and the superheated steam exiting the boiler) is required to be greater than or equal to 20°C. The superheated steam exiting the boiler enters into the steam turbine stages of turbogenerator, where it expands producing mechanical power and driving the electric generator. Thus, electric power is produced, which can partly or entirely cover the ship needs in electricity depending on the installation operating conditions. The steam pressure downstream of the steam turbine must be kept in such levels so that the steam quality is in the region of 90%, otherwise moisture drops may be formed in the final stages of steam turbine, which can cause wear of the steam turbine blades. The steam exiting the steam turbine is advanced to the condenser, where it condenses by the usage of sea water. The condenser absolute pressure is set in the range from 0.05 bar to 0.08 bar, so that the steam turbine expansion ratio is as high as possible and the maximum mechanical work is produced in the steam turbine. The condensate is pumped into the feed water tank (hot well) through the condensate pump. In case where surplus amount of saturated steam is produced, it is also forwarded into the surplus steam condenser where it converts to condensate water, which subsequently is pumped to the feed water tank.

In the case of heating the feed water using the engine air cooler, an additional pump and the appropriate piping is required, so that the feed water is pumped to the high temperature stage of the engine air cooler, where it is heated by the simultaneously cooling of the compressed air. The air is further cooled in the low temperature stage of the engine air cooler using the cooling fresh water. In that case, lower heat is required in the boiler economizer, resulting in greater amount of heat available in the other sections of the boiler. However, as the temperature of feed water (heated in engine air cooler) increases, lower amount of heat is required in the economizer, and therefore the temperature difference at pinch point decreases. In such case, the increase of the exhaust gas temperature at the boiler outlet may be required to ensure the unproblematic function of the boiler.

3 MODELING OF THE WHR INSTALLATION

The governing equations for the mathematical modeling of the WHR installation under steady state conditions are derived by applying the mass and energy conservation equations in the various components of the installation.

The heat transferred to the steam/water in the exhaust gas boiler is given by:

$$\dot{Q}_b = \eta_b mfr_g c_{p\_g}(T_{g\_i} - T_{g\_o}) \tag{1}$$

where mfr_g is the exhaust gas mass flow rate; c_{p_g} is the exhaust gas mean specific heat at constant pressure; T_{g_i} is the temperature of the exhaust gas entering into the boiler; T_{g_o} is the temperature of the exhaust gas exiting the boiler; and η_b is the boiler efficiency that is in the order of 1 2% (1 η_b is the boiler heat transfer losses).

Considering that the water exits the economizer section in saturated condition, the energy balance in the boiler gives:

$$\begin{aligned}\dot{Q}_b = \dot{Q}_{ec} + \dot{Q}_{ev} + \dot{Q}_{sh} = mfr_{cw\_ec}(h_w - h_{cw''\_ec}) \\ + mfr_{cw\_ev}(h_w - h_{cw'\_ev}) + mfr_s(h_s - h_w) \\ + mfr_{sh}(h_{sh} - h_s)\end{aligned} \tag{2}$$

where mfr_{cw_ec} is the economizer circulating water mass flow rate; mfr_{cw_ev} is the evaporator circulating water mass flow rate; mfr_{sh} is the superheated steam mass flow rate; $h_{cw''\_ec}$ is the specific enthalpy of the circulating water entering into the economizer (which is exiting from the external heat exchanger); h_{sh} is the specific enthalpy of the superheated steam exiting the boiler; $h_{cw'\_ev}$ is the specific enthalpy of the circulating water entering into the evaporator (exiting the evaporator circulating water pump); h_w and h_s are the specific enthalpies of saturated water and steam exiting the evaporator, respectively.

The mass balances in the feed water tank and in the boiler drum provide:

$$mfr_{fw} = mfr_s = mfr_{sh} + mfr_{s\_hs} \tag{3}$$

where mfr_{fw} is the feed water mass flow rate; mfr_s is the mass flow rate of the saturated steam produced in the evaporator; and mfr_{s_hs} is mass flow rate of the saturated steam required for covering the ship heating services.

Taking into account the Equation 3 and applying the energy balance in the external heat exchanger, we get:

$$mfr_{cw\_ec}(h_{cw'\_ec} - h_{cw''\_ec}) = mfr_s(h_{fw''} - h_{fw'}) \tag{4}$$

where $h_{cw'\_ec}$ is the specific enthalpy of the economizer circulating water entering into the external heat exchanger; $h_{fw'}$ is the specific enthalpy of the feed water entering into the external heat exchanger (exiting the feed water pump); and $h_{fw''}$ is the specific enthalpy of the feed water exiting the external heat exchanger and entering into the drum.

The specific enthalpy of the feed water entering the drum is derived by manipulating the equation that defines the heat exchanger effectiveness, η_{HE} (Ganapathy, 2003):

$$h_{fw''} = h_{fw'} + \eta_{HE}(h_{cw'\_ec} - h_{fw'}) \tag{5}$$

By applying the energy conservation in the water/steam drum and after some manipulation, we get:

$$(mfr_{cw\_ec} + mfr_{cw\_ev})(h_{cw} - h_w) = mfr_s(h_{fw''} - h_w) \quad (6)$$

where h_{cw} is the specific enthalpy of water contained in the drum.

The required power for each pump of the installation can be calculated using the following equation:

$$P_i = mfr_i(h_{i'} - h_i) = mfr_i \Delta p_i / (\eta_i \rho_i) \quad (7)$$

where mfr_i is the pump mass flow rate; h_i and $h_{i'}$ are the fluid specific enthalpies upstream and downstream the pump, respectively; Δp_i is the pump pressure increase; η_i is the pump efficiency; ρ_i is the fluid density; and $i = fw$ for the feed water pump, $i = cw\_ec$ for the economizer circulating water pump, $i = cw\_ev$ for the evaporator circulating water pump, $i = c$ for the condensate water pump, and $i = c\_sw$ for the condenser sea water pump.

The energy balance in the feed water tank gives:

$$mfr_{fw}h_{fw} = mfr_{sh}h_{c'} + mfr_{s\_hs}h_{w\_hs} + Q_{fw} \quad (8)$$

where h_{w_hs} is the specific enthalpy of the condensate water returning from the ship heating services to the feed water tank; Q_{fw} is the thermal power used for heating the feed water tank.

In the case where the feed water tank is heated by using saturated steam, the thermal power added to the feed water is calculated by:

$$Q_{fw} = mfr_{s\_fw}(h_s - h_w) \quad (9)$$

whereas, in the case where the high temperature stage of the engine air cooler is used for heating the feed water tank, the thermal power added to the feed water is given by:

$$Q_{fw} = \eta_{ac}\, mfr_a\, c_{pa}(T_{ac\_i} - T_{ac\_HT\_o}) \quad (10)$$

where mfr_{s_fw} is mass flow rate used for heating the feed water tank; mfr_a is engine air mass flow rate; c_{pa} is engine air cooler mean specific heat at constant pressure; T_{ac_i} is the temperature of the air entering the engine air cooler; and $T_{ac\_HT\_o}$ is the temperature of the air exiting the high temperature stage of the engine air cooler.

The set of equations (1–10) is solved in order for the produced superheated steam to be calculated providing as input the following parameters: the engine exhaust gas mass flow rate, temperature and air/fuel equivalence ratio, the pressure of the drum water/steam, the pressure and temperature of the feed water tank, the pressure losses in the various boiler sections and the piping of the WHR installation, the ratio of the economizer circulating water to the produced saturated steam mass flow rates, the ratio of the evaporator circulating water to the produced saturated steam mass flow rates, the mass flow rate of saturated steam required for the ship heating services, the boiler efficiency, the pumps efficiency, the external heat exchanger effectiveness, the temperature of superheated steam exiting the boiler, the temperature drops at various sections of the WHR installation, the properties of water/steam, exhaust gas and air.

The first estimation for the mass flow rate of the superheated steam is calculated using the following equation, which is derived considering the ideal Rankine cycle:

$$mfr_{sh} = \frac{\eta_b mfr_g c_{p\_g}(T_{g\_i} - T_{g\_o}) - mfr_{s\_hs}(h_s - h_{fw})}{h_{sh} - h_{fw}} \quad (11)$$

The superheated steam is expanded in the steam turbine of turbogenerator (steam turbine and electric generator), thus providing the required mechanical work to drive the electric generator. The produced electric power is calculated by:

$$P_{el} = mfr_{sh}(h_{sh'} - h_{st,is})\eta_{TG}\, f_b\, f_T\, f_L \quad (12)$$

where $h_{sh'}$ is the specific enthalpy of the superheated steam entering the steam turbine; h_{st_is} is the specific enthalpy of the steam exiting the steam turbine in the case where the steam expands isentropically; η_{TG} is the efficiency of the turbogenerator; and f_b, f_T, f_L are correction factors for the steam turbine back pressure, steam temperature and steam turbine load, respectively. Data for the estimation of turbogenerator efficiency and the correction factors are given in SNAME (1990).

The specific enthalpy of the steam exiting the steam turbine is calculated by:

$$h_{st\_o} = h_{sh'} - \eta_{ST}(h_{sh'} - h_{st,is}) \quad (13)$$

where $\eta_{ST} = \eta_{TG}/(\eta_G\, \eta_m)$ is the steam turbine efficiency; η_G is the generator efficiency; and η_m is the turbogenerator mechanical efficiency.

The heat rate that has to be transferred from the steam to the cooling medium of the condenser (usually sea water) is given by:

$$Q_c = mfr_{sh}(h_{st\_o} - h_{c\_w}) \quad (14)$$

where h_{c_w} is the specific enthalpy of the condensate water exiting the condenser.

The mass flow rate of the condenser sea water pump is calculated by:

$$mfr_{c\_sw} = Q_c / (c_{p\_sw} \Delta T_{sw}) \quad (15)$$

where c_{p_sw} is the condenser sea water specific heat, and ΔT_{sw} is the temperature increase of the sea water in the condenser.

The required input for calculating the produced electric power includes the condenser pressure, as well as the specific enthalpy of steam as function of pressure and specific entropy.

The increase in the ship propulsion installation efficiency is calculated by the following equation:

$$\Delta\eta = [(P_{el} - \sum_{pumps} P) + mfr_{s\_hs}(h_s - h_w)]/(mfr_f\, H_L) \quad (16)$$

where mfr_f is mass flow rate of the engine fuel; and H_L is the fuel lower heating value.

The minimum temperature difference (pinch point) is calculated using the following equation, which is derived using the energy balance in the evaporator and superheater sections of the boiler:

$$\Delta T_{pp} = T_{g\_i} - \frac{\dot{Q}_{ev} + \dot{Q}_{sh}}{\eta_b mfr_g c_{p\_g}} - T_s \quad (17)$$

where T_s is the temperature of saturated steam/water.

4 IMO ENERGY EFFICIENCY DESIGN INDEX

The Marine Environmental Protection Committee (MEPC) of the IMO recognized the need to develop an energy efficiency design index for new ships in order to stimulate innovation and technical development of all elements influencing the energy efficiency of a ship from its design phase (IMO 2009a).

The attained new ship Energy Efficiency Design Index (EEDI) is a measure of ships CO_2 efficiency and in the case of a typical bulk carrier (no ice-class) is calculated by the following formula:

$$EEDI = \frac{P_{ME} \cdot C_{FME} \cdot SFC_{ME} + P_{AE} \cdot C_{FAE} \cdot SFC_{AE}}{Capacity \cdot V_{ref}} - \frac{f_{eff} \cdot P_{AEeff} \cdot C_{FAE} \cdot SFC_{AE}}{Capacity \cdot V_{ref}} \quad (18)$$

where C_F is a non-dimensional conversion factor between fuel consumption (in gr) and CO_2 emission (also in gr), and is based on fuel carbon content; C_F = 3.206 gr CO_2/gr fuel for the case of Diesel Gas/Oil; This value of C_F were used throughout this study since the majority of Diesel engine testing for processing the EIAPP certificates are performed using DMX through DMC grade distillates; *ME* and *AE* refer to the main and auxiliary engine(s), respectively; *Capacity* is defined as the deadweight in case of the dry cargo carrier; P_{ME} is defined as the 75% of the rated installed power (MCR) of the main engine, after having deducted any installed shaft generator; P_{AE} is the required auxiliary engine power to supply normal maximum sea load including necessary power for propulsion/machinery systems, but excluding any other power e.g., ballast pumps, thrusters, cargo gear etc, in the condition where the ship engaged in voyage at the speed V_{ref} under the design loading condition of *Capacity*; P_{AEeff} is the auxiliary power reduction due to innovative electrical energy efficient technology (e.g., WHR) measured at P_{ME}; f_{eff} is the availability factor of each innovative energy efficiency technology, which should be 1.0 for waste energy recovery system according to IMO 2009a; SFC_{ME} and SFC_{AE} are the specific fuel consumptions (in gr/kWh) of the main and auxiliaries engines at the 75% and 50% of MCR, respectively.

It must be also noted, that according to IMO (2009b), a baseline EEDI can be defined, based on regression analysis of data of several merchant ships. For the case of dry cargo bulk carriers, the following expression was proposed:

$$Baseline\ Value = 1354.0 \times Capacity^{-0.5117} \quad (19)$$

5 APPLICATION CASES

The propulsion plant installation of a typical handymax bulk carrier having deadweight of 55,000 tons is examined. Such a ship requires a marine Diesel engine that produces about 8900 kW at maximum continuous rating (MCR) point (MAN Diesel 2010b). At that power range, the following engine models can be selected as the ship propulsion engine: a) the two-stroke engine model 6S50-B8-TII from MAN Diesel (MAN Diesel 2010a) or 6RT-flex50 from Wärtsilä (Wärtsilä 2011), b) the four-stroke engine model 8L48/60 from MAN Diesel (MAN Diesel 2009b) or 9L46 from Wärtsilä (Wärtsilä 2007). The main characteristics of the ship as well as the alternative propulsion plants are given in Table 1.

The required engine data for modeling the WHR system include the exhaust gas mass flow rate, the temperature of the exhaust gas at the turbocharger turbine outlet and the air to fuel equivalence ratio of the exhaust gas. For the case of the heating the

Table 1. Main parameters of ship and examined engines.

Ship parameters—typical values		
Type	bulk carrier	(Handymax)
Size (at scantling draught)	55,000	mt
Length between perpendiculars	185	m
Breadth	32.0	m
Draught (scantling)	12.5	m
Vessel speed	14.5	kn
Engine parameters—typical values		
Engine type	2-s	4-s
Bore	500 mm	460 mm
Stroke	2000 mm	580 mm
Brake power at MCR	8815 kW	8775 kW
Engine speed at MCR	122 rpm	500 rpm
BSFC at MCR	171 ± 5% gr/kWh	183 ± 5% gr/kWh

feed water using the engine air cooler, the temperature of the air exiting the turbocharger compressor is also required as input. All the required parameters are calculated for engine loads in the range from 50% to 100% of MCR, considering that the engines operate according to propeller law and ISO ambient conditions, using the data given in the engine project guides (MAN Diesel 2010a, MAN Diesel 2009b, Wärtsilä 2007, Wärtsilä 2011). The engine fuel and air mass flow rates are derived from the exhaust gas mass flow rate and the air to fuel equivalence ratio.

For both engines the following tolerances were taken into account; for the exhaust gas temperature at turbocharger turbine outlet: +15°C; for the brake specific fuel consumption: +3%; for the exhaust gas mass flow rate: –3%. The values of the temperature and mass flow rate of the exhaust gas exiting the engine for both the examined engines, which are used in the WHR installation simulation cases presented below, are shown in Figure 2.

A set of results including the net produced electric power, the ship powerplant efficiency increase due to the production of the electric power from the WHR system, the steam cycle efficiency and the total powerplant efficiency for the cases of the ship propulsion installations with the two-stroke and the four stroke engines, are presented in Figures 3 and 4, respectively. The following cases for the WHR system parameters were simulated: no heating of the feed water tank, heating of the feed water tank using saturated steam and heating the feed water tank using the engine air cooler. In all the simulated cases, the production

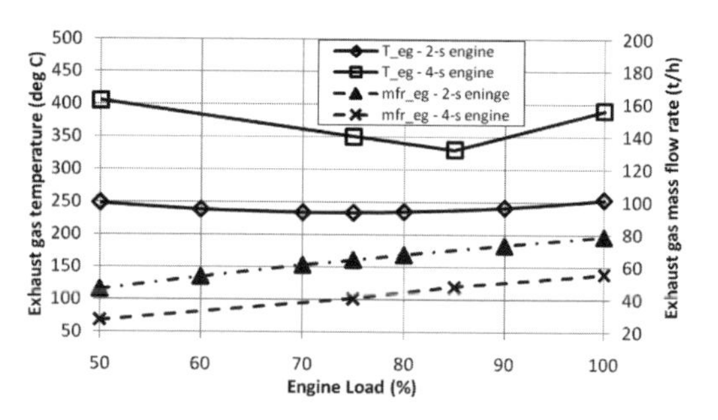

Figure 2. Engine parameters as function of engine load.

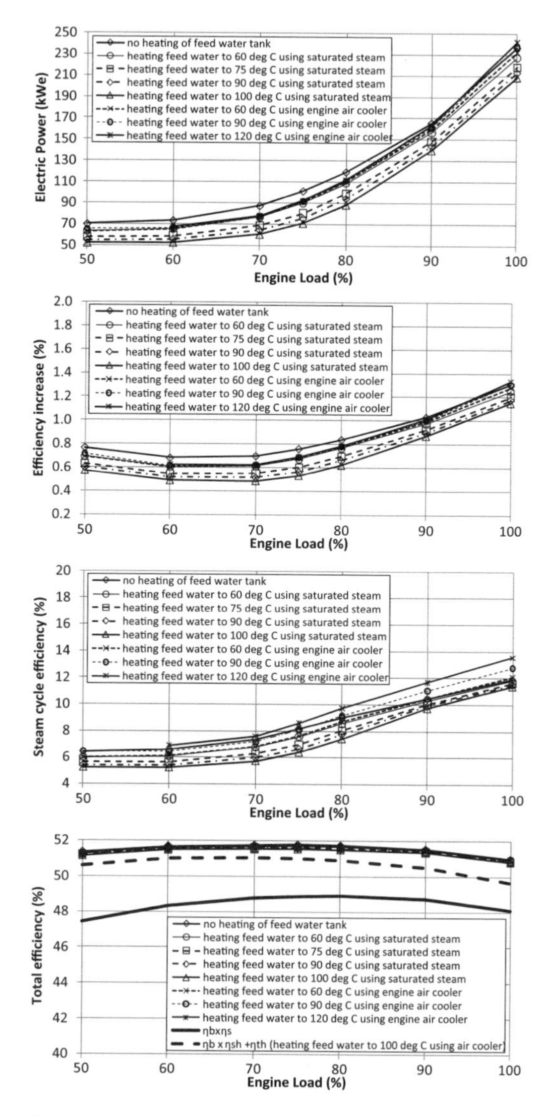

Figure 3. WHR system parameters for the examined two-stroke engine propulsion plant.

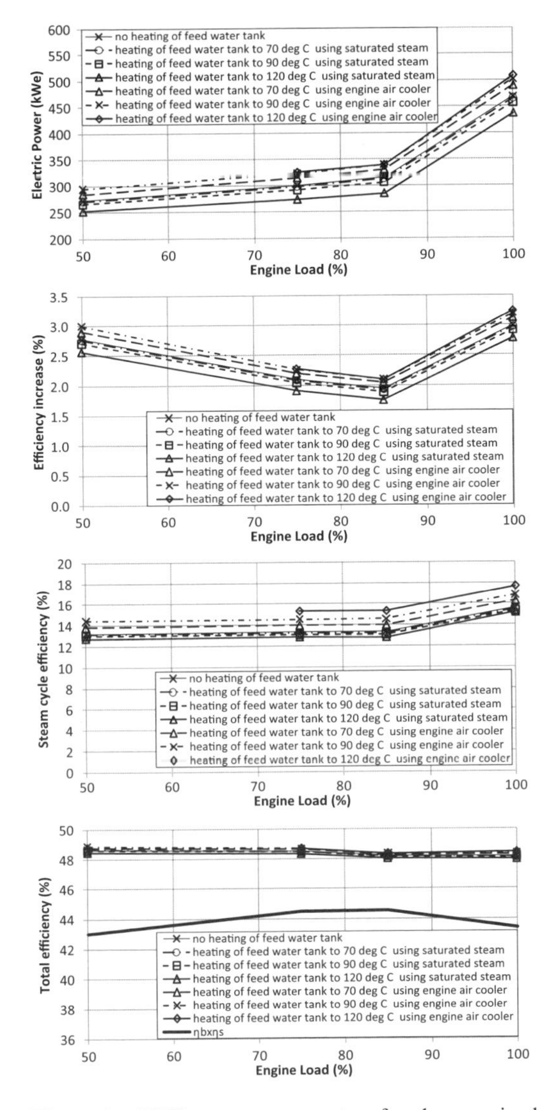

Figure 4. WHR system parameters for the examined four-stroke marine engine propulsion plant.

of saturated steam for the ship heating services is assumed to be 500 kg/h. The used fuel is assumed to have a sulfur weight fraction 3%, which limits the lower values of the temperature of the exhaust gas exiting the boiler to approximately 160°C. The boiler drum pressure is considered to be 8.5 bar for the case of four-stroke engine, whereas it was set at 7 bar for the case of two-stroke engine in order to retain the temperature difference at pinch point above 10°C (due to the lower temperatures of the exhaust gas). The pressure of the condenser is taken as 0.065 bar. The water steam properties are calculated using the equations given in Wagner & Kretzschmar (2008). The average electric power, which is required for the examined ship, is estimated to 450 kWe according to SNAME (1990). The net produced electric power for the two-stroke and four stroke engine installations operating at load 85% of MCR, as function of the feed water temperature, are shown in Figure 5.

As it can be observed from Figure 3, the produced electric power for the case of the examined two-stroke engine propulsion installation using WHR is not enough to cover the engine demands. The ship propulsion installation efficiency increase is in the region from 0.5 to 1.3% depending on the engine load. Due to the relatively low temperature of the engine exhaust gas, the produced superheated steam is also of low temperature, the pressure of the boiler drum is kept to 7 bar, and the obtained steam cycle efficiency is in the region from about 6 to 14%. The total ship propulsion plant efficiency (considering that the produced saturated steam thermal power is entirely consumed in ship heating services) increases in the case of the WHR installation, reaching values in the region of 51 to 52%. Due to the significant amount of electric power that is required to cover the ship demands, the installation of the considered WHR system is sought to be unbeneficial. In the case where a simpler, and thus less expensive, WHR system is installed for producing saturated steam for the ship heating services, the total ship propulsion plant efficiency also improves by 1.8 to 3% depending on the engine load.

For the case of the four-stroke engine ship propulsion installation, the higher temperature of the exhaust gas (approximately 80–150°C in comparison to the two-stroke engine case) results in higher temperature of the superheated steam and higher drum pressure (8.5 bar). Thus, the steam cycle efficiency is in the region of 12.5 to 18% and the produced net electric power reaches 450 kWe at 100% load and over 250 kWe at 50% load. The efficiency increase due to the electric power production

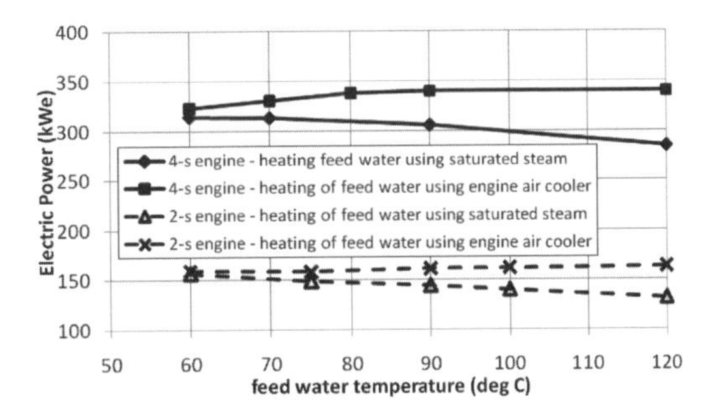

Figure 5. Effect of feed water temperature on produced electric power for the two-stroke and four-stroke engines installations.

is in the region from 2 to 3%, thus resulting in a considerable improvement of the total efficiency of the ship propulsion installation, which reaches 48 to 49%. The best results concerning the net produced electric power and WHR system efficiency increase are obtained using the option of heating the feed water tank to 120°C using the engine air cooler. From the above analysis, it is inferred that the installation of the examined WHR system is beneficial for the substantial improvement of the ship propulsion plant efficiency. The four-stroke engine propulsion installation with the WHR system reaches efficiencies comparable to the ones of the two-stroke engine propulsion system. However, the two-stroke engine with a WHR system for producing saturated steam to cover the ship heating services seems to be the best option for the examined ship propulsion installation, obtaining efficiency values from 49.5 to 51%.

As it is shown in Figures 5, the heating of the feed water in higher temperature levels by using the saturated steam results in lower net electric power production, whereas the net produced electric power from the WHR installation increases in the case of heating the feed water by means of the engine air cooler. In the former case, the observed behavior is attributed to the greater amount of the saturated steam required for the feed water heating and therefore, less energy is available for the superheated steam production. However, in the latter case, the boiler economizer heat transfer rate is lower (for higher feed water temperatures), resulting in lower temperature difference at the boiler pinch point. Therefore, higher temperature of exhaust gas at boiler exit is required to retain the pinch point temperature greater than 10°C. This, in turn, causes the reduction of the recoverable exhaust gas heat in the boiler, and thus the net produced electric power reaches an upper bound.

The attained Energy Efficiency Design Index was calculated using Equation 18 for the examined ship and for the various options for the ship propulsion plant installation (two-stroke engine with and without WHR, four-stroke engine with and without WHR), as well as for some recently build similar bulk carriers, the main characteristics of which are published in RINA (2005). The derived results are presented in Figure 6. As it can be observed from Figure 6, in the case of four-stroke engine propulsion plant, the calculated EEDI is 5.45, whereas in the case of two-stroke engine propulsion plant the EEDI is 5.11. The baseline EEDI for the investigated bulk carrier is 5.08. It is also deduced that the WHR installation can reduce the EEDI by approximately 5.4% (EEDI = 5.16) in case of the four-stroke engine propulsion plant and 1.8% (EEDI = 5.02) in case of the two-stroke

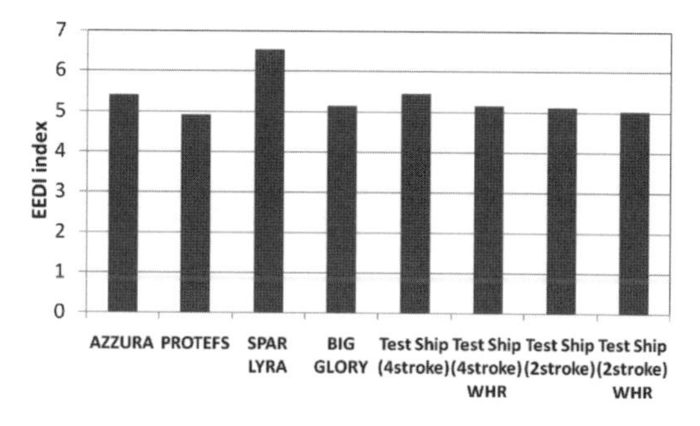

Figure 6. EEDI for the several alternative propulsion plants of the test ship and some recently build similar ships.

engine installation. From EEDI point of view, a two-stroke engine solution is better compared to a four-stroke engine, since in the first case (two-stroke) the EEDI is slightly lower compared to the last case (four-stroke), even when WHR system is used in the last solution. In addition, it can be concluded that only a two-stroke engine solution with WHR can result in EEDI below the baseline. However, in such a case, owing to the relatively low electric power production using the WHR installation of single steam pressure, more complicated WHR installations are required (e.g., with two different steam pressure levels and additional heat recovery from the engine air cooler and/or engine jacket cooling water).

6 CONCLUSIONS

The performance of the WHR system of single steam pressure type comprising an external heat exchanger for heating the feed water entering into the boiler drum was investigated for the cases of a handymax class bulk carrier propulsion plant installation with a two-stroke engine and a four-stroke engine. The options of heating the feed water using saturated steam or alternatively by recovering heat from the engine air cooler were also considered. In addition, the ship EEDI values, which represent the ship CO_2 efficiency, were calculated for the examined cases of the ship propulsion plant installation.

The main findings derived from this work are summarized as follows.

The two-stroke engine in combination with a WHR system for producing saturated steam is the best option for the propulsion plant installation of the examined ship. However, in order to obtain an EEDI below the corresponding baseline value, a two-stroke engine combined with a more complex (and as a result more effective)

WHR system (e.g., with two steam pressure levels and recovery of heat for the engine air cooler and jacket cooling water) has to be used.

In the case of four-stroke engine propulsion installation, the utilization of the WHR system is necessary for considerably increasing the installation efficiency, which however remains lower than the one of the two-stroke engine installation.

The heating of feed water tank using the engine air cooler improves the WHR system efficiency, but an upper bound exists at temperatures about 100 to 120°C. On the other hand, the heating of the feed water tank by means of saturated steam slightly decreases the WHR system efficiency.

REFERENCES

Aklilu, B.T. & Gilani, S.I. 2010. Mathematical modeling and simulation of a cogeneration plant. *Applied Thermal Engineering* 30: 2545–2554.

Dimopoulos, G.G. & Kakalis, N.M.P. 2010. An integrated modeling framework for the design operation and control of marine energy systems, *Proc. CIMAC Congress 2010, Paper No. 15*. Bergen: Norway.

Dzida, M. 2009a. On the possible increasing of efficiency of ship power plant with the system combined of marine diesel engine, gas turbine and steam turbine, at the main engine-steam turbine mode of cooperation. *Polish Maritime Research* 1(59), Vol. 16: 47–52.

Dzida, M. 2009b. On the possible increasing of efficiency of ship power plant with the system combined of marine diesel engine, gas turbine and steam turbine in case of main engne cooperation with the gas turbine fed in parallel and the steam turbine, *Polish Maritime Research* 2(60), Vol. 16: 40–44.

Ganapathy, V. 2003. Industrial Boilers and Heat Recovery Steam Generators-Design, Applications, and Calculations. Marcel Dekker Inc.

Grimmelius, H., Boonen, E.-J., Nicolai, H. & Stapersma, D. 2010. The integration of mean value first principle diesel engine models in dynamic waste heat and cooling load analysis. *Proc. CIMAC Congress 2010, Paper No. 280*. Bergen: Norway.

IMO. 2009a. Interim Guidelines on the Method of Calculation of the Energy Efficiency Design Index for New Ships. MEPC.1/Circ.681.

IMO. 2009b. Recalculation of energy efficiency design index baselines for cargo ships. GHG-WG 2/2/7.

Ioannidis, J. 1984. Waste Heat Recovery from Diesel Engines. *Proc of IMAM 1984 conference*. Athens: Greece.

MAN Diesel. 2005. Thermo Efficiency System (TES) for Reduction of Fuel Consumption and CO2 Emission. Publ. No.: P3339161. Copenhagen: Denmark.

MAN Diesel. 2008. Exhaust Gas Emission Control Today and Tomorrow-Application on MAN B&W Two-stroke Marine Diesel Engines, Publ. No. 5510-0060-00. Denmark.

MAN Diesel. 2009a. Soot Deposits and Fires in Exhaust Gas Boilers. Publ. No. 5510-0065-00. Copenhagen: Denmark.

MAN Diesel. 2009b. Project Guide for Marine Plants, Diesel Engine L+V48/60CR. Augsburg; Germany.

MAN Diesel. 2010a. MAN B&W S50ME-B8-TII Project Guide. 7020-0103-00. Copenhagen: Denmark.

MAN Diesel. 2010b. Propulsion trends in bulk carriers. Publ. No. 5510-0007-02. Copenhagen: Denmark.

RINA. 2005. Significant Ships of 2005. UK.

Rupp M. 2007. Waste heat recovery for lower engine fuel consumption and emissions. *ABB Turbo Systems Ltd.* Baden: Switzerland.

Schmid, H. 2004. Less emissions through waste heat recovery. *Green Ship Technology Conference, London, 28–29 April 2004*. London: UK.

SNAME. 1990. Marine Diesel Power Plant Practices. *T&R Bulletin* 3–49.

Tien, W.-K., Yeh, R.-H. & Hong J.-M. 2007. Theoretical analysis of cogeneration system for ships. *Energy Conversion and Management* 48: 1965–1974.

Valdes, M. & Rapun, J.L. 2001. Optimization of heat recovery steam generators for combined cycle gas turbine power plants, *Applied Thermal Engineering* 21(2001), 1149–1159.

Wagner, W. & Kretzschmar, H.-J. 2008. International Steam Tables. Springer-Verlag.

Wärtsilä. 2007. Wärtsila 46 project guide. 3/2007. Finland.

Wärtsilä. 2011. Ship Power Product Catalogue. Finland.

Sustainable Maritime Transportation and Exploitation of Sea Resources – Rizzuto & Guedes Soares (eds)
 ISBN 978-0-415-62081-9

Amerigo Vespucci Ship: Retrofitting the most beautiful ship in the world

T. Mazzuca
Italian Navy General Staff, Electrical Plants Office, Rome, Italy

ABSTRACT: Once described constructive characteristics and operative requirements of Vespucci Training Ship, it will be reported the criteria adopted to define the new configuration of the distribution and propulsion systems in relation to the global platform retrofitting. Then, will be described the peculiarities of the innovative architecture chosen in relation to ship operative profile, and will be outlined the system engineering activities to be carried out to guarantee a safe and efficient plant operation and to benefit from technical innovations adopted.

1 INTRODUCTION

Vespucci Ship was designed together with Cristoforo Colombo Ship in 1930 by the engineer Francesco Rotundi, commander of the Engineer Officers team and chief of Castellamare di Stabia (Naples) shipyards. It was launched on the 22nd of February 1931, and on 15th of October 1931 it was docked in Genoa, to support Cristoforo Colombo ship until the end of Second World War.

Vespucci Ship was designed as a tall ship with the aim to maintain alive old sea traditions.

The ship was equipped with three masts, the mizzenmast, the mainmast and the foremast, globally holding a tall surface of about 2800 mq. It was provided with eight transverse watertight bulkheads. The hull was proportionated with respect to Cristoforo Colombo ship with a 5% increase, was realized with Martins Siemens mild steel plates, with 12 and 16 mm thickness, and was endowed with a double bottom structure. The keel was more massive, with higher dimensions, leading to an increase in stability of about 20% with 3800 tons displacement.

The original armament consisted in four antiaircraft and anti-ship guns.

At present it is the oldest Italian Navy ship still in service. The ship saying, formalized in 1978 is "Non chi comincia ma quel che persevera" (Not who begins, but who persists), expresses its vocation to the training of future Italian Navy officers. It is memorable the encounter with the U.S. Navy aircraft carrier USS Forestal (CV-59), that, in response to the ship annunciation, declared "You are the Most Beautiful Ship in the World".

In Table 1 are indicated the ship main characteristics:

Table 1. Main characteristics of the ship.

Class and type	Full rigged ship
Displacement	3800 tonnes (9,140,000 lb)
Length	82.4 metres (270 ft) + 16 metres (52 ft) including bowsprit
Beam	7.7 metres (25 ft)
Height	54.0 metres (177.2 ft)
Draught	7.0 metres (23.0 ft)
Propulsion	26 Sails, 2800 square metres (14,600 sq ft) Engine, FIAT B 308 ESS 1 fixed pitch 4 blades propeller
Speed	Sails, 10 knots (19 km/h) Engine, 12 knots (22 km/h)
Standard crew	16 permanent crew 70 passengers 190 sailors

In the first realization of the propulsion plant, the direct current electric propulsion motor, equipped with a Ward-Leonard drive system, was energized by two dinamo generators, driven by two 2-stroke 6-cylinder FIAT Q 426 engines, so that to produce up to about 1471 kW (2000 hp). The primary distribution network dedicated to electrical ship service users was separated from the propulsion system, and was supplied by dinamo generators sets at 110 Vcc voltage level.

In 1964 the platform system was subject to the first global retrofitting, consisting in the complete substitution of primary engines and equipments by similar ones having the same dimensional and operation characteristics. Since then the propulsion subsystem has been provided with two 4-stroke, 8-cylinder FIAT B 308 ESS diesel engines in place of the original ones. The plant configuration remained substantially the same, with the major conversion of ship service distribution to an AC 380 V.- 50 Hz system by means the provision of four identical 500 kVA, 380 V - 50 Hz electrical generators.

At present it counts 80 years of service and the Energy Platform Component shows at present severe criticities of efficiency, supportability, logistic and operation safety, due mainly to the obsolescence reached by plant equipment.

In 2006, after 75 years of service, it was interested by the complete substitution of the foremast, exactly rebuilt with craftmade period old techniques, the substitution of some hull plates riveted by skilful craftmen, reconfigurations of auxiliary rooms with new conception plants. So, in consideration of these recent severe interventions and of the particular and important role played by the ship for the training of Italian Navy officers, it was decided to proceed to another complete platform systems revamping so that to allow the ship to continue its traditional duty.

2 REQUIREMENT ANALYSIS IN VIEW OF NEXT RETROFITTING

Existing platform subsystems, for which it is envisaged complete substitution are:

- electric generation and primary distribution for ship services, composed by the main switchboard, installed in the electrical station room, and divided in two half-busbars by a tie-breaker.
- electric propulsion plant, comprehensive of diesel prime movers, direct current generators, propulsion motor subsystem.

The base idea for platform revamping has been the implementation of the IFEP (Integrated Full Electric Propulsion) concept in consideration of the subsistence of the following conditions:

- recognition of the comparability of entity of ship services electric power request with power required by propulsion plant, at similarity of the situation occurring in cruise ships;
- ascertained advantages in terms of operation flexibility, higher efficiency, global reliability, ease of management.

Starting from load absorption data collected during on-purpose intensive measure campaign conducted during an overall summer training period, considering the final overall request of energy given by the sum of ship services and of the propulsion equipped with a modern propulsion drive, and pursuing following criteria:

- minimization of annual operation rate of the installed prime movers;
- selection of prime movers of the same product series to simplify logistic support;
- reduction of the number of machines simultaneously at work and reduction of watching personnel;
- minimization of operation costs;
- operation of prime movers in an optimal power range in consideration of their operative profile;

it has been individuated as the best choice as regards to number and sizes of the DD.GG. (Diesel Generators), the following consistency:

- n. 2 DD.GG. with rated power in the range 1200 ÷ 1400 KW each; these groups will be indicated in the following as "Navigation DD.GG.", because of their use mainly in navigation trims. Their output voltage characteristics will depend on the selected choice for the plant architecture.
- n. 2 DD.GG. with rated power in the range of 600 ÷ 700 KW; these groups will be indicated in the following as "Harbour DD.GG." because of their main use in harbour trim. Voltage output characteristics are those of the switchboard to which these DD.GG. are connected, that means the so-called "harbour switchboard", and so 400 V–50 Hz three-phase insulated neutral.

All DD.GG. will have rated speed of 1500 rpm.

With the above indicated rating, it descends the following DD.GG. utilization in the various main operative conditions of ship operation:

Table 2. Generators at service.

Operative condition	Required power	Number of G1	Number of G2
Harbour	450	1	0
Sail	510	1	0
Navigation <70%	1440	1	1
Navigation >70%	1790	0	2
Maneuver	2190	0	2

from which descend the following foreseen DD.GG. operative margin, reported in Table 3:

The first conceived electric plant architecture, showed in Figure 1, consisted in two main switchboards, one operated at 690V-50 Hz, called Navigation Switchboard, with the two Navigation DD.GG. connected, and the other at 400V-50 Hz, with the two Harbour DD.GG. connected, called Harbour Switchboard.

Table 3. Operative margins of DD.GG.

Operative condition	DD.GG. operative margin
Harbour	71%
Sail	81%
Navigation <70%	77%
Navigation >70%	72%
Maneuver	87%

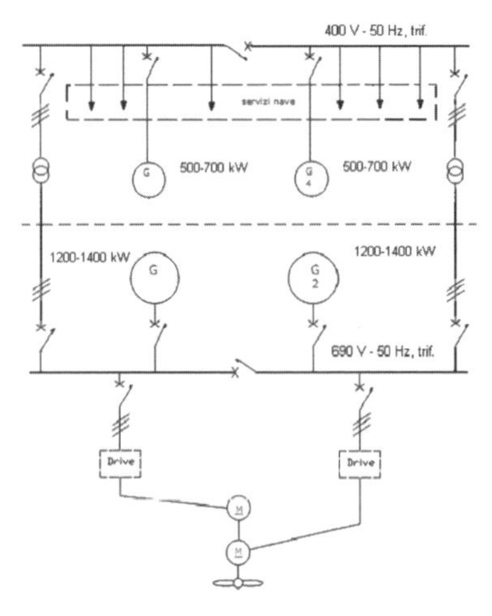

Figure 1. First IFEP architecture.

The two main switchboards are linked by two transformers, to be rated about 500–750 kVA.

Because of initial lack of accurate information about existing and future components and equipment characteristics, the architecture selected resulted as the best choice combining different operation requirements and existing plant constraints:

1. operation of DD.GG. in their optional load percentage even in case unavailability of one machine;
2. avoid the parallel operation of the two busbars connection transformers have not to be operated in parallel, from which descends the necessity to maintain always opened at least one tie-circuit breaker;
3. main switchboards circuit breakers ratings are to be chosen according to uninterrupted absorbed currents, and on the maximum short circuit currents that can occur when at the same time are operated in parallel or two Navigation DD.GG. or 1 Navigation D.G. and two Harbour DD.GG.. With such circuit breakers ratings, it will be verified the compatibility of the energy associated to circuit breaker operation with the let-through energy of existing distribution plant cables, especially of those connecting the secondary switchboards, with the aim to avoid their substitution.
4. there must be the possibility to actuate degraded plant configurations, so that to guarantee service continuity, even in a partial way, in case of unexpected fault to a main plant component as a generator, a connection cable, or half a busbar. It must be then guaranteed, with respect of the first point, a share of power available either immediately or in a short time, to face the occurrence of an unexpected fault on one in service generator, so that to limit as much as possible the level of degradation. It must be preferred the plant configurations able to guarantee the continuity of service, even partial, of ship service and propulsion load, even in case of an unexpected fault on a main plant component as a generator, a switchboards connection line, or an half busbar. At this purpose it is preferable to operate the plant with the two main switchboards connected with respect to criteria n. 3.
5. with both transformers in service it will preferred such plant configurations characterised by a balanced load distribution on the two transformers, so that to uniform their service conditions and to avoid their oversizing. With both transformers in service, must be preferred the plant configurations allowing an as more balanced as possible sharing of powers transferred between the two main switchboards; such a criteria guarantees/assures the equal exploitation of the two machines and avoids an unnecessary oversizing of the two transformers.

During maneuver, and however during restricted waters navigation, it must be assumed a plant configuration allowing ship service continuity of supply in case of abrupt loss of an in service generator.

From the above indicated operation requirements have descended the following preliminary design criteria:

- it must be possible parallel operation and active and reactive load sharing between Navigation and Harbour DD.GG., even if operated at different voltage levels;
- it must be possible the flow of electrical energy in both directions from one main switchboard to the other;
- it must be provided a functionality of PLS Power Limiting System able to actuate an immediate limitation of the propulsion power in case of unexpected D.G. out of service so that to avoid opening of other generators circuit breakers for overload protection intervention.

Installation works will have to be executed without modifying the attitude and the stability conditions of the ship. For what concerns installation of the equipments and their location, it should be taken into consideration the following aspects:

- the possibility to execute easily all maintenance works to equipment subject to intervention, and then accessibility for maintenance purpose, either of new plants or of existing ones.
- the structural impact, in the sense that new basements should not compromise ship structure.

It will be shipyard care the preliminary verification about present ventilation-exhaust system capability in relation to the new thermal requirement of future equipment to be installed. Whenever present ventilation/exhaust system should result insufficient for this purpose, the installation company will have to provide the ventilation system upgrade to satisfy the maximum design temperature increase of 15°C.

For what concerns the electrical primary distribution and generation system power quality, the first requirement was individuated as a mix of Rina and Stanag 1008 Ed. 9 one, attaining to this last relatively to frequency transient recovery time of 2 s instead of the 5 s of RINa. The THDv factor was initially fixed at 8%, as a compromise between the consideration about present electrical appliances harmonic presumable tolerances, and the opportunity to avoid a design of the converter too accurate and complicated in order to contain its harmonic emission contribution.

Even if the ship will not be susceptible to be classified according to a Class Register, e.g. RINa or another standard, new plants will have to comply with in force generic and product IEC, CENELEC and CEI standards.

3 STATE OF THE ART TECHNOLOGIES FOR AN OLD STYLE SHIP

3.1 *Electric primary distribution system*

The definitive network configuration had descended from the results of first electromechanical simulations, showing the possibility to avoid two different voltage rating main switchboards, to be connected necessarily by transformers. In fact such technical studies have showed the assurance of compatibility of shortcircuit currents level with the existing secondary switchboard shortcircuit current withstand capability, by selecting a minimum value of generators subtransient reactance of 16%.

Electrical energy supplied to ship electrical appliances will be of three phase insulated neutral 400 V 50 Hz and 220 V-50 Hz typologies. Electrical distribution will consist of:

- one 400 V - 50 Hz, 3600 A, 50 kA-1 s shortcircuit rated withstand current, switchboard, divided in two sections by one tie-breaker;
- energy cables for the connection to the switchboard of the generators, the propulsion transformers and the electrical users.

Figure 2 below shows the final network configuration:

in which the same DD.GG. defined in the preliminary design stages are connected to the same switchboard, divided in two half busbars, each connecting one Navigation D.G. and one Harbour D.G.

DD.GG. exact ratings are:

- Navigation DD.GG.: 1500 kVA;
- Harbour DD.GG.: 775 kVA.

Depending on the operative condition, the following DD.GG.: combinations can be actuated:

- n. 2 × 1500 kVA DD.GG., busbars connected;
- n. 2 × 775 kVA DD.GG., busbars connected;
- n. 1 × 1500 kVA Navigation D.G. and n. × 2 775 kVA DD.GG.;
- n. 2 × 1500 kVA and n. 2 × 775 kVA DD.GG., separated busbars (transitional phase).

The main switchboard will be provided with an electric interlock preventing parallel operation of all DD.GG. at the same time or of n. 2 Navigation DD.GG. and n. 1 Harbour D.G. except that in transient conditions for a maximum duration of 30 s.

The preliminary network simulations studies highlighted the opportunity to define the quality of power supply in compliance with RINa requirement.

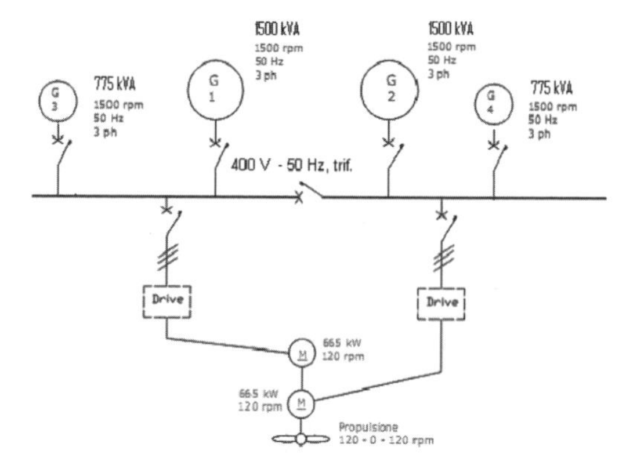

Figure 2. Second IFEP architecture.

The main switchboard will be connected to the ship automation system, and in particular to the SACIEM (Propulsion and Electric Plant Control Automation System) by means of a serial interface in preference of the Ethernet type, Modbus or OPC, or secondarily of RS432 or RSRS485 Modbus standards.

The switchboard will be realized in an individual frame to be installed in place of present 380 V - 50 Hz main switchboard, and it is of interest to note that in the same space and volume it will be possible to install the more powerful switchboard managing and distributing the overall power requested by either propulsion system or ship services.

Further technical elements that will be provided by the switchboard supplier are:

- Shortcircuit analysis;
- Harmonics analysis;
- Dynamic electromechanical system simulations (frequency and voltage transients following abrupt loss of a generator, abrupt load disconnection;
- Signal and energy cables characteristics;
- Selectivity analysis and protection setting tables.

3.2 *The DD.GG.*

DD.GG. will be dimensioned so that to sustain motors' heavy starting and operating conditions changes, included those of the propulsion motor, without determining over-current protection intervention or DD.GG. circuit breakers opening, and without generating network critical conditions with respect to network stability and quality of power supply.

Every alternator will be endowed with an off-line insulation control system able to detect insulation condition when the machine is not in operation.

Every D.G. will be equipped with a local control panel, named DGLCP, and in case, with a further panel exclusively dedicated to the alternator and interfaced with the DGLCP, for the collection and management of signal related to:

- machine control, protection and monitoring;
- interlocks and automatic functions necessary to a D.G. safe and reliable control;
- alternators insulation control system.

DGLCPs will be interfaced with SACIEM, that will be so able to control the DD.GG.

The remote control will require the provision of the following typology of interface to the automation plant:

- Modbus via RS 485 protocol;
- TCP/IP via Ethernet/IP protocol.

Every D.G. will have to be provided with a dedicated software for the automatic evaluation of motor performances.

In particular, the system will correlate alternator active power with the fuel flow rate, and will verify the congruency with the acquired operation parameters with respect to the expected ones, keeping into account environmental conditions.

DD.GG. electrical protection must be set so that to consider alternators operation parameters correspondent to an effective machine abnormal condition, without provoking unexpected interventions, with particular concern to heavy operating conditions such those occurring during maneuvers, during which it can happen that an over-current detection doesn't correspond to an overtemperature condition, that is the physical parameter indicative of the effective thermal state of the machine.

3.3 *The propulsion system*

The propulsion system will assure with continuity torque to the unique propulsion shaft in the speed range 30 ÷ 170 rpm: up to 130 rpm rated speed, operation will be of constant torque type, while from 130 to 170 rpm regulation will be of constant power type. The propulsion system will consist of the following equipment:

- N. 1 electric propulsion motor (EPM), asynchronous type, composed by two active modules within the same casing frame (n. 2 statoric packs and n. 2 independent rotoric cages) such to realize two internal submotors rated 750 kW each at 130 rpm rated speed;
- N. 2 × 24-pulses drives, with VSI (Voltage Source Inverter) converters, PWM modulation and DFE (Direct Front End) network side interface, each with its supply and separation double secondary-windings transformer, interposed between the main distribution and the static converters;
- N. 2 braking resistors;
- Connection cables, complete with support and protection cable trays, between all subsystems above enlisted.

Electric Propulsion System will be monitored and controlled by the same SACIEM preposed to the control of the electric plant. The number and the typology of alarms and signals managed will be in conformity to RINa rules.

EPM must be designed so that to sustain without damage an abrupt shortcircuit event at its input terminals, and the maximum possible overvoltage condition indicated by the manufacturer during speed variations.

The constructive shape of the statoric frame has to be realized so that to be installed by means of self-flushing supports to the existing basement, and so that to have the rotoric centre aligned with the propulsion shaft. The coupling flange will be installed in the same position of the existing electric propulsion motor flange. EPM cooling system will be air forced closed circuit liquid cooled double tube heat exchanger, IC 86 W type.

For what concerns propulsion system drives, each converter will be equipped with:

- fresh water sea water cooled, double tube, heat-exchanger, provided with fresh water electropump, integrated within the converter switchboard;
- anti-vibration elastic dampers;
- power electronics modules, for an easy substitution, during motor operation, in case of damage, in a time less than 2 h;
- power suppliers, PLCs, measurements transformer, and other electric and electronic auxiliary equipments, all provided in a modular construction, so that to assure their easy substitution in a maximum time of 2 h;
- emergency push-buttons, local/remote selectors;
- anti-condensation heaters;
- EPM and converters electrical protections;
- earth fault propulsion system monitoring;
- propulsion power dynamic limitation system PLS (Power Limiting System);
- local control panel to be installed on the front, for the interface with SACIEM, by means of a OPC, or alternatively Modbus, communication protocol;
- UPS to assure continuity of power supply to auxiliary electrical equipments inside the converter board.

The characteristics, defined by now, of the two propulsion transformers are the following:

- Typology: dry-type, cast resin;
- Primary rated voltage: 380 V ± 2,5%;
- Environmental class: E2 class;
- Fire resistance class: F1 class;
- Electrostatic shield.
- Natural air cooled.

The following main characteristics:

- rated power;
- secondary rated voltage;

will be defined in coordination with the EPM design so that to optimize the propulsion system efficiency in the entire speed range. The propulsion system efficiencies requirements are so fixed:

- 90% at rated power;
- 88% at 50% rated power;
- 80% from 10% to 80% of rated power.

The electric braking of the ship will be actuated by means of braking resistors, that have been subject to a preliminary dimensioning, with the assumption to allow the complete stop of the ship in a time not higher than 400 s starting from an initial speed of 11 kn. Such hypothesis led to a dimensioning of 120 kW, and considering this value have been then evaluated all the characteristics of different stop action sequences.

3.4 *The propulsion and electric plant automation system*

The Platform Automation system will be completely renewed in accordance with the consolidated Italian Navy automation systems standard dedicated to second line combatant ships, derived by standards adopted for most recent naval units such as aircraft carrier "Cavour", with some degree of reduction in hardware and software redundancy requirements.

It will be constituted by two independent subsystems:

- SACIEM, described in the previous paragraphs;
- SACSEN, dedicated to hull services management;

sharing the same hardware equipment.

It will be based on a Microsoft platform operative system, in the most updated version, with distributed architecture, composed by intelligent distributed subsystem, and machinery dedicated devices connected to a redundant fiber optic bus.

The architecture is structured in the following levels:

- at the third level (higher level), will be foreseen the operator stations, and so all the hardware and the software, necessary to the coordinated automatic control of all the platform subsystems;
- at the second level, will be foreseen the LOPs (Local Operating Panels), from which it will be possible to control the plants from a subsystem level, either SACIEM or SACSEN, even in case of loss of the fiber optic bus;
- at the lower level (second level), will be foreseen the remote intelligent terminal boxes, for the interfacing of:
  - field sensors and actuators;
  - IFE (Intelligent Front End), equipping specific equipment such for example the circuit breakers of the main distribution switchboard;
  - UAL (Local Automation Unit), for automatic functions implementation at system level or for a limited subsystem area.

The IFE represent the intelligence for local machinery and systems control. Differently from the UAL, accepting a general type configuration, the IFE implements specialistic functions associated to the particular equipment under control. The IFE will be supplied as associated to individual equipments, while UAL will be included in the automation plant supply.

It will then be realized a single fiber optic ring loop, connecting automation units and operator interfaces, assuring the required information transfer speed, and an high security level in the transmission of data between all the units connected to the network. Network LAN will be composed by active switches linked through a backbone with capacity not less than 100 Mbps.

UAL will be realized same typology modular microchips units of the same typology. Such UAL will be able to be programmed either through a suitable local interface (laptop connected through a serial port), or from the MFC (Multi Function Consolles) located in COP (Platform Operative Control Centre) and properly predisposed.

UAL will have the purpose to acquire, directly or by means lower level units, information related to the systems or services monitored, and to implement their direct control by sending the correct commands to the actuators and to the IFE.

In principle it will not be foreseen operator interfaces at a UAL level with the exception of the main distribution switchboards, for which, the operator interfaces associated to the UAL will constitute the IFE of the switchboard.

To assure such a characteristic and, at same time, to respect the principle of distributed intelligence, the equipment databases or the local area services managed by a single UAL, must be resident within it, and available for the control equipment placed at an higher level.

4 FURTHER IMPROVEMENTS FOR AN INNOVATIVE CONCEPT

The choice of equipment technical characteristics will be supported by a more detailed system engineering activity including at present:

- integration and definition of alternator and functional parameters of DG:
- the supplier will provide the electrical characteristics of the alternator to be used in cooperation with the EPM supplier;
  - definition of functional architecture of speed and voltage regulators on DG, in order to be able to synchronize the main switchboard with the shore power supply;
  - definition of voltage/frequency control loop for DG control in accordance with power quality standards defined;
  - definition of interfaces between SACIEM, the DGLCP and the main switchboard controller (protection, alarms, monitoring and control);
  - definition of protections.
- definition of alternator and propulsion transformer characteristics in order to allow the connection of one transformer with one DG running (inrush current);
- short circuit evaluation, protection plan, including selectivity studies and protection settings in all possible operating modes;
- definition of the equipment necessary to synchro nize the main switchboard with the shore panel;

It is intention of Italian Navy General Staff Electrical Plant Office to valorize the use of electromechanical simulations, as a means to obtain a complete analysis of systems and components interaction and to improve plant operation, taking also into account the return of experience deriving from cruise ships IFEP plants, and considering the importance for technical staff to have in advance sufficient knowledge of system behaviour for an innovative plant configuration.

So it has been deemed necessary to carry out further simulations as final verification and adjustment to validate the network behaviour in different operative conditions, so to define the possible conduction procedures without impair system operation, safety and continuity of supply, also in view of the definition of proper training programmes, informing on permitted plant configurations and their management.

With this purpose have been individuated some additional simulation activities defined by initial system configuration, system or operation modification, electrical parameters to analyze, purpose of the study.

The simulations have been divided into two groups:

- phenomena originated from events related to normal operation of the system:
  - maneuvers not originated by undesired circuit breakers intervention or by human errors, causing the passage between two normal plant operation states or from a normal state to a pre-alert or pre-emergency state,
  - power supply request changes due to high power demanding users connections and disconnections. The related transients imply both aspects of static and dynamic stability, and are considered to verify power supply quality contractual requirements and so to design network control systems and regulations, and to evaluate the opportunity to adopt new technologies or additional control actions so to obtain a better system response.
- phenomena originated from events related to abnormal system operation:
  - loss of critical system components, such as generators, main power stations connection and supplies lines, HV/LV transformers, propulsion transformers, converters, motors.
  - high, abrupt and unexpected load variation.

caused by faults in electrical components or in their connections, and/or undesired intervention of protection devices. The related transients concern dynamic stability aspects and are characterized by electrical parameters going beyond contractual requirements ranges for normal network operation.

They are studied with the aim:

- to evaluate ability of control/protection/regulation system to confine the anomaly or the damage on the network and so:
  - to prevent voltage and frequency from reaching intolerable variations such to infer irreparable damages to electrical equipments;
  - to conduct the system in another normal state, even different from the initial one, or in a pre-alert state.

  In this case simulations help to define the characteristics and the performances of such control/protection/regulation systems.
- foresee partial or total blackout potential risks, and then provide to the relative prevention actions, also through implementation of automatic actions as automatic non vital loads disconnection.

For the first group have been individuated the following situations:

- transitions between two normal operative conditions, addition/elimination of one DG in electric and non electric (sail) propulsion. Evaluation of Multiple Input Multiple Output (MIMO) system stability.
- normal variations of ship service and propulsion load. One of the aim of this study is to define EPM acceleration and deceleration slopes.
- crash stop, that will allow to define strategies for backwards energy dissipation
- evaluation of network harmonic content with respect to limits prescribed by RINa rules.

In the second group are included the following:

- loss of one DG caused by undesired machine circuit breaker opening, or by electric system anomaly such as overload, minimum voltage, ...), or by fault occurrence. This studies can help in define blackout prevention strategies such as Propulsion Power Limiting Systems and automatic load disconnection strategies.
- total or partial electric propulsion loss, caused by whatever anomaly or fault in one component of electric propulsion drive. Transition to half-motor operation.
- electric equipment fault. HV busbar fault, distribution transformer fault, LV busbar fault, ecc.
- transition from a configuration with connected HVMSB to another one with separated busbars and viceversa. Evaluation of opportunity to adopt a secondary voltage regulation for each HVMSB.
- Electromechanical simulations allow moreover the study of network reliability comprehensive either of adequacy aspect, and so static stability, and of security aspects, that refer to dynamic stability, to consider for attribution of operative states in normal and degraded conditions of the system.

5 CONCLUSIONS

In this paper have been synthetized the aims, the targets and the consistency of next platform system revamping, that will allow Nave Vespucci to perform another operative life. Following such deep technical intervention, Nave Vespucci will have a further element of distinction in the Italian Fleet, that is the peculiarity to combine at the same time tradition and innovation, being either the oldest ship still in service, or the ship to be endowed with the most innovative IFEP plant, of the overall Italian Navy fleet.

REFERENCES

Enrico Gurioli, 2006. *Vespucci La Nave più Bella del Mondo*. De Agostini.

NATO Nato Standardization Agency (NSA), 2004. *Stanag 1008 Characteristics of Shipboard Low Voltage Electrical Power Systems in Warship of the Nato Navies Ed. 9.*

RINA, 2001. *Rules for the Classification of Ships* Ed 2011.

Stato Maggiore della Marina—SPMM 9° Ufficio, 2010. *Nave Amerigo Vespucci. Progetto Funzionale e di Principio per l'Ammodernamento dell'Impianto Elettrico di Generazione, Distribuzione e Propulsione.*

6 *Safety & operation*

6.1 *Safety*

Sustainable Maritime Transportation and Exploitation of Sea Resources – Rizzuto & Guedes Soares (eds)
© 2012 Taylor & Francis Group, London, ISBN 978-0-415-62081-9

The use of influence diagrams in risk-based inspection

J.R. San Cristóbal
Nautical Sciences Department, University of Cantabria, Spain

M.V. Biezma
Material Sciences Department, University of Cantabria, Spain

ABSTRACT: Inspection and Maintenance planning based on risk minimizes the probability of a system failure and its consequences and it helps management in making correct decisions concerning investment in maintenance and related fields. An interesting tool to model decision problems and implement Bayesian decision theory for risk analysis are Influence Diagrams. This paper shows how the use of Influence Diagrams in Risk Based Inspection (RBI) increases the knowledge we have on equipment condition and help us gain confidence in the planning of future inspections or repair decisions.

1 INTRODUCTION

The essential goal of inspection is to prevent incidents that impair the Safety and Reliability of operating facilities. The inspection process provides a means of gaining confidence in the service reliability of the component being inspected. When an inspection reveals an excessive deterioration, actions are initiated, such as the repair or replacement of the affected component or a change to the operating conditions. In a facility with substantial production revenues, the cost of downtime can be significant. An effective inspection program, centred on knowing when, where and how to inspect, enables the operator to not only control the integrity of the assets, but to control it with a focus on the economic value, while maintaining an acceptable service performance.

In 1990s, Risk-Based Inspection and Maintenance methodology started to emerge and gain popularity beyond 2000. Until 2000, Maintenance and Safety were treated as separate and independent activities (Raouf 2004). An integrated approach incorporating Maintenance and Safety is the appropriate mean of optimizing plant capacity, as both activities are not mutually exclusive functions. Inspection and Maintenance planning based on risk minimizes the probability of a system failure and its consequences and it helps management in making correct decisions concerning investment in Maintenance and related fields (Arunraj & Maiti 2007).

Most operators have reached their current practice by an evolutionary process based upon experience, regulatory and classification society compliance. In the marine and offshore industry, e.g., the inspection frequencies for pressure equipment have traditionally been driven by prescriptive industry practices, usually at time-based or calendar-based intervals (API 2002). This inspection practice, founded mainly on general industry experience for each type of component, has thus far provided an adequate level of reliability. However, such a practice does not explicitly consider the likelihood of failure of a component under its operation and loading conditions, nor the consequences of a failure. Current inspection practices make it difficult to recognize if the same or improved service reliability can be achieved by varying inspection methods, locations or frequencies. Also, current practices do no easily identify if an inspection activity is excessive and provides no measure of increased assurance for the integrity of the component.

Very few operators have developed their existing programs on the basis of a systematic process that seeks to achieve a balance between risk and the level of inspection effort. Certain sectors of industry have recognized that significant benefits may be gained from more informed inspection methods and have begun evolving into inspection program philosophies that combines factors such as satisfaction operating experience, low deterioration rates, minimal consequences of failure and condition-based inspection interval setting. RBI focuses on the optimization programs for pressure retaining equipment and structures. As a risk-based approach, RBI provides an excellent means to evaluate the consequences and likelihoods of component failure from specific degradation mechanisms and develop inspection approaches that will effectively reduce the associated risk of failure.

RBI provides a methodology for determining the optimum combination of inspection methods and frequencies.

In this paper we show how Influence Diagrams can be a useful tool in RBI. With this aim we consider the case of a pipeline suspected of possible corrosion. We must to determine whether to repair or do not repair the pipeline under three different scenarios and also whether to take any information gathering regarding corrosion activity. The paper is organized as follows. In the next section we introduce some aspects of Decision Theory and Influence Diagrams. Next, we show through an application how Influence Diagrams can be used in RBI. Finally, a concluded section with the main findings of the paper is presented.

2 DECISION MAKING PROCESS AND INFLUENCE DIAGRAMS

Very often, at the time the decision has to be made, it is impossible to know with certainty which of the possible strategies will turn out to be the best choice. This typically happens when the degree of success of any particular strategy will depend on external factors which can not be perfectly predicted. In other words, decisions must be made in an environment in which there is uncertainty about the future behaviour of those factors that will determine the consequences following from the possible course of actions. Although scientists and engineers may be willing to make predictions about unknown situations, there is a need to assess the level of uncertainty (Suslick & Schiozer 2004). As the level of information increases, the uncertainties are mitigated. So it is necessary to define the value of information associated with important decisions such as deferring an inspection or maintenance action. Information has value in a decision problem if it results in a change in some actions to be taken by a decision maker. The information is seldom perfectly really and generally it does not eliminate uncertainty, so the value of information depends on both the amount of uncertainty (on the prior knowledge available) and payoffs involved. The value of information can be determined and compared to its actual cost and the natural path to evaluate the incorporation of this new data is by Bayesian analysis. The process of actualisation of uncertain knowledge based on tests (inspections) having themselves a limited reliability is usually addressed by the so-called "Bayesian techniques". Within this approach, it is expected that the performance of inspection, provided effective techniques are used, increases the knowledge we have on equipment condition and help us gain confidence in the planning of future inspections (Giribone & Vallete 2004).

An interesting tool to those who wish to model decision problems and implement Bayesian decision theory for risk analysis are the Influence Diagrams (Singpurwalla 2006). Influence Diagrams have been developed as an alternative to Decision Trees for representing and evaluating decision problems (Howard & Matheson 1981, Shachter 1986). Decision Trees have several drawbacks, such as extremely large tree sizes for complex problems, not allowing explicit representation of independent relations among the variables, and leading the participants to think only in a forward direction when, in fact, decision making often requires thinking in both forward and backward directions (Shachter & Heckerman 1987). The advantages of Influence Diagrams include smaller diagram sizes as compared to Decision Tress, allowing different levels of conceptual modelling (i.e., conceptual relationships and numerical probability distributions), and suitability for use in building automated decision tools.

An Influence Diagram is a network consisting of a directed graph having no loops (a directed graph consists of a finite set of nodes and directed arcs, and a network is a graph in which additional data are stored in the nodes or arcs) with no directed cycles and detailed data stored within the nodes of the graph. Each node in the graph represents a variable in the model. This variable can be a constant or an uncertain quantity, a decision to be made, or an objective. The elements of a decision problem (namely the decision to be made), the uncertain outcomes and the consequences of a decision-outcome combination show up in the Influence Diagram as different shapes. A decision node is indicated by a rectangle and a random node by a circle; the consequence node, also called the value or payoff, is denoted by a triangle. Nodes are then linked up by arrows called arcs. Arcs between nodes pairs indicate influence of two types: i) Informational influences, represented by arrows leading into a decision node. These show exactly which variables will be known by the decision maker at the time that the decision is made representing a basic cause/effect ordering; and ii) Conditional influences represented by arrows leading into a chance node. These show the variables on which the probability assignment to the chance node variable will be conditioned. Represent a somewhat arbitrary order of conditioning that may not correspond to any cause/effect notion and that may be changed by application of the laws of probability (e.g., Baye's Rule).

Influence Diagrams describe the dependencies among aleatory variables and decisions, visualizing the probabilistic dependencies in a decision

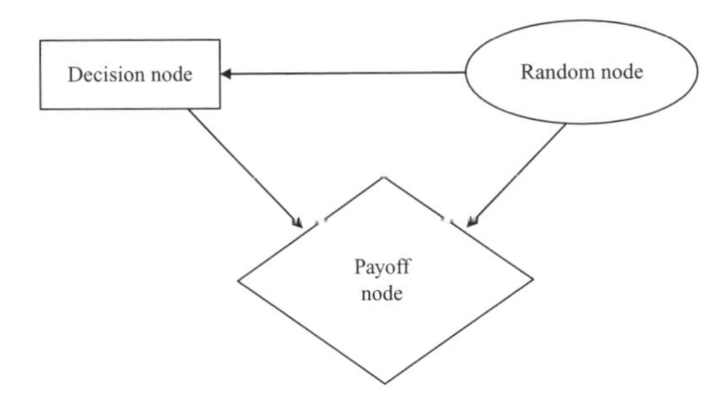

Figure 1. Influence diagram network.

analysis and specifying the states of information for which dependencies can be assumed to exist (Howard & Matheson 1981). Influence Diagrams are considered a generalization of the fault tree technique (Singpurwalla 2006). They despict the decision problem as a compact graph, whose size, as measured by the number of its nodes, grows linearly with the number of variables. In addition, Decision Trees and Influence Diagrams are isomorphic. That is, any properly constructed Influence Diagram can be converted to a Decision Tree and vice versa.

3 APPLICATION

To assess and quantify the threat of corrosion is a prior and essential step to implement an inspection and maintenance plan. To calculate the failure probability of each component or structure, i.e., pipelines, tanks, etc., and, therefore, the integrity of these components, it is necessary to know the probability of detection, the probability of false indication of each survey technique, the time of initiation of corrosion defects and the defect density, the corrosion rate and the defect depth are fundamental. Several methods are available to assess the threat of corrosion. With the use of External Corrosion Direct Assessments (ECDA), proposed by the NACE (NACE 2002), it is possible to adjust the number of coating defects and the number of corrosion defects. Other techniques as Direct Current Voltage Gradient (DCVG) and Close Interval Potential Survey (CIPS) (van O & van Mastright 2006, Webb et al., 2006) allow for an accurate assessment of external coating conditions and Cathodic Protection performance. These techniques, used separately or indeed jointly, may provide sufficient data to assess the threat of corrosion and subsequently to quantify the risk of corrosion. In other cases, such as the large and complex variations that affect chemical and physics variables outside pipeline corrosion this method provides a very useful tool (Hamano et al., 2006).

With the aim to show how Influence Diagrams can be a useful tool in RBI let us consider the case of a pipeline suspected of possible corrosion. Depending on the corrosion activity (Low, Moderate and High) the pipeline may be under three possible damage states. If the deterioration is low the damage state will be A, while if the deterioration is high the corresponding damage state will be B. Finally, if the pipeline is highly deteriorated the damage state will be C. The probabilities associated with the corrosion activity and damage states are shown in Table 1 and the costs associated with the decision whether repair or do not repair depending on the corrosion activity and the damage state of the pipeline are shown in Table 2.

We must to determine whether to repair or do not repair the pipeline and also whether to take any information gathering regarding corrosion activity. Let us consider the three following scenarios: i) to decide whether to repair or do not repair without undertaking any test which permits the decision maker to know the corrosion activity; ii) to decide repair or do not repair knowing the degree corrosion activity before understanding any test; and iii) before to decide repair or do not repair the decision maker performs an inspection test to know the degree of corrosion activity.

The primary decision problem can be formulated by drawing the Influence Diagram shown in Figure 2. In this Influence Diagram, Damage State is dependent upon Corrosion activity but Repair decision in this primary scenario is independent of the Damage state and Corrosion activity

Table 1. Probabilities of corrosion activity and damage state.

	Damage state		
Corrosion activity	A	B	C
Low	0.80	0.15	0.05
Moderate	0.10	0.80	0.10
High	0.05	0.15	0.80

Table 2. Cost ($\$10^6$) associated to the decision whether Repair or Do not repair.

	Repair			Do not repair		
	Damage state			Damage state		
Corrosion activity	A	B	C	A	B	C
Low	1	2	5	0	1	3
Moderate	1.5	3	7	2	5	10
High	2	5	10	5	10	25

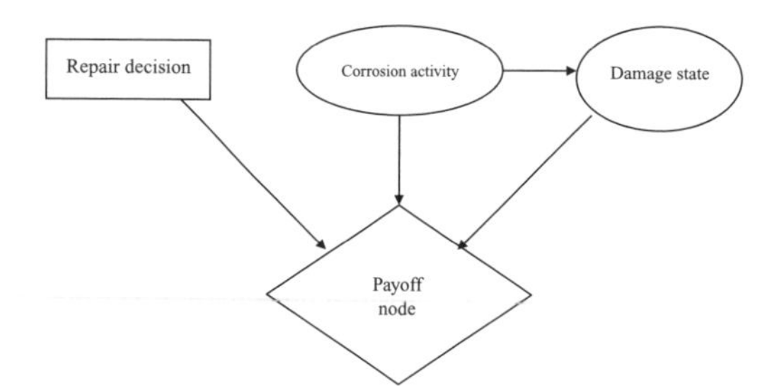

Figure 2. Influence diagram without knowledge on corrosion activity.

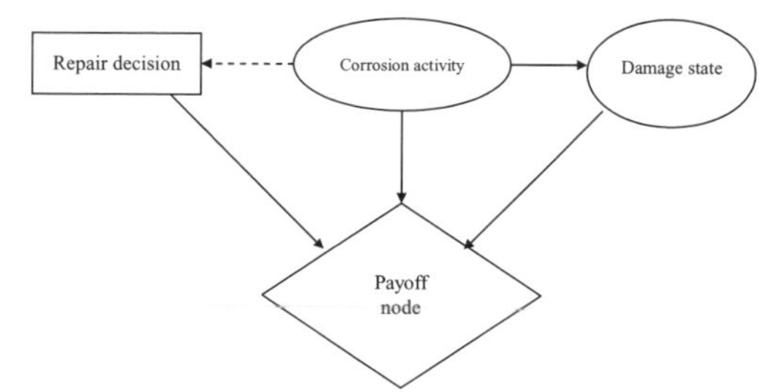

Figure 3. Influence diagram with knowledge on corrosion activity.

because, at the time the decision is made, the Corrosion activity is not known. The Costs (Payoff) are dependent upon the Repair decision as well as on both the Corrosion activity and Damage state. Once all of the information has been captured, the analysis can proceed. Solving this Influence Diagram with the program Precision Tree from Palisade (Precision Tree 2009), the best decision is to repair and the expected value given this decision is a cost of $1.76 millions.

In the second scenario, before understanding any inspection test, the decision maker knows the degree of Corrosion activity when he makes the repair or do not repair decision. This next scenario permits us to calculate the value of the problem with clairvoyance (perfect information) on Corrosion activity. In this situation the usual decision analysis practice is to determine the value of this perfect information on the uncertain variables, in this case the Corrosion activity. To calculate this value we need only add in the Influence Diagram an influence arrow indicated by a dotted line from the Corrosion activity node to the Repair decision node as shown in Figure 3. Solving this Influence Diagram, the expected value of the Costs is $1,235 millions. This means that the expected value of perfect information is the original $1.760 millions minus $1.235 which is $0.525 millions.

To place a value on imperfect information we must model the information source. Let us consider the third scenario in which it might be possible to carry out an inspection test of the Corrosion activity. In this case we begin by adding a chance node to represent the report from the Corrosion activity text. In the Influence Diagram network shown in Figure 4 we have added an activity Test node. We have drawn an arrow to it from the Corrosion activity node showing that the Test results depend on the actual Corrosion activity, and we have also drawn an arrow from the Test results to the Repair decision showing that the decision-maker will know the Test results when he makes the Repair decision. We consider that there are three Test results called "High", "Moderate", and "Low" corresponding

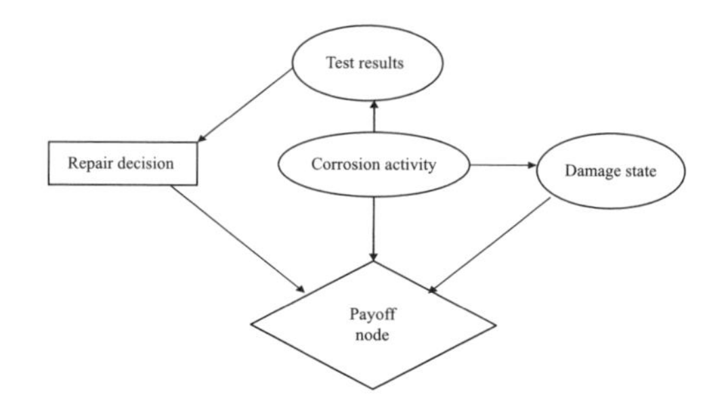

Figure 4. Influence diagram carrying out an inspection test about corrosion activity.

Table 3. Probabilities of Test results depending on Corrosion activity.

	Test results		
Corrosion activity	A	B	C
Low	0.70	0.20	0.10
Moderate	0.20	0.70	0.20
High	0.10	0.10	0.70

to the possibilities for the actual Corrosion activity. Unlike the case of perfect information, these test indications may be misleading so we must to supply the probabilities of these test results for each state of Corrosion activity. Table 3 shows the probabilities that inspection results in High, Moderate, and Low, depending on the Corrosion activity.

To solve this Influence Diagram requires the application of Bayes' Rule to determine from the actual probability assignments new assignments conditional in the opposite order. Evaluation of this network yields an expected cost, given the Corrosion activity test option of $1.500 millions. Subtracting this cost from the original cost of $1.760 millions yields an expected value of $0.260 millions. This is the upper limit on the price the decision maker should pay for the actual test and it is the value of imperfect information for the decision maker.

4 CONCLUSIONS

During Inspection and Maintenance programs, decisions must be made in an environment in which there is uncertainty about the future behaviour of those factors that will determine the consequences following from the possible course of actions. Through the so-called "Bayesian techniques", this uncertain knowledge is actualized, which helps management in making correct decisions concerning inspection or repair. As a risk-based approach, RBI provides an excellent means to evaluate the consequences and likelihoods of component failure from specific degradation mechanisms. In this paper we show how Influence Diagrams are an interesting tool to those who wish to model decision problems and implement Bayesian decision theory for risk analysis. This paper shows how the use of Influence Diagrams in Risk Based Inspection increases the knowledge we have on equipment condition and help us gain confidence in the planning of future inspections or repair decisions.

REFERENCES

API. 2002. Risk-based Inspection. American Petroleum Institute Recommended practice. First Edition.

Arunraj, N.S. & Maiti, J. 2007. Risk-based maintenance-Techniques and applications. Journal of Hazard Materials 142:653–61.

Giribone, R. & Vallete, B. 2004. Principles of failure probability assessment. International Journal of Pressure Vessel and Piping 81: 797–806.

Hamano, K., Kamaga, A., Tateno, S. & Matsuyama, H. 2006. Risk based selection of inspection parts for surface corrosion of piping in chemical plants. SICE-ICASE International Joint Conference Korea.

Howard, R.A. & Matheson, J.E. 1981. Readings on principles and applications of Decision Analysis II. Strategic Decision Group, Menlo Park, CA, 719–762.

NACE, 2002. Pipeline External Corrosion Direct Assessment Methodology, RP 0502-2002, NACE International, Houston, TX.

Precision Tree. 2009. www.palisade.com

Raouf A. 2004. Productivity enhancement using safety and maintenance integration: an overview. International Journal of Systems Cybernetics 33(7):1116–26.

Shachter, R.D. 1986. Evaluating Influence Diagrams. Operations Research 34: 871–882.

Shachter, R.D. & Heckerman, D.E. 1987. Thinking backward for knowledge acquisition. A1 Magazine 8(3): 55–61.

Singpurwalla, N.D. 2006. Reliability and Risk, A Bayesian perspective. Wiley.

Suslick, S.B. & Schiozer, D.J. 2004. Risk analysis applied to petroleum exploration and production: An overview. Journal of Petroleum Science and Engineering 44: 1–9.

Van O, M.T. & van Mastright, P. 2006. A direct assessment module for pipeline integrity management at Gasuine. 23rd World Gas Conference. Amsterdam.

Webb, N., Wyatta, B., Roche, M. & Thirkettle, J. 2006. Combined external coating and cathodic protection assessment technique- A case study for a pipeline in West Africa. 8th International Corrosion Conference.

Sustainable Maritime Transportation and Exploitation of Sea Resources – Rizzuto & Guedes Soares (eds)
© 2012 Taylor & Francis Group, London, ISBN 978-0-415-62081-9

Sensitivity analysis of a fire model used in fire consequence calculations

A.M. Salem & H.W. Leheta
Alexandria University, Alexandria, Egypt

ABSTRACT: Consequence analysis is an important part of the risk assessment technique, where fire consequence models have to be used to predict the consequences of a given fire scenario inside a compartment. As fire models use several input parameters to construct the fire scenario of concern, sensitivity analysis is considered suitable for application to determine how the fire model output depends upon these input parameters. CFAST is one of the most common fire models. To investigate its sensitivity, a number of simulations were conducted considering a single room fire scenario representing a fire in a cabin aboard a Ro-Pax vessel. Four selected input parameters were varied about a base scenario. Both small and large variations were studied. The results indicate that fire growth parameter and peak of heat release rate HRR are the most important input parameters influencing most of CFAST outputs. Simple response-surface correlations were then used to study the sensitivity of CFAST outputs to only the HRR with two door statuses (open/closed). The results demonstrated that CFAST outputs are more sensitive to changes in HRR when the cabin door is closed, i.e., when the fire is ventilation controlled.

1 INTRODUCTION

Regulation II-2/17 of the SOLAS convention, 1974 as amended, allows ship designers to deviate from the current prescriptive rules, which govern the fire safety design of ships provided that risk assessment techniques be used to demonstrate that the level of safety of the alternative design and arrangement is equivalent to that of the design which follows the prescriptive rules. Consequence analysis is an important part of the risk assessment technique, where fire consequence models have to be used to predict the consequences of a given fire scenario inside a compartment [Salem, 2010].

Deterministic computer fire models have been developed and tested to be useful in predicting the consequences of ship's compartment fires. The most important consequences are those related to life safety. If a fire breaks out in any compartment, the environment inside such a compartment becomes fatal and all people inside this compartment and other adjacent compartments may be subjected to untenable conditions, which may lead to their deaths [SFPE, 2005].

CFAST [NIST, 2009] is the fire model that is commonly used in many practical applications within the community of fire protection engineering.

This is due to its ability to deal with multi-connected compartments, its availability to everyone, and its continuous update. It belongs to the family of zone models, where the environment inside each compartment is divided into two zones (layers), an upper hot layer and a lower cold layer. Figure 1 shows a schematic drawing of the two-zone approach.

Sensitivity analysis is a study of how changes in the inputs to a model influence the results of the model [Ferson and Tucker, 2006]. As fire models use several input parameters to construct the fire scenario of concern, sensitivity analysis is considered suitable for application in such a situation to determine how the fire model output depends upon these input parameters. This is an important measure for checking the quality and reliability of the model predictions. If a small change in any of the input parameters results in a large change in the output, the output is said to be sensitive to that input parameter. This means that this particular

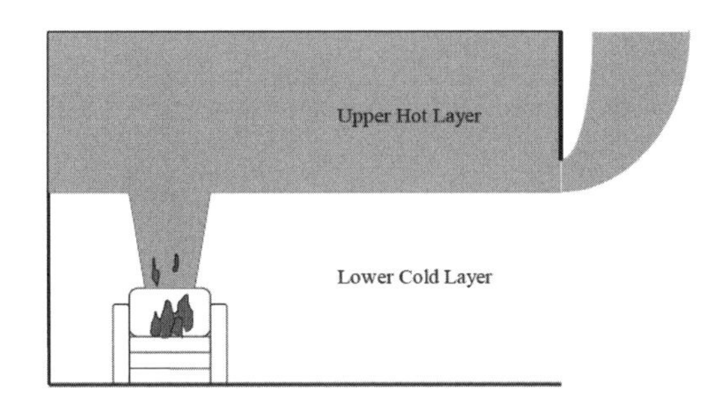

Figure 1. Schematic drawing of a two-zone fire model.

input parameter should be determined precisely in order to achieve a reasonable result.

To investigate the sensitivity of a complex fire model such as CFAST, one should be aware of the fact that obtaining an overall assessment of model sensitivity becomes very complex with numerous model inputs and outputs. Iman and Helton [Iman and Helton, 1988] noted down some of the properties of complex computer models that make the sensitivity analysis difficult to conduct, and the following are just a few of those which are relevant to the fire model CFAST:

- There are many input and output variables;
- it is difficult to reduce the model to a single system of equations;
- correlations may exist among the input variables and the associated marginal probability distributions are often non-normal;
- model predictions are nonlinear, multivariate, time-dependent functions of the input variables, and
- the relative importance of individual input variables is a function of time.

It should be noted here that for a given model input and a given model output, there may be regions of time where the model output is sensitive to the input and also regions where the model output is not sensitive to the same input [Peacock et al., 1998].

The life safety of the crew and passengers aboard ships is of paramount importance. Fire models such as CFAST can be used to predict the Available Safe Egress Time (ASET) for the people to evacuate safely from the ship under consideration. This time is considered in this study to be the minimum of three different times, namely, the time to reach untenable condition due to smoke obscuration (t_{smoke}), the time to reach untenable condition due to convective and radiative heats (t_{heat}), and the time to reach untenable condition due to inhalation of toxic gases ($t_{toxicity}$). Empirical equations for calculating these three times are selected from the literature [Salem, 2007] and used in this study. The equations call some of the outputs of CFAST, namely: upper and lower layer gas temperatures, layer interface height, ambient heat flux, upper optical densities, and upper and lower concentrations of CO, CO_2, as well as O_2.

2 AIM AND OBJECTIVES

Surveying the previous works, which have been carried out in this area showed that there is a lack of information regarding the combined sensitivity of both CFAST outputs and the outputs of the empirical formulas used to estimate the ASET timeline to variation in the input parameters. The aim of this study is to conduct a sensitivity analysis to investigate these combined effects.

To this end, a number of simulations were conducted considering a single room fire scenario representing a fire in a cabin aboard a Ro-Pax vessel. Four selected input parameters, namely, cabin floor area, cabin height, fire growth parameter and peak of heat release rate (HRR) were varied about a base scenario. Both small and large variations for the selected input parameters were studied. Simple response-surface methodology was then used to study the sensitivity of the outputs of concern to changes of the HRR. Finally, the sensitivity of CFAST on single values representing the time to reach untenable conditions inside the cabin has been evaluated.

3 SENSITIVITY TO SMALL CHANGES IN CFAST INPUTS

To investigate the sensitivity of CFAST outputs to small variations (±10% about a base scenario) in its inputs, a number of simulations were conducted. A single room fire scenario representing a fire in a cabin onboard a Ro-Pax ship is considered. Only the variables, which enter in the calculation of the ASET timeline, are studied. Table 1 shows a list of the selected input and output variables for the present analysis, while Figure 2 shows a snapshot of the cabin and the fire source. The fire source is assumed to be located in the middle of the cabin floor, and a t^2-fire growth type is chosen to simulate a piece of furniture fire. Figure 3 shows the different types of t^2-fires.

Table 2, shows the values of the input parameters used to construct the base scenario. Figures 4 to 7 show the results of a time dependent sensitivity analysis of CFAST outputs to a ±10% change in both floor area and cabin height of the cabin for two different door statuses, namely, closed and opened. Considering the effect of changing the floor area while the door is closed, Figure 4 shows that all CFAST outputs have an average relative difference less than 0.1. The upper layer optical

Table 1. List of selected input and output variables.

Input variables	Output variables
Cabin area	Upper/Lower gas temperatures
Cabin height	Smoke layer height
Fire growth parameter	Ambient target heat flux
Peak of HRR	Upper optical density
Door status (opened/closed)	Upper CO, CO_2, and O_2 concentrations

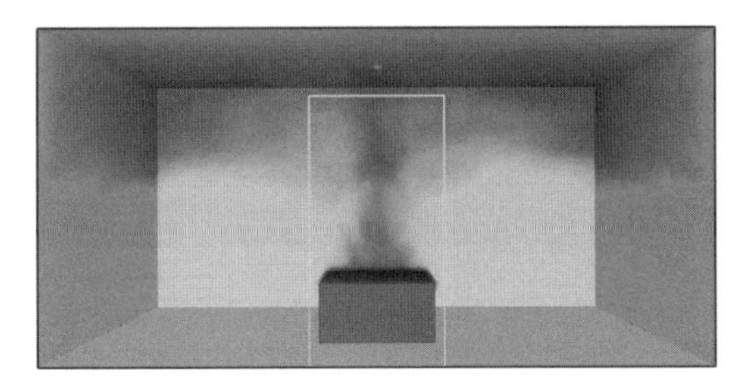

Figure 2. Snapshot of the cabin showing the fire source in red.

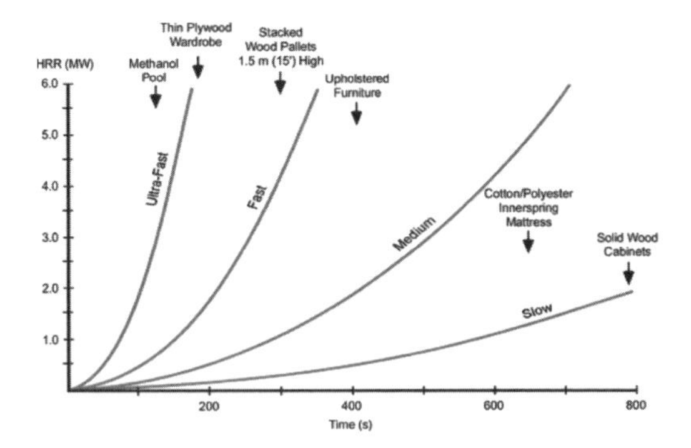

Figure 3. Different types of t^2-fires [Karlsson & Quintiere, 2000].

Table 2. Values of the input parameters used in the base scenario.

Input parameter	Base scenario
Cabin floor area	4.0 m^2
Cabin height	2.2 m
Fire growth parameter "α"	0.004 kW/s^2
Peak of HRR "Q_{max}"	500.0 kW

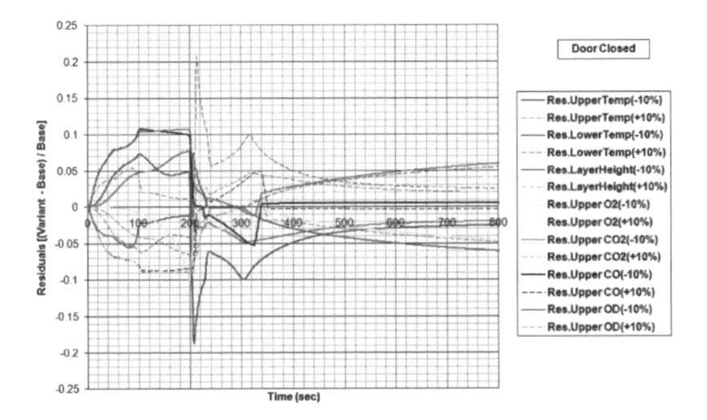

Figure 4. Time dependent sensitivity of some of CFAST outputs to a ±10% change in cabin floor area with door being closed.

density, smoke layer interface height, ambient target heat flux, and upper layer CO concentration are respectively more sensitive to changes in floor area than other outputs. When the fire became steady, i.e., at the end of the fire growth period, the figure showed a fairly constant relative difference for the changes as a function with time.

Considering the effect of changing the floor area while the door is opened, Figure 5 shows that all the outputs of CFAST have an average relative difference less than those that occurred when the door is closed. In this case, the upper layer optical density, the upper layer CO concentration, and the ambient target heat flux respectively have more sensitivity than other outputs.

Figure 6 shows the effect of changing the cabin height while keeping the door closed on CFAST outputs. The figure shows that the relative differences of the model outputs have almost average values less than or equal to 0.1, and that the upper O_2 concentration, the upper optical density, and the ambient target heat flux have respectively more sensitivivity than other outputs.

Figure 7 shows the effect of changing the cabin height while keeping the door opened on CFAST

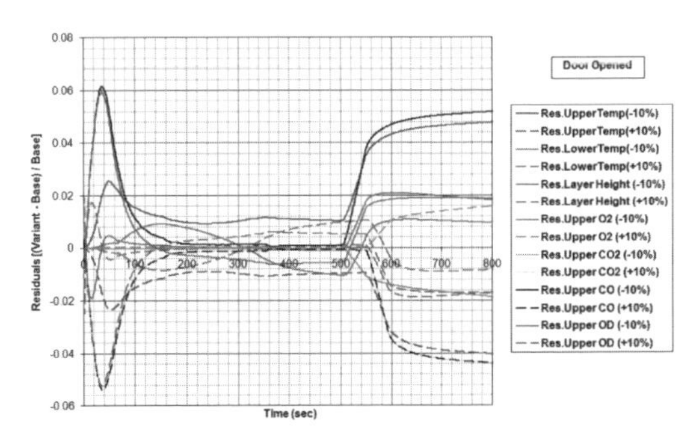

Figure 5. Time dependent sensitivity of some of CFAST outputs to a ±10% change in cabin floor area with door being opened.

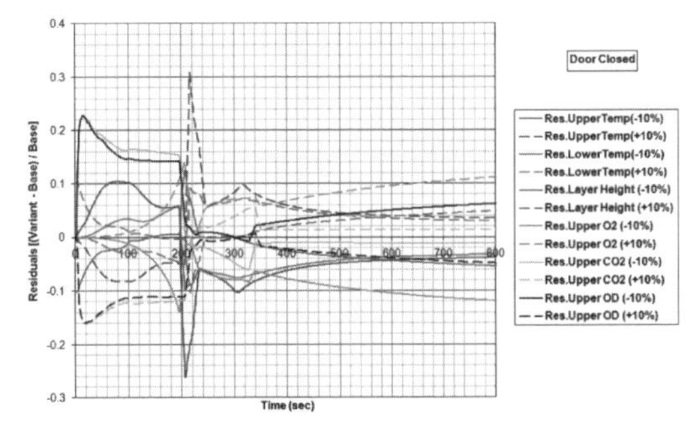

Figure 6. Time dependent sensitivity of some of CFAST outputs to a ±10% change in cabin height with door being closed.

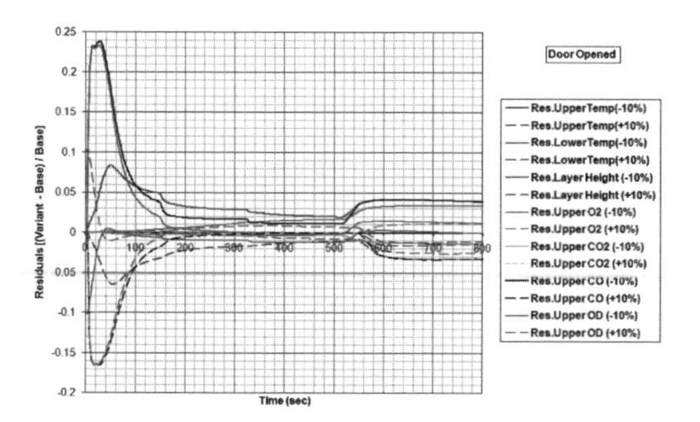

Figure 7. Time dependent sensitivity of some of CFAST outputs to a ±10% change in cabin height with door being opened.

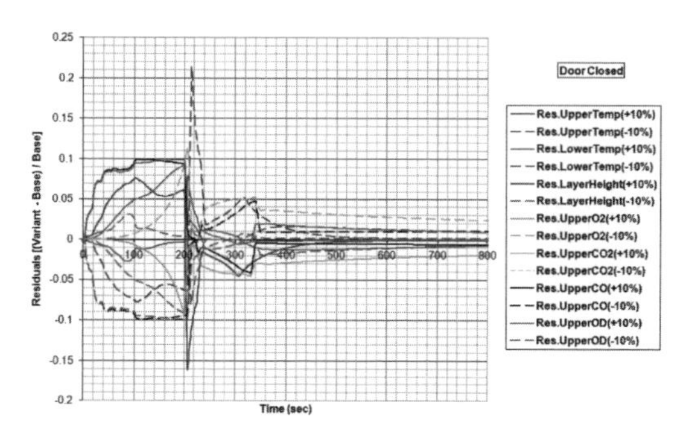

Figure 8. Time dependent sensitivity of some of CFAST outputs to a ±10% change in fire growth parameter with door being closed.

outputs. The graph shows that all the model outputs have almost average values less than 0.1, and that the upper CO_2 concentration, the upper CO concentration, upper layer temperature and the ambient target flux are respectively the most sensitive outputs.

In essence, this implies that reasonable uncertainties in cabin dimensions would have little effect on the results predicted by the model for this scenario. And that the effects are much lesser if the door of the cabin of fire origin is kept open.

Figures 8 to 11 show the results of a time dependent sensitivity of CFAST outputs to a ±10% change in both the fire growth parameter (α), and the peak value of the HRR (Q_{max}) for the two different door statuses. From these figures it is obvious that changing both the fire growth parameter and the peak of the HRR by ±10% have a bit stronger effects on the model outputs than other input parameters, especially, when the door is opened. The effect of changing Q_{max} appears at higher simulation time which corresponds to the time at which the fire growth period ends and the fire starts to level-off. This time differs according to the value of Q_{max}.

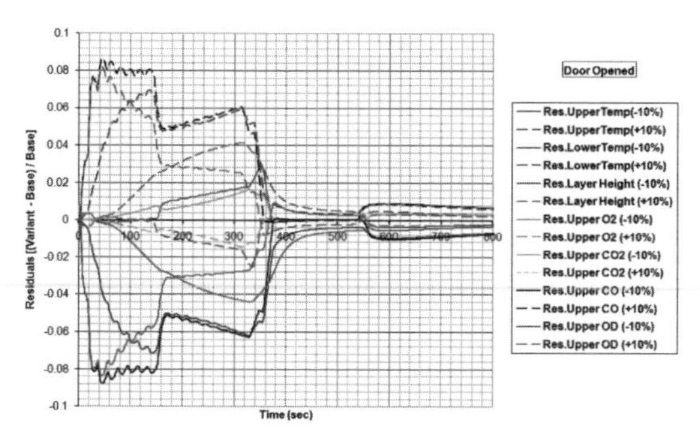

Figure 9. Time dependent sensitivity of some of CFAST outputs to a ±10% change in fire growth parameter with door being opened.

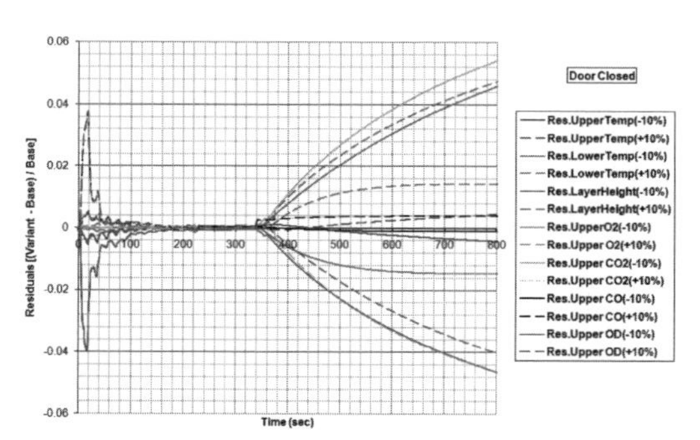

Figure 10. Time dependent sensitivity of some of CFAST outputs to a ±10% change in peak value of HRR with door being closed.

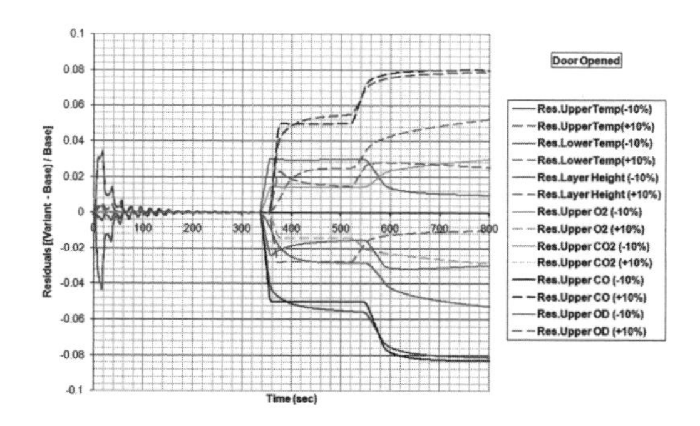

Figure 11. Time dependent sensitivity of some of CFAST outputs to a ±10% change in peak value of HRR with door being opened.

Although the effect of small changes is not too high, it is clear from these figures that both α and Q_{max} appear to have a greater effect on most of the model outputs than other inputs. In other words, the effect of changing the HRR (α and Q_{max}) has the greatest effect on the model outputs.

4 SENSITIVITY TO LARGE CHANGES IN CFAST INPUTS

To investigate the effects of much larger changes in the model inputs, a series of simulations where the inputs varied from 0.1 to 5.0 times the base value was conducted. The cabin fire scenario is also considered here. Only samples of the simulations where the fire growth parameter and the peak value of the HRR were varied are shown in Figures 12 to 15. Each set of the results of the simulations appears as families of curves with similar functional forms.

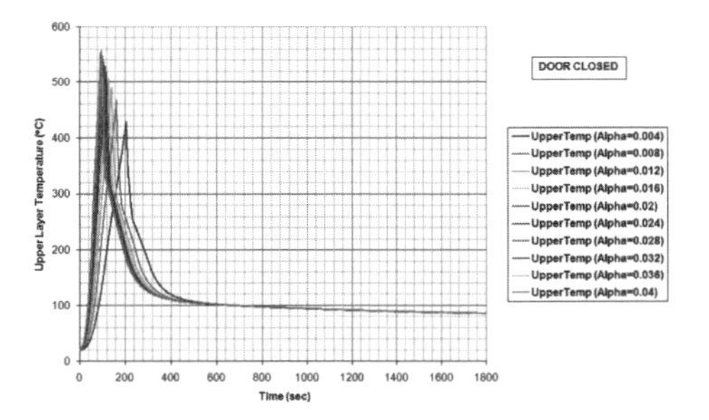

Figure 12. Upper layer temperature resulting from variation of fire growth parameter with door being closed.

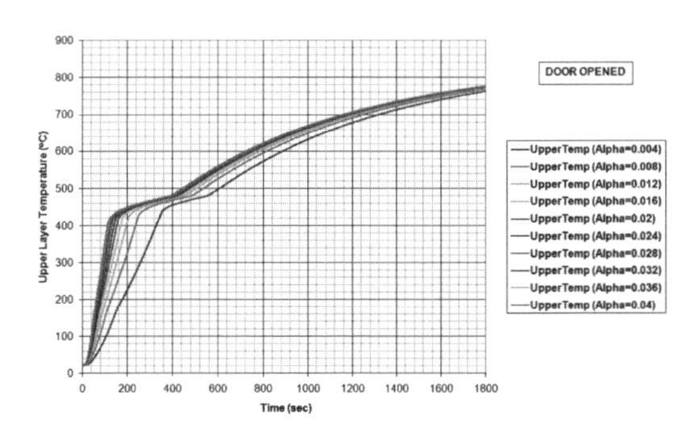

Figure 13. Upper layer temperature resulting from variation of fire growth parameter with door being opened.

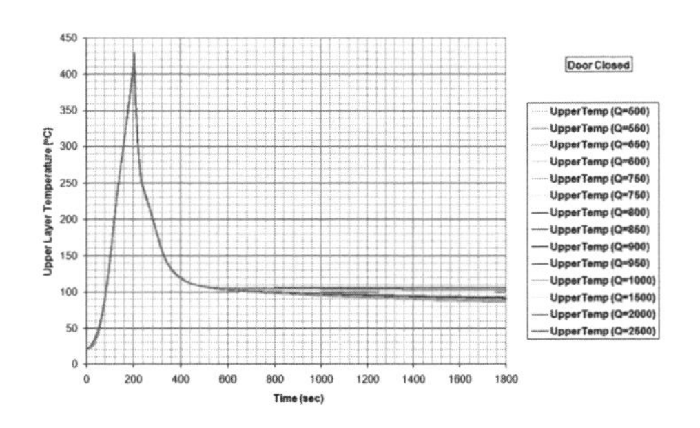

Figure 14. Upper layer temperature resulting from variation of the peak value of the HRR with door being closed.

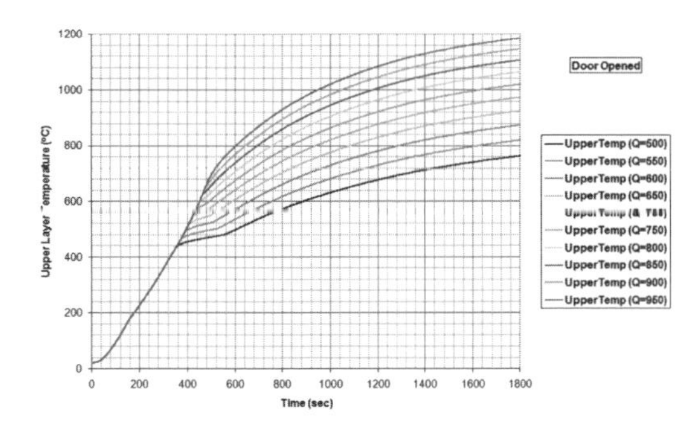

Figure 15. Upper layer temperature resulting from variation of the peak value of the HRR with door being opened.

The results of these simulations indicate that the HRR is the input of CFAST that has greater effect on its outputs than other inputs. This, of course, agrees with the fact that the HRR is the single most important variable that characterises the flammability of products and their consequent fire hazard. Thus, it is decided to use the simple response-surface correlation methodology to examine the sensitivity of CFAST outputs to changes in only the HRR.

5 SIMPLE RESPONSE-SURFACE CORRELATION METHOD

As mentioned above, the HRR is the most important input parameter in any fire scenario; therefore, it will be used as a characteristic parameter to cross-plot the outputs of interest.

Figures 16 to 19 present plots of the upper and lower layer temperatures plotted against the heat release rate for two door statuses. The dotted lines in these figures show the locus of all the individual data points. In this study, only the data up to the time when the fire became ventilation controlled were included. Regression analysis is then used to fit the data. The curves for both upper and lower layer temperatures for both door statuses show a strong functional dependence on HRR. This means that the HRR provides a simple predictor of the temperature in the cabin. Moreover, this relationship allows calculation of the sensitivity of the temperature outputs to the HRR inputs as a simple slope of the resulting correlation between HRR and temperature.

Figure 20, a plot of the slope of the curves shown in Figures 16 through 19, shows the sensitivity of the upper and lower layer temperatures, for the studied cabin fire scenario, to changes in the HRR. Not surprisingly, Figure 20 illustrates that, regardless of the door status, the upper layer

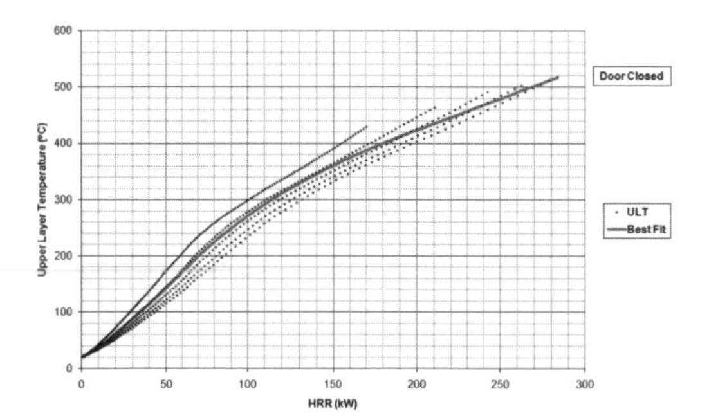

Figure 16. Upper layer temperature against HRR with door being closed.

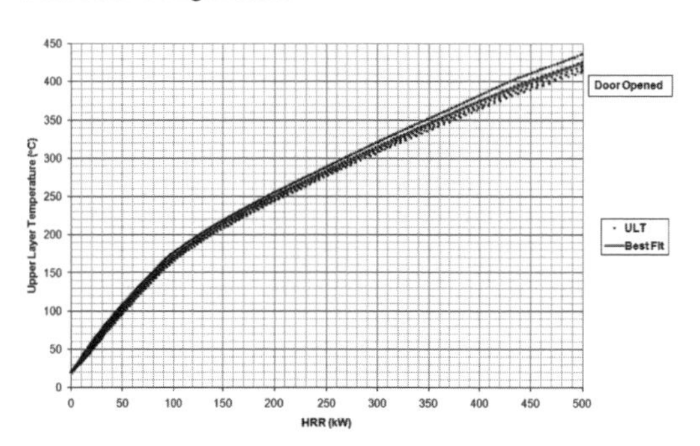

Figure 17. Upper layer temperature against HRR with door being opened.

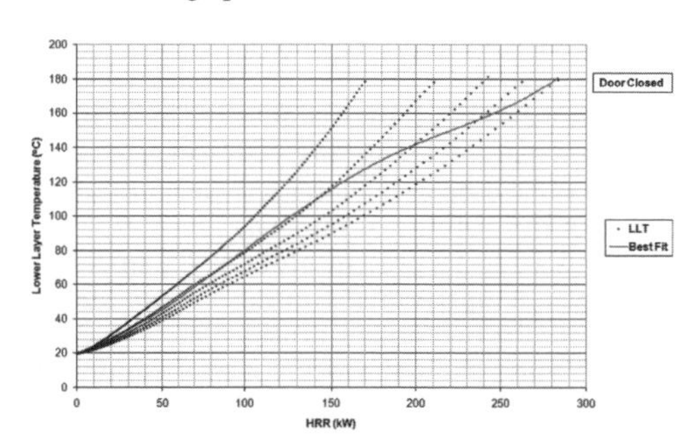

Figure 18. Lower layer temperature against HRR with door being closed.

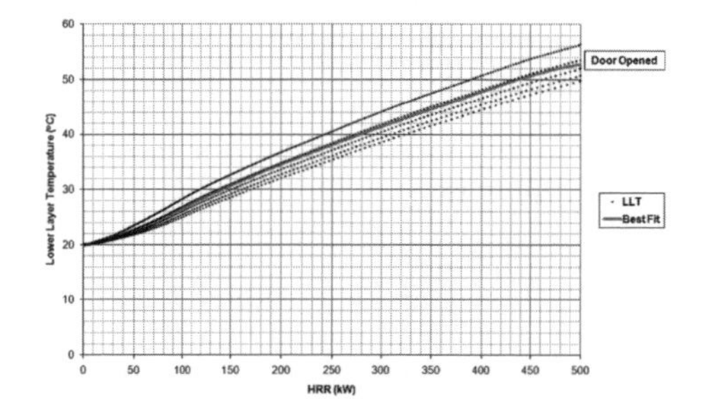

Figure 19. Lower layer temperature against HRR with door being opened.

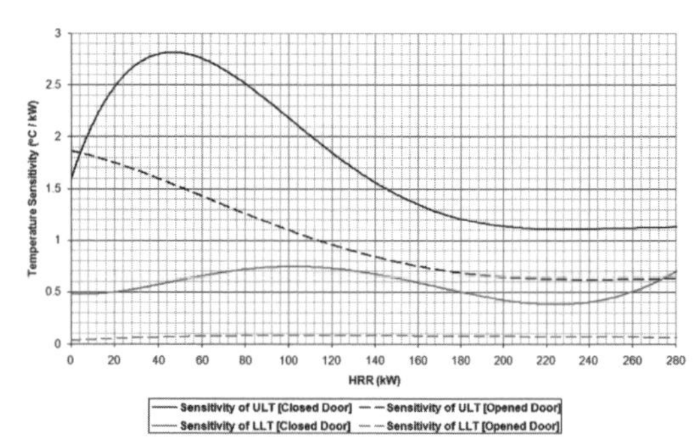

Figure 20. Sensitivity of temperature to heat release rate.

temperature is more sensitive to changes in the HRR than the lower layer temperature. When the door is opened, both layer temperatures show less sensitivity to changes in the HRR than if the door is closed. Except for relatively small HRR (i.e., less than 100 kW), the upper layer temperature sensitivity, when the door is closed, is less than 2.5°C/kW while the sensitivity when the door is opened is less than 1.0°C/kW. This implies, for example, that if the HRR for a 200 kW fire is known to within 50 kW, the resulting uncertainty in the calculation of the upper layer temperature is about ±68°C when the door is closed, and about ±32°C when the door is opened. The lower layer temperature has an average sensitivity of about 0.57°C/kW when the door is closed and an average sensitivity of about 0.06°C/kW when the door is opened. This implies that, for example, if the HRR for a 200 kW fire is known to within 50 kW, the resulting uncertainty in the calculation of the lower-layer temperature is about ±21°C when the door is closed and about ±3.5°C when the door is opened.

The same methodology has been used with the other output parameters of concern and more or less similar features have been recorded. The general conclusion, which could be drawn from the results of this analysis is that CFAST outputs are more sensitive to changes in the HRR if the door of the cabin is kept closed, i.e., when the fire is ventilation controlled.

6 EVALUATING SENSITIVITY OF CFAST ON SINGLE VALUES

In the following, the results of a sensitivity analysis on single values representing the time to reach untenable conditions (t_{smoke}, t_{heat}, and $t_{toxicity}$) will be illustrated. Once again, due to the fact that HRR is the most important input parameter, only the effect of changing both the fire growth parameter and the peak value of the HRR on the single values will be studied.

Considering the effect of changing the fire growth parameter (α) on the single values, Figures 21 to 24 show that regardless of the status of the door, the sensitivity of all the single values decreases as the fire growth parameter increases. As an example, if the fire growth parameter for a

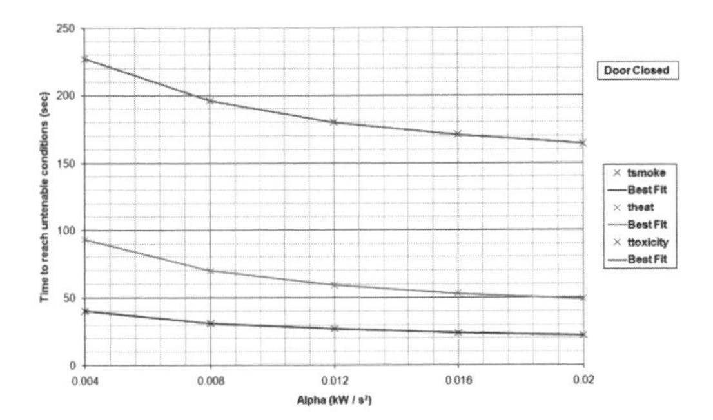

Figure 21. Time to reach untenable conditions against fire growth parameter with door being closed.

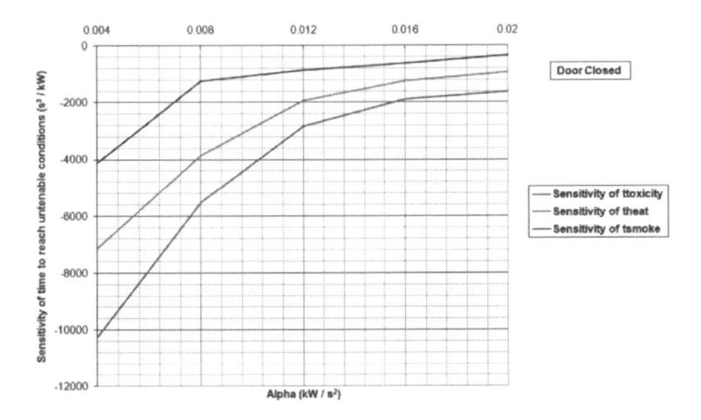

Figure 22. Sensitivity of time to reach untenable conditions to changes in fire growth parameter with door being closed.

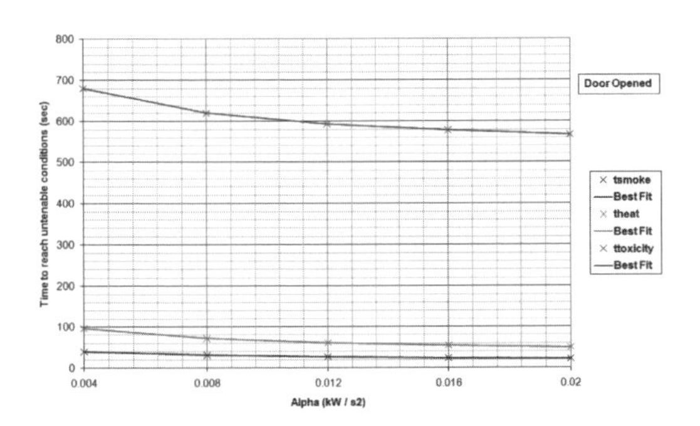

Figure 23. Time to reach untenable conditions against fire growth parameter with door being opened.

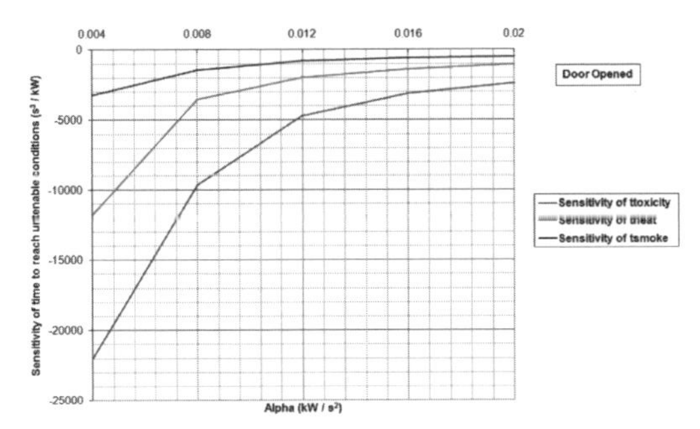

Figure 24. Sensitivity of time to reach untenable conditions to changes in fire growth parameter with door being opened.

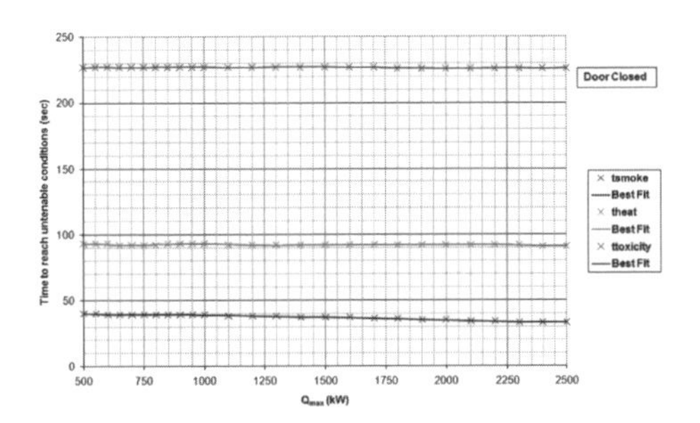

Figure 25. Time to reach untenable conditions against peak of HRR with door being closed.

0.016 kW/s^2 fire is known to within 0.001 kW/s^2; the resulting uncertainty in the calculation of:

1. t_{smoke} is about ±0.630 sec when the door is closed and about ±0.592 sec when the door is opened.
2. t_{heat} is about ±1.231 sec when the door is closed and about ±1.375 sec when the door is opened.
3. $t_{toxicity}$ is about ±1.883 sec when the door is closed and about ±3.112 sec when the door is opened.

Finally, considering the effect of changing the peak value of the HRR (Q_{max}) on the single values, Figures 25 to 28 show that the sensitivity of the critical times to changes in Q_{max} is small when the door is closed while $t_{toxicity}$ showed a much higher sensitivity when the door is opened. As an example, if the peak value of the HRR (Q_{max}) for a 750 kW fire is known to within 100 kW; the resulting uncertainty in the calculation of:

1. t_{smoke} is about ±0.027 sec when the door is closed and about ±0.079 sec when the door is opened.
2. t_{heat} is about ±0.417 sec when the door is closeed and about ±0.000 sec when the door is opened.
3. $t_{toxicity}$ is about ±0.050 sec when the door is closed and about ±30.069 sec when the door is opened.

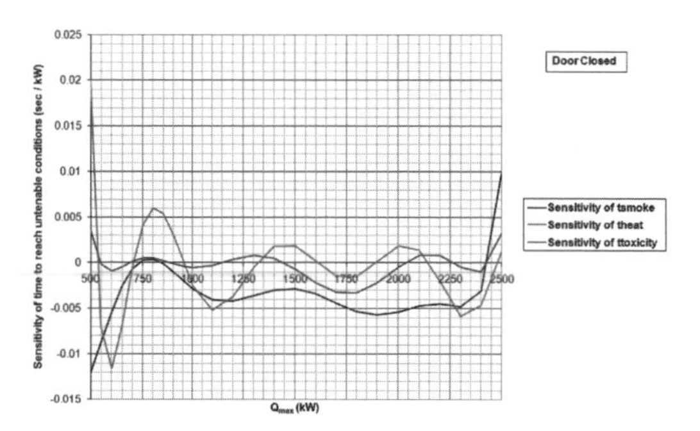

Figure 26. Sensitivity of time to reach untenable conditions to changes in peak of HRR with door being closed.

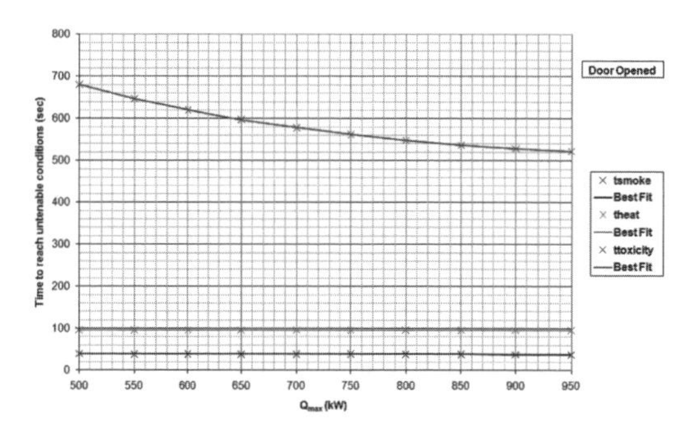

Figure 27. Time to reach untenable conditions against peak of HRR with door being opened.

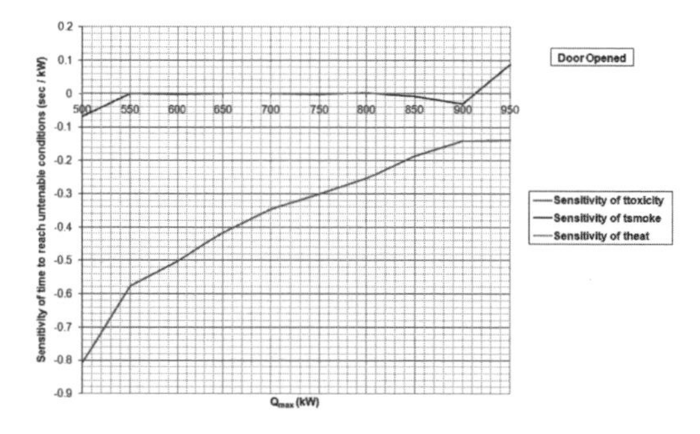

Figure 28. Sensitivity of time to reach untenable conditions to changes in peak of HRR with door being opened.

According to the previous examples, one can conclude that the effect of changing (α) on most of the single values is much greater than the effect of changing Q_{max}, and that the effect of changing Q_{max} appears only on $t_{toxicity}$ when the door is left opened.

7 CONCLUDING REMARKS

The results of the sensitivity analysis obtained in this study using one of the well-known fire simulation programs of the zone model type, CFAST, leads to the following conclusions:

1. For a given fire scenario, i.e., given topology (enclosure dimensions), design fire (fire load and location), and ventilation conditions, the most important input parameters, which affect the outputs of the fire model CFAST are the fire growth parameter (α) and the peak value of the HRR (Q_{max}).
2. In cabin fire scenarios, the effect of changing Q_{max} is much greater when the vent door is opened than when the vent door is closed.
3. Moreover, in cabin fire scenarios, the fire becomes ventilation-controlled, and hence the effect of changing the fire growth parameter (α) on the time to reach untenable conditions is more important than Q_{max} except for $t_{toxicity}$ as it showed much sensitivity to changes in Q_{max} when the door is left opened.

REFERENCES

Ferson, S. & Tucker, W. 2006. Sensitivity in risk analysis with uncertain numbers. Sandia Report SAND2006–2801. Sandia National Laboratories, Albuquerque, NM.

Iman, R. & Helton, J. 1988. An investigation of uncertainty and sensitivity analysis techniques for computer models. *Risk Analysis* 8(1): 71–90.

Jones, W., Peacock, R., Forney, G. & Reneke, P.A. 2009. CFAST—Consolidated Model of Fire Growth and Smoke Transport (Version 6)—Technical reference guide, NIST Special publication 1026, National Institute of Standards and Technology, Gaitherburg, MD.

Karlsson, B. & Quintiere, J. 2000. *Enclosure fire dynamics*, CRC Press LLC, ISBN: 0-8493-1300-7, 2000.

Peacock, R., Reneke, P., Forney, C. & Kostreva, M., 1998. *Issues in evaluation of complex fire models*. Fire Safety Journal, 30:103–136.

Salem A. 2007. Risk-based design for fire safety of ro-ro/passenger ships [PhD Thesis]. [Glasgow, UK]: Glasgow and Strathclyde Universities.

Salem A. 2010. Fire engineering tools used in consequence analysis. Journal of Ships and Offshore Structures, Vol. 5, No. 2: 155–187.

SFPE 2005. *SFPE engineering guide to application of risk assessment in fire protection design*. Society of Fire Protection Engineers. Review draft. Bethesda, MD.

Sustainable Maritime Transportation and Exploitation of Sea Resources – Rizzuto & Guedes Soares (eds)
© 2012 Taylor & Francis Group, London, ISBN 978-0-415-62081-9

Managing uncertainty in performance-based fire safety assessments of ships

Nikos Themelis, Stavros Niotis & Kostas J. Spyrou
School of Naval Architecture and Marine Engineering, National Technical University of Athens, Athens, Greece

ABSTRACT: We address the problem of "uncertainty" that pervades all facets of performance-based studies of ship fire safety. Very often only a limited number of fire scenarios are examined and it is practically unknown whether the result could sufficiently characterize the entire design. Motivated by this, we have carried out several fire simulations referring to a fictitious cruise ship where key factors such as the location of a fire onboard, fire products' yields and safety criteria norms, are treated as uncertain. For each examined scenario we calculated, near to evacuation exits, the time required to reach conditions affecting the efficiency of evacuation, such as visibility loss and toxic incapacitation. The performance of a small group of cabins appears to be quite representative of the aggregate performance of all cabins of the deck, particularly with respect to the visibility criterion. Sensitivity analyses performed reveal notable variations in failure probabilities.

1 INTRODUCTION

Market's demand for innovative solutions in the design of large cruise ships has led to the introduction of alternative assessment procedures for evaluating fire vulnerability on board that deviate from the current prescriptive SOLAS-based practice. Regulation 17 of SOLAS Chapter II-2 in combination with MSC 1002 constitute the platform for performing fire safety engineering analysis in order to ascertain that the alternative designs and arrangement provide, at a minimum, an equal level of fire safety with the one deriving from application of the prescriptive framework. In reality, the introduction of Regulation 17 has opened the door to performance-based fire safety assessments. Their implementation is supported by a number of technical guidelines published by authoritative organisations like the International Organisation for Standardization; the Society of Fire Protection Engineers and the National Fire Protection Association (see for example ISO/TC92 1999; SPFE 2002; NPFA 101 2000). More narrowly to the maritime sector, classification societies have also produced their own guidelines for the implementation of MSC 1002 (e.g. ABS 2004). Even though this step forward in designer's freedom has already impacted upon design evolution in various respects (e.g. from layout and materials to the installed fire safety systems onboard), one should not disregard a few areas of concern, related with the implementation of the associated assessment procedures.

The usual practice in performance-based studies is to evaluate and compare the performance of trial designs and arrangements with respect to certain fire safety objectives, by subjecting them to a limited set of "representative" fire scenarios. Due to the intrinsically vast number of the parameters that affect the process of fire development and the often significant uncertainty about their most appropriate values, the level of confidence to the outcome may not be reasonably specified. In other words, one is not certain that the assumed set of fire scenarios comprises a sufficient basis for decision-making. Furthermore, variations in the input of the scenarios within the uncertainty limits might alter the conclusion about the acceptability of a design. As a matter of fact, it is not clear how to systematically treat the various uncertain parameters that are present in a "performance based" fire safety assessment. Fire scenarios are commonly selected empirically, by "expert judgment". On the other hand, promising works have appeared towards a more holistic treatment of fire safety within a probabilistic framework (Hakkarainen et al., 2009; Vassalos et al., 2010).

We have proposed recently a methodology for the probabilistic generation and analysis of design fires (Themelis & Spyrou 2010). A design fire comprises a core element of a fire scenario as it describes the fire development in the defined space and it is usually expressed by the rate of released heat in time ("HRR curve"). A mathematical model for the generation of a HRR curve has been developed taking into account: the amount and type of available combustible materials (described as "fire load"); various sizes of ignition items ("fuel packages") and their potential in terms of intensity of fire growth, ventilation restrictions, flashover occurrence etc. Even at such an early stage like the definition of design fires however, uncertainties in several parameters values can be identified. These are commonly classified as "epistemic uncertainties", referring to lack of knowledge or data about the values of the various quantities involved. For example, the fire growth characteristics, expressed by the HRR curve, of a specific chair placed in a certain position within an enclosure are not known theoretically in an exact sense and thus, burning tests should be relied upon. Nevertheless, due to the complex underling chemistry that governs the burning of an item as well as the dependence of growth upon many parameters (including the ignition source), repeatability of the result of tests that were carried out with the same specification should not be taken for granted. The use of experimental data is nostrum rather than true remedy for proper parameter selection during fire safety analyses. In the scientific uncertainties one should include also choices accruing from several, implicitly or explicitly made, assumptions in the mathematical model(s). One example is the model of flashover prediction that is often based on semi—empirical formulas. Detailed fire CFD codes are devoid of such uncertainties neither.

Another category of uncertainties, the so called "aleatory uncertainties", are also (tacitly or loudly) present (Notorianni 2002). They arise from the truly random character of several parameters affecting fire growth (for example, an open or closed door, the exact position of some ignition item etc). The treatment of aleatory uncertainties in a probabilistic framework seems almost natural. In fact, all types of uncertainties can be treated in such manner if suitable distributions could be reasonably assumed. However, it is easier to obtain distributions for parameters with respect to aleatory uncertainties (where measurement and statistics can be used) than for the epistemic.

It becomes apparent that our aim in this paper is to address the management of uncertainties associated with ship fire safety assessments. The field is of course vast; but here we are interested specifically on the relation between the location of a fire onboard and the fire products' yields. A cruise ship will be used (in particular a cabin deck area), while numerical simulations based on a well-known CFD code ("FDS") will be employed for predicting the time histories of fire products (McGrattan et al., 2010).

Better understanding about the relationships of uncertain parameters can be very beneficial. This relates also with the fact that, in practice only a limited number of fire scenarios can be studied with detailed CFD models because the associated simulations are CPU time consuming. Therefore, it should be ensured that, the few fire locations selected for performing detailed simulation constitute a good basis for calculating the inherent risk associated with "all possible" locations.

In the first part of the study the location of the fire ignition cabin will be considered as random. This is an aleatory uncertainty, reflecting that ignition might take place in every cabin of the considered deck. In the second part we focus on the variation of fire product yields (specifically smoke and CO yield) whose values could be considered as "epistemically uncertain". These parameters affect directly the risk imposed to passengers during an evacuation process, so differences in this input of fire scenarios could change design characterization from safe to unsafe.

Simulation results are analyzed taking into account passenger hazards faced during an evacuation process. In more detail, we study performance against specific safety targets, expressed in terms of the time to achieve life threatening conditions during evacuation (e.g. reduced visibility, high temperature and high concentration of toxic gases).

In summary, the current work is a first attempt towards elucidating the effect of the uncertain parameters that permeate in a performance-based fire safety assessment; hoping to contribute towards the establishment of a more robust and rational fire safety assessment methodology that would lessen the need for expert judgment.

2 MODEL SET UP

2.1 *Considered space characteristics*

Cabins' layout corresponds to a fictitious but plausible design, based on a cruise ship. The length of the fire vertical zone has been extended to 47 m.

The deck is 2.5 m high and 20 m wide. It is equipped with three longitudinal and one transverse 1.5 m wide corridors, which connect the 40 cabins with a pair of staircases (aft and forward). The staircases are demonstrated as open hatches—3.4 m length by 1.4 m wide—through where air and/or smoke can freely communicate with the external environment. Port and starboard corridors are outfitted with 16 cabins each, evenly distributed, while the middle corridor hosts the other eight.

The deck area of each cabin is 13.4 m^2 with corresponding height 2.5 m, except from the extreme aft cabins, which are 15 m^2. Every cabin door is

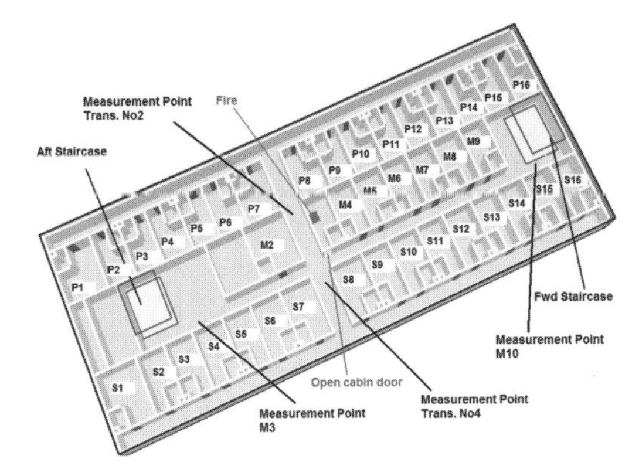

Figure 1. Deck overview showing the cabins and the location of measurement devices.

0.6 m wide and 2.0 m high. The cabin door remains open only when the subject cabin is the place of primary ignition. Figure 1 illustrates deck's general arrangement. For illustration, fire is shown ignited in the middle corridor cabin M3 (they are counted from aft to forward). All interior and exterior boundaries are assumed as B-15 class layered walls consisted of PVC paint, galvanized steel and Rockwool insulation. The exact distribution and properties of the materials used appear in Table No. 1.

2.2 *Fire specifics and simulation parameter settings*

The basic input data required are the HRR curve and the yields of fire effluents (soot and CO). The cabin door has been assumed open all the time and the fire load has been based on experimental data concerning full scale fire tests in a passenger cabin (Arvidson et al., 2009). Other parameters like the incipient time duration, the size and growth characteristics of the "fuel package", have been treated probabilistically. The general framework has been described in Themelis & Spyrou (2010). From the generated set of 200 fires, we select the HRR curve that corresponds to the same approximately maximum HRR and the time to achieve it with the respective experimentally measured HRR (see Figure 2).

Furthermore, the yields for soot and CO have been assumed corresponding to polyurethane and they were taken equal to 0.013 (g/g) and 0.035 (g/g). Specie's yield expresses the specie's mass that emanates from the fire, in terms of fuel mass loss (Karlsson & Quintiere 2000).

The longitudinal corridors are equipped with 10 measurement devices each, equally spaced every 5 m. The transverse corridor has been provided with five measurement points, also spaced every 5 m. The 35 devices in total are placed at 1.5 m height. They record temperature, CO, CO2, O2 volume fraction and smoke obscuration at every time step.

Table 1. Boundary materials (type: "sandwich").

Material	Thick-ness	Density	Thermal Con-ductiv.	Specific heat	Ignition temper-ature
	mm	kg/m^3	W/mK	kJ/kgK	°C
PVC	2 × 0.5	1,380	0.192	1.290	750
Galvanized Steel	2 × 0.6	7,850	51.900	0.483	–
Rockwool	50.0	229	0.041	0.750	750

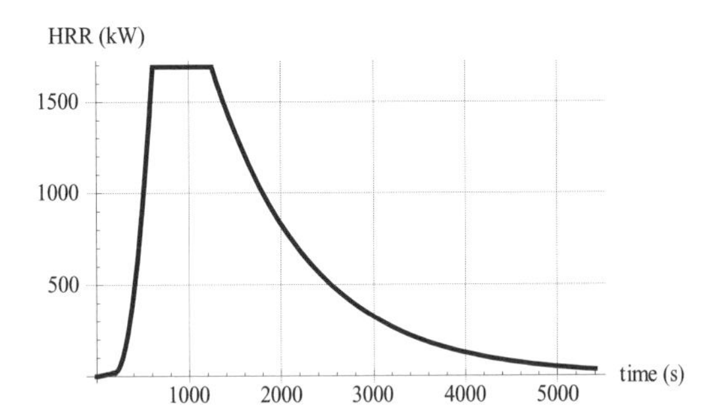

Figure 2. Selected curve of heat release rate.

In total, 320,000 cells have been generated with each cell size (in m) equal to 0.25 × 0.25 × 0.15. Average computational time for 90 minutes of simulation in a standalone PC (3.4 GHz) was 38 hr.

2.3 *Safety objectives*

Trial designs need to be assessed against effective performance criteria (also named here as "safety objectives"). However, even the setting of such criteria entails uncertainty because procedures of criteria development have subjective elements. Different criteria could produce different notions of acceptable design. Since we do not intend to perform evacuation analysis, quantitative criteria can be deduced from the limiting values of fire effluents at specific locations, implying the encounter of threatening conditions to human life during evacuation. The choice of measurement locations brings in another subjective influence that could be handled through a sensitivity analysis.

The following criteria and associated norms have thus been set:

1. Inability to view a light-reflecting-sign at 4 m distance due to smoke obscuration.
2. Temperature should not exceed 60°C.
3. Toxic gases' effect (basically CO) should not reach the incapacitation level.

We examine whether they are satisfied at locations near to evacuation exits and in the transverse middle corridor (see points M3, M10 and Trans. No. 2 in Figure 1). For the third criterion in particular we note that, in an evacuation process the toxic gas dose that a passenger could receive depends on the path that he follows. Here however we are not targeting the received dose by an individual who participates in the evacuation. Instead, we determine the time required to reach the incapacitation level, for a person that stands at one of the three specified locations.

The visibility criterion will be satisfied as soon as the parameter FEC_{smoke} ("fractional effective concentration"), which is calculated according to Equation 1 below, becomes equal to unit. FEC_{smoke} depends on the optical density parameter OD that is calculated in turn by Equation 2 (Mulholand 2002):

$$FEC_{smoke} = \frac{OD}{0.25} \tag{1}$$

$$OD = \frac{C}{2.3 \cdot S} \tag{2}$$

C is a non-dimensional constant characteristic of the type of object viewed through the smoke cloud. For the specific calculations we selected $C = 3$, corresponding to a light-reflecting sign. S is the visibility distance (in m).

The toxic effect of the asphyxiating gases (in fact the increase of CO and CO_2 concentrations and the decrease of O_2) will be considered through the parameter FED_{IN}, which is the "fractional effective dose" for incapacitation (Purser 2002):

$$FED_{IN} = FED_{CO} \cdot V_{CO_2} + FED_{O_2} \tag{3}$$

where:

$$FED_{CO} = \sum_{t_1}^{t_2} \frac{K \cdot [CO]^{1.036}}{D \cdot 60} \cdot \Delta t \tag{4}$$

$$V_{CO_2} = \frac{exp(0.1903 \cdot [CO_2] + 2.0004)}{7.1} \tag{5}$$

$$FED_{O_2} = \sum_{t_1}^{t_2} \frac{1}{60 \cdot exp[8.13 - 0.54(21\% - [O_2])]} \cdot \Delta t \tag{6}$$

$[CO]$, $[CO_2]$ and $[O_2]$ are the average volumetric concentrations (in ppm for CO,% volume for CO_2 and O_2) during a time increment Δt (in s). K and D are parameters related with human activity (for "light work" they take the values 8.2925×10^{-4} and 30 respectively according to Purser). High percentage of CO_2, which in small concentrations (less than 5%) can be considered as non toxic, results in acceleration of breathing (hyperventilation), while the reduction of O_2 causes oxygen hypoxia.

3 RESULTS AND CHARACTERISTIC LOCATIONS

3.1 *Time histories of fire effluents*

Figure 3 shows a screenshot of smoke spreading (produced by the Smokeview program of NIST (Forney 2008) for a fire in a cabin that is located at the middle corridor. Furthermore, Figures 4–5 show the time histories of temperature and CO concentration at specific points on the deck for a variety of fire case scenarios. We can observe the significant variation during a 5 min time period (from 10 to 15 min).

3.2 *Statistical analysis based on all cabins*

Next we calculate the required time for satisfying the criteria at the selected spot locations, for each examined scenario. At first stage the location of the cabin of fire ignition has been considered as uncertain. This means practically that all cabins had to be examined. In Figure 6 is plotted the estimated

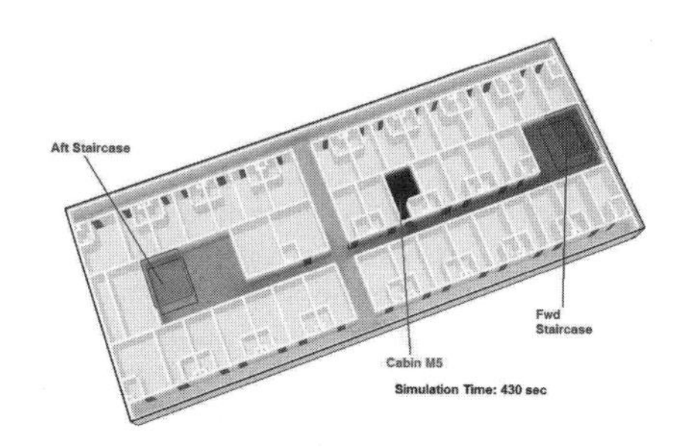

Figure 3. Overview of the deck at time 430 s. Fire case scenario—ignition at cabin M5.

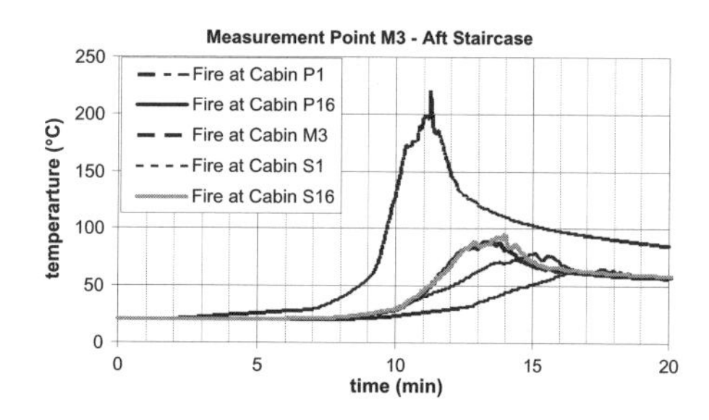

Figure 4. Time series of temperature at aft staircase for 5 different fire scenarios.

critical time required for reaching the visibility threshold due to smoke obscuration, at the three spot locations and for a fire at any of the cabins in the middle row. Furthermore, in Figure 7 is seen the critical time for reaching the incapacitation level due to toxic gases, for the starboard cabins. Lastly, in Figure 8 is shown the critical time with reference to the temperature criterion, considering the port set of cabins.

We have performed statistical analysis of the calculated critical times based on all cabins of the deck. Some key results are summarised in Tables 2, 3 and 4. They will be utilized subsequently in order to determine the probability some criterion to be violated during evacuation, at the three considered locations (see Figures 9–11). Some variability of the critical time is noticeable, depending on the location of fire origin.

Figure 9 contains sufficient information for building an index of evacuation efficiency. One observes that, if the evacuation time lasted for more than 12 minutes, there would be failure due to smoke obscuration. In the worst-case, a fire in cabin M3 seems to incur the lowest average critical time at the evacuation exits (the corresponding time is about 8.12 min). Designing by such an objective would lead to a $P = 0.975$ probability of successful evacuation, for all possible fire locations.

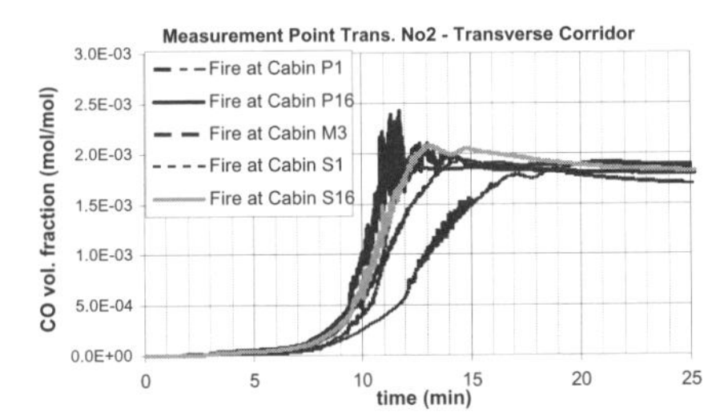

Figure 5. Time series of CO volume fraction at transverse corridor, for different fire scenarios.

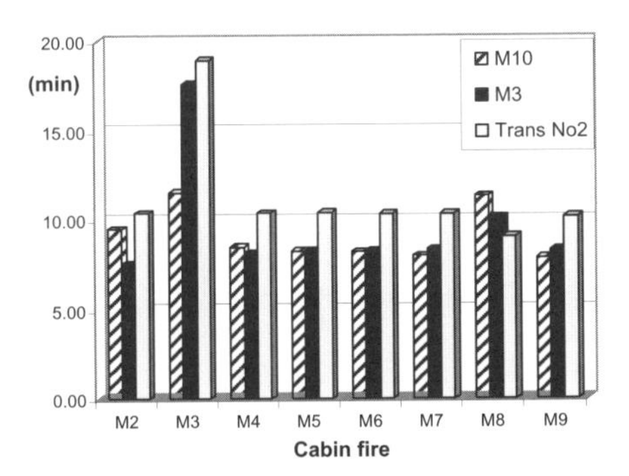

Figure 6. Critical time for smoke (middle set of cabins).

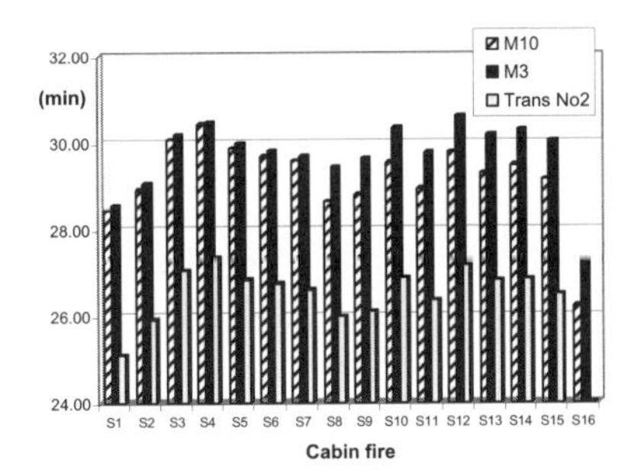

Figure 7. Critical time for toxic gases (starb'd set of cabins).

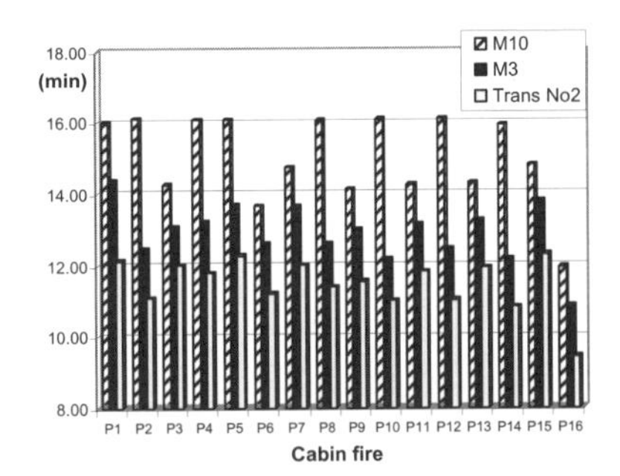

Figure 8. Critical time for temperature rise (port set of cabins).

Table 2. Statistics of time until visibility loss due to smoke.

Measurement point	M10	M3	Trans N2
Mean (min)	11.006	9.846	9.661
Variance (min)	1.383	2.140	2.554
Minimum (min)	7.887	7.560	7.726
Maximum (min)	12.034	17.598	18.908

Table 3. Statistics of time until incapacitation from toxic gases.

Measurement point	M10	M3	Trans N2
Mean (min)	27.810	28.583	26.281
Variance (min)	6.512	4.104	1.070
Minimum (min)	20.851	23.282	23.708
Maximum (min)	30.430	30.605	28.862

Table 4. Statistics of time until critical temperature.

Measurement point	M10 temp	M3 temp	Trans N2
Mean (min)	13.827	12.677	12.094
Variance (min)	6.672	3.507	1.426
Minimum (min)	8.248	8.236	9.232
Maximum (min)	16.679	16.133	14.173

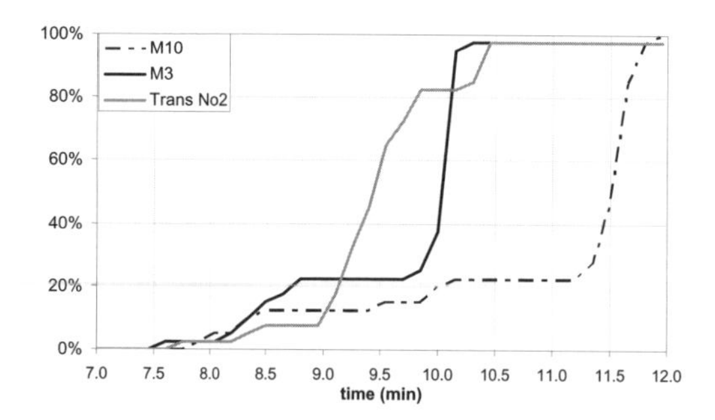

Figure 9. Probability of failure of the safety objective corresponding to visibility incapacity.

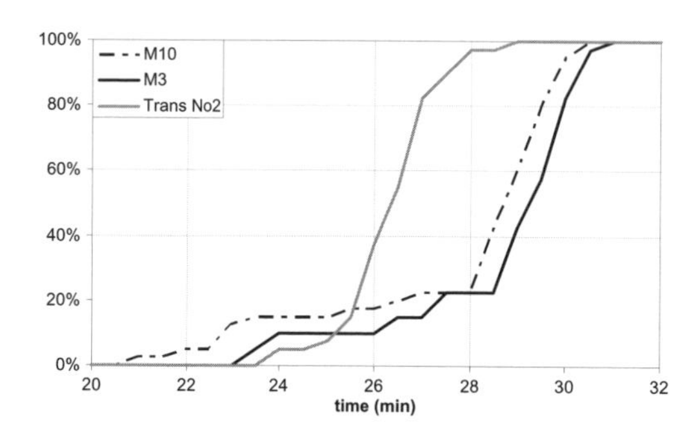

Figure 10. Probability of failure of the safety objective corresponding to toxic gases effect.

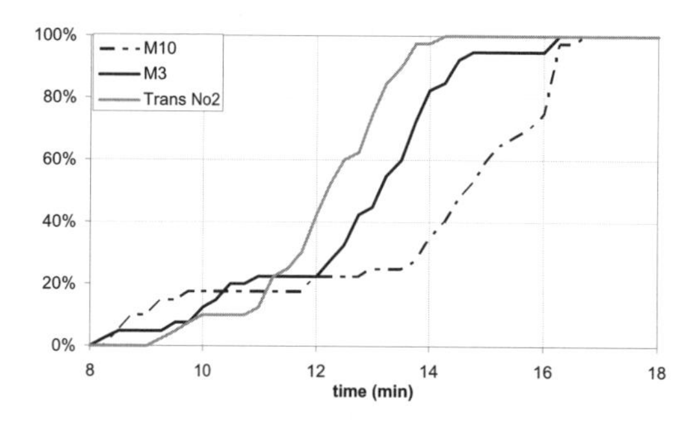

Figure 11. Probability of failure of the safety objective corresponding to temperature.

Nevertheless it is a matter of discussion what would constitute an acceptable probability level.

From the respective results for toxic gas concentration and temperature rise, we observe that toxicity incapacitation requires much more time; while temperature follows smoke with a time lag of 2–3 min. Furthermore, both failure probabilities show to obey a smoother distribution than the one for smoke in which rapid failure could be realized in less than 1 min. It is reminded that the result refers to a person standing at a fixed location and does not include any history of movement on the deck.

3.3 *Selection of representative set*

It becomes evident that it is unlikely to be able to capture the inherent risk from examination of a single cabin fire. Uncertainty of fire location is unavoidable and one wonders whether (and when) the selection of a single cabin could lead to a more or less conservative result. Not being capable to predict variations in risk by varying the fire location will lead to less confidence in the outcome of a performance-based study as different choices could change the acceptability of a trial design. However, carrying out tests for all possible scenarios in terms of location will be impractical due to the long simulation time required by CFD models.

A question naturally arising is thus, whether the required number of simulations could be significantly reduced, by identifying a representative group of cabins that provide similar average behavior, in terms of criteria satisfaction, as the entire set. This reduction process could be producing different representative sets, depending on the examined criterion. Consider for instance the visibility criterion that resulted earlier in smaller critical time than the similar time associated with the other two criteria. Say that we could afford to investigate up to 3 cabins. By empirical judgment and exploiting the privilege of having access to a large set of data referring to all cabins, we found the set comprised by cabins M2, S2 and P14 as the most representative. However we admit that we have not examined meticulously all possible combinations: due to limited time we were unable to perform exhaustively all calculations per criterion and set.

Failure probabilities were determined as functions of the average critical time, for the evacuation exits (M10 and M3). The obtained probabilities are compared against those corresponding to the whole set of cabins of the considered deck. A similar trend is noticed while there is a rather consistent quantitative difference in actual values.

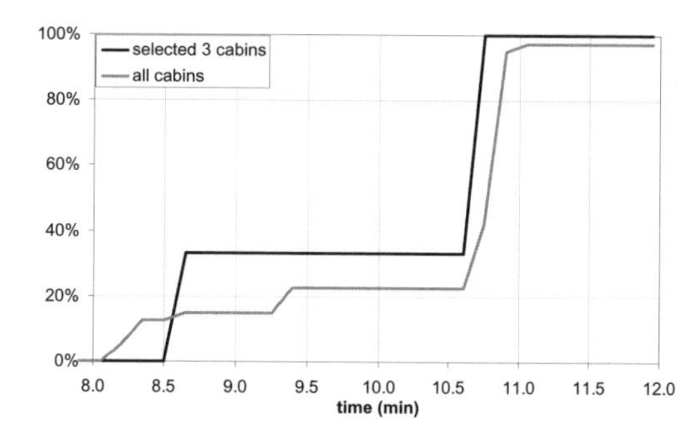

Figure 12. Probability of failure of the safety objective corresponding to visibility near evacuation exits: a) for the selected set of 3 cabins; and b) for all cabins.

4 UNCERTAINTY IN THE NORMS DEFINING VISIBILITY RANGE, SMOKE AND CO YIELD

We turn our focus now to some "epistemic" uncertainties that could affect evacuation efficiency. The uncertain parameters that will be examined are: the distance that defines visibility range in the presence of smoke; the soot and CO fire yields. The respective sensitivity analyses will be carried out for the set of three cabins that we found earlier as most representative.

Inability to see a light-reflecting sign in a distance of 4 m had been considered as a criterion in the previous calculations. We will now perform sensitivity analysis for a range (from to 2.5 m to 4 m visibility, with a step of 0.5 m). Such a range is used in practice in evacuation calculations. The obtained probability values are seen in Figure 13. Next we modify the soot yield by increasing stepwise the initially taken value (0.013 g/g) by 25%, 50% and 75%. The result is seen in Figure 14. Summary results, in terms of critical time, for various combinations of soot yield and visibility, for various percentiles of failure, are presented in Figure 15 and in Figure 16.

Lastly, we perform sensitivity analysis concerning the CO yield, which directly affects the critical time for toxic incapacitation. Figure 17 shows the probability of evacuation failure for a range of CO yields (up to 75% from the initially considered value). The critical time corresponding to various failure percentiles is presented in Figure 18.

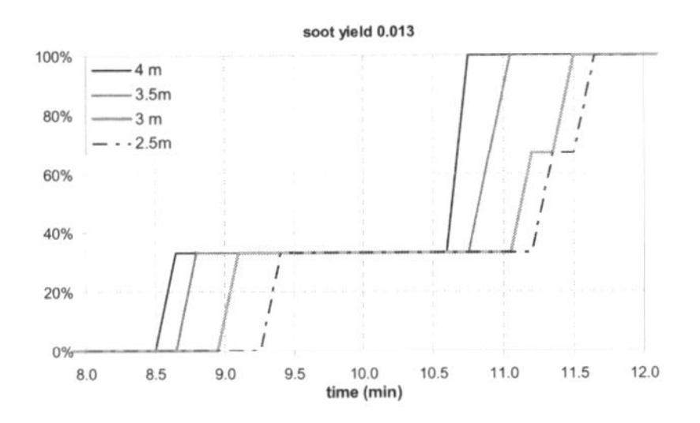

Figure 13. Probability of failure for different norms for smoke [soot yield is fixed at 0.013 (g/g)].

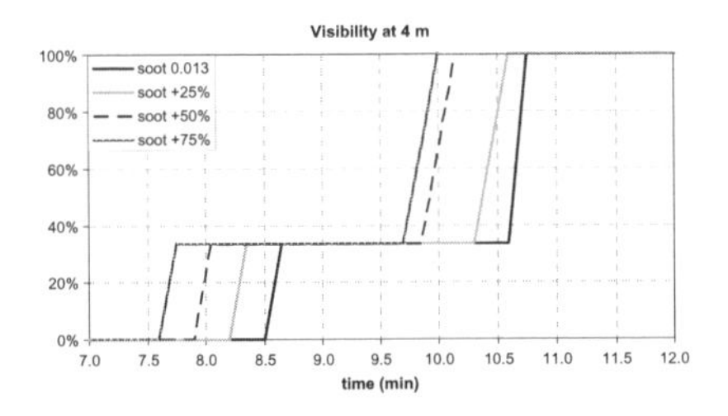

Figure 14. Probability of failure for different soot yields with 4 m visibility.

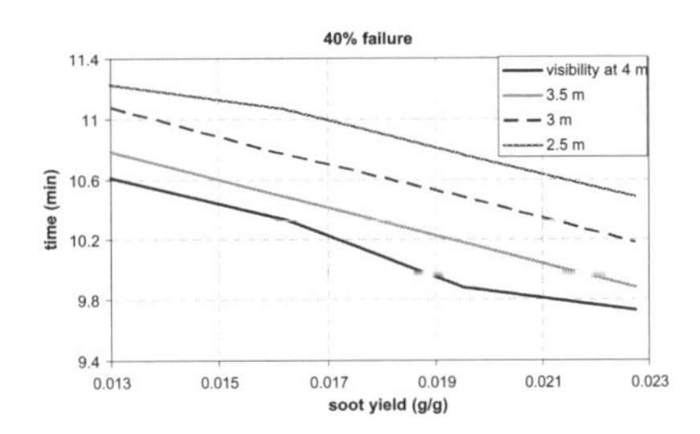

Figure 15. Critical time for soot yield with 40% probability of failure.

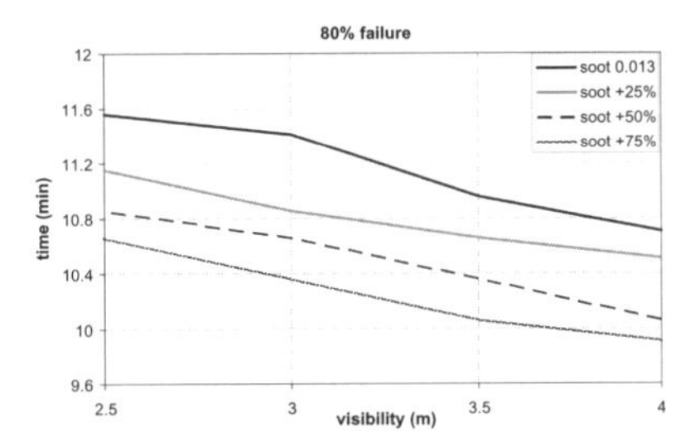

Figure 16. Critical time for visibility at 80% failure probability.

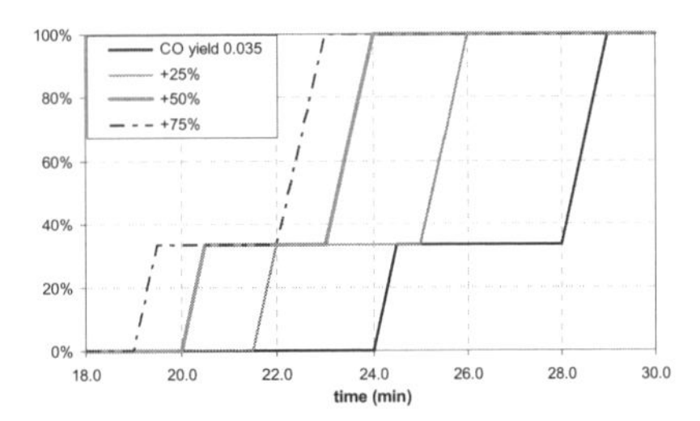

Figure 17. Probability of failure due to toxic incapacitation for different CO yields.

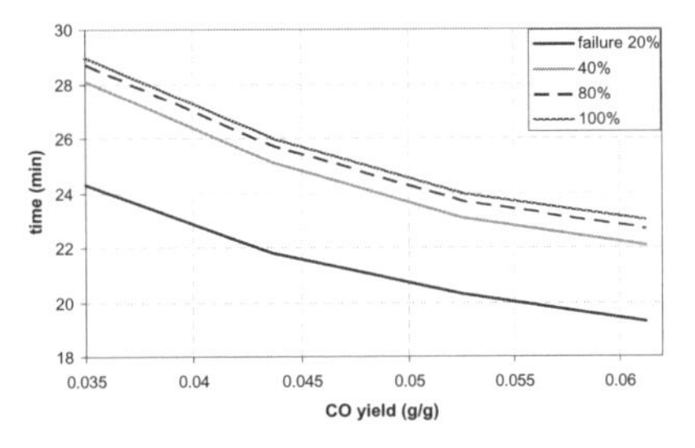

Figure 18. Critical time for toxic incapacitation, for different CO yields and failure percentiles.

From the above results we derive that, by reducing the distance of visibility loss by 0.5 m (12.5% of the initial value) and keeping constant the soot yield, we obtain approximately 3% increase of the critical time (this is natural since more time will be required for reaching a stricter condition).

On the other hand, by increasing 25% the soot yield the respective time will be decreased by 3.5%. Therefore it seems that the critical distance for visibility incapacity affects slightly more the available evacuation time than the soot yield. Also, we can conclude that variation of CO release during a fire, say 25% increase of the yield of CO, will bring about 9% reduction in the critical time for incapacitation. Thus the selection of materials in terms of CO emission should be particularly considered, keeping in mind also the fact that, in smouldering types of fire, the CO yield could be increased by 50–100 times.

5 CONCLUDING REMARKS

In the first part of the current study we have focused on the randomness of the location of the space of fire origin that, in actual fact, could never be exactly anticipated. We have thus considered ignition incidents at any cabin of a deck expanding within a main vertical zone of a fictitious cruiser. We then calculated the probabilities corresponding to safety criteria failure. The key aspect of this part was an heuristic consideration whether a small group of cabins could produce similar tendency in failure probability. If that was true, one could capture inherent risk more efficiently and reliably. Of course the current study is a preliminary one as the issue is major and further research will be required for obtaining a concrete answer.

In the second part of the paper we studied the epistemic uncertainties associated with the proper setting of safety norms in the criteria. Summarizing the results presented in section 4, we can conclude that by reducing the visibility norm by 37.5%, the increase in the estimated critical time was approximately 0.8 min (about 8%). Besides, for the soot, a 50% increase of yield resulted in 1 min less available evacuation time (about 9.1% decrease for 20% success). Variations in the CO yield have also been considered: a fire that emits 50% more CO was found to give, approximately, a 21% decrease of the critical time (of course for the specific conditions assumed). Therefore, notable variations in the failure probabilities have been found and especially, CO yield variations proved to be the most significant.

Uncertainty in performance-based studies is a challenging area in fire safety assessments and it definitely deserves more attention. Modern numerical tools of fire modeling are very useful but they are not panacea since, if used in an unstructured and haphazard manner, they can produce different outcomes and thus lead to different design choices whose true effectiveness is very difficult to evaluate. More studies on these issues are necessary in order to formalize the selection of representative scenarios with respect to the various safety objectives that are relevant to the fire safety analysis of ships.

ACKNOWLEDGMENT

Part of the present study has been carried out within the framework of the EU project FIREPROOF: Probabilistic Framework for Onboard Fire Safety (Grant agreement number: 218761).

REFERENCES

ABS 2004. Guidance notes on alternative design and arrangements for fire safety, American Bureau of Shipping, Houston, USA.

Arvidson, M., Axelsson, J. & Hertzberg, T. 2008. Large-scale fire tests in a passenger cabin. *Fire Technology.* SP Report 2008:33.

Forney, G. 2008. NIST Special Publication 1017-1-User's Guide for Smokeview Version 5—A tool for Visualizing Fire Dynamics Simulation Data. Washington: U.S. Government Printing Office.

Hakkarainen, T., Hietaniemi, J., Hostikka, S., Teemu Karhula, S., Kling, T., Mangs, J., Mikkola, E. & Oksanen, T. 2009. Survivability for ships in case of fire, Final report of SURSHIP-FIRE project. *VTT Research notes* 2497. Finland.

International Maritime Organisation, MSC/Circ. 1002. Guidelines on alternative design and arrangements for fire safety, London, IMO, 2001.

International Organisation for Standardization (ISO), Fire safety Engineering—Part 2: Design Fire Scenarios and Design Fires, ISO/TR 13387-2, 1999.

Karlsson, B. & Quintiere, J.G. 2000. Enclosure fire dynamics. USA: CRC McGrattan, K., McDermott, R., Hostikka, S. & Floyd, J. 2010. NIST Special Publication 1019–5-Fire Dynamics Simulator (Version 5) User's Guide. Washington: U.S. Government Printing Office press, ISBN: 0-8493-1300-7.

Mulholland, G.W. 2002. *SFPE Handbook of Fire Protection Engineering, chapter Smoke Production and Properties, 3rd edition.* Quincy, Massachusetts: National Fire Protection Association.

Notorianni, K. 2002. *SFPE Handbook of Fire Protection Engineering, chapter Uncertainty, 3rd edition.* Quincy, Massachusetts: National Fire Protection Association.

Purser, D.A. 2002. *SFPE Handbook of Fire Protection Engineering chapter Toxicity Assessment of Combustion Products, 3rd edition.* Quincy, Massachusetts: National Fire Protection Association.

Themelis, N. & Spyrou, K. 2010. An efficient methodology for defining probabilistic design fires. *Proceedings, 4th International Maritime Conference on Design For Safety, October 2010.* Trieste.

Vassalos, D., Spyrou, K., Themelis N. & Mermiris, G. 2010. Risk- based design for fire safety—A generic framework. *Proceedings, 4th International Maritime Conference on Design for Safety, October 2010*, Trieste.

Sustainable Maritime Transportation and Exploitation of Sea Resources – Rizzuto & Guedes Soares (eds)
© 2012 Taylor & Francis Group, London, ISBN 978-0-415-62081-9

Detections of potential collision situations by relative motions of vessels under parameter uncertainties

Lokukaluge P. Perera & C. Guedes Soares
Centre for Marine Technology and Engineering (CENTEC), Instituto Superior Técnico, Technical University of Lisbon, Lisbon, Portugal

ABSTRACT: The detection of potential collision situations by relative motions of vessels under parameter uncertainties in vessel manoeuvring is presented in this study. The detection process consists of the observations of the relative navigation trajectory and course-speed vector between two vessels. The proposed detection process is developed as a part of the intelligent navigation system that makes decisions under multi-vessel collision situations. A two vessel collision situation is considered and the extended Kalman filter algorithm is used in this study to estimate the relative navigational trajectory as well as the relative course-speed vector. Finally, prior and posterior collision/near-collision situations are simulated and successful simulation results on the detection of potential collision situations are also presented in this paper.

1 INTRODUCTION

The detections of collision situations are important facilities of transportation systems to improve the safety and security in navigation. However, collision situations could be simplified by assuming that the targets are moving in straight line motions and states/parameters conditions are deterministic. Even though, land and air transportation systems could satisfy these assumptions, maritime transportation systems are often involved with maneuvering trajectories and stochastic state/parameter situations under varying sea conditions.

Furthermore, the navigation constraints and routing schemes in maritime transportation have enforced vessels to execute close quarter navigation, which increases the risk of collisions (Robson, 2006). Therefore, the detections of collision situations under maritime transportation will be a complicated process that needs advanced tools and technologies.

This study proposes a methodology to detect collision and near collision situations by estimating the relative navigation trajectory and the relative course-speed vector between two vessels. Furthermore, the vessels' navigation under maneuvering and stochastic states/parameter conditions is considered. Even though, this study is limited to a two vessels collision situation, this concept can be developed for a multi-vessel collision situation by accumulating multiple two vessel collision situations as proposed by Perera et al. (2011a).

The estimated relative navigation trajectory and relative course-speed vector can use as an evaluation mechanism prior to collisions or near collision situations in maritime transportation. The proposed methodology (i.e. detections of collision situations) is a part of the intelligent navigation system (INS) that is presented in Figure 1 and further described in section 2.

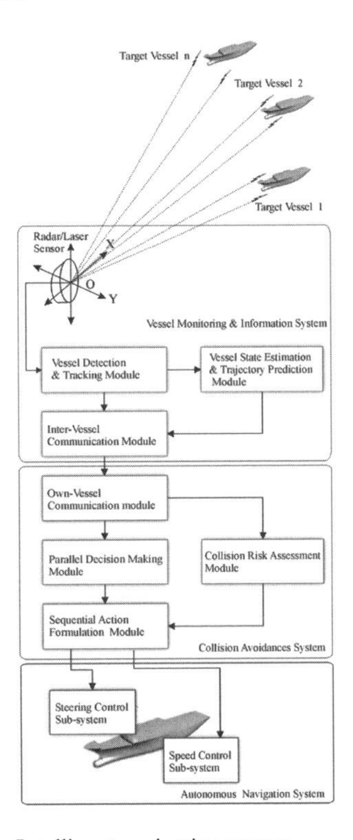

Figure 1. Intelligent navigation system.

1.1 *Collision in maritime transportation*

Human errors are still one of the major causes of maritime accidents (Guedes Soares & Teixeira, 2001) and 75–95% of marine accidents and causalities are caused by some types of human errors (Rothblum et al., 2002, and Antão & Guedes Soares 2008) in accordance with the reported data. Therefore, as illustrated by e-navigation (eNAV 2010), the accumulation of intelligent decision making capabilities into navigational systems will limit the human subjective factors in navigation, which can increase the safety and security of maritime transportation.

The proposed e-navigation concept can be formulated as a collaborated network of traffic information among vessels and shore based stations to improve safety and security in maritime transportation. Furthermore, the e-navigation can decrease navigational errors, increase awareness of vessel situations, improve traffic monitoring facilities, and reduce transportation costs. (Ward and Leighton, 2010).

1.2 *Safety measures and risk assessments*

The safety measures of maritime transportation were influenced by several groups (Wang et al., 2006): ship designer, ship operators and maritime societies. The ship designers influence by safe design of bridge layout, navigational equipments, engine and steering control, maneuverability, and redundancy. The ship operators influence by safe operation of ship speed, manning levels, crew attitude and training, and maintenance. The maritime societies influence by safe aiding and monitoring of vessel traffic systems, pilots, traffic lanes, aids to navigation (i.e. AIS, GPS) and safety inspection procedures.

However, the effectiveness of maritime safety measures are eventually evaluated under rigorous navigation and collision conditions with respect to the vessel operator's decisions. Therefore, the best onboard navigation tools (i.e. intelligent systems and sensors) should be available to influence the ship operation to make better decisions that improve the safety and security conditions within maritime transportation.

The analysis of vessel navigation information will help to detect collision situations and to assess collision risk. The collision risk should be evaluated in real-time by vessels and/or Vessel Traffic Monitoring and Information Systems (VTMIS) in order to guarantee safety and security measures in maritime transportation. As illustrated by Imazu (2006), the mathematical formulation of collision detection between two vessels can be divided in two methods: Closest Point Approach method (CPA) that is a two dimensional method (2D) and Predicted Area of Danger method (PAD) that is a three dimensional method (3D).

The CPA method consists of calculating the shortest distance between two vessels and assessing the collision risk that could be predicted with respect to each vessel domain. However, the CPA method alone cannot be implemented in the evaluation process of collision risk, since it does not consider the vessel size, course and speed variations. An extensive study of the CPA method with respect to a two vessel collision situation is presented by Kwik (1989).

The PAD method consists of modeling one vessel possible trajectories as an inverted cone and the other vessel trajectory as an inverted cylinder, being the region of both object intersections categorized into the Predicted Area of Danger. Both vessels' size, course and speed conditions could be integrated into the geometry of the objects of navigational trajectories in this study.

However, both studies are limited to constant parameter conditions (i.e. fixed vessel's speed and course conditions) that may not always be realistic in maritime transportation. Therefore, a novel method to detect potential collision situations with the parameter uncertainties in maritime transportation (i.e. variation in vessel speed and course conditions) is proposed in this study.

1.3 *Collision risk assessment*

This study formulates a methodology to detect potential collision situations, while vessels are maneuvering in close proximity. The proposed detection process consists of the derivation of relative navigation trajectory and course-speed vector between two vessels that could use to evaluate prior collision/near collision conditions.

Even though, in this study, the collision detection process is derived with respect to a two vessel collision situation that can be developed for a multi-vessel collisions situation by the accumulation of two vessel collision situations. The proposed collision detection process consists of following steps; the observation of both vessels' positions; the estimation of both vessels' velocities, accelerations and navigational trajectories; the calculation of the vessel relative navigational trajectory and relative course-speed vector of a selected vessel with respect to other vessel.

In general, the vessel navigators monitor collision situations by observing the relative bearing of other vessels in open sea; the unchanged relative bearing of a vessel could lead to a collision situation. However, this requirement alone could not predict accurate collision conditions and should not be used in the decision making process under complex navigational conditions; that involve multiple vessels.

Therefore, the observation of relative navigation trajectory and relative course-speed vector of the other vessel could use to improve the detection of collision situations. The relative navigation trajectory could illustrate as a conventional bearing observation situation. However, the relative course-speed vector of the other vessel can be used as an additional tool that could improve the collision detection process.

It is assumed that both vessels' positions are measured by conventional AIS and GPS systems. However, there are many challenges faced by the systems during its position measurements: The first, the AIS and GPS position signals can be associated with sensor noise and/or system errors, therefore the measurements accuracy would be compromised. The second, the vessels are maneuvering under varying sea conditions; the own and target vessel kinematics and dynamics could be associated with time-varying parameter conditions. Therefore, these conditions are identified as parameter uncertainties that have been illustrated in this study.

Hence, a proper mechanism to identify the vessel states (i.e. position, velocity and acceleration) is considered. The extended Kalman filter, one of the well known estimation algorithms, to overcome previous challenges and to estimate accurate vessel states is proposed. One should note that the state estimation based only on both vessels' position measurements is another advantage in this approach.

The main contribution of this study can be summarized as the estimation of vessel's relative navigation trajectory and course-speed vector based on parameter uncertainties in vessel maneuvering that can be used to detect potential collision situations among vessels. The organization of this paper is as follows. Section 2 contains an overview of the INS. Section 3 contains the mathematical formulation of detection of collision situations. Computational simulations are presented in Section 4. Finally, the conclusion is presented in section 5.

2 INTELLIGENT NAVIGATION SYSTEM

The proposed INS that is designed to accumulate intelligent e-navigation facilities into maritime transportation is presented in Figure 1. As indicated in the figure, the system consists of three main sub-systems: Vessel Monitoring & Information System (VMIS), Collision Avoidance System (CAS), and Autonomous Navigation System (ANS).

The main objective of the VMIS is to facilitate the INS by vessel traffic information that consists of vessels' position, course, speed, acceleration and trajectory conditions. The system consists of a scan sensor (i.e. Radar/Laser Sensor) and three main modules: Vessel Detection & Tracking (VDT) Module, Vessel State Estimation and Trajectory Prediction (VSETP) Module and Inter Vessel Communication (IVC) Module.

A Radar/Laser sensor is used as a target detection unit in the VMIS. The VDT module consists of an Artificial Neural Network (ANN) based multi-vessel detection and tracking process. The main objective of the VDT module is to detect and to track vessels that are represented by clusters of data points that have been generated by the Radar/Laser sensor.

The VSETP module consists of an Extended Kalman filter (EKF) based vessel state estimator (i.e. position, velocity and acceleration) and navigational trajectory prediction process. This process is executed by information given by the VDT module. Furthermore, each vessel state conditions (i.e. position, course, speed, etc.) will transfer from the IVC module to the respective vessel OVC module through a wireless network.

The proposed CAS is presented in Figure 1. The main objective of the CAS module is to general collision avoidance decisions/actions in a sequential format that could be executed during vessel navigation. As presented in the figure, the CAS consists of the following modules: Own Vessel Communication (OVC) Module, Parallel Decision Making (PDM) Module, Sequential Action Formation (SAF) module, and Collision Risk Assessment (CRS) Module.

The OVC module is the communication unit between the vessel and the VMIS as mentioned previously. The PDM module consists of a Fuzzy logic based decision making process that generates parallel collision avoidance decisions with respect to each vessel that is under collision risk. The inputs to the PDM module are the range, bearing, relative course and relative speed of the other vessel. The outputs from the PDM module are course and speed change decisions of the vessel. The inputs and outputs are formulated as fuzzy membership functions. The Convention on the International Regulations for Preventing Collisions at Sea (COLREGs) rules and regulations and expert navigational knowledge are considered for the devolvement of Fuzzy rules.

The main objective of the CRA module is to evaluate the collision risk and the expected time until collision of each target vessel with respect to vessel navigation. The tools developed in this study, the relative navigation trajectories and the course-speed vectors, will use in the CRA module to improve the system capabilities.

Furthermore, the CRA module will transfer collision risk information to the SAF module for collision avoidance actions. The main objective of the SAF module is to organize the parallel decisions that were formulated by the PDM module into sequential actions, considering the time until collisions that were formulated by the CRA module. Furthermore, these actions will be executed on vessel navigation.

Finally, the collision avoidance actions formulated by the SAF module will be transferred into the ANS. These actions are further divided into two categories of course and speed controls that will implement on vessel navigation.

The main objective of the ANS is to control the vessel course and speed conditions with respect to collision situations. The proposed ANS is associated with a decentralized control approach where the two control sub-systems are introduced: Steering Control Sub-system (SCS), and Speed Control Sub-system (SPS). The main objective of the SCS is to control the vessel course conditions. The main objective of the SPS is to control the vessel speed conditions. Therefore, the proposed INS is capable of handling multi-vessel collision situations under complex collision conditions. Further details on the INS can be found on Perera et al. (2010a,b, 2011a,b).

3 DETECTIONS OF COLLISION SITUATIONS

The mathematical formulation of detection of collision situations is presented in this section. Therefore, the section is divided into three sections of derivation of system model, formulation of measurement model and Extended Kalman filter. In the system model section, a mathematical model for a two vessel collision situation is derived. In the measurement model section, the observations of available vessel states are formulated. In the extended Kalman filter section, the procedure for the estimation of relative vessel navigation trajectory and course-speed vector is presented.

3.1 *Two vessel collision situation*

A two vessel collision situation is presented in Figure 2. The own vessel, the vessel that is equipped with the INS, is located in point O (x_o, y_o). The target vessel, the vessel that needs to be avoided, is located at point A (x_a, y_a). The own vessel speed and course conditions are represented by V_o and χ_o respectively. The target vessel speed and course conditions are represented by V_a and χ_a respectively. The own and target vessels' instantaneous radius of curvature of maneuvering are presented

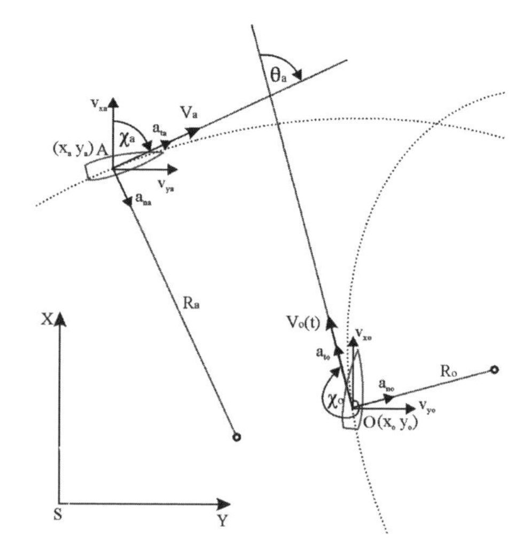

Figure 2. Two vessel collision situation.

by R_o and R_a. The x and y velocity components of the own and target vessels are presented by v_{xo}, v_{yo}, v_{xa} and v_{ya} respectively. The own and target vessels' normal and tangential acceleration components are presented by a_{no}, a_{to}, a_{na} and a_{ta} respectively. The collision encounter angle between vessels is presented by θ_a.

To capture the maneuvering conditions in both vessels a suitable mathematical model is considered. The continuous-time curvilinear motion model that could be formulated for ocean vessel navigation is illustrated in this study. The standard continuous-time curvilinear motion model for the own vessel can be written as:

$$\begin{aligned} \dot{\chi}_o &= \frac{a_{no}}{V_o}; \qquad \dot{V}_o = a_{to} \\ \dot{x}_o &= v_{xo} = V_o \cos\chi_o; \qquad \dot{y}_o = v_{yo} = V_o \sin\chi_o \end{aligned} \tag{1}$$

The standard continuous-time curvilinear motion model for the target vessel can be written as:

$$\begin{aligned} \dot{\chi}_a &= \frac{a_{na}}{V_a}; \qquad \dot{V}_a = a_{ta} \\ \dot{x}_a &= v_{xa} = V_a \cos\chi_a; \qquad \dot{y}_a = v_{ya} = V_a \sin\chi_a \end{aligned} \tag{2}$$

To avoid trigonometrical angle conditions, the following functions are proposed:

$$f^{vyo} = \sin\chi_o = \frac{v_{yo}}{\sqrt{v_{xo}^2 + v_{yo}^2}}; \; f^{vxo} = \cos\chi_o = \frac{v_{xo}}{\sqrt{v_{xo}^2 + v_{yo}^2}}$$

$$f^{vya} = \sin\chi_a = \frac{v_{ya}}{\sqrt{v_{xa}^2 + v_{ya}^2}}; \; f^{vxa} = \cos\chi_a = \frac{v_{xa}}{\sqrt{v_{xa}^2 + v_{ya}^2}}$$

3.2 *System model*

The own and target vessels' continuous-time curvilinear motion models presented in Equations (1) and (2) can be summarized into a system model and that can be written as:

$$\dot{x} = f(x) + w_x \tag{3}$$

where the system states, x, and the system function, $f(x)$, can be further illustrated as:

$$x = \begin{bmatrix} x_o \\ y_o \\ v_{xo} \\ v_{yo} \\ a_{to} \\ a_{no} \\ x_a \\ y_a \\ v_{xa} \\ v_{ya} \\ a_{ta} \\ a_{na} \end{bmatrix}; \qquad f(x) = \begin{bmatrix} v_{xo} \\ v_{yo} \\ a_{to} f^{vxo} - a_{no} f^{vyo} \\ a_{to} f^{vyo} + a_{no} f^{vxo} \\ 0 \\ 0 \\ v_{xa} \\ v_{ya} \\ a_{ta} f^{vxa} - a_{na} f^{vya} \\ a_{ta} f^{vya} + a_{na} f^{vxa} \\ 0 \\ 0 \end{bmatrix}$$

where w_x is white Gaussian process noise with 0 mean and Q covariance. The covariance, Q, can be further written as:

$$Q = diag\begin{bmatrix} Q_{xo} & Q_{yo} & Q_{vxo} & Q_{vyo} & Q_{ato} & Q_{ano} \\ Q_{xa} & Q_{ya} & Q_{vxa} & Q_{vya} & Q_{ata} & Q_{ana} \end{bmatrix}$$

where Q_{xo}, Q_{yo}, Q_{vxo}, Q_{vyo}, Q_{ato}, Q_{ano}, Q_{xa}, Q_{ya}, Q_{vxa}, Q_{vya}, Q_{ata} and Q_{ana} are respective system state covariance values. Furthermore, the own and target vessels tangential and normal acceleration components are presented by a_{to}, a_{no}, a_{ta}, and a_{na} respectively. The respective acceleration derivates can be written as:

$$\dot{a}_{to} = w_{ato} \; ; \dot{a}_{no} = w_{ano}$$
$$\dot{a}_{ta} = w_{ata} \; ; \dot{a}_{na} = w_{ana}$$

where w_{ato}, w_{ano}, w_{ata} and w_{nta} are derivates of tangential and normal accelerations of the own and target vessels that are formulated as white Gaussian distributions with 0 mean and Q_{ato}, Q_{ano}, Q_{ata}, and Q_{ana} respective covariance values.

The Jacobian of the system function, $f(x)$, can be written as:

$$\frac{\partial}{\partial \mathrm{x}}(f(\mathrm{x})) =$$

$$\begin{bmatrix}
0 & 0 & 1 & 0 & 0 & 0 & 0 & 0 & 0 & 0 & 0 & 0 \\
0 & 0 & 0 & 1 & 0 & 0 & 0 & 0 & 0 & 0 & 0 & 0 \\
0 & 0 & a_{to} f^{vxo}_{vxo} - a_{no} f^{vyo}_{vxo} & a_{to} f^{vxo}_{vyo} - a_{no} f^{vyo}_{vyo} & f^{vxo} & -f^{vyo} & 0 & 0 & 0 & 0 & 0 & 0 \\
0 & 0 & a_{to} f^{vyo}_{vxo} + a_{no} f^{vxo}_{vxo} & a_{to} f^{vyo}_{vyo} + a_{no} f^{vxo}_{vyo} & f^{vyo} & f^{vxo} & 0 & 0 & 0 & 0 & 0 & 0 \\
0 & 0 & 0 & 0 & 0 & 0 & 0 & 0 & 0 & 0 & 0 & 0 \\
0 & 0 & 0 & 0 & 0 & 0 & 0 & 0 & 0 & 0 & 0 & 0 \\
0 & 0 & 0 & 0 & 0 & 0 & 0 & 0 & 1 & 0 & 0 & 0 \\
0 & 0 & 0 & 0 & 0 & 0 & 0 & 0 & 0 & 1 & 0 & 0 \\
0 & 0 & 0 & 0 & 0 & 0 & 0 & 0 & a_{ta} f^{vxa}_{vxa} - a_{na} f^{vya}_{vxa} & a_{ta} f^{vxa}_{vya} - a_{na} f^{vya}_{vya} & f^{vxa} & -f^{vya} \\
0 & 0 & 0 & 0 & 0 & 0 & 0 & 0 & a_{ta} f^{vya}_{vxa} + a_{na} f^{vxa}_{vxa} & a_{ta} f^{vya}_{vya} + a_{na} f^{vxa}_{vya} & f^{vya} & f^{vxa} \\
0 & 0 & 0 & 0 & 0 & 0 & 0 & 0 & 0 & 0 & 0 & 0 \\
0 & 0 & 0 & 0 & 0 & 0 & 0 & 0 & 0 & 0 & 0 & 0
\end{bmatrix}$$

where the respective functions are derived as:

$$f^{vxo}_{vxo} = \frac{v_{yo}^2}{\left(v_{xo}^2 + v_{yo}^2\right)^{3/2}}, \qquad f^{vxo}_{vyo} = -\frac{v_{yo} v_{xo}}{\left(v_{xo}^2 + v_{yo}^2\right)^{3/2}},$$

$$f^{vyo}_{vxo} = -\frac{v_{xo} v_{yo}}{\left(v_{xo}^2 + v_{yo}^2\right)^{3/2}}, \qquad f^{vyo}_{vyo} = \frac{v_{xo}^2}{\left(v_{xo}^2 + v_{yo}^2\right)^{3/2}}$$

$$f^{vxa}_{vxa} = \frac{v_{ya}^2}{\left(v_{xa}^2 + v_{ya}^2\right)^{3/2}}, \qquad f^{vxa}_{vya} = -\frac{v_{ya} v_{xa}}{\left(v_{xa}^2 + v_{ya}^2\right)^{3/2}},$$

$$f^{vya}_{vxa} = -\frac{v_{xa} v_{ya}}{\left(v_{xa}^2 + v_{ya}^2\right)^{3/2}}, \qquad f^{vya}_{vya} = \frac{v_{xa}^2}{\left(v_{xa}^2 + v_{ya}^2\right)^{3/2}}$$

3.3 *Measurement model*

The measurement model is formulated to measure the own and target vessel actual positions. The position measurements in discrete-time are considered due to the availability of own and target vessels' positions in discrete time instants. It is assumed that the correlations between vessel position measurements are negligible. The own and target vessels' position measurement model can be written as:

$$z(k) = h(x(k)) + w_z(k) \tag{4}$$

where the system states, $z(k)$, and the function, $h(x(k))$, can be further illustrated as:

$$z(k) = \begin{bmatrix} z_{xo}(k) & z_{yo}(k) & z_{xa}(k) & z_{ya}(k) \end{bmatrix}^T;$$

$$h(x(k)) = \begin{bmatrix} x_o(k) \\ y_o(k) \\ x_a(k) \\ y_a(k) \end{bmatrix}$$

where $z_{xo}(k)$, $z_{yo}(k)$, $z_{xa}(k)$ and $z_{ya}(k)$ are the own and target vessel x and y position measurements respectively, and $w_z(k)$ is white Gaussian measurement noise with 0 mean and covariance $R(k)$. The covariance, $R(k)$, can be further illustrated as:

$$R(k) = diag\begin{bmatrix} R_{xo}(k) & R_{yo}(k) & R_{xa}(k) & R_{ya}(k) \end{bmatrix}$$

where $R_{xo}(k)$, $R_{yo}(k)$, $R_{xa}(k)$, and $R_{ya}(k)$ are presented by the respective system measurement covariance values. The Jacobian matrix of the measurement function, $h(z(k))$, can be written as:

$$\frac{\partial}{\partial x}(h(x(k))) = \begin{bmatrix} 1 & 0 & 0 & 0 & 0 & 0 & 0 & 0 & 0 & 0 & 0 & 0 \\ 0 & 1 & 0 & 0 & 0 & 0 & 0 & 0 & 0 & 0 & 0 & 0 \\ 0 & 0 & 0 & 0 & 0 & 0 & 1 & 0 & 0 & 0 & 0 & 0 \\ 0 & 0 & 0 & 0 & 0 & 0 & 0 & 1 & 0 & 0 & 0 & 0 \end{bmatrix}$$

3.4 *Extended kalman filter*

The Kalman filter is a well known estimation algorithm. However, the Kalman Filter (KF) general algorithm is limited for application of linear systems; therefore the Extended Kalman Filter (EKF) is considered as a standard technique that could be used for a number of non-linear estimation applications. The Extended Kalman filter is proposed in this study as the estimation algorithm for the own and target vessels' states (i.e. position, velocity and acceleration). The estimated vessel states are used to calculate the relative navigation trajectory and course-speed vector of the target vessels.

In general, the own and target vessel positions are measured as noisy position values; therefore, the estimation algorithm is used to increase the position accuracy. In some situations, the own vessel acceleration conditions can also be measured and can be used to improve the state estimation. However, this study is limited to the vessel positions measurements. The EKF algorithm (Gelb et al., 2000) can be summarized as:

- System Model

$$\begin{aligned} &\dot{x}(t) = f(x(t)) + w_x(t) \\ &E[w_x(t)] = 0, \quad E[w_x(t); w_x(t)] = [Q(t)] \end{aligned} \tag{5}$$

- Measurement Model

$$\begin{aligned} &z(k) = h(x(k)) + w_z(k), \quad k = 1, 2, \ldots \\ &E[w_z(k)] = 0, \quad E[w_z(k); w_z(k)] = [R(k)] \end{aligned} \tag{6}$$

- Error Conditions

$$\tilde{x}(k) = \hat{x}(k) - x(k) \tag{7}$$

where $\tilde{x}(t)$ is the state error vector and $\hat{x}(t)$ is the estimated states of the system.

- System Initial States

$$x(0) \sim N\hat{x}(0), P(0) \tag{8}$$

where $P(0)$ is the state initial covariance, describing the uncertainty present on the initial estimates.

- Other Conditions

$$E[\mathrm{w_x}(t); \mathrm{w_z}(k)] = 0 \quad for\ all\ k,t \tag{9}$$

- State Estimation Propagation

$$\frac{d}{dt}\hat{\mathrm{x}}(k) = f(\hat{\mathrm{x}}(k)) \tag{10}$$

- Error Covariance Extrapolation

$$\begin{aligned} &\frac{d}{dt}P(k) = F(\hat{x}(k))P(k) + P(k)F^T(\hat{x}(k)) + Q(k) \\ &F(\hat{x}(k)) = \frac{\partial}{\partial x(k)} f(x(k))\Big|_{x(k)=\hat{x}(k)} \end{aligned} \tag{11}$$

where $P(k)$ is the estimated error covariance with

$$\begin{aligned} P(k) = diag\Big[& P_{xo}(k) \quad P_{yo}(k) \quad P_{vxo}(k) \quad P_{vyo}(k) \\ & P_{ato}(k) \quad P_{ano}(k) \quad P_{xa}(k) \quad P_{ya}(k) \\ & P_{vxa}(k) \quad P_{vya}(k) \quad P_{ata}(k) \quad P_{ana}(k) \Big] \end{aligned}$$

where $P_x(k)$, $P_{vx}(k)$, $P_y(k)$, $P_{vy}(k)$, $P_{at}(k)$ and $P_{an}(k)$ are respective estimated state error covariance values.

- State Estimate Update

When the measurement data is available from the sensors, the system state can be estimated as:

$$\hat{x}(k^+) = \hat{x}(k^-) + K(k)\left[z(k) - h_k\hat{x}((k^-))\right] \tag{12}$$

where $x(k^-)$ and $x(k^+)$ are the prior and posterior estimated system states respectively, and $K(k)$ is the Kalman gain.

- Error Covariance Update

$$P(k^+) = \left[1 - K(k)H(\hat{x}(k^-))\right]\; P(k^-) \tag{13}$$

where $P(k^-)$ and $P(k^+)$ are the prior and posterior error covariance values of the system respectively.

$$H(\hat{x}(k)) = \frac{\partial}{\partial x(k)} h(x(k))\bigg|_{x(k)=\hat{x}(k)}$$

- Kalman Filter Gain

$$K(k) = P(k^-)H(\hat{x}(k^-)) \left[H(\hat{x}(k^-))P(k^-)H(\hat{x}(k^-))^T + R(k)\right]^{-1} \tag{14}$$

3.5 *Relative trajectory and course-speed vector*

The relative trajectory of the target vessel can be calculated by the target vessel's relative x and y positions that can be written as:

$$\hat{x}_{ao} = \hat{x}_a - \hat{x}_o \tag{15}$$

$$\hat{y}_{ao} = \hat{y}_a - \hat{y}_o \tag{16}$$

where $\hat{x}_{ao}$ and $\hat{y}_{ao}$ are estimated relative x and y positions of the target vessel. The relative course-speed vector of target vessel is calculated by the relative x and y velocity components and can be written as:

$$\hat{v}_{xao} = \hat{v}_{xa} - \hat{v}_{xo} \tag{17}$$

$$\hat{v}_{yao} = \hat{v}_{ya} - \hat{v}_{yo} \tag{18}$$

where $\hat{v}_{xao}$ and $\hat{v}_{yao}$ represent the estimated x and y relative velocity components of target vessel respectively. Hence, the target vessel's estimated x and y relative velocity components can be used for the calculations of relative course-speed vector.

4 COMPUTATIONAL SIMULATIONS

A computational simulation of a two vessel near collision situation is presented in Figure 3. The top plot of Figure 3 represents the own and target vessels' actual (Act.), measured (Mea.) and estimated (Est.) navigation trajectories. The vessel position measurements are generated by adding sensor noise into the actual trajectory. As presented in the figure, a near collision situation can be observed. A zoomed view of the same trajectories near the collision point is presented in the top plot of Figure 4. The zoomed view of the relative trajectory of the same situation is presented in the bottom plot of Figure 4.

One should note that the intersection of the two trajectories will not necessarily represent a collision point because each vessel can pass the collision point in different time intervals. However, this confusion can be clarified by the observation of the relative navigational trajectories; where the relative trajectory of target vessel propagation near the own vessel initial position should be in a near collision situation. These conditions are further illustrated in Figures 3 and 4.

Considering the respective zoomed view near the own vessel position in the bottom plot of Figure 4, each estimated position of the target vessel consists of the relative course-speed vector

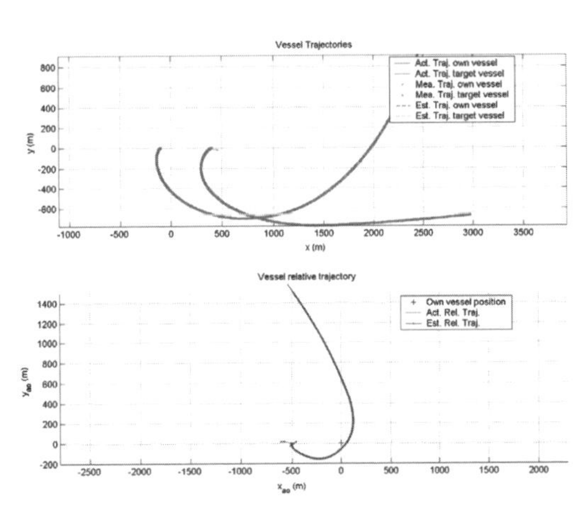

Figure 3. Two vessels near collision situation.

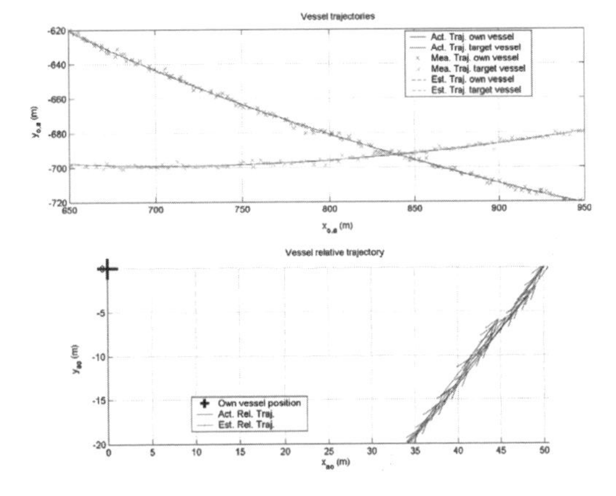

Figure 4. Two vessels collision situation (zoomed view).

that is presented by an arrow and estimated by the extended Kalman filter. This representation is an important factor in the detection of collision conditions even under near collision situations.

The own and target vessels' velocity and relative velocity components that are estimated by the EKF are presented on Figure 5. The own and target vessels' acceleration components that are also estimated by the EKF are presented in Figure 6. These estimated states can be further used for the decision making process of collision avoidance among vessels.

This detection of collision situation is further elaborated in Figure 7, where a prior collision situation is presented. The navigational trajectories of both vessels are presented in the top plot of Figure 7.

The zoomed view of the prior collision situation is presented in the bottom plot of Figure 7. As presented in the plot, the target vessel relative trajectory is heading towards the own vessel initial position with course-speed vectors that are pointed towards the own vessel. Therefore, the target vessel relative navigational trajectory and course-speed vectors can be used for prior collision situations in order to detect the collision risk as proposed in this study.

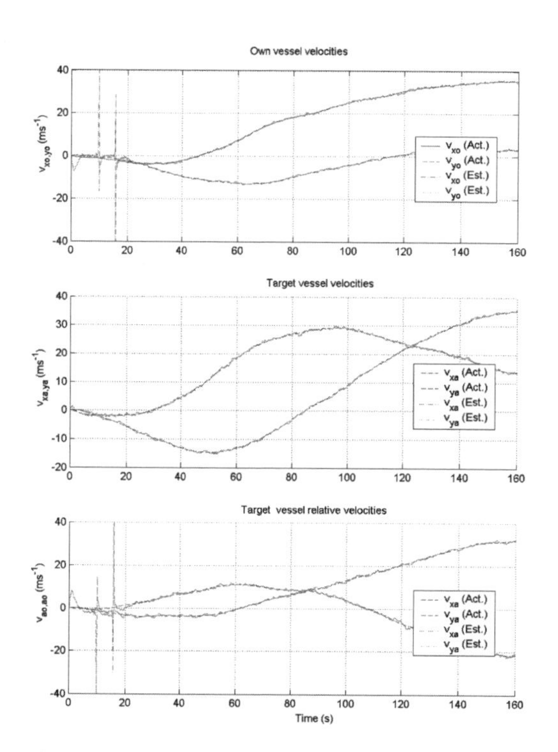

Figure 5. Velocity estimation.

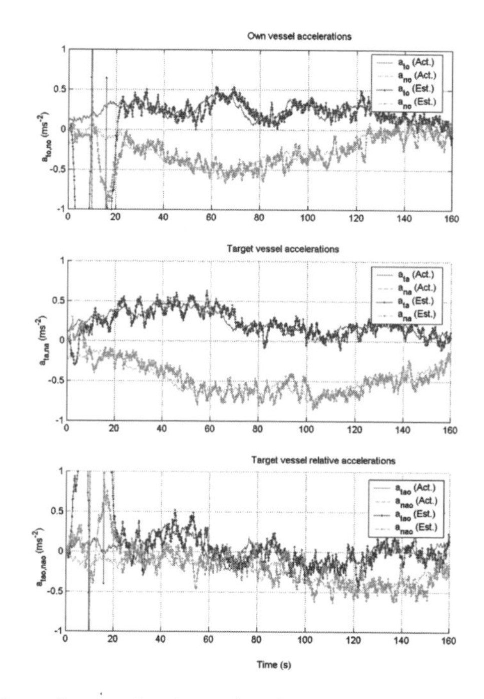

Figure 6. Acceleration estimation.

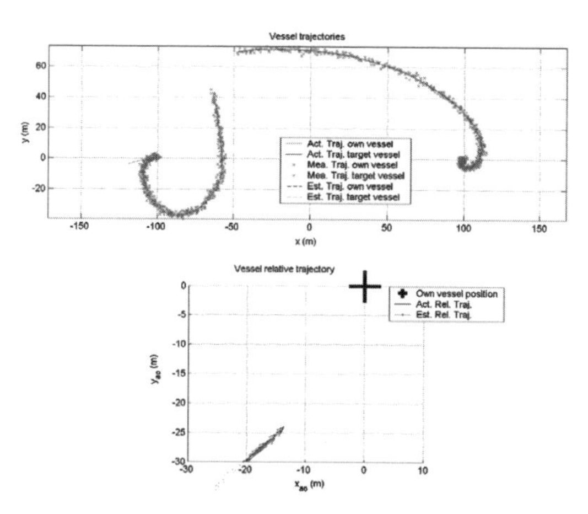

Figure 7. Two vessel prior collision situation.

5 CONCLUSION

A methodology for detection of collision situations that is based on uncertain parameters in vessel maneuvering is presented in this study. As presented in the simulations, the target vessels' relative trajectory as well as the relative course-speed vector could be used for the assessment of prior collision situations. The proposed collision detection process will be used on the INS that is previously described in this paper.

ACKNOWLEDGEMENTS

The first author has been supported by the Doctoral Fellowship of the Portuguese Foundation for Science and Technology (Fundação para a Ciência e a Tecnologia) under contract n.° SFRH/BD/46270/2008. Furthermore, this work contributes to the project on the "Methodology for ships manoeuvrability tests with self-propelled models", which is being funded by the Portuguese Foundation for Science and Technology under contract n.° PTDC/TRA /74332/2006.

REFERENCES

Antão, P. and Guedes Soares, C. 2008, "Causal factors in accidents of high speed craft and conventional ocean going vessels," *Reliability Engineering and System Safety*, vol. 93, pp. 1292–1304.

Cockcroft, A.N. and Lameijer, J.N.F. 2001. A Guide to The Collision Avoidance Rules. Elsevier Butterworth-Heinemann, Burlington, MA. USA.

eNAV, e-navigation 2008. URL http://www.enavigation.org/

Gelb, A., Kasper, Jr., J.F., Nash, Jr., R.A., Price, C.F. and Sutherland, Jr., A.A. 2001. Applied Optimal Estimation. The MIT Press, MA. USA.

Guedes Soares, C. and Teixeira, A.P. 2001, Risk Assessment in Maritime Transportation. *Reliability Engineering and System Safety*, Vol. 74, pp. 299–309.

Imazu, H. 2006. Advanced topics for marine technology and logistics. Lecture Notes on Ship collision and integrated information system, Tokyo University of Marine Science and Technology, Tokyo, Japan.

Kwik, K.H. 1989. Calculations of ship collision avoidance manoeuvres: A simplified approach. Ocean Engineering 16 (5/6), 475–491.

Perera, L.P., Carvalho, J.P. and Guedes Soares, C. 2010a. Bayesian network based sequential collision avoidance action execution for an ocean navigational system. In: In Proc. 8th IFAC Conference on Control Applications in Marine Systems. Rostock, Germany, pp. 301–306.

Perera, L.P., Carvalho, J.P. and Guedes Soares, C. 2010b. Fuzzy-logic based parallel collisions avoidance decision formulation for an ocean navigational system. In: In Proc. 8th IFAC Conference on Control Applications in Marine Systems. Rostock, Germany, pp. 295–300.

Perera, L.P., Carvalho, J.P. and Guedes Soares, C. 2011a. Intelligent Collision Avoidance Facilities for Maritime Transportation. In: Proceedings of the 1st International Conference on Maritime Technology and Engineering (MARTECH2011). Lisbon, Portugal.

Perera, L., Carvalho, J.P. and Guedes Soares, C. 2011b. Fuzzy-logic based decision making system for collision avoidance of ocean navigation under critical collision conditions. Journal of Marine Science and Technology 16, 84–99.

Robson, J.K. 2006. Overview of collision detection in the UKCS. Research Report (RR514), AEA Technology PLC, Oxon, UK.

Rothblum, A.M., Wheal, D.,Withington, S., Shappell, S.A.,Wiegmann, D.A., Boehm,W. and Chaderjian, M. April 2002. Key to successful incident inquiry. In: 2nd International Workshop on Human Factors in Offshore Operations (HFW).

Wang, G., Ji, C., Kujala, P., Lee, S.G., Marino, A., Sirkar, J., Pedersen, P.T., Vredeveldt, A.W. and Yuriy, V. 2006. Collision and grounding. In: 16th International Ship and Offshore Structures Congress, Committee V. 1 Report, Vol. 2, Southapton, UK.

Ward, N. and Leighton, S. 2010. Collision avoidance in the e-navigation environment. In: 17th Conference of the International Association of Marine Aids to Navigation and Lighthouse Authorities. Cape Town, South Africa.

Sustainable Maritime Transportation and Exploitation of Sea Resources – Rizzuto & Guedes Soares (eds)
 ISBN 978-0-415-62081-9

Assessment of safety of ships after the collision and during the ship salvage using the matrix type risk model and uncertainties

M. Gerigk
Gdansk University of Technology, Gdansk, Poland

ABSTRACT: The paper concerns the safety of seaborne transportation and it is directly devoted to safety of ships in damaged conditions. There is a lack of methods enabling an assessment of safety of ships after the collision and during the ships salvage. The existing methods can be based on the expert assessment using the brain storming techniques. Then the safety measure can be based on the qualitative risk assessment. The another possibility can be to use the methods of safety assessment of ships in damage conditions which are based on the requirements included in the SOLAS 2009 regulations. These methods are of prescriptive character and they are more devoted to solve the ship design problems than those existing in operation. Therefore this is very difficult to apply these methods directly to assess the safety of ships in damaged conditions after the collision and for the ship salvage purposes. There is a need to introduce a novel method for assessment of safety of ships in damaged conditions during the accident at sea. Such a method based on the damaged ship performance and risk assessment is briefly presented in the paper. The method is based on implementation of the system approach to safety, performance-oriented approach and risk-based approach to safety using the elements of the Formal Safety Assessment FSA. The holistic approach to safety assessment has been applied. An influence of design, operational, management related and human factors has been taken into account. The holistic matrix type risk model has been applied to enable to take into account all the possible scenarios of an accident. This multi-level risk model is based on the event tree analysis ETA and enables to consider the accident scenarios consisting of the hazards, intermediate events, additional events and consequences. The measure of safety of ships in damaged conditions during the accident is the risk level. The risk acceptance criteria can be either the risk matrix or ALARP concept. The paper shows how to use the method and some examples are presented.

1 INTRODUCTION

The paper presents a few problems regarding a method of risk and safety assessment of ships in damaged conditions where the key issue concerns the risk model. This model should take into account the uncertainties connected with the possible scenarios of an accident. The scenarios consist of combinations of hazards, intermediate events, additional events and consequences. The research is connected with development of a performance-oriented risk-based method for assessment of safety of ships in damaged conditions, Gerigk (2010). The current methods of assessment of safety of ships in damaged conditions are mainly based on the regulations included in the SOLAS convention (Chapter II-1), (IMO 2005, IMO 2008). These methods are more directed towards solving the design problems than those connected with operation or salvage. It follows from many reasons. First of all, these methods are prescriptive in their character. Generally, these methods are based on the probabilistic approach to safety but in some cases the semi-probabilistic components are included. These methods do not take into account all the possible scenarios of an accident. Application of some of these methods to certain types of ships e.g. car-carriers, ro-ro vessels or passenger ships may lead to insufficient level of safety or provide unnecessary design or operational restrictions. The following methods relay on the regulations included in the SOLAS convention and may be considered for application for the design, operational and salvage purposes.

The first method is based on the holistic risk model for the assessment of safety of ships in damaged conditions which can presented as follows, (Jasionowski et al., 2006, Skjong et al., 2006):

$$R = P_c \times P_{c/f} \times P_{c/f/ns} \times P_{c/f/ns/tts} \times C \quad (1)$$

where P_c = probability of collision (hazard); $P_{c/f}$ = probability of flooding having the ship hit from given direction at data position with given extent conditional on collision; $P_{c/f/ns}$ = probability of not surviving conditional on having flooding when the ship is hit from given direction at data position with given extent conditional on collision; $P_{c/f/ns/tts}$ = probability of given time to sink conditional on not surviving the conditional on having flooding when the ship is hit from given direction at data position with given extent conditional on collision; C = consequences regarding the fatalities, property (cargo, ship) and/or environment.

The second method is based on the casualty threshold, time to capsize and return to port concepts, where the basic ship safety objectives have been divided into three categories, Vassalos (2007):

- category I—vessel remains upright and afloat and is able to return to port under own power (RTP—Return To Port);
- category II—vessel remains upright and afloat but unable to return to port under own power and is waiting for assistance (WFA—Waiting For Assistance);
- category III—vessel likely to capsize/sink and abandonment of the ship may be necessary (AS—Abandonment of the Ship).

The third method is based on a concept of an absolute survivability where the Safe Return to Port (SRtP) attained subdivision index A_{SRtP} should be calculated according to the ship residual stability characteristics, IMO (2009):

$$A_{SRtP} = 0{,}4\ A_{SRtP,s} + 0{,}4\ A_{SRtP,p} + 0{,}2\ A_{SRtP,l} \quad (2)$$

where the subdivision indices $A_{SRtP,s}$, $A_{SRtP,p}$ and $A_{SRtP,l}$ regard the subdivision (s), partial (p) and light ship (l) loading conditions. The $A_{SRtP,s}$, $A_{SRtP,p}$ and $A_{SRtP,l}$ indices should be calculated according to the following formula, IMO (2009):

$$A_{SRtP,lc} = \Sigma\ p_i\ s_{SRtP,i} \quad (3)$$

where lc = loading index (lc = s, p, l); p_i = probability that only the compartment or group of compartments under consideration may be flooded, as defined in regulation 7–1, (IMO 2005, IMO 2008); $s_{SRtP,i}$—probability of survival after flooding the compartment or group of compartments under consideration, in the final stage of flooding only, as defined in literature, IMO (2009).

2 A METHOD OF RISK AND SAFETY ASSESSMENT OF SHIPS IN DAMAGED CONDITIONS

Copy The methodology where the risk-based design and a formalized design methodology were integrated together in the design process with the prevention/reduction of risk as a design objective, along the standard design objectives was introduced by Vassalos and the others, (Skjong et al., 2006, SSRC 2009, Vassalos 2006).

The proposed method is a kind of performance-oriented risk-based procedure which enables the risk and safety assessment at the design stage, in operation or during the ship salvage. Within the method the holistic approach to safety assessment of ships is applied. It is based on application of a risk model which includes all the possible scenarios of events during an accident. The method takes into account an influence of design and operational factors on safety and safety management related factors as well. The method is based on implementation of the system integrated approach to safety, elements of Formal Safety Assessment FSA, ship performance-oriented approach and risk-based approach to safety, Gerigk (2010). For the ship performance evaluation the statistics, investigations using the physical models and numerical simulation techniques can be applied. The ship performance evaluation enables to determine the intermediate events, additional events (releases) and consequences and risk of each accident scenario. The aim is to achieve an adequate level of risk by reducing the risk if necessary. Providing a sufficient level of safety based on the risk assessment is the main objective. It is either the design, operational or organizational objective. Then, safety is not a limitation any more, existing in the regulations. It is just the objective. The measure of safety of a ship in damaged conditions is the level of risk. The proposed method may be used at any ship's life circle, including safety assessment of the ship during a catastrophe. Some elements of the method can be used for safety assessment of different means and systems of seaborne transportation. In the future the method can be useful in elaboration of a new methodology of safety assessment of ships, which bases on the application of risk analysis. The method is based on the following steps: setting the objectives, hazard identification, scenarios development, risk assessment, risk control (prevention, reduction), selection of designs (operational procedures) that meet the objectives. The structure of the method is presented in Figure 1, Gerigk (2010).

The safety case considered in the paper regards the safety of a ship in damaged conditions when the ship skin is damaged due to the following hazards: collision, grounding, stranding or another reason. The risk and safety assessment for a ship in damaged conditions starts from the modeling of the risk contribution tree. For each hazard a separate event tree should be modeled using the Event Tree Analysis ETA.

Generally, sixteen safety functions have been used for each event tree from the ship safety in damaged conditions point of view.

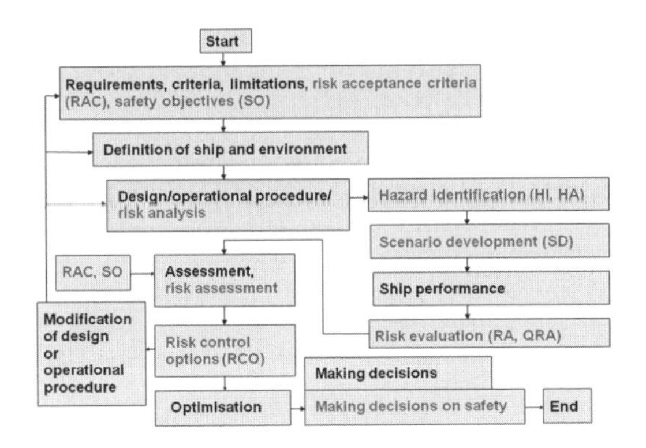

Figure 1. Structure of the method of risk and safety assessment of ships in damaged conditions.

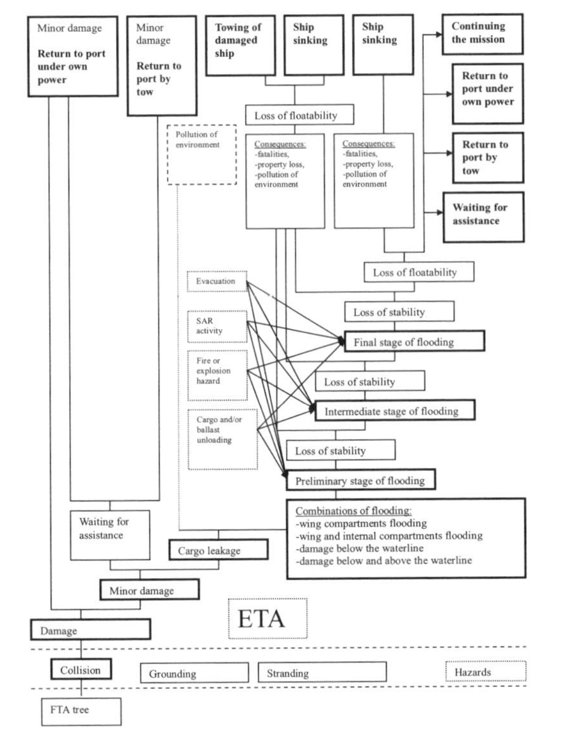

Figure 2. A basic event tree (ETA) for safety assessment of ships in damaged conditions and salvage purposes.

Between them are as follows, Gerigk (2010): function1 (avoiding the hazard), function 2 (hull skin damage (flooding), function 3 (position and extension of damage), function 4 (equalization of the ship heel at the preliminary stage of flooding), function 5 (loss of the ship stability at the preliminary stage of flooding), function 6 (loss of the ship stability during the intermediate stages (and phases) of flooding, function 7 (loss of the ship stability at the final stage of flooding), function 8 (loss of the ship floatability at the final stage of flooding), function 9 (ship is waiting for assistance), function 10 (ship returns to port under own power), function 11 (ship returns to port by tow), function 12 (ship is continuing thc mission), function 13 (mustering and abandonment of the ship (evacuation)), function 14 (SAR action), function 15 (fire and/or explosion), function 16 (emergency cargo unloading (pollution of the environment)). A position of each function can easily be found in Figure 2 where the basic event tree ETA for the safety assessment of ships in damaged conditions and salvage purposes is presented. It is important to notice that each event tree may have a dynamical character.

The holistic approach to ship safety has been applied. According to this approach two major assumptions have been done. First that the system failures can be either the hardware, software, organizational or human failures. The second assumption was that the risk model for assessment of safety of ships in damaged conditions should be the holistic risk model.

3 THE RISK ANALYSIS USING THE MATRIX TYPE RISK MODEL

The risk associated with the different hazards and scenario development was estimated according to the well known general formulae, Gerigk (2010):

$$R_i = P_i \times C_i \tag{4}$$

where P_i = probability of occurrence of a given hazard; C_i = consequences following the occurrence of the data hazard and scenario development, in terms of fatalities, injuries, property losses and damage to the environment.

The risk model (1) may have four different kinds of losses regarding the human fatalities (HF), cargo and ship losses (CS), environment pollution (E) and financial losses ($) and for the ship in damaged conditions can be presented as follows:

$$R = P_c\, P_{c/f}\, PoC_{dam} \times C \tag{5}$$

where P_c = probability of collision (hazard); $P_{c/f}$ = probability of flooding having the ship hit from given direction at data position with given extent conditional on collision; PoC_{dam} = probability of capsizing in damaged conditions in given time conditional on having flooding when the ship is hit from given direction at data position with given extent conditional on collision; C = consequences regarding the fatalities, property (cargo, ship), environment and finance ($C = C_{HF/C}, C_{CS/C}, C_{E/C}, C_{\$/C}$).

The PoC_{dam} probability of capsizing in damaged conditions in given time can be calculated according to the following formulae, Gerigk (2010):

$$PoC_{dam} = P_{c/f/ns} \; P_{c/f/ns/tts} \quad (6)$$

where $P_{c/f/ns}$ = probability of not surviving conditional on having flooding when the ship is hit from given direction at data position with given extent conditional on collision; $P_{c/f/ns/tts}$ = probability of given time to sink conditional on not surviving the conditional on having flooding when the ship is hit from given direction at data position with given extent conditional on collision.

The PoC_{dam} probability can be estimated during the accident at sea using the following methods, Gerigk (2010):

- binary method;
- method based on definition of the probability of surviving a collision;
- method based on definition of the damage stability;
- method based on definition of the ship performance during the accident.

In the case of the last method the roll function in time domain has been anticipated as a major function enabling the risk assessment, Gerigk (2010).

The risk analysis requires to calculate the conditional probabilities regarding the initial events ZI_i, major events (hazards) ZG_j, intermediate events ZP_k and final events ZK_l which can be treated as consequences. The basic mathematical formula are as follows, Gerigk (2010).

First of all the row matrix of initial events is evaluated:

$$P(ZI) = P(ZI_i) \quad \text{for} \quad i = 1 \text{ to } n \quad (7)$$

Then the matrix of major events is calculated:

$$MP_{ZG} = P(ZG_j / ZI_i) \quad \text{for} \quad j = 1 \text{ to } m \quad (8)$$

After that the matrix of intermediate events is calculated:

$$MP_{ZP} = P(ZP_k / ZG_j) \quad \text{for} \quad k = 1 \text{ to } m1 \quad (9)$$

Then the matrix of final events is calculated:

$$MP_{ZK} = P(ZK_l / ZP_k) \text{ dla } l = 1 \text{ do } m2 \quad (10)$$

Finally, the row matrix of final events may be estimated as follows:

$$P(ZK) = P(ZI) \; MP_{ZG} \; MP_{ZP} \; MP_{ZK} \quad (11)$$

Because of the above mathematical model used the entire risk model is called as the matrix type risk model. The risk model enables to consider many possible scenarios of an accident using the event tree like presented in Figure 1. In the case when the additional events occur the $PoC(C_i)$ probability of occurring the C_i given consequences can be calculated according to the formula presented by Gerigk, Gerigk (2010). The typical additional events may concern the water on deck, air cushions, cargo leakage, additional heeling moments or passenger behavior.

4 APPLICATION OF THE METHOD FOR THE SHIP SALVAGE

The ship accidents at sea may lead to the loss of life, loss of properties (ship and cargo) and pollution of environment. The modern approach to safety of seaborne transportation requires to apply the so-called life-circle approach by integrating the design for safety, safe operation and safe salvage methods. Most of the salvage activities have been associated with using the rules of thumb without a support based on the safety assessment process. The effective and safe ship salvage requires to develop the methods, models and tools for the assessment of risk and safety of ships in damaged conditions.

The proposed method of risk and safety assessment of ships in damaged conditions together with the matrix type risk model enables to assess a ship survivability at each stage of flooding including the preliminary, intermediate and final stages of flooding. Such an assessment is necessary to predict any further deterioration of the damage condition. It may be connected with further flooding of compartments and ship listing increase. This assessment also regards predicting how long it would take the ship to capsize or sink. If the ship survives flooding (a ship remains upright (or heeled) and afloat) and is unable to return to port under own power it is necessary to solve the problems associated with the towing when waiting for an assistance. The number of towing points and towing speed in the calm and rough weather conditions should be evaluated. During the towing the risk and safety assessment of the ship should be permanently conducted.

The structure of the procedure for the assessment of safety of ships in damaged conditions during the accident at sea and for the salvage purposes is presented in Figure 3, Gerigk (2010).

An example application of the ship performance data regarding the ship roll function in time domain and ALARP risk evaluation criteria for the assessment of safety of a ship in damaged condition is presented in Figure 4.

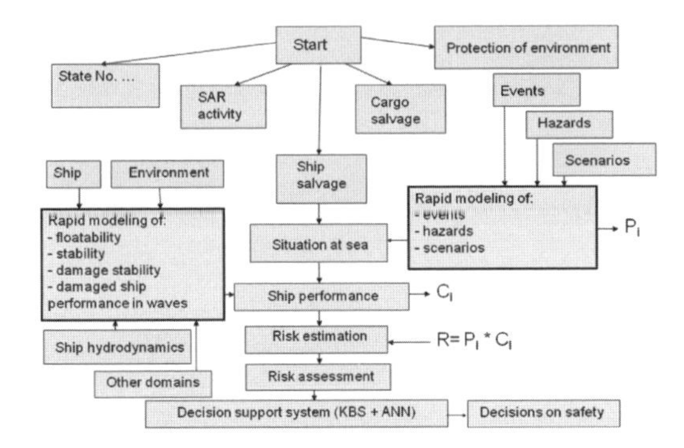

Figure 3. An example of the roll function in time domain and ALARP risk evaluation criteria applied for the assessment of safety of a ship in damaged conditions.

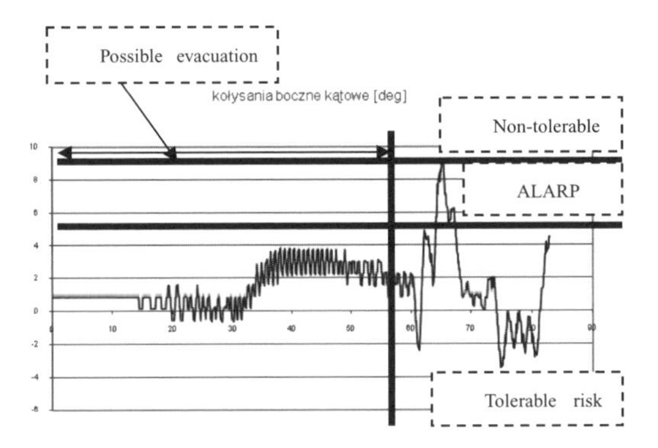

Figure 4. An example of the roll function in time domain and ALARP risk evaluation criteria applied for the assessment of safety of a ship in damaged conditions.

5 CONCLUSIONS

Some elements of the method of risk and safety assessment of ships in damaged conditions together with the matrix type risk model are presented in the paper. The proposed risk model is much more complicated than the model given by Skjong et al. (Jasionowski et al., 2006, Skjong et al., 2006, SSRC 2009, Vassalos 2006). By using (7), (8), (9), (10) and (11) the risk model enables to estimate the risk level for all the possible scenarios of an accident when ship is in damaged conditions during the accident at sea.

The current research is associated with further developing the risk models necessary for the ship performance-oriented and risk-based assessment. From the practical point of view the research should bring a model for the computer simulation of the ship salvage process.

ACKNOWLEDGMENT

The research has been carried out at the Gdansk University of Technology according to support obtained from the Ministry of Science and Higher Education in Warsaw. The author would like to express his sincere gratitude to the above organisations.

REFERENCES

Gerigk, M. 2010. A complex method for safety assessment of ships in damaged conditions using the risk assessment (D.Sc. thesis), *Monography No. 101, Published by the Gdansk University of Technology (in Polish)*, Gdansk 2010.

IMO. 2005. Report of the Maritime Safety Committee on Its Eightieth Session, *MSC 80/24/Add.1*, London 2005.

IMO. 2008. http://www.imo.org

IMO. 2009. Stability and Seakeeping Characteristics of Damaged Passenger Ships in a Seaway When Returning to Port by Own Power or Under Tow, A survey of residual stability margin, *Submitted by Germany, SLF 52/8/1*, London, 26 October 2009.

Jasionowski, A. & Vassalos, D. 2006. Conceptualising Risk, *Proceedings of the 9th International Conference on Stability of Ships and Ocean Vehicles STAB 2006*, Rio de Janeiro, 25–29 September 2006.

Maritime Magazine (Nasze Morze, in Polish). 2007. What may happen to the Baltic Sea (Co z tym Bałtykiem, in Polish), *Volume 12 (24)*, December 2007.

Mohaghegh, Z. & Mosleh, A. 2009. Incorporating organizational factors into probabilistic risk assessment of complex socio-technical systems: Principles and theoretical foundations, *Journal of Safety Science, Vol. 47, 2009*, pp. 1139–1158.

Shipbuilding Magazine (Budownictwo Okrętowe, in Polish). 2005. It can be a catastrophe (Może być strasznie, in Polish), *Volume 3 (548)*, March 2005.

Skjong, R. & Vanem, E. & Rusas, S. & Olufsen, O. 2006. Holistic and Risk Based Approach to Collision Damage Stability of Passenger Ships, *Proceedings of the 9th International Conference on Stability of Ships and Ocean Vehicles STAB 2006*, Rio de Janeiro, 25–29 September 2006.

SSRC. 2009. http://www.ssrc.na-me.ac.uk

Vassalos, D. 2006. Passenger Ship Safety: Containing the Risk, *Marine Technology, Vol. 43, No. 4*, October 2006, pp. 203–213.

Vassalos, D. 2007. Safe return to port, *Seminar for the 50th session of the IMO SLF Sub-Committee*, London, May 2007.

Sustainable Maritime Transportation and Exploitation of Sea Resources – Rizzuto & Guedes Soares (eds)
© 2012 Taylor & Francis Group, London, ISBN 978-0-415-62081-9

A model for consequence evaluation of ship-ship collision based on Bayesian Belief Network

Jakub Montewka, Floris Goerlandt, Soren Ehlers & Pentti Kujala
School of Engineering, Aalto University, Finland

Sandro Erceg, Drazen Polic & Alan Klanac
As2con-alveus ltd. Research, development and applications, Rijeka, Croatia

Tomasz Hinz
The Foundation for Safety of Navigation and Environment Protection, Gdansk, Poland

Kristjan Tabri
Technical University of Tallinn, Estonia

ABSTRACT: In this paper, an attempt is made to define a new, proactive model for estimation the ship-ship collision consequences, assuming a RoPax as a struck ship.

Therefore two major issues are addressed by this paper, one is an identification of the events that follow a collision between two ships at open sea and another is an estimation of the prior probabilities of the events. The latter are obtained in the course of the numerical simulations, observations and literature survey. Then the model is developed by means of Bayesian Belief Network.

Furthermore the sensitivity analysis of the proposed BBN is performed and results are compared with the available models, thus an initial verification of the results is provided.

Finally the probability of ship loss and human loss given a collision is estimated and the obtained results are discussed.

1 INTRODUCTION

Maritime traffic poses various risks in terms of human casualties, environmental pollution or loss of property. In particular, accidents where RoPax ships are involved may pose a high risk in terms of human casualties. In the publications related to modelling the risk of a RoPax ship (Vanem and Skjong 2004), (Antao and Guedes 2006), (Vanem et al., 2007), (Konovessis and Vassalos 2008), (Guarin et al., 2009), one type of models prevails, which is based on an event-tree concept. Moreover the data used to populate the models are based on the accident statistics, thus can hardly be considered proactive.

Therefore this paper introduces a proactive and transferable model for the probability estimation of ship loss and human loss resulting an open sea collision with respect to a selected type of RoPax ship sailing in the selected location. However the modular nature of the model allows for continuous improvement. The model is based on a Bayesian Belief Network (BBN) and utilises series of logically connected events (nodes). The relations among the nodes are given in a probabilistic way, where the prior knowledge is obtained in the course of numerical experiments, observation, analysis and simulation.

In the model presented, a RoPax loss occurs if one of the two accident scenarios is met:

1. The struck RoPax inner hull gets breached and the consecutive flooding is experienced, which can result further in a ship loss. Therefore the critical striking speed and angle for the given mass ratios are obtained with the use of the finite elements simulations;
2. The struck RoPax has no significant hull damage, however the ship is set disable and is experiencing rolling due to wave and wind action, which can result further in ship capsizing thus ship loss. Thus the probability of RoPax capsize is calculated with the use of the six degree of freedom ship motion model.

Moreover numerous variables affecting the consequences are taken into account: maritime traffic composition in the analysed sea area, collision dynamics, ship hydrodynamics, weather conditions, locations of rescue ships with respect to the probable location of an accident, evacuation time from the ship and time of the day at which an accident is probable to happen.

Finally the outcome of the model is the probability of a struck ship loss given an open sea collision. The following boundary conditions should be observed: a given size, type and loading conditions of the struck ship, specific maritime traffic composition and weather conditions corresponding to the ice-free season in the Gulf of Finland.

2 MODEL FRAMEWORK

The presented model is part of the Formal Safety Assessment concept (FSA), which is commonly accepted and approved by the International Maritime Organisation (IMO) rule-making process (IMO 2002). Thus our focus is on the step 2 of the FSA process, see Figure 1.

Field of our interest depends on numerous related and highly uncertain factors, hence the model presented utilises a BBN, which is recognised tools for knowledge representation and efficient reasoning under uncertainty, see (Madsen et al., 2003).

2.1 *Bayesian Belief Network—quantitative and qualitative description*

A classical Bayesian Belief Network is a pair $N = \{G, P\}$, where G = (V,E) is a directed acyclic graph (DAG) with its nodes (V) and edges (E) while P is a set of probability distributions of V. Thus a BBN consists of two parts: a qualitative part (named structure) that is presented as a DAG, and a quantitative part (named parameters) that specifies the dependence relations defined by the structure.

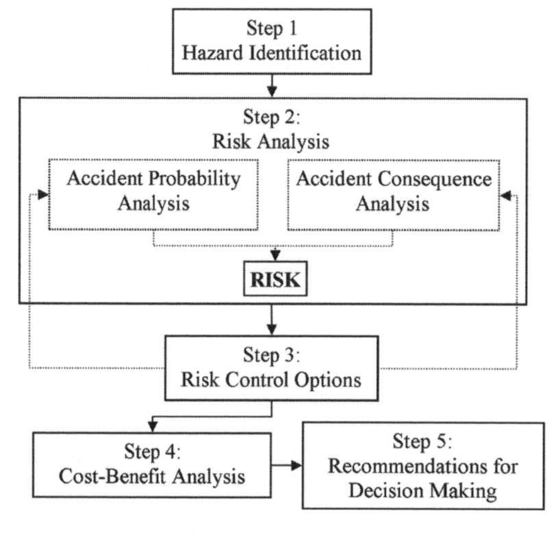

Figure 1. General outline of FSA methodology.

The nodes of a BBN are represented by discrete random variables $X = \{X_1; X_2; X_n\}$ whereas edges reflect the relationships among nodes. Each node is annotated with a conditional probability table (CPT), which represents the conditional probability of the variable given the values of its parents in the graph ($P(X \mid pa(X)) \in P$). The CPT contains all conditional probabilities for all possible combinations of the parent nodes states. If a node does not have parents, its CPT reduces to an unconditional probability table, named also a prior probability of that variable.

Thereby a network $N = \{G, P\}$ is an efficient representation of a joint probability distribution $P(V)$ over V, given the structure of G following the formula, see also (Madsen et al., 2003), (Darwiche 2009):

$$P(V) = \prod_{X \in V} P(X \mid pa(X)) \tag{1}$$

A BBN adopted in our study consists of the CPTs which are obtained in two ways: either by the means of experiment or literature survey. In order to determine, which factors are essential thus should be modelled with greater caution, the sensitivity analysis is carried out at the initial stage of the model development. Once defined, the most vulnerable nodes were tailored to the specific ship type and location by means of experiments and methods described in the previous sections. The CPTs for the remaining nodes that had lower impact on the outcome of the model were based on the generic data available in the literature.

The qualitative and quantitative descriptions of the BBN developed are given in Figure 2.

2.2 *Maritime traffic modelling*

One of the inputs for the model presented is maritime traffic data, in terms of traffic composition, ship types, ship sizes, collision angles, collision speed and time of the day of a hypothetical collision. Most of these, except collision speed and angle are obtained from the dynamic model of maritime traffic, see (Goerlandt and Kujala 2011). The dynamic traffic model simulates the trajectory for each single vessel sailing in the area, while assigning a number of parameters to this vessel as illustrated in Figure 3. The input to this traffic simulation model is taken from the AIS, augmented with the harbours statistics concerning the traded cargo types. The results obtained from the simulation cover the following: time and location

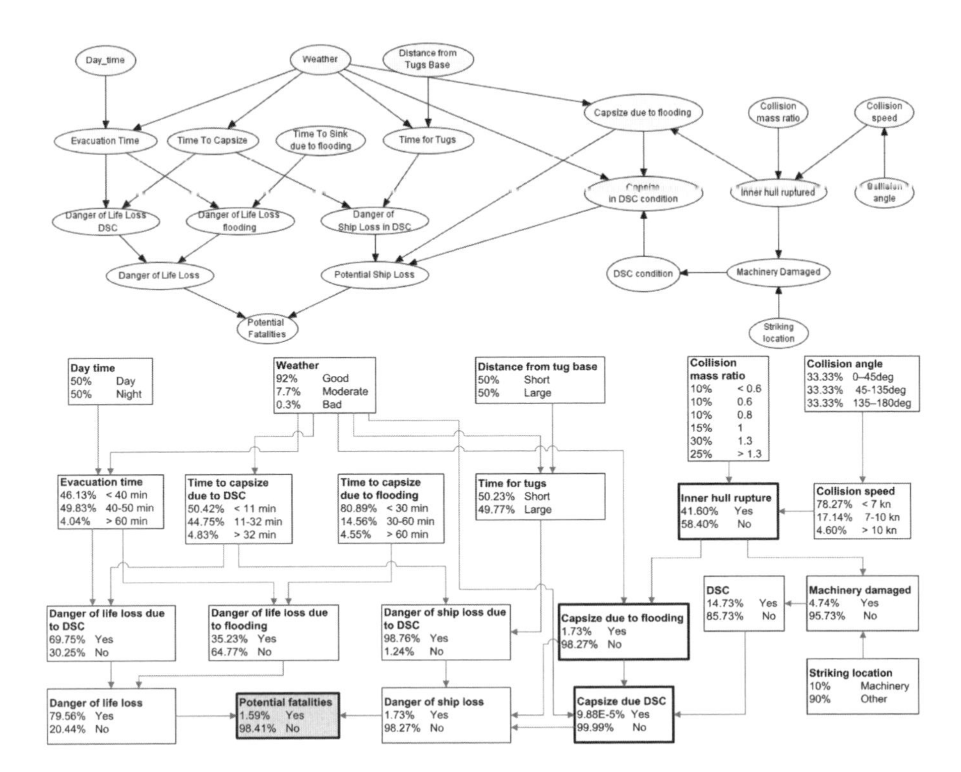

Figure 2. Qualitative description of the model.

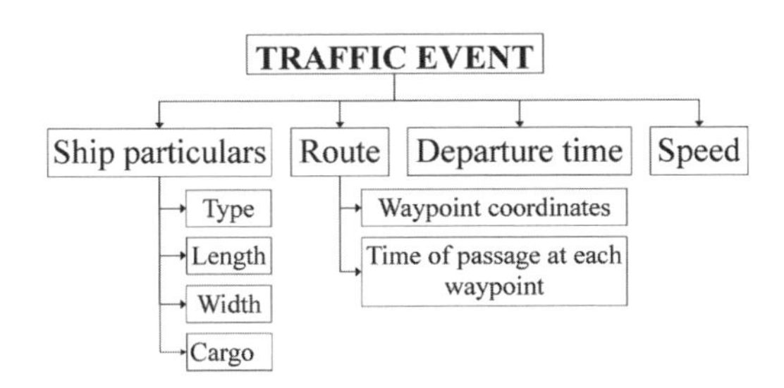

Figure 3. Generated data for each simulated vessel—traffic event.

of a hypothetical collision encounter, types of ship, main dimensions, speeds and courses.

Additionally the collision speed and collision angle are modelled using statistical models taking into account the changes of the initial parameters due to collision evasive action, see (Lützen 2001). Whereas the initial speed and angle of ships are obtained from the traffic simulator.

2.3 *Collision speed and angle modelling*

There are several models to estimate the parameters relevant to a collision scenario, see also (Goerlandt et al., 2011). However most of them are based on the accident statistics and do not take into account the initial speed and angle of colliding ships, except one given by (Lützen 2001). Thereby, this concept is applied in the presented study, as follows:

- The velocity of a striking ship follows a uniform distribution for velocities between zero and 75 percent of an initial speed, then triangularly decreasing to zero;
- The velocity of a struck ship is approximated by a triangular distribution with a most likely value equal to zero and a maximum value equal to initial speed of a ship;
- The collision angle is uniformly distributed between 10 and 170 degrees.

Additionally the actual collision speed is estimated following the adopted five-step procedure: step 1—random sample a striking speed (V_A) of a ship A from an appropriate uniform-triangular distribution; step 2—random sample a struck speed (V_B) of a ship B from an appropriate triangular distribution; step 3—random sample the collision angle—from the uniform distribution; step 4—calculate the relative speed ($V_{(A,B)}$) at which

Table 1. The conditional probability table (CPT) for a *collision speed* variable, given the *collision angle*.

Collision speed [*kn*]	Collision angle [*deg*]	Probability
<7	0–45	0.9987
7–10	0–45	0.0013
>10	0–45	0
<7	45–135	0.3784
7–10	45–135	0.4837
>10	45–135	0.1379
<7	135–180	0.9709
7–10	135–180	0.0291
>10	135–180	0

ship A hits ship B; step 5—calculate a component of $V_{(A,B)}$ that is normal to a hull of struck ship B, finally $(V_{(A,B)\perp})$ is collision speed. For the above procedure Monte Carlo simulations are applied, to get the distributions of collision speed and collision angle.

Finally a CPT is obtained, for model variable *collision speed* given *collision angle*, see Table 1.

2.4 *Probability of inner hull rupture modelling*

To determine the probability of RoPax inner hull rupture given a collision, the critical collision speed and angle is evaluated, using the concept of collision energy. Hence, the collision energy is evaluated for the given collision encounters, where the reference RoPax vessel is struck by: a vessel of similar size (mass ratio 1.0); a vessel smaller by 25 percent (mass ratio 1.33); a vessel larger by 25 percent (mass ratio 0.8) and vessel larger by almost 70 percent (mass ratio 0.6). The available energy for structural deformations is obtained according to the calculation model introduced in (Tabri 2010). This model estimate the dynamics of ship collision and the share in energy available for ship motions and structural deformations. As a result of the combination of this dynamic simulation procedure and the nonlinear finite element method a good estimation of structural damage in various collision scenarios under oblique angles and eccentricity of the contact point can be achieved.

For the purpose of collision simulations the solver LS-DYNA version 971 is used. The ANSYS parametric design language is used to build the finite element model of the reference RoPax vessel. The main characteristics of the ship are gathered in the Table 2.

Three-dimensional model is built between two transverse bulkheads spaced at 26.25 m (see Figure 4) and the translational degrees of freedom are restricted at the plane of the bulkhead locations. The remaining edges are free. The structure is modelled using four nodded, quadrilateral Belytschko-Lin-Tsay shell elements with five integration points through their thickness. The characteristic element-length in the contact region is 50 mm to account for the non-linear structural deformations, such as buckling and folding. The element length dependent material relation and failure criterion according to (Ehlers 2010) is utilised for the simulations. Standard LS-DYNA hourglass control and automatic single surface contact (friction coefficient of 0.3) is used for the simulations. The collision simulations are displacement-controlled.

Table 2. The analysed RoPax vessel characteristics.

Length	188.3 m
Breadth	28.7 m
Draught	6.0 m
Displacement	19610.0 t

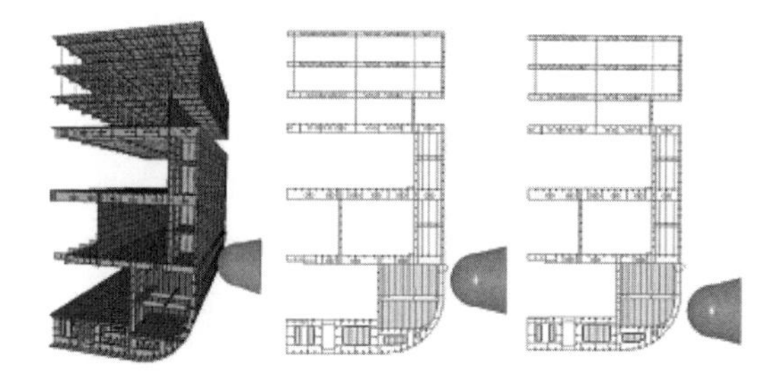

Figure 4. FEM model and vertical striking locations.

The rigid bow is moved into the ship side structure in a quasi-static fashion. Hence, this approach results in the maximum energy absorption of the side structure alone, which is needed for a comparison and can be considered conservative and thereby suitable for a fast prediction.

As a result, the relative energy available for structural deformations as a function of the longitudinal striking location is obtained (see Figure 5) for a mass ratio of 1.0. For a mass ratio of 1.33 and 0.6 these curves are scaled with 0.84 and 1.13 respectively to account for the change in dynamic behaviour. Therefore the graphs for critical striking speeds, for a given mass ratio and striking angle, causing the inner hull breach are obtained. An exemplary graph for a mass ratio 1.0 is presented in Figure 6. The striking speed obtained is a function of striking location along the hull of struck ship.

Finally the CPT for the variable *inner hull rupture* is constructed, see Table 3.

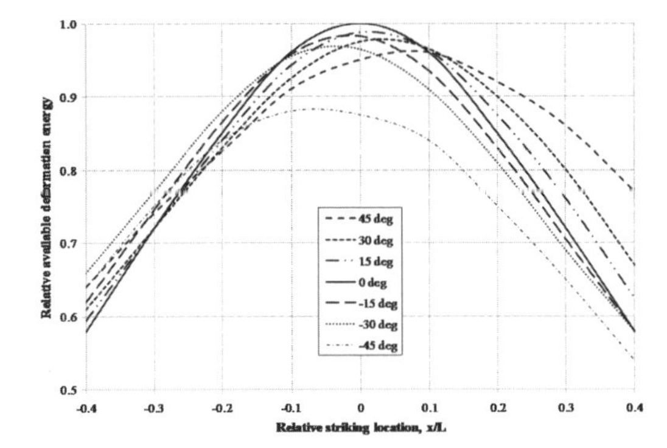

Figure 5. Relative available deformation energy versus relative striking location and striking angle.

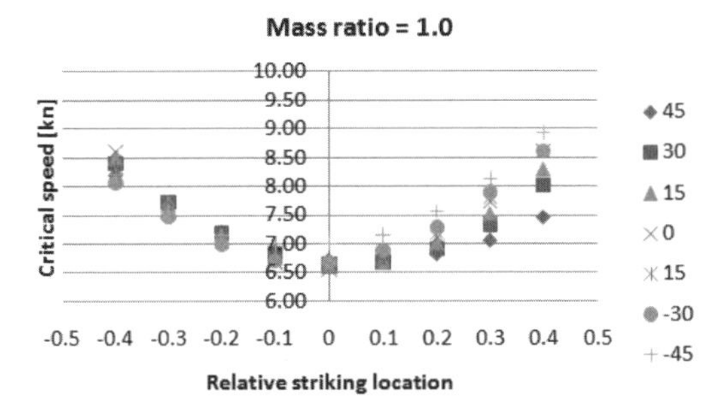

Figure 6. Minimum speeds required for the inner hull breach as a function of the longitudinal striking location and a collision angle.

Table 3. The conditional probability table (CPT) for *inner hull rupture* variable, given two variables *collision mass* and *collision speed.*

Collision mass [-]	Collision speed [*kn*]	Probability
<0.8	<7	0.8
<0.8	7–10	1.0
<0.8	>10	1.0
0.8	<7	0.1
0.8	7–10	1.0
0.8	>10	1.0
1.0	<7	0.0
1.0	7–10	1.0
1.0	>10	1.0
1.3	<7	0.0
1.3	7–10	0.5
1.3	>10	1.0
>1.3	<7	0.0
>1.3	7–10	0.0
>1.3	>10	0.8

2.5 *Estimation of the probability of ship capsize due to flooding in damage condition*

As a result of a ship-ship collision, where the collision speed exceeds the critical speed for breaching the inner hull (see Figure 6), ship flooding can be expected. Moreover, if the wave height is greater than a critical height, a ship can experience flooding which can contribute to a ship loss.

To determine the probability of ship loss due to flooding a concept of "capsize band" has been recently introduced, see (Papanikolaou et al., 2010). The band begins at the wave height that does not cause capsize ($P_{capsize} = 0$) and ends at the wave height where a ship loss is expected always ($P_{capsize} = 1$). The capsize boundaries are symmetrical, around the value of critical wave height, which correspond to $P_{capsize} = 0.5$. Another attribute is that the capsize band gets broader with the increase of the critical wave height. For the purpose of this study we assume the damage stability conditions of a RoPax corresponding to the critical wave height of 5.5 m, with a symmetrical bandwidth of 4 m around it, see (Papanikolaou et al., 2010). Moreover the flooding of a car deck and two compartments underneath is assumed. Additionally the hypothetical damage opening is constant regardless of collision scenario, as defined by SOLAS'95 (B-II Reg. 14).

Time to capsize due to flooding was estimated by means of a simulation model by (Spanos and Papanikolaou 2010).

The weather conditions applied in the model are categorised into three groups (see Table 4). These division corresponds to the adopted capsize band, as follows: *moderate* corresponds to the capsize band, *good* means no capsize at all, whereas *bad* stands for sea conditions in which ship will always capsize (*Weather = Bad*). Table 4 shows the prior probabilities for the weather conditions, which are obtained from the Global Wave Statistics, see (BMT Ltd 1986).

The model provides the following marginal probability of RoPax capsizing due to flooding in damage condition:

$$P_{capsize|flooding|collision} = 1.70 \times 10^{-2} \quad (2)$$

Table 4. The prior probabilities for the variable *weather*.

Variable instance [*m*]	Wave height [*m*]	Probability
Good	0–3	0.920
Moderate	3–6	0.077
Bad	>6	0.003

Furthermore, a complement of this number is the probability of survive given the collision and consecutive flooding.

2.6 *Estimation of the probability of ship capsizing in Dead Ship Condition*

Another type of the consequence arising from a collision addressed by this paper is ship capsizing due to wave and wind action, whereas the ship is in dead ship condition (DSC). DSC means "a condition in which the entire machinery installation, including the power supply, is out of operation and the auxiliary services such as compressed air, starting current from batteries etc., for bringing the main propulsion into operation and for the restoration of the main power supply are not available", see also (BureauVeritas 2005). To determine the probability of ship capsizing and the time to capsize in DSC, the simulations are performed with the use of the state-of-the-art ship dynamics model. Whereas the probability of ship capsizing is assumed equal to the probability of exceeding a particular angle of roll, which in this analysis means 60°. Thus, the ship dynamics in waves is estimated by means of six-degree of freedom ship motion model LAIDYN, which assumes that the overall ship response is a sum of linear and nonlinear parts, see (Matusiak 2011).

To calculate the probability of reaching the intended roll angle ($P_{capsize}$) the Monte Carlo simulations are applied. Hence the probability of ship capsize in DSC is obtained by means of the formula:

$$P_{capsize} = \frac{N_{\phi_c}}{N_s} \tag{3}$$

where N_{ϕ_C} means a number of simulations where the intended angle was reached and N_S is a number of all simulations. The probability of a RoPax capsizing given the analysed conditions yields:

$$P_{capsize|DSC-condition|weather} = 1.20 \times 10^{-4} \tag{4}$$

2.7 *Sensitivity analysis of the model*

To determine how sensitive the results of the model are to variations of the parameter of the model, the parameter sensitivity analysis is carried out. The parameters analysed are the entries of the given CPT. In the analysis presented the influence of all variables on the model outcome, which is a two-state node "Potential Fatalities", is performed. For this purpose the following sensitivity function

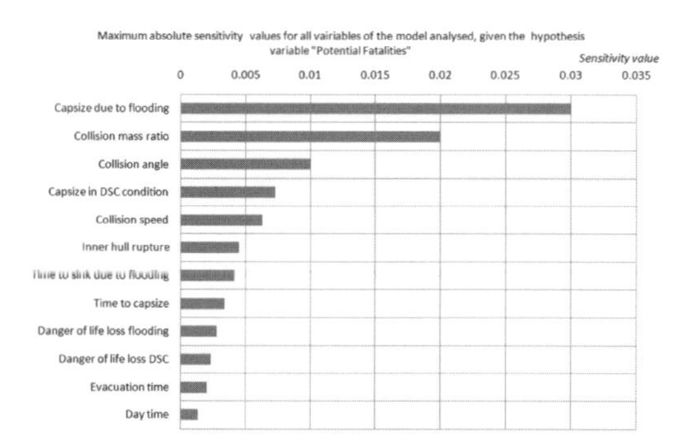

Figure 7. Maximum of absolute sensitivity values for the most relevant variables.

is estimated for each single node, see (Chan and Darwiche 2002):

$$f(t) = \frac{(c_1 t + c_2)}{(c_3 t + c_4)}, \tag{5}$$

where f is an output probability of interest, given observations and $c_1, \ldots, c_4$ are the constants. Whereas the effect of minor changes in the original parameter on the output is given by the sensitivity value, which is computed based on the first derivative of the sensitivity function. Hence, the results of the sensitivity analysis are depicted in Figure 7, where the maximum absolute value of the sensitivity function for the selected nodes are presented. The nodes that obtained the highest values of the sensitivity function are depicted only.

3 RESULTS AND DISCUSSION

In the course of the analysis presented, we assume an accident scenario, where a RoPax ship of a given structure and size, sailing in the Gulf of Finland is struck by another ship of certain mass, with certain collision speed and angle. Then the probability of human loss and/or a ship loss is estimated by means of newly developed model, see Table 5. Additionally, all the marginal probabilities are depicted in Figure 2.

Furthermore the model estimates the probabilities of other variables which are considered important from the viewpoint of maritime traffic risk analysis, for instance: collision speed, ship sinking due to flooding given a collision and the inner hull rupture given a collision, see Table 6.

At the present stage of the model development, a structural analysis is performed for the following mass ratios: 0.6, 0.8, 1.0, 1.3, thus covering almost 80 percent of maritime traffic in the

Table 5. The conditional probabilities of consequences given a collision for a given RoPax.

Consequences given a collision	
Potential ship loss	0.017
Potential fatalities	0.016

Table 6. The selected marginal probabilities obtained from the model.

Collision speed	Probability
<7	0.78
7–10	0.17
>10	0.05
Inner hull rupture	Probability
yes	0.41

Gulf of Finland. The remaining 20 percent belongs mostly to ratios higher than 1.3 (which means that the striking ship becomes smaller and lighter, thus the speed required to rupture the inner hull increases), however some percent of ratios lower than 0.6 is not taken account. The latter is an issue, as the critical collision speed in these cases may be significantly smaller than values adopted in the present analysis, which may increase the probability of ship loss.

The probability of ship capsizing following a collision is estimated using a concept of "capsize band". In the model presented only one band is applied, which corresponds to specific damaged stability conditions. However, the size of the opening which affects flooding is assumed constant regardless of a collision scenario, according to the SOLAS recommendations.

Finally, we compare the obtained results with the statistical data and results obtained from existing models. Thereby the probabilities per collision, of a RoPax loss (capsizing) following a hit by other ship while under way, based on different sources are as follows:

- $P_C = 1.73 \times 10^{-2}$—by the model introduced in this paper, valid within certain boundary conditions,
- $P_C = 4.20 \times 10^{-2}$—valid for an arbitrary passenger RoRo vessel, see (Otto et al., 2002),
- $P_C = 1.89 \times 10^{-2}$—statistics based model, data refers to any RoPax vessel, the boundary conditions unknown, see (Konovesis and Vassalos 2007),
- $P_C = 1.76 \times 10^{-2}$—statistics based model, data refers to any RoPax vessel, the boundary conditions unknown, see (Guarin et al., 2009).

4 CONCLUSIONS

In this paper, we introduced a novel, proactive and transferable model for the probability of ship-ship collision consequences estimation, with respect to a selected type of RoPax ship sailing in the selected location. In the analysis the ship loss and human loss are considered the consequences of a collision scenario. The results obtained form the model are valid within certain, predefined boundary conditions, however due to modular nature of the model it can be used for any ship type and geographical locations.

The presented results are normalised over the whole maritime traffic, according to the prior probabilities of the ship masses ratio, collision speed and collision angle obtained from the AIS data analysis for the Gulf of Finland.

In the light of the sensitivity analysis performed, the most sensitive nodes of the model refer to the weather conditions, the probability of ship sinking due to flooding, the parameters describing a collision scenario (a mass ratio, a collision angle, a collision speed) as well as the probability of intact, however disable ship capsizing.

Notwithstanding all the assumptions and simplifications made, the model results are comparable with the results obtained form the existing models and accident statistics.

ACKNOWLEDGEMENT

The authors appreciate the financial contributions of the following entities: Merenkulun säätiö from Helsinki and the city of Kotka. Risto Jalonen from Aalto University is thanked for his invaluable comments.

REFERENCES

Antao, P. and Guedes, S.C. (2006). Fault-tree models of accident scenarios of ropax vessels. *International Journal of Automation and Computing* 3(2), 107–116.

BMT Ltd. (1986). *Global Wave Statistics*. London: BMT Ltd.

BureauVeritas (2005, April). Rules for the classification of steel ships. Technical report, Bureau Veritas.

Chan, H. and Darwiche, A. (2002). When do numbers really matter? *Journal of artificial intelligence research 17*, 265–287.

Darwiche, A. (2009). *Modeling and Reasoning with Bayesian Networks* (1 ed.). Cambridge University Press.

Ehlers, S. (2010). *Material relation to assess the crashworthiness of ship structures*. Ph.D. thesis, Aalto University, School of Science and Technology, Espoo, Finland.

Goerlandt, F. and P. Kujala (2011). Traffic simulation based ship collision probability modeling. *Reliability Engineering & System Safety* 96(1), 91–107.

Goerlandt, F., Montewka, J., Lammi, H. and Kujala, P. (2011). *Analysis of near collisions in the gulf of finland.* ESRE conference. Manuscript submitted and accepted.

Guarin, L., Konovessis, D., and Vassalos, D. (2009). Safety level of damaged ropax ships: Risk modelling and cost effectiveness analysis. *Ocean Engineering* 36(12–13), 941–951.

IMO (2002, April). Guidelines for formal safety assessment (fsa) for use in the imo rule-making process. MSC/Circ.1023; MEPC/Circ.392.

Konovesis, D. and Vassalos, D. (2007). Risk-based design for damage survivability of passenger ro-ro vessels. *International Shipbuilding Progress* 54, 129–144.

Konovessis, D. and Vassalo, D. (2008). *Risk evaluation for ropax vessels. Proceedings of the Institution of Mechanical Engineers, Part M: Journal of Engineering for the Maritime Environment* 222(1), 13–26.

Lützen, M. (2001). *Ship collision damage.* Ph.D. thesis, Technical University of Denmark, Department of Mechanical Engineering.

Madsen, A., Lang, M., Kjrulff, U. and Jensen, F. (2003). The hugin tool for learning bayesian networks. In Nielsen, T. and Zhang, N. (Eds.), *ECSQARU 2003, LNAI 2711*, pp. 594–605. Springer-Verlag Berlin Heidelberg.

Matusiak, J. (2011). On the non-linearities of ships restoring and the froude-krylov wave load part. *International Journal of Naval Architecture and Ocean Engineering* 3(1).

Otto, S., Pedersen, P.T., Samuelides, M. and Sames, P.C. (2002). Elements of risk analysis for collision and grounding of a roro passenger ferry. *Marine Structures* 15(4–5), 461–474.

Papanikolaou, A., Mains, C., Rusaas, S., Szalek, R., Tsakalakis, N., Vassalos, D. and Zaraphonitis, G. (2010, June). Goalds—goal based damage stability. *In Proceedings of the 11th International Ship Stability Workshop, Wageningen*, pp. 46–57. MARIN.

Spanos, D. and A. Papanikolaou (2010, June). On the time dependent survivability of ropax ships. *In Proceedings of the 11th International Ship Stability Workshop, Wageningen*, pp. 143–147. MARIN.

Tabri, K. (2010). *Dynamics of ship collisions.* Ph.D. thesis, Aalto University, School of Science and Technology, Es-poo, Finland.

Vanem, E., Rusås, S., Skjong, R. and Olufsen, O. (2007). Collision damage stability of passenger ships: Holistic and risk-based approach. *International Shipbuilding Progress* 54(4), 323–337.

Vanem, E. and Skjong, R. (2004). Collision and grounding of passenger ships—risk assessment and emergency evacuations. In SNAJ (Ed.), Proceedings of the 3rd *International Conference on Collision and Grounding of Ships*, pp. 195–202.

Sustainable Maritime Transportation and Exploitation of Sea Resources – Rizzuto & Guedes Soares (eds)
© 2012 Taylor & Francis Group, London, ISBN 978-0-415-62081-9

Use of VELOS platform for modeling and assessing crew assistance and passenger grouping in ship-evacuation analysis

K.V. Kostas
Department of Naval Architecture (NA), Technological Educational Institute of Athens (TEI-A), Athens, Greece

A.-A.I. Ginnis
School of Naval Architecture & Marine Engineering (NAME), National Technical University of Athens (NTUA), Athens, Greece

C.G. Politis
Department of Naval Architecture (NA), Technological Educational Institute of Athens (TEI-A), Athens, Greece

P.D. Kaklis
School of Naval Architecture & Marine Engineering (NAME), National Technical University of Athens (NTUA), Athens, Greece

ABSTRACT: *VELOS*, which stands for *"Virtual Environment for Life On Ships"*, is a multi-user VR system that aims to support designers, early in the design process, to assess passenger and crew activities on ship and improve ship design accordingly. VELOS functionalities provide design aids required for both normal and hectic operational conditions. This has been accomplished by integrating a broad range of software components in VELOS platform which includes tools targeting geometric and VR modeling, crowd microscopic modeling based on steering behaviors technology, as well as communication interfaces with external computational software packages. In the present work, we focus on the evacuation-specific functionality of VELOS by enhancing it with *passenger-grouping* and *crew-assistance* behavior. This is mainly achieved by combining and extending steering behaviors, already used within VELOS, for crowd modeling, as, e.g., *Leader-Follow* and *Cohere* behavior. This enhancement allows to simulate the evacuation process more realistically and compare results acquired for the scenarios prescribed by the IMO, with and without the consideration of grouping and *crew-assistance* behavior.

1 INTRODUCTION

In the wake of a series of events involving large number of fatalities on passenger ships (Vanem and Skjong 2006), evacuation has been a high priority on the International Maritime Organization (IMO) agenda since 1999, when SOLAS (I.M.O. 2001) required evacuation analysis to be carried out early in the design process of new Ro-Ro passenger ships. Following this, the Maritime Safety Committee (MSC) of IMO adopted Circular 1033 (I.M.O. 2002) and recently its new and updated version Circular 1238 (I.M.O. 2007) entitled *"Guidelines for evacuation analysis for new and existing passenger ships"*. In these guidelines two distinct methods of analysis are presented: a simplified and an advanced method of evacuation analysis. As it is pointed out in these guidelines, the assumptions inherent within the simplified method are by their nature limiting. As the complexity of the vessels increases (through the mix of passenger types, accommodation types, number of decks and number of stairways) these assumptions become less representative of reality. In these cases, the use of advanced method would be preferred. However, it is worth noticing that even the advanced method of analysis addresses issues related to the lay out of the main escape routes and passenger demographics, while other operational issues arising in real emergency conditions, such as unavailability of escape arrangements (due to flooding or fire), crew assistance in the evacuation process, group behavior and ship-motion effects are only dealt with by means of safety factors (I.M.O. 2007).

"Virtual Environment for Life On Ships" (VELOS) is a multi-user Virtual Reality (VR)

system that supports designers to assess (early in the design process) passenger and crew activities on a ship for both normal and hectic conditions of operations and to improve the ship design accordingly (Kostas, Ginnis, Kaklis, and Politis 2007), (Ginnis, Kostas, Politis, and Kaklis 2010). VELOS is based on VRsystem (Kostas 2006), a generic multi-user environment with a broad range of functionalities including geometric and VR-modeling, as well as crowd microscopic modeling through a library of over 20 steering behaviors. Additionally, VRsystem can communicate with computational packages, as, e.g., seakeeping software for improving the environment realism and taking into account the ship-motion effect on passengers' movements. VELOS evacuation-specific functionality is greatly enhanced by the VR nature and client-server architecture provided by the VRsystem, namely, the participation and real-time interaction of remote multiple users in the form of avatars. For example, avatars in the evacuation simulation may act as crew members, family-group leaders or just passengers. These VRsystem-inherited features entail a very distinctive approach to evacuation analysis in VELOS, when compared with evacuation tools in pertinent literature.

In this work we focus on the evacuation-specific functionality of VELOS whose main features are summarized as follows:

- An integrated framework for both the IMO simplified and advanced method of analysis.
- An enriched geometrical ship model with topological information in order to improve path-planning procedures.
- Efficient communication through a number of interfaces that enable dynamic specification and handling of the required input data.
- Post-processing of the fundamental output which is agents' trajectories for extracting evacuation-specific information, as travel time, cumulative arrival time and passenger density at specified areas.

More specifically, we herein enrich VELOS by addressing issues as passenger grouping and crew assistance, which can enhance emergency preparedness and management, resulting in this way at a more realistic evacuation analysis. This is mainly accomplished by combining and extending steering behaviors already used within VELOS for crowd modeling, i.e., *Cohere, Leader-Follow, Seek* and *Arrive* behaviors; see § 2. As a result we are able to simulate the evacuation process and compare results with and without consideration of *crew-assistance* and *grouping* behavior for a generic four-room configuration (see § 3.1) and a IMO-based scenario of evacuating the stern zone of a passenger ship; see § 3.2.

2 PASSENGER GROUPING & CREW ASSISTANCE IN VELOS

The motion behavior of an agent is better understood by splitting it into three separate levels, namely *action selection, steering* and *locomotion*. In the first level, goals are set and plans are devised for the action materialization. The steering level determines the actual movement path, while locomotion provides the articulation and animation details.

Agents' autonomy is materialized within the steering level, where the steering behaviors technology is applied. Specifically, agents' autonomy is powered by an artificial intelligence structure, referred to in the pertinent literature as *mind*; see, e.g., (Green 2000), (Reynolds 1987). The mind utilizes a collection of simple kinematic behaviors, called *steering behaviors*, to ultimately compose agent's motion. For each time frame, agent's velocity vector is computed by adding the previous-frame velocity vector to the mind-calculated steering vector. This vector is a combination of the individual steering vectors provided by each associated steering behavior in agent's mind.

In mind modeling we employ two different approaches for the steering vector calculation. The first and rather obvious one, referred to as *simple mind*, produces the steering vector as a weighted average of the individual ones. The second approach that takes into account priorities, called *priority blending*, is an enhanced version of the simple priority mind proposed in (Reynolds 1999). Furthermore, mind is affected, during simulation time, by *Triggers*, scene areas which, when visited by an agent, a prescribed list of actions or property changes are applied to its mind.

Over twenty steering behaviors have been so far implemented within VELOS, including: *Seek, Arrive, Pursuit, Flee, Evade, Offset-{Seek, Flee, Pursuit, Evade, Arrive}, Leader-Follow, Separation, Obstacle-Avoidance & Containment, Inclination, Wander, Path-following and Cohesion & Alignment*.

As the present paper aims towards modeling and assessing the passenger-grouping and crew-assistance functionalities of VELOS for ship-evacuation analysis, the rest of this section is structured in two subsections respectively.

2.1 *Passenger grouping*

Passenger grouping in VELOS is based on the *Enhanced-Cohere* behavior which constitutes an enhancement of the standard *Cohere* behavior, that is responsible for keeping together agents that are geometrically close. Since passengers and crew members on a ship are moving principally on planar surfaces, geometric vicinity is implemented by defining a planar neighborhood with the aid of a radius parameter; see Fig. 1. All vehicles

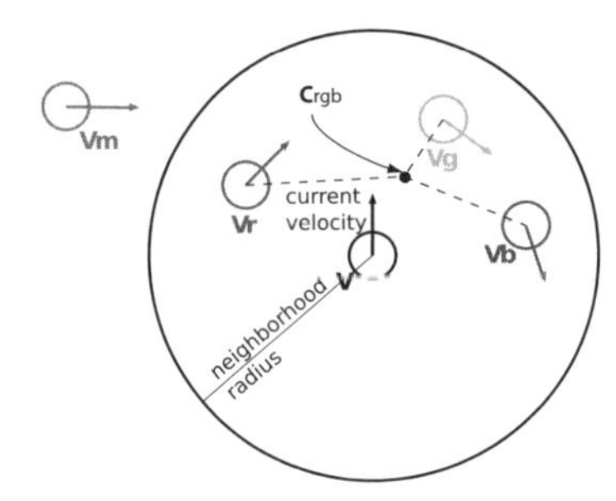

Figure 1. Standard Cohere behavior sketch: $\mathbf{V}_g$, $\mathbf{V}_r$ and $\mathbf{V}_b$ are in **V**'s neighborhood and their centroid C_{rgb} would be **V**'s new target position for application of the basic *Seek behavior*.

found within this neighborhood are then used for calculations. More specifically, the centroid of the positions of all agents neighboring the vehicle in question constitutes the target position for our agent (black circle in Fig. 1). This position is then used as target for the basic *Seek* behavior, described in the ensuing sub-subsection § 2.1.1, for evaluating the resulting steering force and, eventually, the new velocity vector of our agent.

Enhanced-Cohere behavior is responsible for keeping together agents that are not only geometrically close to each other, but also belong to the same group, e.g., a family, a crew guided group, etc. For this purpose, each agent is endowed with an ID in the form of a common length binary representation and the new velocity vector of every agent is obtained by applying the standard *Cohere* calculations on the subset of the neighboring agents that belong to the same group.

In this way, by blending properly the *Cohere* behavior we can produce different grouping levels which can be roughly categorized as follows:

Grouping Level 0: On this level, grouping is formed indirectly, via a common short-term target for the "group" members, as, e.g., followers of the same leader, or through the usage of the standard *Cohere* behavior.

Grouping Level 1: The members of the group are endowed with an *ID* and the *Enhanced-Cohere* behavior described above. Group cohesion is maintained only among nearby agents (within Cohere's neighborhood) sharing a common ID. However, if a member of the group gets out of the *Cohere* behavior's neighborhood, the remaining members will take no action.

Grouping Level 2: The members of the group are endowed with the same properties as in Level 1 and moreover at least one member (e.g., the group leader) has the responsibility of checking group's integrity. In this way, cohesion of the group is maintained, since if a member of the group is lost the group leader will take some corrective action as to wait for the lost member to join the group or to search for finding the lost member.

2.1.1 *Seek behavior*

Seek is a very basic and at the same time a very simple behavior. Its aim is to move a vehicle towards a specified target position with constant speed. *Seek* behavior adjusts the vehicle so that its velocity is radially aligned towards the target (see Fig. 2 (A)). The vehicle will possibly pass through the target and then turn back to approach again. Thus the motion produced is a bit like a moth buzzing around a light bulb. This does not pose any problem if the target moves intermittently but it is rather unnatural if the target is stationery or almost stationery. For handling such cases *Arrive* behavior is more suitable, which is described in the ensuing sub-subsection § 2.1.2.

The implementation of *Seek* behavior in pertinent literature (see Reynolds 1999) and (Green 2000)) has as follows:

1. Compute desired_velocity as:

$$\mathbf{new\_v} = \frac{\overrightarrow{\mathbf{T}-\mathbf{V}}}{\|\mathbf{T}-\mathbf{V}\|} \cdot \max\_speed,$$

where max_*speed* is a parameter referring to the maximum allowable vehicle speed.

2. Compute steering_force as: $\mathbf{new\_v} - \mathbf{v}$

In VELOS we have introduced an alternative approach to *Seek* computation. We don't use the *desired_velocity* notion but instead we try to "position" our vehicle at its target's site under two restrictions:

a. The restriction of vehicle's maximum speed, applied by max_*speed* parameter, and
b. The restriction of the maximum force (steering force) our vehicle is allowed to sustain, applied by a new parameter, namely max_*force*.

Specifically, if we were to position our vehicle at its target site **T** we should add $\overrightarrow{\mathbf{T}-\mathbf{V}}$ to its current position **V** or alternatively add $\mathbf{f}'$ ($\mathbf{f}' = \overrightarrow{\mathbf{T}-\mathbf{V}} - \mathbf{v}$) to its current velocity **v**. In this construction, $\mathbf{f}'$ represents the steering force which is restricted by $\|\mathbf{f}'\| \leq \max\_force$. Thus,

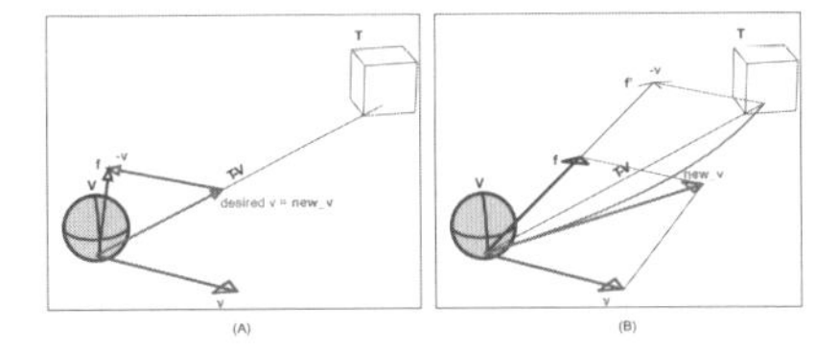

Figure 2. (A) Original *Seek* behavior implementation (B) VRkernel's *Seek* behavior implementation. In (A) our vehicle (sphere) in position **V** follows the **T-V** line while in (B) it will ultimately follow the green trajectory to the target **T**.

if $\|\mathbf{f}'\| > max\_force$, we scale $\mathbf{f}'$ and we get our actual steering $\mathbf{f} = max\_force \cdot \frac{\mathbf{f}'}{\|\mathbf{f}'\|}$. Adding $\mathbf{f}$ to our current velocity $\mathbf{v}$ we get our vehicle's new velocity vector. Finally, if $\|\mathbf{new\_v}\| > max\_speed$ we scale it by $max\_speed / \|\mathbf{new\_v}\|$. In Fig. 3 the details of our proposed *Seek* behavior implementation are sketched.

Furthermore, in *Seek*, as in all steering behaviors implemented in VELOS, we have introduced the notion of the *activation radius* parameter. Specifically, this parameter defines the *activation circular disc* which marks the behavior's active area. This means that if a behavior's target does not lie in that area, the behavior calculations are not performed and essentially the behavior becomes inactive. Vehicle's visual range and auditory field are two of the most common physical interpretations for the activation area in VELOS steering behaviors' implementations.

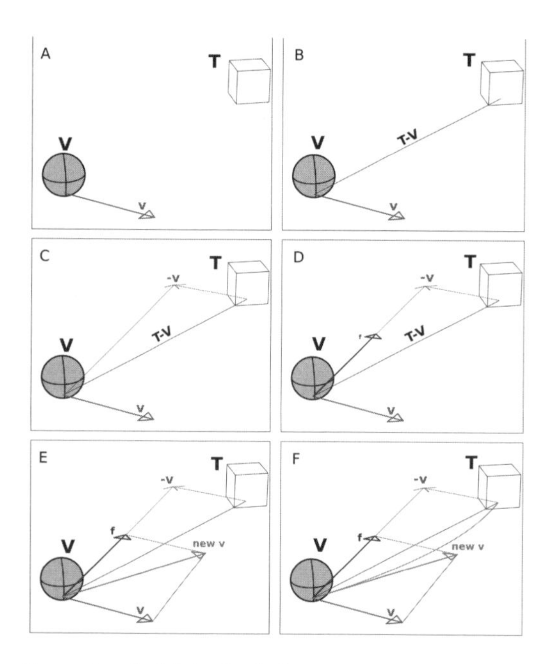

Figure 3. *Seek* behavior in 6 steps.

step 1: Vehicle at position **V** with current velocity **v** is equipped with mind and *Seek* behavior and has as its target the box at position **T**.
step 2: Firstly, vector $\mathbf{T}-\mathbf{V}$ (distance vector) is computed.
step 3: $(\mathbf{T}-\mathbf{V})+(-\mathbf{v})$ is the direction of force vector $\mathbf{f}$.
step 4: Force vector's magnitude is clipped (if needed) to the *maximum allowed*.
step 5: **f** and **v** are added to form vehicle's new velocity **new v**.
step 6: Ultimately the vehicle will follow the green trajectory to its target **T**.

2.1.2 *Arrive behavior*

A major problem with *Seek* is that when the vehicle approaches its target, it is travelling with full speed and this leads to a target "overshooting".

Arrive behavior (see (Reynolds 1999)) is identical to *Seek* as long as the vehicle is far from its target. However, when the vehicle approaches its target, *Arrive* behavior causes the vehicle to slow down and eventually to stop coincident with its target. The distance at which slowing begins is a parameter for the *Arrive* behavior. Its implementation is similar to *Seek*: a desired velocity, with the aid of the force vector, is calculated pointing towards the target. Outside the slowing-down area the velocity is clipped to max_speed, while inside this area the desired velocity is ramped down (linearly, quadratically, etc) to zero.

When implementing *Arrive*, it is rather obvious that for numerical reasons the vehicle isn't going to actually stop moving as reaching the target. The system moves asymptotically towards a stop and the movement will be unnoticeable unless the system does not stop updating vehicle's local space which might lead to a sudden direction flip; here, we must recall that vehicle's speed vector coincides with local $x-axis$ direction. Thus, in the implementation of vehicle's local space update, it is crucial to include an ε-test to terminate local coordinate system updates for small velocity vectors.

2.2 *Crew assistance*

Crew-Assistance behavior in VELOS is offered by affecting the *simple-* or *priority-mind mechanism* in two ways, either by using *Triggers* or via the *Guide Operation*.

A *Trigger* attached to a crew agent is a scene object and at the same time a scene area (*Neighborhood or TN*) that, when visited by a passenger agent, a prescribed list of actions or property changes, the so called *Trigger Actions or TAs*, are applied to the agent. A TA example could be the following: if passenger density at the chosen TN exceeds a prescribed limit, the TA enables the crew agent to redirect passengers towards the closest muster station along a path different from the main escape route; see Test Case 2 in Generic Example of § 3.1.

Guide Operation is materialized through the *Enhanced-Cohere* behavior, described in the previous subsection, and the basic *Leader-Follow* behavior, described below. A *Guide-Operation* example could involve a crew member that is ordered by the officer in charge to guide a group of passengers from a specific site to the closest muster station along a path different from that provided by the evacuation plan; see Test Case 3 in Generic Example of § 3.1.

Furthermore improvement of *Crew Assistance* services could be provided by properly combining *Triggers* with *Guide Operation*.

2.2.1 *Leader-follow behavior*

This behavior causes one or more characters to follow another moving character designated as the leader. Generally, the followers want to stay near the leader, without crowding the leader and taking care to stay out of the leader's way, in case they happen to find themselves in front of the leader. In addition, if there is more than one follower, they want to avoid bumping each other. As a result, the implementation of *Leader-Follow* behavior relies on the *Arrive* behavior, described in the previous sub-subsection § 2.1.2, and the *Separate* behavior, that is needed to prevent followers from crowding each other; see in the ensuing sub-subsection § 2.2.2. The above discussion is better understood if we closely follow the pair of steps summarized in the two configurations, namely *"Follower behind/front Leader"*, illustrated in Fig. 4.

2.2.2 *Separate behavior*

Separate behavior is used to make vehicles keep a certain distance to each other. This stems both from psychological factors preventing especially strangers to exist in close proximity and from an automatic process related to collision avoidance. Essentially, application of *Separate* models this "safe area" people need between each other and also acts as an initial measure in preventing collisions of agents. Human beings do the same thing automatically. A person walking through a crowded place keeps a distance from all other, while at the same time, avoids bumping into other persons without a second thought. In this process only the persons in the immediate distance are considered by the acting person and most of the time this set is reduced to only persons in front of him/her. The same "neighborhood" notion and distinction between "in front of" and "behind" vehicles is modeled within Separation as we'll discuss later.

The basis for *Separate* implementation is the standard theory on forces between electrically charged particles. Fig. 5 provides a four-step illustration of *Separate* implementation. The required repelling forces are calculated by the formula: $\vec{\mathbf{F}} = \vec{\mathbf{r}} / \| \vec{\mathbf{r}} \|^2$, where $\vec{\mathbf{r}}$ is the vector connecting the position of the vehicle in question with each one of its neighboring vehicles in the active area. E.g., $\vec{\mathbf{F}}_1 = \vec{\mathbf{r}}_b / \| \vec{\mathbf{r}}_b \|^2$, where $\vec{r}b = V \quad V_b$.

In the discussion above, all vehicles in our neighborhood are included in the calculations, which may result in unnatural vehicle movements. For handling this, a new attribute `look_ahead` was introduced in VELOS implementation of

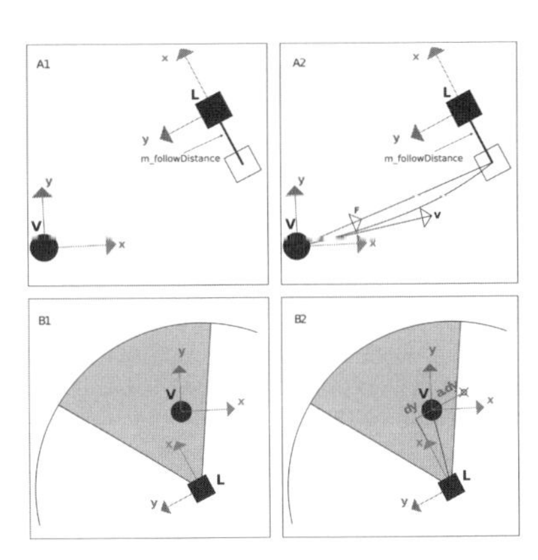

Figure 4. *Leader-follow* behavior in 2 steps.

A. *Follower behind Leader*:
step A.1: Vehicle at position $\mathbf{V}$ computes the virtual position $(0, -m\_followDistance)$ of Leader $\mathbf{L}$ with respect to its local coordinate system.
step A.2: The vehicle behaves as equipped with *Arrive* behavior and a target position the *Leader's virtual position* computed in the previous step.
B. *Follower in front of Leader*:
step B.1: The vehicle is located within the gray circular section, defined by $\mathbf{L}$'s local $x-axis$ and an angular parameter.
step B.2: The vehicle behaves as equipped with the *Arrive* behavior and a new target position ⊗, obtained by computing its distance (dy) from $\mathbf{L}$ local $x-axis$, multiplied by a parameter $a > 1$.

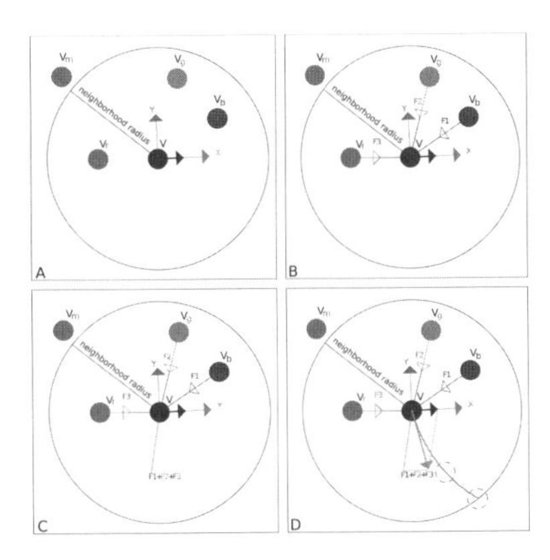

Figure 5. *Separate* behavior in 4 steps.

Step 1: Vehicle at position $\mathbf{V}$ is surrounded by vehicles $\mathbf{V}_b$, $\mathbf{V}_r$, $\mathbf{V}_g$, and $\mathbf{V}_m$. A circle of radius *neighborhood_radius* is *Separate*'s active area.
Step 2: Repelling forces $\vec{\mathbf{F}}_i$, $i = 1, 2, 3$ are calculated for each vehicle within active area.
Step 3: The computed repelling forces are combined to form the steering vector $\vec{\mathbf{F}} = \sum_{i=1}^{i=3} \vec{\mathbf{F}}_i$.
Step 4: Combine vehicle's current velocity (black vector) with the gray steering vector $\vec{\mathbf{F}}$ for updating the velocity vector.

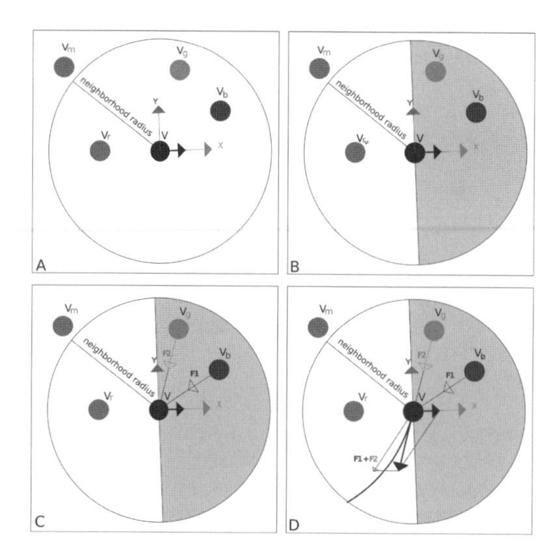

Figure 6. *Separate* behavior in 4 steps with `look_ahead` option activated.

Step 1: As Step 1 in Fig. 5.
Step 2: Repelling forces $\mathbf{F}_i$, $i = 1, 2, 3$ are calculated for each vehicle within the intersection of the active area with the positive x-halfplane, defined by the local coordinate system of the vehicle at position $\mathbf{V}$.
Step 3: As Step 3 in Fig. 5.
Step 4: As Step 4 in Fig. 5.

the *Separate* behavior. Specifically, `look_ahead` controls whether we want to use the complete neighborhood area or only the part in front of our vehicle. Thus, if `look_ahead` is set to true only neighboring objects in front of our vehicle contribute to *Separate* calculation. This neighborhood subdivision is materialized with the aid of our vehicle's local coordinate system as shown in Fig. 6 B.

3 EXAMPLES

This section is structured in two subsections illustrating *crew-assistance* and *grouping* behavior in VELOS for a generic four-room configuration (see § 3.1) and a IMO-based scenario of evacuating the stern zone of a passenger ship; see § 3.2.

3.1 *Generic example*

The proper implementation of group's behavioral modeling and crew assistance in VELOS is demonstrated in the following generic example.

Scenario: 70 persons are located in room A; see Fig 7. Population demographics is as proposed by IMO Guidelines (I.M.O. 2002). The persons must move from room A to D which is considered as the muster station. We calculate the total time required for all persons to reach room D. Three test cases were examined.

Test Case 1: No crew assistance (Fig. 8). The persons move from room A to room D using the main escape route, which is depicted by the solid line in Fig. 7. This is implemented through the assignment of the escape route to agents' *Path-Follow* behavior. In this context, crew assistance functionality is inactive. Significant queue appears in the exit of room A which seriously affects the travel time. Only the main escape route is used in the evacuation process. The total time needed for all persons to reach room D is 203 s.

Test Case 2: Crew assistance via *Trigger* (Fig. 9). In this case crew assistance is active and is implemented as follows: a crew member (represented by a static *Trigger*) is located at the exit of room A and monitors if congestion occurs there. Congestion is considered to occur if persons' density exceeds a prescribed limit (in our case 4 persons/m^2). In the simulation process congestion is demonstrated by getting red a density sensor (circle). Whenever congestion takes place the *Trigger* is activated and the attached to it list of actions consists in directing the persons towards the upper exit of room A until congestion disappears. In this particular implementation persons directed to the upper exit of room A are deprived of their *Path-Follow* behavior and proceed to the room D through successive signs. Both escape routes are used for the evacuation process. The total time needed for all persons to reach room D is 153 s.

Test Case 3: Crew assistance via *guide operation* (Fig. 10). In this case the crew functionality of guidance is activated by exploiting the *Leader-Follow* behavior. A crew member, which is marked by grey color, plays the role of leader and undertakes the mission to guide a subgroup of persons (marked by white color) to room D via the upper (secondary) escape route. These persons (followers) are considered as a group of Level 1, as described in subsection § 2.1. In this aspect, *Leader-Follow* behavior is combined with *Passenger Grouping* functionality. Grouping Level 1 explains also the fact that some members of the group don't follow their guide (crew member) up to the destination place room D. Moreover, the followers are deprived of their *Path-Follow* behavior. The total time needed for all persons to reach room D is 114 s.

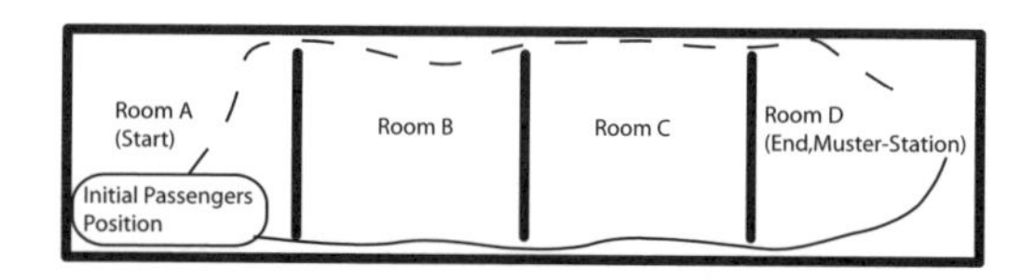

Figure 7. Generic example.

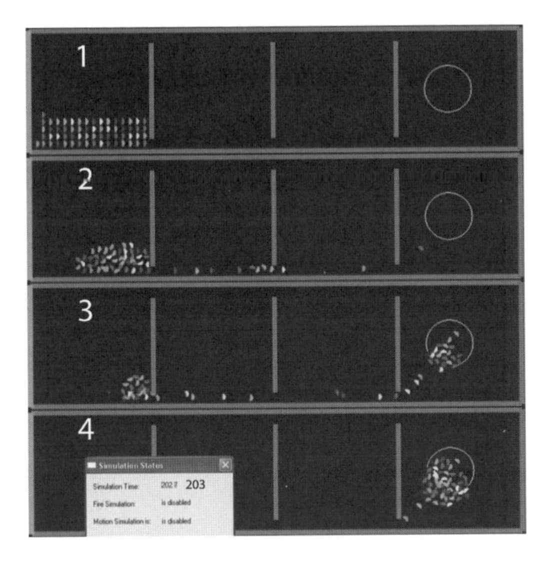

Figure 8. Example, Case 1: All passengers follow the main escape route.

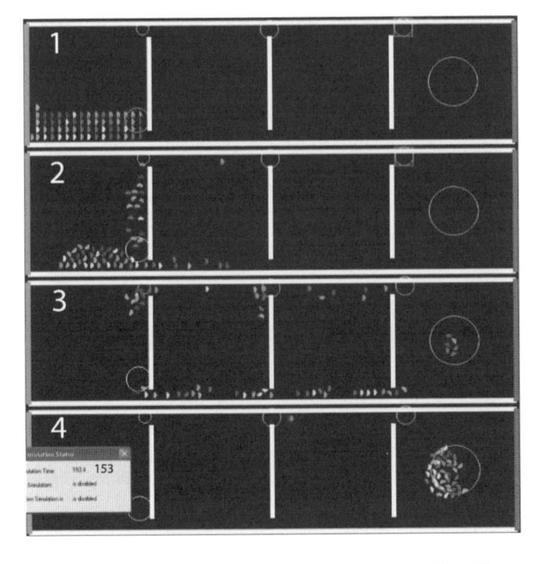

Figure 9. Generic example, case 2: Crew member diverts passengers when queuing at room's A lower exit.

3.2 *Ship example*

In this section we shall use VELOS for performing evacuation analysis for a ro-ro passenger ship with and without crew assistance. One hundred passengers are located in the cabins of Deck 5 of the after vertical zone of the ship, while Muster Station is located on Deck 7. Population demographics are as proposed in (I.M.O. 2007). For every simulation run we distribute randomly the population in the aforementioned areas. Three variations of the above scenario are simulated 3000 times each. For each variation, we compute the *travel time* required for all passengers to reach Muster Station as well as *cumulative arrival time* corresponding to the percentage of passengers reaching Muster Station for each time unit.

In the first variation (Scenario 1), passengers follow the designated escape route without crew assistance; Fig. 11 provides a snapshot of the evacuation process. The other two variations involve crew assistance. In Scenario 2 passengers are directed by two crew members to follow two distinct routes (see Fig. 12), while in Scenario 3 a crew member monitors passengers' density at a specified place and, whenever congestion is likely to arise, he redirects a group of passengers towards a secondary escape route; see Fig. 13 In both cases, crew assistance is materialized through *Triggers*,

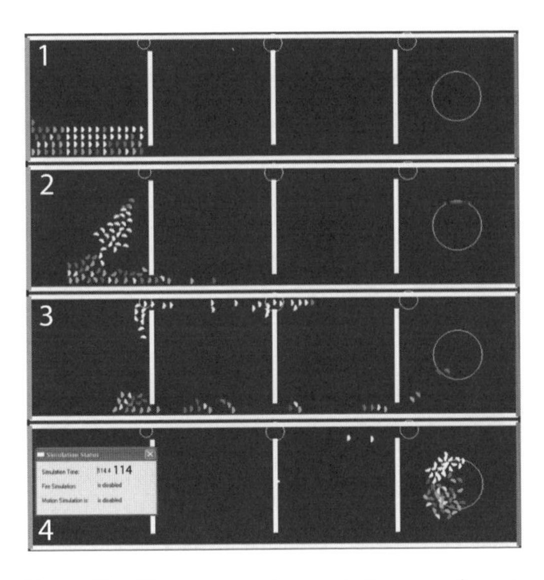

Figure 10. Generic example, case 3: Crew member acts as leader to a group of passengers (white agents).

Figure 11. Ship example, scenario 1: Snapshot from VELOS.

Figure 12. Ship example, scenario 2: Snapshot from VELOS.

Figure 13. Ship example, scenario 3: Snapshot from VELOS.

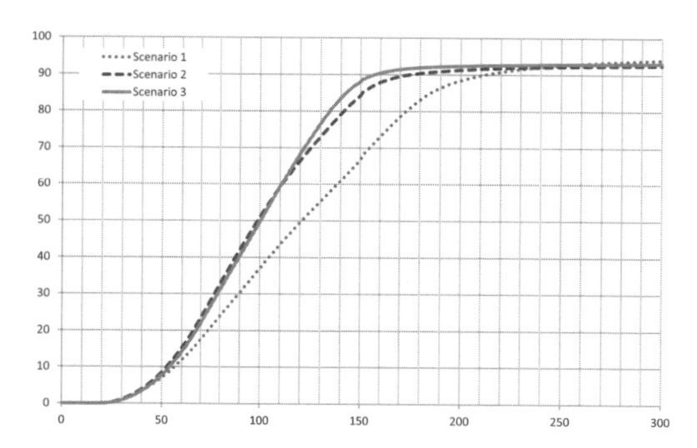

Figure 14. Average cumulative arrival time for Scenarios 1, 2 and 3.

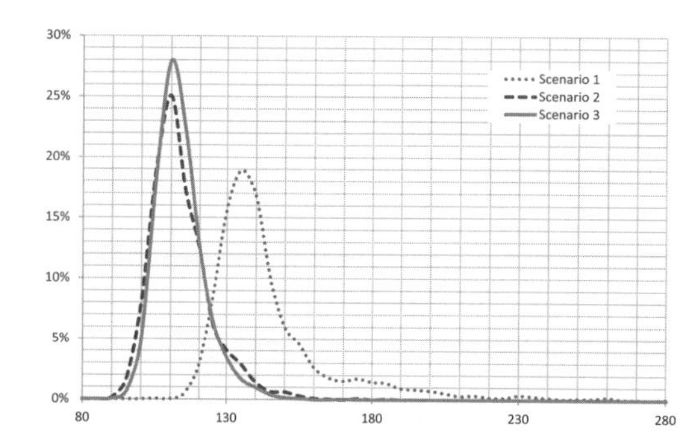

Figure 15. Travel time distribution for scenarios 1, 2 and 3.

which in Scenario 2 involves TAs applied to all passengers passing through the corresponding TN, while in Scenario 3 TAs are of dynamic character as a result of the attached density sensor.

Fig. 14 depicts the average of the cumulative arrival time for each scenario. As it can easily be seen from this figure, Scenarios 2 and 3, based on crew-assistance, achieve a considerably better performance compared to Scenario 1. Among Scenarios 2 and 3, the latter is marginally better as a result of the dynamic crew-assistance policy adopted. Analogous conclusions can be drawn from Fig. 15, where the distributions of travel-time of the three scenarios are depicted. Average travel time for Scenarios 1, 2 and 3 are equal to 147 s, 112 s and 113 s, respectively. Moreover, in Scenarios 2 and 3 travel-time distribution is narrow-banded, which reflects the effectiveness of the adopted evacuation processes versus that of Scenario 1.

REFERENCES

Ginnis, A.-A.I., Kostas, K.V., Politis, C.G. and Kaklis, P.D. (2010). Velos: A vr platform for ship-evacuation analysis. *CAD, special issue on Computer Aided Ship Design* *42*(11), 1045–1058.

Green, R. (2000). Steering behaviors. In *SIGGRAPH 2000 Conference Proceedings.*

I.M.O. (2001). IMO. *SOLAS Amendments 2000. International Maritime Organization, London*, UK. I.M.O.

I.M.O. (2002, June). *Interim Guidelines for evacuation analyses for new and existing passenger ships* (MSC/Circ. 1033 ed.). I.M.O.

I.M.O. (2007, 30 October). *Guidelines for evacuation analyses for new and existing passenger ships* (MSC.1/Circ. 1238 ed.). I.M.O.

Kostas, K. (2006). *Virtual Virtual Reality Kernel with Support for Ship Life-cycle Modeling*. Ph. D. thesis, National Technical University of Athens, School of NAME.

Kostas, K.V., Ginnis, A.-A.I., Kaklis, P.D. and Politis, C.G. (2007). Velos: Virtual environment for life on ships. In *Proc. of the 3rd International Maritime Conference on Design for Safety, September 26-28, 2007 Berkeley, California* USA.

Reynolds, C.W. (1987). Flocks, herds, and schools: A distributed behavioral model, in computer graphics. In *SIGGRAPH '87 Conference Proceedings*, pp. 25–34.

Reynolds, C.W. (1999). Steering behaviors for autonomous characters. In *GDC'99 (Game Developers Conference)*.

Vanem, E. and Skjong, R. (2006). Designing for safety in passenger ships utilizing advanced evacuation analyses—a risk based approach. *Safety Science 44*(1), 111–135.

Sustainable Maritime Transportation and Exploitation of Sea Resources – Rizzuto & Guedes Soares (eds)
© 2012 Taylor & Francis Group, London, ISBN 978-0-415-62081-9

High-flexible multi-technique acquisition system for human postural stability tests onboard ships

E. Nocerino
DIN—Dipartimento di Ingegneria Navale, "Federico II" University of Naples, Italy
LTF—Dipartimento di Scienze Applicate, "Parthenope" University of Naples, Italy

F. Menna
3DOM Research Group, FBK—Fondazione Bruno Kessler, Trento, Italy
LTF—Dipartimento di Scienze Applicate, "Parthenope" University of Naples, Italy

S. Ackermann, S. Del Pizzo & A. Scamardella
LTF—Dipartimento di Scienze Applicate, "Parthenope" University of Naples, Italy

ABSTRACT: For validating theoretical models aiming to simulate the postural behaviour of working personnel one of the possible methods is the execution of trials onboard full-scale ships. This kind of tests requires measuring ship and human movement simultaneously. Up to now, international guidelines for standardizing human postural stability tests onboard have not been established. Scientific works focused on this topic do not make use of human motion capture techniques commonly employed for postural stability studies indoor. This paper presents an innovative motion acquisition system usable onboard ships while accomplishing the daily mission. The system integrates different techniques (photogrammetry, inertial measurements, global positioning system) for acquiring both ship and human motions. Its core is an own-developed low-cost motion capture system fundamental in analysing and understanding the measurements from the inertial sensors. Preliminary laboratory tests and results from measurement campaigns onboard are presented and discussed.

1 INTRODUCTION

Postural stability is a key topic for the maritime sector as wave induced ship motions make the maintenance of upright stance demanding and moving in a controlled manner very difficult, negatively affecting safety of personnel working onboard. Mariners have to concentrate on standing upright while performing the allotted task, avoiding risk of potential injury. Crewmembers of fishing vessels, navy craft, supply vessels, experience conditions of works that are different from those faced by workers in other sectors. The fatality rate for fishers is typically several times higher than for other employees, making fishing a very hazardous activity (ILO 2007).

Several models have been proposed for predicting loosing of balance events, called Motion Induced Interruptions (MIIs), or simulating the postural behaviour of a crewmember onboard sailing vessels (Nocerino 2011). The commonly used model, proposed by Graham et al. (1992) assumes the individual to react as a rigid body (a dummy) and underestimates the human capacity to counteract external disturbances (i.e. ship motions). More complex postural models have been proposed but necessitate to be verified (Nocerino 2011). Validating theoretical models requires many experiments to be conducted, in order to obtain relevant (statistically) data and infer significant results. Because of the peculiarities of the issue, the only practicable solution is to execute experiments in motion simulators or full-scale sea trials. Trials in motion simulators have the advantage that the same sequence of motions may be reproduced to obtain controlled experiments on a large number of subjects. However, some simulators cannot reproduce all six degrees of freedom or very large ship motions (in particular, translational motions). Practical experiments are limited since it is not possible to simulate all the tasks or phenomena occurring on a real ship. The results of tests conducted in motion simulators are likely more conservative than measures taken on actual ships since the harsh environmental conditions likely encountered onboard (unfavourable temperature, weather and slippery platform conditions) cannot be reproduced. The costs for using such facilities can be very high, even prohibitive if the research is not supported by appropriate funding. Onboard experiments allow studying postural stability in the field and correlating impaired balance with ship motions and onboard conditions in the real environment. Two main problems are related to the execution of tests onboard. First of all, standard procedures for conducting experiments onboard and validating human postural model have not been established yet. Few studies focused on this topic have been conducted and published.

The most part of them was interested in experimentally determining empirical coefficients for improving the prediction of MIIs occurrence employing the standard model by Graham (Crossland & Rich 1998). In the case of more complex human stability models, full-scale trials had principally the aim of tuning the proposed models in order to reproduce the MII rate and time of occurrence observed during the tests (Langlois et al., 2009). The investigators were not focused in verifying the agreement between the simulated response of the balance control model and the measured behaviour of individual working onboard vessels. A general procedure should require executing a huge number of trials onboard several ships, testing many individuals involved in several tasks and in different environmental conditions, resulting in unbearable costs. This contribution presents a procedure for measuring human motion onboard ships while accomplishing their daily mission. The designed acquisition system results from the collaboration among researchers involved in many scientific fields. Diverse techniques (photogrammetry, inertial measurements, global positioning system) and instruments (digital cameras, inertial and positioning sensors) have been integrated with the ultimate goal of acquiring both ship and human motions. The whole process of planning, designing, calibrating, and assessing the accuracy of the motion capture system is reported and discussed. The motivations that are at the basis of setting up this "ad hoc" acquisition system are analysed. Laboratory tests were conducted for testing and validating the entire measurement system before executing preliminary trials onboard a fishing vessel. Results from the first measurement campaigns are presented and discussed.

2 HUMAN POSTURAL STABILITY

Human upright stance is inherently unstable. Even in the absence of additional disturbances (movements of the supporting surface, external forces, etc.), the gravitational component causes the COM (Centre of Mass) to move away from the equilibrium position (continuous sway movement), around a pivot point (the ankle joint) at a certain height from the ground. Posture means the orientation of the body relative to the gravity, that is an angular measure from the vertical (Winter 2005). The postural motion of an individual is classically studied separately into two perpendicular planes defined relative to the human body. Figure 1a shows the sagittal plane (in red) that bisects the body in the antero/posterior direction (plane of symmetry), and the coronal plane (in blue) that halves the body into front/back parts. In the sagittal plane, the simplest biomechanical model for reproducing and analysing the postural behaviour of a person standing upright is a single-segment inverted pendulum, whose parameters are chosen for matching mass and inertial characteristics of the individual. This model is able to embody the intrinsic instability of human posture, characterized by continuous oscillation of the COM around the equilibrium position (Nocerino 2011). Representing the human body as an inverted pendulum is common in literature for modelling the postural

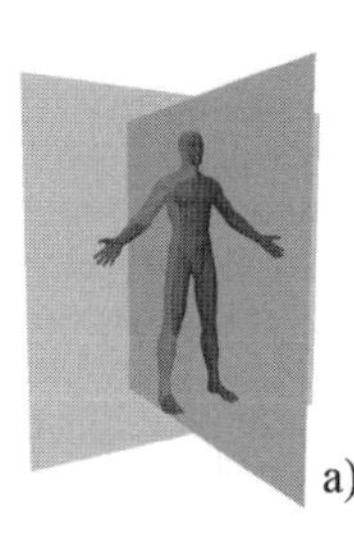
a)

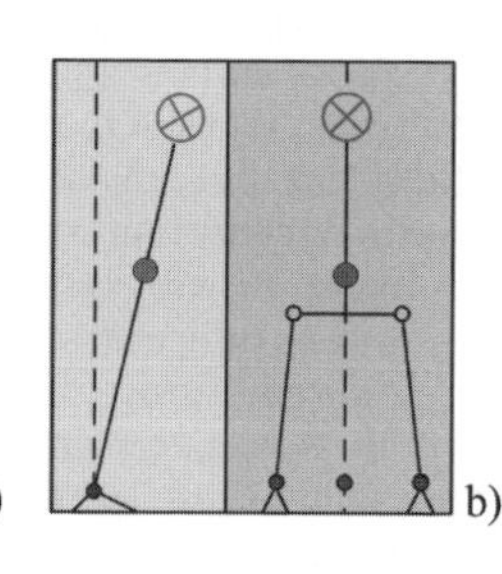
b)

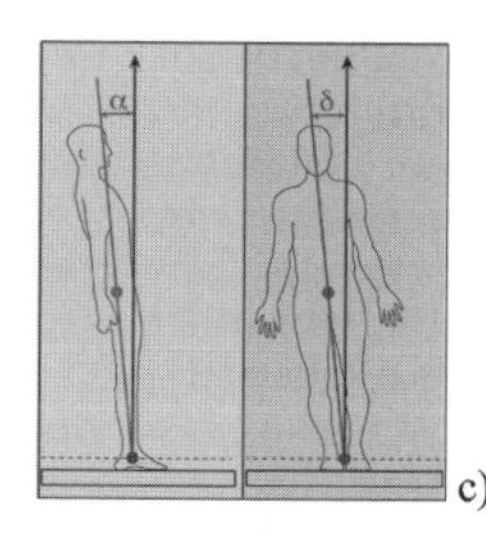

c)

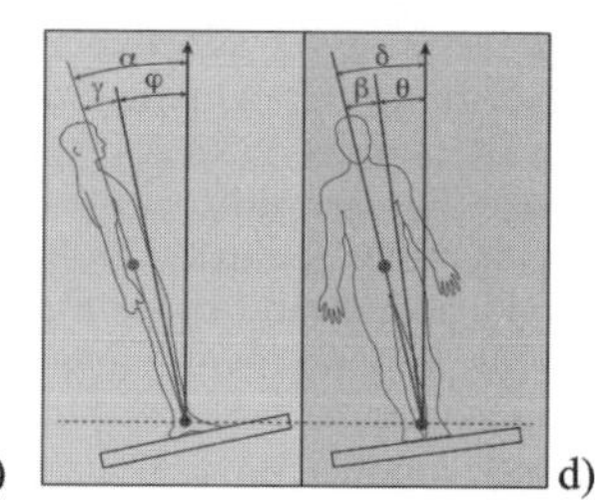

d)

Figure 1. a) Anatomical planes: the sagittal (red) and coronal (blue) planes. b) Biomechanical models for human postural stability in the sagittal (red) and coronal (blue) planes. c) Leaning angle α in the antero/posterior direction (sagittal plane, in red) and δ in the lateral direction (coronal plane, in blue). d) Human postural sway on a moving ship deck: the person is facing laterally. Left: in the sagittal plane (human stability influenced by the deck roll angle φ) postural sway angle in the antero/posterior direction α relative to the vertical and γ to the normal to ship deck. Right: in the coronal plane (human stability influenced by the deck pitch angle θ) the postural sway angle in the lateral direction δ relative to the vertical and β to the normal to the deck.

behaviour in the sagittal plane (Fig. 1b, in red). Winter et al. (1998) extended the use of the inverted pendulum model also in the coronal plane (Fig. 1b, in red), even if in this plane more complex models are usually employed (Nocerino 2011). A measure for quantifying the postural stability of a person standing on a stationary (quiet standing) or moving (perturbed or dynamic standing) is provided by the sway or leaning angles of the person's COM from the ankle joint. In Figure 1c, the angles in the sagittal (α) and coronal plane (δ) are shown in the case of quiet standing. In perturbed conditions, the postural sway is influenced by the movements of the platform (e.g., ship motion). Figure 1d shows the sway angles for a person facing laterally on a ship's deck: (I) φ and θ are the ship roll and pitch angles; (II) α and δ are the human sway angles in the antero/posterior and lateral directions relative to the vertical (inertial frame); (III) γ and β are the human sway angles in the antero/posterior and lateral directions relative to the normal to ship deck (ship-fixed frame).

3 HUMAN POSTURAL STABILITY TESTS ONBOARD SHIPS

Full-scale sea experiments for studying human stability combine elements characterising the classical seakeeping trials with procedures of human motion analysis. While the main goal of full-scale seakeeping tests is to assess and compare the performance of ships in rough weather (Lloyd 1989), the tests conducted in the present study aim at observing and recording the postural stability behaviour of sailors in the real working environment. The crucial interest is to examine the actual movement of subjects trying to maintain balance, rather than to assess only the seakeeping qualities of the tested ship. These requirements demand for a multidisciplinary approach, involving the classical seakeeping method, together with human motion capture techniques.

3.1 *Seakeeping full-scale trials*

Seakeeping trials, by their nature, are highly-costly and time consuming. They require careful planning, even if the final result is strongly related to the environmental conditions. During the execution, the sea state and weather in general can significantly change, causing the tests to be not practicable or the result not reliable. Evidently, the ship is subtracted from the daily service with heavy costs for the owner. Each trial run must be of sufficient duration in order to ensure a reasonably estimate of the physical parameters to be assessed (Lloyd 1989).

Until some years ago, instruments used for seakeeping tests were very expensive. With the advent of new technologies, flexible and low-cost instrumentations for measuring the motion of objects are now available. Advances in Micro-Electro-Mechanical System (MEMS) technology have permitted the development of low-cost inertial sensors (composed of a cluster of accelerometers and gyroscopes), that can be used for evaluating both ship's angular motions and linear accelerations (Koning 2009). Global Navigation Satellite Systems (GNSS), such as the American Global Positioning System (GPS) or the Russian GLObal Navigation Satellite System (GLONASS) together with augmentation systems (e.g. the European Geostationary Navigation Overlay Service—EGNOS), provide satisfactory accuracy in determining craft position even during high speed dynamic manoeuvres (Pacifico et al., 2009).

3.2 *Human motion capture*

Motion capture is the technique for recording the movement of a subject in order to obtain measurements of kinematic variables (linear and angular displacements, velocities, and accelerations), acquire information for animating digital character models, recognise and track individuals in areas that need to be monitored (banks, airports, military installations, etc.). Considering the different applications, motion capture can be divided into the main areas of body movement, facial capture, and hand gestures. For studying postural stability and human movements in general, different measurement methodologies are used. According to the sensors employed, these techniques can be classified in direct measurement methods, based on mechanical, electromagnetic, acoustic, inertial trackers, and indirect optical (imaged-based) methods (D'Apuzzo 2003, Roetenberg 2006). Nowadays, portable and light sensors have been developed; in particular, inertial sensors, often integrated with magnetic sensors or GPS receivers, have become quite common for motion capture and measurement (Roetenberg 2006). Image-based sensors are also very popular in both human body modelling and movement analysis (Chiari et al., 2005, D'Apuzzo 2003). Among the wide range of existing methodologies and technologies, the choice of instrumentations and sensors depends on the requirements of the specific experimentation. Studying human stability and postural movements onboard ships means to conduct experiments in locations that differs substantially from any conventional ashore laboratory. The need of capturing the motion of an individual standing on a ship during its daily service does not permit the use of optical systems commonly employed for human motion analysis in indoor, motionless laboratories. Commercial optical systems (e.g., Motion Analysis™, Vicon™), made up of many cameras, provide real time 3D data, but are costly and not sufficiently flexible. They have to be installed in specific locations and connected to a dedicated computer; these sensors are not designed to operate in disadvantageous environmental conditions (water, wetness, saltiness) and with not optimal lightning.

4 REQUIREMENTS FOR A MOTION ACQUISITION SYSTEM ONBOARD

In order to encompass the strong practical limitations related to the execution of postural stability tests onboard ships, an ad-hoc motion capture system was developed. The motion acquisition system has been designed for being usable onboard ships while accomplishing their daily mission and handled even by only one person in order to minimise the encumbrance onboard. The peculiarities of the proposed motion measurement system can be summarised as followings:

- affordable cost for operating in potentially "dangerous" environment (water, wetness, saltiness);
- ease of use and calibration even in disadvantageous conditions, as on a ship's deck;
- speed in setting up in different configurations and removing in order to not hinder the normal execution of onboard tasks;
- flexibility for being employed on a movable platform, in cramped spaces, in unfavourable lighting conditions (also in the dark);
- taking up a minimal amount of space;
- to be portable on different platforms (diverse ships), without interfering with onboard work.

The designed system is the core of a wider "mobile laboratory" for the assessment of manoeuvrability, seakeeping and personnel safety onboard developed in cooperation with the Laboratory of Topography and Photogrammetry of the "Parthenope" University of Naples (http://lft.uniparthenope.it) and the 3DOM Research Group of FBK (Fondazione Bruno Kessler) of Trento (http://3dom.fbk.eu).

5 THE MOTION ACQUISITION SYSTEM FOR HUMAN POSTURAL ANALYSIS ONBOARD

Taking into account the requirements for human postural tests onboard, the measurement system has been realised into two main components (Fig. 2): (i) ship motion measurement unit, and (i) human motion capture system.

5.1 *Ship motion measurement*

For tracking and measuring ship motions in the six degrees of freedom an integrated GPS and Inertial Measurement Unit (IMU), based on MEMS technology, has been used. The Motion Tracker MTi-G (www.xsens.com) consists of three accelerometers, three gyros, three magnetometers, magnetic and static pressure sensors. The sensor provides three linear accelerations, rates of turn and orientation data (Euler angles) referred to an inertial reference frame. When the GPS signal is available, the MTi-G also supplies position and velocity data.

5.2 *Motion capture system for human postural measurement onboard*

For measuring the postural behaviour of an individual onboard, two methodologies have been employed: motion measurement with inertial sensor and motion capture with videogrammetry. The integration of different techniques assures redundancy in the achieved measurements. Moreover, the use of videogrammetry allows the control (reliability) in the estimation of the parameters of interest: videos can be revised for better understanding and reconstructing the recorded phenomena and, also, identifying measurement errors.

5.2.1 *Inertial measurement of human motion*

A portable device (Xbus Master) controlling a lightweight MEMS inertial sensor (Xsens Xbus kit, www.xsense.com) is used for acquiring 3D orientation as well as kinematic data of individuals onboard ships referred to an inertial reference frame. The master can be attached to a belt and

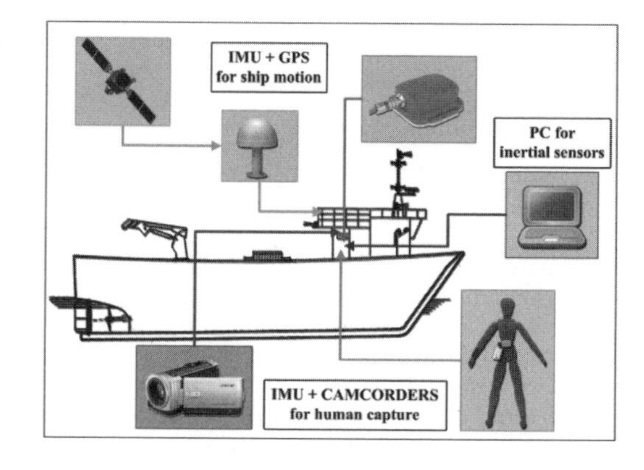

Figure 2. The proposed motion acquisition system for human postural measurement onboard.

fasten around the waist, while the inertial unit can be fixed to the body.

5.2.2 *Videogrammetry for human motion capture*

Videogrammetry is an extension of photogrammetry, a well established method for making precise and reliable measurements and performing three-dimensional (3D) reconstruction of an object from images. Over the years the photogrammetric technique has been successfully applied in various sectors (aerial mapping, cartography, remote sensing, industrial measurement, quality control, cultural heritage), including the maritime field (Ackermann et al., 2008, Menna et al., 2009). The major difference between photogrammetry and videogrammetry is that the latter makes use of video components (such as CCD-cameras, video recorders and frame grabber) for image acquisition (Gruen 1997). If the body is moving and the aim of the survey is to retrieve its dynamics, images from different points of view have to be taken simultaneously. This is achieved by using synchronised cameras that record the moving object from different positions. The movement is sampled according to the camcorder frame rate and reconstructed by tracking points of interest across the epochs (frames).

The proposed image-based motion capture system is made up of three low-cost, light and compact video cameras (Sony™ HDR-CX106E) and different kind of supports (three solid tripods or clamps with respective ball heads). The camcorders employ a CMOS (Complimentary Metal-Oxide Semiconductor) image sensor with a resolution of 1920 × 1080 pixels and a frame rate of 50 fps (frames per second). The cameras do not require a computer controller (no cables are necessary) and operate autonomously storing data on their internal memory.

6 SYSTEM SET UP AND CALIBRATION

The simultaneous employment of different sensors and acquisition techniques has required the development of specific procedures for calibrating and validating the single system components, synchronizing the whole acquisition system and making different data comparable.

6.1 *Videogrammetric system calibration procedure*

For retrieving 3D precise measurement of a moving object employing imaged-based systems, it is necessary to compute parameters which describe both (i) the internal characteristics of the cameras and (ii) their relative position and orientation in the space (Ackermann et al., 2008, Menna et al., 2009). For the proposed system, the first step (i), called camera calibration is performed in laboratory, the second (ii), named exterior orientation, is computed on site every time the system is set up (Nocerino et al., 2011). The precision in determining calibration and orientation parameters for the whole set of cameras influences the precision in measuring 3D coordinates.

A fundamental requirement of a videogrammetric system is the synchronization among the cameras: for the present study, this operation is performed by means of a LED (Light Emitting Diode). The LED is switched on (off) within the Field of View (FOV) of all the recording cameras: the first frame in which it changes its state (from off to on or vice-versa) is taken as the synchronization event. The misalignment error between the three video tracks can be one frame at most. This systematic synchronization error affects the accuracy of point coordinates as much as the recorded object or the recording system moves with high speed.

For tracking automatically one or more moving points, a specific algorithm has been developed: well-distinguishable multiple points (e.g., LEDs) are detected and tracked through the frame sequence of each camera (Nocerino et al., 2011, in press).

The proposed system was tested and compared with a commercial motion capture system in order to verify its accuracy and robustness in surveying objects moving at different speeds. For low movements, the comparison showed that employing low-cost optical systems is an acceptable compromise between cost, flexibly and achievable accuracy (Nocerino et al., 2011, in press).

6.2 *Combining IMU and videogrammetric sensors*

In order to synchronise videos and IMU data, a suitable system was assembled. A mechanical switch simultaneously hits the inertial sensor and turns on a LED fixed to the IMU housing. The inertial sensor measures the hit as a peak in the value of the acceleration along the sensitive axis and, at same time, the videogrammetric system captures the LED switching on: this point is selected as the synchronization moment for the two different measurement systems.

A suitable base of support was built of non-magnetic rigid material (Fig. 3a, b). The IMU for human motion measurement is fastened on the plate, with four coloured high-power LEDs fixed at the corners. The LEDs are tracked with the own-developed algorithm for retrieving both the linear and angular movements of the plate and, consequently, the motion of the inertial sensor. The plate, where a recognisable path with

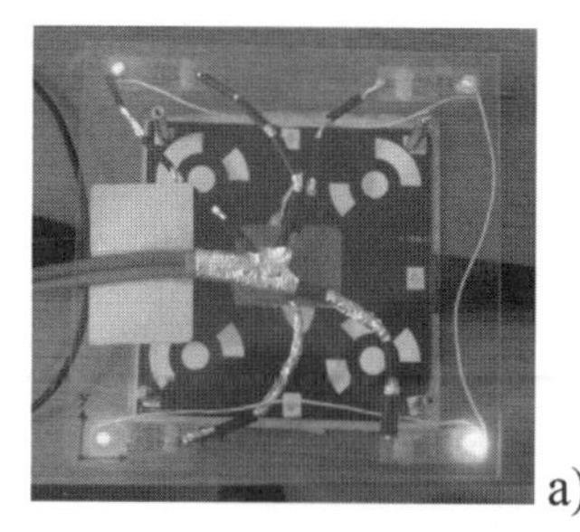
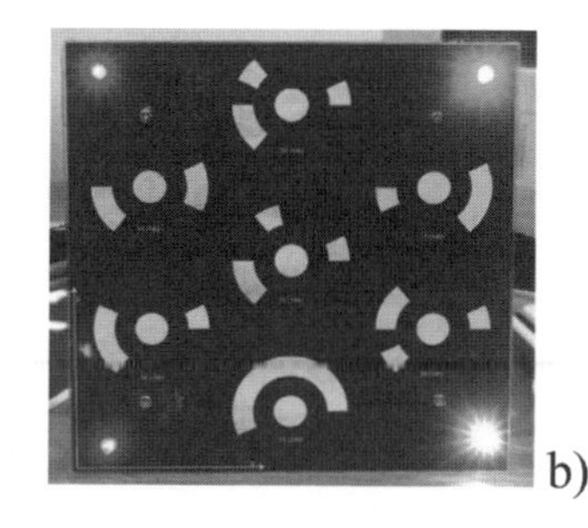
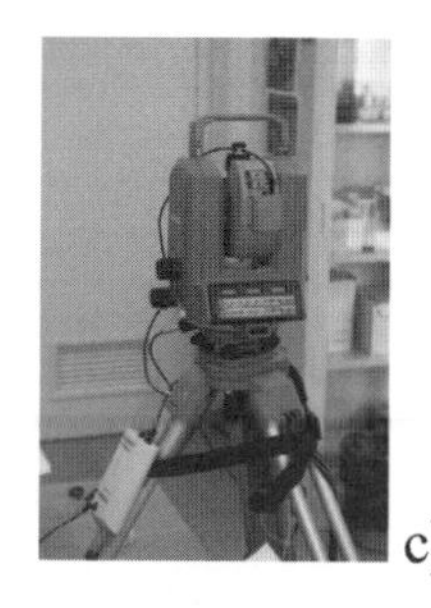

Figure 3. a) and b) Photogrammetric plate for the MTx sensor equipped with coloured LEDs. c) IMU fastened on the telescope of the WILD T2000 theodolite for static tests.

photogrammetric coded targets was drawn, is also use for defining a reference frame parallel to the IMU-fixed coordinate system.

6.3 *Reliability of the inertial sensors*

As widely discussed and motivated in literature (e.g., Roetenberg 2006) there are important limitations in the current motion measurement systems based on MEMS IMU. These arise mainly because the inherent drift of the orientation and position estimates limits long-term stability. In order to verify the reliability of the inertial sensors for the application of interest, several static and dynamic tests have been conducted.

6.3.1 *Static tests*

Laboratory indoor tests have been conducted in order to verify the orientation output provided by the inertial sensors (Mti-G without the GPS antenna and MTx) in static conditions. The units were fastened to a high-precision electronic WILD T2000 theodolite (Fig. 3c). By turning the instrument in different positions, calibrated data from the sensors were acquired using the factory software and recorded for at least 60 seconds. The tests showed that the inertial sensors, after short transients, were able to measure attitude (roll and pitch) with an accuracy of about 0.15°. Moreover, during the time interval of each single trial, the drift was not significant. Troubles were found in determining the yaw (heading), since the magnetometers were strongly influenced by the magnetic field generated by the circuitry of the theodolite. It was concluded that in static conditions the inertial sensors work well as inclinometers. Unfortunately, they cannot be used for heading assessment if strong and variable magnetic fields are expected. As far as the MTi-G is concerned, improvements are presumed if data from GPS are integrated.

6.3.2 *Dynamic tests*

The videogrammetric system was used as reference for testing the reliability of the inertial units in dynamic conditions. Two different sets of trials have been conducted. The videogrammetric system was arranged into the laboratory (indoor) for executing the experiments with the MTx, the IMU for measuring human motion. In the case of the MTi-G employed for recording the ship motion, the experiments were carried on outside (on a terrace for heaving the skyline free from obstruction), in order to test the integration with the GPS. The inertial sensors were fastened on the photogrammetric plate, which during the tests was manually moved by an operator, gradually increasing the dynamics of movement. The motion of the plate was recorded by the three camcorders while calibrated data from the inertial sensors were acquired and sampled at 100 Hz using the factory software. Each trial lasted about 90 seconds. The laboratory testes showed that roll and pitch angles from the inertial sensor are not characterised by a significant drift compared to the videogrammetric output. The maximum differences between the two measurement techniques are observed when the movement dynamics increase (rate of turn greater than 4-degrees/sec ca.). The yaw angle provided by the MTx is not sufficiently reliable, since a non-negligible drift was present at the end of the test. The reason is that the magnetometers, that provide the measurement of the yaw angle, are no longer able to compensate the local magnetic field. On the contrary, the yaw angle provided by the MTi-G was proved to be more stable thanks to the integration with the GPS. The acceleration are obtained with videogrammetry by deriving twice the tracked plate motion. For comparing them with the accelerations from the inertial units, the components due to gravity are calculated and added. The comparison showed the accordance between the data from the different techniques. Figure 4 reports a representative comparison between the roll angle and the acceleration along the X-axis measured by the MTx (in red) and calculated with videogrammetry (in blue).

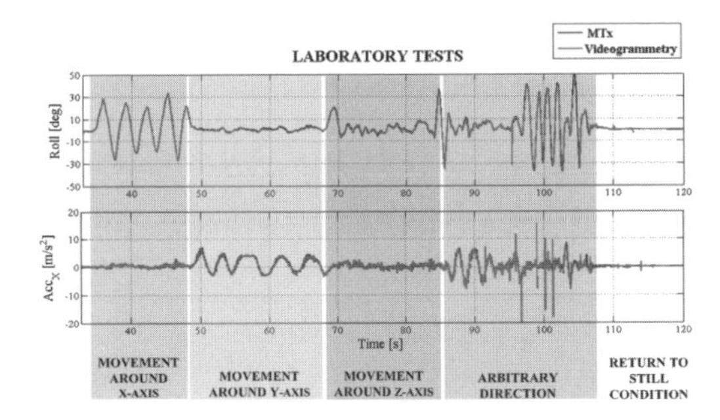

Figure 4. Comparison between the roll angle (up) and acceleration along the X-axis (down) measured with videogrammetry and inertial sensor (MTx).

7 HUMAN CENTRE OF MASS CALCULATION

For the present study, evaluating the COM's height was necessary in order to obtain an approximated location for placing the inertial sensor and videogrammetric targets. For this purpose, the model proposed by Winter (2005) has been used: the human body is divided into 14 segments, identified by 21 markers. Measuring the 3D coordinates of the 21 markers with a videogrammetric system, the COM position in 3D is determined according to the general formula for barycentre calculation (Nocerino 2011, Winter 2005). In a suitable space (indoor laboratory), the proposed capture system was set up and calibrated. A set of 21 adhesive bandage markers were prepared for each individual to be measured. The markers were drawn as black circles on a white background, in order to ensure a high contrast; this made easier the point measurement phase with the Photomodeler™ photogrammetric software (Fig. 5). Once calculated the COM's height, each subject was also instrumented with the IMU placed on the reference plate. Assuming that the person body is rigid, the COM motion can be tracked recording the displacement of the photogrammetric reference plate positioned at the measured COM height.

Figure 6a reports the two-dimensional displacement of the COM for an 80-second record in quiet standing. The trend shows that the main motion component is in the anterior/posterior direction, in agreement with the hypothesis of modelling a person as an inverted pendulum. In Figure 6b, the motion component outside of the coronal plane is illustrated; the trend gives information about the validity of the "inverted pendulum" hypothesis. If the magnitude of the out-of-plane component increased significantly, the "inverted pendulum" model would be no longer suitable for describing the individual postural stability. The final part of the displayed trace shows that the out-of-plane COM component is increasing: this corresponds to the moment where the recorded subject is moving at the end of the postural stability test.

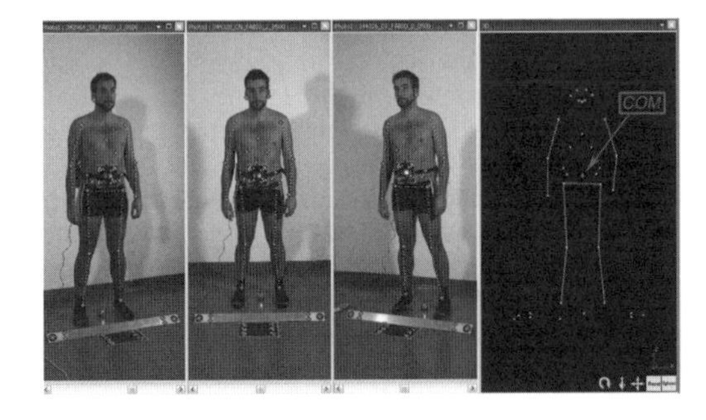

Figure 5. COM calculation. Left: Synchronised image frames from the three video cameras. Right: 3D positions of body COM (red dot).

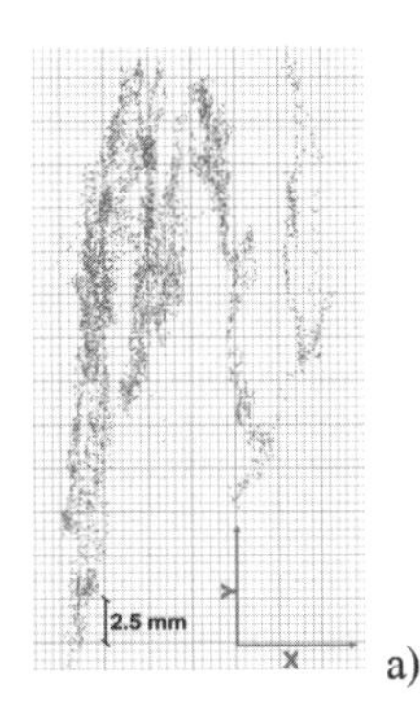

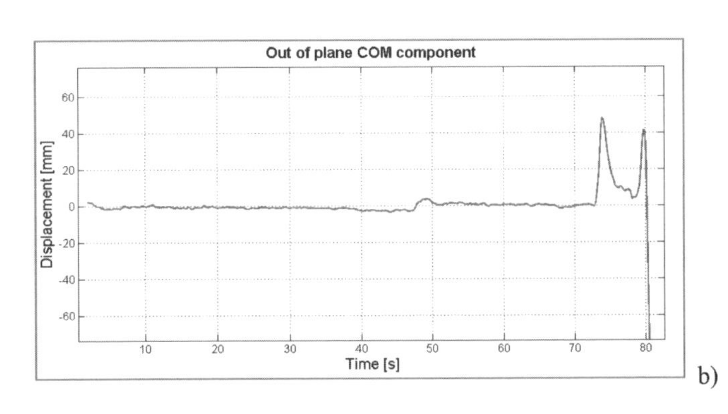

Figure 6. a) 2D displacement of the COM for an 80-second record in quiet standing. b) Out-of-plane COM component from videogrammetry.

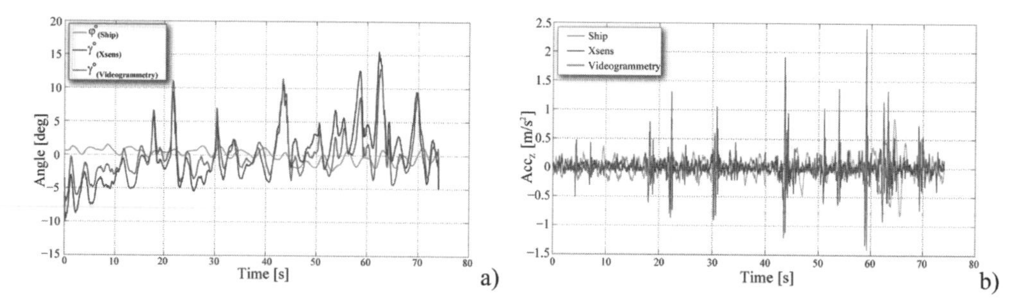

Figure 7. Results of postural test onboard (the person is facing laterally). a) Postural sway angle in the antero/posterior direction γ (relative to the normal to ship deck) from videogrammetry (red) and inertial sensor (blue); ship roll angle φ in green. b) Z-component acceleration from videogrammetry (red) and inertial sensor (blue); ship Z-component acceleration without the g-component in green.

8 POSTURAL STABILITY TEST ONBOARD SHIPS

Onboard trials were conducted to test the proposed motion capture system in the real environment, characterized by challenging conditions (cramped spaces, adverse lighting conditions, vibrations, etc.). The ship employed to this purpose is a 16.30-meter long wooden ship. The "San Gaetano" is a purse seiner inshore fisheries vessel operating in the Bay of Naples. In order to not interfere with the normal activities of the crew members, the trials were performed inside a cabin located behind the pilot bridge. The camcorders were secured to the ship structures using suitable clamps. Two video cameras were placed for viewing simultaneously an individual facing forward. The third camera was arranged in order to form a base with one of the other two devices and record an individual facing laterally. The tested subjects were instrumented with the inertial sensor MTx positioned at the COM height estimated during preliminary laboratory tests (Section 7). The MTi-G for ship motions measurement was positioned on the cabin deck close to the vertical projection of the individual's COM and the GPS antenna was located on the top of the pilot bridge. The two separate inertial sensors were connected to the same PC and the data from both the MTi-G and MTx were collected via the factory software at a 120 Hz acquisition rate. This allowed to reduce the hardware required onboard and, at the same time, to synchronise the data from the two inertial units. The method explained in Section 6.2 was applied to synchronize the inertial sensors with the videogrammetric system. Several trials were executed: the postural motion of the tested subjects was measured while maintaining the upright stance facing forward and laterally. The inertial sensors provide the orientation of human body and ship relative to an inertial frame, while the optical-based system measures the movement of the observed individual in the ship-fixed frame. Both the orientation output and accelerations for the MTx were opportunely transformed and compared with the measurements from videogrammetry. Figure 7a shows the postural sway measured both by the inertial device and with videogrammetry for an individual facing laterally. The deck roll angle is also illustrated.

In Figure 7b, the Z-component accelerations for the same subject are reported, together with the deck accelerations without the g-component. The results show a good correlation between the two measurement systems.

9 CONCLUSIONS

This contribution presented a multi-technique system for human motion capture onboard ships, designed for being flexible, light and manageable. It is made up of two main components, for both human and ship motion measurements, that can be installed in different configurations and settings onboard ships in ordinary operation. Laboratory tests were conducted for testing the reliability of low-cost inertial units by using the videogrammetric system as reference. For 120-second tests at most, the inertial data provided satisfactory results, in terms of stability and accuracy. Several trials were conducted onboard a fishing vessel to test the proposed motion system in the real environment. Despite the practical difficulties related to the specific setting (cramped spaces, adverse lighting conditions, vibrations, etc.) the motion acquisition system provided satisfactory results. The two human capture systems are able to provide the main parameters (sway angles and accelerations) for studying human postural stability onboard.

REFERENCES

Ackermann, S., Menna, F., Scamardella, A. & Troisi, S. 2008. Digital Photogrammetry for High Precision 3D Measurements in Shipbuilding Field. *6th CIRP International Conference on ICME—Intelligent Computation in Manufacturing Engineering*, Naples, Italy, 23–25 July.

Chiari, L., Croce, U.D., Leardini, A. & Cappozzo, A. 2005. Human movement analysis using stereophotogrammetry: Part 2: Instrumental errors. *Gait & posture* 21(2): 197–211.

Crossland, P. & Rich, K.J. 1998. Validating a Model of the Effects of Ship Motion on Postural Stability. *The 8th International Conference on Environmental Ergonomics*, San Diego, USA: 385–388.

D'Apuzzo, N. 2003. *Surface Measurement and Tracking of Human Body Part from Multi Station Video Sequences*. Ph.D. Thesis, ETH—Swiss Federal Institute of Technology, Zurich, Switzerland.

Graham, R., Baitis, A.E. & Meyers, W.G. 1992. On the Development of Seakeeping Criteria. *Naval Engineers Journal* 104(3): 259–275.

Gruen, A. 1997. Fundamentals of videogrammetry—A review. *Human Movement Science* 16(2–3): 155–187.

ILO—International Labour Organization 2007. *Work in fishing*. Convention No. 188, Recommendation No. 199. Geneva.

Koning, J. 2009. *Lashing@Sea. Executive Summary*. Report No. 19717-20-TM. Wageningen, the Netherlands: MARIN—MAritime Research Institute Netherlands.

Langlois, R.G., MacKinnon, S.N. & Duncan, C.A. 2009. Modelling Sea Trial Motion Induced Interruption Data Using an Inverted Pendulum Articulated Postural Stability. *The Transactions of The Royal Institution of Naval Architects, International Journal of Maritime Engineering* 151(A1): 1–9.

Lloyd, A.R.J.M. 1989. *Seakeeping: Ship Behaviour in Rough Weather*. Chichester, West Sussex, England: Ellis Horwood Ltd.

Menna, F., Ackermann, S., Scamardella, A., Troisi, S. & Nocerino, E. 2009. Digital photogrammetry: a useful tool for shipbuilding applications. *13th Congress of Intl. Maritime Assoc. of Mediterranean (IMAM2009)—Towards the Sustainable Marine Technology and Transportation*, Vol. I, Istanbul, Turkey, 12–15 October: 607–614.

Nocerino, E. 2011. *Human postural stability onboard ship as seakeeping criterion. Stance control model and procedure for validating it: a proposal*. Ph.D. Thesis, "Federico II" University of Naples, Naples, Italy.

Nocerino, E., Ackermann, S., Del Pizzo, S., Menna, F. & Troisi, S. 2011. Low-cost human motion capture system for postural analysis onboard ships. *Proc. SPIE*, Vol. 8085, Munich, Germany, 23–26 May, (in press).

Pacifico, A., Nocerino, E., Scamardella, A. & Vultaggio, M. 2009. Implementation and testing of EGNOS full-scale dynamic trials onboard of HSC. *Atti dell'Istituto Italiano di Navigazione* 190: 142–152.

Roetenberg, D. 2006. *Inertial and Magnetic Sensing of Human Motion*. Ph.D. Thesis, University of Twente, Enschede, The Netherland.

Winter, D.A. 2005. Biomechanics and Motor Control of Human Movement. John Wiley and Sons Inc.

Winter, D.A., Patla, A.E., Prince, F., Ishac, M. & Gielo-Perczak, K. 1998. Stiffness control of balance in quiet standing. *Journal of Neurophysiology* 80(3): 1211–1221.

Sustainable Maritime Transportation and Exploitation of Sea Resources – Rizzuto & Guedes Soares (eds)
 ISBN 978-0-415-62081-9

Investigating the contributors of fishing vessel incidents

R.E. Kurt, I. Lazakis & O. Turan
Department of Naval Architecture & Marine Engineering, University of Strathclyde, UK

ABSTRACT: The present research paper demonstrates an in-depth analysis of fishing vessel incidents occurring in UK territorial waters as registered in the MAIB database. The aim is to investigate the main factors and sub-factors in the incidents involving fishing vessels and the role of human element in them. In order to achieve this aim, fishing vessel incidents recorded in the UK MAIB database for the last 19 years are thoroughly examined. The objective through which the main aim is achieved includes the examination of the incident location, the various types of the incident as well as the identification of the factors which contribute to the above mentioned incidents. Furthermore, the present paper compares the outcomes of the analysis with current rules, guidelines and regulations from national and international bodies and moreover suggests measures to overcome the above mentioned concern.

1 INTRODUCTION

Fishing vessel incidents (including accidents and near misses) are considered as the most important contributor to fatalities in the maritime sector worldwide (FAO 2001). An estimated number of more than 24,000 lives being lost worldwide during fishing operations is a substantial number, which overshadows any other related occupational accidents and fatalities in onshore industries (ILO 1999). The later prompted the International Maritime Organization (IMO) to initiate various efforts in order to address this issue. However, fishermen and workers onboard fishing vessels are still high up in the list of being employed in one of the most dangerous jobs in the world. This observation refers not only to the harshness of the actual working environment but also to a number of other reasons, which are not in-depth examined and analyzed. This is performed in the present study by investigating the UK MAIB database incidents in order to establish the key human and technical factors contributing to fishing vessel accidents and near misses.

Moreover, the present paper is structured as follows: the first section provides an in-depth overview of the legislative regime regarding commercial fishing at international level as well as includes relevant research studies carried out addressing fishing vessel accidents. The next section presents the analysis of the statistical data of the fishing vessel fleet and the incidents they are involved with in UK territorial waters as presented in the MAIB database over a period of 19 years. The last section of the paper presents the discussion and conclusions deriving from the analysis performed herein, providing further suggestions on how to tackle the identified problems as well.

2 REVIEW OF FISHING VESSEL INCIDENTS

As early as the 70's, IMO originated the effort for introducing guidelines and regulations regarding commercial fishing safety with the Torremolinos International Convention in 1977, applying to fishing vessels of 24 m in length and over. Eventually, it was not until 1993 that the Torremolinos Protocol for the safety of fishing vessels was adopted (IMO 1993). It was decided that it will enter into force one year after it is ratified by 15 member states with an aggregate number of 14,000 vessels. Bearing in mind that, by the time this paper was being prepared, a total number of 17 states but of only a cumulative number of 3,000 vessels has approved the Protocol, the standards for the specific category of fishing vessels have still a long way to go until their full implementation is achieved.

Nevertheless, IMO (1995) continued and intensified its efforts highlighting the need for better standards regarding fishermen education and training. This was achieved by introducing the International Convention on Training, Certification and Watchkeeping for fishing vessel personnel (STCW-F) which supplements the original IMO STCW code. The STCW-F convention also applies to ocean-going fishing vessels greater than 24 m in length and consists of four chapters including general provisions, certification guidelines for skippers, officers, engineer officers and radio operators; addresses basic safety training for all fishing vessel personnel and finally considers watchkeeping as well.

Apart from the above, IMO has been comprehensively working with other international bodies such as the International Labour Organization (ILO) and the Food and Agriculture Organization (FAO) in order to introduce non-mandatory requirements for fishing vessels. This is the case of the International Code for fishermen and fishing vessels divided into two parts. Part A refers to the safety and health practices (IMO 2005a) while Part B mentions the safety and health requirements for the construction and equipment of fishing vessels (IMO 2005b). Moreover, the legislative efforts of international organizations are complemented by the IMO voluntary guidelines regarding the design, construction and equipment of small fishing vessels, addressing the standards on vessels smaller than 24 m in length (IMO 2005c).

In addition to the above, fishing vessel accidents have also been of concern worldwide as shown in various studies performed by different researchers. Roberts (2010) examines the fatality rate of fishermen compared to other industrial sectors in the UK and concludes that fishing-related accidents are the highest in terms of worker-years. He suggests that more attention should be placed on personal safety devices as well as not properly maintained vessels. This comes in addition to the MAIB report on the analysis of fishing vessel accidents underlining the main causes of fatal accidents (MAIB 2008). These include among others the automatic bilge alarm not working properly, pipework failures leading to vessel flooding, watertight doors, hatch and other exposed openings left open at sea and last but certainly not least fatigue and lack of sleep.

In another paper by Perez-Labajos et al. (2006), they examine accidents occurring in the Spanish fishing fleet over a period of years and highlight the type of fishing vessels at risk more than others including trawlers and long-line fishing vessels. They also emphasize on the fact that even though international policies are in place, they have not yet been enforced but are only applied on a local and voluntary basis. This is also related to relevant studies carried out at various other countries including among others the work of Jin et al. (2001) for the US commercial fishing industry, Antao et al. (2008) regarding the occupational fishing accidents in Portugal and the investigation of Wiseman & Burge (2000) for the fishing vessels less than 20 m in the Newfoundland region of Canada. All the above conclude that national and international rules and regulations should be made mandatory and that the education and training of seafarers should be enhanced together with allowing for better maintenance and sea-worthiness of the fishing fleet.

Wang et al. (2005) also address the problem of fishing vessel accidents. They initially examine the relevant national and international regulations regarding fishing vessels as well as other relevant studies conducted in various countries worldwide. They then proceed with the investigation of fishing vessel accident statistics in terms of the vessels being lost per year as well as examining the contributing factors to these accidents. They conclude by identifying a number of dominant causal factors that lead to the mentioned accidents such as machinery damage, foundering and flooding, grounding as well as collisions and contacts among others. Machinery failure is also identified as the most common factor for vessels less than 12 meters in length by the MCA report regarding small fishing vessels (MCA 2004). However, to the best of the authors' knowledge, grounding and collisions are not by themselves contributing features to accidents as there exist other underlying factors, which are directly related to fishing vessel accidents including lack of machinery maintenance, lack of crew training and others as is shown in the next section of the present paper.

The decreasing trend of fishing vessel casualties is also confirmed by the study of Branagan and Turan (2010), in which they tackle the issue of accidents of fishing vessels in relation to the culture in the fishing industry and the way this is associated with the health and safety of crew and workers onboard the vessels. More specifically, they examine the accident statistics of the UK fishing fleet over a period of 16 years and reach the conclusion that although there is a constant decrease in the fishermen fatality trend, there is an alarming increasing trend in the accidents of fishing vessels since 2002. The above result is also confirmed by a survey and field trip carried out by one of the authors, during which the fishermen opinion is also taken into account revealing some more in-depth causes of accidents apart from the obvious harsh and hazardous working environment. These include among others communication problems while on duty (especially in the case of foreign workers being employed), national and international legislation restrictions, legislative gaps regarding specific vessel sizes as well as the social aspect of commercial fishing itself with crew and workers being exposed to a prolonged working schedule away from family and friends.

In another paper by Turan et al. (2003), an effort to investigate the loss of life regarding fishing vessels is carried out by employing a combination of reliability and Fuzzy Set Theory ((FST) tools). In this case, the Fault Tree Analysis (FTA) tool is used to model the possibility of loss of life at sea when onboard a fishing vessel while FST is used to compliment the analysis of the accidents occurring with fishing vessels. At first, the FTA is populated with numerical values already existing in the literature (e.g., available database) whereas any missing values are derived from the fuzzy set assessment assigning linguistic terms in the various end-events of the Fault Tree structure responsible for the loss of life onboard fishing vessels. Consequently, the paper provides a priority ranking of the major critical events contributing to the loss of life onboard fishing vessels. In this way, the design and operational issues emerging from this analysis can be enhanced so as to avoid the high number of casualties in this industrial sector.

3 ANALYSIS OF MAIB DATABASE

The data regarding the fishing vessels analyzed in the present paper derive from the UK Marine Accident Investigation Branch (MAIB), which is a separate division in the UK Department for Transport. Overall, the accessible data span a period of 19 years (1991 to 2009) and provide a thorough insight for the incidents and accidents occurring onboard UK fishing vessels as well as on accidents occurring in UK territorial waters.

At first, Figures 1–3 show the general trends and information about the fishing vessel fleet registered in the UK. More specifically, the total number of fishing vessels together with their total GT capacity is demonstrated. Furthermore, the fishing vessel loses and total accidents occurring for the UK registered fleet are shown as well as the injuries and fatalities for the same regarding crew and workers for the same number of vessels.

As is shown in Figure 1, there is a declining trend in the total number of fishing vessels registered in the UK over the years as well in the GT capacity of these vessels in the same period. This denotes the overall diminution of the professional fishermen as well as of the workers employed onboard the vessels. It is also explained by the decrease in the number of bigger fishing vessels employed in distant areas far away from shore with smaller ones mainly operating in waters around the UK (Roberts 2010).

As is observed in Figure 2, the fishing vessel loses decline through the years apart from fluctuations shown at specific years, which can be attributed in cumulative accidents occurring at that time. The later is moreover confirmed by the reduction of the total number of accidents, which has also significantly decreased from 600 in 1995 to around 200 in 2009. This clearly shows the improvement in training of fishermen and workers onboard fishing vessels as well as the development regarding the awareness about safety culture in general. Other contributing factors also include the enhancement in the maintenance regime concerning the subject vessels together with improving the overall design in terms of stability issues.

As can be seen in Figure 3, the overall trend of injuries and fatalities is highly related to the decrease in the total number of accidents and vessel loses depicted in Figure 2. This can be attributed to the initial introduction of national and international guidelines and regulations, which although are not mandatory, they have laid the foundations for the enhanced understanding of the fishing vessel accidents and thus improve the overall fatality and injury rate.

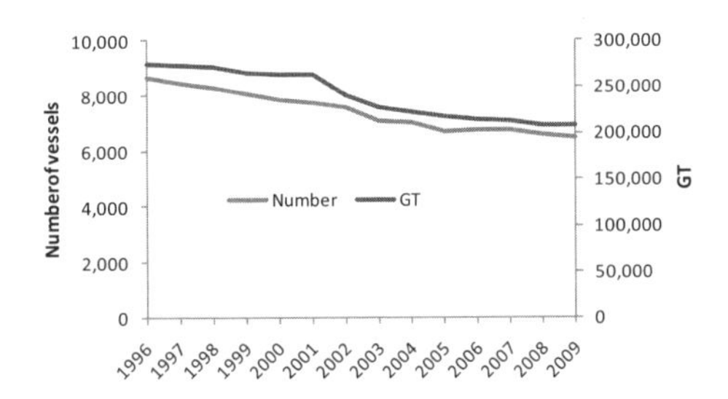

Figure 1. Cumulative number and GT of UK fishing vessels per year (UK MMO 2011).

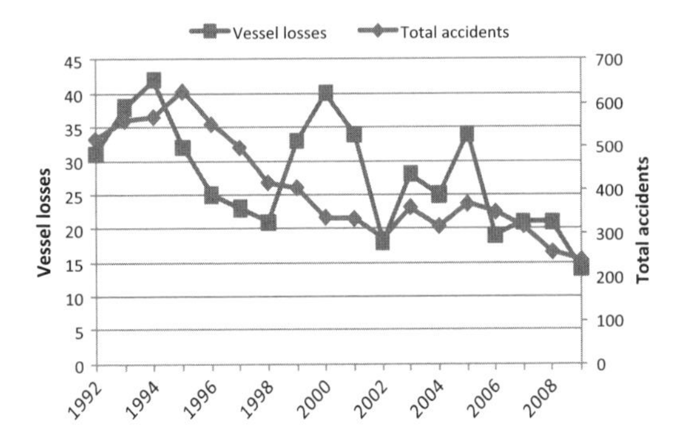

Figure 2. Fishing vessel loses and total accidents occurring for the UK registered fleet between 1992–2009.

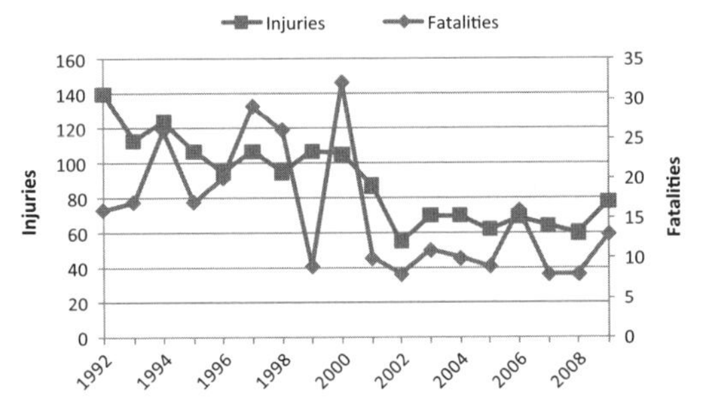

Figure 3. Fishing vessel injuries and fatalities for the UK registered fleet between 1992–2009.

Moreover, in order to investigate the key contributors of the fishing vessel accidents, the MAIB database is analyzed in depth. First of all, the database is analysed in terms of the nember of accidents per vessel type for the last 19 years. In this respect, Figure 4 shows the distribution of the occurence of fishing vessel incidents per vessel type.

As can be seen, out of 21,871 vessels which have been involved in near misses or accidents, fishing vessels include a total number of 6,832 incidents, which represents 31.2% of the entire database. This clearly shows the need for further investigation of the fishing vessels incidents in order to identify the underlying causes, which can then be utilized to derive further recommendations aiming at achieving safer operations.

Figure 5 shows the types of fishing vessels (including fish catching and processing vessels) in the MAIB database. It can be seen that most of the vessels are trawlers even though the majority of the vessel type remains uncategorized. This is due to the fact that either in the accident investigation procedure the relevant field for "vessel type" was left blank or the old accident investigation forms did not have this "vessel type" field category. Subsequently, this type of data was not collected in a consistent way. For consistency reasons, further analysis carried out in the present paper excludes the uncategorized data mentioned above.

In addition to the above, another result stemming from the analysis of the MAIB databse is the "vessel size" data in terms of GRT (Gross register tonnage). In this respect, the size data are grouped in line with the "International Standard Statistical Classification of Fishery Vessels (ISSCFV) by GRT Categories" Figure 6 shows the grouping of fish catching and processing vessels according to their corresponding GRT.

By filtering the results, it is shown that 41.2% of fishing vessel incidents belongs to the group

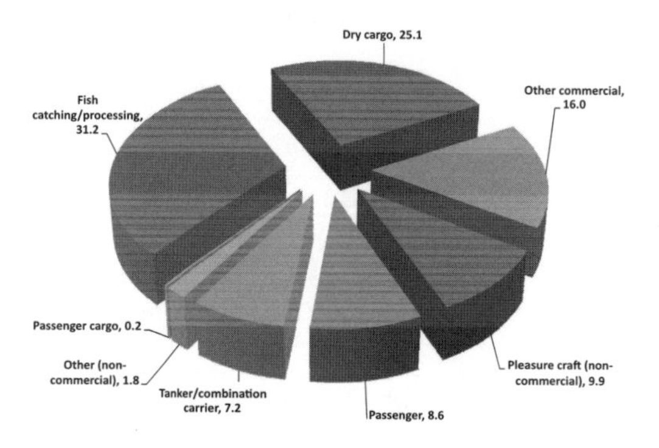

Figure 4. Accidents per vessel type between 1991 and 2009.

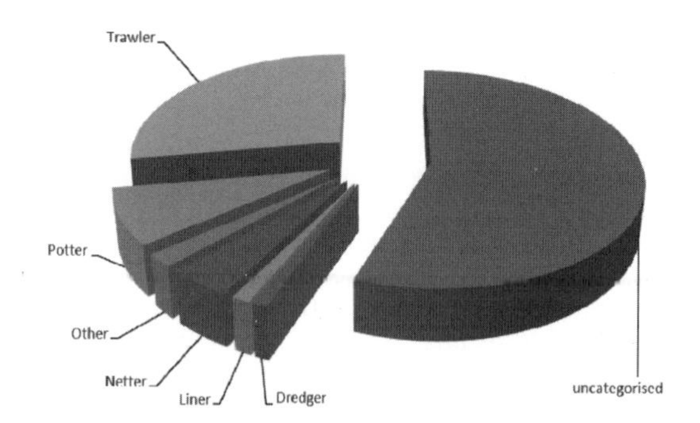

Figure 5. Fishing catching and processing vessel types.

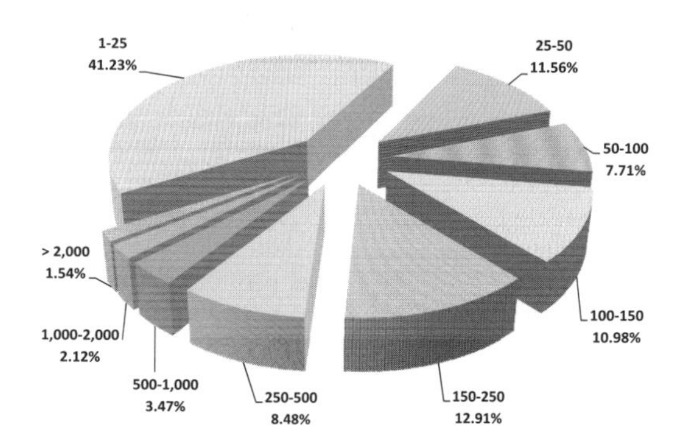

Figure 6. Fish catching and processing vessels grouped by GRT.

of smaller vessels (1–25 GRT). Moreover, almost 53% of the fishing vessel incidents belong to those, which are smaller than 50 GRT.

When dealing with the fishing vessel incidents the Authors considered the location as of the incidents of a great importance to the analysis. Therefore, Figure 7 shows the location of incidents over the years.

From Figure 7 it can be seen that the majority of fishing vessel accidents occur in coastal waters. According to the aforementioned gross tonnage distribution, most of the accidents in the database belong to vessels smaller than 50 GRT. Therefore, the majority of accidents happening in the coastal waters can be explained due to the fact that most of these vessels are smaller and do not operate in long distance open sea areas.

Moreover, Figure 8 shows that the majority of incidents including al vessels (apart from netters) occur in coastal waters. Netters, on the other hand, have slightly more accidents in high seas than coastal waters. The port/harbor or river/channel area does not seem to be a critical location for fishing vessel incidents.

Each incident in the MAIB database is attributed an Underlying Accident Factor Category (UAFC), which defines the exact reason for which

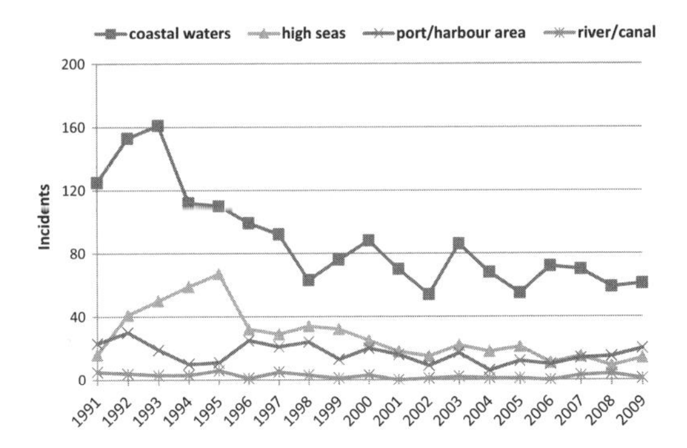

Figure 7. Location of fishing vessel incidents through years.

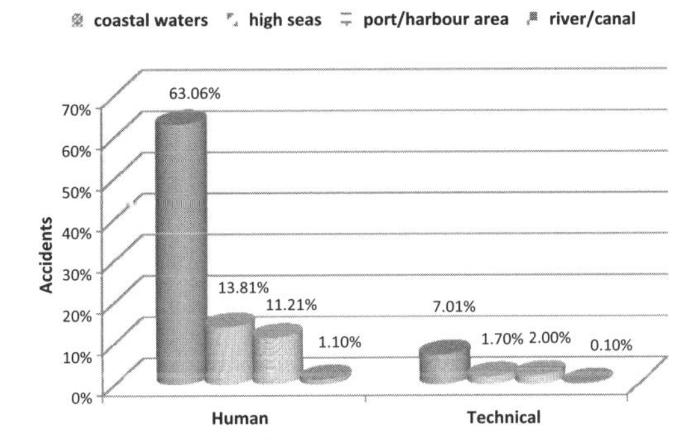

Figure 9. Location of accidents vs. UAFC.

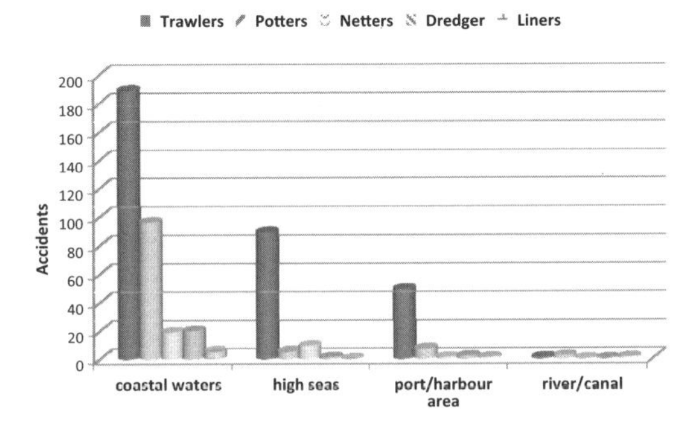

Figure 8. Types of fishing vessels vs. location of incidents.

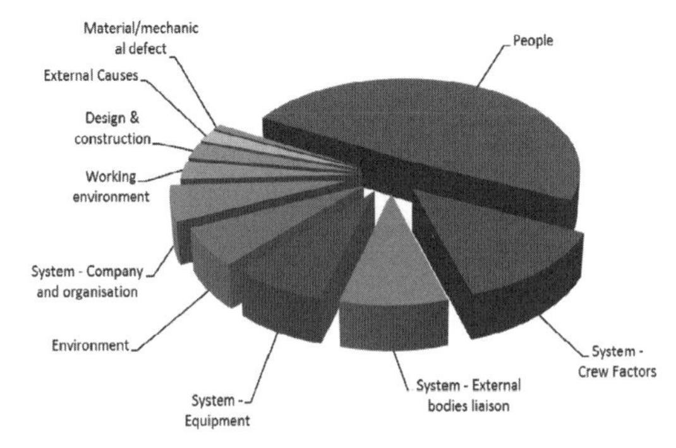

Figure 10. Fishing vessel underlying accident factors (UAF).

the incident occurred. In this respect, UAFC is divided into either technical or human factors as the category for underlying chain of factors leading to an incident. Figure 9 shows the UAFC category according to the location of incidents.

As a further result of the investigation procedure regarding fishing vessel incidents, the majority of these are attributed to human factors (87.5% human factors, 12.5% technical factors). Another interesting result is that 63.8% of the mentioned incidents having occurred in coastal waters are attributed to human factors. This is due to the fact that these vessels are mostly small size vessels with an inadequate human factors practice.

As can be seen, the incidents attributed to human factors constitute a large number when compared to the overall size of the fishing fleet. The demand for decreasing the crew numbers on board ships has caused the shipping industry to concentrate its effort on the human element side of the industry and accordingly address the related human factors. Due to the installation of intelligent systems and tools onboard ships, human capabilities and behavior may become the bottleneck in the entire system. That is the reason behind an increasing trend in the number of incidents attributed to human factors. Therefore, it was decided to expand the investigation of the incident database in order to identify the underlying accident factors (UAF) and underlying accident sub-factors (UASF).

In this respect, Figure 10 shows the UAF for fishing vessel incidents. The analysis shows that 48.8% of the UAF belongs to the group categorized as "people". "People" are defined in the MAIB database as "*an individual related to the investigation*" and will have further sub factors, which can be found in the UASF.

The second major category is the "System—Crew" factors, which is defined in the MAIB database as, "*the interaction of the crew, the internal organization and the way in which individuals work together as a team, all impact on the likelihood of a human error on board ship*". Crew factors constitute a portion of 13.4% of all UAF in fishing vessel accidents. The rest of the major factors identified are as follows: initially the "System-external bodies' liaison" (8.7%), which is defined as "*External bodies*" influence shipboard operation, which leads to incident. The next factor is the "System-Equipment" (7.6%) which is related to the equipment of the vessel. The "Environment" (6.5%) is next in the list, followed by the "System—company and organization" (6.0%) which is defined as "*management failures contributing to the occurrence of the incident event to be classified*". The "Working environment" (3.1%), the "Design & construction" (2.8%)' as well as the "External Causes" (2.1%) and finally "Material/mechanical defect" (1.0%) conclude the list with the UAF.

In the MAIB database, the underlying incident factors have been further sub-categorized to identify the initial reasons leading to the incidents. At this point, it should be mentioned that MAIB upgraded their data collection procedure in 2005. Therefore, there have been some changes in the way the data are recorded since then. In order to overcome the duplicated values in the database, the authors of this paper combined several UASF together before identifying the top ten UASF listed in Table 1. Out of 93 different UASF, the authors filtered the top ten which represent the 38.9% of all UASF.

Noncompliance appears to be the most common underlying accident sub-factor in fishing vessels. This is considered mostly human factors related. Unfortunately, not following or fulfilling the relevant regulations is quite a common case in the fishing vessels which can easily be overcome by drawing more attention together with enforcing regulations as well as introducing the awareness trainings.

"Fatigue & vigilance" is one of the key problems that need to be addressed, not only in fishing vessels but for the entire shipping industry.

Especially work-rest hours of crewmembers play a important role in crew fatigue. However, fatigue and vigilance has many interdependent contributors, which researchers have been trying to investigate further. In the report prepared by Houtman et al. (2005) it was mentioned that "Research is needed to address the unique combinations of potential stressors, which may interact in various ways to produce fatigue, poor health and increased accident risk". However, fatigue and vigilance, similar to many other human factors, are interdepended with other environmental and psychological factors, which makes the problem very multidimensional to overcome For example based on the findings of recent research carried out on crewmembers in ship bridge simulators (Kurt et al., 2010), it was concluded that noise has significant relationship on crew's vigilance performance. Moreover, crewmembers also reported that it is more annoying, more tiring and harder to concentrate on their tasks in high noise conditions. Similar to noise other contributors of vigilance and fatigue needs to be minimized to achieve a decrease in fatigue and vigilance.

Table 1. Top ten UASF in fishing vessel incidents.

Ranking top 10 UASF in fishing vessels accidents		Number	Percentage
1	Non compliance	130	5.6%
2	Fatigue & vigilance	115	5.0%
3	Training	111	4.8%
4	Equipment related	104	4.5%
5	Perception of risk	98	4.2%
6	Weather	79	3.4%
7	Competence	71	3.1%
8	Poor decision making/ information use	67	2.9%
9	Inadequate resources	62	2.7%
10	Inattention	61	2.6%

"Training" is one of the key solutions to many problems as well as it can be a reason for an accident in its absence. Many accidents have happened due to lack of or inadequate training given to crewmembers. Since the ships become more complex day-by-day, effective training becomes a hot topic for maritime safety, security and efficiency. Furthermore, maritime training experts have initiated further research for innovative trainings resulting to safer maritime operations. An example of the above can be EU funded Leonardo Lifelong Programme (LLP) M'Aider project which aims to develop simulator scenarios of key accidents in the maritime sector, in order to help the cadets and existing crew members by learning from past accidents (Bosma et al., 2010).

"Equipment related" category is the combination of several equipment related categories such as; equipment not available, equipment badly maintained, equipment misuse, equipment poorly designed, etc. It can be clearly observed, that even though it is generally mentioned as "equipment category", it still contains the human element at the core of it. In this case, the human element (crewmembers) can be liable in case of badly maintained equipment in contrast to the same crewmembers, who can be the victims in the rest of the categories (i.e., equipment poorly designed).

"Perception of risk", in simple terms, can be defined as the subjective judgment that people make about the severity of a risk. However, in most cases this individual-dependent judgment signifies an important part in the safe operation of the vessel. By proper trainings, crewmembers' risk perception can be improved.

"Weather" is an element, which is hard to control and affects the safety of ships. Human on board vessels are forced to the extremes in heavy weather conditions therefore human error is likely to occur in conditions too. Being prepared for these conditions by having predefined safety procedures as well as personal safety equipment will help to maintain the crewmembers' safety. However, extreme weather conditions and the subsequent safety issues arising from them are in some cases unavoidable as crewmembers need to carry on with their daily routine work.

"Competence" suggests that crewmember is not competent enough to carry out the duties assigned to him/her. Crewmembers may be certificated but not actually competent to the level prescribed in the certificate. It might also be that the crewmembers lose their competence on the relevant subject as time goes by. Therefore, it is necessary to keep a good record and investigation on crew members' competence levels as well as update the crewmembers' competence level when necessary.

"Poor decision making/information use" poses a threat to the overall system safety. Decision-making is the study of identifying and choosing alternatives based on the preferences of the decision-maker. Mental fatigue and stress are very important in decision making by creating slower and low quality responses in decisions in an intricate environment. Therefore, it is important to develop relevant procedures to deal with mental fatigue as well.

"Inadequate resources" category can be defined as the resources needed to complete a job effectively and safely is not available (such as time, finance and personnel). This category is mainly concerned with manning procedures. In the case of not adequate time allowed for crew hand-over to a new crewmember, this may result in the new crewmember not completely updated with the requirements of the task and will create a threat for ship's safety.

"Inattention" refers to the loss of attention and may include failing to monitor displays; not maintaining a proper lookout; forgetting to perform an assigned duty. Similar to all human factors, inattention may also originate from other related human factor causes such as a personal problem, fatigue, drugs, boredom, or hearing problems. As mentioned before keeping good work rest patterns will help in solving all fatigue as well as inattention problems.

4 CONCLUSIONS AND DISCUSSIONS

In the context of this paper, the authors have presented a thorough review of the legislative efforts to address the issue of fishing vessel incidents at an international level. It is observed that there are still no mandatory rules and regulations in place which may be one of the biggest contributing factors for the high rates of fishing vessels incidents. In addition to the above, a review of past research activities regarding fishing vessel accidents and near misses has been also performed, showing the continuous interest of the maritime community towards these type of vessels.

Furthermore, the incidents regarding fishing vessels recorded over a period of 19 years have been analysed by using the UK MAIB database. The authors were faced with some limitations during the analysis performed, such as missing data in underlying accident factors fields. In this case, the incidents with missing data have been excluded from the analysis carried out in this paper. However, the number of data filtered out is not expected to have a big effect on the quality of the results.

In the results section of this paper, the underlying accident factors for the fishing vessels has been identified and explained. It is important to draw the attention of maritime and fishing industry to the fact that most of the underlying factors can be prevented by delivering effective trainings. Introducing an effective training programme for fishing vessels' crewmembers and enforcing these trainings through obligatory regulations will dramatically reduce the fishing vessel incidents overall.

ACKNOWLEDGEMENTS

The Authors of this paper would like to thank the UK MAIB (Marine Accident Investigation Branch) for providing the accident database, which was necessary to carry out the present research.

REFERENCES

Antao, P., Almeida, T., Jacinto, C. & Soares, C.G., 2008. Causes of occupational accidents in the fishing sector in Portugal. *Safety Science*, 46: 885–899.

Bosma, T., Lazakis, I., Turan, O. & Kurt, R.E., 2010. Developing Maritime Aids for Emergency Responses. *Proceedings of the 4th PAAMES and AMEC conference*, 6–8 December, Singapore.

Branagan, O. & Turan, O., 2010. The culture of the fishing industry vs. health & safety: can there only be one winner?. *International Conference on Human Performance at Sea* (HPAS), 16–18 June, Glasgow, UK.

FAO, 2001. *Safety at sea as an integral part of fisheries management*. FAO Fisheries Circular No. 966, Food and agriculture organization (FAO) of the United Nations, Rome.

Houtman, I. et al. 2005. Fatigue in the shipping industry. TNO Quality of Life. Report 20834/11353.

ILO, 1999. *Report on Safety and Health in the Fishing Industry*. International Labour Organization (ILO), Geneva, 13–17 December 1999, Geneva.

IMO, 1993. *Torremolinos Protocol for the Safety of Fishing Vessels*. Torremolinos International Convention for the safety of fishing vessels, International Maritime Organization (IMO), London.
IMO, 1995. International Convention on Training, Certification and Watchkeeping for Fishing Vessel Personnel (STCW-F). International Maritime Organization (IMO), London.
IMO, 2005a. *Code of safety for fishermen and fishing vessels, Part A: Safety and Health Practice*. International Maritime Organization (IMO), London.
IMO, 2005b. *Code of safety for fishermen and fishing vessels, Part B: safety and health requirements for the construction and equipment of fishing vessels*. International Maritime Organization (IMO), London.
IMO, 2005c. *Voluntary Guidelines for the Design, Construction and Equipment of Small Fishing Vessels*. International Maritime Organization (IMO), London.
Jin, D., Kite-Powell, H. & Talley, W., 2001. The safety of commercial fishing: determinants of vessel total losses and injuries. *Journal of Safety Research*, 32: 209–228.
Kurt, R.E. et al., 2010 "An Experimental Study Investigating the Effects of Noise on Seafarers' Performance Onboard Ships", *International Conference on Human Performance at Sea (HPAS)*, 16–18 June, Glasgow, UK.
MAIB, 2008. *Analysis of UK Fishing Vessel Safety 1992 to 2006*. Marine Accident Investigation Branch (MAIB), Department of Transport, Southampton.
MCA, 2004. A study into incidents involving Under 12 m Fishing Vessels. Maritime Coastguard Agency (MCA) prevention Branch, prevention report 2.
Perez-Labajos, C, et al., 2006. Analysis of accident inequality of the Spanish fishing fleet. *Accident Analysis and Prevention*, 38: 1168–1175.
Roberts, S.E., 2010. Britain's most hazardous occupation: Commercial fishing. *Accident Analysis and Prevention*, 42: 44–49.
Turan, O., Olcer, A.I. & Martin, P.L., 2003. Risk assessment of loss of life for fishing vessels in fuzzy environment. *Safety and reliability*, 23(2): 19–39.
UK MMO, 2011. http://www.marinemanagement.org.uk/fisheries/statistics/annual2009.htm#ch1, UK Marine Management Organization (accessed 31/03/2011).
Wang, J. et al., 2005. An analysis of fishing vessel accidents. *Accident Analysis and Prevention*, 37: 1019–1024.
Wiseman, M. & Burge, H., 2000. *Fishing Vessel Safety Review (less than 65 feet)*. Maritime Search and Rescue, Newfoundland Region, Canada.

Sustainable Maritime Transportation and Exploitation of Sea Resources – Rizzuto & Guedes Soares (eds)
 ISBN 978-0-415-62081-9

Simulation method for risk assessment in LNG terminal design

L. Gucma & M. Gucma
Maritime University of Szczecin, Szczecin, Poland

M. Perkovič & P. Vidmar
Faculty of Maritime Studies and Transport, University of Ljubljana, Portoroz, Slovenia

ABSTRACT: Localization of sea terminal, especially when it is designed to handle potentially dangerous cargo like LNG gas, requires extremely demanding and complete study. Although appropriate tools exists, each study demands special adaptations of risk models. Article presents risk calculation methods with determination of all stages parameters and variables used in formal assessment and special respect to simulation methods. Article copes with localization of LNG (Liquefied Natural Gas) terminal from point of navigational advantages. Full mission simulations are describing very well the environment of vessel and provide full interactions, while limited task simulations along with Monte Carlo methods can in certain aspects speed up the development process. Authors copes with design of new terminal and case studies for Slovenia and Poland are presented.

1 INTRODUCTION

1.1 *The LNG projects in Poland and Slovenia*

Liquefied Natural Gas is in interest of many countries as a alternative, clean and efficient fuel. Europe countries are looking for reliable and diverse solution, not based on one fuel source—mainly pipe. In Poland aim is a project of building LNG terminal in north western Poland (Port of Swinoujscie), whilst in Slovenia LNG port will be outside of Koper Port. Although different types of localizations are described many similarities can be outlined. In both ports, concept for navigation risk based detailed study has been prepared. Vessels that will be operating in ports are in ranges of capacities from 75000 to 216000 m^3 of LNG cargo.

Currently there are no coherent and complex methods of LNG port and infrastructure design from point of navigational risk view. Many studies are and has been undertaken in this field to mention most important: (Vanem et al., 2008) as well as general methodologies: (IMO, 2007), or (Fang et al., 2005). Article is continuity of the work undertaken in recent work, and preliminary results were presented in: (Gucma & Gucma, 2007).

1.2 *Mathematical model used in simulators*

Hydrodynamic models of two classes are run on limited tasks simulators utilized in Maritime University of Szczecin and University of Ljubljana. One class models are used only when limited parameters are known (usually when non-existing ships or general class ships are modeled). The other class models are used when precise characteristics of hulls, propellers and steering devices are known from field trials. Additionally, real maneuvering characteristics are often utilized for model validation.

The model used is based on methodology where all influences like hull hydrodynamic forces, propeller drag and steering gear forces as well as certain external influences are modeled as separate forces and at the end summed up as longitudinal, transverse and rotational forces—so called forces model.

The model operates in the loop where, for known input variables, the system instantly calculates the forces and moments generated on the hull, from which instantaneous accelerations and speeds of surge, sway and yaw are determined. The most important forces acting on the model are (Gucma S., 2009):

- thrust of propellers;
- side force of propellers;
- sway and resistant force of propellers;
- bow and stern thruster forces;
- current, wind and ice effects;
- moment and force of bank effect;
- shallow water forces;
- mooring and anchor forces;
- reaction of the fenders and friction between fender and ship's hull;

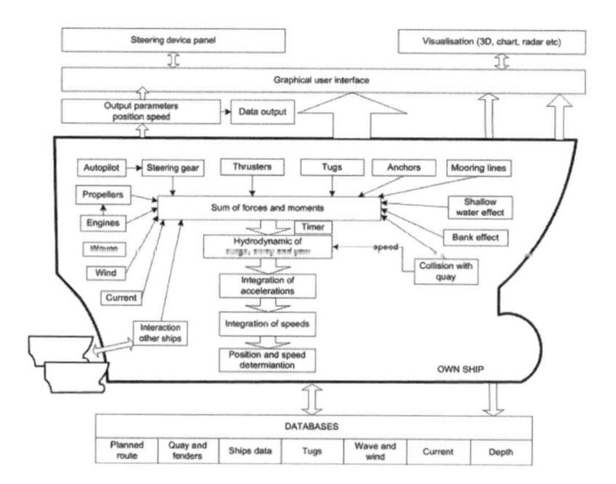

Figure 1. Functional diagram of the ship maneuvering simulation model.

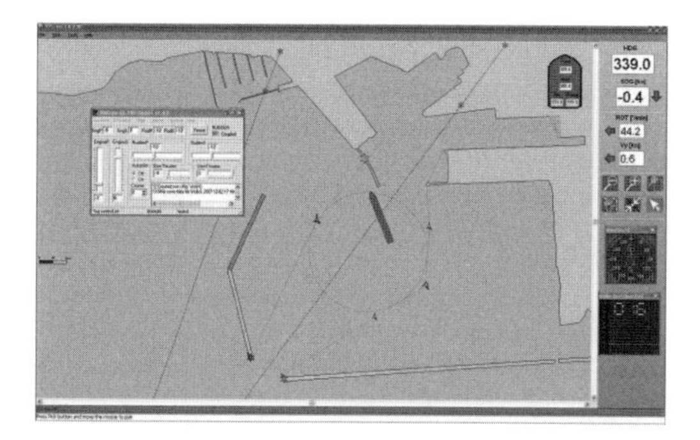

Figure 2. 2D Chart view during maneuvering in designed port model.

Figure 3. 3D full mission simulator view of LNG port in Swinoujscie.

- tugs' forces;
- other forces depending on the characteristics of ship's propulsion and steering units.

The functional diagram of the ship manoeuvring simulation model is presented in Figure 1.

The model interface is a typical 2D chart interface (Fig. 2) and 3D world as seen on real vessel (Fig. 3) The interface brings information on the ships state (position, course, speed, rot. etc.), berth and shore localization, navigational marks, soundings, external conditions, tug and line control and controls on the model. The model is implemented in integrated environment Visual C™ using C++ language.

2 DESIGN OF AN EXPERIMENTAL SYSTEM IN SIMULATION RESEARCH

Based on the studies of the theory of modeling and simulation (Zeigler, 1984), simulation-based research can be summarized as the following procedure:

- formulation of the problem,
- modeling of the object to be examined,
- development of computer programs,
- model validation,
- design of an experimental system,
- experiment,
- analysis of the results.

While designing an experimental system we have to define the scope of research that in our considerations depends on the following (Gucma & Gucma, 2008):

- examined variants of the area infrastructure,
- examined types of ship (their maximum parameters),
- hydrometeorological conditions under which operations in the examined area will be performed (maneuvers, types and number of tugs in maneuvers, etc.).

Having defined the data specifying the experiment range, we can establish specific research conditions (Gucma & Gucma, Planowanie eksperymantu w badaniach symulacyjnych inżynierii ruchu morskiego, 1998):

- number of series of planned manoeuvres,
- conditions prevailing in particular manoeuvre series.

3 VERIFICATION OF SIMULATION MODELS

The acceptable level of model conformity with reality in marine traffic engineering studies aimed at the determination of waterway parameters is assumed at 90% (Gucma S. 1990). Model verification is a procedure applied before each experiment. The repeated verification of optimum results in simulation tests was valid, which proves that the human-machine-environment system was well designed (Gucma S. 2001). There are three levels of simulated model verification:

1. verification of the mathematical model,
2. verification of pilot or navigator behaviour,
3. verification of the simulated system.

To verify the model, researcher have to answer three questions:

1. Will the ship proceed along the same or similar route during an identical maneuver in the same waters and under identical conditions?
2. Will the pilot perform the same maneuvers in the same time period?
3. Will the simulated system suit real world?

The first answer can be given during process of the mathematical model verification. To answer the second question is far more difficult due to various ways and results of maneuvering process, even with the same pilot. This may result from the fact that the navigator, while planning a maneuver, predicts future positions of the ship. Predictions are often burdened with errors, and in addition, there exist various ways of making the same maneuver. (Gucma S. 2009). On the other hand, complete adequacy of the model to reality does not have to be greater than designed tolerance of fairways, i.e. approximately three meters (Webster, 1992). Higher accuracy of simulations for waterway designs would be an academic discussion rather than practical nature need. The model can be verified on the basis of (Gucma S. 2001):

- data from real tests,
- physical model tests,
- theoretical relationships—analytical calculations,
- expert opinions.

3.1 *Verification of mathematical models*

The verification of a mathematical model, i.e. quantitative assessment of its adequacy, is always carried out after the model is built and based on conformity indicators of ship's manoeuvring characteristics (Gucma S. 2009). The verification makes use of comparative parameters to compare ship's maneuvers (Gucma S. 2009):

- acceleration—distance and time for various operating modes of the main engine;
- stopping—distance, time, deviation from the original course for various ME working modes;
- turning ability (turning circle tests)—diameter, linear velocity and rate of turn in turning;
- course stability (zigzag tests)—time of rudder response, maximum angle of deviation, yaw time, non-dimensional yaw period;
- turning ability with thrusters or engines—time to alter course by a specific value.

Verification is a long period and usually iterative process. The verification of a mathematical model and pilot's behavior involves teams of pilots and professionals in the field of ship movement computer simulation. To eliminate systematic errors, at least two groups working independently are arranged. Then the results are compared. The verification by pilot experts is a subjective process, so the results should be approached with reserve (Gucma S. 2009).

3.2 *Mathematical model and verification of navigator behavior*

With replacing the real navigator's work by an algorithm reflecting his decision process, we obtain an autonomous model of ship control. Human navigator adjusts his method of control to the type of object that is steered and controlled. Research on ship and aircraft autopilots unveils that the human can control objects similarly to a PID controller burdened with inertia and delay, thus it is possible to build a human function as a controller transmittance equation (Gucma S. 2009):

$$G(s) = \frac{u(s)}{e(s)} = \frac{e^{-sT_0}}{(1+sT_{nm})(1+sT_r)}\left(k_p + sk_D + \frac{k_I}{s}\right)$$

where: T_0–delay time, T_{nm}–time constant characterizing the inertia of human nervous and muscular systems, T_r–another time constant of inertia dependent on object control, k_P, k_D, k_I–the proportional, derivative and integral gains.

As the gains: k_P, k_D, k_I vary from and may even assume zero values, this variability of gains can be modeled in a fuzzy controller instead of a conventional linear PID controller. Therefore, two types of algorithms have been designed to represent navigator's decision process: one by using experts' knowledge contained in 'reference passages' and the other using a fast time prediction system based on the hydrodynamic model of examined LNG carriers and an analytical model of maneuver choice (Gucma S. 2009). Verification in both fuzzy and PID version is done in following manner:

1. Model functional tests,
2. Preparation of comparable testbed and area model for fast time and real time
3. Performance of fast time and real time (by experienced pilots & captains) simulations on same conditions,
4. Comparison of results,
5. Fast time model tune—up.

Also experts' opinions are very valuable at this stage. During LNG for Poland project as well as Slovenian LNG study fast time simulations were performed.

4 MONTE CARLO SIMULATION METHODS

The safety assessment of complex marine systems requests from models a number of parameters, such as: vessels traffic, hydrometeorological conditions, area parameters and many others. These parameters are mostly random; therefore analytical methods are not suitable for constructing their models, particularly if these models are to include the human factor. Such systems can be modeled by Monte Carlo (MC) simulations methods. These well known instrument, is still underestimated by researchers, consist in generating random numbers aimed at estimating their distributions. The generation of numbers is done by special numerical programs implemented in high speed computers. The method is extremely valuable when relations between distributions are described by sophisticated, often non solvable functions and depends on random elements (like human behavior). The MC method offers several advantages, enabling (Vose, 2000), (Brandt, 1998):

- examination of a large number of scenarios,
- modeling of correlations and internal relations,
- application of a simplified mathematical level,
- application of a computer for calculations,
- fast introduction of active changes in the model,
- sensitivity analysis by changing distributions and their parameters,
- incorporation of a large quantity of data from a long time interval,
- evaluation of the method error.

The disadvantages of the MC methods are as follows:

- weak convergence of the method, which can be compensated by a large number of iterations;
- only approximate solutions can be obtained;
- high computing power demand (some very complex problems require long simulation time).

This article presents methodology of creating MC models for the determination of marine structures safety in view of their possible damage by passing LNG vessels. These models often have to take into account possible human errors and natural human behavior (e.g. decision to execute emergency anchor dropping). The presented models can be used for assessing the probability of ship's collision with marine structures, such as: wind turbines, drilling rigs and other objects that may be considered dangerous for surface navigation or the probability of indirect damage to submarine pipelines or cables by a ship's anchor (Gucma S., 2009).

In cases we have to know the function of probability distribution changes or their parameters in time. Generally there are four groups of issues solved by marine traffic engineering using MC simulation methods:

1. stochastic models of vessel traffic streams with elements of MC simulation (Gucma L. 2003), (Hansen & Simonsen, 2000);
2. MC methods based on generalized simulation and real models (Iribarren, 1999);
3. simulation models of ship movement in fast time with random external excitations (Hutchison, Gray & Mathai, 2003);
4. methods of measurement error analysis and uncertainties using MC simulation (Sand, Nielsen & Jakobsen, 1994).

The first group of methods comprises simulations of large marine traffic engineering systems, mostly in offshore areas. The subject of simulation are vessel traffic streams, but the objective of simulation is the evaluation of system safety. Microscopic methods of simulating vessel traffic streams are utilized, thus single ship movement is simulated. These methods utilize very general models of the navigator, while the ship is regarded as a theoretical point. These models contain analytical or empirical subsystems, whose task is to define probabilities of some modeled events. Time flow is accelerated in order to create adequate large number of scenarios, as simulated events are rare (groundings, collisions, indirect damage etc.). Designed magnitudes of random variables can be introduced in this type of models.

The second group of methods distinguishes MC tests, using distributions of random variables and their parameters obtained from previously conducted real or simulation tests. These methods are generally aimed at getting a larger number of accident scenarios that may be obtained by simulation methods. The models are often combined with analytical or empirical models (Gucma L., 2005).

The third group of methods is modifying simulation models of ship movement operating in fast time with a navigator's model. Random disturbances, often dependent on time, are introduced. Also, random models of navigator's behavior can be implemented. Methods of this type are mostly used for port areas, basically in the designing and optimization of waterway parameters (Gucma S., 2009).

Last group is helpful in the determination of measurement errors in complex measurement systems especially where this cannot be done by analytical methods. These models can be combined with analytical or empirical models in conjunction with uncertainty analysis. Accidents happen due to navigator mistake, often enhanced by ship technical failure and variable conditions.

The discussed models are based on an assumption that parameters of vessel traffic stream are the most important factor affecting safety (Gucma S. 2009). These models can also include empirical or analytical algorithms of ship's behavior in various conditions, such as models of drift, surge or leeway.

The above methods can be used for determining the probability of ship's collision with:

- open sea drilling rigs and oil platforms,
- offshore wind farms,
- fixed aids to navigation,
- submarine pipelines or cables,
- bridges and their supports,
- other stationary marine structures.

In most cases randomizers used in simulations were of polynomial normal type, where route determination and density of traffic were relayed on random variables generated by Poisson distribution. All randomizers were tuned in order to have high stability output.

4.1 *Model for the assessment of LNG carrier safety—case study*

Safety assessment is based on the original probabilistic model created at the Maritime University of Szczecin. The model is capable of assessing the risk of a large complex system with consideration of human (navigator) behavior models, ship dynamics model, real traffic streams parameters and external conditions such as wind, current, visibility, etc. The model works in fast time and can simulate a large number of scenarios. The output from the model such as a place of collision, ships involved, and navigational conditions can be useful for risk assessment of the proposed LNG terminal locations using FSA method. The results were used for the determination of the optimal location of the LNG terminal based on the navigational risk criterion. Of the several simplifications assumed, one of the most important was considering an LNG carrier as a conventional ship.

The developed model, presented in Figure 4, can be used for assessing almost all navigational accidents, such as collisions, groundings, collision with a fixed object (Gucma L. 2003), indirect accidents involving anchors or accidents caused by ship generated waves (Gucma & Zalewski

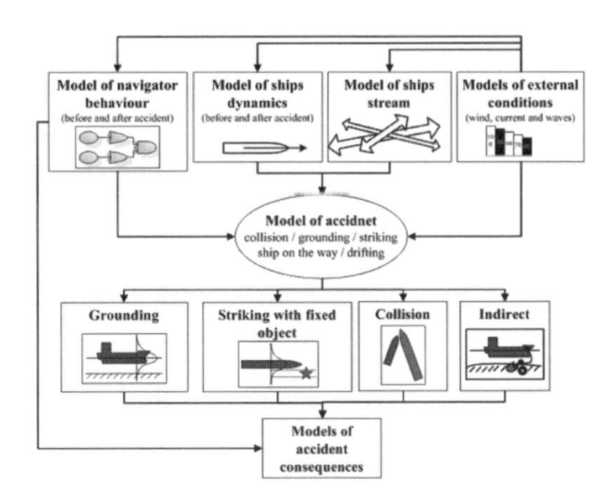

Figure 4. Diagram of a fully developed stochastic model of navigation safety assessment (Gucma L., 2005).

2003). The model can comprise several modules responsible for different navigational accidents. This methodology has already been used by several authors with different effect (Merrick et al., 2001), (Hansen & Simonsen, 2000), (Otay & Tan, 1998.). In these studies the model is used to assess the safety of different variants of LNG terminal location.

Mathematical model used in MC consist of model of navigator behavior, implemented here as Bayesian network program, ship dynamics model—presented in ch. 1.2, model of ships streams—where date were obtained by real systems (AIS) and scaled for constant factors to meet future traffic and model of external conditions—where various factors like wind, current, visibility etc. are set of functions randomly taken from database. Details of each component can be found in (Gucma, L. 2008) & (Gucma, L. 2005).

In the following case LNG carrier was passing in the loop routes to 2 possible locations of LNG terminal in Poland (Gdansk-eastern and Swinoujscie-western). The presented models of collisions, fire and groundings were adapted to the stochastic model of safety determination. Several experiments were performed. The long time of experiments is necessary to achieve statistically stable results. Vessel traffic was estimated at the level observed in 2005. The LNG traffic was estimated as two LNG carrier passages per week—one entry and one departure from the Baltic Sea (96 per year).

The simulated places of groundings of LNG carriers in Baltic Sea are presented in Fig. 5. The routes of highest grounding probability are the S3, G2 and G3.

For assessment of passage safety other types of accidents shall be respectively assessed. Worst type of accident apart from grounding, where cargo tank rapture can happen—are collisions of loaded LNGC with other vessels. In many literature studies collisions were calculated (Vanem et al., 2008), although social risk criteria's were treated only as function of total loss of human lives between crews of both vessels. At Baltic as well as Adriatic there is a large number of ferries and cruising ships with over thousand passengers on board. Hence risk based social criteria shall be updated and treated as far higher. As the researches showed (Gucma et al., 2010), that mean risk (at 0.95 confidence level) of collision with sea ferry of length over 170 m, varies from 0,0022 to 0,0043 collisions per year per LNG ship. Thus special kind of system for traffic regulation must be applied as it is done in many terminals over the world. Some of results—collisions of LNGC at Baltic Sea with ferry of L_{pp} over 170 m—from this studies are presented in Fig. 6.

This tool is under adoption to Slovenia LNG case where denser traffic occurs in Gulf of Trieste leading to many ports, existing Italian LNG terminal and planned Slovenian LNG terminal.

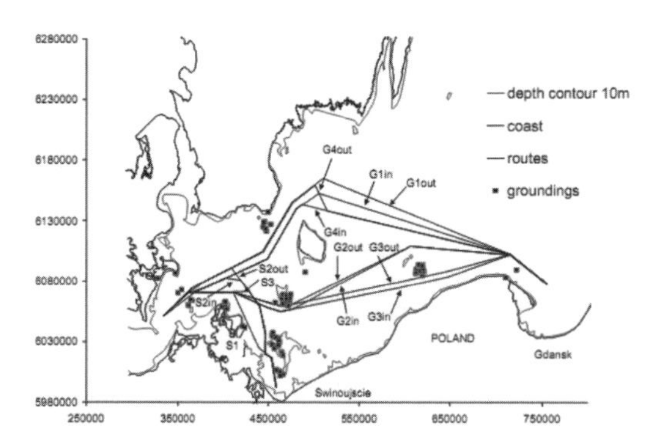

Figure 5. Results from MC simulations in Baltic Sea area for LNGC passage—groundings.

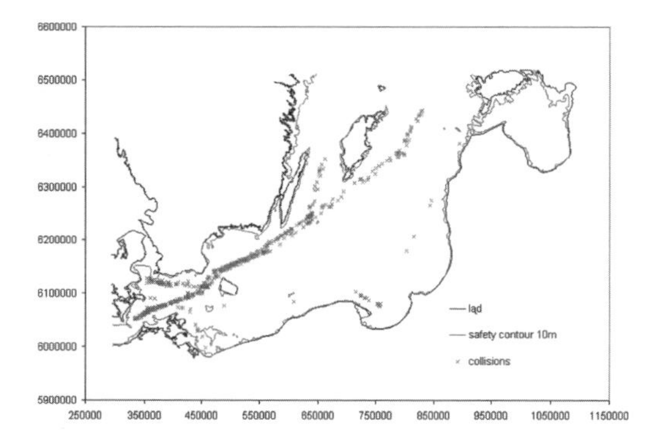

Figure 6. Results from MC simulations in Baltic Sea for LNGC passage—collisions with other vessels.

First results will be compared to general traffic model at Adriatic Sea.

5 CONCLUSIONS

Sophisticated simulation systems require extremely careful planning, design, and validation of assumptions for researches. Often, trying to achieve fast results to big simplifications or wrong assumptions can in fact extend period of time to achieve valid results. LNG terminal design is rapidly evolving part of hydro-structural design branch and without having really well working models it's quite difficult to obtain valuable results.

Existing methods are covering only some sectors of design process and detailed studies must be conducted to describe all phenomena in this field.

Methods of fast simulation let to choose best solution of LNG terminal localization in Poland (Swinoujscie port), prepare safe transit route to the port and design approach.

REFERENCES

Brandt, S. (1998). *Analiza danych.* Warszawa: PWN.

Fang, Q., Yang, Z., Hu, S. & Wang, J. (2005). Formal safety assessment and application of the navigation simulators for preventing human error in ship operations. *Journal of Maritime Science Applied 2005; 4(3),*. Rotterdam: Balkema.

Gucma, L. (2003). Models of ship collision with offshore navigational obstacles. *KONBIN Confernce.* Gdynia.

Gucma, L. (2005). *Risk factors modeling of vessels collisions with port and sea constructions.* Szczecin: Studies 44; Maritime University in Szczecin Publish House.

Gucma, L. *Complex model of navigational accident probability assessment based on real time simulation and manoeuvring cycle concept*, Safety, Reliability and Risk Analysis, Esrel 2008, 21–27.09.2008.

Gucma, L. & Gucma, S. (1998). *Planowanie eksperymantu w badaniach symulacyjnych inżynierii ruchu morskiego*. Szczecin: WSM.

Gucma, L. & Zalewski, P. (2003). *Damage probability of off-shorepipelines due to anchoring ships*. Polish maritime research.

Gucma, M. & Gucma, S. (2007.). Simulation method of navigational risk assessment in optimization of LNG terminal parameters. *Proc. intern. symp., ESREL 2007.* (pp. In T. Aven, J.E. Vinnem (eds.), Risk, Reliability and Societal Safety (vol. 3).). Rotterdam: Balkema.

Gucma, S. (1990). *Methods of waterway determination and design*. Szczecin: WSM.

Gucma, S. (2001). *Marine Traffic Engineering.* Gdańsk: Wydawnictwo Okrętownictwo i Żegluga.

Gucma, S. (2008). *Simulation methods in marine traffic engineering.* Szczecin: Maritime University of Szczecin.

Gucma, S. (2009). *LNG terminals design and operation.* Szczecin: Maritime university.

Gucma, S. et al. (2010). Quantitive risk for LNG sea terminal in Swinoujscie Poland, report from reserche, (in polish), Szczecin: Maritime Univeristy.

Hansen, P. & Simonsen, B. (2000). *GRACAT: software for grounding and collision risk analisys.* Copenhagen: Proc. 2nd Int. Conf. on Collision and Grounding of Ships.

Hutchison, B., Gray, D. & Mathai, T. (2003). *Maneuvering simulations—an application to waterway navigability.* San Francisco: Proc. of SNAME world maritime technology conference.

IMO. (2007). Guidelines for Formal Safety Assessment (FSA) for use in the IMO rule-making process. (p. MSC83/INF.2 (Consolidated version)). London: International Maritime Organization.

Iribarren, J. (1999). *Determining of horizontal dimension of ship meanuvering areas.* Bruxells: Pianc bulletin no 100.

Merrick, J. et al. (2001). *Modelling risk in dynamic eniironent of maritime transportation.* Washington: Proc. of Winter Simulation Conference.

Otay & Tan. (1998). Stochastic Modeling of Tanker Traffic through Narrow Waterways. *Proc. 1st International Conference on Oil Spills in the Mediterranean and Black Sea Regions,* (pp. 85–96). Istanbul.

Sand, S., Nielsen, D. & Jakobsen, V. (1994). *Risk analisys of simulated approach to ports.* Seville: Proc. of PIANC Congress.

Vanem, E., Antao, P., Ostvik, I. & Castillo de Comas, F. (2008). Analysing the risk of LNG carrier operations,. *Safety in Maritime Transportation,* (pp. Pages 1328–1344). Reliability Engineering & System Safety, Volume 93, Issue 9,.

Vose, D. (2000). *Risk Analisys. A quantitive guide.* New York: John Willey and Sons.

Webster, W. (1992). *Shiphandling simulation, Application of Waterway Design.* Washington D.C.: National Academy Press.

Zeigler, B. (1984). *Teoria modelowania i symulacji.* Warszawa: PWN.

Sustainable Maritime Transportation and Exploitation of Sea Resources – Rizzuto & Guedes Soares (eds)
© 2012 Taylor & Francis Group, London, ISBN 978-0-415-62081-9

The influence of large accidents on risk assessment for LNG terminals

P. Vidmar, S. Petelin & M. Perkovič
Faculty of Maritime Studies and Transport, University of Ljubljana, Portorož, Slovenia

L. Gucma & M. Gucma
Akademia Morska Szczecin, Poland

ABSTRACT: The investigation is focused on the risk that an LNG terminal located in a port area poses to the port itself and to the nearby populated area. Experience shows that the acceptance of the risk for new installations in an already existing industrial area quickly becomes a matter of public discussion. Although the additional risk could be negligible for the nearby population in the sense of an industrial risk, the fear of the population from industrial plants exists. The study is oriented toward answering the actual questions of a state policy and civilian community about the safety and risks of an LNG terminal located close to a populated area. The approach of risk assessment is focused on the analysis of LNG spill consequences followed by an evaluation of individual and societal risks that arise from those consequences. The LNG jetty is located inside a port area but close to a populated area on two sides. The consequences analysis is done with two models, depending on the magnitude of accident. The dispersion of natural gas is simulated using an expansion model or lumped model. For larger spills where the gas cloud could reach a populated area the dispersion is analysed with a CFD model to obtain a detailed view of the concentration field. Next is the evaluation of individual and societal risk depending on population distribution, statistics of weather conditions and a scenarios based approach. The conclusion summarizes the obtained results of the analysis and provides guidelines for a comprehensive study of risk assessment in practice.

Keywords: LNG safety, risk assessment, CFD model, lumped model

1 INTRODUCTION

1.1 *LNG hazards in a transport process*

Hazard from LNG cargo begins in the first processing stage of natural gas liquefaction and loading the substance into LNG tankers. The transport itself over the sea is the safest part of the distribution process, as is demonstrated by the statistic of nautical accidents in the past 40 years (DNV 2007, Perkovič et al., 2010 and Gucma 2007).

A review of a Rand Corporation document published in 1976 indicates a high level of safety for LNG tankers. The document indicates that in the initial 16-year history (from 1959 up to 1974) there had been no significant accidents. It should be noted, though, that in 1974 the world LNG fleet included only 14 vessels; by November, 2009, there were 327 vessels, a figure expected to increase to 350 vessels sometime in 2010.

The DNV (Det Norske Veritas) counts 185 nautical accidents involving LNG tankers, all without severe consequences for the crew. The frequency of LNG tanker accidents is therefore 5.6×10^{-2} per ship year. The findings of the DNV (2007) furthermore demonstrate that the potential loss of life for the LNG crew member is 9.74×10^{-3} or less, considering the occupational fatality rate onboard gas tankers is 4.9×10^{-4}. The analysis of the northern Adriatic Sea (Petelin et al., 2009), or, precisely, the gulf of Trieste, demonstrates that nautical accidents should occur with a frequency of 1.25×10^{-2} per year, assuming the current traffic density, and increases to 2.62×10^{-2} if the ship traffic increases by 100%.

The hazard associated with LNG is mainly in its potential to cause severe fires resulting in heat radiation. If a large quantity of LNG is spilled into a pool, the cloud that is formed as it evaporates is a mixture of natural gas, water vapour, and air. Initially the cloud is heavier than air (due to its low storage temperature) and remains close to the ground. The buoyancy moves the natural gas upward at a gas temperature of around 170 K (−103°C), as experimentally demonstrated by ioMosaic (2007). The major influences on natural gas diffusion are environmental conditions. The cloud moves in the direction of the wind and the wind causes the cloud to mix with more air. If the concentration of gas in the air is between 5% and 15% it is flammable and burns if it contacts any ignition source. A concentration of gas smaller than 5% will not ignite and if the concentration is over 15% the air becomes saturated. The explosion of natural gas is not possible in open spaces because the low velocity of flame spread, around 0.4 m/s, is not enough to produce a pressure wave (Fells 1969). The burnout of gas/air mixture in open air could result only in a flash fire. Explosions could only result in enclosed spaces where gas is going to accumulate.

2 RISK DEFINITION AND ACCEPTANCE CRITERIA

Activities involving hazardous substances like LNG are often risk assessed for their impact on external safety by using two criteria, one for the individual risk and one for the societal risk.

2.1 *Individual risk*

The individual risk in regard to LNG is calculated as location specific, to a person exposed outside 24 hours per day. In several countries the authorities have defined criteria which have to be met in order to assess a level of societal risk as acceptable. Criteria for some countries were analysed and discussed by Trbojević (2005). Some of these are used internationally. In most countries an individual risk of $1E10^{-3}$ per year is taken as the upper bound criteria to assess the acceptability of an activity for employers working in an industrial installation. The upper risk criterion for the public is therefore $1*10^{-4}$ per year. Also the risk at locations where vulnerable objects are situated is taken into account. Vulnerable objects are those where people are present who react in a different way to a threat posed upon them as ordinary people. This difference can be caused by differences in state of health or the possibility of evacuating the location in case of danger. The calculation of the individual risk for a specific failure event is influenced by three main parameters: the two coordinate location x, y and weather conditions (wind speed and stability). The individual risk $IR_{xy/w}$ is therefore the function of the frequency F_{fe} of the failure event occurring in a year, the probability of the event in a particular direction θ (influenced by weather) and the probability of people dying due to exposure (DNV 2009).

$$IR_{x,y/w} = F_{fe} \int_{\theta_1}^{\theta_2} \left[P_{\theta/w} P_{d/\theta w} \right] d\theta \qquad (1)$$

θ is the direction of the release,
θ_1 is the lower value of θ that influences the computation point,
θ_2 is the upper value of θ that influences the computation point,
$P_{\theta/w}$ is the probability of the release occurring in the direction of the wind and
$P_{d/\theta w}$ is the probability of death considering the direction of the release and weather.

This is the contribution to the individual risk from a single event under specific weather conditions. To obtain the overall individual risk at a given point all possible events must be taken into account. However in the sense of the order of magnitude the worst events (low probability/high consequences) have the most influence on individual risk. In the calculation process, low consequence events increase the size of high risk isolines, but usually do not influence the low risk isolines, risk lower than $1*10^{-6}$ that reaches greatest distances. The strong dependence of the individual risk on weather conditions influences the total risk, calculated using the following equations:

$$IR_{point\_x,y} = \sum_{Weathers} P_w \cdot IR_{x,y/w} \qquad (2)$$

and

$$IR_{total\_x,y} = \sum_{all\_x,y} IR_{point\_x,y} \qquad (3)$$

where

$IR_{point\_x,y}$ is the individual risk over all weather conditions for a specific event

P_w is the probability of specific weather and

$IR_{total\_x,y}$ is the individual risk as a sum of all events under all weather conditions.

2.2 *Societal risk*

Societal risk posed by an LNG terminal facility or hazardous activity is measured by the probability that a group of persons would be exposed to a hazardous level of harm (fatality) due to all types of accidents at the facility or its hazardous activity. The societal risk is dependent on both the density of people in the vicinity of an LNG terminal and the location of the population with respect to the facility. The societal risk is generally presented in the form of a curve, expressing the relationship between the annual probability (F) of exceeding a given number of fatalities and the number N (Trbojević 2005). In most countries the risk assessment is performed on the basis of potential fatalities to the exposed population. Different countries use slightly different criteria for risk acceptability. In the UK, the Health and Safety Executive (HSE)

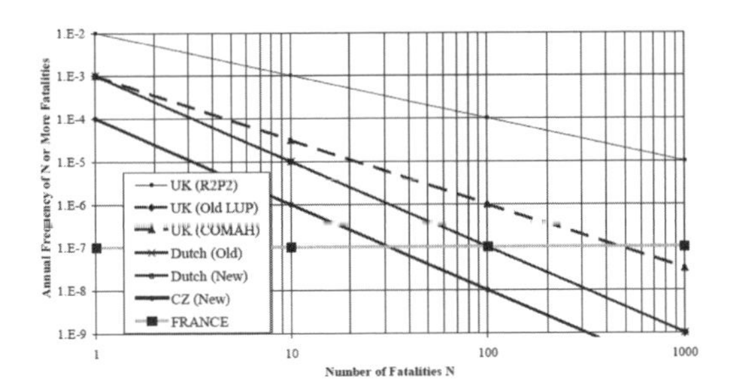

Figure 1. Comparison of F-N Criteria (Trbojvić, 2005).

guidelines are to use the individual risk as the principal measure, but also to use the societal risk criteria for land use planning. The acceptability criteria levels for risks for facilities in the UK are specified by HSE (1989). Facilities are permitted only when these (published) criteria are met. In the Netherlands, however, both the individual risk criteria and the societal risk criteria have to be met when considering (in risk assessment) those events whose hazardous effects extend to such distances at which the conditional probability for lethality is higher than 1%. The risk tolerability criteria for fatalities established in various countries for societal risks are summarized in Figure 1 (Trbojević 2005).

The risk calculations have to determine the number of people killed by a particular event and attach this to the associated frequency of the event to form the F-N curve. Probability of death is calculated from the consequence model and so it only remains for the risk model to integrate the probability of death for each event over the specified population. $N_{f/fe}$ represents the number of people killed by a given event, particular considering the weather category/wind direction combination. It is calculated for the population on a grid according to Equation (4) (DNV 2007).

$$N_{f/fe} = \iint_{area} P_{d,xy} n_{xy} dxdy \tag{4}$$

3 CONSEQUENCES ANALYSES BACKGROUND

3.1 *Rupture size*

The estimation of the rupture size caused by an accident involving an LNG carrier is not easy, because of the variables: a complex structure, the type of accident and the location of the primary rupture. The LNG carrier has four to five barriers that would have to fail before the LNG cargo is spilled. The shield of the LNG containment allows deformation before it ruptures. The characteristic of this material is that it remains elastic at a cargo storage temperature of −162°C. The second fact is that the reservoir is filled up to 96–97% and the rest is gas, about 800 m³. Because of this gas space the containment structure allows slow deformations that occur during an accident. The accident of the El Paso Paul Kayser described by Bubbico (2009) is a good example. The accident caused significant deformations of the ship's hull and ribbed construction yet the deformation of the cargo containment was limited to one meter and there was no release. Pitpablo (2004) suggests that the maximum possible rupture size is 250 mm in diameter when caused by grounding or collision with the shore.

Considered and analyzed ruptures on LNG vessels are only caused by accidents from the traffic point of view. Terroristic attacks and attacks with weapons are not considered. However, the interpretation of DNV (2009) and Sandia National Laboratory (2004) is widely acceptable. Major damage to LNG tankers by weapons is less probable because the vessel's structural stiffness is much greater than any building, bridge or other land structure. In case a projectile breaks through the primary reservoir, there is a high probability of immediate ignition, a local fire or even the destruction of the ship, but not the formation of a flammable gas cloud and subsequent flash fire that is a danger to the neighbour population. On this basis DNV suggests the consideration of the largest rupture size of 1500 mm for a terroristic attack.

Therefore the further scenarios analysis would consider the following LNG reservoir rupture size:

- 250 mm Maximum possible rupture size caused by grounding
- 750 mm Maximum possible rupture size caused by collision
- 1500 mm Maximum possible rupture size caused by terrorist attack
- 7,000 m³/h Maximum possible leakage rate for 10 min.
- 10,000 m³/h Maximum possible leakage rate caused by sabotage for 60 min.

3.1.1 *Tank rupture model*

The tank model is a quasi-steady-state model, which calculates the discharge rate of gases and/or liquids from a tank or pipe system caused by a rupture in the tank or shearing of the attached pipe. It takes into account chemical properties, environmental variables (atmospheric pressure and ambient temperature), rupture geometry (circular, rectangular, smooth or jagged edges), and the containment variables (pressure, temperature). In contrast to a pipe leak model, that is steady state, the quasi-steady-state model means that variables like pressure and temperature slightly change in time depending on the leak release rate.

Three discharge models exist: one to describe the release of liquid from a hole (including two-phase flow), one to describe the release of vapour from a rupture above a boiling liquid, and one to describe the release of a compressed gas. The logic for passing parameters between these models is illustrated in Figure 2 (Safer System, 1996).

3.2 *Pool size related to rupture size*

The pool size is related to the rupture size through which the LNG is leaking as well as with the condition that the gas is starting to burn during the process of leaking. The rapidity of the ignition means a smaller pool and fewer consequences to the surroundings. The pool size is also conditioned by the surface roughness and the heat transfer from the environment to the pool. LNG spilled on water is a particular phenomena wherein the heat transfer depends on water surface conditions, especially waves and temperature. The evaporation rate is also the first step in the vapour cloud formation that is moved downwind. The size of the cloud and its location strongly depends on wind speed. Weaker wind (about 2 m/s) speed keeps the cloud homogeneous for a longer period, but the buoyancy moves it higher, stronger wind speed (5 m/s and more) dilutes gases and the cloud reaches greater distances. Investigations conducted by

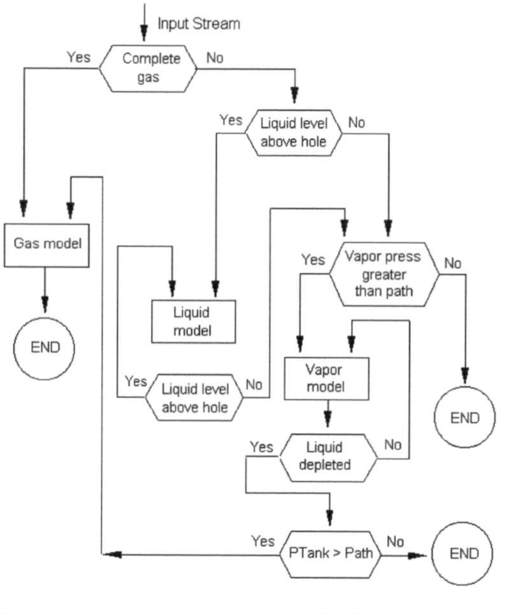

Figure 2. Logic for algorithms within the Tank rupture model.

DNV (2009) and Sandia (2004) differ for wind speeds from 2 to 5 m/s.

Studies conducted by ioMosaic (2007) demonstrate the relation of flow rate, spilling time and the size of pool formed as presented in Figure 3. The basis is a typical LNG tanker with a single reservoir capacity of 20,000 m³ that is spilling through different rupture sizes.

The pool is modelled as an upright cylinder of volume V, radius R and height H as seen in Figure 4.

At any time t the volume of the pool can be written as:

$$V_P(t) = V_0 + \int_{t=0}^{t} \left(\dot{V}_{IN}\right) dt - \frac{m}{\rho_L} \tag{5}$$

m—Mass of liquid vaporized from pool starting at t = 0 to time t.

$\dot{V}_{IN}$—Volumetric flow of liquid into the pool

The above Equation (5) represents a volumetric balance of the liquid pool. A heat transfer model is also included in the pool evaporation model. The main purpose is to compute the evaporation rate due to radiation, conduction and natural or forced convection heat transfer. If the pool temperature is below the boiling point, as would be with LNG

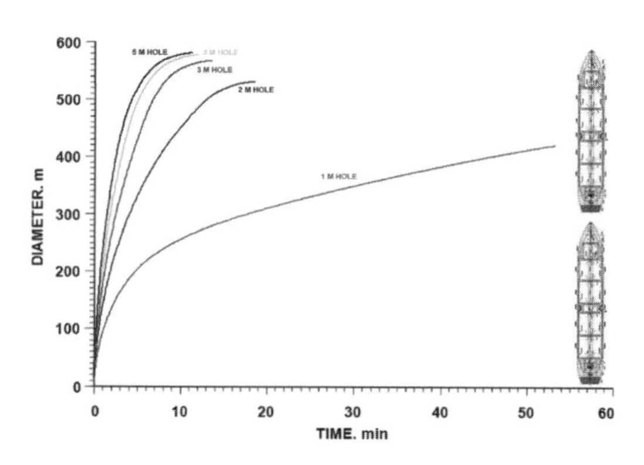

Figure 3. Pool spread on water surface (ioMosaic 2007).

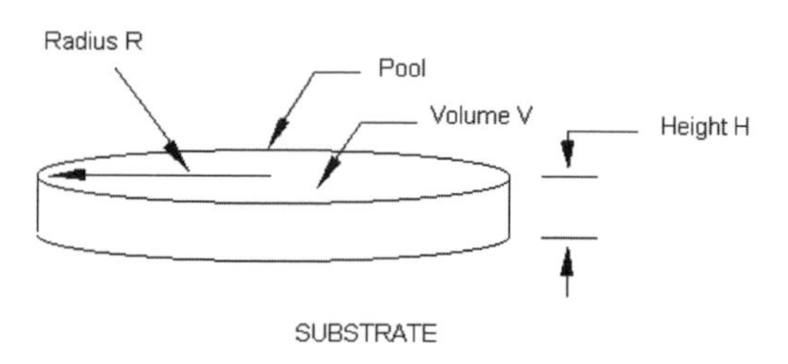

Figure 4. Liquid pool modelled as an upright cylinder.

spills, then the pool is assumed to be evaporating at that temperature at a rate defined by:

$$E_{vap}(t) = \frac{Q_{evap}}{\Delta h(T_{pool})} \tag{6}$$

As mentioned above, the spreading of the LNG pool is a dynamic process that is alimented by the spilling from the tank slowed down by evaporation or even vapours burning. Research conducted by Hissong (2007) and ioMosaic (2007) presented in Table 1 explain that the evaporation rate could vary depending on water conditions.

3.3 *Source dynamics*

The algorithms mentioned in the tank rupture section 3.1.1 determine the state of the released stream from containment due to a failure/rupture. This released chemical should, in general, consist of three release "streams": escaping gas, liquid that remains airborne (which will break up into a flashing fraction of gas and an aerosol fraction), and liquid that falls to the ground to form a pool, which then evaporates. The various "streams" and the initial source dynamics are involved in the formation of a vapour cloud.

The estimation of the flash and aerosol fraction is defined in [15] and [23] as:

$$x = c_L(T_o - T_B)/\Delta h_V \tag{7}$$

The estimation of the airborne liquid (or aerosol) stream is much more difficult to formulate. The magnitude of the aerosol stream depends upon the containment variables (pressure, temperature, etc.) and is also affected by the rupture characteristics (orientation, geometry of hole).

Hence, the present approach consists of achieving the capability of defining the aerosol fraction. In addition, the computer code has a default aerosol estimation technique. The aerosol content of the stream is assumed to be a function of the flashing fraction, defined above, in Equation (7).

Table 1. Evaporation rates depending on water conditions.

	Evaporation rate Max. value kg/m²s	Evaporation rate 20 sec. avg. kg/m²s
Water	0.303	0.147
Water	0.245	0.191
Water	0.230	0.196
Ice	0.513	0.191
Ice	0.333	0.171
Brine	0.254	0.186

An enthalpy balance for the system produces:

$$(m_{V1} - m_{V0})L_a + m_{V1}\int_{T_0}^{T_1} c_V dT + m_{L_1}\int_{T_0}^{T_1} c_L L dT + m_A \int_{T_A}^{T_1} c_A dT + m_{A_0}\int_{T0}^{T_1} c_A dT - 0 \tag{8}$$

The ideal gas law is assumed in the vapour phase,

$$p_{V_1} = \frac{p \cdot \frac{m_{V_1}}{MW}}{\frac{m_{V_1}}{MW} + \frac{(m_{A_0} + m_A)}{MW_A}} \tag{9}$$

and an Antoine vapour pressure equation is used (SFPE Handbook, 1995).

$$\ln p_{V_1} = C_8 + \frac{C_9}{C_{10} + T} \tag{10}$$

3.3.1 *Atmospheric dispersion of a vapour cloud*

Leak rates from a tank rupture were calculated; initial gas, "flashing" and an aerosol are formed. An aerosol is a cloud of tiny liquids droplets or fine solid droplets suspended in air. Calculating the droplet evaporation along the cloud trajectory, the overall vapour generation rate is obtained. In this section the dispersion model of the vapour cloud is described. The model considers different types of release: instantaneous, steady continuous, and transient for dense (active) and lean (passive) gases.

Because of a very fast transient and changes of variables, it is difficult to predict the course of events, especially close to the source of dispersion. Once the cloud, modelled as a cylinder, is formed, it begins to slump under the effect of gravity. The velocity of the edge of the cloud can be described as [15]:

$$\frac{dR}{dt} = k_1 \left[\left(\frac{\rho_{CLOUD} - \rho_A}{\rho_{CLOUD}} \right) \cdot g \cdot h \right]^{1/2} \tag{11}$$

k_1 is a slumping constant that depends on the characteristic of released gases and weather conditions. The most important factor is wind speed. It is important to note that the dispersion model does not assume turbulent flow.

Dispersion of the cloud is therefore is a good candidate to be computed by the CFD model (McGrattan, 2001). The rate at which liquid fuel evaporates when burning is a function of the liquid temperature and the concentration of fuel vapour above the pool surface. According to the Clausius-Clapeyron relation, the volume fraction of the fuel vapour above the surface is a function of the liquid boiling temperature;

$$V_f = \exp\left[-\frac{h_v W_f}{R_f}\left(\frac{1}{T_s} - \frac{1}{T_b}\right)\right], \quad (12)$$

where h_v is the heat of vaporization, W_f is the molecular weight, T_s is the surface temperature, and T_b is the boiling temperature of the fuel.

4 LNG FIRE EFFECT

For reference only, this chapter provides some basic understanding of the potential effect distances that may occur if LNG releases are set free from large containments of liquefied natural gas. An example of a release scenario including breach of cargo containment is given below.

Here are some explanations of terms used in tables and figures.

- LFL = Lower Flammable Limit: The gas will need a certain concentration in order to be flammable. LFL is the point where the gas becomes too thin to ignite. The gas cannot ignite outside this point.
- Pool Fire is the combustion of vapour evaporating from a layer of liquid at the base of the fire.
- kW/m^2—Escape shall be possible. Heat radiation level leading to 2nd degree burns after 1 minute 30 seconds exposure.

For intentional acts it is noted that a release of LNG is very likely to be ignited directly, which will lead to a pool fire effect. From earlier studies of Sandia (2004) it can be assumed that LNG pools up to about 230 meter radius might be formed, which provide pool fire effects of maximum a several hundreds of meters.

It is important to understand that there are significant uncertainties related to large releases of LNG. This applies for the release scenario as well as the calculations. In general, fire radiation calculations performed by Phast Risk (DNV, 2007) as above are assessed as conservative as they do not account for the soot effect for large fires. Scenarios comparable to the one presented above have been used as a basis to establish the Sandia Zoning as presented in the next Section. DNV follows this zoning as a first and conservative approach.

4.1 *Hazard zones*

As shown in previous chapters, the calculation of precise effect distances for potential fire effects (exposure to certain levels of significant heat radiation) is difficult unless the specifics of any given actual context are considered to derive the appropriate representative event scenarios. Calculated effect distances are directly influenced by assumptions about possible hole sizes, weather conditions, ignition probabilities etc. As these types of assumptions are difficult/impossible to predict in the context of intentional acts, it makes it also difficult to assess the potential affected area from selected fire scenarios.

Because of this it has become a best practice in the LNG industry to assess the hazard zoning around LNGCs in the context of security mitigating measures related to them. As the vulnerability is directly influenced by the effectiveness of deterrent measures and restrictions a pro active approach is advocated based on a geographical analysis and an in depth risk assessment. Sandia National Laboratories is a US-based government-owned/contractor operated (GoCo) facility that has developed science-based technologies that support national security. Sandia advocates the following security zones to take into account for LNG carriers/activities (Sandia 2004):

Note that the in case of a pool fire the application of zone 1 restrictions will prevent severe domino effects. Zone 2 and 3 are aimed at maximizing the chances for rescue in case of vapour cloud scenarios. But more important all zones are also intended to identify pro active measures to prevent such intentional acts from happening. The appropriate restrictions and measures should be based on an in depth threat analysis done by local experts.

In the specific context of the Port of Koper, the indicative hazard zones (see Figure 5) would encompass the population of Koper who are located at approximately 1300 meters of the jetty and the LNGC. Closer to the jetty and the LNG

Table 2. Ship cargo tank release, 1500 mm.

Scenarios	Results
LNG carrier cargo tank Hole size: 1500 mm Sustainable burning rate: 8130 kg/s Released onto the water surface	Maximum burning pool radius = 230 m At 1 m height above ground
Weather condition for radiation and dispersion calculations:	Sustainable burning pool radius = 180 m
Wind velocity: 5 m/s (prevailing wind condition)	Distance to 5 kW/m^2 (with sustainable pool radius): 750 m
Stability class: D	Distance to LFL: 1800 m
Surface Roughness: 3 mm	
Humidity: 70%	
Temperature: 20°C	

Table 3. Security zones around LNG terminal.

	Distance to LNGC	Dominating hazards	Risk management strategies
Zone 1	500 m	Thermal radiation (>37,5 kW/m^2) and vapor dispersion can pose a severe public safety and property hazard and can damage or significantly disrupt critical infrastructure located in this area.	• No critical or vulnerable infrastructure is allowed in zone 1. • Risk management strategies should address vapor dispersion and fire hazards. • Most rigorous deterrent measures, such as vessel safety, waterway traffic management schemes, and establishing positive control over the vessel. • Incident management and emergency response measures should be carefully evaluated to ensure adequate resources are available for consequence and risk mitigation.
Zone 2	500 m –1600 m	Thermal radiation (>5 kW/m^2) and vapour dispersion transitions to less severe hazard levels to public safety and property.	• Infrastructure can be present in Zone 2 but emer gency preparedness must be guaranteed. • Risk management strategies should address vapour dispersion and fire hazards • Emergency response measures such as ensuring areas of refuge and development of community warning signals to ensure persons know what precautions to take.
Zone 3	1600 m –2500 m	Risks and consequences to people and property are minimal. Thermal radiation poses lesser risks to public safety and property. The dominating hazards are flammable vapor clouds.	• Nor restriction for infrastructure in Zone 3. • Risk management strategies should focus on dealing with vapour cloud dispersion and can be less complicated or extensive than Zones 1 and 2. • Ensure that areas of refuge are available and measures should be implemented to ensure that persons know what to do in the unlikely event of a vapour cloud.

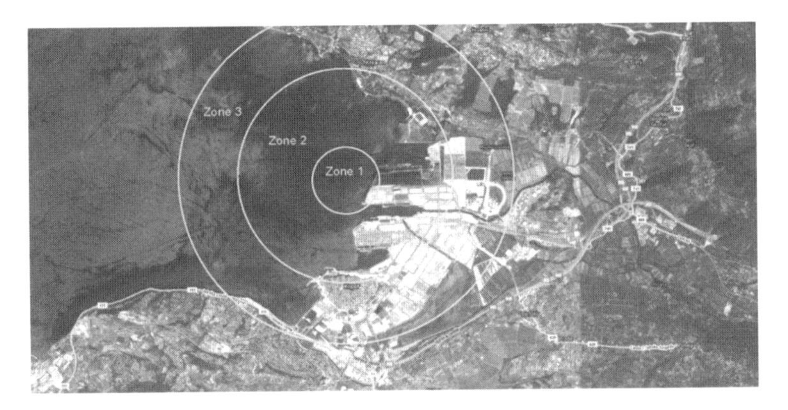

Figure 5. Hazard zones for Port of Koper (DNV 2009).

pipeline various activities and infrastructure is present that could be characterised as vulnerable.

5 SCENARIOS AND RESULTS

The simulations conducted are used to compare results obtained from the lumped model (fast computation) and CFD model (FDS). The diameter of the spilled pool is related with the diameter of the hole on a reservoir shield. The following scenarios assume the hole at the bottom of the reservoir and 10^5 kg of LNG content is spilled in 10 seconds and forms the pool. The evaporation rate of natural gas from the pool surface is calculated by the lumped model and assumed to be 0.16 kg/m^2s (Vidmar 2006) for the CFD model in a stable sea and stable weather conditions.

1. D = 90 m Calculated pool diameter for the hole diameter 500 mm.
2. D = 450 m Calculated pool diameter for the hole diameter 1500 mm.

5.1 *Scenario 1: Medium release with pool formation*

The first scenario assumes the evaporation of an LNG pool of radius 45 m (D = 90 m) as a consequence of a larger spill. The amount spilled is 100 tons of LNG and is assumed to be the largest spill that can occur during unloading operations of an LNG carrier. This and other scenarios are described in the risk assessment in paragraph 6.

The size of the computational domain is of 1500 m length (x), 300 m width (y) and 150 m height (z). The spatial numerical discretisation is divided into 7.8 E6 cells compressed in two dimensions (x and z) in the neighbourhood of the pool. The wind speed is 2 m/s in x direction. The evaporation dynamics are computed separately in a lumped model and introduced as an input in the CFD calculation.

Figure 6 shows the pool evaporation dynamics and the pool size. The evaporation rate per square meter is 0.155 kg/m^{2}s and is 3% less than in a CFD assumption. The reason is that heat transfer is lower in a thicker pool. Results obtained using two models differ, but the differences could be accepted at this stage.

The LFL Methane concentration for both models is plotted in Figure 7. The Lumped model results show the cloud height as a function of time. The cloud area covers concentrations between 5% and 15% of Methane in the air. CFD results are plotted as concentration fields observed at different times. The field labelled CFD 500 s almost matches the downwind distance of the lumped model results, although the shape of the cloud doesn't match. The reason of this difference is the distribution of concentration assumed in a lumped model. This is a Gaussian distribution, and clouds normally take typical circular shapes. Assuming higher wind speed would not result in different cloud shape, but rather in different sizes. Therefore the downwind range of the cloud, calculated by the lumped model, matches well with the CFD results, but cloud heights could not. The recognition of this difference provides the reason that larger releases need to be analysed using different methods or different models to make sure that the results reflect the real scenario. Results from the lumped model are used in the next paragraph to model individual and social risk. The risk zones are influenced by downwind cloud range on a ground

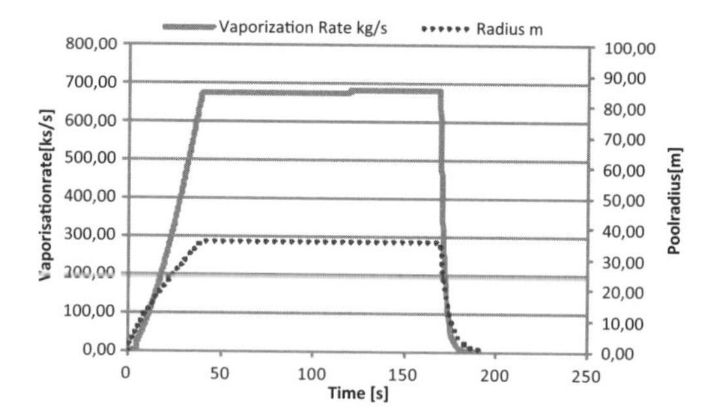

Figure 6. Pool evaporation and size dynamics for scenario 2.

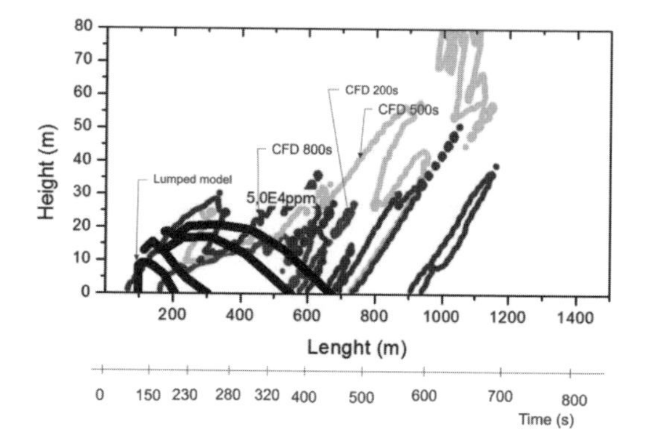

Figure 7. Methane concentration of 5% computed with a lumped model and CFD model for scenario 2.

level that is assumed to reach the largest distance. As presented in Figure 7 the flammable concentration cloud could reach 50% greater distances 30 to 50 meters above ground and therefore influence the risk level.

The event under scenario 2 matches the end of hazard zone 1 and the beginning of zone 2.

5.2 *Scenario 2: Large release with pool formation*

The second scenario assumes the evaporation of an LNG pool of diameter 450 m as a consequence of a 1.5 m hole in a primary containment of LNGC. The amount spilled is 12.000 m^3 of LNG. The environmental conditions are assumed to be the same as in scenario 1 and the wind speed is 1,5 m/s and second 4 m/s in x direction. The evaporation dynamic is computed in a lumped model as the dispersion dynamics. The CFD model is not used at this time because of the excessive domain discretisation.

Figure 8 shows the cloud size in two directions for concentration area between 5% and 15% gas concentration. Results show that that at low wind

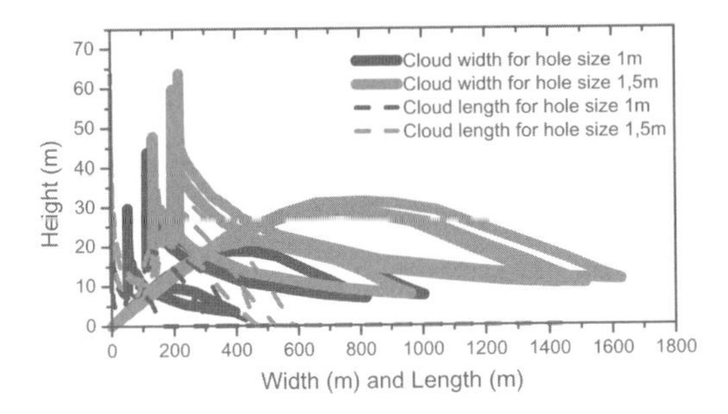

Figure 8. Cloud width and length between 5% and 15% natural gas concentration for different hole sizes on primary containment and wind velocity 1,5 m/s.

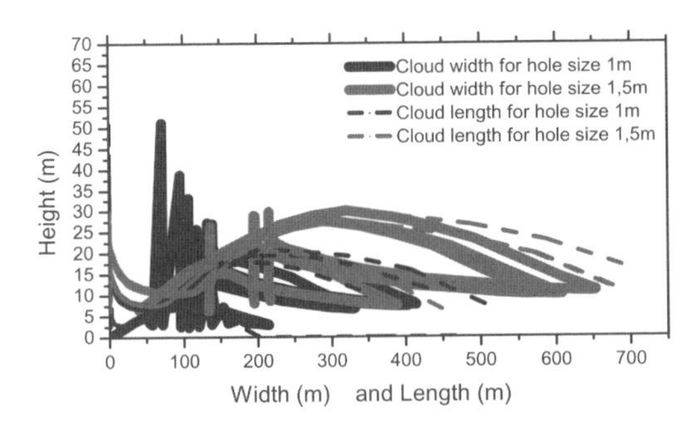

Figure 9. Cloud width and length between 5% and 15% natural gas concentration for different hole sizes on primary containment and wind velocity 4 m/s.

speed (1,5 m/s) the gas cloud could reach about 1600 m area from spill source, when the direction depends on wind direction. The difference in hole size of 0.5 m changes substantially the affected area. In sense of hazard zones listed on Figure 5 the gas cloud reach the outside bound of the second zone.

At higher wind speed the gas is diluted faster in air and the gas cloud reach the area about 700 m from the source. The affected zone is enclosed in the inside the port area and about 200 m inside the second hazard zone.

6 RISK ASSESSMENT OF AN LNG TERMINAL

The risk assessment of an LNG terminal identifies and quantifies the risks from the perspective of safety associated with the arrival of an LNG carrier into the port and the unloading activities at the terminal jetty. Depending on the requirement of accuracy a qualitative or quantitative approach is used. The overall process of risk assessment includes the following activities:

- LNGC at sea,
- LNGC entry into the port until the vessel is moored at the jetty
- LNGC unloading operation

The assessment of the installation presented is focused on the accidental events of the LNGC approaching territorial waters though the northern Adriatic separation zone, entering the port and the unloading operation in the sense of individual risk and social risk. Therefore the appropriate approach is to discover and analyse processes and installation elements that could lead to undesired consequences. There are several widely approved methods, like HAZID (Hazard Identification) and HAZOP (Hazard and Operability), described in (Macdonald et al., 2004 and Center for Chemical Process et al., 2008) that are used for structured and systematic examination of a planned or existing process or operation in order to identify and evaluate problems that may represent risks to personnel or equipment. The HAZOP study and report are not presented in the paper, but the evaluated accident events frequencies are used for the evaluation of individual and social risk.

The population is divided into employed in the port and the population outside the installation or the 'civilian' population. Of great importance is the risk estimation for a civilian population that is not acquainted with emergency situations and safety procedures.

Figure 10 shows the distribution of the population within and in the neighbourhood of the port based on the national geographic system (SFPE, 1995). Particular attention is paid to the population density and migration inside a port close to the jetty and to the dense populated living areas in the neighbourhood.

Applying the risk model on terrorist scenarios defined above we should evaluate individual and societal risk. Depending on the assumption of the intentional act event likelihood we should find risk zones with evaluated risk levels. If the intentional

Figure 10. Population density in neighbour of the installation.

act event resulting in LNG spill as a consequence of 1,5 m hole in primary LNGC reservoir has a probability of 1E-7/AvgeYear the individual risk is as presented on Figure 11. Assuming the upper individual risk level allowance for population of 1E-6, as proposed by HSE, the risk in enclosed in Zone 1 (Figure 5) that is inside the port protected area.

The societal risk curve (F-N) on Figure 12 is located within the ALARP (As Low as Reasonably Practicable) area delimited by the Max and Min Risk Criteria straight lines. The summary of risk assessment shows that the 1E-6/AvgeYear individual risk contours do not reach any vulnerable location outside the port area and therefore the level of societal risk is not unacceptable, since the presence of people in the direct vicinity of the jetty is limited and the distance to populated areas like the town centre is too far to pose a societal risk.

Applying a different probability of 1E-6 to the same intentional act event as before the risk area increase. On Figure 13 the area with 1E-6/AvgeYear correspond to the Zone 2 of Figure 5. The societal risk is analysed for both, employed people and neighbour population. The analysis shows that the F-N curve exceed upper bound of the ALARP region and therefore outside the allowed risk range.

After several simulations we found the allowed probability of intentional act event, when the F-N curve not crosses the maximum risk criteria. This value is around 7E-7/year and should be the reference for security planning.

In the unlikely event that a terrorist intervention does not give a direct ignition of the flammable release a vapour cloud may be formed that drifts off with the wind. A vapour cloud may be ignited at a remote location if an ignition source with sufficient energy potential is present within

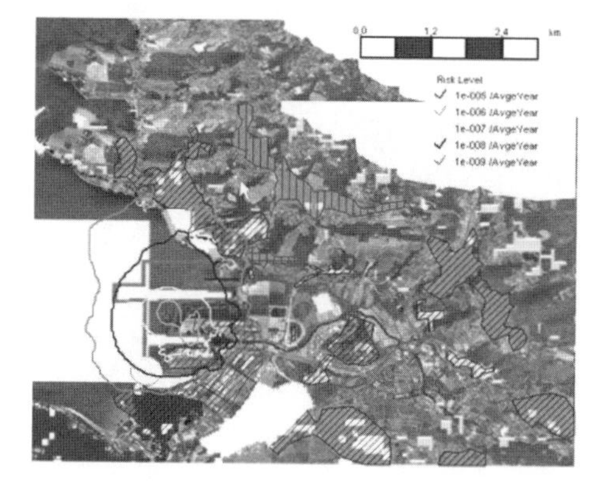

Figure 11. Individual risk for intentional act with probability 1E-7.

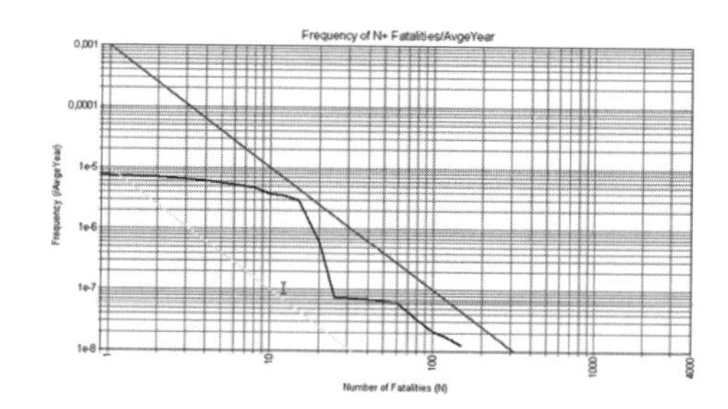

Figure 12. Societal risk for intentional act with probability 1E-7.

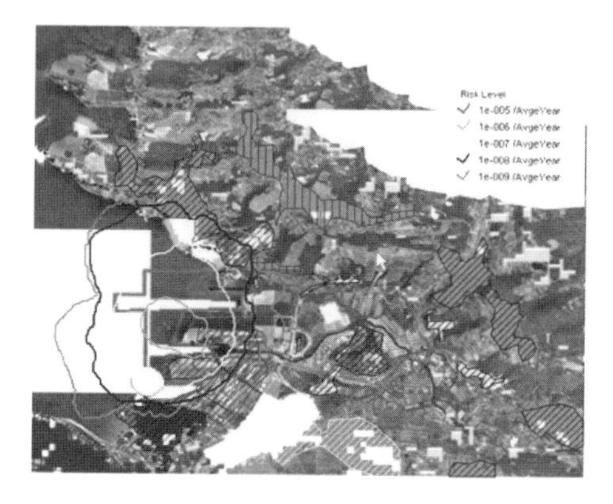

Figure 13. Individual risk for intentional act with probability 1E-6.

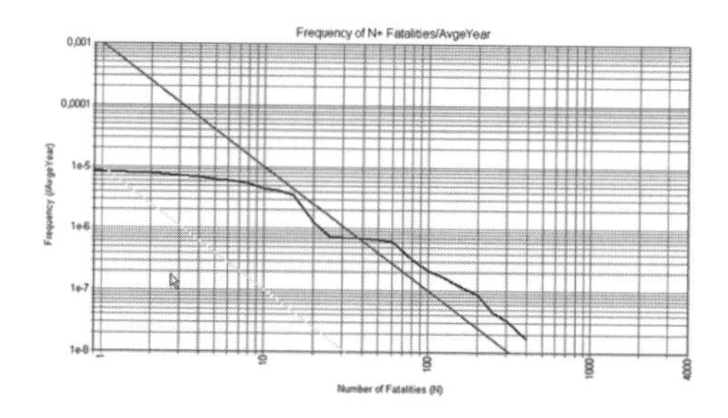

Figure 14. Societal risk for intentional act with probability 1E-6.

the developing flammable cloud (e.g., a flare or furnace in adjacent industries, or traffic movements on land or water).

Because of the unpredictability of terrorist interventions and their potential impact on LNGC it has become best practice in the LNG industry to assess hazard zoning around LNGCs in the context of security mitigating measures related to them. Sandia/DNV recommends applying 3 different zones (500, 1600 and 2500 meters) around LNGCs with different levels of security associated with them.

7 CONCLUSION

The consequence analysis and risk assessment conducted show that the intentional act could lead to a rupture size of not larger than 1.5 m on primary containment of LNGC. The consequence of such rupture on a single reservoir could lead to a spill of 12.000 m^3 of LNG into the water and that 450 m large poll is going to form. At low wind speed the gas cloud formed could reach a zone up to 1600 m downwind that is the area that covers a hazard zone 2 that is on the limit of the populated area. The risk evaluation for individual and collective threat is analysed backward to find the allowed probability of intentional act where the individual and societal risk do not exceed the acceptable levels. The probability of the intentional act on LNGC could therefore be less than 7E-7/AvgeYear. If the security system for the LNG and the nation itself could keep the probability of intentional act below this level, the societal risk is kept inside the ALARP area.

The focus of any intentional act management plan needs to be directed towards the prevention and deterrence of such acts. A thorough and detailed security risk evaluation can provide a good foundation for a suitable risk management plan. As discussed in this report the vulnerability is one of the most important aspects when dealing with intentional acts. To lower the vulnerability in general several options are available: one of the most important is to obey the zoning as described in the paper. The zoning should ensure the presence of no vulnerable objects near the LNG installation.

REFERENCES

Bubbico, R., Di Cave, S. & Mazzarotta, B. Preliminary risk analysis for LNG tankers approaching a maritime terminal, Journal of Loss Prevention in the Process Industries, Volume 22, Issue 5, September 2009, pp. 634–638.

Center for Chemical Process Safety. Guidelines for the management of change for process safety, John Wiley & Sons, Inc., Hoboken, New Jersey, 2008.

Det Norske Veritas. MPACT Theory manual, Internal publication, DNV software, June 2007.

Det Norske Veritas. Risk Assessment LNG import Koper: nautical and unloading operations, Report no/ DNV Reg No.: / 124UI0A-4, 2009.

Drysdale Dougal. An introduction to Fire dinamics, John Wiley & Sons Ltd – 1998.

Fells, I. & Rutherford, A.G. Burning velocity of methane-air flames, Combustion and Flame, Volume 13, Issue 2, April 1969, Pages 130–138. Country Reports on Terrorism 2008. http://www.aon.com

GRM, A. Wind rose in the southern half of the Gulf of Trieste, Marine Biology Station, National Institute of Biology, April 2002.

Gucma, L. "Evaluation of oil spills in the Baltic Sea be means of simulation model and statistical data. International Maritime Association of Mediterranean", Kolev and Soares editors), Balkema 2007.

Handleiding risicoberekeningen Bevi: inleiding, Module A/B/C – Versie 3.0", RIVM, January 2008.

Hissong, D.W. Keys to modeling LNG spills on water, Journal of Hazardous Materials 140 (2007) 465–477. http://www.hse.gov.uk/http://antoine.fsu.umd.edu/chem/senese/101/liquids/faq/antoine-vapor-pressure.shtml

Kevin, B. Mc Grattan, Howard R. Baum, Smoke Plume Trajectory from In Suit Burning of Crude oil in Alaska, NIST-USA-1997.

Macdonald, D. Practical Hazops, Trips and Alarms, IDC Technologies, imprint of Elsevier 2004, ISBN: 0750662743.

McGrattan, K., Baum, H., Rehm, R., Hamins, A., Forney, G.P., Floyd, J.E. and Hostikka, S., 2001. Fire Dynamics Simulator—Technical reference guide, National Institute of Standard and Technology, NISTIR 6783, 2001.

"Nautical Risk assessment LNG transport Rostock". DNV Energy, December 2007.

Perkovic, M., Gucma, L., Przywarty, M., Gucma, M., Petelin, S. & Vidmar, P. Nautical risk assessment for LNG operations at the Port of Koper, Inernational conference on traffic science, Portorož 2010.

Petelin, S., Vidmar, P., Perkovič, M., Luin, B. & Kožuh, M. Predlog prometno-varnostnih analiz za plinski terminal (Sovenian only), Portorož 2009.

Pitblado, R.M., Baik, J., Hughes, G.J., Ferro, C. & Shaw, S.J. Consequences of LNG Marine Incidents, Det Norske Veritas (USA) Inc., Houston, 2004.

Population Census 2002, Statistical Office of the Republic of Slovenia, www.stat.si.

Richard, A. Clarke. LNG Facilities in Urban Areas, A Security Risk Management Analysis for ATTORNEY GENERAL PATRICK LYNCH RHODE ISLAND, May 2005. ioMosaic, Modelling LNG Spreading and Vaporisation, ioMosaic Corporation 2007.

Roger, M. Cooke. A Brief History of Quantitative Risk Assessment, Rosources Summer 2009 (http://www.rff.org/Cooke.cfm).

Sandia Natl. Lab. "Guidance on Risk Analysis and Safety Implications of a Large Liquefied Natural Gas (LNG) Spill on Water", SAND2004-6258, Dec. 2004. ioMosaic documents, http://www.iomosaic.com/, 2006.

Safer System LLC. User's Guide Trace 8-Description of modelling algorithms, Westlake Village, California, USA-1996.

SFPE Handbook. Fire protection engineering, 2nd edition, National Fire Protection Association 1995.

Trbojevic, V.M. Risk criteria in EU, ESREL'05, Poland, 27–30 June 2005.

Vidmar, P. & Petelin, S. An analysis of a fire resulting from a traffic accident, Journal of Mechanical Engineering 49(2003), ISSN 0039-2480, pp. 1–13.

Vidmar, P. & Petelin, S. Analysis of the effect of an external fire on the safety operation of a power plant, Fire Safety Journal 41 (2006) 486–490.

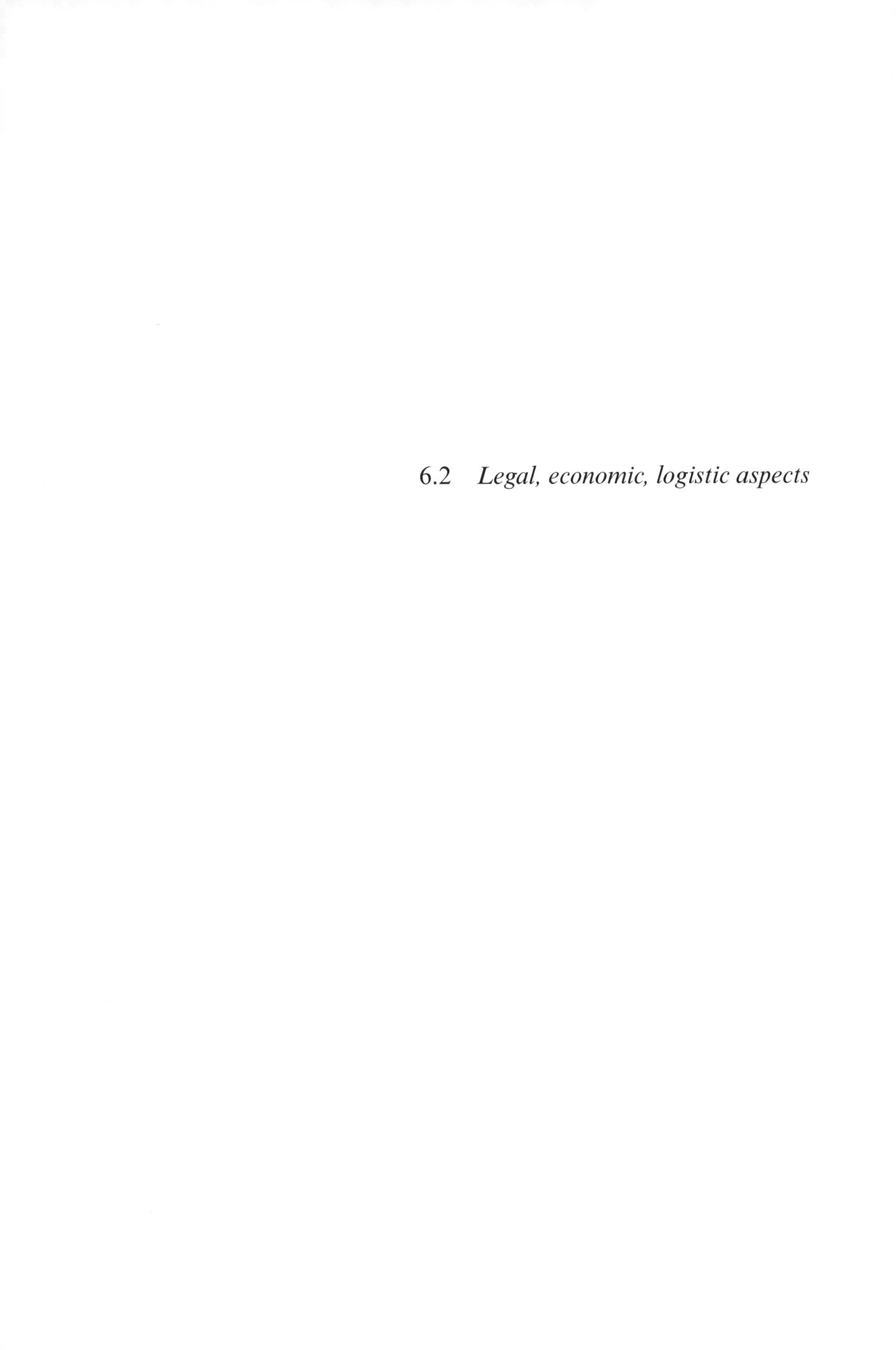

6.2 *Legal, economic, logistic aspects*

Sustainable Maritime Transportation and Exploitation of Sea Resources – Rizzuto & Guedes Soares (eds)
 ISBN 978-0-415-62081-9

The regime of the contiguous zone in the context of international law

E. Xhelilaj
University of Vlora "Ismail Qemali", Albania

O. Metalla
Parliamentary Member, Albania

ABSTRACT: The legal regime of the contiguous zone, as articulated in article 33 of UNCLOS 1982, has brought a situation wherein its legislative and enforcement status continues to be vague and ambiguous. The contents of article 33 regarding the contiguous zone, lacks the clarification on the issues such as the enforcement and jurisdictional rights of a coastal state. Moreover, the ambiguous formulation of this article creates potential premises for misunderstanding and/or misinterpretation. Many coastal states have applied the legal regime of the contiguous zone in their waters, others have not established yet, and some states seems to have problems in appreciating the exact rights and powers to exert in their established contiguous zone. The authors, are of the opinion that there may be a need currently to seriously take under consideration these challenges of the legal nature, and that a revision of the legal regime of the contiguous zone in the international law may be necessary in order to find an appropriate solution for the aforementioned issues.

1 INTRODUCTION

Since in the eighteenth century the legislative and enforcement rights against smuggling ships in the British coastal waters were regarded as universal courtesy of nations for their convenience, and it was rationalized on the same principles as other powers of the State's authority over these waters. This concept may be considered the genesis of the contiguous zone, formally established in the regime of the international law by the 1958 Geneva Convention. It was UNCLOS 82, however, that consolidated the contiguous zone regime, providing to coastal States limited enforcement jurisdiction over "this buffer and checking zone" (Larson, 1976, p. 38) necessary to prevent and punish infringements of its customs, fiscal, immigration, and sanitary laws that occurs within its territory or territorial sea (UNCLOS, 1983). In light of these considerations, contiguous zone regime is considered significant since "it serves as safety valve from the rigidities of the territorial sea, satisfying therefore important demands of coastal States through the exercise of limited authority" (McDougal et al., 1987, p. 76).

For about a century, the legal status of the contiguous zone has been on the focus of studies carried out by many authors such as George Schwarzenberger "International Law" in 1957; S. Oda "The concept of the contiguous zone" in 1962; O'Connell "The international law of the sea" in 1984; Churchill & Lowe "Law of the Sea" in 1999 and so forth. All these notable authors have shed light in many issues with regard to the genesis and development of the contiguous zone; the enforcement jurisdiction of coastal States in this zone; and immigration concerns. In light of current developments, however, there are many issues and challenges pertaining to the contiguous zone which need to be tackled. Accordingly, the evolution of maritime technology, the adoption of new maritime regulations, the prevalence of security concerns, and the increased traffic of the narcotic drugs at sea exposed many issues such as the enforcement authority in this zone; the relevance of the contiguous zone in nowadays; as well as the possible establishment of legislative jurisdiction and the adoption of the anti-narcotic laws in this area.

Therefore, it is the purpose of these authors to provide an analytical discussion concerning the contiguous zone in the context of enforcement and legislative jurisdiction by discussing first the notion of this zone; then, examining its legislative and enforcement jurisdiction; and finally analyzing the challenges that contiguous zone is facing today—trying from the authors' perspective to address the important issues previously stated.

2 THE NOTION OF THE CONTIGUOUS ZONE

2.1 *Genesis and development*

The origin of the contiguous zone appears to derive from the eighteenth century British Hovering

Acts, initially enacted in 1750 (Frommer, 1981–2) as well as in 1765 (Improvement of Revenues and Customs Act), and revoked by the Customs Consolidation Act in 1876, which was adopted as an instrument against foreign smuggling ships hovering within the coastal waters of Great Britain in a distance up to 24 miles (Churchill & Lowe, 1999). The root of these Acts originates from the House of Lords' formal discussion in 1739, held in response to the continuing evil of smuggling at sea, particularly, from the Spanish interference with British shipping in Caribbean waters (O'Connell, 1984). According to the House of Lords' philosophy, these anti-smuggling rights were accorded by the law of nations, because has been found that all the safety measures taken on land could not thwart smuggling (The Parliamentary History of England from the Earliest Period to the Year, 1803). In parallel, U.S Congress in 1790 (Customs Duties Act) enacted also legislation against smuggling at sea, requiring from all vessels to carry an anti-smuggling manifest, which was demanded by the officers of the customs within the limits of four leagues of the U.S.A coastal waters with the purpose to prevent illicit trade.

As the establishment of the three-mile territorial sea based in the Anglo-American practice in the nineteenth century became inefficient for the States' interests (Larson, 1976), several States such as Chile in 1855, Argentina in 1869, Canada in 1877, and Mexico in 1902 through agreements and unilateral decisions claimed a zone adjacent to the territorial waters to enforce customs and security laws (Lowe, 1982). In this respect, the most notorious laws were advanced by the U.S.A in its 1919 Volstead Act and 1922 Tariff Act, enacted in response to alcohol smuggling within 12 miles of its coast (Churchill & Lowe, 1999). Nevertheless, the real notion of using a distinct contiguous zone adjacent to territorial waters was put forward by the French lawyer Renault in a formal discussion in the Institute of International Law in 1894 (Churchill & Lowe, 1999). This proposal was further discussed in the 1930 Hague Conference, but except recognising the existence of the concept of the continuous zone as an autonomous institution in international legislation, no respective law was enacted. The adoption of the contiguous zone as an international regime was finally established in the Convention on the Territorial Sea and the Contiguous Zone in Geneva on April 29, 1958.

In light of these considerations, the development of the contiguous zone seems to be unique because it has been internationally recognized not only through unilateral claims, but also by international treaties and multilateral agreements (Larson, 1976). Be that as it may, it is submitted that identifying the genesis of the contiguous zone exclusively from the British Hovering Acts it might be partially realistic; yet, credits in the development of this concept should be given also to the U.S. maritime anti-smuggling legislation from the eighteenth and nineteenth century, which was incorporated in different treaties and agreements between the U.S and other States, becoming therefore prevalent among nations and more compatible with the international law at that time (Frommer, 1981-2). Lowe (1982, p. 110) in this respect, notes that, "the enforcement of the U.S. liquor treaty beyond the three-mile limit in 1920 focused the attention of the international community on the benefits of this significant regime".

2.2 *UNCLOS 1982 vs. Geneva Convention 1958*

According to article 24 of 1958 Geneva Convention, the coastal States have "the rights in a zone of the high seas, up to 12 miles from the baseline, to exercise control necessary to prevent and punish infringement of its customs, fiscal, immigration or sanitary regulation within its territory or territorial sea" (Convention on Territorial Sea and Contiguous Zone, 1958, p. 8). This provision was revised by the UNCLOS III, adopted in 1982 in Montego Bay, Jamaica, establishing perhaps a better approach of the contiguous zone. Although, the article 33 of the UNCLOS III pertaining to the contiguous zone shares similar principles with the article 24 of the 1958 Geneva Convention, there are several substantial differences between them.

First, in contrast to article 24 of Geneva Convention, in article 33 of UNCLOS III were deleted the words *on the high seas*, establishing the contiguous zone as something separate from the territorial sea, while taken into account the juridical status of the E.E.Z. Secondly, in article 24 of the Geneva Convention the contiguous zone was extended up to 12 miles from the baseline, whereas in the UNCLOS III the same zone was provided for a maximum extend of 24 miles. In addition, the paragraph 3 of article 24 regarding the delimitation of the contiguous zone boundary between opposite coastal States was omitted from the text of article 33 of UNCLOS III, due to the insertion of E.E.Z and territorial sea delimitation in this Convention (Nandan & Rossene, 1993). Moreover, in contrast to Geneva Convention, article 303 in relation to article 33 of UNCLOSS III provides additional rights to the coastal States in order to exercise control over archaeological and historical objects out to 24 miles (UNCLOS, 1983). Finally, UNCLOS' definition of Hot Pursuit in article 111 is more comprehensive, allowing coastal States in the contiguous zone to exercise their authority when the foreign vessel or one of its boats is violating the rights for the protection of which this zone was established (UNCLOS, 1983).

3 LEGISLATIVE JURISDICTION

3.1 *International Law*

The legal status of the contiguous zone is provided by the international regime of the 1982 UNCLOS—considered the constitution of the oceans, which in article 33 stipulates that;

"In a zone contiguous to its territorial sea, described as the contiguous zone, the coastal State up to 24 miles from the baseline, may exercise control necessary to prevent and punish infringement of its customs, fiscal, immigration or sanitary laws and regulations within its territory or territorial sea" (UNCLOS, 1983).

In this respect, the objectives of a coastal State in establishing a contiguous zone according to UNCLOS III are diverse from those applied to the establishment of the territorial sea, because contiguous zone is not part of the territorial sea, but is within E.E.Z, and the high seas freedom of navigation is applicable in it (Nandan & Rossene, 1993). Additionally, it is important to note that the nature of control and the enforcement powers in the contiguous zone does not create any sovereignty rights over this zone or its resources.

3.2 *Legislative jurisdiction vacuum*

Considering the importance of the matters it addresses, the brevity of article 33 raises doubts whether the coastal State may or may not have the power to adopt national legislation (Johnson, 2004) i.e. legislative jurisdiction, with the purpose of making foreign vessels subject to its customs, immigration, fiscal and health laws in the contiguous zone (O'Connell, 1984). This issue is important since it determines the exact authority exercised by the coastal States in the contiguous zone, not only as a separate regime, but also in the context of the E.E.Z which is "considered a Sui Generis area" (Klein, 2005, p. 144). This concern was brought up in the controversial case of the Japanese vessel *Taiyo Maru* arrested in 1974 within the U.S.A contiguous fisheries zone, in which the U.S. court held that "article 24 of 1958 Geneva Convention was permissive, not exhaustive, and in the contiguous zone a coastal State can established both enforcement and legislative jurisdiction" (Churchill & Lowe, 1999, pp. 138–139). Similarly, the Italian Court of Criminal Cassation in 1963, adjudicating an analogue case held that "article 24 could not limit the right of a State to enact laws, i.e. to establish legislative jurisdiction in the contiguous zone" (O'Connell, 1984, p. 1059).

Despite of the aforementioned legal cases, scholars such as Churchill and Lowe (1999) observe that, the authority exercised by coastal States in the contiguous zone is enforcement jurisdiction only. According to Oda (1962), this was the purpose of the 1958 Geneva Convention in the first place, which had been considered to create a zone of enforcement and not legislative jurisdiction in the contiguous zone. The same theory is suggested by Johnson (2004), where she recognizes the enforcement jurisdiction in the contiguous zone, but seriously doubts the prescriptive jurisdiction existence in the contiguous zone. The main reason against the establishment of the legislative jurisdiction in the contiguous zone was that, these rights may endanger the freedoms of navigations in this area (Churchill & Lowe, 1999), which according to Roach & Smith (1996) this is a zone comprised of international waters, and vessels of all nations enjoy these international rights. Larson (1976) as well notes that, the contiguous zone is governed by the principle of the freedoms of the seas, and coastal States in this zone has limited control and rights regulated by international provisions—a theory which might indicate a lack of legislative jurisdiction in this zone. Although, the contiguous zone is included within E.E.Z, yet, it is still governed by the general principles of the high seas i.e. international law, which characterizes the E.E.Z itself (Tsarev, 1987), and since article 33 in contrast to article 56, provides only control and not jurisdiction rights (UNCLOS, 1983), the coastal States can arguably establish legislative jurisdiction in the contiguous zone.

In light of these considerations, the absence of the legislative jurisdiction by the coastal States in the contiguous zone seems to be apparent, and in this regard, it may be submitted that both article 24 and article 33 of Geneva and UNCLOS Conventions, allow only enforcement and not legislative jurisdiction in the contiguous zone. In other words, both provisions allow to the coastal State to exercise control only in respect of offences committed within territorial sea and not in respect of violations occurred in the contiguous zone per se (Churchill & Lowe, 1999). Consequently, a legislative jurisdiction vacuum appears to prevail in the contiguous zone, obscuring as a result the exact authority as well as rights and powers of the coastal States in this particular sea area.

4 ENFORCEMENT JURISDICTION

4.1 *Control vs. jurisdiction*

The enforcement authority of the coastal States is clearly stipulated in article 24 and article 33 of both Conventions, which provides the rights to these States to control and punish in the contiguous zone the infringements of the laws and regulations pertaining to the territorial sea. In light of many concerns such as the enhancement of security issues, and the increased volume of illicit trade and drugs, determining the level of the authority exercised in the contiguous zone is a vital matter. In this regard, the term *control* stipulated in the provisions of both Conventions brings doubts whether *control* or *jurisdiction* is exercised in this zone. The article 24 of 1958 Geneva Convention according to Sir Gerald Fitzmaurice (1959, p. 113), ascribes only *control* not *jurisdiction* rights, and as he further notes "it is control, not jurisdiction that is exercised, and that the power is primarily that of policeman rather than of a judge". This theory is advocated also by O'Connell (1984, p. 1058), who highlights that the "enumeration of rights in article 24, and article 33 of both Conventions indicate the absence of any jurisdictional rights". Johnson (2004) as well following the same philosophy argues that the term *control* stipulated in the article 33 of UNCLOS III questions the level of the authority exercised by the coastal State.

On the other hand, Oda (1962) in his renowned article argues that the theory proposed by Fitzmaurice does not reflect precisely the practice exercised by many coastal States in the recent past, and that the 1958 Geneva Convention on contiguous zone represents somehow the coastal States practice over the years which have exercised jurisdiction and not control in their coastal waters. Substantial information of these jurisdictional practices, according to Lowe (1982), is provided from the works of Dickinson "Jurisdiction at Maritime Frontier" in 1926 and Masterson "Jurisdiction in Marginal Sea. Similarly, Churchill and Law (1999, p. 137) note that "the literal wording of the article 24 and article 33 of both Conventions ascribes only enforcement jurisdiction to the coastal State, highlighting perhaps their partiality of using *jurisdiction* rather than *control* phrase.

It is the view of these authors, nonetheless, that costal States in respect to article 33 of UNCLOS III are exercising *control* more than *jurisdiction*. This might be because in UNCLOS III, article 33, is clearly stipulated the word *control*, in contrast for instance to *jurisdiction* exercised in E.E.Z explicitly stated in the article 55 of UNCLOS III (Nandan & Rossene, 1993). This fact indicates that *control* as a term is established to emphasize for another level of authority distinct from the *jurisdiction*. Secondly, contiguous zone appears to be characterized by a legislative jurisdiction vacuum, which forbids a sovereign State to properly exercise its jurisdiction in it. In this regard, international law prevails in the contiguous zone (Larson, 1976), and this regime limits dramatically the jurisdiction exercised by the coastal State. Mukherjee (2002, p. 151) in his study Maritime Legislation notes that, "national courts in respect to the validity of the jurisdiction .... may have to contend with more than one international instrument and the domestic legislation". In this respect, the domestic legislation in the contiguous zone is absent; therefore, the jurisdiction in this zone is essentially limited. Finally, the main enforcement mechanism of the jurisdiction is the power of arrest. Nevertheless, the power of arrest can doubtingly be exercised against foreign vessels in the contiguous zone (O'Connell, 1984), and this is clearly stipulated in the article 27 of UNCLOS III, which forbids the arrest of vessels for offences committed before they enter the territorial sea (Churchill & Lowe, 1999).

4.2 *Power of arrest*

Under the article 33, the coastal State has the authority to *control* the contiguous zone as to prevent the infringement of the appropriate laws from occurring in territorial sea not in the contiguous zone, and to *punish* the infringements committed within territorial sea. This statement raises the question—what enforcement mechanisms are exercised by the coastal State to control and punish the foreign vessels in the contiguous zone? Since contiguous zone is considered a separate regime, and the high seas freedom of navigation is applicable in it (Nandan & Rossene, 1993), may a coastal State arrest in this zone foreign vessels only for violations of its laws committed in the territorial sea? In this respect, it may be submitted that foreign vessels arguably can be subject of arrest by the coastal State in the contiguous zone, unless they are committing high seas international crimes namely piracy, illegal broadcasting and slavery, as well as hot pursuit and constructive presence. This is because the powers exercised in the contiguous zone are limited and mainly regulated by the international law (Prescott & Victor, 2005) which is perhaps incomprehensible in this respect. Moreover, the municipal law of the coastal States which serves as a legal platform for the enforcement rights of the coastal State seems to be nonexistent in the contiguous zone.

This theory is also supported by O'Connell (1984, p. 1058) in his study, stressing that "a State exercises in the contiguous zone only limited power of police and foreign ships are not bound to conform to the laws of coastal State in the contiguous zone". Churchill and Lowe (1999, p. 139) similarly observe that "in a case of dispute arises concerning a claim by a coastal State regarding to jurisdictional rights in the contiguous zone not expressly granted under the Convention the question is to be resolved on the 'basis of equity' and 'importance of interests'" highlighting perhaps their doubts on the applicability of the enforcement powers in this zone. Consequently, it appears that article

33 in respect to the power of arrest is ambiguous, and legal improvements or amendments perhaps should be made in the relevant provisions in order to clearly define the enforcement powers of the coastal States in the contiguous zone.

5 CHALLENGES OF THE CONTIGUOUS ZONE

5.1 *The relevance of the contiguous zone*

This question was set forth for the first time at the second session of UNCLOS in 1974, indicating as a consequence reservations by the Second Committee as to whether the retention of the concept of the contiguous zone would be justified in the new concept of the new Law of the Sea, and it was argued by Israel that this zone would no longer be needed due to the consensus on the 12 mile limit to the territorial sea (Nandan & Rossene, 1993). Other States such as Mexico, Cameron, and Indonesia maintained also that the "establishment of the E.E.Z by the UNCLOS would render the contiguous zone superfluous and unnecessary" (Nandan & Rossene, 1993, p. 269). This uncertainty regarding the relevance of the contiguous zone after the adoption of the 1982 UNCLOS is still argued currently—revealed also by the fact that "only 75 coastal States globally claim a contiguous zone beyond the territorial sea" (UKHO, 2004, as cited in Prescott & Victor, 2005, p. 35) and most of the other coastal States seem to consider unnecessary the establishment of this zone. Hence, in light of these events the question raises whether the contiguous zone is still relevant, particularly, in light of current legal and technological developments in the maritime field.

It is the perspective of these authors, in this regard, that contiguous zone is not only relevant, but also in light of the new technological developments, the increased volume of maritime trade and transport, and the recent illegal narcotic concerns, its existence becomes paramount today. Firstly, contiguous zone existence is vital since it relates to certain exact powers and controls which are not part of the regime of the coastal States rights and jurisdiction in the E.E.Z (Nandan & Rossene, 1993). Secondly, the customs, immigration, fiscal and immigration matters regulated in this zone are considered crucial for the coastal States' progress and prosperity (McDougal et al., 1987). Thirdly, global maritime trade recently is growing very fast and seaborne goods reached a record high of 7.11 billion tons (UNCTAD, 2006), a figure which indicates that foreign vessels traffic visiting coastal States ports is increasing annually, and the need for the contiguous zone authoritative powers are inevitability essential for the States. Moreover, advances in techniques of underwater recovery, leading States to exercise further control under the article 303 over the recovery of archaeological and historical objects out to 24 miles has increased the relevance of the contiguous zone for many States (Nandan & Rossene, 1993) such as Greece and France. Finally, the volume of illicit drugs moved by the sea transport which in many States is subject of national customs authority, continue to increase (International Chamber of Drug Trafficking and Abuse, 1994) and the necessity to prevent and deter these acts in the contiguous zone, as a buffer area, is more relevant than ever.

5.2 *The potential establishment of the legislative jurisdiction*

As discussed in the previous chapters, the legislative jurisdiction vacuum in the contiguous zone appears to bring confusions and uncertainty regarding the exact powers, rights and jurisdiction exercised by the coastal States in this zone. Noticing the significance of this issue, in the 1930 Hague Conference many States were prepared to admit the existence of both sorts of jurisdiction in the contiguous zone based on their practices (Lowe, 1982). In addition, efforts were made by the Italian and Polish Delegation in 1958 Geneva Convention to delete the words "within its territory or territorial sea" in the article 24 in order to establish a legal regime in the contiguous zone (Oda, 1962, p. 141). Further vague efforts were made in the 1982 UNCLOS to establish legislative jurisdiction over the contiguous zone, but both attempts were opposed by States which observed in such claims danger to free navigation in this sea area (Churchill & Lowe, 1999). These events appear to illustrate the importance that legislative jurisdiction in the contiguous zone represented for many coastal States.

Many years have passed since last efforts were made to establish legislative jurisdiction in the contiguous zone and, in light of current developments, perhaps is time to consider the establishment of the legislative jurisdiction in the contiguous zone. This concern for certain States perhaps is not relevant, since it has been opposed long time ago; yet, the maritime industry and the international legal regime is changing continuously, and what was best strategy or philosophy for yesterday is not at present. This might bring as a result a re-examination of this important issue, with the final outcome to allow coastal States to establish legislative jurisdiction in the contiguous zone—directed of course at preventing infringement of customs, immigration, fiscal, sanitary and perhaps even anti-narcotic laws, in order to protect the territorial sea. The fact that many coastal States such as India, Sri Lanka and Pakistan has adopted national legislation in the contiguous zone (Territorial Waters, Continental Shelf, E.E.Z and Other Maritime Zones Act, 1976) indicates perhaps the need for the establishment of legislation in this zone currently.

O'Connell (1984, p. 1060) in this regard takes the view that "to say that the coastal State may not enact laws at all to apply in the contiguous zone would be to contract the very notion which has underlain article 24" and consequently the article 33 of UNCLOS III. Johnson (2004) also claims that the authority in the contiguous zone must include at least the ability from the coastal State to prescribe laws and regulations. Churchill & Lowe (1999, pp. 138–139), in this respect, observe that the claims regarding the legislative jurisdiction in the contiguous zone have not evoked significant international opposition in practice and that "… the inclusion of both jurisdictions are more easy to defend now than formally" indicating perhaps that many States currently may support the adoption of the legislative jurisdiction in this significant zone.

5.3 *The approach regarding the adoption of anti-narcotic laws*

The maritime drug traffic poses an increasingly threat to coastal States' security and stability, and consequently to the international peace and security. Recently, the high value of drugs when smuggled in large quantities have attracted the attention of terrorist and criminal groups to penetrate in the shipping transport, resulting thus to a no safe shipping routes, where national authorities can be certain that there are no illicit drugs in the vessels (International Chamber of Drug Trafficking and Abuse, 1994). The national authorities have recognised that maritime transport is the most vulnerable of all modes of transportation in carrying large amounts of drugs, and the growth of containerization have increased the opportunities for drug concealment. The fact that in 1991 over 327 tonnes of illicit drugs were reportedly seized by customs authority in maritime transport, and that the figure has remained high till recently (International Chamber of Drug Trafficking and Abuse, 1994) shows that the drug volume transported by ships has increased in the last decades.

In this respect, the prevention and punishment of illicit drug traffic regulated by national legislation in the contiguous zone might be imperative. This concern was illustrated in the case of *U.S. vs. Gonzales* in 1985 in which the US court held that the protective principles of the international law justifies the exercise of narcotic legislative jurisdiction in the high seas (O'Connell, 1984) i.e. contiguous zone. Similarly, the increased popularity in the establishment of the contiguous zone by some 38 States in 1992 was ascribed by the UN to the problem of narcotics traffic and to the need by coastal States to take additional measures to prevent drug smuggling in this zone (UN Doc, 1992). Also, the indented coasts of countries such as Croatia, Greece and archipelagic States might require that drug control is more efficient when occurs in the contiguous zone, because within territorial sea smugglers might lose track. Consequently, perhaps presently is time to establish anti-narcotic laws outside the customs jurisdiction where it is so, and thoughts might be given to the possible adoption of anti-narcotic laws in the contiguous zone, normally without creating obstacles to the freedoms of navigation.

6 CONCLUSIONS

As a result, it appears that the genesis of the contiguous zone is derived from the British Hovering Acts back in the eighteenth century; however, significant contribution to the origin and development of this concept has provided also the U.S.A anti-smuggling legislation enacted initially in 1790. Although, the 1958 Geneva Convention established the contiguous zone as an international regime, the 1982 UNCLOS consolidated furthermore this concept and presented a more comprehensive view of the contiguous zone in relation to the delimitation of this zone and its establishment as a separate regime.

The analysis of the contiguous zone legal status in this paper has revealed a potential legislative jurisdiction vacuum in this zone; and moreover, the coastal State enforcement jurisdiction appears to be ambiguous. In addition, the authority exercised by the States in the contiguous zone it appears to be *control* more than *jurisdiction*, and the *power of arrest* can be arguably exercised in this zone. In this respect, improvements might be required in the respective provision to further clarify this issue. What's more, the status of the contiguous zone within E.E.Z appears to be controversial because of the vague relationship between the two respective provisions in 1982 UNCLOS, and a possible revision of these articles might result in a comprehensive view of this issue in the future.

After examining the past and current developments in relation to the contiguous zone, this paper has revealed that although contiguous zone existence was threaten by 1982 UNCLOS, and perhaps is still in doubt at present, the significant issues that it addresses and might tackle currently make this zone more relevant than ever. Therefore, in light of many issues concerning the coastal States presently, it might be relevant to consider the adoption of the legislative jurisdiction by the coastal States in the contiguous zone, aiming of course on the prevention of customs, immigration, fiscal, and sanitary issues. In this regard, taken into account the crucial concern that drug traffics represents at present for the coastal States, the enactment of preventive laws against this phenomena in the contiguous zone might be reasonable. Overall, the contiguous zone is an important international regime which tackles issues with great interests for the coastal States, and as such, further studies should aim in the possible improvement and amendments of the legal framework of this zone for the benefit of the whole maritime community.

REFERENCES

Brown, E.D. & Churchill, R.R. (ed.) 1987. *The UNCLOS: Impact and Implementation Proceedings*: Law of the Sea Institute Nineteenth Annual Conference, Hawaii, University of Hawaii.

Churchill, R.R. & Lowe, A.V. 1999. *The law of the sea*, 3d Edition, Manchester, Manchester University Press.

Cornell University Law School,. *Definition of Jurisdiction*, Retrieved October 22, 2007 from http://www.law.cornell.edu/

Convention on Territorial Sea and Contiguous Zone, 1958, article 24.

Customs Duties Act, 1790, 1 Stat. at large, 2nd. Sess., c.35, s. 9.

Frommer, A.M. 1981-2. "The British Hovering Acts: a contribution to the study of the contiguous zone", 16 Revue Belge de droit international 434–58.

Improvement of Revenues and Customs Act, 1765, 5 Geo. 3, c. 43, s. 27.

India, Territorial Waters, Continental Shelf, E.E.Z and Other Maritime Zones Act, 1976, UN Leg, Ser. B/19.

International Chamber of Drug Trafficking and Abuse, 1994. *Drug Trafficking and Drug Abuse: Guidelines for Owners and Masters on Prevention, Detection and Recognition*, Published by Witherby & Co. LTD, London.

Johnson, L. 2004. *Coastal State Regulation of International Shipping*, New York, Oceana Publications.

Klein, N. 2005. *Dispute settlements in the UNCLOS*, Cambridge, Cambridge University Press.

Larson, D.L. (ed.) 1976. *Major Issues of the Law of the Sea*, New Hampshire, University of New Hampshire Publication.

Lowe, A.V. 1982. "The development of the concept of the contiguous zone", in Jennings, R.Y. & Brownlie, I (ed.), *The British yearbook of International Law 1981*, Oxford, Clarendon Press.

Nandan, S.N. & Rosenne, S. (ed.) 1993. *United Nations Convention on the Law of the Sea, A Commentary*, Volume II, Virginia, Martinus Nijhoff Publishers.

McDougal, M.S. & Burke, W.T. 1987. *The public order of the oceans: a contemporary international law of the sea*, New Heaven, New Heaven Press.

Mukherjee, P.K. 2002. *Maritime Legislation*, Malmo, WMU Publications O'Connell, P.D. 1984. *The international law of the sea*, Volume II, Oxford, Clarendon Press.

O'Connell, P.D. *The international law of the sea*, Volume II, Oxford, Clarendon Press, 1984.

Oda, S. 1962. "*The concept of Contiguous Zone*", 11 ICLQ 131–53.

Prescott, J.R. 2005. *The maritime political boundaries of the world,* Second Edition, Boston, Martinus Nijhoff Publishers.

Progress made in the implementation of the comprehensive legal regime in the UNCLOS, UN Doc, 1992, A/47/512.

Review of Maritime Transport in 2006, UNCTAD Doc, 2006.

Roach, J.A. & Smith, R.W. 1996. *United States Response to Excessive Maritime Claims*, Boston, Martinus Nijhoff Publication.

Sir Gerald Fitzmaurice. 1959. "Some Results of the Geneva Conference on the Law of the Sea", 8 Int. & Comp. L.Q. 73, 113.

The Parliamentary History of England from the Earliest Period to the Year 1803, Hansard, Vol. 10, col 1232.

Tsarev, V.F. 1987. "The juridical nature of E.E.Z" in Brown, E.D, (ed.). *The UNCLOS: Impact and Implementation Proceedings*: Law of the sea Institute, Nineteenth Annual Conference, Hawai, University of Hawai.

United Nations Convention on the Law of the Sea, 1983.

Valenzuela, M. 1999. "*Enforcing Rules against Vessel Source Degradation of the Marine Environment*".

Sustainable Maritime Transportation and Exploitation of Sea Resources – Rizzuto & Guedes Soares (eds)
 ISBN 978-0-415-62081-9

Manning crisis in the international shipping: Fiction vs reality

E. Xhelilaj, K. Lapa & L. Prifti
University of Vlora "Ismail Qemali", Albania

ABSTRACT: Among the most controversial matters regarding the manning of ships recently is related to the issue whether there is a lack of the total number of seafarers globally, or is there a shortage of quality seafarers to man the world fleet in nowadays. In this regard, BIMCO/ISF survey of 2000/2005 revealed a shortage of officers and a surplus of ratings on global bases, tendency which is likely to increase in the future. On the other hand, ILO research in 2001 confirms similar trends, but with the exception that the total supply level of seafarers is higher than the BIMCO/ISF study. In a survey of 2010, BIMCO/ISF reveals that still there is a modest shortage of officers and a balance between supply/demand of ratings. In light of these considerations, the authors are of the opinion that although there is a certain level of shortage of seafarers, yet this is more in terms of quality officers employed in international shipping today rather than in total number of seafarers globally. Perhaps inclusive and detailed surveys may be carried out in the future by joint efforts from BIMCO/ISF, ILO, Maritime Administrations as well as other international organization in order to reveal comprehensive figures in connection to this important matter for the shipping industry.

1 INTRODUCTION

Perceived as an important component in the context of the global shipping industry, crew members manning the ships, according to several sources recently, appear to be inadequate in numbers. As many international maritime organizations have underscored lately, the matter of employment of the personnel working onboard is important not only in terms of efficiency of maritime transport, but as well as in relations to safe and secure shipping in cleaner oceans. Although maritime administrations, shipping companies and international maritime organizations have made essential efforts in respect to recruitment, retention and training of seafarers worldwide, yet statistics of 2010 stressed from certain international surveys indicate that there is an overall shortage of officers' numbers employed onboard ships.

In this regard, this is an important matter to analyse since human element, even though taken into consideration the high technological level and automation of merchant ships in nowadays, is found to be responsible for a considerable percentage of accidents at sea worldwide. The accidents occurred from human error, caused from both fatigue and lack of competence, have concluded in the loss of human lives, severe pollution of coastal and ocean environment as well as to damage of state and private property.

In light of these considerations, the authors' purpose in this paper is to analyse whether these studies reflect the reality of the manpower situation in global shipping industry. Following this line, the authors will first take under consideration an analytical review regarding the background of this potential problematic situation and, then examining this particular matter from the present situation perspective. Finally, the authors will reflect some concluding remarks, which probably will shed light toward the aforementioned issue.

2 BACKGROUND OF THE SITUATION

According to the survey carried out by BIMCO/ISF in 2005, as well as by the International Labour Organization in 2001, it appeared that there was a shortage of seafarers working onboard ships at international level back then. This trend appears to follow even in the survey carried out by BIMCO/ISF in 2010, with the exception that there is a balance between supply/demand in terms of ratings and a modest overall shortage of 2% of officers working onboard. From the above figures, it seems that the manpower situation has been inherited from diverse problems occurring internationally in the past. These problematic situations appeared to emanate mainly from economic recession or political crises in different parts of the world.

The genesis of the recent manpower situation in the international merchant fleet probably goes back from 1973 oil crises, when OPEC decided to raise oil prices significantly by cutting back on world supply, leading as a result to high unemployment level in most industrial nations, including the shipping industry which was significantly affected (Mejia, 2008).

Similarly, a serious worldwide recession in shipping in 1980s, resulting from a major overcapacity of tonnage, imposed the need to reduce manning costs, forcing the ship-owners to increase the employment of seafarers from lower cost labour supply countries (India and Philippines) at the expense of seafarers from TMC within the OECD (BIMCO/ISF, 2000). Thus, a major part of the shipping industry remained dependant on a progressively ageing stock of senior officers from OECD countries. BIMCO/ISF in 2000, reported that the severity of the recession in the 1980s forced many shipping companies to reduce radically on their training funds, aggravating further the manpower situation in terms of quality and quantity.

During the early 1990s, an increasing numbers of seafarers from the Eastern and Central Europe supplemented this seafarers' stock in the world fleet (BIMCO/ISF, 2000), which had declined at that time to a critical degree due to low recruitment levels as well as to inefficient retention policies (Leggate, 2004). However, doubt remained over the skills and competence of officers from several of these new labour sources. In this respect, a 1990 and 1995 report of BIMCO/ISF revealed that the outmost concern of the industry was the lack of adequate recruitment of youngsters, as well as to ensure that the training and qualifications of the seafarers would be sufficient and of the highest achievable common minimum international standards.

The manpower situation was further affected from the economic crises in Asia in 1997, followed in South America and Central and Eastern Europe, resulting thus in a collapse in the value of many national currencies against the US dollar, the currency in which many seafarers are paid (BIMCO/ISF, 2000). Apparently, these economic problems had a major impact on world trade, including shipping industry, and subsequently influenced both the global supply and demand of the seafarers. Whereas the world freight rates recovered from the 1990s global recession, the EU freight rate remained stained, causing thus oversupply, which in turn exerted financial pressure on the ship-owners to reduce their costs by decreasing their fleet capacity or registering their ship to other flags (Commission on European Communities, 2001). This impacted to some extent the demand level, particularly of the EU seafarers, because the EU ship-owners controled 40.8% of the world fleet measured in gross tonnage (Mitroussi, 2008).

The strong world economic growth in 2000s (mainly in Asia), combined with a rapid world trade growth, produced a stable increase in demand for shipping services, which in turn increased the world trading fleet as well as the average vessel size. Subsequently, there was a considerable demand of crew members working onboard ships on global basis (BIMCO/ISF, 2005), although to some extent the ship automation had some reverse effect. From 1990 to 2001, the annual growth rate of the world active fleet has been 2.8% (UNCTAD, 2002), and from 2000 to 2005 this rate has been over 4%, considerable figures which suggest that the number of seafarers would need to grow respectively to sustain the same level of trade. On the other hand, the adoption of ISPS Code in 2002, caused by the global terrorism, and the full and final implementation of STCW 95 (including white list), in effect from 2002, impacted faintly the supply level of seafarers due to the security, and to the seafarers' qualification and certification strict requirements that many countries fail to meet.

At the same time, whereas the political unrest and wars in Middle East slightly affected the tanker's seafarers' demand, the expansion of the EU towards East Europe, China's and India's economic boom (BIMCO/ISF, 2005) as well as Philippines seafaring capacity (Leggate, 2004) increased the supply level of seafarers, including junior officers. Nevertheless, this particular seafarers' population, to some extent, appeared to reduce the average manpower quality level due to certain educational and training inefficiencies in these countries as well as the absence of experience of these junior officers onboard ships. In light of these considerations, it seems that the high instability of the global economy, world trade, shipping financial capacity, political situation, and seafarers' employment structure, combined with legislative changes and hardships of seafaring profession over the last three decades caused the present potential manpower alarming situation.

3 THE ISSUE OF MANPOWER CURRENTLY

The most contentious issues surrounding the manning of ships today is related to the question whether there is a lack of a global number of seafarers, or is there a shortage of quality seafarers to man the world fleet today. In light of these considerations, BIMCO/ISF survey of 2000/2005 revealed a shortage of officers and a surplus of ratings on global bases. The same survey highlights that the tendency regarding the level of officers' shortage and ratings' surplus is likely to increase in the future.

On the other hand, ILO research in 2001 confirms similar trends, but with the exception that the total supply level of seafarers is higher than the BIMCO/ISF study. The last mentioned organization, in the latest survey reveals that despite the global economic downturn, and the dramatic reduction in demand for shipping services in 2009, the data suggests that while the supply and demand for ratings are more or less balanced, there are still some shortages for officers, particularly for certain grades and for ship types such as tankers and offshore support vessels (BIMCO/ISF 2010).

Hence, the worldwide supply of seafarers in 2010 is now estimated to be 624,000 officers and 747,000 ratings (BIMCO/ISF 2010). This is based on the numbers holding STCW certificates and is therefore somewhat broader and not directly comparable to estimates in previous studies. It reflects significant increases in seafarer supply in some countries, notably in China, India and the Philippines, as well as in several European nations. Based on the figures and data laid down by the 2010 Manpower Update, the current estimate of worldwide demand for seafarers in 2010 is 637,000 officers and 747,000 ratings (BIMCO/ISF, 2010).

The aforementioned results suggest that the situation in 2010 is one of approximate balance between demand and supply for ratings with a modest overall shortage of officers (about 2%); the implication being there is currently not a serious shortage problem for officers in aggregate. This does not, of course, mean that individual shipping companies are not experiencing serious recruitment problems, but simply that overall supply and demand are currently more or less in balance (BIMCO/ISF, 2010). This new Update highlights that the industry is likely to face a challenging future for crewing. There are many uncertainties, but the results indicate that the industry will most probably face a continuing tight labour market, with recurrent shortages for some officers, particularly if shipping markets recover (BIMCO/ISF, 2010).

Be that as it may, perhaps the most substantial point is that all the aforementioned sources indicate in general terms that there is shortage of officers manning the world fleet for the past 20 years, and this trend will follow in the future. Although these surveys are consider significant contribution towards the study of supply/demand level of the seafarers, since they are based on comprehensive information from relevant maritime institution worldwide; and involve factors such as the global trade, world fleet growth, technological improvements, and international standards; yet, scholars maintain that these research studies highlight the fact that the baseline or current position on seafarers numbers in not known with any degree of accuracy (Leggate, 2004).

This theory is based on several reasons. Firstly, it appears that, the process of data collection suffers somehow deficiencies in the survey "since it is manly based on questionnaires sent to governments and shipping companies as well as on the perception of senior executives in shipping companies" (BIMCO/ISF, 2010) without indicating how many replies or feedback were received by BIMCO on the matter as well as whether these feedback was sufficient to come up with concluding figures. For instance, in 1995 survey only 60 replies out of 120 applications worldwide were obtained (BIMCO/ISF, 1995). Moreover, these studies are ambiguous in terms of the concept of 'number of seafarers', which in the eyes of diverse people may have different connotation such as; only the number of seafarers working on board, seafarers employed both on board and at shore, or solely qualified seafarers with an appropriate official seamen's discharge book (Li & Wonham, 1999).

Secondly, attempting to quantify the seafarers is extremely difficult as many countries have no established system to achieve this, and the numbers are often based on employment statistics which leaves the problem of determining the number available for employment or active population (Leggate, 2004). Again, BIMCO/ISF data of 2005 according to Li (1999) it appears to underestimate the supply level of Chinese and other Asian countries seafarers which subsequently may impact the total supply figure. In 2010 Update, this issue was somehow tackled since Dalian Maritime University has helped obtain input from Asian countries (BIMCO/ISF, 2010). Yet, Philippines, Indian and other Asian countries' seafarers occupy a considerable part of ship manning worldwide currently and exact figures of supply/demand on these seafarers are difficult to reveal even from a reputable Chinese organization. A positive approach, in this regard, might have been that several leading universities in Asia, particularly in India, Philippines and China would have cooperated together in order to conduct a comprehensive survey to precisely evaluate the Asian manpower working onboard ships.

Furthermore, the differences in ILO and BIMCO/ISF results, though both assimilated information from the same sources, indicate once again that their data may probably lack accuracy (Leggate, 2004). Another potential issue is that the latest results from BIMCO/ISF cannot be directly compared with previous studies which in turn may impact a comprehensive evaluation of present situation as well as the future trends regarding manpower in the shipping industry.

Finally, there is no explanation from the BIMCO/ISF study on how the figures of officers/ratings

Table 1. Supply levels of officers and ratings manning the international merchant fleet.

	Current supply			
Area	Officers (1000's)	%	Ratings (1000's)	%
OECD	184	29.4	143	19.2
Eastern Europe	127	20.3	109	14.6
Africa/Latin America	50	8.0	112	15.0
Far East	184	29.5	275	36.7
Indian Sub-Continent	80	12.8	108	14.5
All National Groups	624	100	747	100.0

*Source: 2010 BIMCO/ISF estimates.

global supply came from the partial response; and on what bases it can be judged that the partial responses account for 90% of all seafarers available worldwide (Li & Wonham, 1999). Therefore, taken under considerations the above analysis, the results of these surveys or studies somehow may probably lack genuine figures.

In contrast to BIMCO/ISF and ILO outcomes, many scholars and organizations support the theory that perhaps there is no shortage of seafarers in numbers, but instead there is an absence in qualitative officers and ratings. Thus, while studies reveal a decline of OECD officers working on board ships during the last three decades (Mitroussi, 2008), it appears that many relevant international organizations are concern more with the possible lack of qualified and well trained seafarers (Commission on European Communities, 2001). The 2010 BIMCO/ISF Manpower Update, as indicated below, shows that the officers and ratings from OECD countries currently comprise only a small proportion of the number of seafarers employed onboard, with the rest of it originated from Indian Sub-Continent, Eastern Europe, Far East, Africa and Latin America.

Moreover, the shipping industry reveals that the weakest point currently and possibly in the future is the number of quality seafarers, particularly officers, rather than their shortage in numbers (Leggate, 2004). Seafarers who can competently man diverse ships, whatever their size or management, with the good command in English and skills in communicating with colleagues from different cultures and backgrounds appears to be in great demand from the ship-owners and shipping industry (Li & Wonham, 1999). In this regard, the 2010 Update of BIMCO/ISF indicate that there is absence of officers in tankers and offshore support vessels which normally require a higher level of competence than most of other types of ships. There is also particular concern over the current and future availability of senior management level officers, especially specialists and engineers (BIMCO/ISF, 2010). From the abovementioned analysis, therefore, may emanate the controversial question; Is there a manning crises today?

4 CONCLUSIONS

In view of these authors, it appears that there is a certain manning crisis currently; but this is more in terms of the shortage of quality officers and ratings, rather than in the total number of seafarers. Although there has been e relentless decline in the number of seafarers and officers coming from OECD countries due to an appreciable reduction in recruitment and retention, this has being counteracting by an increase in those seafarers from the labour supply countries, that is both at junior officers as well as at ratings levels deriving from China, India, Eastern Europe and Philippines.

Moreover, results from the shipping companies survey indicate problems with the supply of particular grades of seafarers, such as senior officers and engineers in some labour markets worldwide (BIMCO/ISF, 2010) which are probably more experienced and well trained—an indication perhaps of quality absence among the population of seafarers on global bases. Therefore, the counteracting supply of junior officers coming from labour supply countries may not tackle the issue of qualitative manpower situation currently.

The seafarers stock in China and Eastern Europe after 2000s (Zhao & Amante, 2005) as well as the Philippines, Indian and Indonesian seafaring supply-capacity have provided the grounds in the last decade for a considerable increase in the numbers of junior officers and ratings available to work on board (BIMCO/ISF, 2005), which currently counts for more than 60% of the total number of officers (BIMCO/ISF, 2010). Nonetheless, it appears that a certain percentage of this manpower group, considering also their possible absence of experience onboard, probably lacks quality in terms of international maritime standards, which subsequently may impact the overall performance of the shipping industry globally.

This increase in seafarers' total number from aforementioned countries just recently may be supported by the 2010 Manpower Update, figures of which reflect significant increases in seafarers' supply in some countries, notably in China, India, Africa/Latin America and the Philippines. Furthermore, the same sources highlights that some evidence have been found of continuing recruitment and retention problems of seafarers, mainly in OECD countries but not as severe as envisaged by the 2005 Update, probably an effect of the downturn (BIMCO/ISF, 2010).

At the end, perhaps another important concern for the future is that according to some estimations an increase in the number of ships in the world fleet of 2.3% per annum will occur (BIMCO/ISF, 2010), and with current problems in recruitment and retention, manning levels are assumed to decline slightly on average. Unless measures are taken to ensure a continued rapid growth in terms of qualified seafarers' numbers, especially for officers, and/or to reduce wastage from the industry, existing shortages are likely to intensify over the next decade. Therefore, comprehensive studies should aim in the future on how to tackle this essential issue. These studies or surveys may be best carried out by joint efforts from BIMCO/ISF, ILO, maritime administrations as well as other international maritime organization, in order to reveal comprehensive figures of supply/demand of seafarers as well as how to face these challenges in connection to this imperative matter for the shipping industry.

REFERENCES

BIMCO/ISF (2010). *Manpower update: The worldwide demand for and supply of seafarers.* Warwick Institute for Employment Research, Dalian Maritime University. United Kingdom.

BIMCO/ISF (2005). *Manpower update: The worldwide demand for and supply of seafarers.* Warwick Institute for Employment Research, United Kingdom.

BIMCO/ISF (2000). *Manpower update: The worldwide demand for and supply of seafarers.* Warwick Institute for Employment Research, United Kingdom.

Commission of the European Communities (2001). *Communication from the Commission to the Council and the European Parliament on the Training and Recruitment of Seafarers.* Brussels.

Leggate, H. (2004). The future shortage of seafarers: Will it become reality? *Maritime Policy and Management, 2004*, 31–1.

Mejia. M. (2007). *Issues in International Ship-manning.* MLP-401. Unpublished Lecture, WMU. Malmo, Sweden. *Key Economic Events 1973—The OPEC Oil Crises: Forcing up world oil prices.* Government of Canada.

Mejia, M. (2008). *Issues in International Ship-manning.* MLP-401. Unpublished Lecture, WMU, Malmo, Sweden.

Mitroussi, K. (2008). Employment of seafarers in the EU context: challenges and opportunities. *Marine Policy, 2008.*

Li, K.X. & Wonham, J. (1999). Who mans the world fleet: will it become reality. *Maritime Policy and Management, 1999,* 26–3.

SIRC Global Seafarers Database (2003). as cited in Zhao, M.H & Amante, M.V (2005). Chinese and Filipino Seafarers: A race to the top or the Bottom? *Modern Asian Studies, 2005,* 39–3.

Sustainable Maritime Transportation and Exploitation of Sea Resources – Rizzuto & Guedes Soares (eds)
 ISBN 978-0-415-62081-9

Regression and probability analysis of dry bulk indices

Ch.N. Stefanakos
Department of Naval Architecture, Technological Educational Institute of Athens, Aegaleo, Greece

O. Schinas
Maritime Business School, Hamburg School of Business Administration, Hamburg, Germany

J. Barberakis
Department of Naval Architecture, Technological Educational Institute of Athens, Aegaleo, Greece

ABSTRACT: Maritime transport of dry commodities in bulk is one the two major bulk markets of the shipping industry. Baltic Exchange has developed a series of indices for the various sizes of bulk carriers' world fleet in order to monitor daily evolution of relevant freight rates. In the present work, an extensive regression analysis is performed between indices related to Capesizes (BCI, 1999–2010), Panamaxes (BPI, 1999–2010), Supramaxes (BSI, 2005–2010) and Handysizes (BHSI, 2006–2010) bulk carriers. The analysis covers the pairs (BCI vs BPI), (BPI vs BSI) and (BSI vs BHSI) and the corresponding data have been examined on a yearly basis. Apart from some exemptions, most of the results give a very good coefficient of correlation. Moreover, the univariate probability structure of the indices is also studied. Histograms are derived using both yearly segments and the whole series of data for each index. These results help us better understand and visualize the probabilities of occurrence of the value of the various indices for each examined time period. On the whole, both regression and probability analysis are proved to be versatile tools in examining the behavior of the dry bulk indices in order to prepare reliable market researches.

1 INTRODUCTION

The bulk carriers are considered to be the 'workhorses' of the maritime industry. Dry bulk trades are also very significant to the international trade. "The share of dry cargo in the total volume of goods loaded has been growing over the years, and continues to account for the lion's share of the total (66.2 per cent)" is stated in the last UNCTAD report (p. 9, 2010a). Therefore, the study of indices that provide better understanding of the market developments, as well as supports analytical models and managerial decisions is indispensable, and in many cases tests the validity of intuitive solutions.

The indices developed and maintained by the Baltic Exchange, and precisely the BCI, BPI, BSI, and BHSI that refer to capers, panamaxes, supramaxes and handysize bulk carries respectively, provide a good attestation of the respective market. Moreover, the indices serve day-to-day operations and decisions, as many charter parties include freight rate and hire terms dependent on the value of one of the above indices, futures and other derivatives are based on them and last but not least ship financing proposals are tested vis-à-vis to the historical data of the indices. Therefore the current paper aims to provide substantial statistical information for the indices under scrutiny.

The paper is structured as following: the next section presents the data and provides the basic statistics of the indices. The analysis is broken down per annum of monthly available data. The next section, provides the regression analysis results; interesting statistical outcomes are highlighted and offer a solid base for practioners— mainly- to support decisions or to justify market forecast intuition. In the fourth section, the multipurpose probability analysis of the historical index data is presented. This is of particular interest to ship financiers, who normally take into consideration the assessment of the probability of market levels into their decision-making process. Finally, the paper concludes with remarks and suggestions for further work.

All statistical calculations are conducted on MA-TALB ® and the data are drawn out the SIN Clark-sons Database ®. The results of the analysis rivet interest on global shipping markets and are not focused on a specific region or trade.

2 DATA USED

The data source used in the present study is the monthly time series of the following indices:

- BCI (Baltic Capesize Index),
- BPI (Baltic Panamax Index),
- BSI (Baltic Supramax Index), and
- BHSI (Baltic Handymax Index)

from the Baltic Exchange.

The available data at hand covers the following time period for each index:

- BCI: 1999–2010,
- BPI: 1999–2010,
- BSI: 2005–2010, and
- BHSI: 2006–2010.

The results in the next two sections are based in annual segments of our data.

In Tables 1–4, basic statistics for all these segments are presented.

Table 1. Basic statistics for index BCI.

	Count	Mean	Min	Max	St.Dev.	Skewn.	Kurt.
1999	166	1318.01	831	1901	395.39	0.28	1.42
2000	252	2187.10	1602	2557	233.18	–0.77	3.23
2001	251	1472.77	898	2186	412.23	–0.06	1.38
2002	249	1394.40	991	2381	347.75	1.12	3.19
2003	250	3662.59	2016	6911	1570.70	1.09	2.52
2004	251	6011.20	3575	8911	1275.16	0.12	2.36
2005	249	4602.85	2319	6801	1323.49	0.06	1.64
2006	249	4288.83	2711	6104	1076.99	0.20	1.42
2007	250	9924.00	5943	16256	3237.20	0.62	1.95
2008	251	9363.39	830	19687	5267.52	–0.27	2.10
2009	250	4171.55	1361	8243	1791.77	0.45	2.16
2010	215	3540.27	1640	5455	833.10	–0.35	3.07
TOTAL	2883	4426.14	830	19687	3451.99	1.63	5.70

Table 2. Basic statistics for index BPI.

	Count	Mean	Min	Max	St.Dev.	Skewn.	Kurt.
1999	250	1064.86	709	1361	169.10	–0.23	2.34
2000	252	1540.21	1247	1677	90.18	–1.24	4.17
2001	251	1248.33	822	1649	311.99	–0.11	1.28
2002	249	1129.96	892	1731	198.35	1.39	4.14
2003	250	2543.97	1496	4539	976.66	1.12	2.60
2004	251	4383.01	2329	6110	889.72	–0.34	2.40
2005	249	3128.07	1488	4956	1008.42	0.41	1.79
2006	249	3020.75	1841	4394	786.45	0.31	1.54
2007	250	7031.95	3923	11713	2278.43	0.49	1.98
2008	251	6089.78	440	11425	3268.07	–0.61	2.01
2009	250	2405.36	492	4453	980.83	–0.21	2.33
2010	215	3259.85	1941	4622	719.99	–0.02	1.84
TOTAL	2967	3068.74	440	11713	2278.42	1.50	4.98

Table 3. Basic statistics for index BSI.

	Count	Mean	Min	Max	St.Dev.	Skewn.	Kurt.
2005	125	1869.18	1391	2129	191.79	–1.23	3.57
2006	249	2248.45	1442	2989	499.15	0.03	1.52
2007	250	4537.88	2732	6956	1277.65	0.37	1.97
2008	251	3973.29	421	6743	1970.39	–0.70	2.14
2009	250	1658.12	389	2487	515.90	–0.97	3.67
2010	215	2246.15	1628	3111	409.97	0.45	2.11
TOTAL	1340	2852.79	389	6956	1550.37	0.92	2.98

Table 4. Basic statistics for index BHSI.

	Count	Mean	Min	Max	St.Dev.	Skewn.	Kurt.
2006	110	1321.77	1000	1535	155.82	–0.63	2.28
2007	250	2219.42	1324	3232	579.13	0.20	1.77
2008	251	2006.21	281	3407	963.70	–0.72	2.25
2009	250	788.37	268	1240	232.06	–0.59	3.29
2010	215	1174.07	856	1520	178.21	0.37	1.92
TOTAL	1076	1536.55	268	3407	797.53	0.54	2.31

3 REGRESSION ANALYSIS

3.1 *Analysis procedure*

Assume two series of data: Sample X (e.g., BCI) $X = \{X_1, X_2, \ldots, X_n, \ldots, X_N\}$ and Sample Y (e.g., BPI) $Y = \{Y_1, Y_2, \ldots, Y_n, \ldots, Y_N\}$.

Ordinary Least Squares (OLS) regression theory suggests the relationship of Y and X, in form of a straight line, as below:

$$y = ax + b \tag{1}$$

by minimizing the square of the distances of the actual values Y_n from the estimated values y_n (Spanos 2003, Davidson & MacKinnon 2004), i.e.,

$$\sum_{n=1}^{N} (Y_n - y_n)^2 = \min. \tag{2}$$

The parameters of the line are then given by the following formulas:

$$b = \bar{Y} - a\bar{X}, \text{ and} \tag{3}$$

$$a = \frac{N^{-1}\sum_{n=1}^{N} X_n Y_n - \bar{X}\bar{Y}}{N^{-1}\sum_{n=1}^{N} X_n^2 - \bar{X}^2}, \tag{4}$$

where

$$\bar{Y} = \frac{1}{N}\sum_{n=1}^{N} Y_n \quad \text{and} \quad \bar{X} = \frac{1}{N}\sum_{n=1}^{N} X_n. \tag{5}$$

The goodness-of-fit is measured by means of the Pearson's correlation coefficient

$$R^2 = \frac{\sum_{n=1}^{N}(y_n - \bar{Y})^2}{\sum_{n=1}^{N}(Y_n - \bar{Y})^2}, \tag{6}$$

the range of which is *[0,1]*. As the value of R^2 approaches unity, the fit becomes better and better.

3.2 *BCI vs BPI results*

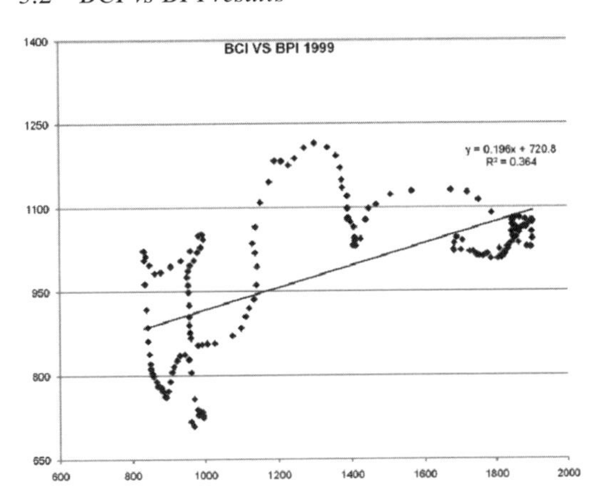

Figure 1. Regression analysis of BCI vs BPI for the year 1999.

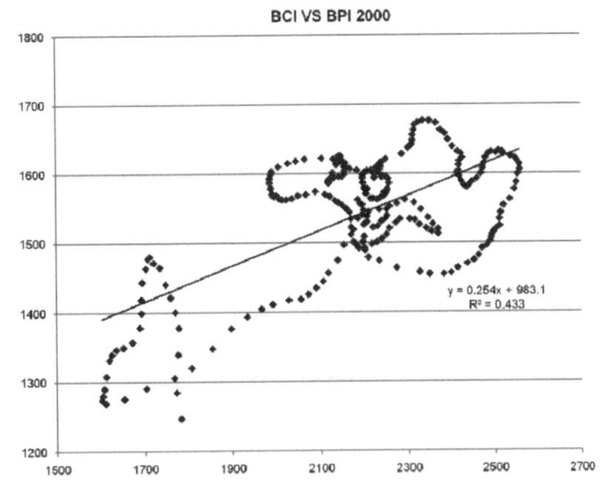

Figure 2. Regression analysis of BCI vs BPI for the year 2000.

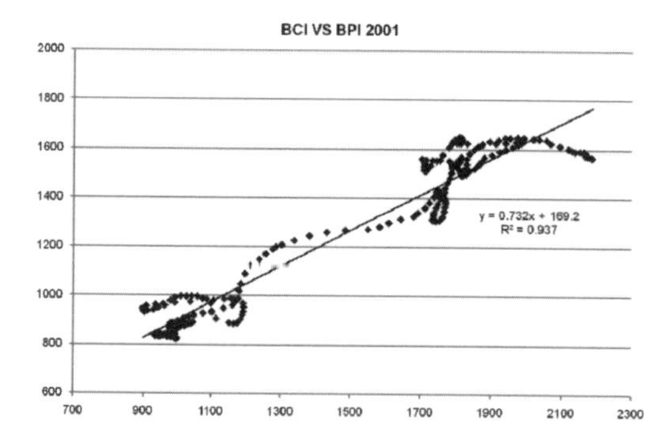

Figure 3. Regression analysis of BCI vs BPI for the year 2001.

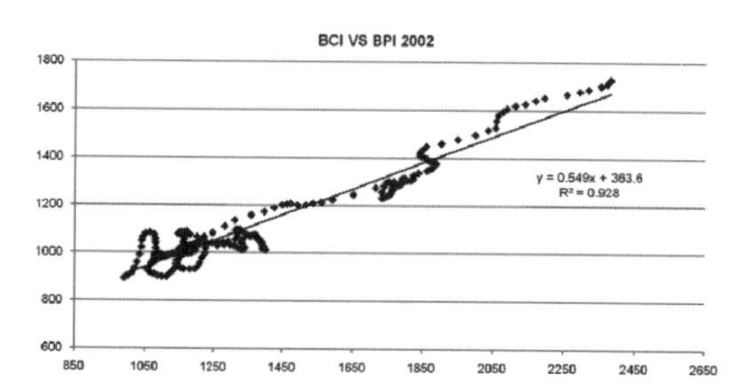

Figure 4. Regression analysis of BCI vs BPI for the year 2002.

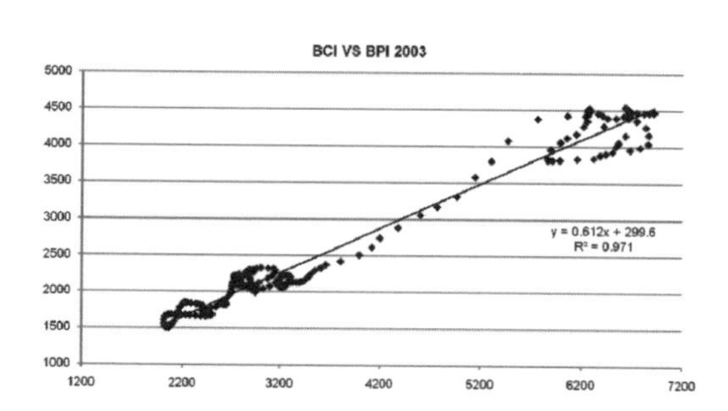

Figure 5. Regression analysis of BCI vs BPI for the year 2003.

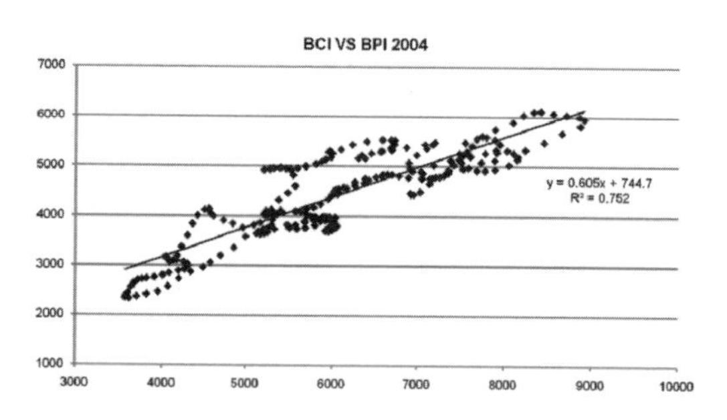

Figure 6. Regression analysis of BCI vs BPI for the year 2004.

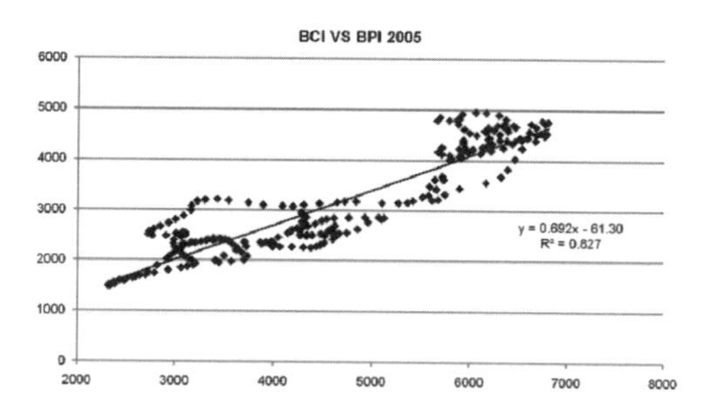

Figure 7. Regression analysis of BCI vs BPI for the year 2005.

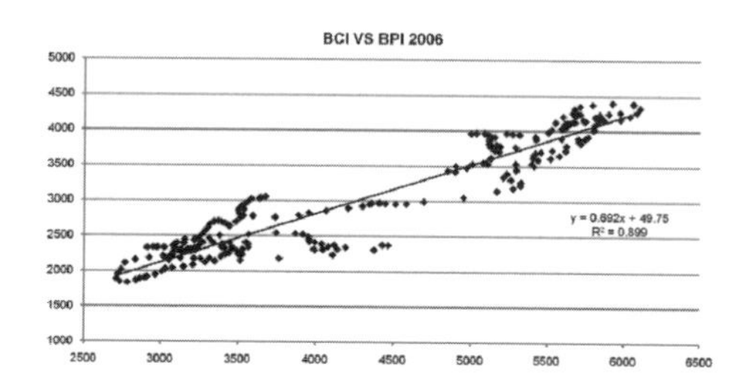

Figure 8. Regression analysis of BCI vs BPI for the year 2006.

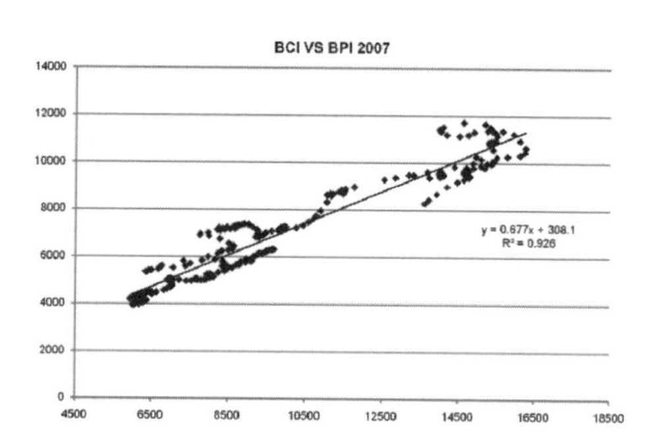

Figure 9. Regression analysis of BCI vs BPI for the year 2007.

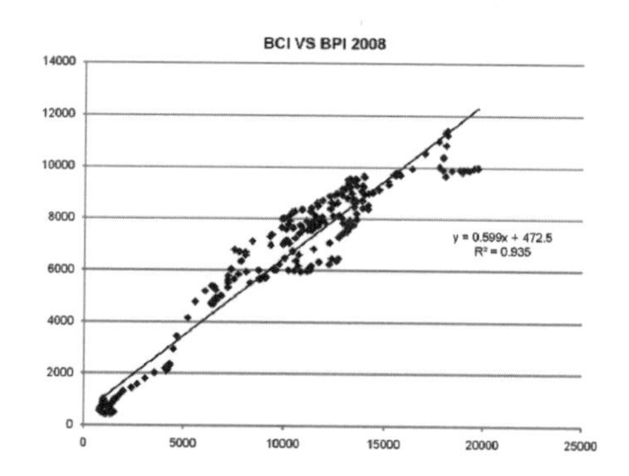

Figure 10. Regression analysis of BCI vs BPI for the year 2008.

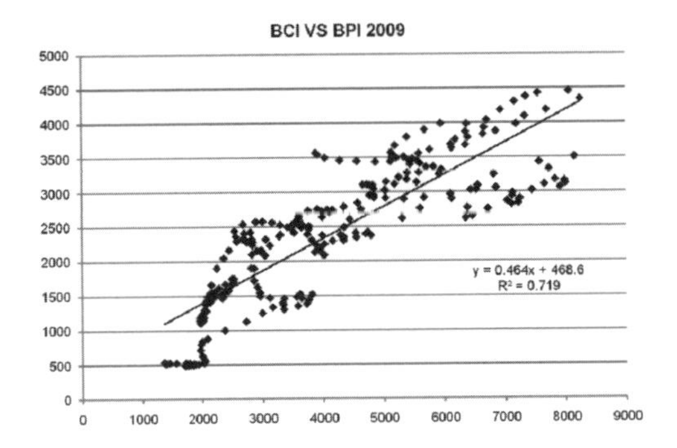

Figure 11. Regression analysis of BCI vs BPI for the year 2009.

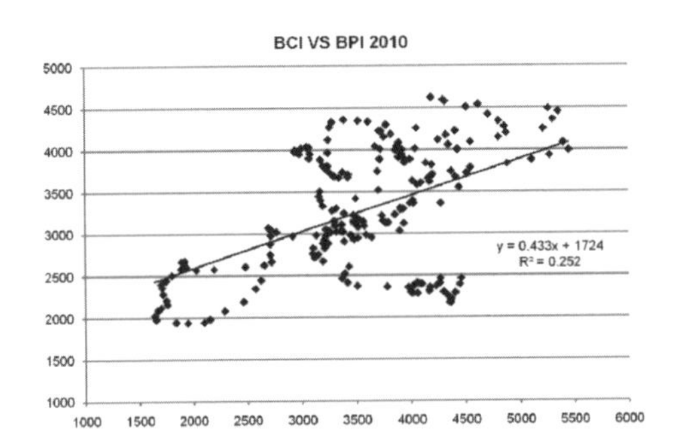

Figure 12. Regression analysis of BCI vs BPI for the year 2010.

3.3 *BPI vs BSI results*

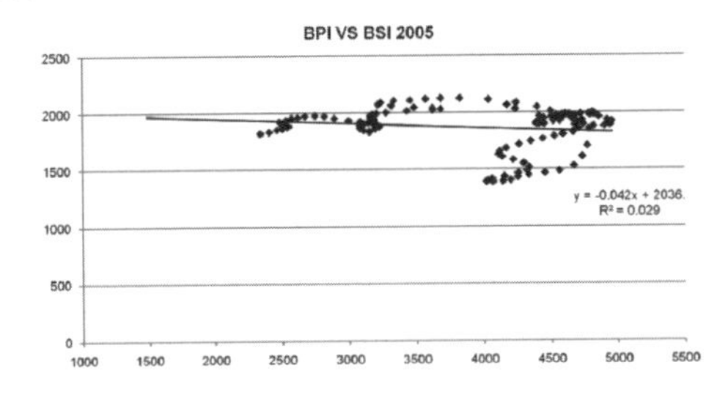

Figure 13. Regression analysis of BPI vs BSI for the year 2005.

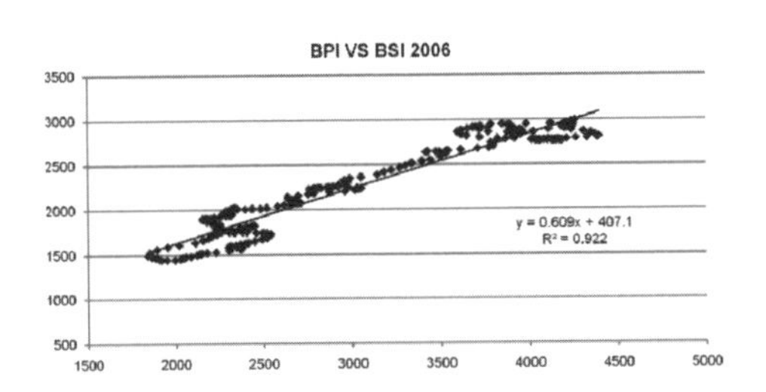

Figure 14. Regression analysis of BPI vs BSI for the year 2006.

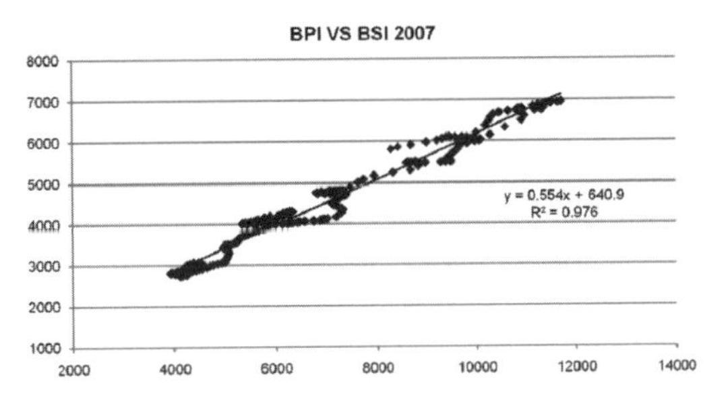

Figure 15. Regression analysis of BPI vs BSI for the year 2007.

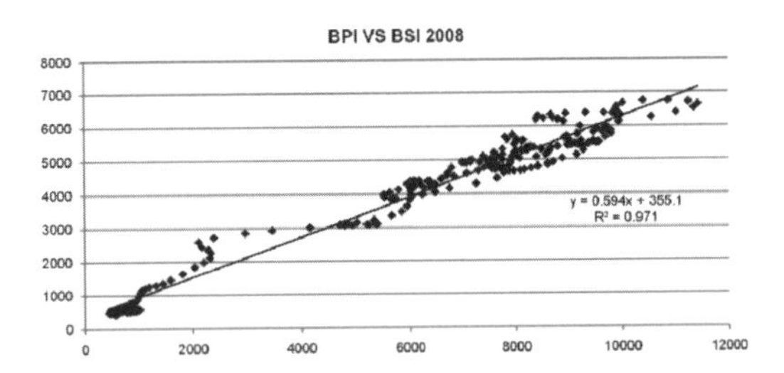

Figure 16. Regression analysis of BPI vs BSI for the year 2008.

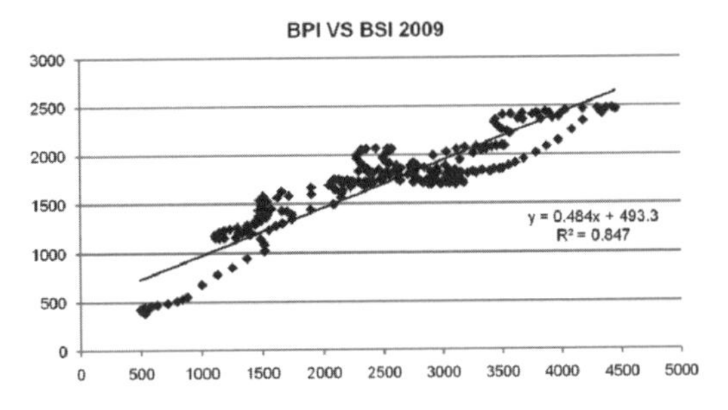

Figure 17. Regression analysis of BPI vs BSI for the year 2009.

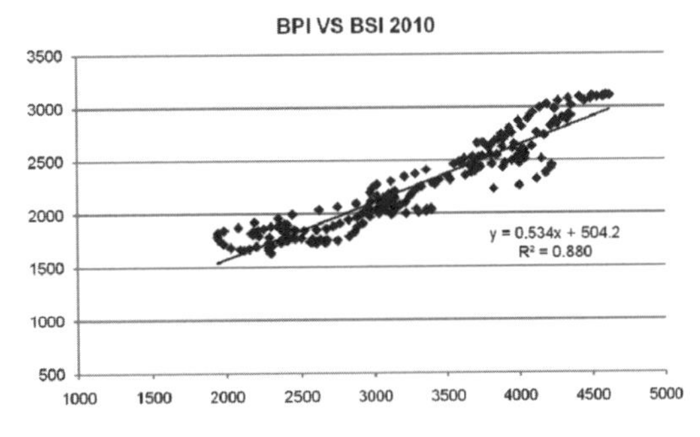

Figure 18. Regression analysis of BPI vs BSI for the year 2010.

3.4 *BSI vs BHSI results*

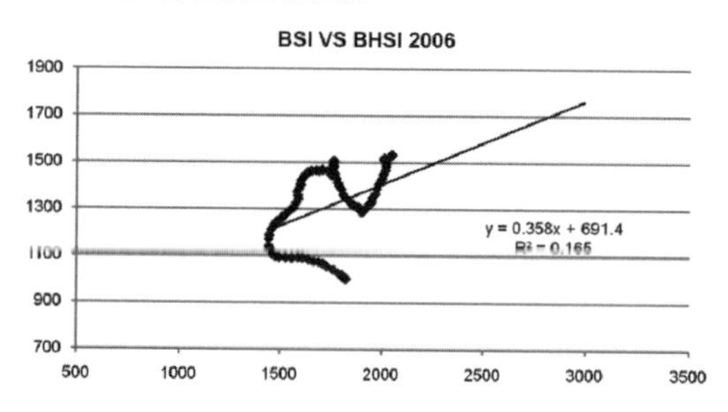

Figure 19. Regression analysis of BSI vs BHSI for the year 2006.

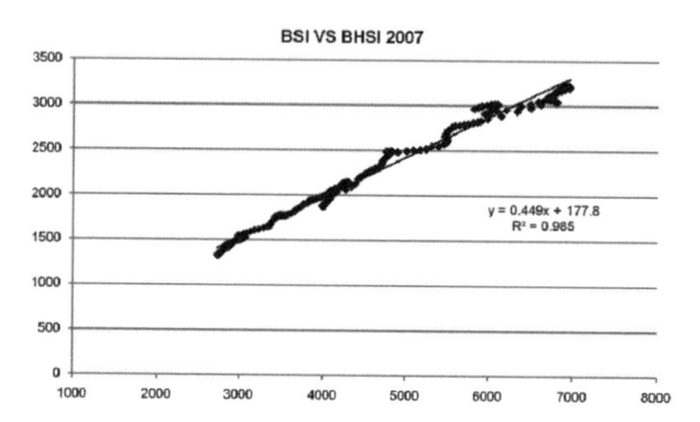

Figure 20. Regression analysis of BSI vs BHSI for the year 2007.

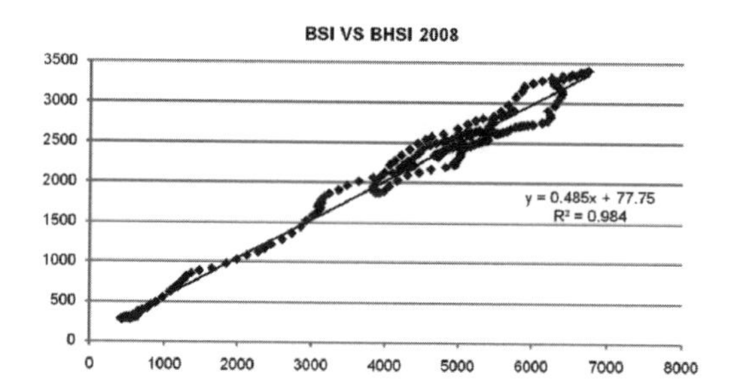

Figure 21. Regression analysis of BSI vs BHSI for the year 2008.

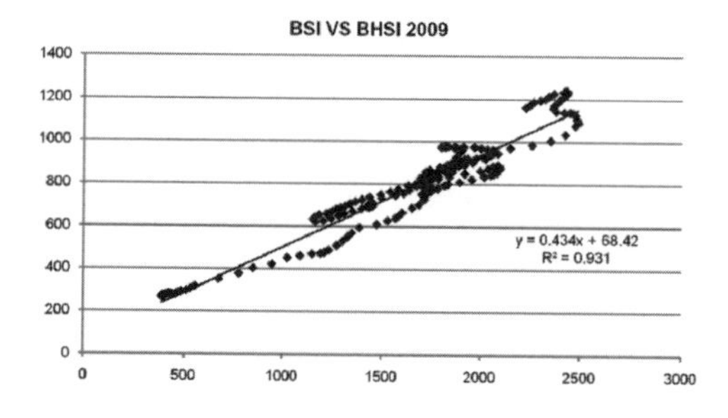

Figure 22. Regression analysis of BSI vs BHSI for the year 2009.

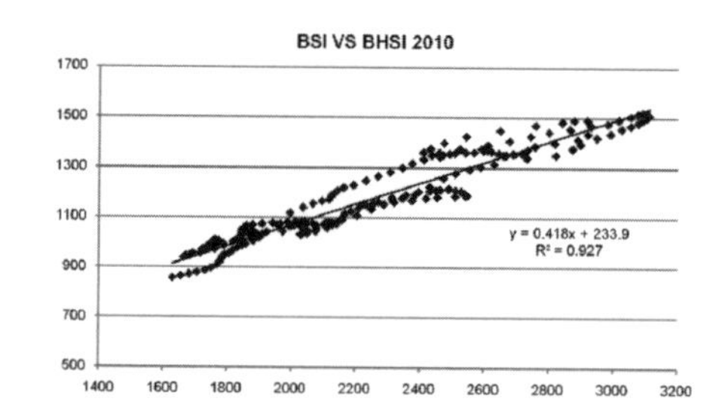

Figure 23. Regression analysis of BSI vs BHSI for the year 2010.

Table 5. Correlation coefficient R^2.

Year	BCI vc BPI	BPI vs BSI	BSI vs BHSI
1999	0.3645	n/a	n/a
2000	0.4337	n/a	n/a
2001	0.93726	n/a	n/a
2002	0.92827	n/a	n/a
2003	0.97114	n/a	n/a
2004	0.75247	n/a	n/a
2005	0.82701	0.02238	n/a
2006	0.89993	0.92238	0.16566
2007	0.92665	0.97666	0.98513
2008	0.935	0.97107	0.98478
2009	0.71933	0.84766	0.93173
2010	0.25199	0.88069	0.9272

3.5 *Analysis results*

The analysis procedure presented in section 3.1 has been applied to the following three pairs of indices:

i. BCI vs BPI,
ii. BPI vs BSI, and
iii. BSI vs BHSI.

By examining the results given in Figures 1–23 of sections 3.2–3.4, the following conclusions can be drawn.

Indices BCI and BPI seem to be positively correlated for the period 2001–2009, although in years 2004, 2005 and 2009 the correlation is slightly decreased .

Accordingly, indices BPI and BSI seem to be positively correlated for the period 2006–2010, with a small decrease in correlation in years 2009 and 2010.

Finally, indices BSI and BHSI seem to be positively correlated for the period 2007–2010.

In Table 5, a summary of the Pearson's correlation coefficient R^2 of the above results is given.

4 PROBABILITY ANALYSIS

Probability analysis is a versatile tool for the analysis of historical data of the indices of the dry market (Stopford 2009).

For each dry index, the empirical histogram is calculated as follows (Spanos 2003).

First, a particular partition is defined of the form

$$\{\xi_1, \xi_2, \ldots, \xi_i, \ldots, \xi_{I+1}\} \tag{7}$$

in order to appropriately segment the range of possible values of the index. Then, the table of relative frequencies of occurrence (histogram) is calculated as

$$\nu_i = \frac{k_i}{N}, \quad i = 1,2, \ldots, I, \tag{8}$$

where

$$k_i = \{\# of\ X_n's : \xi_i \le X_n < \xi_{i+1}, n = 1,2, \ldots, N\} \tag{9}$$

and N is the total number of observations (measurements).

The partitions used for the present probability analysis are given in Table 6.

In this way, the probability that "the index falls within a and b values" can be calculated.

In Figures 24–27, the histograms of BCI, BPI, BSI and BHSI are presented, respectively, based on the whole amount of data available.

In index BCI, for example, the following areas with concentrated probability mass can be distinguished. There is a 9% of its values falling in the range 970–1220, a 19% in 1720–2470, and a 12% in 2720–3470 and in 5220–6470.

So, these areas of BCI index can be considered that they have greater probability to occur in the future.

Working similarly with the other indices, areas of greater probability can be defined.

For BPI, we mention areas 3770–4520 (12%), 1520–1670 (10%), 1220–1520 (9.2%), 920–1070 (9%), 2120–2420 (9%).

For BSI, 1720–2120 (29%), 3920–6920 (24.5%), 2720–3120 (13%), 320–720 (5%).

For BHSI, 1700–3100 (31%), 800–1100 (23%), 1300–1500 (13.5%), 200–400 (6%).

Furthermore, in order to examine the seasonal variability of these probabilities, the annual histograms of the indices have been produced and the results are shown in Figures 28–31.

Table 6. Partitions used in probability analysis for indices BCI, BPI, BSI and BHSI.

Index	Partition
BCI	[0 720:250:20000]
BPI	[0 320:150:12000]
BSI	[0 320:200:7000]
BSHI	[0 200:100:3500]

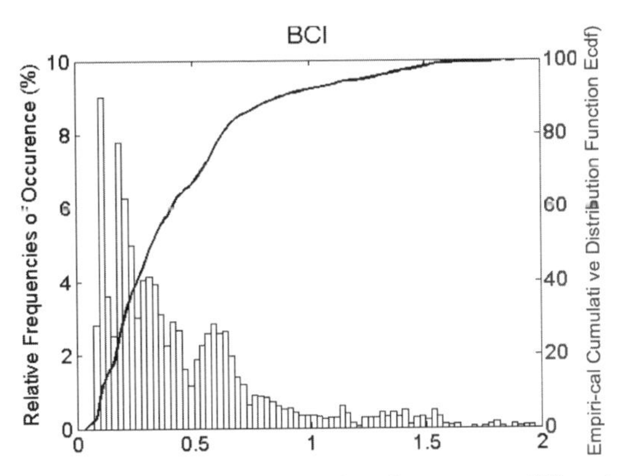

Figure 24. Relative frequencies of occurrence (%) and empirical cumulative distribution function (ecdf) of BCI for all years.

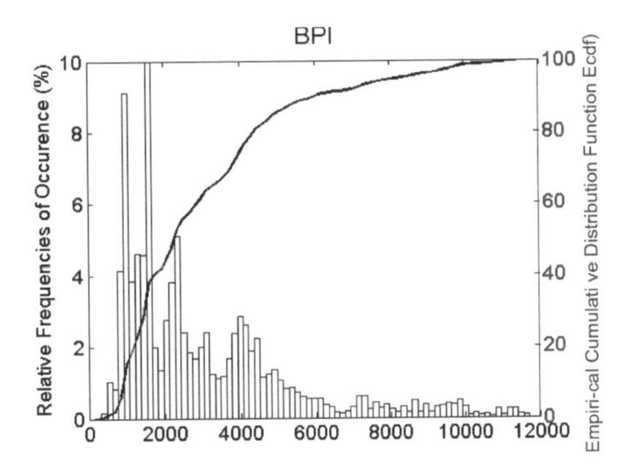

Figure 25. Relative frequencies of occurrence (%) and empirical cumulative distribution function (ecdf) of BPI for all years.

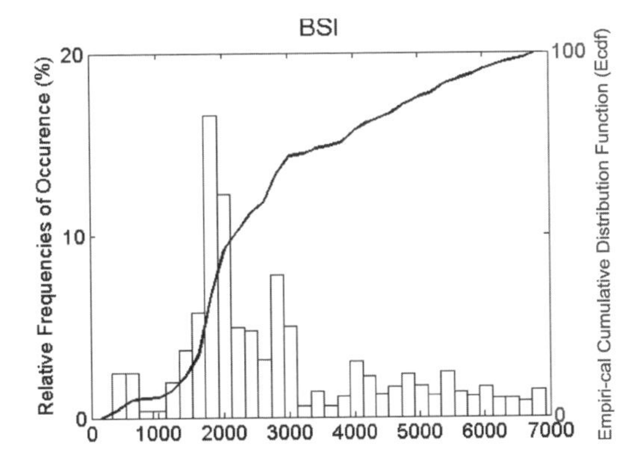

Figure 26. Relative frequencies of occurrence (%) and empirical cumulative distribution function (ecdf) of BSI for all years.

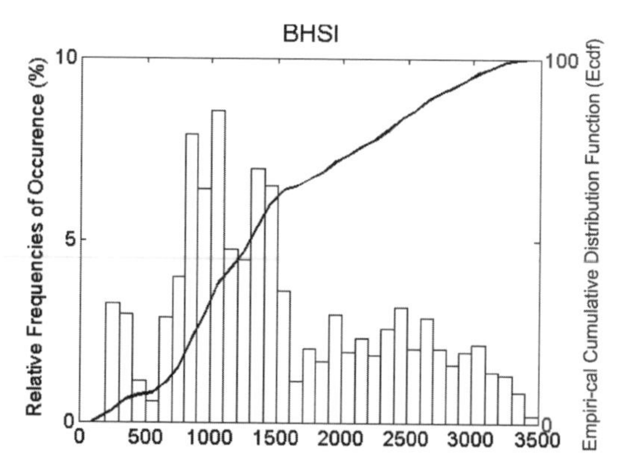

Figure 27. Relative frequencies of occurrence (%) and empirical cumulative distribution function (ecdf) of BHSI for all years.

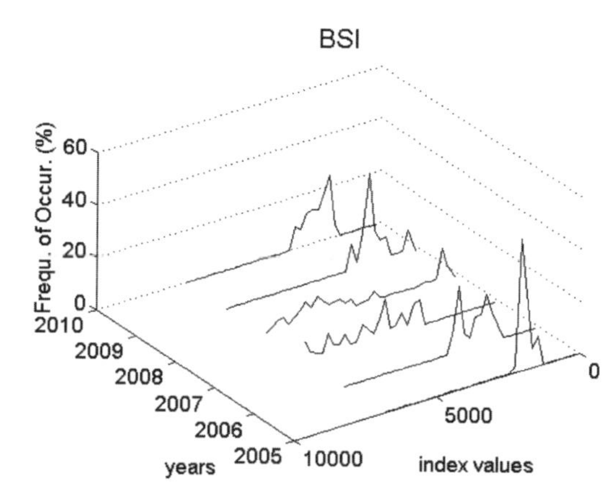

Figure 30. Time history of relative frequencies of occurrence (%) for BSI.

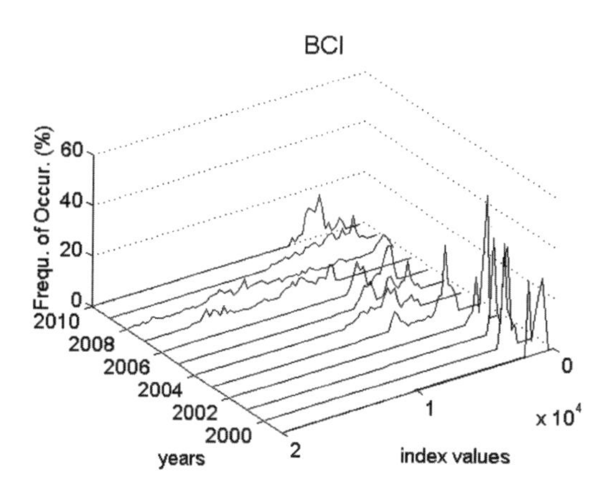

Figure 28. Time history of relative frequencies of occurrence (%) for BCI.

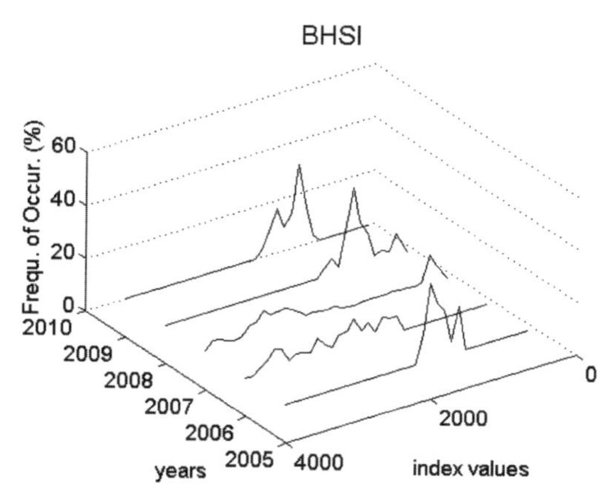

Figure 31. Time history of relative frequencies of occurrence (%) for BHSI.

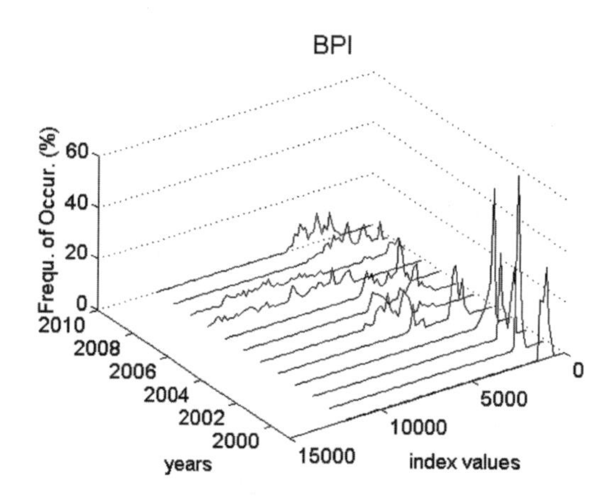

Figure 29. Time history of relative frequencies of occurrence (%) for BPI.

For indices BCI and BPI, one can observe that in the first years (1999–2002), the main probability mass is concentrated in the low values of the range (BCI: 720–1970, BPI: 770–1670). On the contrary, in years like 2008, probabilities of index values are well scattered in the whole range.

Similar situation can be observed in BSI and BHSI Figures for the years 2005–2006 and 2009–2010. Probabilities in years 2007–2008 are again scattered in the almost the whole range of index values.

5 CONCLUDING REMARKS

In the present work, an extensive regression analysis has been performed between indices related to Capesizes (BCI, 1999–2010), Panamaxes (BPI, 1999–2010), Supramaxes (BSI, 2005–2010) and Handysizes (BHSI, 2006–2010) bulk carriers.

The analysis covers the pairs (BCI vs BPI), (BPI vs BSI) and (BSI vs BHSI) on a yearly basis.

Most of the results give a very good coefficient of correlation (97%–84%).

Moreover, the univariate probability structure of the indices is also studied. Histograms are derived using both yearly segments and the whole series of data for each index.

These results help us better understand and visualize the probabilities of occurrence of the various index values for each examined time period.

On the whole, both regression and probability analysis are proved to be versatile tools in examining the behavior of the dry bulk indices in order to prepare reliable market researches.

It is expected that future work will also test the relationship of the above indices with other indices and time-series, especially those of oil prices (UNCTAD 2010b) and indices of major bulk trades, such as the CRB (Commodity Research Bureau) Index. See also Kim (2011).

REFERENCES

Davidson, R., MacKinnon, J.G. 2004. *Econometric Theory and Methods.* New York: Oxford University Press.

Kim, H.-G. 2011. Study about how the Chinese Economic Status affects to the Baltic Dry Index. *International Journal of Business and Management* 6(3): 116–123.

Stopford, M. 2009. *Maritime Economics. 3rd. ed.* London: Routledge.

Spanos, A. 2003. *Probability Theory and Statistical Inference. Econometric Modeling with Observational Data.* Cambridge: Cambridge University Press.

UNCTAD 2010a. *Review of Maritime Transport.* New York: United Nations Publication, UN-CTAD/RMT/2010.

UNCTAD 2010b. *Oil Prices and Maritime Freight Rates: An Empirical Investigation.* New York: United Nations Publication, UNCTAD/DTL/TLB/2009/2.

Sustainable Maritime Transportation and Exploitation of Sea Resources – Rizzuto & Guedes Soares (eds)

Potential of short sea shipping in Brazil

Nayara A.L. de Valois, Afonso Celso Medina & Rui Carlos Botter
University of São Paulo, São Paulo, Brazil

ABSTRACT: Brazil has a transport system with little presence of intermodality and use of Short-Sea Shipping. Nevertheless, these are goals for Brazilian waterborne transportation matrix, which is expected to exceed 13% going to 25% in 2025. Currently, Brazilian transportation matrix isn't balanced surpassing that of other countries in the use of the highway, where approximately 63% of the total cargo is transported. The aim of this paper is to make a diagnosis on the investigative potential use and improvement of Short-Sea Shipping in Brazil. To base this effect, we present a new concept of "fast coastal terminals", its conceptual model and potential gains in an intermodal transportation network.

1 INTRODUCTION

Traditionally, the concept of Brazilian Short-Sea Shipping (SSS) or Coastal Shipping is relieved of the logistics chain and intermodality. The under-representation of intermodality in Brazil exposes the low use of waterways despite of the dissemination and awareness of the benefits to the Brazilian public transport network.

According to Jones *et al.* (2000), intermodal transportation should be generally defined as "the shipment of cargo and the movement of people involving more than one mode of transportation during a single, seamless journey". As Yevdokimov (2000) argued that "intermodal transport requires the physical transfer of cargo between different modes, in a systemic perspective of the transport chain, from collection to distribution, minimizing downtime of goods in their handling between origin and destination". Even with the prospects for growth in cargo handling in Brazilian ports, although some companies have adopted logistics solutions that address the use of waterways, the country is still undergoing political and regulatory reforms, and what it is seen is a slow progress in the dissemination of intermodality, even with a low use of Short-sea Shipping, which represented less than 23% of cargo handling in the country in 2010.

The use of intermodality associated with Coastal or Short-sea Shipping, as featured on Botter *et al.* (2007) "is a necessity for the development of transportation sector in Brazil" and, in a sense, a necessity for achieving the goals of the National Plan of Logistics and Transport—PNLT. The concept also presented on Botter *et al* (2007) for Short-Sea Shipping is understood as "a freight transport logistics chain that relies on one of its links in the maritime transport between points in the Brazilian coast". For authors, SSS should not be relieved of its intermodal nature, since it is dependent on the maritime routes between the terminals and the points of origin and destination and could bring great opportunity for companies to become more competitive by reducing logistics costs and generating greater reliability in service.

Thus, it is suggested a new concept in this article, which considers the SSS inserted into an intermodal transport chain for movement of goods. It is also presented the concept of "fast coastal terminal", with some suggestions to make it competitive. This discussion does not yield a simple improvement to the current Brazilian SSS, but can generate an increasing process of workload attraction to the transportation sector.

This article is divided into four main parts: 1) primary diagnosis of Short-sea Shipping in Brazil; 2) the legal, economic and environmental aspects to implement coastal terminals in Brazil; 3) comparative analysis with the experiences of the European Union, mainly with projects such as the Motorways of the Sea (MoS) and 4) recommendations and suggestions.

2 PRIMARY DIAGNOSIS OF SHORT-SEA SHIPPING IN BRAZIL

According to Brazilian law, it is considered as Short-Sea Shipping (Law 10,893/04): "a coasting navigation that is held between Brazilian ports, using exclusively the sea or the sea and the interior". A broader concept, however, is presented by CGEE (2009) for merchant shipping, based on Regulation for Maritime Traffic (RTM, 1992), classifying SSS as:

- Great Cabotage—held at the merchant shipping between Brazilian ports and or harbors of the

Atlantic coast of South America, West Indies (the Caribbean) and the East coast of Central America, excluding the ports of Puerto Rico and the Virgin Islands;

- Small Cabotage—held between Brazilian ports, the vessel not moving away for more than 20 nautical miles from the coast, or 37.04 km and making large-scale ports whose distance does not exceed 400 nautical miles (740.8 km).

Since the early 1930s the SSS in Brazil is stunted by investments in new roads, new types of vehicles for heavy loads and providing direct service to the final customer. The greatest period of industrial growth in Brazil, between 1940 and 1980, was also marked by high growth of road transportation, which now competes strongly with the sea transportation of goods. This growth took place by the technological development of vehicles, maintenance and construction of new roads and fuel subsidies, which were happening successively in those years.

Many researchers have already developed some studies with the objective of diagnosing the Coastal Shipping in Brazil. Some of these studies, as Moura *et al.* (2008) and CGEE (2009), present major obstacles and difficulties for SSS implementation in Brazil and a diagnosis of the waterway industry and shipbuilding in Brazil, respectively. Year after year, the Brazilian SSS has registered some growth, but in short steps. It may be noted (Fig. 1), as shown by the National Waterborne Transport Agency—ANTAQ statistical report, the small growth in the use of SSS in Brazil, between 1998 and 2010.

In fact, the percentage share of SSS in the cargo movement in Brazil fails to reach the mark of 23%, and by contrast, often show declines, as the 4% between 1999 and 2009 (Fig. 2).

Thus, the barriers of the SSS in the maritime sector may be listed if observed the gaps in regulatory issues, lack of Government incentives, increased rates in ports, fleets with aged vessels, the need for more modern equipment in various Brazilian ports and new investments for the integration of the transport logistics chain.

One of the problems faced by SSS, according to Moura and Botter (2010), is the imbalance that exists in the flow of cargo between the regions of the country. The irregular distribution of goods is also caused by the disparity between modes of transportation, especially because most cargo is transported by road, with roughly 63% stake in array of cargo transportation. Studies for the National Logistics and Transport Plan—PNLT claim that the goal of participation of waterborne transport in Brazil must overcome the mark of 25% in the transport matrix, in 2025. One way to improve this situation would be to provide a different vision for

Figure 1. Cargo movement in Brazil between 1998 and 2010.
Source: ANTAQ annual statistical report of ports, 2010.

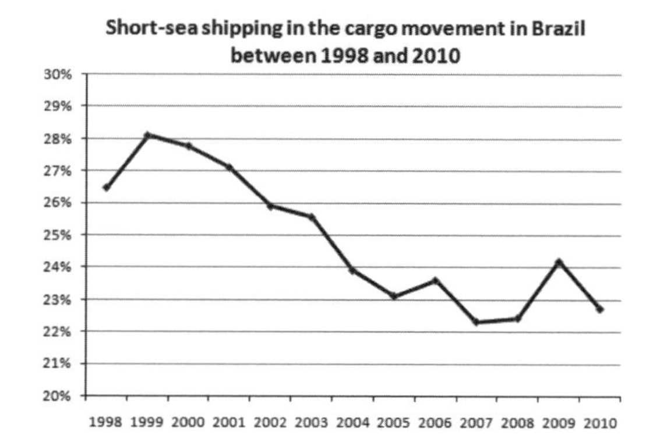

Figure 2. SSS share in the cargo movement in Brazil between 1998 and 2010.
Source: ANTAQ annual statistical report of ports, 2010.

the use of Short-sea Shipping, with appreciation of the various spheres of society and, therefore, a new concept of SSS as part of intermodal transport chains for cargo movement. The concept of SSS can be expanded, not only the one that has its origin and destination within the national territory (Small Cabotage), it can become the integration of transport logistics chain.

Among the main advantages of developing a new model for Brazilian SSS, there are: managing door-to-door cargo integrity, security, agility in customer delivery, competitive cost, integration across all Brazilian regions, using containers, based on a predictable transportation with weekly departures and arrivals. In addition, reduced use of road transportation, reduced congestion in the port access, lower pollutant emissions in port areas and the consequent decrease in the levels of greenhouse gas emissions in the country. Also, with the entry of new vessels to serve domestic market, through the growth of the Brazilian shipbuilding industry, the trend is that the volume of cargo transported (demand) in SSS earns greater impetus in the coming years.

In this context of low competitiveness of Brazilian SSS and low supply of intermodal systems covering coastal shipping on one of its portions, the concept of an *intermodal transport network using "fast" coastal terminals* can expedite the process of transferring cargo between modals, increase speed of operation, making a competitive coastal shipping under the aspect of the service level and enabling the door-to-door carriage.

Knowing this, Figure 3 shows the schematic representation of this new concept (Botter *et al.*, 2007).

So, the *"fast" short-sea terminal* concept is a terminal that is part of an intermodal network, in which the SSS as "a cabotage traditional transport" is the basis of it. It adds the possibility of creating specialized handling for SSS, since the process becomes integrated with other modals and facilitates the transfer of cargo operation at the points of origin and destination. These "fast" terminals for SSS must meet certain proposed criteria, such as:

- To have a different treatment of cargo, with simplified rules for loading, unloading and clearance;
- To own in its surrounding a cargo consolidation center (dry-port) to facilitate the transshipment of containers;
- To facilitate the operation of the Multimodal Transport Operator (OTM);
- To have competitive prices and rates, for better utilization of port infrastructure;
- To have a plan for continuous improvements to modernize equipment, maintenance and control systems;
- To reduce port expenses, travel and operational costs;
- To invest in qualification and training of port operators;
- To schedule regular routes for containers;
- To invest in specialized ships and information technology;

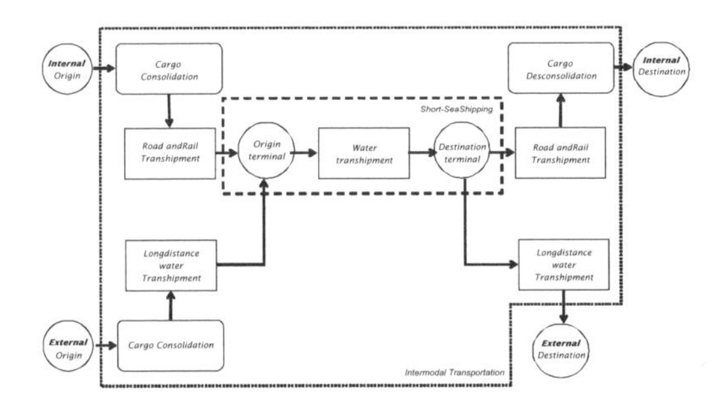

Figure 3. Schematic representation of intermodal transport involving Short-sea Shipping.

- To integrate with other transport modes, with appropriate access to rail and road;
- To improve management of operations, to avoid delays or rescheduling;
- To open market to foreign flagged ships and concessions;
- To adequate legal and regulatory procedures for faster transfers, etc.

The concept of "fast terminals" can turn the Short-Sea Shipping in a most competitive transport of goods in Brazil as it attracts more cargoes to the terminals, in a viable process and with operational, documentary and legal features for the differentiated movement of cargo with the use of exclusive vessels.

3 LEGAL, ECONOMIC AND ENVIRONMENTAL ASPECTS FOR IMPLEMENTING FAST COASTAL TERMINALS

Few studies have been conducted on the potential of SSS cargo attraction in Brazil. An initial estimate can be constructed by the size of the transport market. According to the Brazilian transportation sector in 2010, about 1.2 billion tons of various types of cargo were moved on the Brazilian highways. Excluding bulk handling from this volume, this means that nearly 36 million tons of cargo were handled during the year in trucks, considering flows with scope for Short-sea Shipping, of intrastate and interstate, non-competitive for the maritime. This corresponds to approximately 2.2 million TEUs.

According to CGEE (2009), Brazilian Short-sea Shipping has a great potential to achieve a change in the transportation sector and increase its competitive advantage. "A favorable scenario for Brazil is the development of the port terminals to establish more container operations and consequently coastal lines in the country, complementing long-distance services and *Mercosul*" (CGEE, 2009). It is estimated that companies providing SSS in Brazil handled 348,000 TEUs in 2010, so 23% of the total potential SSS market for the country. This indicates that there is a large market, about three times the current level, not yet met, or yet undeveloped, which presents opportunities to capture market share from road transportation.

However, some improvements should be set out in the maritime transport sector to enable the widespread use of SSS as: adequacy of the national fleet and remodeling in the laws of restrictions on foreign vessels. In recent years, Brazil has seen economic growth and positioning in the ranking of Nations with financial stability and prospects for increasing income and decreasing social differences. New port laws and actions in the transportation sector have shown positive results with investments in the areas of transportation, energy, sanitation, housing and water resources, for example, PAC—a growth acceleration program designed for Brazil, with infrastructure investments of R$ 503.9 billion, near US$ 318 billion, for four years. For the logistics infrastructure, the estimated investments from 2007 to 2010 was R$ 58.3 billion, near US$ 37 billion, however in a report of completed actions, one can observe that many of the planned projects for the transport area have not been completed yet, according to the Institute of Applied Researchs—IPEA (2010), forecasted until December, 31st.

In greater detail, the Table 1 shows the status of projects for the transportation—logistic infrastructure in late 2010.

These initiatives still face known deficiencies, including: lack of an appropriate logistics infrastructure for receiving and distributing goods; few investments for re-rigging of port terminals; lack of harmonization for intermodality (terminals for integration between modals); regulation for transport operation with more than one modal; excessive bureaucracy in customs clearance; regulatory, technological and legal bottlenecks; lack of concern with environmental issues; among others.

Other weaknesses are the lack of adequate road access to the port terminals, the need for modernization of loading and unloading equipment at terminals, low technological level on cargo unitization, making it difficult to shift between modals and lack of regular supply transport service.

Some studies, such as Fernandes (2001), Andrade (2003), Botter *et al.* (2007) and Gonçalves (2008), emphasize regulatory aspects that still hamper the development of SSS, among them, for the transport of general cargo, the Merchant Marine Fund—MMF creates bureaucracy affecting the competitiveness of industry and legislation concerning new shipping companies operating in the sector, which shall have a Brazilian vessel or they shall have at least one vessel being built for operation permit. These and other aspects create barriers for new competitors.

Table 1. Status of the logistic infrastructure projects: 69% running in 2010.

Modal Transportation	Designing, licensing or bidding	Under construction	Number of projects	% Running
Road	253	612	865	70,75%
Rail	9	4	13	30,77%
Ports	17	6	23	26,09%
Waterways	4	39	43	90,70%
Airports	16	10	26	38,46%
Total	299	671	970	69,18%

Source: IPEA, 2010.

Regarding the navigation itself, it falls into two problems: low operating speed compared to other modes of transportation and few routes supply, as measured by the frequency of trips per week. When adding the excessive port times, arising from tax and customs clearance and port restrictions on coastal accesses, there is a huge loss of competitiveness for the Brazilian Short-Sea Shipping.

Therefore, it is clear the need for change in the Brazilian regulation, with regards to improvements to the terminals. Furthermore, it is known the dependency on other modes for the correct operation of water transportation, evidencing the need for modernization of Brazilian ports, development of new terminals, expansion and refurbishment of railways and roads.

4 COMPARATIVE ANALYSIS WITH THE EXPERIENCES OF THE EUROPEAN UNION, MAINLY PROJECTS AS THE MOTORWAYS OF THE SEA (MOS)

International experience in the field of SSS is fundamental to understanding the situation of Brazil and to conduct a comparative analysis appropriate to the country's needs, in aspects of regulation, economy and incentives. Unlike the concept of Short-Sea Shipping used by the European Community, the greatest divergence of Brazilian SSS is presented in the door-to-door managing, verified in cargo handling operations in these countries, which may also be justified by wide use of railroad to move their goods. Undoubtedly, the European Union (EU) is the economic bloc of countries that have the most developed SSS system. What favors their water transportation is mainly the possibility of network connections between countries, making the system more than 67,000 kilometers of coastline and over 25,000 kilometers of inland waterways, with greater turnover of 430 million tons per year, according to ANTAQ (2007).

Inland direct connections between seaports and interior ports of the European Union gives access to several other ports and still rely on vessels that allow this access, increasing the possibilities of coastal navigation. Moreover, the standards for SSS in the EU facilitate the freedom of services provision between member countries and the adoption of measures which highlight the proposals for simplification of operation with single administrative offices and the integration of the logistics chain between agents (chargers, shipowners, shippers, etc).

Among the measures adopted, the Motorways of the Sea (MoS) are the latest and which strengthen the logistics integration strategies of Short-Sea Shipping. The "Motorways of the Sea" was a project defined between EU countries to help achieve goals such as alleviate the major land bottlenecks of the European transport system (replacing road transport), setting up an efficient intermodal system "where the goods are quickly transferred between modes through optimization of port operations, overcoming natural boundaries and sensitive areas as well as other geographical barriers" ("letter of Naples" of EU Transport Ministers, July 2003). The proposed concept for MoS provides an integrated service and system (BAIRD, 2007).

The four main corridors were planned for the MoS in the Baltic Sea, Western Europe, Southeast and Southwest Europe. One can, for example, use the project PORTMOS (Portuguese Motorways of the Sea) as benchmarking. The concepts developed for Portugal, which provides seamlessly services and operational, administrative, bureaucratic and informational systems, as well as logistical infrastructure, enable door-to-door transport of goods (maritime corridors) as an alternative to road transport, showing an effective, affordable and competitive process (PORTMOS, 2008).

In recent years, the European Commission and other countries have promoted extended research activities within research and development framework programs, such as Motorways of Sea, which focused on making short-sea shipping lanes viable, where it is possible to compete with road transport. Until now the obtained results have identified opportunities and constraints for the development in these countries.

For the Brazilian SSS market it can be observed firstly that the adjustment system suffers bureaucracy in clearance of SSS cargo with high taxes and deficiency to demand and lack of vessels. The standard rules for reserve flag, which guarantees to national flag vessels the right to carry cargo and passengers between domestic ports, do not lead to discussions that can develop the sector. Other differences are the high cost of vessels produced in national shipyards and harbors maintenance fee, seen by operators as a reducer of competitiveness.

5 RECOMMENDATIONS AND SUGGESTIONS FOR IMPROVEMENTS IN BRAZIL SHORT-SEA SHIPPING

Bureaucracy, costs and ports inefficiencies are the most significant factors that affect the attractiveness of SSS in Brazil, according to the vision of industry users. However, some stimulus measures can be developed for larger subsidies for construction and differential tax treatment.

One important aspect for SSS in Brazil is the increase of transport supply and cargo demand, which does not necessarily depend on public funding, but access to foreign vessels at an affordable price. Decisions to use tax incentives—additional freight for renewal of the Merchant Navy and a special Brazilian registry for ships could have been effective, but not to become a general rule for all fleets of ships and craft. Another important aspect is the freight of imported and exported cargo which can improve competitiveness, but needs to be evaluated as a reduction of freight charges.

The development of a conceptual model for the intermodal network with participation of Short-Sea Shipping can open the discussions for the establishment of "fast" terminals with characteristics to leverage the change in structure of intermodal transportation, particularly in the transport of containers in the country. With these terminals, cargo transfer process would be facilitated, also ensuring fast port processes, allowing the door-to-door service and attracting new players to the water transportation network.

REFERENCES

ANTAQ—Agência Nacional de Transportes Aquaviários. 2007. Relatório Final do projeto "Diagnóstico da navegação de cabotagem visando à regulação do setor". Universidade de São Paulo—USP. Apoio FINEP—Financiadora de Estudos e Projetos. 185p.

BAIRD, A.J. (2007) The economics of Motorways of the Sea. Maritime Policy & Management, 34(4), 287–310.

BARAT, Josef (org.). Logística e transporte no processo de globalização: oportunidades para o Brasil. São Paulo: Editora UNESP: IEEI, 2007. ISBN 978-85-7139-758-3.

BOTTER, R.C.; MOURA, D.; LOBO, G.; MEDINA, A.C. Diagnóstico da navegação de cabotagem visando à regulação do setor. Relatório Final Projeto FINEP/ANTAQ/USP. 2007.

CGEE—Centro de Gestão e Estudos Estratégicos. Tópicos estratégicos para investimentos em CT&I nos setores de transporte aquaviário e de construção naval. Brasília—DF. 2009. 190 p. ISBN 978-85-60755-17-2.

FERNANDEZ, Enrique L., J. and J. Manuel Fernández N., 1998, Privatization of Ports in Developing countries. The case of container terminals, Paper presented at the 8WCTR, Antwerp, July, 1998.

JONES, W. Brad; CASSADY, C. Richard; BOWDEN, Royce O. Jr. Developing a Standard Definition of Intermodal Transportation Symposium on Intermodal Transportation. 2000. Transportation Law Journal. v. 27. no. 3. p. 8 (345 to 352). Available on <http://heinonline.org/HOL/Page?handle=hein.journals/tportl27&id=447&collection=journals&index=>.

MOURA, D.A.; BOTTER, R.C.; MEDINA, A.C. Diagnosis of Brazilian short-sea shipping and it main obstacles. In: The 12th International Congress of the International Maritime Association of Mediterranean- IMAM: Taylor & Francis, v. 1. p. 593-592, Varna, Bulgary, 2007. ISBN 978-0-415-45523-7.

NAZÁRIO, Paulo. Intermodalidade: Importância para a Logística e Estágio Atual no Brasil. Artigo logística ILOS. 2000. Disponível em http://www.ilos.com.br/site/index.php?option=com_content&task=view&id=1012&Itemid=74.

PAIXÃO, A.C. and MARLOW, P.B. (2002) Strengths and weaknesses of short sea shipping. Marine Policy, 26(3), 167–178.

PAIXÃO, A.C. and MARLOW, P.B. (2005) The competitiveness of short sea shipping in multimodal logistics chains: Service attributes. Maritime Policy & Management, 34(2), 363–382.

PERAKIS, A.N. and DENISIS, A. (2008) A survey of short sea shipping and its prospects in the USA. Maritime Policy & Management, 35(6), 591–614.

PORTMOS—Integração do sistema marítimo-portuário nas auto-estradas do mar. 2008. Available on<http://www.planotecnologico.pt/InnerPage aspx?idCat=73&idMasterCat=30&idLang= 1&idContent=1519&idLayout=4&site=planotecnologico>.

YEVDOKIMOV, Yuri V. Measuring Economic Benefits of Intermodal Transportation. 2000. Transportation Law Journal. v. 27. no. 3. p. 14 (439 to 452). Available on <http://heinonline.org/HOL/Page?handle=hein.journals/tportl27&id=447&collection=journals&index=>.

Sustainable Maritime Transportation and Exploitation of Sea Resources – Rizzuto & Guedes Soares (eds)
© 2012 Taylor & Francis Group, London, ISBN 978-0-415-62081-9

The effects of regulatory changes on green freight corridors

G.P. Panagakos & H.N. Psaraftis
Laboratory for Maritime Transport, School of Naval Architecture & Marine Engineering, National Technical University of Athens (NTUA), Athens, Greece

ABSTRACT: "Green corridors" in freight transportation is a new concept introduced in 2007 as an action of the "Freight Transport Logistics Action Plan" of the European Commission. It pursues a corridor approach in developing integrated, efficient and environmentally friendly transportation of freight between major hubs and by relative long distances. NTUA leads the EU-financed SuperGreen project, which aims at assisting the European Commission in defining green corridors through the use of Key Performance Indicators (KPIs). A set of 9 European corridors have been selected by the project to be used as testing ground for the KPIs and the related methodology. The purpose of this paper is to present the KPIs that have been selected for benchmarking the SuperGreen corridors and to use them for assessing the potential effects of changes in EU and international regulations on green corridor development.

1 INTRODUCTION

What is really a green corridor? In a strict sense, a precise definition of the term is still elusive, and in fact one of the most important contributions of ongoing research on the topic would be to develop an explicit and workable definition of the term.

According to the "Freight Transport Logistics Action Plan" of the European Commission (EU 2007), which introduced the concept, "... industry will be encouraged along these corridors to rely on co-modality and on advanced technology in order to accommodate rising traffic volumes, while promoting environmental sustainability and energy efficiency. Green transport corridors will... be equipped with adequate transhipment facilities at strategic locations ... and with supply points initially for bio-fuels and, later, for other forms of green propulsion. Green corridors could be used to experiment with environmentally-friendly, innovative transport units, and with advanced Intelligent Transport Systems (ITS) applications... Fair and non-discriminatory access to corridors and transhipment facilities should be ensured in accordance with the rules of the Treaty."

The EU-financed SuperGreen project aims at assisting the European Commission in further defining green corridors through a corridor benchmarking exercise using Key Performance Indicators (KPIs). More details about the project can be found at: http://www.supergreenproject.eu.

The purpose of this paper is to present the KPIs selected for benchmarking the SuperGreen corridors and use them for assessing the potential effects of changes in the EU and international regulatory environment on green corridor development. Emphasis will be placed on maritime transport.

The paper is based on the work performed under the SuperGreen project, as it has been reported in Salanne et al. (2010), Pålsson et al. (2010), Ilves et al. (2010) and Panagakos et al. (2011).

2 THE SUPERGREEN KPI'S

2.1 *The project*

SuperGreen is a Coordination and Support Action, in the context of the European Commission's 7th Framework Programme of Research and Technological Development. The objectives of the SuperGreen project concern supporting the development of sustainable transport networks by fulfilling requirements covering environmental, technical, economical, social and spatial planning aspects. This will be achieved by:

- giving overall support and recommendations on green corridors to EU's Freight Transport Logistics Action Plan,
- conducting a programme of networking activities between stakeholders (public and private),
- providing a schematic for overall benchmarking of green corridors based on selected KPIs,
- delivering policy recommendations at a European level for the further development of green corridors, and
- providing the Commission with recommendations concerning new calls for R&D proposals to support development of green corridors.

Project work is organised in 7 work packages. The most relevant one to the present paper is the package concerning corridor benchmarking. It involves: the selection of corridors; definition of the benchmarking methodology and indicators; identification of changes in the operational and regulatory environment that may enhance or hamper green corridor development, the actual corridor benchmarking; and definition of areas for improvement. It is noted that the benchmarking of corridors in relation to green technologies and smart Information and Communication Technology (ICT) applications comprises the subject of different work packages and will not be covered here.

2.2 *The SuperGreen corridors*

An initial list of 60 potential corridors was compiled on the basis of the TEN-T priority projects, the Pan European Transport Network and proposals made by the project's industrial partners. After two consolidation rounds the number of candidate corridors was reduced to 30. A survey was carried out to gather information on these 30 corridors. Based on the information gathered, a pre-selection of 15 corridors was made. A geographic and modal balance was ensured among these pre-selected corridors. The aim at this stage was to select the ones with the highest "greening potential" rate.

Further information was collected on these 15 pre-selected corridors and a deeper analysis was performed taking into consideration land use aspects. The analysis resulted in a recommendation of 9 corridors for final selection, which was presented to a stakeholder workshop especially arranged for this purpose. In line with comments received during the workshop, the selected corridors were modified by adding segments that exhibit advanced "greening" characteristics.

It should be made clear that the selection of these corridors was made only for the purposes of the SuperGreen project and by no means this implies any endorsement, direct or indirect, either by the SuperGreen consortium or by the European Commission, of these corridors vis-à-vis any other corridor, with respect to any criteria, environmental, economic, or other.

Table 1. Initial set of KPIs.

KPI area	Indicator	Unit
Efficiency	Absolute cost	€/tone
	Relative cost	€/ton-km
Service quality	Transport time	Hours
	Reliability (time precision)	% of shipments delivered within acceptable window
	Frequency of service	number per week
	ICT applications	graded scale (1–5)
	–cargo tracking	
	–other ICT services	
	Cargo security	number of incidents per total number of shipments
	Cargo safety	number of incidents per total number of shipments
Environ. sustainability (*)	CO2-eq	g/ton-km
	SOx	g/1000 ton-km
	NOx	g/1000 ton-km
	PM10	g/1000 ton-km
Infrastr. sufficiency	Congestion	av. delay (h)/ton-km
	Bottlenecks	graded scale (1–5)
	–geography	
	–capacity of infrastr.	
	–condition of infrastr.	
	–administration	
Social issues	Corridor land use	
	–urban areas	% of buffer zone (**)
	–sensitive areas	% of buffer zone (**)
	Traffic safety	fatalities & ser. injuries per year per mio ton-km
	Noise	% of corridor length above 50/55 dB

(*) well-to-wheel approach; (**) shaped by a radius of 20 km around the median line of the corridor.

Of particular interest to maritime transport are the two mainly short-sea-shipping corridors covering the Mediterranean and Baltic seas and the mainly deep-sea-shipping corridor concerning the Asia—Europe sea and land connections. Inland waterway transport is represented by the Rhine-Main-Danube corridor, while the remaining corridors are related basically to road and rail transport. More details on SuperGreen corridor selection can be found at Salanne et al. (2010).

2.3 *The initial set of KPIs*

No corridor benchmarking exercise was identified in the literature. The closest case concerns benchmarking of transport chains and was studied by the BE Logic project (Kramer et al. 2009). Based on this experience, the project developed a methodology that consisted of decomposing the corridor under examination into transport chains, benchmarking these chains using a set of KPIs, and then aggregating the chain-level KPIs to corridor-level ones using proper weights for the averaging.

The initial set of KPIs resulted from a process that included the compilation of a gross list of performance indicators, their categorisation into different groups and their filtering during detailed discussions. These KPIs, grouped in five areas (efficiency, service quality, environmental sustainability, infrastructural sufficiency, and social issues), are presented in Table 1 below along with their respective definition.

2.4 *KPI revision*

With the aim of soliciting feedback, the methodology and initial set of KPIs were presented in three events: two regional stakeholder workshops and a meeting of the project's Advisory Committee. The general consensus was that the methodology was in broad terms acceptable and that the KPIs proposed by the project cover all basic facets of the problem. However, there was also a general sense that KPIs as proposed were too ambitious and there was a need to simplify them so that the set be useful. In that sense, reducing the set of KPIs to a more manageable one was considered as a desirable outcome.

Following an internal round of KPI screening, a revised set was presented to a third regional SuperGreen workshop, organised in Malmö, Sweden and hosted by the Swedish Transport Administration. The aim was to set a basis for collaboration with the numerous green corridor initiatives in the Baltic region and take advantage of an audience directly or indirectly exposed to the green corridor concept. The KPI set that resulted from this process is the one of Table 2. This set was reaffirmed at the fourth regional stakeholder workshop of the project in Sines, Portugal.

Table 2. Revised set of KPIs.

Indicator	Unit
Relative transport cost (to the user)	€/ton-km
CO_2 emissions	g/ton-km
SOx emissions	g/1000 ton-km
Average speed	km/h
Reliability (on time delivery)	% of shipments delivered within acceptable window
Frequency of services	Number per year

3 EFFECTS OF REGULATORY CHANGES

Aiming at identifying factors that might promote or hinder green corridor development, the SuperGreen project undertook an extensive literature survey that resulted in 79 changes in the operational and regulatory environment. These changes were grouped in 7 themes (Business environment, Trends in logistics, Public policies, Operations, Infrastructure development, Technology development, and International regulations) and their effects on green corridor development were assessed through the use of the SuperGreen KPIs. For more details on the subject refer to Panagakos et al. (2011).

For the purposes of this paper, the changes concerning regulatory and policy related issues were isolated from the others and regrouped into two categories: the ones concerning EU policy making, and those related to international (non-EU) regulations (refer to Tables 3 & 4 respectively).

In these tables, the direction and level of significance of the effects of each change are depicted through symbols, which have the following meaning:

+ Moderate increase
++ Significant increase
+++ Very significant increase
– Moderate decrease
–– Significant decrease
––– Very significant decrease
+/– Two different forces work in opposite directions
(+) Potential effects
+ (–) Moderate increase but potential decrease under specific conditions described per case.

No symbol means that no effects are expected.

In order to avoid confusion, the definitions of the KPIs used in the analysis are those of Table 2, with the exception of emission KPIs, which are defined in absolute (mass) rather than relative

Table 3. Effects of changes in EU public policies.

No.	Change	Cost	Speed	Reliability	Frequency	CO_2–eq.	SOx
1	EU enlargement	++/—	–	–		++	++
2	EU integration	++	–	–	+	++	++
3	Liberalise transport operations	— (–)	++ (+)	++ (+)	++	— (–)	— (–)
4	Internalise external costs	++				—	—
5	Set energy consumption/emission/ noise standards & other regulatory measures	+				—	—
6	Tighten up and harmonise safety standards	+		+			
7	Tighten up security standards	+/–	–	+			
8	Standardise transport units and vehicles	+/—	++	++		– (+)	
9	Harmonise infrastructure (interoperability)	+/—	++	++		—	—
10	Harmonise rules and enforcement	–		+		–	–
11	Standardise liability and documentation for multi-modal transport	–				(–)	(–)
12	Simplify administration	—	++				
13	Create freight-oriented corridors	–	+++	+++		(—)	(—)
14	Develop green corridors	–		++		—	—
15	Create a core network of high EU added value	— (–)	++ (+)	++ (+)	+	— (–)	— (–)
16	Employ a spectrum of instruments to fund infrastructure and other actions	— (–)	++ (+)	++ (+)		— (–)	— (–)
17	Bring ICT applications to market (ITS, ERTMS, RIS, e-maritime, e-freight, e-customs)	—	++	++		—	—
18	Enhance education and training	–	+	+		–	–
19	Ensure satisfactory working conditions	–		+		–	–
20	Support research & development	—	+	+		–––	–––
21	Educate, inform and involve the greater public in transport policies (incl. labelling)	–	+	+		–	–
22	Monitor and publish service quality indicators	–	++	++		–	–
23	Promote international cooperation with EU neighbouring countries	(–)	(+)	(+)			
24	Green public procurement					–	–

+/– = moderate increase/decrease; ++/— = significant increase/decrease; +++/––– = very significant increase/decrease (and combinations thereof).

(mass/ton-km) terms. The above symbols should be considered in conjunction with the KPI definitions. It is mentioned as an example that the symbol '+' in the CO_2-eq column signifies a moderate increase in GHG emissions and not a positive development in this respect.

It is also noted that in assessing the effects of a particular change, this change is considered independently from all other factors, which are kept unchanged. As an example it is mentioned that in projecting significant increase ('++') of CO_2-eq emissions due to EU enlargement, the capacity of transport infrastructure is kept at today's level, which does not need to be the case in reality. In most cases this assumption places more emphasis on the short term effects of a change. Significant long term effects not captured by this assessment are presented separately in the text.

3.1 *Changes concerning EU policy issues*

A total of 24 changes in EU public policies, as listed in Table 3, have been identified and assessed against the SuperGreen KPIs. Space limitations do not allow justification of these assessments here; this can be found in Panagakos et al. (2011). Instead, we will focus on the following three important issues.

The first one concerns the liberalisation of transport operations. Following the efficiency gains achieved by the market opening in air transport, which have resulted in a significant reduction of

Table 4. Effects of changes in international regulations.

No.	Change	Cost	Speed	Reliability	Frequency	CO_2–eq.	SOx
1	Support fair international trade	–		+		–	–
2	Adopt EEDI	+	–		–	– (+)	– (+)
3	Internalise the external costs of GHG emissions from ships	+				–	–
4	Strengthen restrictions on NOx and SOx	++				(+)	– (+)
5	Establish a mandatory polar code	—	++	+	++	—	—
6	Enhance international security	+/– (++)	– (—)	+			
7	Establish global standards for ICT applications in shipping	—	++	++		—	—
8	Establish global standards for IWT–engines	–				—	—
9	Upgrade EU status in IMO	(+)				(–)	(–)

+/– = moderate increase/decrease; ++/— = significant increase/decrease; +++/—— = very significant increase/decrease (and combinations thereof).

user costs, the European Commission has set the liberalisation of road and rail transport operations as one of its main objectives. With the so-called Third Railway Package for rail and Regulation No 1072/2009 for road haulage, the legal framework of market opening is almost complete. Some issues such as opening up competition in the provision of intermodal terminal and port services, as well as existing differences in taxation and subsidies still need to be addressed. More effort is needed, however, in enforcing the competition rules (EU 2009).

The effects of liberalisation are significant reduction of user costs and emissions, and significant increase of speed, reliability and frequency of service. These gains are achieved basically through better utilisation of infrastructure and vehicles/vessels (higher load factors and lower empty trip factors) and more intensive use of ICT applications. It is noted, however, that the lower transport costs will have a positive impact on transport demand, and for most KPIs the above gains will be mitigated but not reversed.

The internalisation of transport related external costs, an issue that was raised in the 1990s and gained momentum in the last three years with numerous studies and policy papers, is the second one. Prices reflecting all costs—internal and external—convey the right signal to economic actors, who have economic incentives to use safer, more silent and environmentally-friendly vehicles or transport modes and, to plan their trips according to expected traffic conditions, leading to efficiency gains (seen from the welfare economics point of view). The principle applies to all modes. In all cases, it is suggested that revenues generated by internalisation should be used by Member States for making transport more sustainable through projects such as research and development on cleaner and more energy efficient vehicles, mitigating the effect of transport pollution or providing alternative infrastructure capacity for users (EU 2008b).

The expected effects of externality internalisation are significant gains in terms of emissions, at the expense of increased user costs. The role of ICT applications is crucial in making the internalisation possible and in reducing the operating and management costs of the relevant schemes.

It is recommended that the Commission assesses the possibility of including the fair and non-discriminatory access requirement, and the internalisation of external costs as prerequisites for labelling a particular corridor as "green". In this way, green corridors, in addition to being a field for experimenting with environmentally-friendly, innovative transport units, and with advanced ITS applications, can become a laboratory for transport policies, too.

The third point of interest concerns the creation of freight-oriented corridors, as they have been introduced by Regulation No. 913/2010 (EU 2010). The regulation designates 9 European corridors as initial freight corridors, where sufficient priority is given to international freight trains. In addition, it makes it mandatory for each Member State (excluding Cyprus and Malta) to participate in the establishment of at least one freight corridor.

The effects of the freight-oriented corridors on cargoes already transported by rail are very significant improvements in terms of speed and reliability. Improvements are also expected in terms of costs through better coordination. If the scheme succeeds to attract road cargoes, significant gains in emissions will also materialise.

Four valuable lessons can be drawn from Regulation No 913/2010. Firstly, the Regulation separates the criteria for establishing a freight-oriented corridor from the indicators monitored after its establishment. In fact, while the establishment criteria are defined by the Regulation, the indicators to be monitored are left for the corridor's management to decide with only broad directions given. This is a logic that can be followed for the green corridors, too.

Secondly, one of the establishment criteria is the definition of a freight-oriented corridor: "A corridor crossing the territory of at least three Member States or of two Member States if the distance between the terminals served by the freight corridor is greater than 500 km." Although there is no need to expand this definition to the green corridors, it certainly provides a guideline to this end.

Thirdly, in recognition of the multiplicity of entities involved, the Regulation sets up a detailed governance structure, including representatives of the Member State authorities, Infrastructure Managers, Railway Undertakings and terminal owners/managers. To simplify communication with applicants and other interested parties, the establishment of a one-stop-shop is foreseen. Both the international governance structure and the one-stop-shop provided for by the Regulation can be features for the green corridor governance, with minor adjustments where needed.

Fourthly, the Regulation prescribes a number of implementation measures including:

a. a market study,
b. an implementation plan describing the characteristics of the freight corridor, including:
   - bottlenecks,
   - the programme of measures necessary for creating the freight corridor, and
   - the objectives for the freight corridor, in particular in terms of service quality and its capacity,
c. an investment plan including financial requirements and sources of finance,
d. a deployment plan relating to the interoperable systems along the freight corridor,
e. a performance monitoring mechanism,
f. a user satisfaction survey, and
g. the requirement to update all the above periodically.

All these requirements tie very well with the green corridor concept and should be retained.

3.2 *Changes concerning international regulations*

All 9 changes of Table 4 concern shipping (8 apply on marine shipping and 1 on inland navigation). This is not surprising given the international nature of merchant shipping, which makes the international legal context particularly important for this industry.

The IMO's activities to combat GHG emissions from ships are very extensive. Its objective is to finalise soon a mandatory Energy Efficiency Design Index (EEDI) covering the environmental performance of new ships above 400 GRT.

Without going into technical details, the EEDI value is calculated by a formula, the numerator of which is a function of all power generated by the ship (main engine and auxiliaries), and the denominator is a product of the ship's deadweight (or payload) and the ship's 'reference speed', appropriately defined as the speed corresponding to 75% of MCR, the Maximum Continuous Rating of the ship's main engine. The units of EEDI are grams of CO_2 per tonne mile. The way this index will work is as follows: The EEDI of a new ship is to be compared with the so-called "EEDI (baseline)," which is a function of the ship's deadweight. If a ship's EEDI is above the equivalent baseline, the ship would not be allowed to operate until and unless measures to fix the problem are taken (IMO 2010).

However, there have been numerous concerns on EEDI's future use. For instance, an important caveat concerns the accuracy of the speed data that have been used in calculating the EEDI (baseline). Another concern is that the combination of formulae for EEDI and EEDI (baseline) essentially imposes a speed limit, and, in turn, an upper bound on the ship's MCR, shifting the focus from developing the most efficient hull forms, engines or propellers to reduce CO_2, to achieving the same objective just by reducing power and service speed.

The adoption of EEDI, therefore, is expected to reduce GHG emissions (CO_2-eq) directly, and air pollutant emissions (SOx) indirectly through reduced consumption of fuel oil. On the other hand, the reduced speed implied by the EEDI formula will negatively affect the frequency of service. Transport costs will go up in periods of high freight rates, when speed is an important factor, but will remain unaffected in low rate periods, when slow steaming is a usual practice. In the event the measure induces a back-shift from sea to road, the gains in environmental performance will be reversed.

In addition to EEDI, the IMO is considering the adoption of a Market-Based Measure (MBM), as a means to internalise the external costs of GHG emissions from ships. Such a measure would provide economic incentives to ship owners to build ships that are more energy efficient and/or adopt operational measures (for instance, slow steaming, or other) that would reduce GHG emissions. However, utmost care should be exercised on the choice of the instrument and on its implementation scheme, so as to avoid carbon leakage, evasion/fraud and cargo shifts to land-based modes that could produce more GHGs. Another effect of an MBM system is to raise money to purchase offsets for other sectors, i.e. invest in wind farms, photovoltaic parks, or other technologies that would reduce GHG emissions elsewhere. The effect of a possible market-based measure would be reduced emissions of GHGs (directly) and the other air pollutants (indirectly through lower fuel consumption) at the expense of a moderate increase in transport costs.

IMO has also strengthened restrictions on NOx and SOx. NOx emissions from new diesel engines over a certain size have been limited since 1 January 2000. For SOx, a 4.5% maximum worldwide level of sulphur content in fuel oil burned by ships has been established by IMO. This organization has also set up SOx Emission Control Areas (SECAs) where more stringent specifications for fuel burned by ships apply. The Baltic and North Seas (including the English Channel) are currently designated as SECAs. In April 2008 the IMO agreed in principle to further reduce the sulphur content of fuel used both within SECAs and worldwide. In SECAs, maximum sulphur levels have been reduced to 1% from 1 January 2010 and would be further reduced to 0.1% from 1 January 2015. The global limits would be reduced to 3.5% from 1 January 2012, with a further reduction to 0.5% from 1 January 2020 or 2025 if sufficient fuel is not available (EU 2008a).

However, the suggested reduction of maximum sulphur content of fuel oil to 0.1% as from 1 January 2015 in SECAs receives a lot of criticism lately. A recent study commissioned by the Union of German shipowners and the Union of German Ports (Lemper et al. 2010) concludes that such a reduction can only be achieved by using diesel oil in place of heavy fuel oil. This, in turn will increase the operational cost of shipping within SECAs in relation to trucks and trains and may force some shipping lines and ports to exit the market.

The implementation of the said amendments to MARPOL Annex VI concerning SOx emissions from ships is expected to result in improvements in the relevant KPI, albeit at a significant cost increase. In the event that the 'back-shift' from sea to road is not avoided, all environmental sustainability indicators will see a deterioration, which will be extended to other sensitive factors like congestion, accidents and noise. The possibility of amending the new Marco Polo programme to include financial instruments aimed at avoidance of 'back-shift' from SSS to road transport is suggested as an option that should be assessed by the European Commission.

Security has become a high profile policy issue since 2001. Transport is both a target and an instrument of terrorism, a fact that creates obvious overlaps between transport policy and security policy (Petersen et al. 2009). However, the danger of over-regulating is present. An impact assessment study on the US suggested requirement for 100% scanning of US-bound containers (Policy Research Corporation 2009) concluded that such a measure will intensify the inherent congestion problem of several European ports and will raise direct costs for all containers due to reduced throughput capacity, while the security gains are questionable.

Enhancing international security will reduce significantly the relevant risks and improve, at the expense of transport time and cost (although it is possible to have lower insurance premiums). However, due care needs to be given to avoid over-regulation, which can create bottlenecks and have significant adverse effects on transport time and cost.

4 CONCLUSIONS

4.1 *EU transport policy*

All identified barriers to green corridor development have been adequately addressed by EU policies. Of particular importance are the administrative barriers addressed by the Freight Transport Logistics Action Plan. In general, the legal framework is pretty much in place. Special attention should be given to the enforcement of existing legislation

The corridor approach is an effective way to address the fragmented nature of European transport networks, especially in the rail sector.

The effectiveness of transport policy is enhanced by employing packages of complementary instruments. Very important is the role of technology (in particular commercially viable alternative fuels) for the long run, and of ICT applications for the immediate future. The significance of educating, informing and involving the greater public in transport policies is a precondition for their effectiveness.

Over-regulating is an issue that should not be overlooked, since improvements in one aspect might create problems in another. Three such cases were identified, all concerning maritime transport and non-EU institutions. A possibility worth assessing by the European Commission is the amendment of the new Marco Polo programme to include financial instruments aimed at avoiding 'back-shift' from more environmentally-friendly modes to road transport.

4.2 *Green corridors*

The concept, consistent with the corridor approach mentioned above, is by far more complicated than the recently introduced freight-oriented corridors, which can be viewed as a subset of green corridors. Nevertheless, valuable lessons can be drawn from Regulation No 913/2010, which introduced the freight-oriented corridors in relation to:

- separation of the criteria establishing a freight-oriented corridor from the indicators monitored after its establishment,

- the definition of a freight-oriented corridor,
- the detailed governance structure fostering international cooperation among a multiplicity of actors involved, and the introduction of a one-stop-shop for communication with third parties,
- the implementation measures foreseen, including a market study, an implementation plan, an investment plan, a deployment plan relating to the interoperable systems, a performance monitoring mechanism, and a user satisfaction survey, all updated periodically.

In relation to the criteria for labelling a particular corridor as "green", it is suggested that the Commission assesses the possibility of including as prerequisites:

- the fair and non-discriminatory access requirement of the Freight Transport Logistics Action Plan, and
- the internalisation of external costs, which for the time being remains voluntary.

In this way, green corridors in addition to being a field for experimenting with innovative transport technologies and advanced ICT applications, can become a field for experimenting with EU transport policies, too. This is in line with the core network concept proposed for the new TEN-T guidelines, which by placing emphasis on the European added value of the transport networks and their integration, in a way that combines efficiency targets with the sustainable development goals of the EU, basically extends the green corridor concept across all Europe.

Another conclusion concerns the role of intermodal terminals and freight villages in the development of green corridors. The shift of competition from among individual enterprises to among supply chains necessitates optimising performance at the chain level and this is impossible without nodes permitting the effective and efficient modal interconnection.

ACKNOWLEDGEMENTS

As already mentioned above, the work reported herein was supported in part from EU project SuperGreen (grant agreement TREN/FP7TR/233573/ "SUPERGREEN"). The assistance of Rein Jüriado and Fleur Breuillin, both Project Officers at the European Commission (DG-MOVE), for their technical and administrative support and for their advice in general is gratefully acknowledged. We are also thankful to (alphabetically) Sergio Barbarino, Niklas Bengtsson, Bianca Byring, Chara Georgopoulou, Even Ambros Holte, Indrek Ilves, Atle Minsaas, Christopher Pålsson, Konrad Pütz, Sanni Rönkkö, Ilkka Salanne, Anders Sjöbris, Andrea Schön, Panos Tsilingiris, Aud Marit Wahl, the members of the project's Advisory Committee and numerous other individuals, perhaps too many to mention by name, for their help.

REFERENCES

EU 2010. *Regulation (EU) No 913/2010 of the European Parliament and of the Council of 22 September 2010 concerning a European rail network for competitive freight*. Strasbourg, 22.9.2010.

EU 2009. *A sustainable future for transport: Towards an integrated, technology-led and user friendly system*. Communication from the Commission, COM(2009) 279, Brussels, 17.6.2009.

EU 2008a. *Greening Transport*. Communication from the Commission, COM(2008) 433, Brussels, 8.7.2008.

European Commission 2008b. *Strategy for the internalisation of external costs*. Communication from the Commission, COM(2008) 435, Brussels, 8.7.2008.

EU 2007. *Freight Transport Logistics Action Plan*, Communication from the Commission, COM(2007) 607, Brussels, 18.10.2007.

Ilves, I., Panagakos, G., Wahl, A.M., Georgopoulou, C., Rönkkö, S., Vanaale, E., Vio, F. 2010. *Benchmarking of green corridors*, SuperGreen Deliverable D2.4 (Version 1), 02-40-RD-2010-14-01-0, 15.12.2010.

IMO 2010. *Report of the Working Group on Energy Efficiency Measures for Ships*. MEPC 61/WP.10, London, 30.9.2010.

Kramer, H., Sedlacek, N., Jorna, R., van der Laak, R., Bozuwa, J., Gille, J., Ossevoort, R., Magliolo, M. 2009. *Report on overall benchmarking framework*, BE LOGIC Deliverable D2.1, 10.6.2009.

Lemper et al. 2010. *Die weitere Reduzierung des Schwefelgehalts in Schiffsbrennstoffen auf 0,1% in Nord- und Ostsee im Jahr 2015: Folgen für die Schifffahrt in diesem Fahrtgebiet*. ISL study 2411, Bremen, September 2010.

Pålsson, C., Bengtsson, N., Salanne, I., Pütz, K., Tsilingiris, P., Georgopoulou, C., Schön, A., Minsaas, A., Wahl, A.M., Holte, E.A., Panagakos, G. 2010. *Definition of benchmark indicators and methodology*, SuperGreen Deliverable D2.2, 02-22-RD-2010-16-01-6, 23.11.2010.

Panagakos, G., Rönkkö, S., Georgopoulou, C., Varvate, M., Kiliç, R., Chadjinikolaou, S., Psaraftis, H., Salanne, I., Ülkü, K. 2011. *Effects of changes in operational and regulatory environment*, SuperGreen Deliverable D2.3 (draft), 02-30-RD-2011-01-01-3, 22.2.2011.

Petersen M.S. et al. 2009. *Report on Transport Scenarios with a 20 and 40 year Horizon*, TRANSvisions Final report, Copenhagen, Denmark, 23.3.2009.

Policy Research Corporation 2009. *The impact of 100% scanning of U.S.-bound containers on maritime transport*. Antwerp, 24.4.2009.

Salanne, I., Rönkkö, S. and Byring B. 2010. *Selection of Corridors*, SuperGreen Deliverable D2.1, 02-21-RD-2010-03-01-5, 15.7.2010.

6.3 *Weather routing*

Sustainable Maritime Transportation and Exploitation of Sea Resources – Rizzuto & Guedes Soares (eds)
 ISBN 978-0-415-62081-9

Numerical weather and wave prediction models for weather routing, operation planning and ship design: The relevance of multimodal wave spectra

A. Orlandi
DINAEL, University of Genova, Italy & Consorzio LAMMA, Firenze, Italy

D. Bruzzone
DINAEL, University of Genova, Italy

ABSTRACT: From the developments in meteorological and oceanographic sciences, a relatively new field of application for seakeeping computational techniques emerged in the context of operational services for ship operations at sea. Ship responses are traditionally computed by employing parametric waves spectra, that represent only approximations of real seaways. Frequently, directional spectra of real seaways are composed of more than one peak at different directions. Wave spectra in output from state of the art third generation wave prediction models are reliable and well represent such multimodality of real seaways, if fed with reliable wind data. This work is aimed to study and quantify the relevance of wave spectra multimodality, both in frequency and direction, on the computation of seakeeping responses. The study is performed by using numerical wave spectra, discretized on direction frequency grids, so as they come out from wave forecasting models. This approach is considered by the authors as applicable for weather routing and operation planning services.

1 INTRODUCTION

Modern techniques for numerical weather (Kalnay 2004) and wave (Komen et al. 1994) prediction, opportunely coupled with numerical models for the computation of ship motions, have a potential for the improvement of operational services for ship operations at sea. Such a potential has not still been completely exploited, despite the development of computational techniques for weather routing and operation planning commenced years ago (Haltiner et al. 1962, Faulkner 1963, Zoppoli 1972). The high resolution and reliability of modern numerical weather prediction systems and of third generation spectral wave models allow today a very detailed description of wind and wave conditions at sea. Such a description is usually available some days (three or four) in advance, thanks to the high forecasting skills of global and regional prediction systems, capable of ingesting a huge amount of observational initialization data, from many sources such as satellites, ground stations, radiosondes, and so on, by the use of data assimilation techniques. The implementation of ensemble forecasting techniques allows a growing improvement of forecasting skills also for longer leads (five or seven days), for which probabilistic forecasts are more advisable (Hoffschildt et al. 1999, Saetra & Bidlot 2004, Saetra 2004).

Third generation spectral wave models allow the forecasting of wave conditions by the computation of the time evolution of the complete directional spectrum on the whole set of grid points of the computational domain. As a result a huge amount of predicted spectral data is potentially available every day at operational meteo-marine forecasting centers. Nevertheless, traditional weather routing systems are generally based on a very reduced amount of the information present in such data. Frequently, only spectrally averaged quantities such as significant wave height, mean wave direction and mean or peak period are used (Chen et al. 1998, Standing 2005, Padhy et al. 2008, Delitala et al. 2010). This was originally justified by the need of reducing the amount of data to be stored and eventually transmitted and also by the fact that spectral information generated by wave models of first generations was not as reliable (Peterson & Bales 1980) as that from more recent wave prediction systems, due to both a pourer physical description and computational demand limitations. Modern information and telecommunication technology and its growth trends allow much confidence in the capabilities of exploiting more and more huge amount of data, also in an operational context at sea (Lee et al. 2002, Geiszler et al. 2003).

Waves in real seaways are always multi frequency and multi directional phenomena and as such they must be described. The need and relevance of such a complete approach for ship seakeeping has been long studied e.g. by Guedes Soares (Rodriguez & Guedes Soares 1999, Boukhanovsky & Guedes Soares 2009) and also in connection with wave prediction models (Claessens et al. 2005, Rusu et al. 2008) and wave measurement from ship motions in the context of operator guidance systems (Iseki & Ohtsu 2000, Nielsen 2006, Pascoal & Guedes Soares 2009).

This paper reports the first results of a study performed, following the above mentioned ideas, as a collaboration between Consorzio LAMMA (in charge of the Meteorological Service for Tuscany Region) and the Department of Naval and Electric Engineering (DINAEL) of the University of Genoa. The main goal of such a study is the design and preliminary implementation of a seakeeping forecasting system. It is planned to be made by the integration of seakeeping computational algorithms with meteo-marine forecasting models so as to generate a comprehensive and reliable system to be employed for operation planning at sea and in the context of innovative weather routing systems.

The first part of the study has been focused on ocean waves effects on ship motions, that are the main forcing for ships in many situations at sea. Also wind and currents will be considered in the prosecution of the study. In this paper we report the results obtained, adopting a strip theory approximated approach to seakeeping. In particular we have modified the PDSTRIP program (Bertram et al. 2005) in order to render it able to compute ship motions in seaways described by spectral data computed by the third generation wave spectral model Wavewatch III (Tolman 2009). As a first case we have considered the S175 containership (ITTC 1987) and we have investigated several aspects emerging from seakeeping computations.

2 WEATHER AND WAVE PREDICTION

Numerical prediction of meteo-marine conditions is performed at operational centers by the use of a chain of models (e.g. see Orlandi et al. 2009, Ortolani et al. 2007). Letting aside for the moment marine currents, the models chain is composed of a meteorological model and a wave model. They are run in cascade mode, i.e. the meteorological model is run first and its output furnish the atmospheric forcing (namely the wind at a height 10 meters) for the subsequent run of the wave model.

2.1 *Numerical weather prediction*

Atmospheric models, also told Numerical Weather Prediction (NWP) models, solve an initial values fluid dynamical problem comprising a set of numerical equations derived from the thermodynamics and fluid dynamics equations. They are discretized on a grid of points covering the whole computational domain, and are coupled to several parameterization schemas, for the many sub-grid physical phenomena (e.g. interactions with and amongst radiation, water in all its phases, soil, vegetation) that are not explicitly described by the main set of equations.

Global atmospheric models are run daily (four runs each day, see ECMWF and NCEP) and each run is initialized by the complex process of data assimilation, that optimally blends model data of the precedent run with the huge amount of data coming from the world observing system, composed of satellites, meteorological stations, soundings and so on.

Higher resolution, local area atmospheric models are run in cascade mode, using global models data as initial and boundary conditions (Pielke 2002).

Global models forecasts span usually ten days, with the best skills over the first four or six days. High resolution local area models forecasts span usually three or five days, covering restricted areas over the globe. As an example we cite the regional forecasts produced daily at Consorzio LAMMA (see LAMMA), by the local area mesoscale model WRF (Michalakes 2004, WRF).

2.2 *Spectral wave modeling*

If wind data forecasted by the NWP model are reliable, modern third generation spectral wave models are able to predict waves with very high forecasting skills (Komen et al. 1984, Wise group, 2007). Each run is initialized by using restart data produced by the preceding wave model run. This is possible due to the fact that wave dynamics are not affected by the strong nonlinear instabilities affecting atmospheric dynamics and are strongly driven by wind data.

As an example of wave forecasts we cite the ones issued daily by Consorzio LAMMA (see LAMMA), based on the Wavewatch III third generation spectral wave model, developed at NOAA centers (Tolman 2009).

Wave spectral modeling treats sea waves as a stochastic process, due to the impossibility of predicting the phase relationships among the various wave components (Komen et al. 1994 and Massel 1996). What is predictable is the directional wave spectrum, i.e. (Wiener-Khinchine theorem) the Fourier transform of the autocorrelation function of the wave heights (Massel 1996, Price & Bishop 1974). The time scale of the evolution of the directional wave spectrum is much slower than the time scale of the underling stochastic process. Hence it is possible to adopt a sort of adiabatic approximation in which the waves are approximated as a quasi stationary stochastic process, "locally" stationary, but affected by a relatively "slow" temporal drift, caused by: wind, nonlinear interactions among wave components, and dissipation mechanisms.

In this framework the prediction of the space time evolution of the directional wave spectrum is performed by the balance equation (Komen et al. 1994, Wise Group 2007, Massel 1996):

$$\frac{\partial S_{\varsigma}}{\partial t} + \underline{\nabla} \cdot \left(\underline{c}_g S_{\varsigma} \right) = S_{in} + S_{nl} + S_{ds} \tag{1}$$

where $S_{\varsigma} = S_{\varsigma}(\omega,\theta)$ is the directional wave spectrum, ω is wave angular frequency, θ is wave direction, $\underline{c}_g$ is the wave group speed, while the source terms $S_{in} = S_{in}(\omega,\theta)$, $S_{nl} = S_{nl}(\omega,\theta)$, $S_{ds} = S_{ds}(\omega,\theta)$ represent the input from the wind, the non linear interactions among wave components, dissipation mechanisms (mainly wave breaking on the open sea, also interactions with the bottom at coast). Equation 1 represents a whole set of equations, one equation for each wave component (ω,θ), coupled via the non linear term S_{nl}, whose numerical solution in wave prediction models is accomplished by a discretization both in space-time and in frequency-direction. Third generation spectral wave models have been developed by implementing efficient algorithm to solve the integrals involved in the computation of the non linear interactions term S_{nl}, which is the key to correctly reproduce the spectral evolution (Massel 1996).

Under the variable evolution of the wind, third generation spectral wave models generate, on each point of the space computational grid, a time series (one for each time-step) of complex spectral structures, composed of several peaks. Usually one peak is due to the local wind and the others represent the swell components coming, along various directions of propagation, from distant generation areas. When only one wind direction is present for long times and wide fetches are involved, wave models generate spectra that tend to the known saturations spectrum, i.e. the Pierson-Moskowitz one (Komen et al. 1996).

In Figure 1, left side, we report the polar plot of a directional wave spectrum computed, by the wave model in operational use at Consorzio LAMMA.

The meteorological convention for the waves directions is used, i.e. we represent the direction *from* which waves come. The radial coordinate of the plot is referred to the wave period in seconds (upper left scale) but we report also the

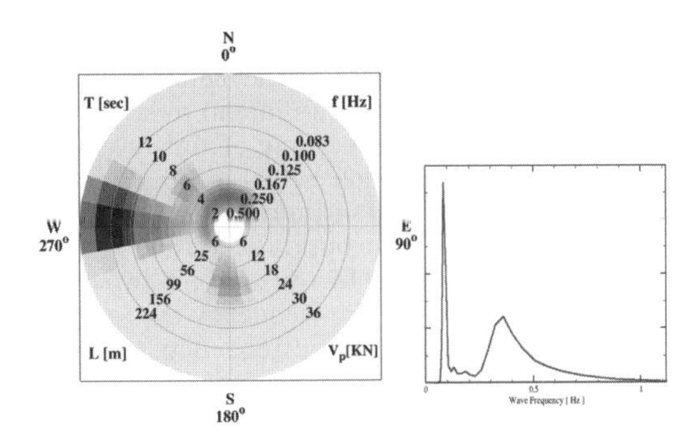

Figure 1. Left: directional wave spectrum computed, by the third generation spectral wave model Wavewatch III in operational use at Consorzio LAMMA, for 18/02/2011 14:00, for a point at North of Capo Peloro near Messina, Sicily. Right: corresponding frequency spectrum obtained by direction integration.

corresponding wave frequency values in Hertz (upper right scale) and, using the dispersion relation for sea waves, we report also the corresponding wavelengths in meters (lower left scale) and the corresponding phase speed of the waves in Knots (lower right scale). For completeness, in right side of Figure 1, the corresponding frequency spectrum (i.e. obtained by the integrating over the directions) is reported. By comparison of the two diagrams, we notice that (due to the radially growing geometrical dilation) the polar plot tends to overemphasize wave components with greater values of the period (i.e. those nearer to the external rim of the diagram). Almost four peaks are distinctly recognizable: a long swell from W (the component with the highest peak period around 12 s), two decaying very weak swell components, one from S and another from NW (peak periods around 6 sec), a wind component from NNW (the second in peak height, with peak period around 3 s). Swell components are always characterized by longer periods and a narrow angular spreading, while wind components have lower periods and are much more directionally spread (as in Figure 1).

In our seakeeping analyses we will adopt the following factorization of the directional spectrum:

$$S_{\varsigma}(\omega,\theta) = H_{1/3}^2 \, \Sigma_{\varsigma}(\omega,\theta) \tag{2}$$

The first factor is the significant wave height squared, and is proportional to the total energy density (Massel 1996). It is well known that significant wave height squared is proportional to the zero-order moment of the wave spectrum (Massel 1996, Price & Bishop 1974). The second factor, that we call *unitary wave directional spectrum*, contains the information on how the energy is distributed in frequency and direction. The unitary wave directional spectrum Σ_ς is a spectrum normalized to a significant wave height of one meter. This kind of factorization is the one we intend to substitute to the usual synthesis consisting of only averaged quantities: significant wave height, mean direction and period, that drop all the information about the frequency direction distribution of wave energy. It is more detailed than the factorization in two separate functions, one for the frequency dependence (usually a standard parametric frequency spectrum) and the other for direction (spreading function). In this paper we intend to investigate the relevance of a detailed description of the frequency direction dependence of multimodal spectra (i.e. with several contemporary peaks) for a correct numerical prediction of ship motions.

Spectral partitioning algorithms (Boukhanovsky & Guedes Soares 2009, Tracy et al. 2007) are another way of approaching this latter point. The information contained in the wave directional spectrum can be compressed by applying an algorithm that recognizes each spectral peak and associates to it a direction, a peak frequency and a partial significant wave height. Then the complete spectrum could be regained by combining parametric spectra, one for each one of the spectral peaks. It must be considered that such operations have a computational cost and are prone to computational errors, that could add numerical noise to the original data. In an operational context, in which computation time is always a relevant constraint, could be better to archive the whole directional spectrum, also if it is more memory demanding.

3 SEAKEEPING FORECASTING

3.1 *Ship motions in multimodal seaways*

The prediction of ship motions in realistic seaways can be performed in a straightforward way approximating the ship as a linear system and applying the well established formalism for the analysis of linear stochastic systems (St. Denis & Pierson 1953, Ochi & Bolton 1973, Price & Bishop 1974). Such an approach is also completely compliant with the framework on which spectral wave modeling has been developed (Massel 1996, Lewis 1990). The linear approximation, both for waves and for ship dynamics is in several cases a crude approximation, but there is also a wide variety of cases in which the results obtained are sufficiently reliable. We limit our analysis to the latter cases, as is common use in seakeeping applications, and demand to specialized literature for a through analysis of non linear effects (e.g. see Newman 1978, 1986, Lewis 1990 for an introductive discussion and literature cited therein). Within this framework, many seakeeping characteristics of ships may be statistically described and predicted by performing an analysis of spectral moments at various orders (Ochi & Bolton 1973, Price & Bishop 1974). In particular we will focus on the computation of significant values of ship motions, that can be computed from zero order spectral moments, analogously to the significant wave height. The response spectra for ship motions are given by the wave spectrum multiplied by the square modulus of the frequency response function for each of the six degrees of freedom of the ship (Ochi Bolton 1973, Lewis 1990). In order to avoid the many complications due to the multivalued nature of the encounter frequency relation, we will compute such integrals in the domain of wave frequencies (e.g. see Lewis 1990):

$$m_{0j} = \int_0^\infty \int_0^{2\pi} Y_j^2\left(\omega_e(\omega,\theta_r),\theta_r,U_s\right) S_\varsigma(\omega,\theta_r)\, d\theta_r d\omega \quad (3)$$

Where: j = 1,2, ..., 6, U_s is the ship speed of advance, Y_j is the response amplitude operator (i.e. the real modulus of the complex valued frequency response function) of the *j*-th ship motion degree of freedom, and the encounter frequency ω_e is related to the wave frequency ω and to the relative (to the ship bow) direction θ_r by the relation:

$$\omega_e = \omega_e\left(\omega,\theta_r,U_s\right) = \omega\left[1 + \omega\frac{U_s}{g}\cos(\theta_r)\right] \quad (4)$$

Notice that, unlike the common convention in ship hydrodynamics, θ_r direction is that of wave provenience. The integrand of Equation 3, as a function of ω and θ_r is not a proper response spectrum, because it is not written in terms of the encounter frequency. Nevertheless it can give important clues on the contributions of the various wave components to the ship responses and we will refer to it as *j–n*th *pseudo-response spectrum*.

In Figure 2 we show a polar plot of the values of the ω_e/ω, for U_s = 25 Knots (as marked by the large dot dashed circle), as a function of ω and θ_r. We can see that the values inside the little circle centered on the stern side are negative, meaning that for frequencies and directions in that area the waves are overtaken by the ship, while outside quartering and following waves overtake the ship. The little circle marks the locus of the zeroes of the encounter frequency. Encounter and wave frequencies are equal ($\omega_e/\omega = 1$) only for "beam sea" wave components, while $|\omega_e/\omega| > 1$ for waves coming from head and bow quarters and for low period (i.e. high frequency) stern and quartering waves.

In Figure 3 we plot the unitary wave directional spectrum of Figure 1, but with the relative direc-

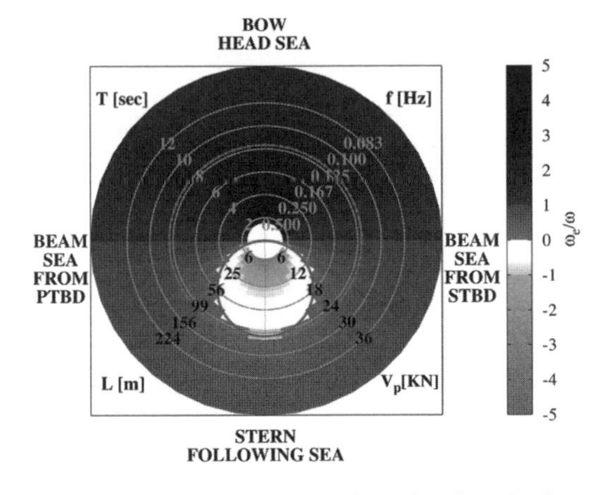

Figure 2. Ploar plot of the ω_e/ω ratio, discretized on the frequency direction grid used by the wave model, for U_s = 25 Knots (as marked by the large dot dashed circle, that serves as a guide to the correspondence with the wave phase speed, length, frequency, period). The little circle on the stern side marks the zeroes of ω_e

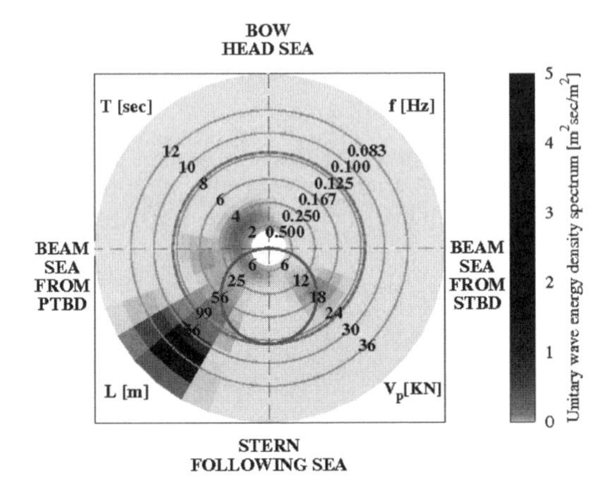

Figure 3. Unitary wave directional spectrum of Figure 1 referred to the bow of a ship heading 45° NE. Also the speed circle and the $\omega_e = 0$ circle of Figure 2 are reported, for a ship speed of 25 Knots.

tion coordinate θ_r referred to the bow of a ship heading 45 ° NE. Note that we have rotated the wave spectrum according to the ship heading, but we have plotted it versus wave frequency, not encounter frequency. We have also plotted the large dot dashed and the little circles of Figure 2 for a ship speed of 25 Knots.

Adopting the factorization defined in Equation 2, the significant values of ship responses may be written as:

$$m_{0j} = H_{1/3}^2 \, M_{0j} \tag{5}$$

where we have introduced the *unitary significant response* M_{0j} induced by the unitary wave directional spectrum Σ_ζ defined in Equation 2. In our analysis we will consider these significant responses to unitary wave directional spectra in order to focus the attention on the frequency directional structure of the analyzed seaways, putting all of them on the same energetic baseline. The values of the significant responses really corresponding to each spectrum can be regained simply by multiplication for the respective significant wave height squared as in Equation 5. In further developments of this study the analysis will be extended also to also higher order moments, useful in computing the statistics of maxima and other quantities needed to quantify derived responses like accelerations, structural loads, and also other seakeeping indices specific for features connected with ship motions like added resistance, motion sickness, slamming and deck wetness.

3.1.1 *Computation of ship responses*

In order to compute spectral moments of ship responses, Response Amplitude Operators (RAO's) must be known as function of ω and θ_r for various ship speeds. The determination of such functions is performed usually either by hydrodynamic computations or by laboratory experiments in towing tanks. In the work here described a purely theoretical approach has been adopted and response amplitude operators have been computed by adopting a strip theory approximation (Salvesen et al. 1970, Söding 1969), which is a standard in many seakeeping studies, at least at preliminary stages of ship design. Being our ultimate goal the implementation of an operational system for seakeeping forecasting in conjunction with meteomarine models, we considered the strip theory as a good compromise between accuracy and reliability on one side, and ease of implementation and computational speed on the other. In our work, we adopted the PDSTRIP program, which is a quite complete strip theory seakeeping program, which allows also for corrections for someone of the main limitations of strip theory. It also allows the computation of drift forces, along both transverse and longitudinal directions. Longitudinal drift force corresponds to the wave added resistance, which is very important in the powering computations needed also in weather routing algorithms. The algorithm implemented in PDSTRIP is based on the Boese approach (Boese 1970).

We use PDSTRIP to compute response amplitude operators for the values of ω and θ_r used in the spectral grid of LAMMA wave model. Moreover we modified routines originally aimed at computing significant responses with standard parametric wave spectra in order to make it able to compute them by using numerical directional wave spectra generated by the wave model.

We also modified the treatment of the contributions of wave components near the zeroes of encounter frequency (surfriding conditions). In PDSTRIP the sum of such contributions in the integrals for significant values is simply skipped, in order to avoid inconsistent results due to limitations of strip theory and consequent RAO's divergences. This problem regards mainly the modes without restoring forces (surge, sway, yaw). Well aware of the potential unreliability of strip theory for very low values of encounter frequency, we eliminated the divergences of RAO's, by applying empirical cut-off values similarly to what is implemented in other approximations (Meyers et al., 1981, Meyers et al. 1985) and eliminated the skipping away of their contributions. With this approach, potentially unreliable contributions could be generated and summed, for higher Froude number values, from spectral regions near the zeroes of encounter frequency (the small circle in Figures 2 and 3). In this initial phase of the study we are more interested in pointing out the roles of the various spectral regions, so we assume that the RAO's so obtained can be an acceptable approximation to highlight the differences resulting from the different spectral model.

4 NUMERICAL RESULTS

In our study we begun to make numerical tests on the S175 containership, because it is a standard benchmark vessel for seakeeping computations.

We used the modified PDSTRIP program to compute the S175 RAO's for values of the ship speed ranging from 0 to 25 KT, with steps of 1 KT (e.g. see Fig. 4).

We used the archived RAO's data to compute and analyze pseudo-response spectra, significant values of the responses along all the six degrees of freedom, transverse and longitudinal (added resistance) drift forces of the S175, for all the above values of the speed of advance. We performed such calculations for various wave spectra generated by LAMMA operational wave model. For comparison we computed the same quantities also for a the parametric spectrum implemented in PDSTRIP, in particular by setting the parameters for the PDSTRIP modified Pierson-Moskowitz spectrum (a two parameters Bretschneider-like form depending on $H_{1/3}$ and mean period, see Söding & Bertram 1998), with $\cos^2$ spreading function, below indicated with PMC. For consistency checks, such computations have been performed both, via the original PDSTRIP algorithm, that skips the zeroes of encounter frequency, and via our renormalized

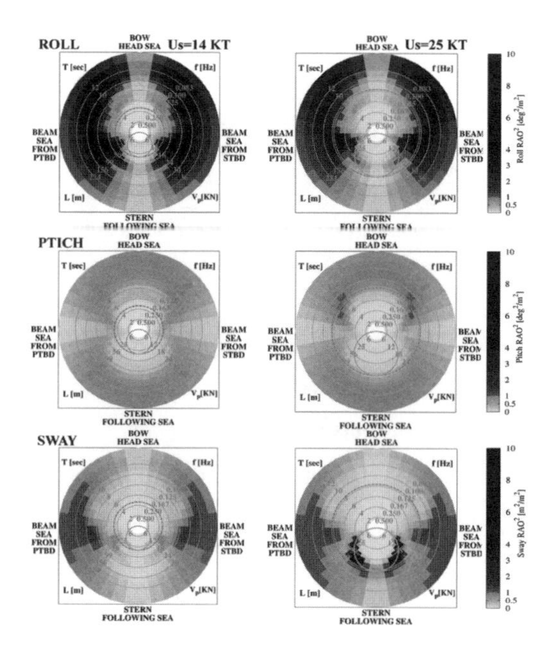

Figure 4. Polar plots of S175 squared RAO's for Roll, Pitch, Sway for two ship speeds: left 14 Knots, right 25 Knots. Also speed and $\omega_e = 0$ circles are drawn.

algorithm, performing a discretization of the PMC spectrum on the frequency directional grid of the wave model. This grid comprises 30 frequencies, from 0.04 to 0.6 Hz (1,6–24 s of wave period), and 36 directions with 10° constant spacing, from 0° to 360°.

The differences in results obtained from this two ways of accounting the parametric spectrum are practically null for heave, pitch and roll, while for surge, sway, and yaw, they give a measure of the contributions due to the renormalized peaks in correspondence of the zeroes of encounter frequency. These differences are negligible for S175 speeds below 20 KT, while are still small but not negligible for speeds greater than that (and obviously only with quartering or stern waves).

For each multimodal spectrum generated by the wave model (from now indicated with numeric spectrum or NMS) we defined a PMC equivalent spectrum, by equating the tree parameters: significant wave height, mean direction and mean period.

This is consistent with the operational practice in common use in routing systems.

We performed responses computations for both by using unitary wave directional spectra defined by Equation 2 and labeling them with PMC1 and NMS1. Comparing the results obtained with a NMS1 and with the equivalent PMC1 we evidenced many relevant differences.

In particular in this paper we report results obtained by performing computations with the spectrum NMS1 of Figure 1 and with its equivalent PMC1 of Figure 5, characterized by: average wave direction (of provenience, rel. to N clockwise) about 314°, average wave period about 4.3 s. In Figure 6 we report polar plots of some of S175 pseudospectra for roll, pitch, sway, computed for ship speed of 25 Knots, in the case of 45° NE heading, with PMC1 (left) and NMS1 (right). They clearly show how different spectral components interact with different portions of

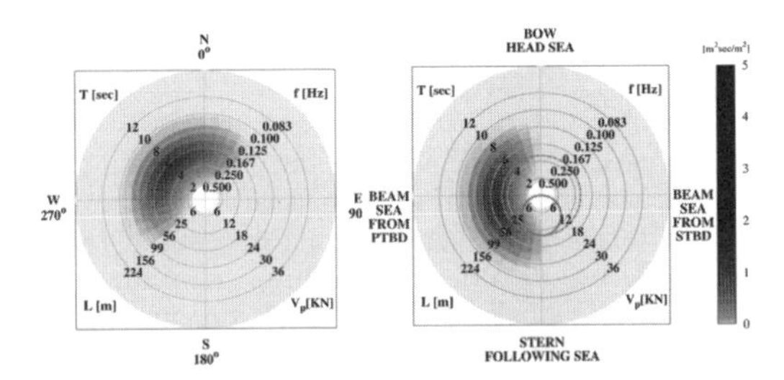

Figure 5. Left: PMC1 modified Pierson-Moskowitz spectrum of PDSTRIP with cos^2 directional spreading. Right: PMC1 with respect to the ship bow, with ship heading 45° NE. Also speed and $\omega_e = 0$ circles like Figure 3 are plotted, but for a speed of 14 Knots.

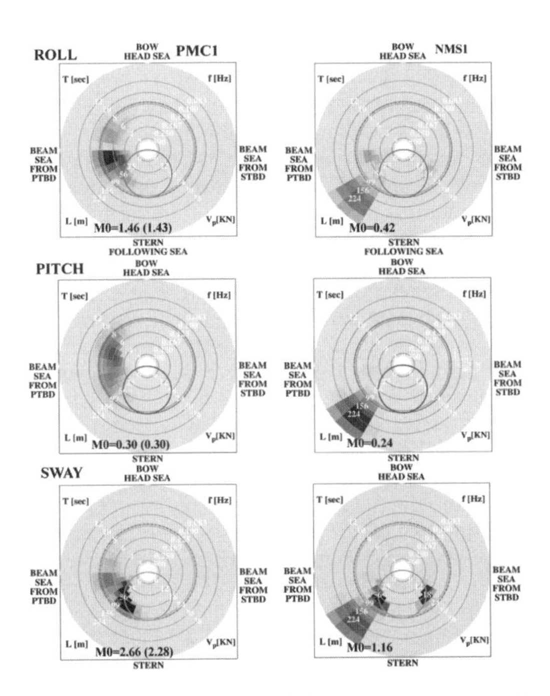

Figure 6. Polar plots of S175 pseudospectra for Roll, Pitch, Sway, for Us = 25 Knots. Computed with 45° NE heading, and PMC1 left, and NMS1 right.

RAOs to give the integral value for (unitary) significant responses M_{0j}, reported in the lower part of each polar plot. In parenthesis near PMC1 values, we report the values computed by the original PDSTRIP (skipping surfriding peaks). It is apparent that only for sway surfriding peaks contribute in this case (with both PMC1 and NMS1), as can be also seen in sway squared RAO for 25 Knots in Figure 4.

To better appreciate the differences in the results obtained from several computations (executed for 36 values of ship heading and speed from 0 to 25 Knots) we built polar plots in which the radial coordinate is the ship speed of advance, the angular coordinate is the ship heading relative to North, and the variables plotted are the significant values of the six main ship responses and of the drift forces.

In Figure 7 the significant responses obtained with PMC1 are reported, while in Figure 8 those with NMS1. The former show a clear symmetry around the mean wave direction, due to the symmetry of the spreading function used for PMC1 and of the response operators. For roll, two symmetric peaks emerge for quartering seas. For surge, sway, yaw modes, surfriding peaks originate a rapid (and probably unreliable) growth of the responses at higher values of ship speed.

The responses computed with NMS1 do not show a defined symmetry and appear globally less intense, mainly for roll. Most of the responses

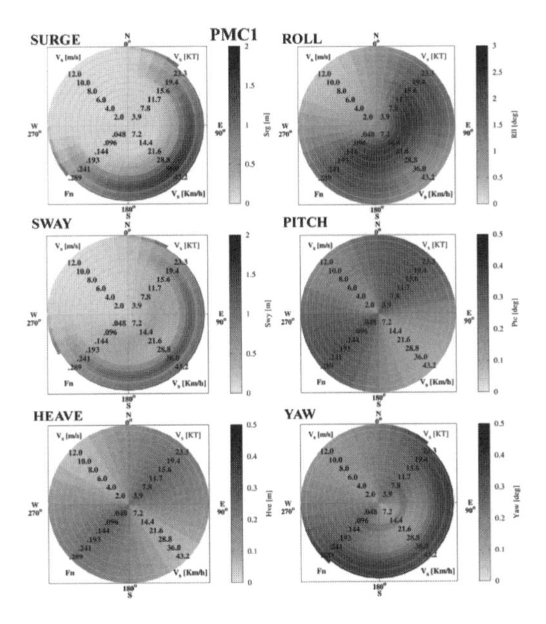

Figure 7. S175 significant responses computed for various ship headings and ship speed from 0 to 25 Knots, with PMC1.

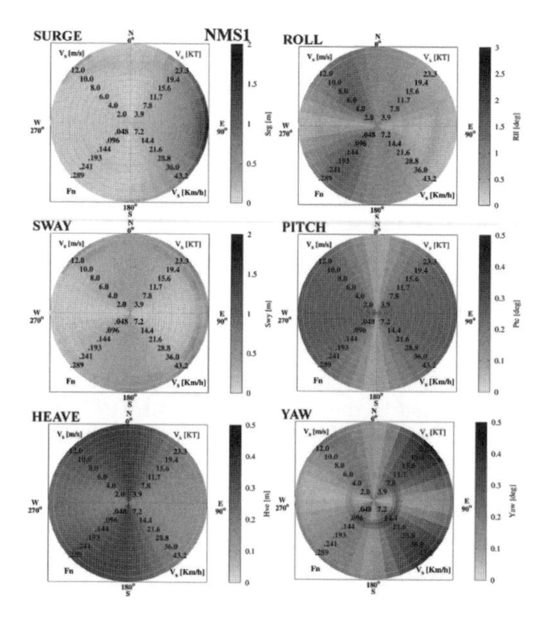

Figure 8. S175 significant responses computed for various ship headings and ship speed from 0 to 25 Knots, with NMS1.

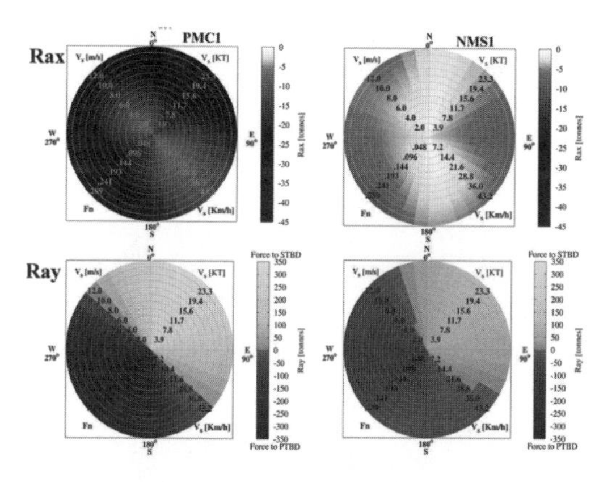

Figure 9. Top: S175 longitudinal drift force Rax (wave added resistance) and transverse drift force Ray, computed with PMC1 (left) and NMS1 (right) as for Figures 7 and 8.

seem to be due to the long period swell component from W (see also Figure 8), but several features can also be traced to the other minor peaks. The more relevant roll peaks with NMS1 are mainly due to bow quartering waves, while they are due to stern quartering waves with PMC1.

The use of the mean period for NMS-PMC equivalence generate a PMC1 peak of too high frequency with respect to the structure of NMS1. Better strategies should be used for adopting equivalent parametric spectra (Boukhanowsky & Guedes Soares 2009).

In Figure 9 we report the results obtained for longitudinal *Rax* (added resistance) and transverse *Ray* drift forces, with PMC1 and NMS1. The symmetry of the drift forces around the mean wave direction for PMC1 is evident both for *Rax* and for *Ray*, and it absent for NMS1, in which case the values are significantly lower, mainly for *Rax*. Another relevant feature is the existence of several minima of *Rax*, corresponding to directions of low added resistance, or of low fuel consumption, very useful in many applications in the fields of weather routing and operations planning.

5 CONCLUSIONS

The work described has been developed as a first step of an investigation aimed at developing a system for detailed seakeeping forecasting, to be obtained by optimally coupling meteo-marine forecasting models with ship motion simulation algorithms. In this first study, by applying strip theory to the S175 containership, we have evidenced that relevant differences emerge when computing ship responses in short crested seas described in two different ways: *i*) a realistic multimodal directional spectrum, with several peaks of wind sea and swell, generated by a third generation wave model; *ii*) a Bretschneider-like parametric spectrum, with $\cos^2$ spreading function, with the same values of three spectrum averaged quantities: significant wave height, mean direction and mean period. The results shown, and others analogous obtained with other multimodal spectra, suggest that, instead of parametric spectra coming from spectrum averaged quantities, a more complete description of directional spectrum in each seaway should be used to correctly describe the angular and speed behaviour of the responses and drift forces of a ship. To accomplish such a more detailed treatment of ship responses in real seaways, an improvement of the treatment of RAO behaviour near the zeroes of encounter frequency is also needed. This point has been investigated in an heuristic way, by introducing a novel diagram for evidencing the zeroes of encounter frequency.

A more detailed treatment of multimodality of real seaways calls for new approaches to be applied to the statistical characterization of the wave climate in a given region of sea. This need emerges also in other fields of marine forecasting and coastal engineering (Bradbury et al. 2007, Orlandi et al. 2008). In particular a frequency directional wave climate should be developed, with approaches similar to that described in Loupatokin et al. (2002), and Stefanakos (2008).

Such an approach will be developed also in our study. We have commenced to develop a spectral climatologic analysis technique, to be applied to time series of directional spectra generated by a wave model. The analysis is based on two steps. A first partition of all spectra, in a given homogeneous area, is based on the sea severity, i.e. on intervals of values of the significant wave height. Then in each of the above defined sea severity classes, unitary wave spectra are computed via Equation 2 and then a spectral classification algorithms (e.g. cluster analysis, principal component analysis) is applied. The output of this analysis will be a definition of the relevant frequency-directional spectral structures (a generalization of the spectral signatures in remote sensing), in each sea severity class, also partitioned for seasonal period.

Such an approach for the characterization of wave climate will be useful also in the ship design process. Many computational resources are devoted to the optimization of ship hydrodynamic properties, based on wave climate data. But if in the region of operation of the ship a relevant fraction of the encountered seaways are of multimodal kind, much of the optimization work with too simple parametric spectra may be questionable. A spectral climate like the above described, associated with scenario simulations like those described in Lewis (1990) and in Claessens et al. (2005), will be the correct answer, also suggested by the results presented in this paper.

REFERENCES

Bertram V., Veelo B., Söding H., Graf K., 2005. Developement of a freely available strip method for Seakeeping.

Boese, P., 1970. Eine einfache methode zur berechnung der widerstandsderhohung eines schiffes im seegang, *JournalSchiffstechnik-ShipTechnologyResearch*, 17 (86).

Boukhanovsky, A.V., Guedes Soares, C., 2009. Modelling of multipeaked directional wave spectra, *Appl. Oc. Res.*, 31, 132–141.

Bradbury, A.P., Mason, T.E., and Poate, T., 2007. Implications of the spectral shape of wave conditions for engineering design and coastal hazard assessment—evidence from the English channel, *10th International workshop on wave hindcasting and forecasting and coastal hazard symposium,* North Shore, Oahu, Hawaii.

Chen, H., Cardone, V., Lacey, P., 1998, Use of Operation Support Information Tecnology to Increase Ship Safety and Efficiency, *SNAME Transacfions*, 106, 105–127.

Claessens, E.J., Pinkster, H.J.M., Dallinga, R.P., 2005. On the use of a wond-wave model in the prediction of operational performance of marine structures, *Int. Shipbuild. Progr.*, 52 (4), 297–323.

Delitala, A.M.S., Gallino, S., Villa, L., Lagouvardos, K., Drago, A., 2010. Weather routing in long-distance mediterranean routes, *Theor. Appl. Climatol.*, 102, 125–137.

ECMWF, web page: http://www.ecmwf.int.

Faulkner, F.D., 1963. Numerical methods for determining optimum ship routes, *Navigation*, 10, 351–367.

Geiszler, D.A., Kent, J., Strahl, J.L.S., Cook, J., Love, G., Phegley, L., Schmidt, J., Zhao, Q., Franco, F., Frost, L., Frost, M., Grant, D., Lowder, S., Martinez, D., McDermid, L.M., 2003. The navy on-scene weather prediction system, COAMPS-OS™, *American Meteorological Society Conference*.

Haltiner, G.J., Hamilton, H.D., 'Arnason, G., 1962. Minimal-time ship routing, *J. Appl. Meteor.*, 1 (1), 1–7.

Hoffschildt, M., Bidlot, J-R., Hansen, B., Janssen, P.A.E.M., 1999. Potential benefit of ensemble forecast for ship routing, *ECMWF Technical Memorandum*, 287, Reading, UK.

Iseki, T., Ohtsu, K., 2000. Bayesian estimation of directional wave spectra based on ship motions, *Control Eng. Practice*, 8, 215–219.

ITTC, 1987. S-175 comparative model experiments—Report of the Seakeeping Committee, Proceedings, 18th ITTC, International Towing Tank Conference, The Society of Naval Architects of Japan, October, Kobe, Japan, 415–427.

Kalnay, E, 2004. *Atmospheric modelling, data assimilation and predictability*, Cambridge University Press, Cambridge, United Kingdom.

Komen, G.J., Cavaleri, L., Donelan, M., Hasselman, K., Hasselman, S., Janssen, P.A.E.M., 1994. *Dynamics and modelling of ocean waves*, Cambridge University Press, Cambridge, United Kingdom.

LAMMA, meteo-marine models web pages:
http://www.lamma.rete.toscana.it/wrf-web/index.html
http://www.lamma.rete.toscana.it/ww3/ww3.html

Lee, H., Kong, G., Kim, S., Kim, C., Lee, J., 2002. Optimum Ship Routing and It's Implementation on the Web, In Chang W. (ed.) Advanced Internet Services and Applications, *Lecture Notes in Computer Science*, 2402, 125–136.

Lewis, E.V. (ed.) 1990. *Principles of naval architecture: volume III motions in waves and controllability*, SNAME Editions.

Lopatoukhin, L., Rozhkov, V., Boukhanovsky, A., Degtyarev, A., Sas'kov, K., Athanassoulis, G., Stefanakos, C., Krogstad, H., The spectral wave climate in the Barents sea. *Proc. OMAE 2002.* Oslo, Norway.

NCEP, global model web pages:
http://www.nco.ncep.noaa.gov/pmb/nwpara/analysis/.

Newman, J.N., 1978. The theory of ship motions. *Advances in Applied Mechanics* 18, 221–285.

Newman, J.N., 1986. *Marine Hydrodynamics*. The Massachusetts Institute of Technology (MIT) Press, Cambridge, MA.

Nielsen, U.D., 2006. Estimation of on-site directional wave spectra from measured ship responses, *Mar. Struct.*, 19, 33–69.

Massel, S.R., 1996. *Ocean surface waves: their physics and prediction,* World Scientific publishing, London, UK.

Meyers, W.G., Applebee, T.R. Baitis, A.E., (1981). *User's manual for the standard ship motion program SMP*, DTRNSWC Technical Report SPD-0936-01.

Meyers, W.G., Baitis, A.E., (1985). SMP84: *Improvements to the accuracy of the standard ship sotion program SMP81*, DTNSWC Technical Report SPD-0936-04.
Michalakes, J., J. Dudhia, D. Gill, T. Henderson, J. Klemp, W. Skamarock, and W. Wang, 2004. The Weather Reseach and Forecast Model: Software Architecture and Performance, in George Mozdzynski (ed.) *Proceedings of the 11th ECMWF Workshop on the Use of High Performance Computing In Meteorology*, 25–29 October 2004, Reading.U.K.
Ochi, M.K., Bolton, W.E., 1973. Statistics for Predictlon of Ship Performance in a Seaway, 1, *International Shipbuilding Progress*, February, April and September 1973.
Orlandi, A., Pasi, F., Onorato, L.F., Gallino, S., 2008. An observational and numerical case study of a flash sea storm over the Gulf of Genoa, Proceedings of 7th European Meteorlogical Society Annual Meeting and 8th European Conference on Applications of Meteorology 2007, Advances Science and Research, 2.
Orlandi, A. et al. 2009. Implementation of a meteomarine forecasting chain and comparison between modeled and observed data in the Ligurian and Tyrrhenian seas, *Volume sulle attività di ricerca scientifica e tecnologica del CNR nell'ambito del mare e delle sue risorse*, CNR, Italy, Rome.
Ortolani, A., A. Antonini, G. Giuliani, S. Melani, F. Meneguzzo, G. Messeri, A. Orlandi, and M. Pasqui, 2007. Implementing an operational chain: the Florence LaMMA laboratory, in V. Levizzani, P. Bauer, and F.J. Turk, (eds) *Measuring precipitation from space—EURAINSAT and the future. Advances in Global Change Research*, Vol. 28, Kluwer Acad. Publ., ISBN: 978-1-4020-5834-9.
Padhy, C.P., Sen, D., Bhaskaran, P.K., 2008. Application of wave model for weather routing of ships in the North Indian Ocean, *Natural Hazards*, 44 (3), 373–385.
Pascoal, R., Guedes Soares, C., 2009. Kalman filtering of vessel motions for ocean wave directional spectrum estimation, *Oc. Eng.*, 36, 477–488.
Peterson, R.S., Bales, S.L., 1980. Ship response sensitivity to forecast and measured wave spectra, *Internal Report, David W. Taylor Naval Ship Research and Development Center*, Bethesda, Maryland.
Pielke, R.A., 2002. *Mesoscale meteorological modelling*, Academic press, San Diego, California.
Price, W.G., Bishop, R.E.D., 1974. *Probabilistic theory of ship dynamics*, Chapman and Hall, London, UK.
Rodrıguez, G., Guedes Soares, C., 1999. A criterion for the automatic identification of multimodal sea wave spectra, *Appl. Oc. Res.*, 21, 329–333.
Rusu, E., Pilar, P., Guedes Soares, C., 2008. Evaluation of the wave conditions in Madeira Archipelago with spectral models, *Oc. Eng.*, 35, 1357–1371.
Saetra, Ø., 2004. Ensemble ship routing, *ECMWF technical memorandum*, 435, Reading, UK.
Saetra, Ø., Bidlot, J.R., 2004. Potential benefits of using probabilistic forecasts for waves and marine winds based on the ECMWF Ensemble Prediction System, *Weather and Forecasting*, 19 (4): 673–689.
Salvesen, N., Tuck, E.O., Faltinsen, O., 1970. Ship motions and sea loads, *Transactions SNAME*, New York, 78, 250–287.
Söding, H., 1969. Eine modifikation der streifenmethode, Schffstechnik, 16, 15–18.
Söding, H., Bertram, V., 1998. Shiffe im seegang, Handbuch der werfen XXIV, Hansa-Verlagh, 151–189.
Standing, R.G., 2005. Review of the role of response forecasting in decision making for weather-sensitive offshore operations, *HSE Research Report*, 347, Prepared by BMT Fluid Mechanics Limited for the Health and Safety Executive.
St Denis, M. and W.J. Pierson 1953. On the motion of ships in confused seas, SNAME Transactions 61, 280–332.
Stefanakos, C.N., 2008. Investigation of the long-term wind and wave spectral climate of the Mediterranean Sea, *Proceedings of the Institution of Mechanical Engineers, Part M: Journal of Engineering for the Maritime Environment*, 222 (1), 27–39.
Tolman, H.L., 2009. User manual and system documentation of WAVEWATCH III version 3.14. *NOAA NWS NCEP MMAB Technical Note* 276. See also: polar.ncep.noaa.gov/waves/wavewatch/wavewatch.shtml.
Tracy, B., E.-M. Devaliere, T. Nicolini, H.L. Tolman and J.L. Hanson, 2007. Wind sea and swell delineation for numerical wave modelling, *in10th international workshop on wave hindcasting and forecasting & coastal hazard symposium*. Paper P12.
Wise Group, Cavaleri, L. et al., 2007. Wave modelling The state of the art, Progress in Oceanography, 75 (4), 603–674.
WRF, model web pages: http://www.wrf-model.org.
Zoppoli R., 1972. Minimum-time routing as an N-stage decision process, *J. Appl. Meteor.*, 11, 429–435.

Sustainable Maritime Transportation and Exploitation of Sea Resources – Rizzuto & Guedes Soares (eds)
 ISBN 978-0-415-62081-9

Optimization of routing with uncertainties

G.I. Papatzanakis, A.D. Papanikolaou & S. Liu

Ship Design Laboratory, National Technical University of Athens, Athens, Greece

ABSTRACT: A route optimization methodology in the frame of an onboard Decision Support/Guidance to the Master System has been developed and is presented in this paper. The method aims at the minimization of the fuel voyage cost and the risks related to the ship's seakeeping performance within acceptable limits of voyage duration. Parts of this methodology were implemented by interfacing alternative probability assessment methods, like Monte Carlo, FORM and SORM, a 3D seakeeping code (Papanikolaou 1989), including a software tool for the calculation of the added resistance in waves (Liu et al. 2010). The entire system is integrated within the probabilistic analysis software PROBAN (Det Norske Veritas 2003). The herein elaborated methodology parts and software modules deal with the estimation of ship's added resistance in waves, the analysis of a variety of seakeeping events and the calculation of the corresponding probability distributions or exceedence levels of predefined threshold values.

1 INTRODUCTION

Ships should always be prepared to encounter extreme weather during their voyage. The captain and his crew maybe well trained to recognize dangerous situations and take measures. Their decisions are in general based on their experience and the information available. Unfortunately, information about the weather and its likely impact on ship's performance is limited and uncertain. Thus, the captain needs to predict the consequences of his mastering decisions based on the available information presented on the various displays at the navigational bridge and his previous experience.

In recent years ship routing and decision support systems have become important issues in relation to the safety and the optimization of the economy of sea transportation. Safety of ship, crew and cargo, minimization of fuel consumption and the release of toxic exhaust gases, time restrictions for the delivery of the cargo, etc. form a complex multi-objective optimization problem with opposing targets.

The assessment of seakeeping events and of structural integrity and added resistance calculations are necessary ingredients of such decision support systems, which are employed for the evaluation of the appropriate route. In the deterministic seakeeping problem these are based on specific ship inherent and environmental data and assumptions. However, in realistic conditions each parameter of the seakeeping problem, even sensitive to seakeeping ship inherent data like GM and radius of gyration, is related to a degree of uncertainty, whereas other parameters, like environmental data, are inherently random. Above uncertain parameters and random variables define an intricate probabilistic problem, the solution of which is beyond complexity very time consuming for onboard applications.

In this paper the methodology of an onboard Decision Support Route Optimization System for the evaluation of the seakeeping parameters and the conditions affecting the ship's navigation is introduced.

Having such a system installed onboard, alternative navigational conditions may be evaluated and the optimum route control option can be given to the master for the minimization of the emerging risk and the fuel cost within the acceptable route time frame.

The modules of the system for the calculation of added resistance in waves and the estimation of the probabilities of exceedence for various seakeeping events are herein elaborated and results of their application to the operation of sample ships are presented and discussed.

2 ROUTE OPTIMIZATION METHODOLOGY

A Route Optimization Decision Support System should focus on optimizing the safety of the ship and the reduction of the journey cost by sensing the environment for actual situation data, and predicting the ship motions accordingly, thereby ensuring optimal operational performance, relying on computer based decision support tool creating an interface to be used in ship operation.

Such system must provide information and guidance by offering an evaluation of consequences, giving insight in the uncertainty of the environmental parameters (wind, waves and currents), ship loading condition, heading against the main wave direction and course and of other data by use of appropriate deterministic or probabilistic models, as applicable.

2.1 *System structure*

The basic structure of the route optimization system methodology of NTUA-SDL is introduced in this paper and sketched in Figure 1.

The system is comprised of four main modules, namely

- The added resistance module for the calculations of the added resistance in waves,
- the seakeeping hazards module for the calculations of seakeeping events affecting the structural integrity of the ship and the safety of the crew and the cargo,
- the risk calculation module which by having the probability of emergence of an event and the corresponding consequences is calculating the risk and the
- Route evaluation module for the evaluation of the alternative navigation conditions based on the users criteria.

The main inputs of the system should be ship's main characteristics, namely ship's geom-

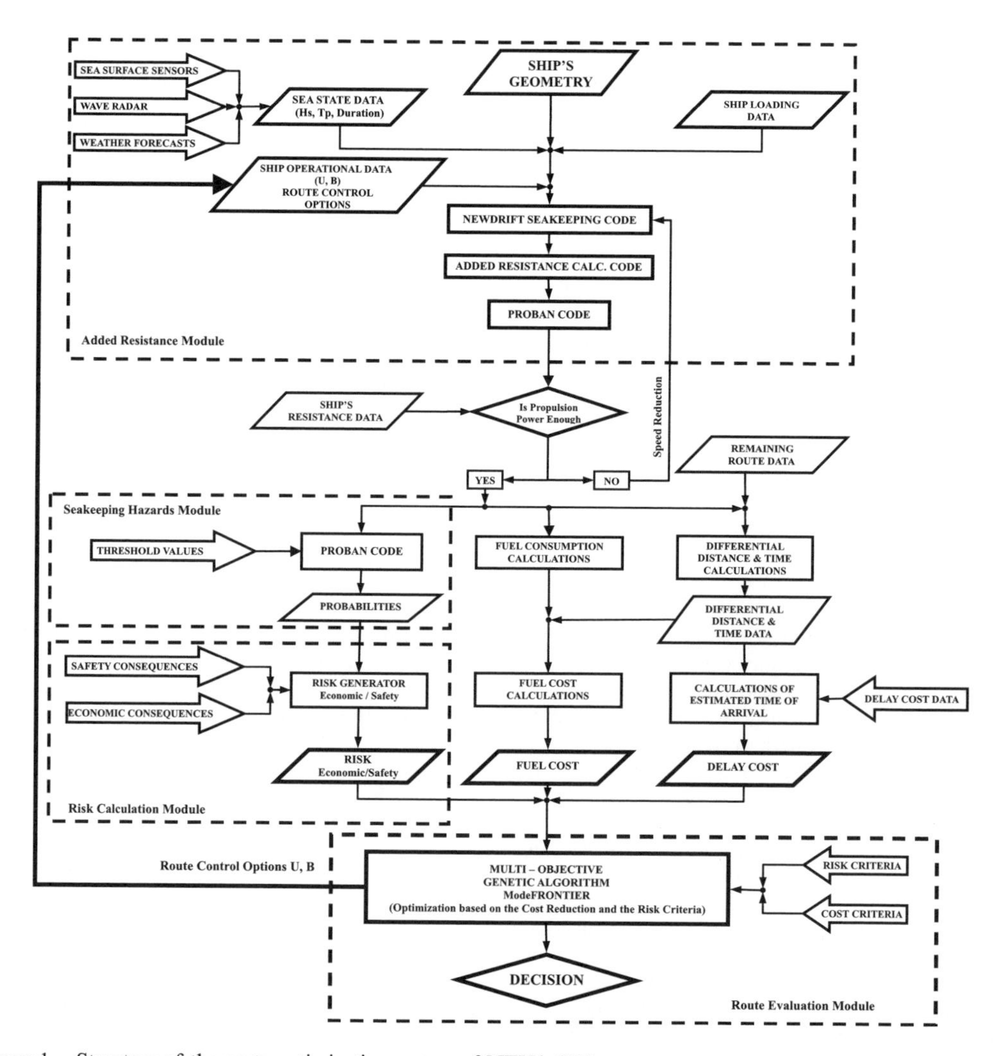

Figure 1. Structure of the route optimization system of NTUA-SDL.

etry and weight distribution, propulsion system characteristics, resistance in calm water and weather data for the specified route as waves, winds and currents. The weather data should be available continuously onboard, to allow predictions in 3 hrs windows.

The output should be the information given to the master for selecting the optimum route that will be minimizing the emerging risks due to weather, the fuel cost and CO_2 emissions, within an acceptable route time frame.

Ship's operational data (speed and heading to the sea) should be utilized as decision parameters for producing different navigation conditions for evaluation.

For the evaluation of the alternative routes, cost criteria relative to the reduction of the fuel and the cost of the delay should be used. Risk can be assessed depending on its category that can be classified as follows:

- Economic risk (damage to ship structure, equipment and cargo, including total loss of cargo and ship, loss of reputation).
- Safety of people (loss of life, injuries and health).
- Risk to the maritime environment (especially oil spillage for damaged tanker ships).

The first of the above categories can be expressed in monetary units and be assessed by cost criteria while the other two categories must be treated by other risk criteria and restrictions. The risk evaluation with respect to the optimum route is governed by several constraint parameters related to ship's stability, structural integrity, specified loading condition and characteristic parameters related to the specified voyage (route scenario).

2.2 *Main software modules and tools*

The calculation of ship responses, including the added resistance and powering in waves, the probabilistic analysis of the seakeeping hazards and the optimization procedure should be conducted by state of the art computational tools, allowing the execution of calculations by an onboard computer. The main software tools employed in the route optimization system of NTUA-SDL are:

- Newdrift developed by Prof. Papanikolaou between 1983 to 1989 (Papanikolaou 1989), is a 6DOF, 3D panel code for the seakeeping and wave induced loads analysis of ships and arbitrarily shaped floating structures, including multi-body arrangements. The code enables the evaluation of 6 DOF first—and quasi second—order motions and wave—induced loads, including drift deviations, forces and moments and is applicable to arbitrarily shaped 3D floating or submerged bodies (like ships, floating structures, or underwater vehicles), operating at zero or nonzero forward speed, finite or infinite water depth and being excited by sinusoidal linear waves or arbitrary frequency or headings. The consideration of natural seaway excitation is enabled through a spectral analysis postprocessor SPECTRA, given the incident spectral characteristics.
- Added Resistance Code developed by Shukui Liu in 2009 (Liu et al. 2010) calculates Added Resistance Response Functions in regular seas by Maruo's far-field theory and uses a semi-analytical correction for short wave lengths. NEWDRIFT or a new Time Domain Green Function Method developed in the PhD thesis of Shukui Liu may be used to calculate the velocity potential and the linear ship responses.
- PROBAN developed by DNV in 1989 (Det Norske Veritas 2003) is a tool for general purpose probabilistic analysis. The main objective of PROBAN is to provide a variety of methods aimed at different types of probabilistic analysis. This includes probability analysis, first passage probability analysis and crossing rate analysis. It can deal with a broad class of probabilistic and statistical problems. PROBAN allows efficient modeling of random variables and events.
- ModeFRONTIER developed by ES.TEC.O in 1999 is a multi-objective optimization and design environment that features the most recent optimization techniques as mono-objectives optimization algorithms (SIMPLEX algorithms, Gradient-based methods), Multi-Objective Algorithms (MOGA), optimization algorithms and robust design tools, statistical analysis tools, Design Of Experiments (DOE) and data-mining tools.

3 ADDED RESISTANCE IMPLEMENTATION

For addressing the reduction of the fuel cost and the CO_2 emissions within a route optimization system it is important to dispose accurate calculations of the ship's added resistance in waves. The added resistance in waves is the increase of a ship's calm water resistance that is caused by the encountered waves. Calculations of the added resistance can be used as a correction to the calm water resistance to predict the total resistance of a ship in a seaway. A ship may experience in common seaways a 20–40% resistance and powering increase, for which the added resistance is the main reason. Thus, being able to predict accurately the added resistance in waves is a vital part of a route optimization system; it is noted that this aspect is greatly neglected in many common commercial route assistance systems.

3.1 *Added resistance in regular seas*

The Added Resistance Code of NTUA-SDL was introduced in 2009 (Liu et al. 2010). This code calculates Added Resistance Response Functions in regular seas by using Maruo's far-field approach theory (Maruo 1963) and the Kochin functions approach and uses a correction formula (Kuroda et al. 2008) for short wave lengths. The velocity potential and the linear ship responses for the calculation of added resistance can be given by the seakeeping code NEWDRIFT (Papanikolaou 1989) or the new Hybrid Time Domain Green Function Method developed in the PhD thesis of Shukui Liu (Liu 2011). The results of this code were compared with those of other well-established authors and experimental data and a reasonable agreement was observed, suggesting that the implemented procedure appears a reliable and robust method for the routine prediction of the added resistance of a ship in waves.

For this module of the optimization methodology, this code and the NEWDRIFT were interfaced and the resulted Response Functions where used for the calculations of the added resistance in irregular seas.

3.2 *Added resistance in irregular seas*

The added resistance in irregular waves, expressed by a sea spectrum, may be expressed by superposition of the ship responses of the constituent regular waves, namely:

$$\overline{R}_{AW} = 2\int_0^\infty R(\omega) \cdot S_\zeta(\omega) \cdot d\omega \quad (1)$$

$$R(\omega) = \frac{R_{AW}(\omega)}{\zeta_a^2} \quad (2)$$

where, $S_\zeta(\omega)$ is the wave energy spectrum, R_{AW} the added-resistance response function in regular waves and ζ_a the regular wave amplitude.

3.3 *Application of the added resistance module*

A validation study for the added resistance of an AFRAMAX oil tanker design was elaborated for sea state conditions of the Caribbean Sea.

The main characteristics of the ship are:

L_{BP} = 242.0 m
B = 44.0 m
T = 13.7 m
Vs = 15.5 kn
P_B = 13560 KW (at 100% MCR)

For the calculations in irregular waves a Bretschneider spectrum was used. In Figure 2 the added resistance response functions for a speed of 16 kn are shown.

Tables 1–3 illustrate the wave scatter for the specified sea. Shaded areas represent the sea states where the added resistance is less that the values of the corresponding curves of 95%, 88% and 84% confidence at Figures 3 and 4. The bolded areas indicate the sea state where the added resistance calculations have been executed for each respective curve and divide the scatter to safe (shaded) and unsafe weather conditions.

Added resistance results compared to the calm water resistance are shown in Figure 3. In 84% of the probable sea states in the area of operation, the ship sustains an added resistance increase of up to 19% of the calm water resistance, while there are 11% of seaways with a resistance increase from 19% to 35% and a 5% with added resistance increase of more than 35%.

The study ship, with a sea margin of 10%, can maintain her speed at 84% of the likely seaways in the specified area by operating at 85% of MCR as Figure 4 illustrates. With the selected main engine of 13560 kW brake power, the ship can operate at the specified area with a confidence of 95%, by going up to about 98% of MCR or by having a minor speed loss of 0.5 kn.

4 SEAKEEPING IMPLEMENTATION

The seakeeping module of the route optimization method has been developed by NTUA-SDL within the EU funded ADOPT project by interfacing the seakeeping code NEWDRIFT and the probabilistic tool PROBAN (see, Spanos et al. 2008).

Seakeeping hazards can be formulated with a limit state function g(X), where $X = (X_1, X_2, \ldots, X_N)$ is a vector of random variables and takes negative

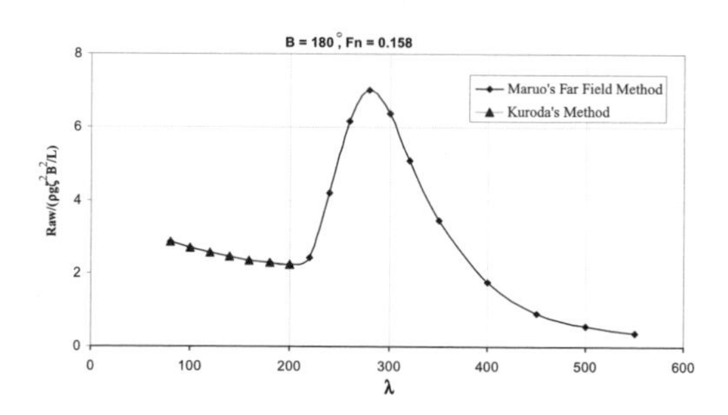

Figure 2. Added resistance response functions.

Table 1. Caribbean Sea wave scatter areas with 95% confidence for added resistance calculations.

Hs\Tz	3.5	4.5	5.5	6.5	7.5	8.5	9.5	10.5	11.5	12.5	13.5	
0.5	0.007493	0.035131	0.045571	0.023275	0.006238	0.001092	0.000145	0.000016	0.000002	0.000000	0.000000	0.118963
1.5	0.001862	0.037208	0.124421	0.134869	0.069947	0.022224	0.005085	0.000932	0.000147	0.000021	0.000003	0.396719
2.5	0.000243	0.009925	0.058954	0.104139	0.082259	0.037695	0.011892	0.002900	0.000591	0.000106	0.000018	0.308722
3.5	0.000025	0.001641	0.014342	0.035414	0.037573	0.022370	0.008912	0.002679	0.000660	0.000141	0.000027	0.123784
4.5	0.000003	0.000234	0.002734	0.008685	0.011517	0.008369	0.003989	0.001409	0.000402	0.000098	0.000022	0.037462
5.5	0.000000	0.000034	0.000502	0.001948	0.003088	0.002633	0.001449	0.000583	0.000187	0.000051	0.000012	0.010487
6.5	0.000000	0.000005	0.000098	0.000448	0.000824	0.000802	0.000498	0.000224	0.000079	0.000024	0.000006	0.003008
7.5	0.000000	0.000001	0.000021	0.000111	0.000232	0.000253	0.000174	0.000086	0.000033	0.000011	0.000003	0.000925
8.5	0.000000	0.000000	0.000005	0.000030	0.000071	0.000085	0.000064	0.000034	0.000014	0.000005	0.000001	0.000309
9.5	0.000000	0.000000	0.000001	0.000009	0.000023	0.000031	0.000025	0.000014	0.000006	0.000002	0.000001	0.000112
10.5	0.000000	0.000000	0.000000	0.000003	0.000009	0.000012	0.000011	0.000006	0.000003	0.000001	0.000000	0.000045
11.5	0.000000	0.000000	0.000000	0.000001	0.000003	0.000005	0.000005	0.000003	0.000002	0.000001	0.000000	0.000020
12.5	0.000000	0.000000	0.000000	0.000000	0.000002	0.000002	0.000002	0.000002	0.000001	0.000000	0.000000	0.000009
13.5	0.000000	0.000000	0.000000	0.000000	0.000001	0.000001	0.000001	0.000001	0.000001	0.000000	0.000000	0.000005
14.5	0.000000	0.000000	0.000000	0.000000	0.000001	0.000002	0.000002	0.000002	0.000001	0.000001	0.000000	0.000009
	0.009626	**0.084179**	**0.246649**	**0.308932**	**0.211788**	**0.095576**	**0.032254**	**0.008891**	**0.002129**	**0.000462**	**0.000093**	**100.06%**

95%	**Area below 95% Curve**
5%	**Area above 95% Curve**

Table 2. Caribbean Sea wave scatter areas with 88% confidence for added resistance calculations.

Hs\Tz	3.5	4.5	5.5	6.5	7.5	8.5	9.5	10.5	11.5	12.5	13.5	
0.5	0.007493	0.035131	0.045571	0.023275	0.006238	0.001092	0.000145	0.000016	0.000002	0.000000	0.000000	0.118963
1.5	0.001862	0.037208	0.124421	0.134869	0.069947	0.022224	0.005085	0.000932	0.000147	0.000021	0.000003	0.396719
2.5	0.000243	0.009925	0.058954	0.104139	0.082259	0.037695	0.011892	0.002900	0.000591	0.000106	0.000018	0.308722
3.5	0.000025	0.001641	0.014342	0.035414	0.037573	0.022370	0.008912	0.002679	0.000660	0.000141	0.000027	0.123784
4.5	0.000003	0.000234	0.002734	0.008685	0.011517	0.008369	0.003989	0.001409	0.000402	0.000098	0.000022	0.037462
5.5	0.000000	0.000034	0.000502	0.001948	0.003088	0.002633	0.001449	0.000583	0.000187	0.000051	0.000012	0.010487
6.5	0.000000	0.000005	0.000098	0.000448	0.000824	0.000802	0.000498	0.000224	0.000079	0.000024	0.000006	0.003008
7.5	0.000000	0.000001	0.000021	0.000111	0.000232	0.000253	0.000174	0.000086	0.000033	0.000011	0.000003	0.000925
8.5	0.000000	0.000000	0.000005	0.000030	0.000071	0.000085	0.000064	0.000034	0.000014	0.000005	0.000001	0.000309
9.5	0.000000	0.000000	0.000001	0.000009	0.000023	0.000031	0.000025	0.000014	0.000006	0.000002	0.000001	0.000112
10.5	0.000000	0.000000	0.000000	0.000003	0.000009	0.000012	0.000011	0.000006	0.000003	0.000001	0.000000	0.000045
11.5	0.000000	0.000000	0.000000	0.000001	0.000003	0.000005	0.000005	0.000003	0.000002	0.000001	0.000000	0.000020
12.5	0.000000	0.000000	0.000000	0.000000	0.000002	0.000002	0.000002	0.000002	0.000001	0.000000	0.000000	0.000009
13.5	0.000000	0.000000	0.000000	0.000000	0.000001	0.000001	0.000001	0.000001	0.000001	0.000000	0.000000	0.000005
14.5	0.000000	0.000000	0.000000	0.000000	0.000001	0.000002	0.000002	0.000002	0.000001	0.000001	0.000000	0.000009
	0.009626	**0.084179**	**0.246649**	**0.308932**	**0.211788**	**0.095576**	**0.032254**	**0.008891**	**0.002129**	**0.000462**	**0.000093**	**100.06%**

88%	**Area below 88% Curve**
12%	**Area above 88% Curve**

Table 3. Caribbean Sea wave scatter areas with 84% confidence for added resistance calculations.

Hs\Tz	3.5	4.5	5.5	6.5	7.5	8.5	9.5	10.5	11.5	12.5	13.5	
0.5	0.007493	0.035131	0.045571	0.023275	0.006238	0.001092	0.000145	0.000016	0.000002	0.000000	0.000000	0.118963
1.5	0.001862	0.037208	0.124421	0.134869	0.069947	0.022224	0.005085	0.000932	0.000147	0.000021	0.000003	0.396719
2.5	0.000243	0.009925	0.058954	0.104139	0.082259	0.037695	0.011892	0.002900	0.000591	0.000106	0.000018	0.308722
3.5	0.000025	0.001641	0.014342	0.035414	0.037573	0.022370	0.008912	0.002679	0.000660	0.000141	0.000027	0.123784
4.5	0.000003	0.000234	0.002734	0.008685	0.011517	0.008369	0.003989	0.001409	0.000402	0.000098	0.000022	0.037462
5.5	0.000000	0.000034	0.000502	0.001948	0.003088	0.002633	0.001449	0.000583	0.000187	0.000051	0.000012	0.010487
6.5	0.000000	0.000005	0.000098	0.000448	0.000824	0.000802	0.000498	0.000224	0.000079	0.000024	0.000006	0.003008
7.5	0.000000	0.000001	0.000021	0.000111	0.000232	0.000253	0.000174	0.000086	0.000033	0.000011	0.000003	0.000925
8.5	0.000000	0.000000	0.000005	0.000030	0.000071	0.000085	0.000064	0.000034	0.000014	0.000005	0.000001	0.000309
9.5	0.000000	0.000000	0.000001	0.000009	0.000023	0.000031	0.000025	0.000014	0.000006	0.000002	0.000001	0.000112
10.5	0.000000	0.000000	0.000000	0.000003	0.000009	0.000012	0.000011	0.000006	0.000003	0.000001	0.000000	0.000045
11.5	0.000000	0.000000	0.000000	0.000001	0.000003	0.000005	0.000005	0.000003	0.000002	0.000001	0.000000	0.000020
12.5	0.000000	0.000000	0.000000	0.000000	0.000002	0.000002	0.000002	0.000002	0.000001	0.000000	0.000000	0.000009
13.5	0.000000	0.000000	0.000000	0.000000	0.000001	0.000001	0.000001	0.000001	0.000001	0.000000	0.000000	0.000005
14.5	0.000000	0.000000	0.000000	0.000000	0.000001	0.000002	0.000002	0.000002	0.000001	0.000001	0.000000	0.000009
	0.009626	**0.084179**	**0.246649**	**0.308932**	**0.211788**	**0.095576**	**0.032254**	**0.008891**	**0.002129**	**0.000462**	**0.000093**	**100.06%**

84%	**Area below 84% Curve**
16%	**Area above 84% Curve**

values when a hazard occurs. Therefore the probability is calculated for the cases where:

$$g(X_1, X_2, \ldots, X_N) < 0 \tag{3}$$

The presented limit state g-functions are mainly functions of ship responses in waves. By having the function S of the employed ship motion model, the ship responses Y, the wave and loading parameters X and some given response control parameters C, then g-function can be defined.

$$Y = S(X|C) \tag{4}$$

$$g = g(X,Y|C) \tag{5}$$

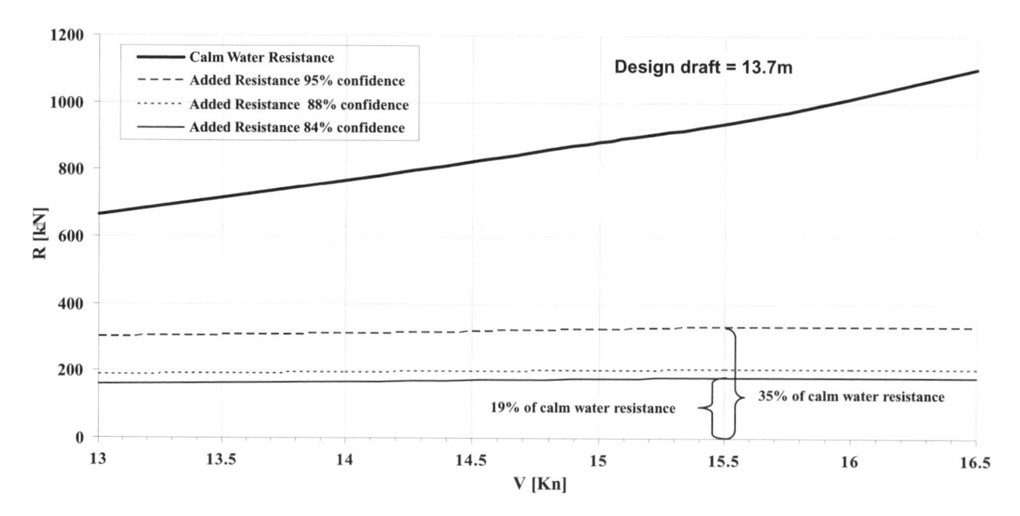

Figure 3. Added resistance results with different percentages of confidence.

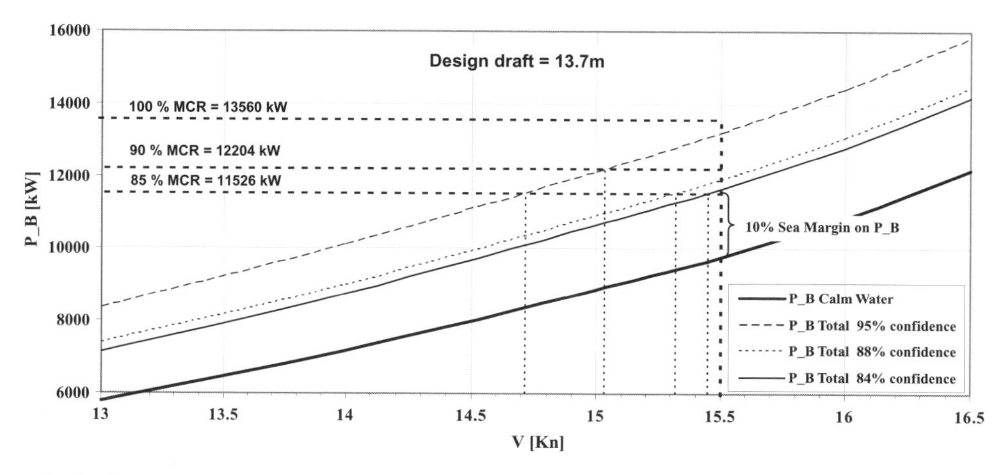

Figure 4. Brake power results with different percentages of confidence.

In realistic conditions each parameter of the seakeeping problem, even sensitive to seakeeping ship inherent data like GM and radius of gyration, is related to a degree of uncertainty, whereas other parameters, like environmental data, are inherently random. The uncertainty of them is not of the same importance for the calculations of the ship's responses.

The seakeeping module can be applied to both ship's operational assessment and design procedure.

In the operational mode parameters relevant to the ship's loading condition can be assumed determined at the beginning of the voyage. In this mode the random variables are the environmental parameters. The speed and the heading to the wave constitute the control parameters. The short time of the calculations enables this mode for onboard probabilistic evaluation.

In the design mode, all of the variables are considered uncertain. This mode can be useful at the design stage for the risk assessment during ship's life. The requested time for the probabilistic calculation does not allow this mode to be used on an onboard route optimization system.

4.1 *Seakeeping hazards*

In the development of the seakeeping module, five limit states (hazards) have been implemented, namely: the vertical acceleration at the bow, the total acceleration at the bridge, the bow slamming, the propeller racing and the deck immersion (green water). Hazards are defined either as excessive acceleration (exceeding threshold values) or high number of occurrences of seakeeping events. They are defined for several locations along the ship, which may be readily modified in a straight forward way. Other hazards (e.g. bending moments etc.) may be also added in the presently implemented system.

Formulation of limit states was used to result to the evaluation of the mean up-crossing rate of some variables. For Gaussian, zero-mean, narrow-band processes, the mean up-crossing rate v^+ of a level α can be approached by

$$v^+ = \frac{1}{2\pi}\sqrt{\frac{m_2}{m_0}}\exp\left(-\frac{\alpha^2}{2m_0}\right) \tag{6}$$

where m_0, m_2 are the zero and second order moment of the variable's spectrum S_R in consideration. For linear ship responses, S_R is calculated from the transformation of the wave spectrums according to the response operator H, both functions of frequency ω.

$$S_R(\omega) = |H(\omega)|^2 S(\omega) \tag{7}$$

4.2 *Threshold values*

The hazards are defined through a set of characteristic threshold values for the involved variables. Suppose the hazard of the frequent propeller racing that occur during severe pitching and subsequent propeller emergences. Such an event is undesired for the propulsion system. So, analyzing the capacity of the propulsion system and its tolerance to the racings, independently of the ship motions, then the threshold value for the racing rate is determined and correlated consequences are attributed. If the frequency of propeller racing is higher than the determined threshold value then the ship will encounter the related consequences. Apparently for a hazard on a top level description several threshold values may be derived each one correlated to a different level of consequences.

4.3 *Probabilistic methods*

For the evaluation of the probabilities of the seakeeping hazards First Order Reliability Method (FORM), Second Order Reliability Method (SORM) and Monte Carlo method have been employed and investigated.

FORM method initially transforms the basic X-variable space of a formulated limit state function g(X) into a u-space, where variables U_i are independent and standard normal variables. The mapping does not change the distribution of the g-variable and preserves all relevant probability information. In this way, the transformed g-function divides the u-space into safe and failure domain, Figure 5, where $g > 0$ and $g < 0$ correspondingly. Then, if the g-function is substituted by a linear function which is passing through a point u^*, the so-called design point, which is the point of the g-function closest to the space origin, a first order approximation is defined, namely the FORM method. Thus, the failure probability is the corresponding to the sub-domain defined by the linear approximation instead of the actual g-function (the shaded set in Fig. 5). Applying the same concept, but implementing a second order approximation then the SORM (Second Order Reliability) method is defined.

Obviously if the limit surface g of a hazard is not strongly non-linear then the approximation defined by FORM and corresponding probabilities could be satisfactory in view of the accuracy for the set problem.

Monte Carlo method that is based on sampling of the evaluated function can be proved efficient for the calculation of the central part of the distribution. Nevertheless for low probability events it suffers from the large number of simulations required to achieve a level of accuracy.

In Figure 6 the u-space mapping for the significant wave height Hs and the Peak Period Tp is shown for a slamming rate > 4 per hour. The straight line is the linear approximation of the g-function according to FORM.

4.4 *Computation time*

A fast computational performance in order to achieve practical application times onboard is a basic requirement for the developed computer-based probabilistic approach. Although the

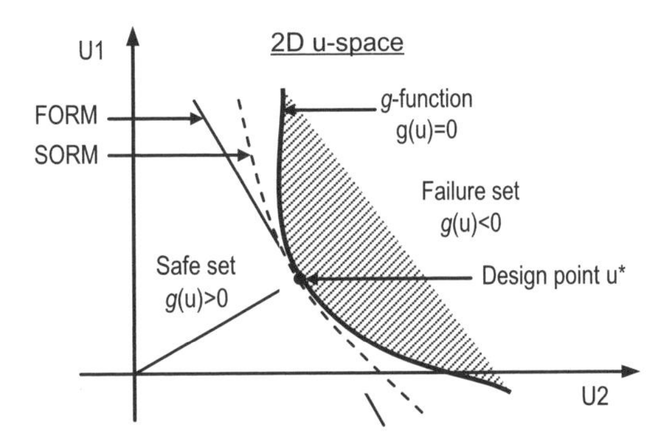

Figure 5. Two dimension g-function approach with FORM.

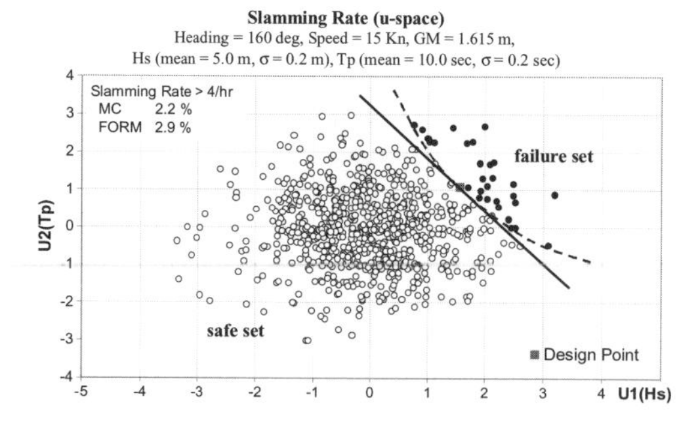

Figure 6. U-space for Hs and Tp.

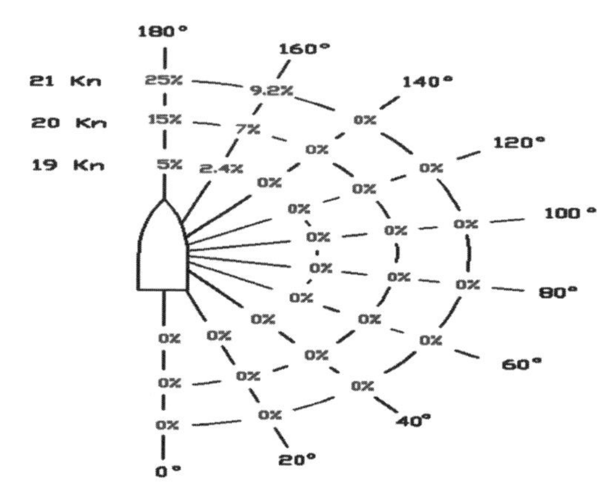

Figure 7. Probabilities for propeller emergence rate > (1/min).

computational time to complete a full set of calculations and evaluations strongly depends on the employed computer machine, the time recorded and provided herein enables a representative view of the current performance achieved at laboratory's environment.

With reference to a single PC computer, Intel Core2 CPU 6600 @ 2.40 GHz, 2 GB Ram and for a dense hull representation (2 × 500 panels), the computational times are:

- 35 sec per Limit State evaluation, when using Monte Carlo
- 5 sec per Limit State evaluation, when using FORM.

An evaluation of the 5 limit states takes about 12.5 min for the calculations of the probabilities for 30 alternative sailing conditions within a range of speed-heading combinations, as shown in Figure 7. For this assessment the wave spectrum parameters have been assumed to be uncertain parameters. The required time for the evaluation of five limit states can be considered short enough for the use of this module in an onboard Route optimization system.

5 CONCLUSIONS

A route optimization methodology aiming the safety of the ship, the crew and the cargo and the reduction of the total cost of the voyage has been briefly presented. Two of the main modules for the calculation of added resistance and the probabilistic assessment for the considered seakeeping hazards have been developed and validation studies proved their efficiency in view of their implementation into an onboard optimization system.

More seakeeping hazard limit states focusing on the structural integrity of the ship and the safety of the crew can be added and will be presented in planned future publications.

ACKNOWLEDGMENTS

The presented work is part of the PhD work of the first author. This was partly financially supported by DNV in the framework of the GIFT strategic R&D collaboration agreement between DNV and the School of Naval Architecture and Marine Engineering of NTUA-Ship Design Laboratory. Finally, another part of the presented work was developed within the FP6 research project ADOPT (2005–2008) of the European Commission, Contract No FP6-TST4-CT-2005-516359. The above financial supports are acknowledged by the authors.

REFERENCES

ADOPT (2005–2008). E.C. research project ADOPT (Advanced Decision Support System for Ship Design, Operation and Training), Contract No FP6-TST4-CT-2005-516359, FP6, Sustainable Surface Transport Programme. http://adopt.rtdproject.net/

Det Norske Veritas, (2003). *Proban Theory, General Purpose Probabilistic Analysis Program*. User Manual, DNV Software Report No. 92-7049/Rev.4, Nov. 1st 2003, program version 4.4.

Kuroda, M., Tsujimoto, M., Fujiwara, T., Ohmatsu, S., Tagaki, M., (2008). Investigation on components of added resistance in short waves, *Journal of the Japan Society of Naval Architects and Ocean Engineers*, Vol. 8.

Liu, S., (2011). *Numerical Simulation of Large Amplitude Ship Motions & Application to Ship Design & Safe Operation*, PhD thesis, NTUA, Athens.

Liu, S., Papanikolaou, A., Zaraphonitis, G., (2011). Prediction of Added Resistance of Ships in waves, *Ocean Engineering Journal*.

Maruo, H., (1957). The excess resistance of a ship in rough seas, *International Shipbuilding Progress*, Vol.4, No. 35.

Maruo, H., (1960). The drift of a body floating on waves. *Journal Ship Research*, Vol.4, No. 3.

Maruo, H., (1963). Resistance in waves, 60th anniversary Series, *The Society of Naval Architects of Japan*, vol. 8.

Papanikolaou, A. (1989). NEWDRIFT V.6: *The six DOF three-dimensional diffraction theory program of NTUA-SDL for the calculation of motions and loads of arbitrarily shaped 3D bodies in regular waves*, Internal Report, National Technical University of Athens, NTUA-SDL.

Papanikolaou, A., Gratsos, G., Boulougouris, E., Eliopoulou, E., (2000), Operational Measures of Avoiding Dangerous Situations in Extreme Weather Conditions, Proc. of the 7th International Conference on Stability of Ships and Ocean Vehicles, STAB2000, Feb. 2000, Tasmania, Australia.

Spanos, D., Papanikolaou, A., Papatzanakis, G., (2008). Risk-based Onboard Guidance to the Master for Avoiding Dangerous Seaways, *Proc. 6th Int. Osaka Colloquium on Seakeeping and Stability of Ships*, March 26–28, 2008, Osaka.

Spanos, D., Papanikolaou, A., Papatzanakis, G., Tellkamp, J., (2008). On Board Assessment of Seakeeping for Risk-based Decision Support to the Master, *Proc. 7th Int. Conf. on Computer Applications and Information Technologies in Maritime Industries*, COMPIT 08, Liege.

Tsujimoto, M., Shibata, K., Kuroda, M., Tagaki, K., (2008). A practical correction method for added resistance in waves, *Journal of the Japan Society of Naval Architects and Ocean Engineers*, Vol. 8.

Sustainable Maritime Transportation and Exploitation of Sea Resources – Rizzuto & Guedes Soares (eds)
© 2012 Taylor & Francis Group, London, ISBN 978-0-415-62081-9

Development of an onboard decision support system for ship navigation under rough weather conditions

L.P. Perera, J.M. Rodrigues, R. Pascoal & C. Guedes Soares
Centre for Marine Technology and Engineering (CENTEC), Instituto Superior Tecnico, Technical University of Lisbon, Lisbon, Portugal

ABSTRACT: The paper describes the development of an onboard decision support system to support ship operation, in particular on decisions about ship handling in waves, which will contribute to vessel safety. The prototype system monitors several motion related parameters, and, by processing these data, provides the ship master with the information about the consequences of the different ship handling decisions. The paper describes the decision criteria and the approaches adopted for the calculation of the parameters that govern the master's decisions. It describes the software that was developed to perform those calculations and to display in a user interface the advice to the master as well as the data acquisition and processing hardware that has been organized for the on board monitoring of motions and strains in the structure.

1 INTRODUCTION

The effect of waves in rough weather is one of the factors that most degrade a ship's operational efficiency. Therefore, the tactical judgment involved in the ship handling decision process takes an essential part in navigation. Rothblum et al. (2002) and Antão & Guedes Soares (2008) have shown that 75% to 96% of marine casualties have their origins in some kind of human errors, where human errors are still one of the major causes of maritime accidents (Guedes Soares, & Teixeira, (2001)). Therefore whenever the navigators can be helped with monitoring and decision support systems, a contribution is being given to safety.

The initial developments of onboard systems to aid the navigation in rough weather had been mainly concerned with structural integrity and equipment safety. Lindemann et al. (1977) developed one such system by measuring the accelerations in six degrees of freedom and the stresses at a cross section. Hoffman (1980), considered a system using ship to shore communications along with charts for routing in heavy weather.

The work of Koyama et al. (1982) consisted of a computer based system capable of computing the mean period and the root mean square prediction of roll motion. The input component was a pendulum for measuring the ship motions and, given a pre-determined criterion, an alarm would fire in case of danger. Unfortunately the pendulum system proved unsatisfactory especially for high speeds so the results were shown to be unreliable.

Later, Huss & Olander (1994) formulated a prototype of an on board based guidance and surveillance system for wave induced effects on ships, where a rate gyro and an accelerometer were used towards the local real-time estimation of the sea state in the form of a spectrum. Köse et al. (1995) proposed a scheme based on low cost equipments which, with special purpose developed software, resulted in an encompassing reliable system for stability monitoring and advising applicable to any ship. Payer, and Rathje, (2004) presented a onboard system thought for containership operation in rough seas.

The present work deals with the development of an onboard decision support system for tactical decisions of ship handling in waves, which enables the master to improve ship performance while minimizing the likelihood of structural damage. The system now reported has been formulated in 2005, before the start of the EU Handling Waves project (http://www.mar.ist.utl.pt/handlingwaves/). It has some similar principles and solutions to the decision support system that has been developed for the operation of fishing vessels, as described by Rodrigues et al (2011).

More recently, Bitner-Gregersen, & Skjong (2009) and Nielsen & Jensen (2010) presented concepts of risk-based guidance of ships, which have some ideas that can be incorporated in future systems.

The system now reported besides monitoring in real time the actual ship responses, also predicts the near term motions and structural loads due both to weather changes and to possible changes in course and speed by the shipmaster. It also includes a component of hull monitoring by strain gauges, which corresponds to a more established technology as described by Slaughter et al. (1997), which however is here integrated with the decision support system.

More specifically this paper deals with the development of monitoring devices able to accurately measure the motions of the ship and the implementation of a Decision Support System (DSS) integrating the various elements required. The system's architecture, working principles, dataflow, calculation procedures and equipment are herein described.

2 DECISION SUPPORT SYSTEM

Figure 1 shows the logical architecture of the DSS. The main modules are: Data Processing and Analysis module, Structural Loads Estimator module, and the Sea Estimator module.

The Data Processing and Analysis module allows the monitoring in real-time of the ship motions and accelerations for arbitrary positions in the ship. The Structural Loads Estimator provides estimation of the loads on the structure, assessed by different approaches.

The Sea Estimator Module uses the filtered and digitized accelerations and velocities associated with ship motions, measured during the previous minutes, and these are used to estimate the directional wave spectrum. This is achieved by the implementation of a *Kalman filter* based algorithm described by Pascoal & Guedes Soares (2009), which proved a better option than the approach considered initially, which is described in Pascoal & Guedes Soares (2008).

From this estimation of the spectrum, the system predicts the near term motions and structural loads due both to weather changes and to possible changes in course, taking for this a probabilistic approach in the form of root mean square values of key motion amplitude and acceleration levels.

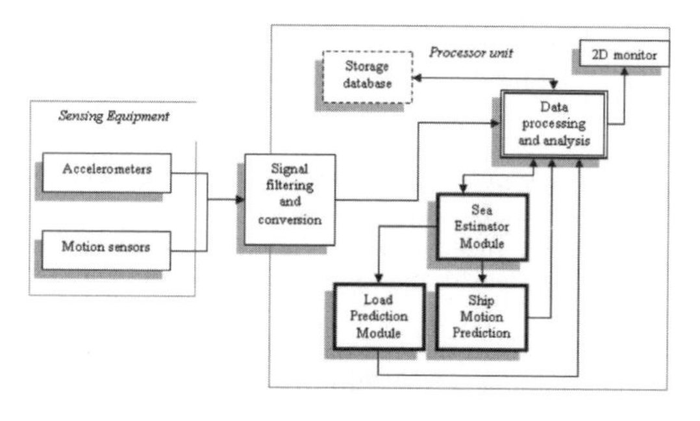

Figure 1. DSS logical architecture.

These parameters are then checked against predefined operational safety criteria, which results in the construction of a polar plot on which the areas with dangerous combinations of ship course and speed are indicated. The prediction of motions and accelerations is done from the estimated directional wave spectra and the existence of pre-calculated motion transfer functions using a strip theory code. With these values, the criteria for operability and seasickness in NORDFORSK (1987) and in O'Halon and McCauley (1974) are assessed.

It has also been implemented the capability to check, for a given the sea state, the probability of occurrence of parametric rolling based on experimental and numerical results. A simple query to a database containing the results for the different combinations of ship speed and course was the methodology chosen.

The Structural Loads Estimator also uses the estimated spectrum and transfer functions for shear forces and bending moments to predict what would be the loads in the structure in the different options of course and speed decisions. It also includes a neural network model (Moreira and Guedes Soares, 2011) that uses as input the measured accelerations and motions at the various locations of the ship and produces as output the shear stresses and bending moments at selected locations, so that the ship master can have on-line information of the loads that the structure is being subjected.

In addition to the motion measurement equipment, a set of strain-gauge units are installed onboard, so as to provide a direct measurement of the strains in the structure, which can be compared with the predicted strains from the neural network model of the Structural Loads Estimator.

3 CRITERIA FOR SHIP OPERATIONAL SAFETY

The operability criterion correlates the type of work to be performed at a given location on the vessel with its maximum *rms* (root mean square) values of lateral acceleration, vertical acceleration and roll amplitude. The limit values according to NORDFORSK (1987) are listed in (Table 1).

The O'Hanlon and McCauley (1974) criterion for MSI (Motion Sickness Incidence) is defined as the percentage of people to experience seasickness during a period of two hours. It is governed by the following expressions:

$$\mathrm{MSI} = 100\left(0.5 \pm \mathrm{erf}\left(\frac{\pm\log_{10}(a_z/g) \mp \mu_{MSI}}{0.4}\right)\right)\% \quad (1)$$

Table 1. RMS criterion.

	$\ddot{x}_{3,P}$	$\ddot{x}_{2,P}$	$x_{4,P}$
Description of work	m/s²	m/s²	deg
Light manual work	0.20 g	0.10 g	6.0
Heavy manual work	0.15 g	0.07 g	4.0
Intellectual work	0.10 g	0.05 g	3.0
Transit passengers	0.05 g	0.04 g	2.5
Cruise liner	0.02 g	0.03 g	2.0

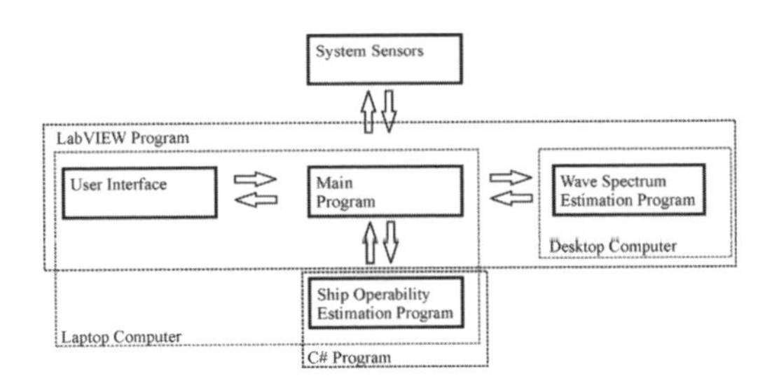

Figure 2. Software architecture.

$$\mu_{MSI} = -0.819 + 2.32(\log_{10}\omega_e)^2 \tag{2}$$

$$erf(x) = erf(-x) = \frac{1}{\sqrt{2\pi}}\int_0^x \exp\left(-\frac{z^2}{2}\right)dz \tag{3}$$

where a_z stands for the vertical acceleration at some point in the ship where one wishes to assess the incidence of motion sickness and, in (Eq. 2), ω_e is the frequency of encounter.

It is crucial to point out that the above defined quantities are evaluated at some point in the structure of the ship where it is significant to assess the motions. Therefore, there is a need to *expand* the linear motions from the centre of gravity, which are given by the raw transfer functions, into the absolute motions at the specific point of interest.

4 IMPLEMENTATION OF THE DECISION SUPPORT SYSTEM

Regarding the actual implementation, the system may be decomposed into 5 units as presented in Figure 2, where the software architecture is shown and the implementation tools are referred (LABVIEW and C#).

The *System Sensors* are responsible for all motion measurements and the *Wave Spectrum Estimation Program* takes into action the task of estimating the real-time local sea state. The *User Interface* allows access to the polar plot as described in Section 1, it permits the graphical monitoring of all variable values that govern the calculations taking place, and constitutes the tool by which the developer may perform analysis and troubleshooting.

The *Ship Operability Estimation Program* is an application responsible for the computation of the predicted near term motions for the current and other possible combinations of speed and course, from the knowledge of the estimated spectrum main parameters.

Finally, the *Main Program* constitutes the kernel of the system's dataflow. It collects and stores all real-time sensor and estimated spectrum data, it is also responsible for communicating with the Ship Operability Estimation Program by sending the spectrum parameters and receiving a set of $m \times n$ matrices to be forwarded to the *User Interface*. Here m is the number of discretized possible speeds of advance and n the number of discretized possible ship course directions. These matrices are populated with *zeros* and *ones*, thus providing a mapping to implement on a polar plot, where a *one* classifies a combination of speed and course which results in an undesirable behaviour of the ship for the current sea state, in light of the criteria defined in Section 2.

4.1 *Wave spectrum estimation program*

The *Wave Spectrum Estimation Program* runs under the LABVIEW Real-Time (RT) operating system. This unit implements a high-speed iterative procedure for estimating the ocean wave directional spectrum from the vessel motion data. It uses as input the measurements from motion sensors and provides spectral updates under quickly changing weather conditions.

The *Kalman filtering* algorithm, for iterative harmonic detection, and frequency domain vessel response data in the form of transfer functions, are used in the estimation process. The output is the estimated directional spectrum parameters: significant wave height, mean period and mean direction. More details on this subject may be found in the references given in Section 1 concerning the real-time spectrum estimation.

4.2 *User interface*

The view of the *User Interface* on the laptop computer is presented in Figure 3. The interface consists of 9 parts: data management, application management, GPS display, channel test, danger zones display, statistics data, estimated spectrum display, loads conditions display, and motion signals display. These sections are presented by separate tabs:

- Data management: The data management section is responsible for collecting sensor and GPS data (accelerometer, wave height measurement

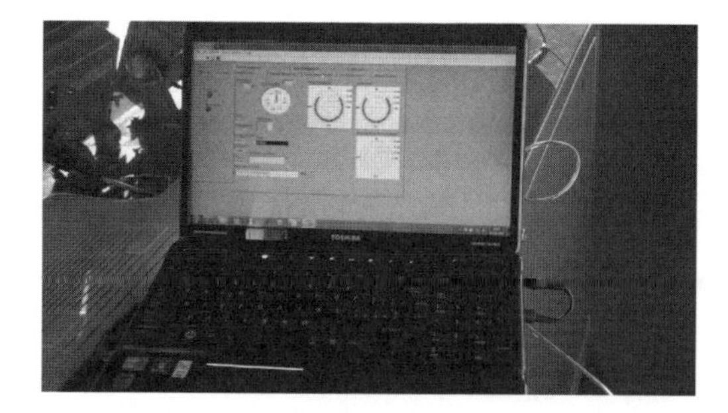

Figure 3. Laptop on the bridge with user interface.

sensor, inclinometer, angular rate sensor, strain gauge sensors). The sensor data is then saved to an external hard drive.

- Application management: The application management section enables the operator to stop, reboot and restart individual processes and programs. These processes and programs are: *User Interface*, *CompactRIO* and *EtherCAT* (see Section 4) units, and *Wave Spectrum Estimator*.
- GPS display: The GPS display is responsible for read and display services of data collected from the GPS sensor.
- Channel test: This section is designed to read and write uncelebrated data from the different channels in the data acquisition system. Any problems with read and write operations with Data Acquisition (DAQ) channels can be evaluated in this unit.
- Danger zones: The main objective of the danger zones section is to display the results calculated by the *Operability Assessment Program*. A polar plot constructed by mapping the matrices previously defined is presented.
- Statistics display: This section consists of statistical information about the accumulated data that has been collected by the ship sensors. However, currently this unit is under development.
- Estimated spectrum display: Relates to the wave spectrum estimation which has been done by the *Kalman filter* algorithm that is running under the LabVIEW RT platform on the *Wave Spectrum Estimation Program*. An isometric view of the bi-dimensional spectrum is displayed in this window.
- Load conditions display: The ship's hull stress conditions that are measured by the strain gauge sensors are displayed in this section. High strain values are monitored by the program and a warning light will also be displayed on this section.
- Motion signal display: The calibrated ship motions that are measured by the sensors are displayed in this section. The ship surge, sway and heave accelerations, pitch and roll angles

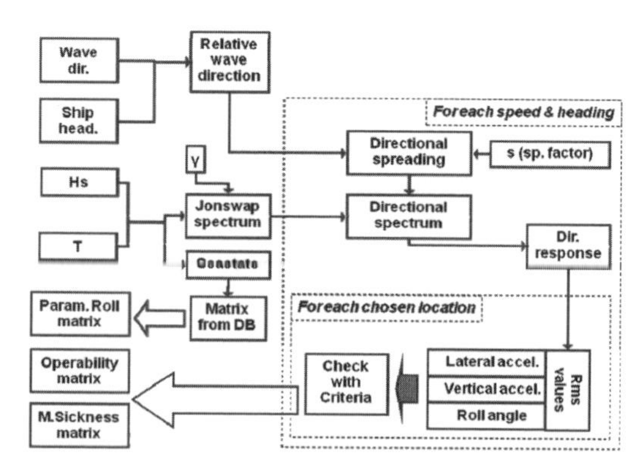

Figure 4. Ship operability estimation program architecture.

and yaw rate are displayed in the top area of this section. The measured wave height is displayed on the bottom area of this section.

4.3 *Ship operability estimation program*

The *Ship Operability Estimation Program* implements the criteria checking defined in Section 2. Based on the estimated spectrum parameter it constructs a JONSWAP spectrum and computes the necessary quantities previously defined. Contrary to the remainder of the system, this tool is not part of the LABVIEW en vironment (see Fig. 2), but is rather a *.net* standalone application. The communication with the Main Program is done through continuously updated input and output files.

In Figure 4 the architecture of this application is presented, where it can be seen the inclusion of a parametric rolling occurrence check. This check has yet to be correctly implemented, although the logic has already been set to work. It consists on a query which is done to a database of simulations/tests done on the particular vessel for which the occurrence of this phenomenon is likely to be expected. These simulations/tests are not yet available and this fact constitutes the cause to which this subject is not discussed further in this paper.

5 EXPERIMENTAL PLATFORM

The DSS prototype has been installed on a Ro-Ro ship with $L_{pp} = 214.0$ m and $B = 32.0$ m. Another prototype is currently being set up on a container vessel with $L_{pp} = 117.6$ m and $B = 20.2$ m.

The system has been put to work on the Ro-Ro ship and it has been collecting data from the motions and calculations results. Once enough data have been collected the same will then be used to calibrate and validate the system. In this section, a brief presentation of the individual components of the experimental platform and its operating logic is carried out.

5.1 *Functional structure*

Considering the assigned tasks, the DSS can be divided into three main sub-systems: *Motion monitoring sub-system*, *Stress monitoring sub-system* and *Wave condition monitoring sub-system*.

The main objective of the *Motion monitoring sub-system* is to evaluate the vessel motions on the seaway. It consists of the midship accelerometer, the midship angular rate sensor, and the inclinometer. The sensor locations are presented in Figure 5. The accelerometer measures the surge, sway, and heave accelerations. The angular rate sensor measures the yaw angular velocity and the inclinometer measures the roll and pitch angles.

The purpose of the *Stress monitoring sub-system* is to evaluate the hull stress condition. It consists of four strain gauges that are oriented: Two strain gauges located starboard and portside of the midship and two strain gauges located fore and aft of starboard. The strain gauges locations are also presented in Figure 5.

The function of the *Wave condition monitoring sub-system* is to evaluate the wave spectrum. It consists of the wave height measurement sensor and the bow accelerometer. Their locations are shown in Figure 5. The bow mounted, down-looking, wave measurement sensor measures the relative wave height and the bow accelerometer compensates for the vessel motions.

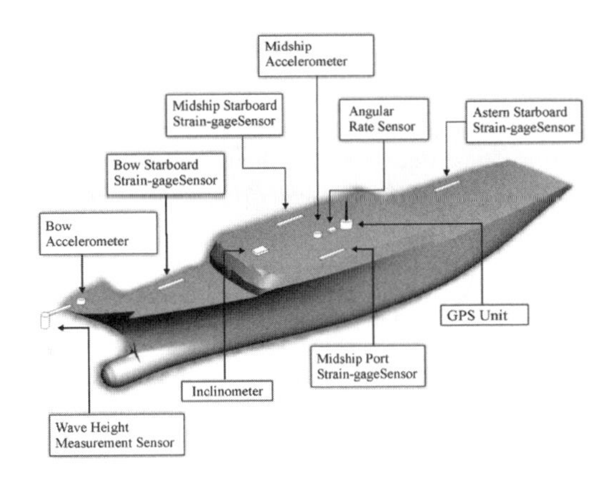

Figure 5. Sensor locations.

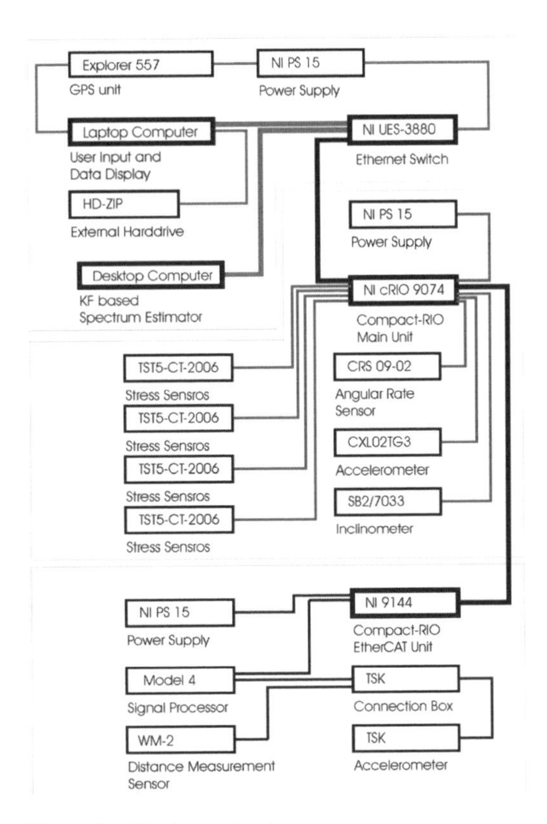

Figure 6. Hardware structure.

5.2 *Hardware structure*

The hardware structure, shown in Figure 6, mainly consists of the real-time digital data acquisition system (DAQ), sensors, computers and power supply units. It is composed of the following units: Laptop computer with external hard-drive, Desktop computer, GPS unit, sensors, *Compact-RIO* and *EtherCAT*, Ethernet switch and Power suppliers.

5.2.1 *Laptop computer*

The laptop computer acts as the main control equipment of the system. There are three software components that run on the laptop: *User Interface*, *Main Program* and *Ship Operability Assessment Program*. The first two are coded with LABVIEW, whereas the third is a standalone .net application developed in C#. An external hard-drive, with the purpose of saving real-time data for further analysis, and the GPS unit, are both connected directly to the laptop.

5.2.2 *Desktop computer*

The desktop computer runs the LABVIEW real-time operating system which enables high-speed calculation for *Kalman filter* based wave spectrum estimation. The inputs to the wave spectrum estimation program are the motion related sensor measurements (midship accelerometers, wave height sensor, inclinometer, and angular rate sensor), the vessel course and speed. The output of is the estimated wave spectrum including the wave direction, significant wave height and mean period. These quantities are then forwarded to the *Ship Operability Estimation Program* for further analysis.

5.2.3 *GPS unit*

The GPS NORTHSTAR Explorer 557 unit that has been installed on the experimental platform is presented in Figure 7. The unit comprises an external antenna and a RS-232 data communication cable. This unit is connected to the laptop through a combined USB/RS-232 communication port. The system parameters fetched by the GPS unit—vessel position, speed and course—are forwarded to the system by the laptop.

5.2.4 *CompactRIO and EtherCAT*

CompactRIO and *EtherCAT* units from NATIONAL INSTRUMENTS are used as the main data acquisition components for the DSS, and are pictured in Figures 8 and 9. The *CompactRIO* unit is located at midhsips and it is the central data acquisition hardware, whereas the *EtherCAT* unit is installed closer to the bow and acts as an extension of the first.

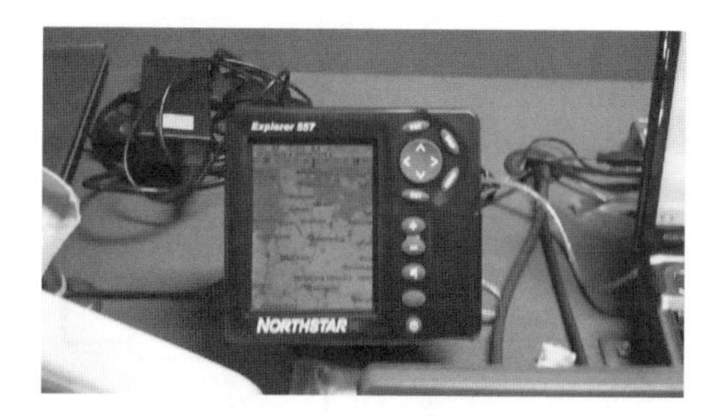

Figure 7. GPS unit.

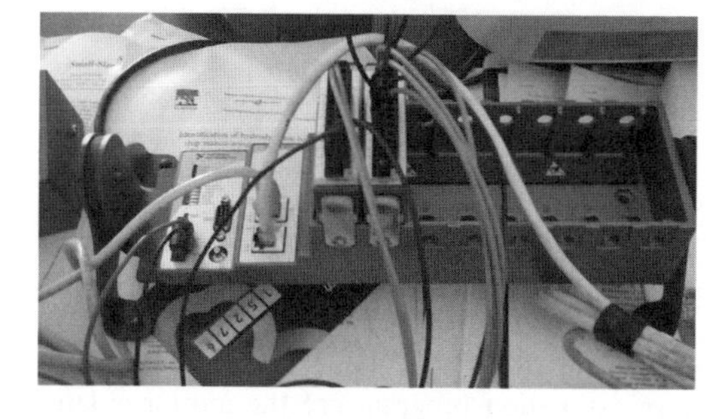

Figure 8. National instruments Compact-RIO unit.

Figure 9. National instruments EhterCAT unit.

5.2.5 *Ethernet switch*

The Unmanaged Ethernet Switch that enables communication between the *CompactRIO* unit, laptop, and desktop computer is presented in Figure 10. This unit incorporates an automated bandwidth management process that can secure the network from overloading among other errors. Furthermore, the unit is capable of online debugging and automatic recovering of IP addresses of other Ethernet linked units.

5.2.6 *Other sensors*

A picture of the compact system sensor configuration is presented in Figure 11. The *CompactRIO* is connected with several sensors: Angular rate sensor, accelerometer, inclinometer and four strain gauge sensors. The *EtherCAT* unit is connected with the wave height measurement sensor and its associated accelerometer, connection box and signal processors. The installation of the signal processor with the *EtherCAT* unit is shown in Figure 12.

5.2.7 *Strain gauge sensors*

As can be inferred from Figure 13, the sensor comprises a steel rod which is allowed to displace longitudinally at one end. The measurement of this displacement at the four distinct locations depicted in Figure 4 and consequent conversion to strain values, gives way to the assessment of the ship's structural loads such as the vertical bending moment.

5.2.8 *Wave height sensor*

The bow mounted, down-looking, wave height measurement sensor is presented in Figure 14. This TSURUMI SEIKI wave height sensor consists of four components: sensor unit, accelerometer, connection box and single processing unit. The sensor component measures the relative height of the bow wave by resorting to a microwave Doppler method, while the accelerometer measures the vessel's acceleration so as to eliminate the relative motion effect.

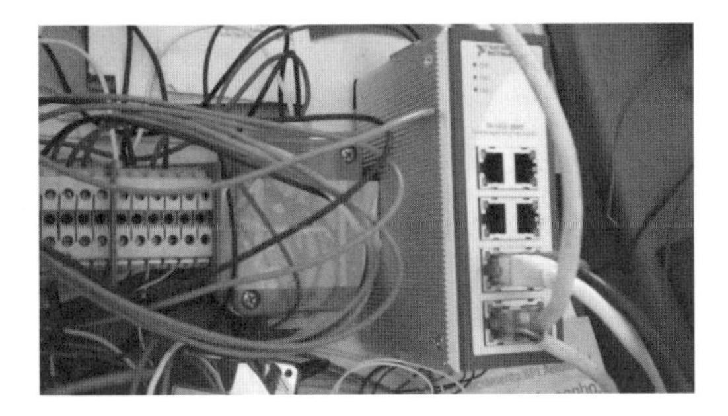

Figure 10. National instrument Ethernet switch.

Figure 11. The compact system with sensors.

Figure 12. Signal processor with EtherCAT unit.

The connection box enables the linkage between both these components and the signal processor unit, from which the output is taken. The analogue output of the signal processor unit which is made available to the DSS is composed by the quantities: ship displacement, relative wave height, wave height, significant wave height and average wave period.

5.2.9 *Power supplies*

Three power supply units are used to power the *CompactRIO*, *EtherCAT*, associated sensors and the GPS unit. One such power supply unit and its wiring connections are presented in Figure 15.

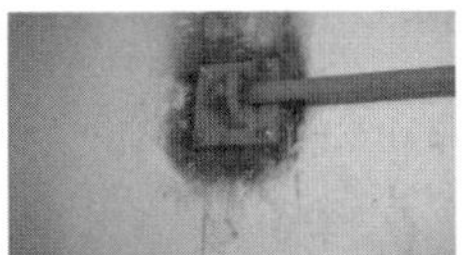

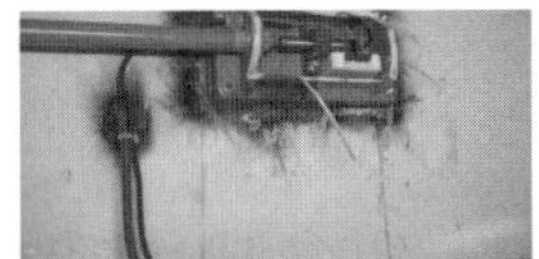

Figure 13. Strain gauge sensor.

Figure 14. Wave height Senso.

Figure 15. Power supply unit with wiring connections.

6 CONCLUSIONS

An prototype of an onboard decision support system for ship navigation under rough sea and weather conditions has been set up and installed on board of a Ro-Ro ship.

The system is currently collecting data as the ship operates, which will be used in its calibration and validation. Besides the motion measurement related instruments, strain gauges have also been installed. The time series of the values of the strains at these sensors locations, will serve as the basis for the training of a neural network to be implemented, capable of quickly and accurately giving the expected loads given the present sea state and possible ship's courses and speeds. The system has been successfully tested in terms of hardware and software integration. Another prototype installation is being done on a container vessel, and the chronological steps will be the same as regarding the first ship.

ACKNOWLEDGEMENTS

This work is done within the project of "Handling Waves: Decision Support System for Ship Operation in Rough Weather", which is being funded by the European Commission, under contract TST5-CT-2006-031489.

The work of the first and second authors has been supported by research fellowship of the Portuguese Foundation for Science and Technology (Fundação para a Ciência e a Tecnologia) under contract SFRH/BD/46270/2008 and SFRH/BD/64242/2009, respectively.

REFERENCES

Antão, P. and Guedes Soares, C. 2008, "Causal factors in accidents of high speed craft and conventional ocean going vessels," *Reliability Engineering and System Safety*, vol. 93, pp. 1292–1304.

Bitner-Gregersen, E.M. Skjong R (2009) Concept for a risk based Navigation Decision Assistant *Marine Structures,* Vol. 22, 2, pp 275–286.

Guedes Soares, C. and Teixeira, A.P. 2001, Risk Assessment in Maritime Transportation. *Reliability Engineering and System Safety*, Vol. 74, pp 299–309.

Hoffman, D. 1980. The integration of shipboard and shore based systems for operation in heavy weather. *Computers in Industry. Volume 1, Issue 4*: 251–262.

Huss, M & Olander, A. 1994. Theoretical seakeeping predictions on board ships—a system for operational guidance and real time surveillance. *Technical* Report, Naval Architecture, Department of Vehicle Engineering, Royal Institute of Technology. Stockholm.

Kose E., Gosine R.G., Dunwoody A.B., Calisal S.M. 1995. An expert system for monitoring dynamic stability of small craft. *Ocean Engineering, Vol. 20, No. 1.*

Koyama, T., Susumu H. & Okumoto, K. 1982. On a micro-computer based capsize alarm system. *Second International Conference on Stability of Ship and Ocean Vehicles*: 329–339.

Lindemann, K., Odland, J. and Strenghangen, J. 1977. On the application of the hull surveillance system for increased safety and improved utilization in rough weather. *SNAME, Transactions volume 85.*

Moreira, L. and Guedes Soares, C. (2011) "Estimation of Wave-Induced Hull Bending Moment and Shear Force from Ship Motion Simulations using Artificial Neural Networks", *Maritime Technology and Engineering*, C. Guedes Soares et al. (Eds), Taylor and Francis (in press).

Nielsen, U.D. & Jensen, J.J. 2010. A novel approach for navigational guidance of ships using onboard monitoring systems. *Ocean Engineering 38*: 444–455.

NORDFORSK. 1987. Seakeeping performance of ships, Assessment of a ship performance in a seaway. *Nordic Cooperative Organization for Applied Research.*

Operation and Maintenance Manual (2010), "TSK Wave Height Meter", Tsurumi Seiki Co. LTD, Japan.

O'Hanlon, J.F. and M.E. McCauley 1974. Motion Sickness Incidence as a Function of Vertical Sinusoidal Motion. *Aerospace Medicine AM-45(4)*, 366–369.

Pascoal, R., Guedes Soares, C. 2008. Non-parametric wave spectral estimation using vessel motions. *Appl. Ocean Res., 30(1)*: 46–53.

Pascoal, R., Guedes Soares, C. 2009. Kalman Filtering of Vessel Motions for Ocean Wave Directional Spectrum Estimation. *Ocean Engineering, 36(6-7)*: 477–488.

Payer, H.G., Rathje, H., 2004. Shipboard routing assistance decision making support for operation of container ships in heavy seas. Transactions of SNAME 112, 1–12.

Rodrigues, J.M., Perera, L.P. and Guedes Soares, C. (2011) "Decision support system for the safe operation of fishing vessels in waves", *Maritime Technology and Engineering*, C. Guedes Soares et al. (Eds), Taylor and Francis (in press).

Rothblum, A.M., Wheal, D., Withington, S., Shappell, S.A., Wiegmann, D.A., Boehm, W., and Chaderjian, M. 2002. Key to successful incident inquiry. *Proc. 2nd International Workshop on Human Factors in Offshore Operations (HFW)*: 1–6.

Slaughter, S.B. Cheung, M.C. Sucharski, D. & Cowper B., (1997) "State of the Art in Hull Monitoring Systems", Ship Structure Committee Report No. SSC-401, Washington, DC.

7 *Environment*

7.1 *Sea waves, wind*

Sustainable Maritime Transportation and Exploitation of Sea Resources – Rizzuto & Guedes Soares (eds)
© 2012 Taylor & Francis Group, London, ISBN 978-0-415-62081-9

A new numerical scheme for improved Businnesque equations with surface pressure

Deniz Bayraktar & Serdar Beji
Faculty of Naval Architecture and Ocean Engineering, Istanbul Technical University, Maslak, Istanbul, Turkey

In this work an improved Boussinesq model with a surface pressure term is discretized by a new approach. By specifying a single parameter the proposed discretization enables the user to run the program either in the long wave mode without dispersion terms or in the Boussinesq mode. Furthermore, the Boussinesq mode may be run either in the classical Boussinesq mode or in the improved Boussinesq mode by setting the dispersion parameter appropriately. In any one of these modes it is possible to specify a fixed or a moving surface pressure for simulating a moving object on the surface. The numerical model developed here is first tested by comparing the numerically simulated solitary waves with their analytical counterparts. The second test case concerns the comparison of the numerical solutions of moving surface pressures with the analytical solutions of the long wave equations for all possible modes (long wave, classical, and improved Boussinesq).

1 INTRODUCTION

The earliest depth-averaged wave model that included weakly dispersive and nonlinear effects was derived by Boussinesq (1871), in which the non-hydrostatic pressure was linearized and included in the momentum equations. The original equations were derived for constant depth only. Later, Mei and LeMeháute [1], Peregrine [2] derived Boussinesq equations for variable depth. While Mei and LeMeháute used the velocity at the bottom as the dependent variable, Peregrine used the depth-averaged velocity and assumed the vertical velocity varying linearly over the depth. Due to wide popularity of the equations derived by Peregrine, these equations are often referred to as the standard Boussinesq equations for variable depth in the coastal engineering community. The standard Boussinesq equations are valid only for relatively small kh and H/h values where kh and H/h represents the parameters indicating the relative depth (dispersion) and the wave steepness (nonlinearity), respectively. Madsen et al [3] and Madsen and Sørensen [4] included higher order terms with adjustable coefficients into the standard Boussinesq equations for constant and variable water depth, respectively. Beji and Nadaoka [5] gave an alternative derivation of Madsen et al's [4] improved Boussinesq equations. Liu & Wu [7] presented a model with specific applications to ship waves generated by a moving pressure distribution in a rectangular and trapezoidal channel by using boundary integral method. Torsvik [9] presented a numerical investigation on waves generated by a pressure disturbance moving at constant speed in a channel with a variable cross-channel depth profile by using Lynett et al [8] and Liu & Wu [7]'s COULWAVE long wave model. The surface disturbance may come from a moving free surface object, bottom movement, or a moving object in between. The first case is associated with a moving surface pressure which is the main problem to be investigated in this study using Beji and Nadaoka's [5] alternative derivation. First of all the numerical model is verified for different test cases, such as comparing the numerically simulated solitary waves with their analytical counterparts.Then numerical solutions of all possible modes (long wave, classical, and improved Boussinesq) for moving pressures are compared with the analytical solutions. The work is currently being carried out to extend the scheme to 2-D case with realistic surface pressure forms so that waves generated by ship-like objects may be simulated.

2 IMPROVED BOUSSINESQ EQUATIONS

Dispersion relation of Peregrine's system [2] is an accurate approximation to Stokes first order wave theory for very small values of the dispersion parameter μ. Madsen et al [3] improved dispersion characteristics of this system by adding extra dispersive terms to the momentum equations as expressed in terms of depth integrated velocities $P=(h+\eta)\bar{u}$ and $Q=(h+\eta)\bar{v}$. The form of the dispersion relation is determined by matching the dispersion characteristics to linear wave theory. Later, this procedure has been extended to the case of variable depth by Madsen and Sørensen [4]. Alternatively, Beji and Nadaoka [5] introduced a slightly different method to improve the dispersion characteristics by a simple algebraic manipulation of Peregrine's work for variable depth.

2.1 *Derivation of Beji and Nadaoka's improved Boussinesq equations*

Following the procedure given by Peregrine [2] the continuity and momentum equations are,

$$\frac{\partial \bar{\mathbf{u}}}{\partial t} + (\bar{\mathbf{u}}\cdot\nabla)\bar{\mathbf{u}} + g\nabla\eta = \frac{1}{2}h\frac{\partial}{\partial t}\nabla[\nabla\cdot(h\bar{\mathbf{u}})] - \frac{1}{6}h^2\frac{\partial}{\partial t}\nabla(\nabla\cdot\bar{\mathbf{u}}) \tag{1}$$

$$\frac{\partial \eta}{\partial t} + \nabla\cdot\left[(h+\eta)\bar{\mathbf{u}}\right] = 0 \tag{2}$$

According to Beji and Nadaoka the second order terms are replaced with their equivalents in the Boussinesq type equations as these equations are the result of an ordering process with respect to two parameters, which are ε and μ^2. As given by Beji and Nadaoka [5] a simple addition and substraction in equation (2) gives

$$\mathbf{u}_t + (\mathbf{u}\cdot\nabla)\mathbf{u} + g\nabla\eta = (1+\beta)\frac{h}{2}\nabla[\nabla\cdot(h\mathbf{u}_t)] - \beta\frac{h}{2}\nabla[\nabla\cdot(h\mathbf{u}_t)] - (1+\beta)\frac{h^2}{6}\nabla(\nabla\cdot\mathbf{u}_t) + \beta\frac{h^2}{6}\nabla(\nabla\cdot\mathbf{u}_t) \tag{3}$$

where β is a scalar to be determined from the dispersion relation. Instead of a full replacement, a partial replacement of the dispersion terms are made so a form with better dispersion characteristics is obtained. Using $\mathbf{u}_t = -g\nabla\eta$ for replacing the terms proportional to β gives

$$\mathbf{u}_t + (\mathbf{u}\cdot\nabla)\mathbf{u} + g\nabla\eta = (1+\beta)\frac{h}{2}\nabla[\nabla\cdot(h\mathbf{u}_t)] + \beta g\frac{h}{2}\nabla[\nabla\cdot(h\nabla\eta)] - (1+\beta)\frac{h^2}{6}\nabla(\nabla\cdot\mathbf{u}_t) - \beta g\frac{h^2}{6}\nabla(\nabla^2\eta) \tag{4}$$

which is a momentum equation with mixed dispersion terms. Setting $\beta = 0$ recovers the original equation, while $\beta = -1$ corresponds to replacing $\mathbf{u}_t$ with $-g\nabla\eta$ in equation (2). Equations (2) and (4) constitute the improved Boussinesq Equations.

2.2 *Specification of dispersion parameter*

Linearized 1-D Boussinesq Equations for mildly varying depth is formulated as follows. The continuity equation in expanded form

$$\frac{\partial \eta}{\partial t} + \frac{\partial h}{\partial x}u + h\frac{\partial u}{\partial x} = 0 \tag{5}$$

The momentum equation can be expanded as

$$\frac{\partial u}{\partial t} + g\frac{\partial \eta}{\partial x} = h\frac{\partial h}{\partial x}\frac{\partial^2 u}{\partial x\partial t} + \frac{h^2}{3}\frac{\partial^3 u}{\partial x^2\partial t} \tag{6}$$

where $h\,\partial h/\partial x\;\partial^2 u/\partial x\partial t$ is the linear shoaling term while $h^2/3\;\partial^3 u/\partial x^2\partial t$ is the linear dispersing term. Linearized 1-D Boussinesq Equations for constant depth simplify to the following equations.

$$\frac{\partial \eta}{\partial t} + h\frac{\partial u}{\partial x} = 0 \tag{7}$$

$$\frac{\partial u}{\partial t} + g\frac{\partial \eta}{\partial x} = \frac{h^2}{3}\frac{\partial^3 u}{\partial x^2\partial t} \tag{8}$$

Combining equations (6) and (7) by cross-differentiation the 1-D Boussinesq equations for constant depth is obtained as

$$\frac{\partial^2 \eta}{\partial t^2} - gh\frac{\partial^2 \eta}{\partial x^2} = \frac{h^2}{3}\frac{\partial^4 \eta}{\partial x^2\partial t^2} \tag{9}$$

where $h^2/3\;\partial^4\eta/\partial x^2\partial t^2$ is the linear dispersion depth. Water waves of different wave lengths travel with different phase speeds, a phenomenon known as frequency dispersion. For the case of infinitesimal wave amplitude, the terminology is linear frequency dispersion. The frequency dispersion characteristics of a Boussinesq-type of equation can be used to determine the range of wavelengths for which it is a valid approximation. Assume $\eta = \eta_0 e^{i(kx\pm\omega t)}$ so that $\eta_{tt} = -\omega^2\eta_0 e^{i(kx\pm\omega t)}$, $\eta_{xx} = -k^2\eta_0 e^{i(kx\pm\omega t)}$ and $\eta_{xxtt} = k^2\omega^2\eta_0 e^{i(kx\pm\omega t)}$. Substituting these expressions into equation (9) gives

$$\omega^2\left(1+\frac{k^2h^2}{3}\right) = ghk^2 \tag{10}$$

which can be rewritten as,

$$\omega^2 = \frac{k^2}{1+\frac{k^2h^2}{3}}gh \simeq k^2\left(1-\frac{k^2h^2}{3}\right)gh \tag{11}$$

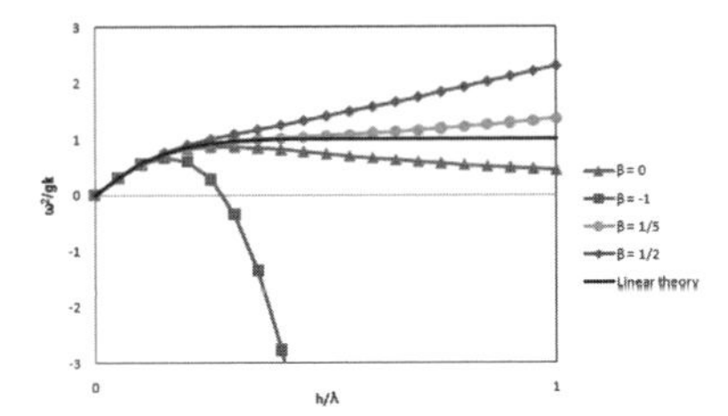

Figure 1. Dispersion curves for various values of dispersion parameter β compared with linear theory.

since $(1+\mu^2)^{-1} \approx 1-\mu^2$ for small values of μ^2. Since $\omega = kc$,

$$c^2 = gh\left(1 - \frac{k^2h^2}{3}\right) \tag{12}$$

Here $k^2h^2/3$ shows the correction to the wave celerity due to the inclusion of the weak dispersion effect. Considering the improved Boussinesq equations, in linearized forms equations (2) and (4) yield the following dispersion relation evaluated by Beji and Nadaoka [5]:

$$\frac{\omega^2}{gk} = \frac{kh(1+\beta k^2h^2/3)}{[1+(1+\beta)k^2h^2/3]} \tag{13}$$

where ω is the wave frequency, $k^2 = k_x^2 + k_y^2$ and k_x, k_y are the components of the wave number vector. Equation (13) is specified according to matching the resulting dispersion relation with a second order Padé expansion of the linear theory dispersion and β is determined from this second order Padé expansion of the linear theory dispersion relation $\omega^2/gk = \tanh kh$:

$$\frac{\omega^2}{gk} = \frac{kh + k^3h^3/15}{1+2k^2h^2/5} \tag{14}$$

In order that Equation (13) be identical with Equation (14) β should be set to 1/5. Figure 1 compares various values of dispersion parameters with the exact expression of linear theory. Among these asymptotic expansions, the one corresponding the Padé type expansion is the best. Thus, when $\beta = 1/5$, the model may propagate relatively shorter waves ($h/\lambda = 1$) with acceptable errors in amplitude and celerity.

3 A NEW DISCRETIZATION SCHEME FOR 1-D IMPROVED BOUSSINESQ EQUATIONS

The finite difference method is the most natural way of solving a PDE directly in an approximate manner. The idea behind this is to discretize the continuous time and space into a finite number of discrete grid points and then to approximate the local derivatives at these grid points with finite difference schemes. For numerical modeling, the discretization of the variables u, v and η are necessary in order to solve momentum and continuity equations. Arakawa C grid which is shown in Figure 2, is the most appropriate system since it enables the discretization of the continuity equation in the most accurate manner. Here, u and η represent the horizontal velocity and the free surface displacement respectively. The surface displacement is obtained from an semi-explicit discretization of the continuity equation which is,

$$\eta_t + \frac{\partial}{\partial x}[(h+\eta)u] = 0 \tag{15}$$

Multiplying both sides of the continuity equation by Δt and differentiating with respect to x gives:

$$\begin{aligned}\left(\frac{\partial \eta}{\partial x}\right)_i^{k+1} &= \left(\frac{\partial \eta}{\partial x}\right)_i^{k} \\ &\quad -\frac{1}{2}h\left[\left(\frac{\partial^2 u}{\partial x^2}\right)_{i-\frac{1}{2}}^{k+1} + \left(\frac{\partial^2 u}{\partial x^2}\right)_{i-\frac{1}{2}}^{k}\right]\Delta t \\ &\quad -2h_x\left(\frac{\partial u}{\partial x}\right)_{i-\frac{1}{2}}^{k+\frac{1}{2}}\Delta t - \frac{\partial^2}{\partial x^2}\left(\eta_i u_{i-\frac{1}{2}}\right)^{k+\frac{1}{2}}\Delta t\end{aligned} \tag{16}$$

where k denotes the time level. It should be noted that this equation is centered at $\eta_i^{k+1/2}$.

The momentum equation which is solved for u is

$$\begin{aligned}u_t + uu_x + g\eta_x &= (1+\beta)\frac{h^2}{3}u_{xxt} \\ &\quad + (1+\beta)hh_x u_{xt} \\ &\quad + \beta g\frac{h^2}{3}\eta_{xxx} + \beta ghh_x\eta_{xx}\end{aligned} \tag{17}$$

Discretization of the momentum equation is given as follows noting that all spatial derivatives are centered at the grid point where u_i^k is located.

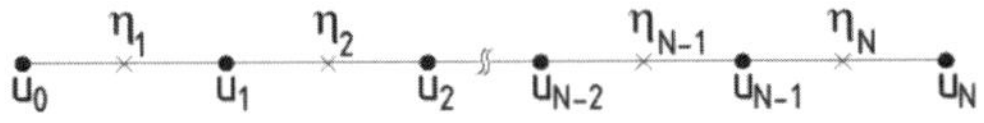

η_1 η_2 η_{N-1} η_N

u_0 u_1 u_2 u_{N-2} u_{N-1} u_N

Figure 2. The Arakawa-C grid.

$$\frac{u_i^{k+1}-u_i^k}{\Delta t}+\frac{1}{2}g\left[\left(\frac{\partial \eta}{\partial x}\right)_{i+\frac{1}{2},j}^{k+1}+\left(\frac{\partial \eta}{\partial x}\right)_{i+\frac{1}{2},j}^{k}\right]$$

$$=(1+\beta)\frac{h^2}{3}$$

$$\left[\frac{(u_{i+1}^{k+1}-2u_i^{k+1}+u_{i-1}^{k+1})-(u_{i+1}^{k}-2u_i^{k}+u_{i-1}^{k})}{\Delta x^2\Delta t}\right]$$

$$+(1+\beta)hh_x\left[\frac{(u_{i+1}^{k+1}-u_{i-1}^{k+1})-(u_{i+1}^{k}-u_{i-1}^{k})}{2\Delta x\Delta t}\right]$$

$$-u_i^{k+\frac{1}{2}}\left(\frac{\partial u_i}{\partial x}\right)^{k+\frac{1}{2}}$$

$$+\beta g\frac{h^2}{3}\left[\frac{\eta_{i+1}^{k+\frac{1}{2}}-3\eta_i^{k+\frac{1}{2}}+3\eta_{i-1}^{k+\frac{1}{2}}-\eta_{i-2}^{k+\frac{1}{2}}}{\Delta x^3}\right]$$

$$+\beta ghh_x\left[\frac{\eta_{i-2}^{k+\frac{1}{2}}-\eta_{i+1}^{k+\frac{1}{2}}-\eta_i^{k+\frac{1}{2}}-\eta_{i-1}^{k+\frac{1}{2}}}{2\Delta x^2}\right] \quad (18)$$

Substituting $(\partial\eta/\partial x)_{i,j}^{k+1}$ from equation (16) into the discretized x-momentum equation (18) and multiplying by Δt gives the expression which is essentially a tridiagonal matrix system for u_{i-1}^{k+1}, u_i^{k+1} and u_{i+1}^{k+1}.

4 TEST CASES FOR THE VERIFICATION OF THE BOUSSINESQ MODEL

4.1 *An analytical solution of Boussinesq equations: Solitary waves*

The most elementary analytical solution of Boussinesq equations is a solitary wave. A solitary wave is a wave with only crest and a surface profile lying entirely above the still water level. It is neither oscillatory nor does it exhibit a trough. The solitary wave can be defined as a wave of translation since the water particles are displaced at a distance in the direction of wave propagation as the wave passes. A true solitary wave cannot be formed in nature because there are usually small dispersive waves at the trailing edge of the wave. On the other hand, long waves such as tsunamis and waves resulting from large displacements of water caused by such phenomena as landslides and earthquakes sometimes behave approximately like solitary waves. Also, when an oscillatory wave moves into shallow water, it may often be approximated by a solitary wave. In this situation, the wave amplitude becomes progressively higher, the crests become shorter and more pointed, and the trough becomes longer and flatter. Only one parameter, wave steepness, $\varepsilon = H/d$ is needed to specify a solitary wave because both wavelength and period of solitary waves are infinite. To the lowest order, the solitary wave profile varies as $sech^2 q$, where $q=(3H/d)^{1/2}(x-Ct)/2d$. The free-surface elevation, particle velocities, and pressure may be expressed respectively as follows

$$\frac{\eta}{h}=\frac{u}{\sqrt{gd}\,\frac{H}{d}} \quad (19)$$

$$\frac{u}{\sqrt{gd}}\frac{H}{d}=\frac{\Delta p}{\rho g H} \quad (20)$$

$$\frac{\Delta p}{\rho g H}=\operatorname{sech}^2 q \quad (21)$$

where Δp is the difference in pressure at a point under the wave due to the presence of the solitary wave. To second approximation, this pressure difference is given by

$$\frac{\Delta p}{\rho g H}=1-\frac{3}{4}\frac{H}{d}\left[1-\left(\frac{Y_s}{d}\right)^2\right] \quad (22)$$

where Y_s is the height of the surface profile above the bottom. Since the solitary wave has horizontal particle velocities only in the direction of wave advance, there is a net displacement of fluid in the direction of wave propagation. The solitary wave is a limiting case of the cnoidal wave. Cnoidal waves may be viewed as the nonlinear counterparts of the sinusoidal waves in shallow water. When $k^2=1$, $K(k)=K(1)=\infty$, and the elliptic cosine reduces to the hyperbolic secant function and the water surface Y_s measured above the bottom reduces to

$$Y_s = d + H\operatorname{sech}^2\left[\sqrt{\frac{3}{4}\frac{H}{d^3}}(x-Ct)\right] \quad (23)$$

and the free surface is given by,

$$\eta = H\operatorname{sech}^2\left[\sqrt{\frac{3}{4}\frac{H}{d^3}}(x-Ct)\right] \quad (24)$$

The numerical tests presented here are done using the numerical scheme developed in Section 3. The water depth is constant and both the original Boussinesq equations ($\beta=0$) and the improved Boussinesq equations ($\beta=1/5$) are used for simulations. As it can be seen in Figures 3 and 4 the analytical and computational results agree very well for both $\beta=0$ and $\beta=1/5$, although from analytical point of view, the solitary waves corresponding to the improved Boussinesq equations should be slightly different. Differences in height between analytical and computational results are shown in Figure 5 for $\beta=0$

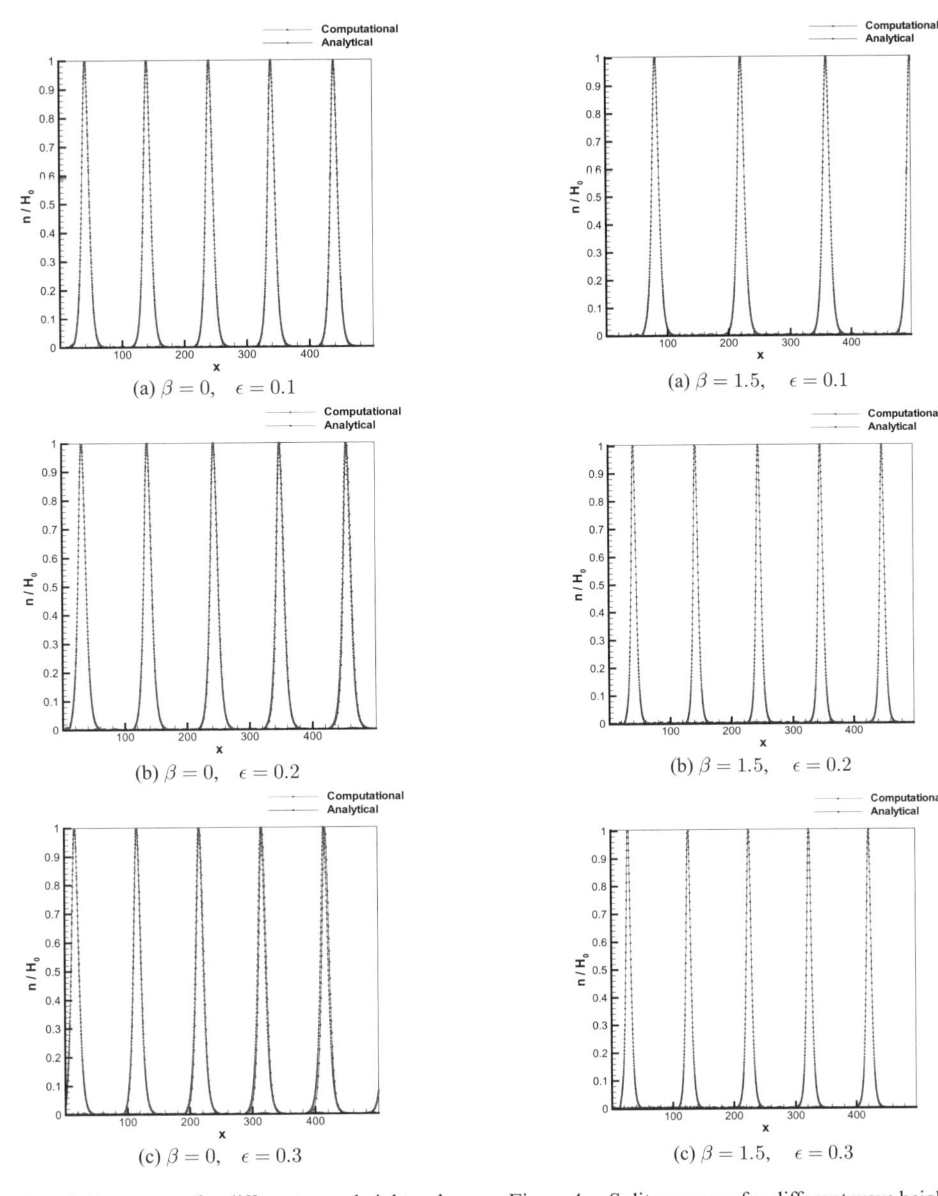

Figure 3. Solitary waves for different wave heights when $\beta = 0$.

Figure 4. Solitary waves for different wave heights when $\beta = 1/5$.

by calculating the relative error percentage for different ε values. It is observed that as nonlinearity parameter, ε, increases, the relative error percentage increases linearly up to 13% for $\varepsilon = 0.4$.

4.2 *Comparison of analytical solution to linear shallow water wave equations*

The linearized long-wave equations in 1-D in presence of a pressure term are given by,

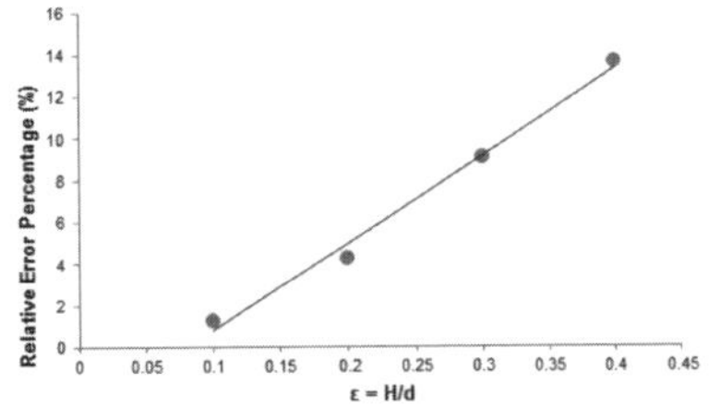

Figure 5. Relative error of the calculated wave height versus nonlinearity parameter ε.

$$\eta_t + hu_x = 0 \quad (25)$$

$$u_t + g\eta_x = -\frac{1}{\rho}\nabla P \quad (26)$$

where η is free surface elevation, u horizontal velocity component, P surface pressure, h constant water depth and g gravitational acceleration. Due to the moving pressure, two different free surface waves moving in opposite directions are generated. The mathematical problem can be separated as the free surface wave problem and the pressure wave problem. Assume that the moving pressure is $P = P_0F(x - Vt)$ and for the two free waves the profiles are $\eta_1 = a_1F(x - c_0t)$ and $\eta_2 = a_2F(x + c_0t)$. Besides these free waves the forced wave profile is $\eta_3 = a_3F(x - Vt)$ and the velocity of the forced wave is $u_3 = b_3F(x - Vt)$. For the right moving sinusoidal wave, wave surface profile can be described as,

$$\eta = a_0 \sin(kx - \omega t) \quad (27)$$

where a_0 is the wave amplitude, k is the wave number. Differentiating the above equation with respect to x gives

$$\eta_x = ka_0 \cos(kx - \omega t) \quad (28)$$

From the momentum equation (26) for the unforced free wave case (no pressure) $u_t = -g\eta_x$. Substituting η_x into this expression

$$u_t = -gka_0 \cos(kx - \omega t) \quad (29)$$

Integrating over time, the horizontal velocity u is found as,

$$u = \frac{gk}{\omega} a_0 \sin(kx - \omega t) \quad (30)$$

where $a_0 \sin(kx - \omega t)$ represents η itself and $\omega = kc$. Substituting these expressions, the horizontal velocity is found as, $u = c\eta / h$. Therefore, right moving wave velocity is $u_1 = c_0\eta_1 / h$ and the left moving wave velocity is $u_2 = -c_0\eta_2 / h$. Now considering the forced wave case with pressure gradient and substituting η_3 into the continuity equation (25) by taking the time derivative, $u_3 = Va_3F(x - Vt)/h$. Noting that $u_3 = b_3F(x - Vt)$ gives

$$a_3 = -\frac{hP_0}{\rho\left(gh - V^2\right)} \quad (31)$$

Substituting $u_3 = b_3F(x - Vt)$ and $P = P_0F(x - Vt)$ into the momentum equation (26) by taking the derivative with respect to t and x respectively, $u_3 = \rho g V a_3 F(x - Vt) + PF(x - Vt)/\rho V$. Noting that $u_3 = b_3F(x - Vt)$ gives

$$b_3 = -\frac{P_0V}{\rho\left(gh - V^2\right)} \quad (32)$$

Substituting a_3 into the expression $\eta_3 = a_3F(x - Vt)$ results in $\eta_3 = hP_0F(x - Vt)/\rho(gh - V^2)$. u_1 and u_2 are found by substituting η_1 and η_2 into the continuity equation (25) respectively. After these substitution $u_1 = a_1c_0F(x - c_0t)/h$ and $u_2 = -a_2c_0F(x + c_0t)/h$. The boundary conditions are,

$$u_1 + u_2 + u_3 = 0 \quad (33)$$

$$\eta_1 + \eta_2 + \eta_3 = 0 \quad (34)$$

Substituting free and forced solutions into (33) and (34) for $t = 0$ and solving for a_1 and a_2 gives

$$a_1 = \frac{hP_0(c_0 + V)}{2c_0\rho\left(gh - V^2\right)} \quad (35)$$

$$a_2 = \frac{hP_0(c_0 - V)}{2c_0\rho\left(gh - V^2\right)} \quad (36)$$

Finally, for three different wave profiles the following expressions for η are obtained.

$$\eta_1 = \frac{hP_0(c_0 + V)}{2c_0\rho\left(gh - V^2\right)} F(x - c_0t) \quad (37)$$

$$\eta_2 = \frac{hP_0(c_0 - V)}{2c_0\rho\left(gh - V^2\right)} F(x + c_0t) \quad (38)$$

$$\eta_3 = -\frac{hP_0}{\rho\left(gh - V^2\right)} F(x - Vt) \quad (39)$$

The corresponding velocities are computed likewise. Let's assume that the moving pressure field is represented by $F(\chi) = Exp[-(x/250)^2]$ where $\chi = x - Vt$. In this case we choose, $h = 20\ m$, $P_0 = -4905$, $g = 9.81\ m/s^2$ and $\rho = 1000\ kg/m^3$. The length of the computational domain is $20000\ m$, grid size is $20\ m$ and time step is $1\ s$. Solutions for linear shallow water waves at $t = 50$ and $100\ s$ and for the velocities, $V = 10$ and $18\ m/s$ are in Figure 6.

The same analytical solution is compared with one dimensional Boussinesq model when $\beta = 1/5$ for the velocities $V = 0{,}10$ and $18\ m/s$ at $t = 50$ and $100\ s$. As it can be seen from Figure 7, the analytical and numerical solutions are again in agreement.

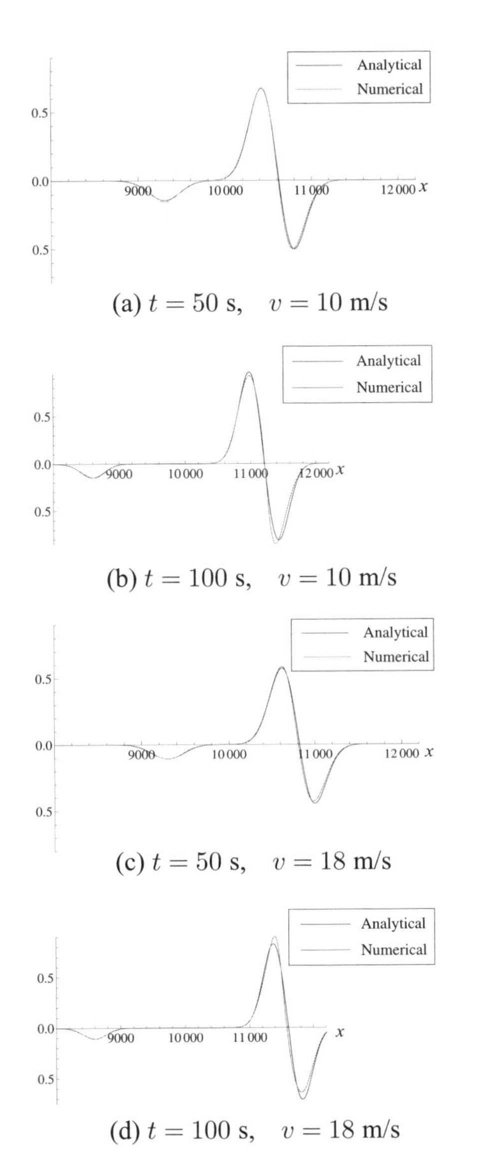

(a) $t = 50$ s, $v = 10$ m/s

(b) $t = 100$ s, $v = 10$ m/s

(c) $t = 50$ s, $v = 18$ m/s

(d) $t = 100$ s, $v = 18$ m/s

Figure 6. Comparison of numerical and analytical solutions of linear shallow water waves generated by a moving pressure at $t = 50\ s$ and $t = 100\ s$.

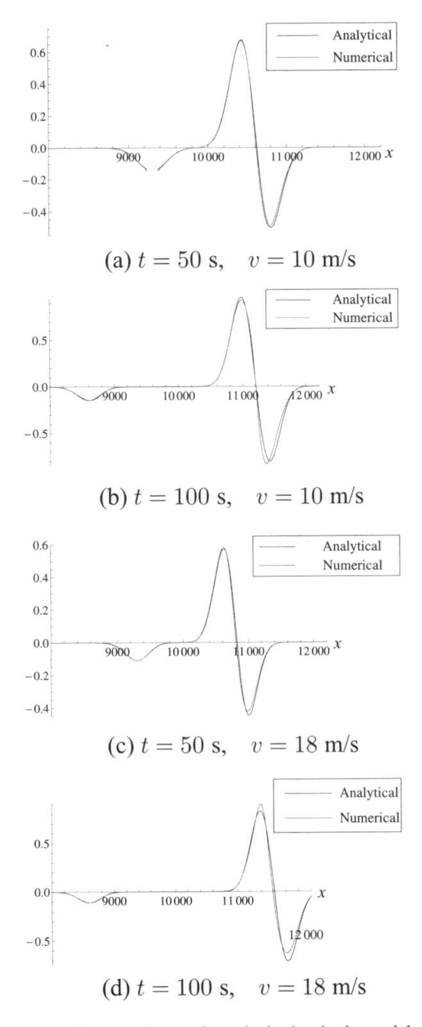

(a) $t = 50$ s, $v = 10$ m/s

(b) $t = 100$ s, $v = 10$ m/s

(c) $t = 50$ s, $v = 18$ m/s

(d) $t = 100$ s, $v = 18$ m/s

Figure 7. Comparison of analytical solution with linear shallow water waves and 1-D Boussinesq solution generated by a moving pressure when $\beta = 1/5$ at $t = 50\ s$ and $t = 100\ s$.

In Figure 8 the average (using 500 points) error percentages between analytical and computational surface elevations are shown for five different pressure field speeds $V = 5, 10, 15, 20$ and 25 m/s which corresponds to depth Froude numbers, 0.4, 0.7, 1.1, 1.4 and 1.8 at $t = 50$ s when $\beta = 1/5$. Results show that around Froude number 1, the relative error percentage takes its maximum value and as Froude number exceeds 1, the error percentage decreases.

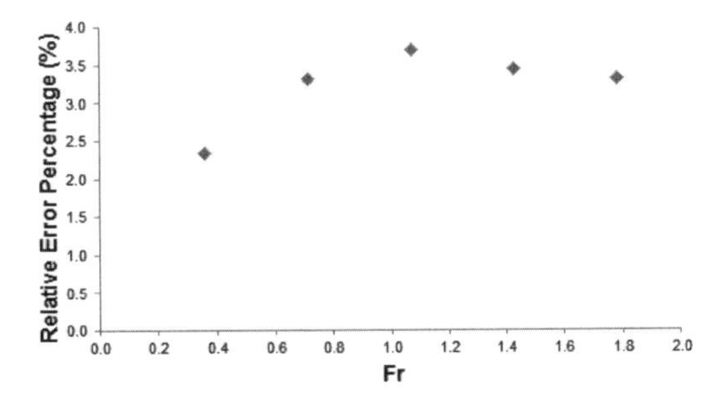

Figure 8. Average relative error of the calculated and analytical surface elevation versus Froude number.

5 CONCLUSIONS

The numerical model developed here is first tested by comparing the numerically simulated solitary waves with their analytical counterparts. These simulations serve to a twofold purpose; namely, testing of the nonlinear and the dispersive performance of the numerical model. The comparisons and the relative error percentage plots give quite satisfactory results hence establish the reliability of the numerical approach adopted. The second test case concerns the application of fixed and moving surface pressures. For the long wave equations the analytical solutions for fixed and moving surface pressures of Gaussian type are recapitulated first. Then, for all possible modes (long wave, classical, and improved Boussinesq) numerical simulations with the Gaussian surface pressure are performed and compared with the analytical results. Comparisons and relative error percentage plots clearly show that all the modes work well with the specified surface pressure and compare quite satisfactorily with the analytical solutions. Also, it is observed that around critical Froude number, the error increases but for Froude number greater than unity, the error percentage decreases again.

REFERENCES

[1] Mei, C.C. and LeMehaute, B. Note on the Equations of Long Waves Over an Uneven Bottom, J. Geophys. Res. 71:393 (1966).

[2] Peregrine, D.H. Long Waves on a Beach, J. Fluid Mech. 27:815.(1967).

[3] Madsen, P.A., Murray, R. and Sørensen, O.R.A New Form of the Boussineq Equations with Improved Linear Dispersion Characteristics, Coastal Engineering. 15:371.(1991).

[4] Madsen, P.A. and Sørensen, O.R. A New Form of the Boussineq Equations with Improved Linear Dispersion Characteristics .2. A Slowly Varying Bathmetry, Coastal Engineering. 18:183.(1992).

[5] Beji, S. and Nadaoka, K.A. Formal Derivation and Numerical Modelling of the Improved Boussinesq Equations for Varying Depth, Ocean Engineering. 23:691.(1996).

[6] Nwogu, O. Alternative Form of Boussinesq Equations for Nearshore Wave Propagation, Journal of Waterway, Port, Coastal and Ocean Engineering. 119:618. (1993).

[7] Liu, P.L.F and Wu, T.R. Waves Generated by Moving Pressure Disturbances in Rectangular and Trapezoidal Channels, J. Hydraul. Res., 42:163 (2004).

[8] Lynett, P., Wu, T.-R., and Liu, P.L.-F.Modeling Wave Runup with Depth-Integrated Equations, Coastal Eng., 46(2), 89–107 (2002).

[9] Torsvik, T., Pedersen, G. and Dysthe, K. Waves Generated by a Pressure Disturbance Moving in a Channel with a Variable Cross-Sectional Topography, Journal of Waterway, Port, Coastal and Ocean Engineering, (2009).

Sustainable Maritime Transportation and Exploitation of Sea Resources – Rizzuto & Guedes Soares (eds)
© 2012 Taylor & Francis Group, London, ISBN 978-0-415-62081-9

A coupled-mode model for water-wave induced groundwater pressure and flow in variable bathymetry regions and beaches

K.A. Belibassakis
School of Naval Architecture & Marine Engineering, NTU of Athens, Greece

ABSTRACT: Beach groundwater-swash interaction is a subject of interest in coastal engineering, sediment transport and groundwater circulation. Focusing on the region where water depths are greater than the wave breaking depth, a phase-resolving coupled-mode model is developed for modelling water wave propagation in variable bathymetry and coastal regions and its interaction with porous flow in the porous layer under the penetrable seabed of general shape, characterized by sloping parts and undulations. Theoretical analysis and experimental results in this direction have been presented by Massel (2004, 2005) who developed a closed-form solution, in constant depth, for the pore-water pressure component and velocity circulation pattern induced by surface waves. The present model is also based on Biot's theory, taking into account volume change and pore-water flow, and extends previous work (Athanassoulis & Belibassakis 1999, Belibassakis et al 2007) to variable bottom topography in the presence of penetrable bottom, enabling prediction of wave-induced groundwater dynamic pressure and flow in the porous medium.

1 INTRODUCTION

Sandy beaches consist of unconsolidated sediment and are permeable. In this case, the changes of pressure associated with both the mean and the oscillatory wave flows produce a groundwater flow of sea water within the porous medium. The mean flow component plays a significant role concerning water table formation and groundwater flow. Furthermore, this component percolates through the permeable bottom and influences the wave forces on structures supported by or extending into the bottom. On the other hand, the oscillatory component of the wave field contributes to damping of the waves over a porous beach. Knowledge of both the above components is important concerning the interaction of physical processes, biodiversity and productivity of sandy beaches, sediment transport and coastal structure stability (see, e.g., Massel 2001, 2004, 2005).

As concerns the mean flow component, useful information can be obtained from the solution of the slow-scale mean-flow equations, forced by the radiation stresses induced by wave propagation in variable bathymetry regions, as well as by the free-surface and the bottom stresses. The radiation stresses can be calculated from the fast-scale wave flow properties over the beach region; see, e.g., Mei (1983), Dingemans (1997). Finally, the resulting phase-averaged mean-pressure on the bottom can be used to calculate induced groundwater circulation in permeable beach in the set-up region, assuming that the groundwater flow is in the Darcy law regime (see, e.g., Belibasakis et al 2007).

For the interaction of free-surface gravity waves with variable bottom topography, in water of intermediate depth and in shallow water, a broad class of approximation techniques has been developed; see, e.g., Dingemans (1997). Most one-equation *mild-slope models* suffer from the fact that the vertical structure of the wave field given by a specific, preselected function is poorly represented. In order to better describe the wave field when the bottom topography is not slowly varying and the depth is sufficiently small so that the wave strongly interacts with the bottom, Massel (1993) and Porter & Staziker (1995) derived extended mild-slope models, in which the vertical profile of the wave potential at any horizontal position is represented by a local-mode series involving the propagating and the evanescent modes. However, this expansion has been found to be inconsistent with the Neumann boundary condition on a sloping bottom, since each of the vertical modes involved in the local-mode series violates it and, thus, the solution, being a linear superposition of modes, behaves the same. This fact has two important consequences. First, the velocity field in the vicinity of the bottom is poorly represented and, secondly, wave energy is not generally conserved. This problem has been remedied by the coupled-mode model developed by Athanassoulis & Belibassakis (1999). In the latter model the standard local-mode representation is enhanced by including an additional term, called the sloping-bottom mode, leading to a consistent coupled-mode system of equations. This model is free of any assumptions concerning the smallness of bottom slope and curvature, and it is consistent since it enables the exact satisfaction of the bottom boundary condition and the correct calculation of the bottom velocities. A key feature of the above coupled-mode technique is that the rate of decay of the modal amplitudes is very fast (see Athanassoulis & Belibassakis 1999), and thus, only a small number of modes is sufficient for a very accurate calculation. These facts facilitate its extension to treat wave propagation and diffraction in general 3D environments, Belibassakis et al (2001), and to predict second-order and fully non-linear waves in variable bathymetry, Belibassakis & Athanassoulis (2002, 2011).

In previous works (Belibassakis et al 2007a) the consistent coupled-mode model has been further extended to include the effects of bottom friction and wave breaking, which are important for the more accurate calculation of the radiation stresses, especially on decreasing depth and in a beach region. Then, this model has been used, in conjunction with an iterative solver of the mean-flow equations, permitting an accurate calculation of wave-induced set-up and flow in open and closed environments. Finally, the resulting phase-averaged mean-pressure has been applied to calculate the induced groundwater circulation on a permeable beach, in the set-up region (Belibassakis et al 2007b). Under the assumption that the groundwater flow is in the Darcy law regime, in the case of a stationary, phase-averaged flow, the velocities can be obtained in terms of the pressure gradient. In this case, the groundwater circulation is governed by the Laplace's equation on the pressure, forced by Dirichlet data specified by the excess pressure on the sea bottom that is induced by the mean flow; see also Massel (2001).

Focusing on the region where water depths are greater than the wave breaking depth, in the present work a phase-resolving coupled-mode model is developed for modelling water wave propagation in variable bathymetry and coastal regions and its interaction with porous flow in the porous layer under the penetrable seabed of general shape, characterized by sloping parts and undulations. Theoretical analysis and experimental results in this direction have been presented by Massel (2004, 2005) who developed a closed-form solution, in constant depth, for the pore-water pressure component and velocity circulation pattern induced by surface waves. The present model is also based on Biot's theory, taking into account volume change and pore-water flow, enabling prediction of wave-induced groundwater dynamic pressure and flow in the porous medium.

2 FORMULATION OF THE PROBLEM

The marine environment consists of a water layer bounded above by the free surface and below by the sea bottom, separating water from porous medium, which is terminated by a horizontally flat, impermeable bottom boundary. The variable water depth is denoted by h and thus, the thickness of the permeable layer is $h_B - h$. The coefficients of permeability (or filtration) and porosity of the sandy bottom are denoted by K_f and n_K, respectively; see Figure 1.

For simplicity we will restrict ourselves to a 2D problem, however the present theory can be extended to general 3D environments; see, e.g. Belibassakis et al (2001), Gerostathis et al (2008) as concerns the water-wave part. It is assumed that the bottom surface exhibits a general variation, i.e. the bathymetry $h(x)$ is characterised by parallel, straight bottom contours lying between two regions of different depth: the deeper water region or region of incidence ($x < a$, where the depth is constant and equal to $h = h_1$) and the shallow water region ($x > b$, where the depth is constant and equal to $h = h_3$). A Cartesian coordinate system is intro-duced, with its origin at the intersection between the plane beach topography and the mean water level, and the z-axis pointing upwards.

The liquid domain D is decomposed in three parts $D^{(m)}$, $m = 1,2,3$, where $D^{(2)}$ is the variable bathymetry subdomain ($a < x < b$). The free-surface ∂D_F and bottom ∂D_B boundaries are similarly decomposed. The same domain decomposition in three parts $D_p^{(m)}$, $m = 1,2,3$, is also introduced in the porous layer; see Figure 1.

The wave field in the region D is excited by a harmonic incident wave of small amplitude, with direction of propagation normal to the depth-contours. In the framework of linearised water wave equations (Mei 1983), the fluid motion is described by the 2D potential

$$\Phi(x,z,t) = \mathrm{Re}\left\{-\frac{igH}{2\omega}\varphi(x,z;\mu)\cdot\exp(-i\omega t)\right\}, \quad (1)$$

where H is the waveheight, g is the acceleration due to gravity, $\mu = \omega^2/g$ is the frequency parameter, and $i = \sqrt{-1}$. The free-surface elevation is obtained in terms of the wave potential on the free surface as $\eta(x,t) = (-1/g)\partial\Phi(x, z = 0, t)/\partial t$.

In the water region, the complex wave potential satisfies the Laplace's equation

$$\frac{\partial^2\varphi(x,z)}{\partial x^2} + \frac{\partial^2\varphi(x,z)}{\partial z^2} = 0, \quad -h(x) < z < 0, \quad (2a)$$

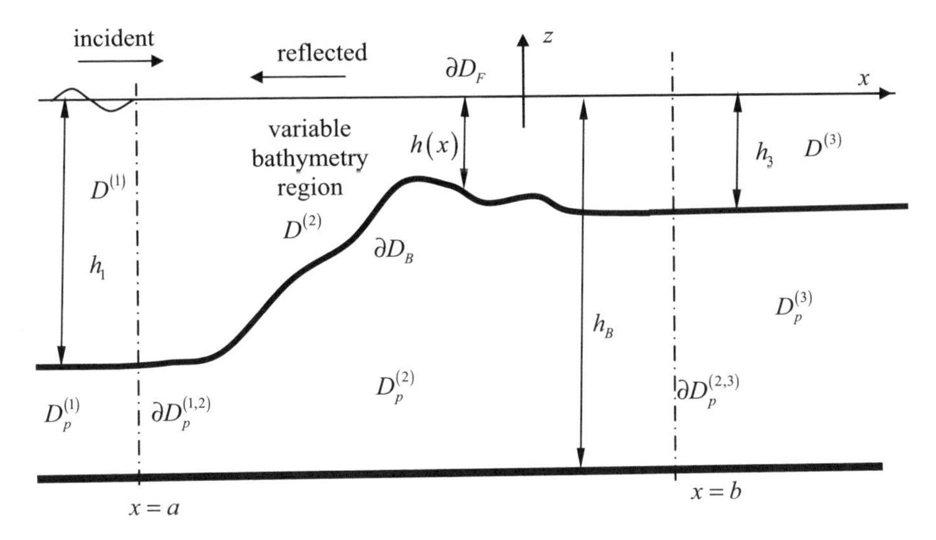

Figure 1. The marine environment (definition of basic quantities).

supplemented by the free-surface boundary condition,

$$\frac{\partial \varphi(x,z)}{\partial z} - \mu\varphi(x,z) = 0, \quad z = 0 \tag{2b}$$

Extended theoretical and experimental evidence (Massel 2004, 2005) indicates that the influence of the bottom permeability on sea surface elevation is small. This influence increases when the point under consideration approaches the permeable sea bottom $z = -h(x)$, where the following boundary conditions are satisfied:

$$p = i\omega\rho\varphi(x, z = -h(x)) = p_K(x, z = -h(x)), \tag{3a}$$

where ρ is the water density, and

$$\frac{\partial \varphi(x, z = -h(x) + 0)}{\partial n} = n_K u_n(x, z = -h(x) - 0), \tag{3b}$$

where u_n denotes the normal component of the pore water velocity and p_K the dynamic pressure in the porous medium. Considering materials finer than gravel, the normal velocity u_n is very small, and the above equation (3b) is simplified by setting $\partial\varphi/\partial n = 0$, i.e.

$$\frac{\partial \varphi(x,z)}{\partial z} + \frac{dh}{dx}\frac{\partial \varphi(x,z)}{\partial x} = 0, \quad z = -h(x) \tag{4}$$

In this case, the wave fields in the water column and in the porous medium are essentially coupled only by the continuity of the dynamic pressure, Eq. (3a), and thus, the water wave propagation problem can be decoupled from the porous flow subproblem. To this connection, the water wave equations (2) and (4) are supplemented by the following conditions at infinity, as $x \to \mp\infty$,

$$\varphi(x,z) \approx \left[e^{ik_0^{(1)}x} + A_R e^{-ik_0^{(1)}x}\right]\frac{\cosh\left(k_0^{(1)}(z+h_1)\right)}{\cosh\left(k_0^{(1)}h_1\right)}, \tag{5a}$$

$$\varphi(x,z) \approx A_T e^{ik_0^{(3)}x}\frac{\cosh\left(k_0^{(3)}(z+h_3)\right)}{\cosh\left(k_0^{(3)}h_3\right)}, \tag{5b}$$

where A_R and A_T denote the reflection and transmission coefficients, and the wavenumbers $k_0^{(1)}, k_0^{(3)}$ are obtained as the positive real root of the dispersion relations, formulated at the depth s h_1 and h_3, respectively,

$$\mu h_m = k_0^{(m)}h_m \tanh\left(k_0^{(m)}h_m\right), \quad m = 1,3. \tag{6}$$

3 EQUATIONS IN POROUS MEDIUM

In shallow water, due to possible wave breaking and the entrance of gas into the porous medium, as well as production of gases by the organisms living in the sand, the apparent bulk modulus of the pore water $\hat{E}_w$ depends on the degree S of saturation by water. Following Massel (2005), we use here the relationship proposed by Verruijt (1969)

$$\hat{E}_w^{-1} = SE_w^{-1} + (1-S)/p_0, \tag{7}$$

where $E_w = 1.9\,10^9$ Nm^{-2} is the bulk modulus of pore water without air, $(1-S)$ is the degree of saturation by air, usually less than one, and p_0 is the absolute pressure.

Assuming that the velocity components of the soil matrix are very small, the dynamic equations of fluid motion in the porous medium can be simplified, and are written in the following form

$$u = -\left(n_K \frac{\partial P_K}{\partial x} + n_K \rho \frac{\partial u}{\partial t}\right)\frac{K_f}{n_K^2 \gamma}, \tag{8a}$$

$$w = -\left(n_K \frac{\partial P_K}{\partial z} + n_K \rho \frac{\partial w}{\partial t}\right)\frac{K_f}{n_K^2 \gamma}, \tag{8b}$$

where (u,w) denote the velocity components of the flow in the porous medium and P_K is the pore pressure. Furthermore, mass conservation, in the form of storage equation, takes the form

$$\frac{\partial u}{\partial x} + \frac{\partial w}{\partial z} = -\frac{1}{\hat{E}_w}\frac{\partial P_K}{\partial t}, \tag{9}$$

see, e.g., Mei (1983, Sec 13) and Massel (2005).

Using Eqs. (8) in Eq. (9) to eliminate the fluid velocities (u,w) we finally obtain the following equation concerning the pore dynamic pressure

$$\frac{\partial^2 P_K}{\partial x^2} + \frac{\partial^2 P_K}{\partial z^2} - \frac{\rho}{\hat{E}_w}\frac{\partial^2 P_K}{\partial t^2} - \frac{n\gamma}{\hat{E}_w K_f}\frac{\partial P_K}{\partial t}. \tag{10}$$

The above equation considered in the frequency domain

$$P_K(x,z,t) = \mathrm{Re}\left\{-\frac{igH}{2\omega} p_K(x,z)\cdot \exp(-i\omega t)\right\}, \tag{11}$$

finally reduces to the Helmholtz equation

$$\frac{\partial^2 p_K}{\partial x^2} + \frac{\partial^2 p_K}{\partial z^2} + \psi^2 p_K = 0, \tag{12}$$

characterised by a complex-valued wavenumber parameter,

$$\psi^2 = \frac{\rho\omega^2}{\hat{E}_w} + i\frac{n\gamma\omega}{\hat{E}_w K_f}, \tag{13}$$

where $\gamma = \rho g$. The imaginary part of ψ indicates the dissipative nature of the examined flow in the porous medium.

Furthermore, since the horizontal bed rock at $z = -h_B$ is considered impermeable, no normal flow is permitted, and from Eq. (8b) we obtain the following Neumann condition for the pore pressure

$$\frac{\partial p_K}{\partial z}(x, z = -h(x)) = 0. \tag{14}$$

Thus, the equation governing the harmonic porous flow is Eq. (12) forced by the water-wave applied (dynamic) pressure on the interface $z = -h(x)$, as obtained from Eq. (3a).

In the case of horizontally flat interface, as e.g. in the two half strips $D_\ell^{(m)}$, $m = 1,3$, periodic solutions of the above system, Eqs. (11), (14) and (3a), can be obtained by separation of variables of the Helmholtz equation. The latter permit us to derive expressions of the pore-pressure field, which far away from the depth inhomogeneity in the regions of incidence and transmission, are as follows:

$$p_K^{(1)}(x,z) = \frac{\cosh\left[\lambda_0^{(1)}(z+h_B)\right]}{\cosh\left[\lambda_0^{(1)}(h_B - h_1)\right]}\frac{i\omega\rho}{\cosh\left(k_0^{(1)}h_1\right)} \times\left(\exp\left(ik_0^{(1)}x\right) + A_R \exp\left(-ik_0^{(1)}x\right)\right) \tag{15}$$

$$p_K^{(3)}(x,z) = \frac{\cosh\left[\lambda_0^{(3)}(z+h_B)\right]}{\cosh\left[\lambda_0^{(3)}(h_B - h_3)\right]}\frac{i\omega\rho A_T \exp\left(ik_0^{(3)}x\right)}{\cosh\left(k_0^{(3)}h_3\right)}, \tag{16}$$

where

$$\lambda_0^{(m)} = \sqrt{\left(k_0^{(m)}\right)^2 - \psi^2} = \sqrt{\left(k_0^{(m)}\right)^2 - \frac{\rho\omega^2}{\hat{E}_w} - i\frac{n\gamma\omega}{\hat{E}_w K_f}}, \quad m = 1,3. \tag{17}$$

Propagating harmonic solutions in constant depth have been derived by (Massel 2005, Sec. 2.3) for the same problem including the dynamic equations of soil momentum, which in the present analysis have been neglected on the basis of very small soil matrix velocity assumption.

As an example, we present in Figure 2 a comparison between present model predictions and experimental data (from Massel 2005), in the case of short waves of period $T = 5$ s and wave height $H = 0.3$ m, propagating in a water layer of constant depth $h = 2$ m overlying

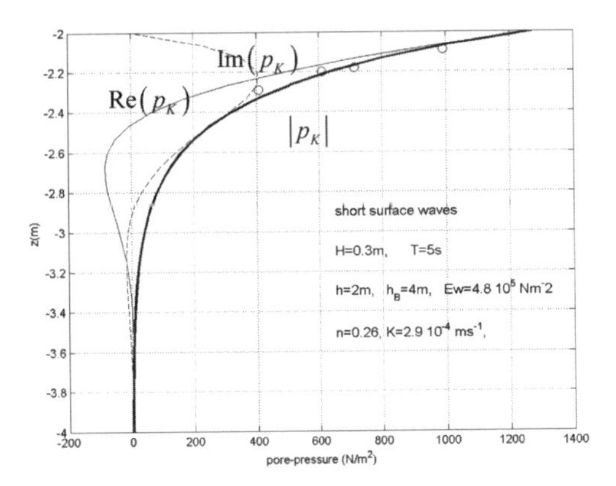

Figure 2. Vertical distribution of dynamic pore-pressure for short waves in constant depth. Comparison between theoretical predictions (lines) and experimental data (circles, from Massel 2005).

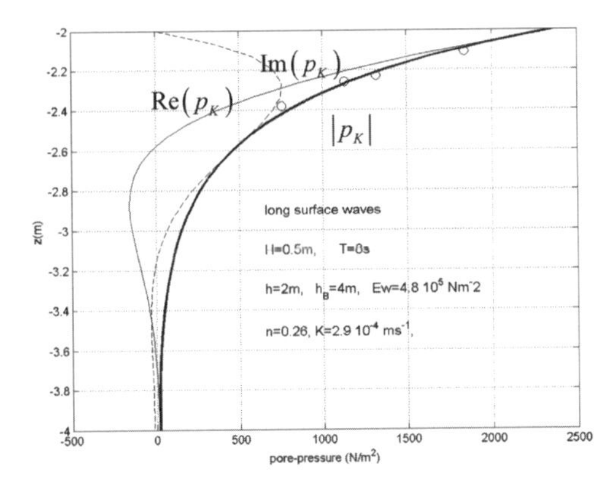

Figure 3. Vertical distribution of dynamic pore-pressure for long waves in constant depth. Comparison between theoretical predictions (lines) and experimental data (circles, from Massel 2005).

a sandy bottom of thickness 2 m (so that $h_B = 4$m. In this case, the permeability of the porous medium is $K_f = 2.9\,10^{-4}\,\mathrm{ms}^{-1}$ and the porosity $n_K = 0.26$. High degree of saturation of water by air is reported so that in the examined case $\hat{E}_w = 4.8\,10^5\ \mathrm{Nm}^{-2}$. We see in Figure 2 that the present results, obtained by neglecting the soil skeleton velocities, still compare quite well with the measured data. The corresponding result, in the case of long waves of period $T = 8$ s and wave height $H = 0.5$ m, propagating the same environment, is presented in Figure 3.

4 WATER-WAVE PROPAGATION

The monochromatic wave propagation-diffraction problem in the variable bathymetry region is reformulated as a matching boundary value problem on the complex wave potential $\varphi^{(2)}(x,z)$, $(x,z) \in D^{(2)}$ This is obtained by using the following modal-type expansions of the wave potential in the constant-depth half strips $D^{(1)}$ and $D^{(3)}$:

$$\varphi^{(1)}(x,z) = \left(\exp\left(-ik_0^{(1)}x\right) + A_R \exp\left(ik_0^{(1)}x\right)\right) Z_0^{(1)}(z) + \sum_{n=1}^{\infty} C_n^{(1)} Z_n^{(1)}(z) \exp\left(-k_n^{(1)}(x-a)\right) \quad (x,z) \in D^{(1)} \tag{18a}$$

$$\varphi^{(3)}(x,z) = A_T \exp\left(ik_0^{(3)}x\right) Z_0^{(3)}(z) + \sum_{n=1}^{\infty} C_n^{(3)} Z_n^{(3)}(z) \exp\left(k_n^{(3)}(b-x)\right) \quad (x,z) \in D^{(3)} \tag{18b}$$

In the above expansions the sets of numbers $\{ik_0^{(i)}, k_n^{(i)}, n = 1,2,\ldots\}$, $i = 1,3$, and the sets of vertical functions $\{Z_n^{(i)}(z), n = 0,1,2,\ldots\}$, $i = 1,3$, are the eigenvalues and the corresponding eigenfunctions of regular Sturm-Liouville problems obtained from the Laplace equation, by separation of variables, in the constant-depth half-strips $D^{(i)}$, $i = 1,3$. Then, the coefficients $A_R, \{C_n^{(1)}\}_{n\in N}$ and $A_T, \{C_n^{(3)}\}_{n\in N}$, and the wave potential $\varphi^{(2)}(x,z)$ in $D^{(2)}$ can be obtained as the solution of the Laplace equation on the vertical plane in $D^{(2)}$, satisfying the free-surface condition on $\partial D_F^{(2)}$, the bottom boundary condition on $\partial D_\Pi^{(2)}$, and matching conditions requiring continuity of wave velocity and pressure on the vertical interfaces, $\partial D_I^{(12)}$ $(x = a)$ and $\partial D_I^{(23)}$ $(x = b)$, separating the variable bathymetry region $D^{(2)}$ from $D^{(1)}$ and $D^{(3)}$.

The wave potential $\varphi^{(2)}(x,z)$ associated with the propagation/diffraction of water waves in the variable bathymetry region can be very conveniently treated by means of the consistent coupled-mode model (Athanassoulis & Belibassakis 1999). This model is based on the following enhanced local-mode representation:

$$\varphi^{(2)}(x,z) = \varphi_{-1}(x) Z_{-1}(z;x) + \varphi_0(x) Z_0(z;x) + \sum_{n=1}^{\infty} \varphi_n(x) Z_n(z;x). \tag{19}$$

In the above expansion the mode $n = 0$ denotes the propagating mode and the remaining terms $n = 1,2,\ldots,$ are the evanescent modes. The additional term $\varphi_{-1}(x)Z_{-1}(z;x)$ is a correction term, called the sloping-bottom mode, which properly accounts for the satisfaction of the Neumann bottom boundary condition on the non-horizontal parts of the bottom. The functions

$Z_n(z;x)$, $n=0,1,2...$, appearing in Eq. (19) are obtained as the eigenfunctions of local vertical Sturm-Liouville problems,

$$Z_0(z;x) = \operatorname{sech}(k_0 h)\cosh[k_0(z+h)],$$
$$Z_n(z;x) = \sec(k_n h)\cos[k_n(z+h)],\ n=1,2,... \quad (20)$$

where the x-dependent eigenvalues $\{ik_0(x), k_n(x)\}$ are obtained as the roots of the dispersion relation, formulated at the local depth,

$$\mu h(x) = -k(x)h(x)\tan[k(x)h(x)],\ a \le x \le b \quad (21)$$

A specific convenient form of the function $Z_{-1}(z;x)$ is $Z_{-1}(z;x) = z^3h^{-2} + z^2h^{-1}$, however, other choices are also possible; see Athanassoulis & Belibassakis (1999). By introducing the above expansion in a variational principle, the following coupled-mode system of horizontal equations for the modal amplitudes of the wave potential is obtained:

$$\sum_{n=-1}^{\infty} a_{mn}(x)\varphi_n''(x) + b_{mn}(x)\varphi_n'(x) + c_{mn}(x)\varphi_n(x) = 0,\ m=-1,0,1,...., \quad (22)$$

in $a<x<b$, where a prime denotes differentiation with respect to x. The coefficients a_{mn}, b_{mn}, c_{mn} of the above system are dependent on x through $h(x)$ and can be found in Table 1 of Athanassoulis & Belibassakis (1999). The coupled-mode system is supplemented by appropriate boundary conditions on the mode amplitudes, at the ends $x=a$ and $x=b$ of the variable bathymetry region, ensuring complete matching on $\partial D_I^{(12)}$ and $\partial D_I^{(23)}$. Finally, from the solution of the coupled-mode system, the reflection and transmission coefficients can be calculated; see Athanassoulis & Belibassakis (1999). An important feature of the calculation of the wave field by means of the enhanced representation (11) is that it exhibits an improved convergence, since the rate of decay of the modal amplitudes $|\varphi_n|$ is of the order $O(n^{-4})$. Thus, only a few number of modes suffice to obtain a convergent solution in this region, even for bottom slopes of the order of 1:1, or higher. More details about the above technique can be found in Athanassoulis & Belibassakis (1999), and as concerns its application to realistic 3D seabed topographies in Belibassakis et al (2001), Gerostathis et al (2008). Moreover, details about the extension of the above theory to nonlinear waves in variable bathymetry can be found in Belibassakis & Athanassoulis (2002, 2011).

5 WAVE FIELD IN THE POROUS MEDIUM

Next, the present model is applied to calculation of groundwater field in the porous medium (like e.g. sandy beaches) with application to the prediction of phase resolving component due to wave propagation in variable bathymetry. The numerical solution to the problem, Eqs. (12),(14), forced by Diriclet data concerning the water-wave dynamic pressure on the sea bottom $z=-h(x)$, defined by Eq. (3a), is obtained by using a boundary integral representation based on source-sink distribution on the bottom boundary. Specifically, for $\mathbf{x}, \xi \in \partial D_p^{(2)}$:

$$p_K(\mathbf{x}) = \int_{\partial D_p^{(2)}} \sigma_\ell(\xi) G(\mathbf{x}\,|\,\xi) dS, \quad (23a)$$

$$\nabla p_K(\mathbf{x}) = -\frac{\sigma_\ell(\mathbf{x})\mathbf{n}(\mathbf{x})}{2} + \int_{\partial D_p^{(2)}} \sigma(\xi)\nabla_x G(\mathbf{x}\,|\,\xi) dS, \quad (23b)$$

where $\mathbf{x} = (x,z)$ and

$$G(\mathbf{x}\,|\,\xi) = \frac{i}{4} H_0^{(1)}(\psi|\mathbf{x}-\xi|), \quad (24)$$

is the free-space Green's function of the Helmholtz equation in 2D (see, e.g., Colton & Kress 1983), and the unit normal vector $\mathbf{n}(\mathbf{x})$ in the above equations is taken to be directed to the interior of $D_p^{(2)}$. The boundary integral in Eq. (23b) is considered in the sense of Cauchy principal value. On the open parts of the boundary $\partial D_p^{(2)}$, which are the vertical interfaces $\partial D_p^{(1,2)}$ and $\partial D_p^{(2,3)}$, at $x=a$ and $x=b$ (see Figure 1), respectively, appropriate radiation conditions are specified ensuring that the pore-pressure field and its normal derivative are continuous, as approaching these interfaces from inside and outside the computational domain. Based on the previous analysis, Eqs.(15), (16), and assuming that the above vertical interfaces are located far from the inhomogeneity the above radiation conditions can be approximated by he following conditions

$$\frac{\partial p_K}{\partial n} + ik_0^{(1)} p_K = 2ik_0^{(1)} \exp\left(ik_0^{(1)} a\right) P^{(1)}(z),\ x=a, \quad (25a)$$

where

$$P^{(1)}(z) = \frac{\cosh\left[\lambda_0^{(1)}(z+h_B)\right]}{\cosh\left[\lambda_0^{(1)}(h_B-h_1)\right]} \frac{i\omega\rho}{\cosh\left(k_0^{(1)} h_1\right)}, \quad (25b)$$

and

$$\frac{\partial p_K}{\partial n} + ik_0^{(3)} p_K = 0, \ x = b, \tag{26}$$

where $\partial p_K / \partial n = \mathbf{n}\nabla p_K$. On the basis of the above considerations the induced wave pressure in the porous layer is found as a solution to the following system of boundary integral equations, $\mathbf{x} \in \partial D_p^{(2)}$:

$$\int_{\partial D_p^{(2)}} \sigma_\ell(\xi) G(\mathbf{x} \mid \xi) dS = p_K(\mathbf{x}) \ , a < x < b, \ z = -h(x), \tag{27a}$$

where $p_K(\mathbf{x})$ in the right-hand side of the above equation is obtained from Eq. (3a),

$$-\frac{\sigma_\ell(\mathbf{x})}{2} + \int_{\partial D_p^{(2)}} \sigma_\ell(\xi) \mathbf{n}(\mathbf{x}) \nabla_x G(\mathbf{x} \mid \xi) dS = 0, \quad a < x < b, \ \ z = -h_B, \tag{27b}$$

$$-\frac{\sigma_\ell(\mathbf{x})}{2} + \int_{\partial D_p^{(2)}} \sigma_\ell(\xi) \left[\mathbf{n}(\mathbf{x}) \nabla_x G(\mathbf{x} \mid \xi) + ik_0^{(1)} G(\mathbf{x} \mid \xi) \right] dS = 2ik_0^{(1)} P^{(1)}(z), \quad x = a, \ \ -h_B < z < -h_1, \tag{27c}$$

$$-\frac{\sigma_\ell(\mathbf{x})}{2} + \int_{\partial D_p^{(2)}} \sigma_\ell(\xi) \left[\mathbf{n}(\mathbf{x}) \nabla_x G(\mathbf{x} \mid \xi) + ik_0^{(3)} G(\mathbf{x} \mid \xi) \right] dS = 0 \quad x = b, \ \ -h_B < z < -h_3 \tag{27d}$$

The numerical solution scheme is based on discretizing the above equation by means of a low-order panel method using linear elements with constant source-sink distributions (see, e.g., Grilli 1998). In Figures 4 and 5, we present results obtained by the present model concerning the calculation of wave pressure fiels in the water and porous media in a shoaling region. In this case, the depth function presents a variation from 4m to 2m over 40m horizontal distance, corresponding to an upslope environment in relatively shallow water. The mean bottom slope is 5% and the maximum bottom slope, appearing in the middle of the domain, is 25%. The hard impermeable botom is located at a depth $h_B = 6$ m. As before, short and longer incident waves have been considered, and the wave data and medium parameters are the same as the ones presented and discussed in Figures 2 and 3 respectively. In the same plots the calculated free-surface elevation is shown by using thick lines.

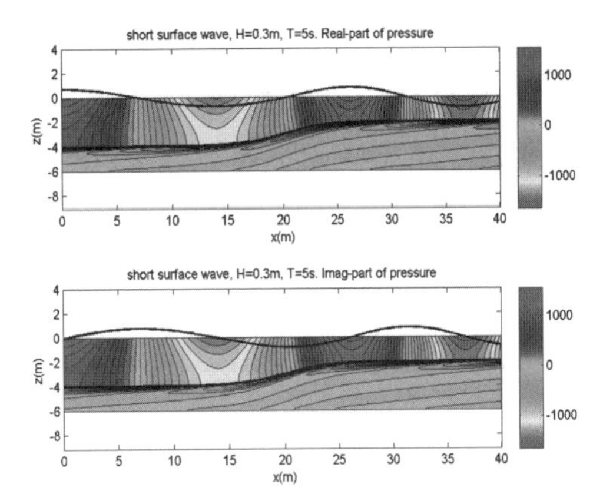

Figure 4. Pressure distribution in the water and in the porous bottom in the case of short waves propagating over a smooth shoal. The free-surface elevation is also shown (5 times exaggerated).

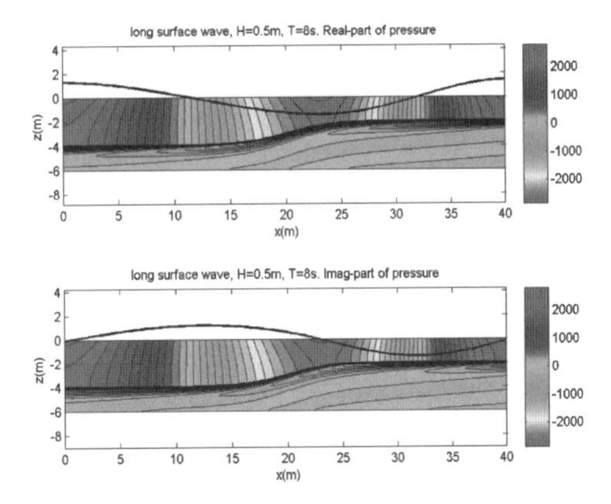

Figure 5. Pressure distribution in the water and in the porous bottom in the case of long waves propagating over a smooth shoal. Free-surface elevation is also shown (5 times exaggerated).

We observe in Figures 4 and 5 that the present nethod is able to accurately model the details of the flow both in the water and in the porous-medium column, where the field exhibits fast attenuation. Groundwater flow velocities are obtained from pressure through Eqs. (8), and, in the above examples, they exhibit a fast decay in depth and their maximum values are a small fraction of the corresponding wave velocities in the water column near the sea bed.

6 CONCLUSIONS

In the present work, a new phase-resolving coupled-mode model is developed for modelling water wave propagation in variable bathymetry and coastal regions and its interaction with porous flow in the layer under the penetrable seabed. The sea-bottom has general shape, characterized by sloping parts and undulations. Our method is based on Biot's theory, taking into account pore-water flow effects, and extends previous work by the authors concerning water-wave propagation in variable bathymetry domains in cases with penetrable bottom. Focusing on the region where water depths are greater than the wave breaking depth, the present model provides reasonable predictions, in agreement with measured data, and provides useful information concerning wave-induced groundwater dynamic pressure and flow in the porous medium. Future work in planned towards the detailed investigation and verification of the present method, and its application to engineering and environmental studies in coastal regions and sandy beaches.

REFERENCES

Athanassoulis G., Belibassakis K., 1999. A consistent coupled-mode theory for the propagation of small-amplitude water waves over variable bathymetry regions. *J. of Fluid Mechanics* 389, pp. 275–301.

Belibassakis K., Athanassoulis G., Gerostathis, T., 2001. A coupled-mode model for the refraction-diffraction of linear waves over steep 3D bathymetry. *Applied Ocean Research* 23, 319–336.

Belibassakis K.A., Athanassoulis G.A., 2002. Extension of second-order Stokes theory to variable bathymetry. *Journal of Fluid Mechanics,* 464, pp. 35–80.

Belibassakis K.A., Gerostathis T., Athanassoulis G.A., 2007. Calculation of wave-induced set-up in variable bathymetry regions and groundwater flow in permeable beaches by a coupled-mode method. Proc. 8*th Int. Congress on Mechanics*, HSTAM, Patras, Greece.

Belibassakis K.A., Gerostathis T.P., Athanassoulis G.A., 2007. A coupled-mode technique for the prediction of wave-induced set-up and mean flow in variable bathymetry domains. Proc. 26*th Int. Conf on Offshore Mechanics and Arctic Engin. OMAE2007,* San Diego.

Belibassakis K.A., Athanassoulis G.A., 2011. A coupled-mode system with application to nonlinear water waves propagating in finite water depth and in variable bathymetry regions. *Coastal Eng.* 58, 337 350.

Colton D., Kress R., 983. *Integral Equation Methods in Scattering Theory*. John Wiley & Sons.

Dingemans M., 1997. *Water wave propagation over uneven bottoms*. World Scientific.

Gerostathis T., Belibassakis K.A., Athanassoulis G.A., 2008. A coupled-mode model for the transformation of wave spectrum over steep 3d topography. A Parallel-Architecture Implementation. *Journal of Offshore Mechanics and Arctic Engineering, JOMAE*, Vol. 130.

Grilli S., Pedersen T., Stepanishen P., 1998. A hybrid boundary element method for shallow water acoustic propagation over an irregular bottom. *Engineering Analysis with Boundary Elements* 21, pp. 131–145.

Mei C.C., 1983. *The applied dynamics of ocean surface waves*. Second Edition 1996, World Scientific.

Massel S., 1993. Extended refraction-diffraction equations for surface waves. *Coastal Engin.* 19, pp. 97–126.

Massel S.R., 2001. Circulation of groundwater due to wave set-up on a permeable beach. *Oceanologia*, 43 (3), pp. 279–290.

Massel S.R., Przyborska A., Przyborski M., 2004. Attenuation of wave-induced groundwater pressure in shallow water. *Oceanologia*, 46 (3), pp. 383–404.

Massel S.R., Przyborska A., Przyborski M., 2005. Attenuation of wave-induced groundwater pressure in shallow water (part2), *Oceanologia*, 47(3), pp. 291–323.

Porter D., Staziker D.J., 1995. Extension of the mild-slope equation, *Jour. Fluid Mech.* 300, pp. 367–382.

Verruijt A., 1969. *Elastic storage of aquifers*. In *Flow through porous media*, R.J.M. Deweist (ed.), Acad. Press, New York, 331–376.

Sustainable Maritime Transportation and Exploitation of Sea Resources – Rizzuto & Guedes Soares (eds)
© 2012 Taylor & Francis Group, London, ISBN 978-0-415-62081-9

On wave height distribution in the space domain and in the time domain

P. Boccotti, F. Arena & V. Fiamma
NOEL laboratory, "Mediterranea" University, Reggio Calabria, Italy

ABSTRACT: A small-scale field experiment has been carried out in the Natural Ocean Engineering Laboratory of Reggio Calabria (NOEL, www.noel.unirc.it), to record waves in the space-time domain. An array of gauges was placed orthogonal to the coastline, so as to cover at least 1.5 wave length. During the experiment 1,404,000 waves were recorded (by considering waves in time domain). It is shown as a theoretical asymptotic wave height distribution well represents the distributions of the wave heights both on the time domain and on the space domain, as α tends to infinity (α being the quotient between an individual wave height and the root-mean-square wave elevation of the sea state). The convergence is faster for the waves on the time domain. The prediction of the quasi-determinism theory that wave heights must be greater on the time domain than on the space domain gets a full confirmation.

1 INTRODUCTION

Longuet-Higgins [1952] was the first who realized that the distribution of crest-to-trough wave heights of a stationary sea state is close to the distribution of wave heights of a stationary random Gaussian process. *Cartwright and Longuet-Higgins* [1956] showed that the probability of exceedance of the crest-to-trough heights tends to the form

$$P(\alpha)=\exp(-\alpha^2/8), \tag{1}$$

as the bandwidth approaches zero, where α is the quotient between an individual wave height and the root-mean-square wave elevation of the sea state. Equation (1) was obtained as a corollary of the solution by *Rice* [1944,1945] for the distribution of local maxima of a stationary random Gaussian process with an arbitrary bandwidth.

Forristall [1978] showed that the actual probability of exceedance of the wave heights in a stationary sea state was systematically somewhat smaller than the probability given by equation (1). Specifically, it was found that the inverse function $\alpha(P)$, that is the function giving the dimensionless height that has a given probability P to be exceeded in a stationary sea state was about a 10% smaller than

$$\alpha(P)=\sqrt{8\ln(1/P)}, \tag{2}$$

which is the inverse of (1).

Since equation (2) is based on two assumptions: (i) bandwidth approaching zero; (ii) linearity (a stationary Gaussian process represents a linear model of a stationary sea state). A question of those years, concerning wave statistics was: is the difference between the actual $\alpha(P)$ and function (2) an effect of finite bandwidth or an effect of nonlinearity?

Longuet-Higgins [1980] concluded that it was an effect of finite bandwidth. *Boccotti* [1982] as a corollary of the quasi-determinism theory for waves on the time domain showed that

$$P(\alpha)\ \mathrm{O}\ \exp\left[-\frac{\alpha^2}{4(1+\psi^*)}\right] \quad \text{with} \quad \psi^*=\frac{|\psi(T^*)|}{\psi(0)}, \tag{3}$$

where O means "is of the same order as". Equation (3) is exact in the limit as α tends to infinity, for stationary Gaussian processes with an arbitrary bandwidth. Equation (3) may be rewritten in the form

$$P(\alpha)=K\exp\left[-\frac{\alpha^2}{4(1+\psi^*)}\right], \tag{4}$$

where K is a coefficient independent from α. The inverse of (4) is

$$\alpha(P)=\sqrt{[4(1+\psi^*)][\ln(K)+\ln(1/P)]}, \tag{5}$$

where the term depending on K can be formally cancelled. This is because in the limit as P approaches zero

$$\ln(K)/\ln(1/P)\to 0. \tag{6}$$

Hence equation (5) may be rewritten in the form

$$\alpha(P) = \sqrt{[4(1+\psi^*)]\ln(1/P)}. \tag{7}$$

With the characteristic values of ψ^* ($\psi^* = 0.73$ for the mean JONSWAP spectrum, $\psi^* = 0.65$ for the Pierson and Moskowitz spectrum) equation (7) gives α (P) of 7%–9% smaller than the α (P) given by (2). That is equation (7) is consistent with the finding of *Forristall* [1978]. This fact suggested that the K (being still unknown in the early eighties) should be rather close to 1. Therefore *Boccotti* [1984] tested the function (4) with $K = 1$, that is

$$P(\alpha) = \exp\left[-\frac{\alpha^2}{4(1+\psi^*)}\right], \tag{8}$$

with data of crest-to-trough heights of waves being subject to reflection at the vertical breakwater of Genoa port.

Boccotti [1989] obtained the closed form solution for the probability $P(\alpha)$ as α tends to infinity, for an arbitrary bandwidth. Indeed he arrived at the relationship between K and the frequency spectrum of the random process:

$$K = \frac{1+\ddot{\psi}^*}{\sqrt{2\ddot{\psi}^*(1+\psi^*)}}, \text{ where } \ddot{\psi}^* = |\ddot{\psi}(T^*)| / \ddot{\psi}(0), \tag{9}$$

so that the expression of $P(\alpha)$ is given by:

$$P(\alpha) = \frac{1+\ddot{\psi}^*}{\sqrt{2\ddot{\psi}^*(1+\psi^*)}} \exp\left[-\frac{\alpha^2}{4(1+\psi^*)}\right] \text{ for } \alpha \to \infty. \tag{10}$$

Equation (10) is a corollary of one of the theorems of the QD theory. This is the so called theorem of the necessary condition [see *Boccotti*, 2000, sections 9.2,9.6–10].

Finally, the following model:

$$\begin{aligned} P(\alpha) &= \sqrt{\frac{1+r}{2r}}\left(1+\frac{1-r^2}{4r\alpha^2}\right)\exp\left(-\frac{\alpha^2}{4(1+r)}\right) \\ r^2 &= \frac{1}{m_0}\left[\left(\int_0^\infty E(\omega)[\cos(\omega-\bar{\omega})]\frac{T_m}{2}\mathrm{d}\omega\right)^2 \right. \\ &\quad \left. + \left(\int_0^\infty E(\omega)[\sin(\omega-\bar{\omega})]\frac{T_m}{2}\mathrm{d}\omega\right)^2\right] \\ T_m &= 2\pi\, m_0 / m_1 \end{aligned} \tag{11}$$

was given by *Tayfun* [1990] for $\alpha > \sqrt{2\pi}$ approximately.

The aforementioned theoretical distributions are expected to be valid for both surface waves and pressure head waves beneath the water surface. This is because both surface waves and pressure head waves with the linear theory of stationary wind-generated waves, represent stationary Gaussian random processes of time at every given point.

More recently, the wave height distribution has been investigated by many authors, including nonlinear effects, also by considering field data or numerical simulations [*Mori and Janssen*, 2006; *Stansell*, 2004; *Cherneva et al.* 2009, *Shemer and Sergeeva*, 2009; *Arena and Guedes Soares*, 2010; *Casas-Prat and Holthuijsen*, 2010]. A complete review was given in *Tayfun and Fedele* [2007], where theoretical models were compared with field data: they concluded that nonlinearity does not affect the wave height distribution [see also *Forristall*, 2005] and that Boccotti's model well represent the distribution of high waves. Nonlinear effects modify both the crest and the trough distribution, with respect to the Rayleigh law, which represents both distribution if a linear (Gaussian) model is considered [*Tayfun*, 1994; *Forristall*, 2000; *Fedele and Arena*, 2005; *Fedele and Tayfun* 2009; *Arena and Ascanelli*, 2010]: crest and troughs amplitudes are increased and reduced respectively with respect to the Rayleigh law. To the second order, *Arena* [2005] showed that the height of the largest waves is not modified by nonlinearity, in agreement with *Tayfun and Fedele* [2007] and *Casas-Prat and Holthuijsen* [2010] conclusions that wave height distribution are not affected by nonlinear effects.

2 DESCRIPTION OF THE FIELD EXPERIMENT

On May, 2010 a horizontal beam was placed 1.2 m beneath the mean water level, in the Natural Ocean Engineering Laboratory (NOEL, www.noel.unirc.it) off the beach at Reggio Calabria (Italy). This beam was basically orthogonal to the shoreline, had a length of 15 m, and supported 26 pressure transducers for measuring the pressure head waves. Each pressure transducer was connected to a small vertical tube (0.40 m long) with a bending section at the top (like a small periscope).The opening at the top of this small tube was in a vertical plane orthogonal to the shoreline. This tube served only to measure pressure head waves in the undisturbed wave field. Thanks to the small tubes, the pressure head waves were measured 0.40 m above the horizontal beam which consisted of a truss of high stiffness and small section.

As to angle $\bar{\theta}$ of the dominant wave direction, it was estimated from the relative phase of point 26 and point 27. Pressure transducer 27 (not seen in Figure 1) was 0.75 m from transducer 26, and these two transducers were aligned with the shoreline.

The water depth ranged from 5 m at the seaward end of the beam to 2 m at the landward end (see Figure 1). Variations of the water depth due to tide were within $\pm 0.15\mathrm{m}$.

As said, the distributions of wave heights in the stationary Gaussian processes are expected to hold both for surface waves and for pressure head waves beneath the water surface. Hereafter, it will be understood that 'wave' stands for 'pressure head wave'.

Ealing with pressure head waves is preferable to dealing with surface waves. This is because pressure head waves are measured by pressure transducers which, usually, are more reliable and precise and less expensive than the gauges used for measuring surface waves.

A rather homogeneous data set was constituted with the records wherein

$$\psi^* > 0.7, \qquad T_p < 2.6s, \quad |\bar{\theta}| < 20° . \qquad (12)$$

Here ψ^* is the narrow bandedness parameter being defined as the absolute value of the ratio between the minimum and the maximum of the autocovariance of the pressure head waves. The condition on the narrow-bandedness parameter ψ^* serves to select wind waves, that is waves in the generating area. Indeed wind waves have some characteristic narrow spectra, like the *Pierson and Moskowitz* [1967] spectrum or the JONSWAP spectrum [*Hasselmann et al.*,1973], whose ψ^* is greater than 0.65 for the surface waves, and the ψ^* of the pressure head waves is greater than the ψ^* of the surface waves. This is because the spectrum shrinks gradually from the water surface to the seabed, since it sheds its high frequency tail.

The second of conditions (12), that dealing with T_p guarantees that the length (15 m) of the horizontal beam is at least 1.5 times the wave length, what will serve in the next section dealing with the distribution of wave heights on the space domain. Finally, the condition on $\bar{\theta}$ guarantees that the direction of wave advance is close to the alignment of the gauges.

These wind waves with a H_s between 0.2 m and 0.4 m and a T_p between 1.8 s and 2.6 s are typical at the location of the experiment, where they

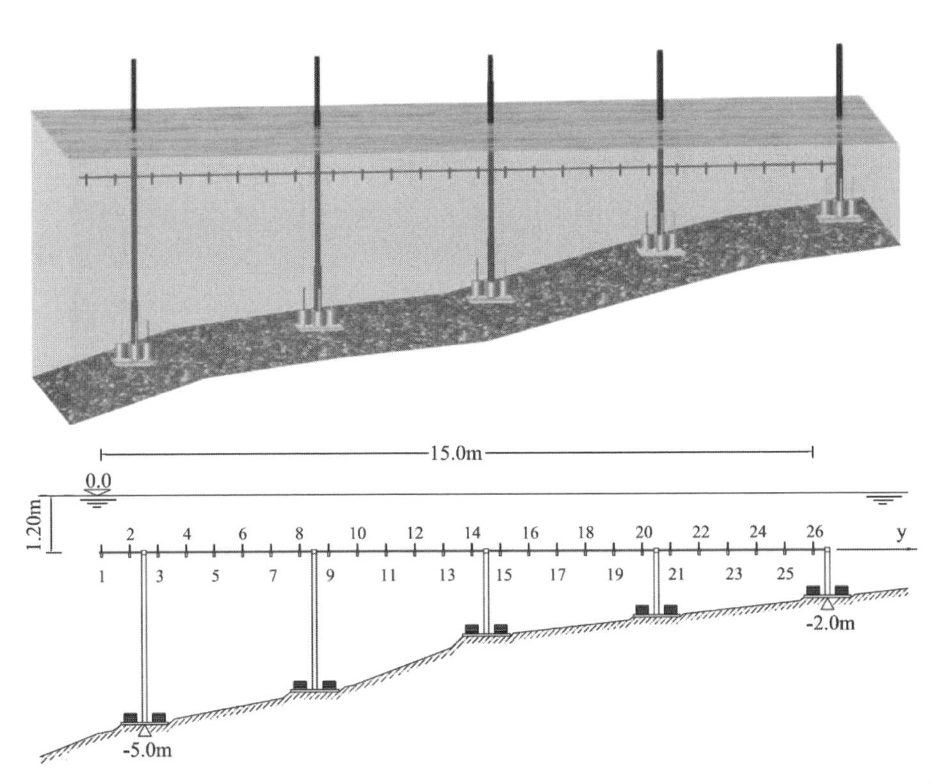

Figure 1. Reference frame; the horizontal beam of 15 m length, 1.2 m beneath the mean water level, orthogonal to the shoreline. The beam supported 26 pressure transducers.

are generated by a local wind of the Straits of Messina. These wind waves are very interesting for ocean engineering in that they represent some natural small scale models of storm seas in the Froude dynamic similarity [see *Boccotti*, 2000, sects. 4.5 and 4.6].

There were 410 records fulfilling conditions (12). Each of these records was of five minutes, and the sampling rate was of 10 Hz per each gauge.

3 PROBABILITY OF WAVE HEIGHTS ON THE TIME DOMAIN: DATA VS THEORY

The set of 410 five-minute records fulfilling conditions (12) consisted of about 1,404,000 individual waves, which make an average of about 54000 individual waves per each of the 26 transducers on the horizontal beam.

Because of the stochastic dependence between wave heights measured at some points close to one another, a confidence interval at the probability level 1/1,404,000 (say the 90% confidence interval) will be wider than this confidence interval would be if the 1,404,000 waves were stochastically independent of one another. On the opposite, the confidence interval at the probability level 1/54000 (or at any probability level greater than 1/54000) will be narrower than this confidence interval would be if there was a single set of 54000 waves stochastically independent from one another.

The matter is essentially the same as for the probability of exceedance $P(H_s > h)$ of the significant wave height at a given location, obtained from satellite data. The multiple records of an individual wave made by the array of gauges in our experiment are like the multiple observations per each pass of the satellite over the sampling area. Re the relationship between number of passes, number of observations per each pass, and width of the confidence interval of $P(H_s > h)$ [see *Boccotti*, 2000, pp. 463–464].

In the 410 records fulfilling conditions (12) the two parameters (ψ^* and K) of equation (10) proved to be in the following ranges

$$0.700 < \psi^* < 0.882, \quad 1.036 < K < 1.160. \tag{13}$$

with the following average values

$$\psi^* = 0.773, \quad K = 1.080. \tag{14}$$

Points in Figure 2 represent the probability of exceedance of the wave height, obtained from our data; and the continuous lines represent respectively equation (1) and equation (10) with the values (14) of ψ^* and K.

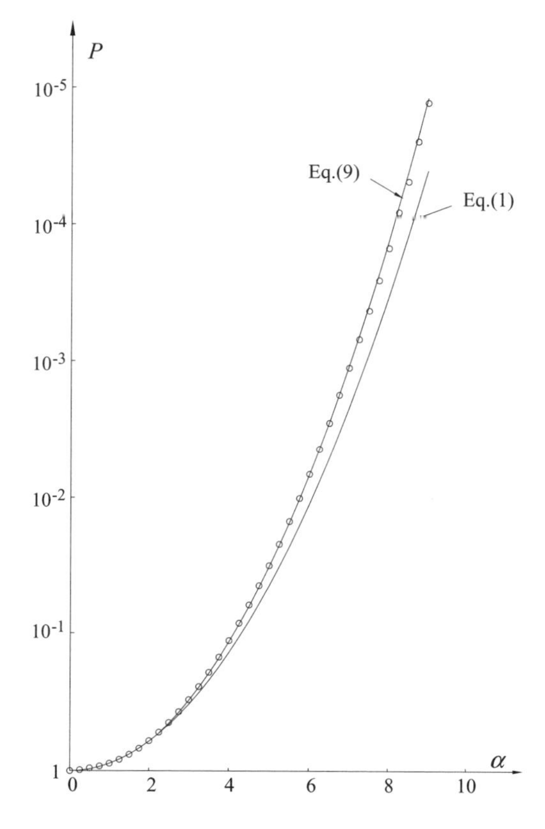

Figure 2. Probability of exceedance $P(\alpha)$ for the waves on the time domain.

Equation (1) fits well the data points up to $\alpha = 2.5$. Hereafter there is an impressive closeness between data points and equation (10). This means that equation (10) being exact as α tends to infinity, holds very good for $\alpha > 2.5$. Or in other words the convergence of the actual probability function to the asymptotic form (10) occurs as α exceeds 2.5.

The $P(\alpha)$ (11) of *Tayfun* [1990] in its turn is nearly coincident with equation (10). The average value of the unique parameter r of equation (11), in the 410 records proves to be

$$r = 0.766 \tag{15}$$

and with this average value of r, equation (11) proves to be practically coincident with equation (10) with the average values (14) of ψ^* and K.

In order to appreciate better the impressive agreement between equations (10) and (11) and the data, Table 1 gives the values of the actual probability and of the probability calculated with equations (1), (8), (10), (11). The difference between equations (8) and (10) is due to parameter K. This difference is small, given that K is close to 1. However it is

Table 1. 1000 $P\alpha$ time domain: (a) data zero up-crossing waves; (b) data zero down-crossing waves; (c) equation (1); (d) equation (8); (e) equation (10); (f) equation (11).

α	(a)	(b)	(c)	(d)	(e)	(f)
0	1000	1000	1000	1000	–	–
0.5	956	955	969	965	–	–
1	880	881	882	869	938	–
1.5	758	759	755	728	787	828
2	603	603	607	569	615	630
2.5	442	442	458	414	448	453
3	300	300	325	281	304	305
3.5	190	189	216	178	192	192
4	111	111	135	105	113	112
4.5	60.9	60.9	79.6	57.6	62.2	61.5
5	31.3	31.3	43.9	29.5	31.8	31.3
5.5	14.9	14.9	22.8	14.1	15.2	14.9
6	6.68	6.75	11.1	6.25	6.75	6.59
6.5	2.84	2.84	5.09	2.59	2.8	2.72
7	1.13	1.11	2.19	1	1.08	1.05
7.5	0.44	0.41	0.88	0.36	0.39	0.37
8	0.14	0.16	0.33	0.12	0.13	0.12

thanks to this small difference that the full agreement between theory and data is achieved.

The convergence onto the asymptotic function (10) occurs for $P < 0.5$. *Boccotti* [1989, 2000], using the data of numerical simulations of *Forristall* (1984) showed that the actual probability converges onto the asymptotic function (10), the faster the narrower the spectrum is. In particular, Figure 9.3 of the book of *Boccotti* [2000] shows that for a very large bandwidth ($\psi^* = 0.4$) the convergence occurs for $P < 6 \cdot 10^{-4}$, for a large bandwidth ($\psi^* = 0.55$) the convergence occurs for $P < 3 \cdot 10^{-3}$, for a narrow band spectrum ($\psi^* = 0.65$) the convergence occurs for $P < 3 \cdot 10^{-1}$. Here we deal with a narrower spectrum, a spectrum of the pressure head waves wherein ψ^* is at 0.77. So, we may expect that the conver gence onto the asymptotic function (10) occurs for some even greater value of the probability of exceedance, and in fact we find that the convergence occurs for $P < 5 \cdot 10^{-1}$. The table gives one column for the zero up-crossing waves, and one column for the zero down-crossing waves. On comparing to each other these two columns, we may see that the distribution of the zero up-crossing wave heights is coincident or very close to the distribution of the zero down-crossing wave heights.

4 THEORETICAL APPROACH TO THE PROBABILITY OF WAVE HEIGHTS ON THE SPACE DOMAIN

According to the linear theory of stationary wind-generated waves, the free surface displacement is given by

$$\eta(y,t) = \sum_{i=1}^{N} a_i \cos(k_i y - \omega_i t + \varepsilon_i) \tag{16}$$

where for simplicity it has been assumed that the waves are long-crested. As usually it is assumed that N tends to infinity, phase angles ε_i are uniformly distributed in $(0, 2\pi)$ and are stochastically independent from one another, frequencies ω_i are all different from one another, and amplitudes a_i are all of the same order. Under these assumptions η at some fixed point (x,y) represents a stationary random Gaussian process of time, whose spectrum is such that

$$E(\omega)\delta\omega = \sum_j 0.5 a_j^2 \quad \text{for } j \text{ such that } \omega < \omega_j < \omega + \delta\omega. \tag{17}$$

Also η as a function of y, at some fixed time instant represents a Gaussian random process of space, whose spectrum $E_s(k)$ is related to the spectrum of waves on the time domain by

$$E_s(k)\mathrm{d}k = E(\omega)\mathrm{d}\omega \tag{18}$$

where k is related to ω by the linear dispersion rule. From equation (18), it may be verified that E_s is broader than E, and hence ψ^* of waves on the space domain if smaller than ψ^* of waves on the time domain. What implies that also the probability of exceedance of wave heights on the space domain is smaller than the probability of exceedance of wave heights on the time domain.

On shallow water, if the water depth is not constant, $\eta(y,t)$ for a fixed t, no longer is a stationary random process. In part this is the case of our experiment wherein d/L_{p0} ranges from 0.2 to 1.0. Therefore, we shall deal with the normalized wave elevation

$$\tilde{\eta}(y,t) = \eta(y,t)/\sigma(y) \tag{19}$$

where $\sigma(y)$ is the root mean square wave elevation of the sea state, which, generally, depends on the location on the horizontal beam.

5 OBTAINING THE AUTOCOVARIANCE ON THE SPACE DOMAIN

For applying equation (10) to pressure head waves on the space domain we must know the values of ψ^* and K on the space domain. That is we must obtain the normalized autocovariance of pressure head waves on the space domain:

$$\psi_s(Y) = \langle \tilde{\eta}(y,t) \cdot \tilde{\eta}(y+Y,t) \rangle . \tag{20}$$

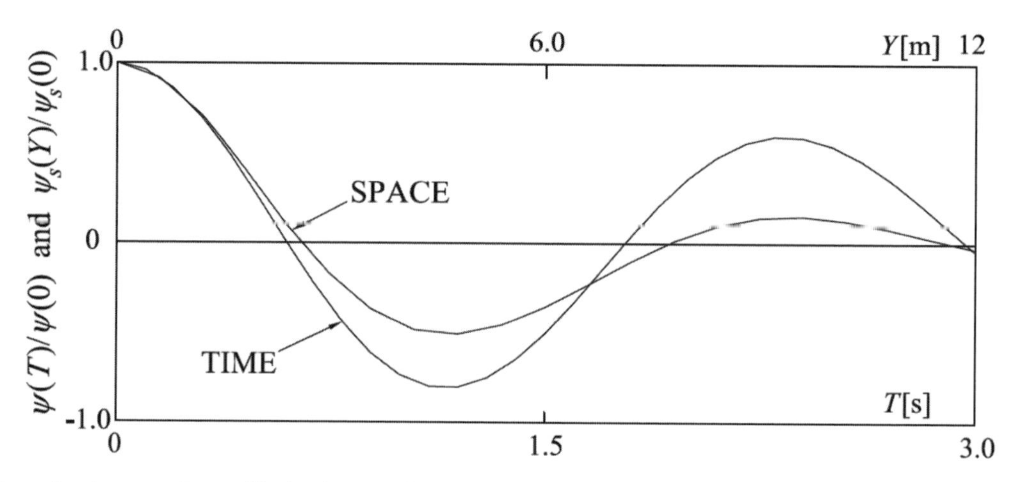

Figure 3. Autocovariance with time lag T; and autocovariance with space lag Y.

For a given space lag Y, $\psi_s(Y)$ is the average of products of $\tilde{\eta}$ at a location y and $\tilde{\eta}$ at the location $y+Y$, with the two $\tilde{\eta}$ being taken at the same time instant. Let us call $\tilde{\eta}_{i,j}$ the $\tilde{\eta}$ at the ith transducer ($i = 1,26$) at the jth time instant of a record ($j = 1,3000$). The given lag Y must be a multiple of the interval $\Delta y = 0.6$ m between the transducers. For $Y = n \cdot \Delta y$ with n being a given entire we have

$$\psi_s(n \cdot \Delta y) = \left(\sum_{j=1}^{3000} \sum_{i=1}^{26-n} \tilde{\eta}_{i,j} \cdot \tilde{\eta}_{i+n,j} \right) \Big/ N, \quad N = 3000 \cdot (26-n) \tag{21}$$

This is the average of the total number N of products of $\tilde{\eta}(y,t)$ and $\tilde{\eta}(y+Y,t)$ that we obtain from one five minute record.

Figure 3 shows $\psi(T)$ and $\psi_s(Y)$ for one of the records. We see that ψ^* on the time domain is greater than ψ^* on the space domain. This is because, as said, the spectrum $E_s(k)$ of waves on the space domain is broader than the spectrum of waves on the time domain. Then, looking at Figure 3, we see that the minimum of the autocovariance on the space domain is flatter than the minimum of the autocovariance on the time domain. As a consequence K on the space domain is greater than K on the time domain. (As the ratio between the curvature of the minimum of the autocovariance and the curvature of the maximum of the autocovariance approaches zero, K tends to infinity, what follows from equation (9a) of K.) The ranges of ψ^* and K for the waves on the space domain for the 410 records fulfilling conditions (12) proved to be

$$0.12 < \psi^* < 0.73, \quad 1.10 < K < 2.15. \tag{22}$$

with the following averages:

$$\psi^* = 0.50, \quad K = 1.26. \tag{23}$$

6 PROBABILITY OF THE WAVE HEIGHT ON THE SPACE DOMAIN: DATA VS THEORY

Figure 4 represents $\tilde{\eta}$ as a function of y at a given time instant. We have got

$$3000 \cdot 410 = 1230000. \tag{24}$$

pictures like that of Figure 4: 3000 being the number of pictures per record (the duration of a record being of 300s and the sampling rate being of 10 Hz), and 410 being the number of records fulfilling inequalities (12). We filed the height of each individual zero up-crossing wave and of each zero down- crossing wave being present in each picture. As an example in the picture of Figure 4 there are one zero up-crossing wave and one zero down-crossing wave. The total number of waves was 2,696,000, one half of which were zero up- crossing waves and one half zero down-crossing waves.

The probability of exceedance of the wave heights on the space domain is represented in Figure 5 (data points). The continuous lines represent equations (1) and (10), with the average values ($\psi^* = 0.50$, $K = 1.26$ of coefficients ψ^* and K). The convergence of the probability of the heights of pressure head waves onto equation (10) is evident from Figure 5. The convergence is slower than in Figure 2 relevant to the waves on the time domain.

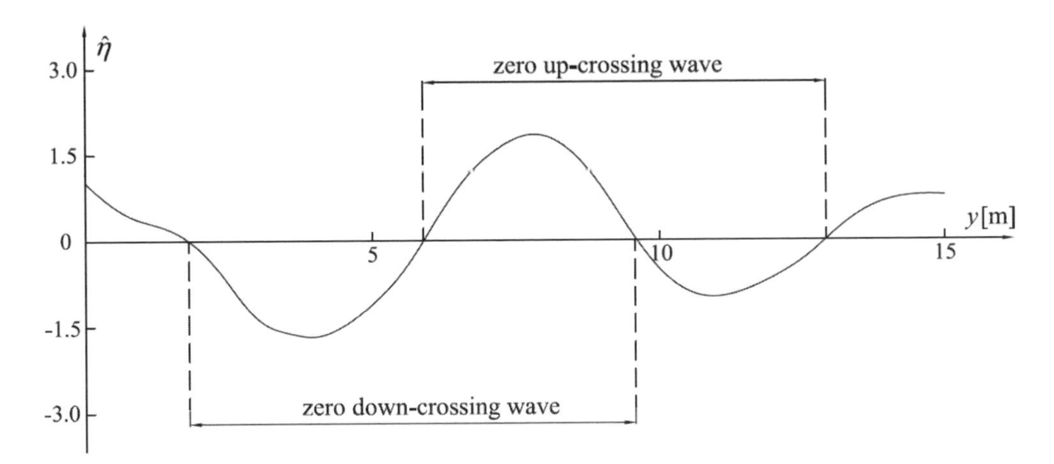

Figure 4. Example of waves on the space domain (the y-axis is parallel to the beam).

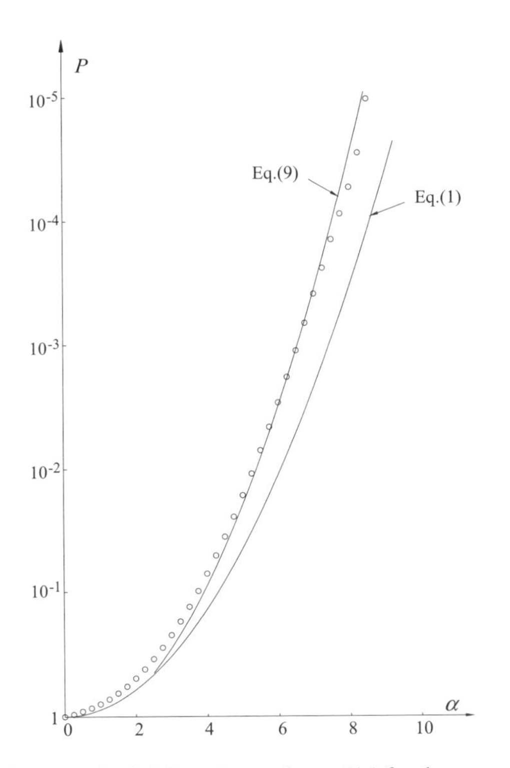

Figure 5. Probability of exceedance $P(\alpha)$ for the waves on the space domain.

This is consistent with the fact that the spectrum of the waves on the space domain is broader than the spectrum of the waves on the time domain. On the time domain (Fig. 2) the convergence occurs for $\alpha > 2.5$, on the space domain the convergence occurs for $\alpha > 6$.

The same comment re the confidence interval done for the waves on the time domain applies here too. Specifically, our data base consists of about 2,696,000 waves on the space domain (counting all the zero up- crossing waves and all the zero down-crossing waves). What we can say is that the $P(\alpha)$ at probability levels greater than or equal to 1:54000, obtained with the 2,696,000 waves, will be statistically more reliable than the $P(\alpha)$ obtained from 54,000 waves recorded by a single gauge.

7 WAVES ARE HIGHER ON THE TIME DOMAIN THAN ON THE SPACE DOMAIN!

The difference between the largest waves on the time domain and the largest waves on the space domain, is foreseen by the quasi-determinism theory [see *Boccotti*, 2000, section 10.3]. Specifically, the QD theory shows that the largest wave height on the space domain during the evolution of a wave group is *necessarily* smaller than the largest wave height on the time domain.

Let us see why. Let the continuous line in Figure 6 represent the largest wave on the space domain during the evolution of a group of exceptionally large waves. According to the QD theory, at the time instant when the largest wave height occurs on the space domain, the envelope centre falls at point (x_o, y_o) wherein there is the central zero of the largest wave. The crest of this wave comes before the envelope centre and hence this wave crest is decreasing, what in turn implies that this wave crest had a greater height when transited at (x_o, y_o). The trough of the largest wave on the space domain comes after the envelope centre and hence this wave trough is growing, what in turn implies that this wave trough will be greater when it shall transit at (x_o, y_o). The conclusion is that the height of the wave on the time domain at point (x_o, y_o) will be greater than the height of the largest wave on the space domain.

Here below, we show the result of a comparison wave by wave, that is a comparison between a wave on the time domain (like (a) in Fig. 7) and the same wave on the space domain (like (b) in Fig. 7). The time instant t_z of the zero at point 13 is obtained by linear interpolation of the pressure head wave at this point. The pressure head wave as a function of y is obtained by means of the Fourier series from the 26 values of the pressure head at the locations of the transducers at time instant t_z.

The results are given in Table 2. The first column of these table gives the threshold of the dimensionless wave height, the second (fourth) column gives the number of waves on the time domain, at the centre of the beam, which exceed this threshold. The third (fifth) column deals with the same waves on the space domain, and gives the number of these waves which exceed the threshold. The difference between the column concerns the wave definition: in columns 2 and 3 the waves on the time domain are zero up-crossing waves; while in columns 4–5 the waves on the time domain are

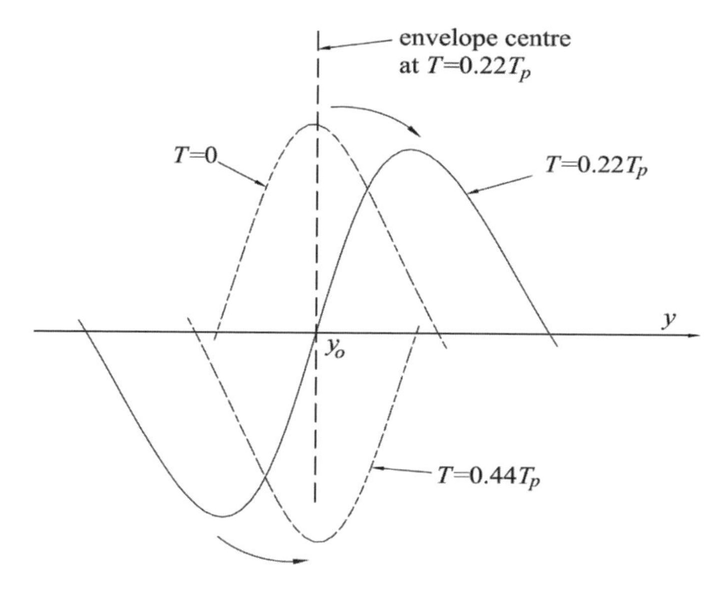

Figure 6. QD theory: The largest wave height on the space domain is necessarily smaller than the largest wave height on the time domain.

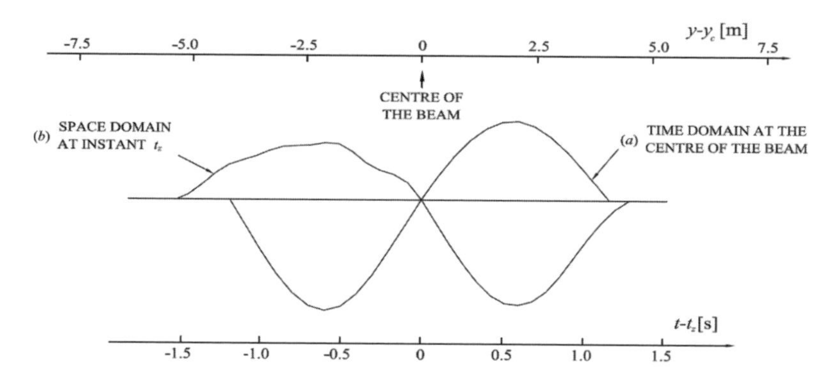

Figure 7. (a) A wave vs time at point 13 at the centre of the horizontal beam; (b) the same wave vs space (y-axis) at the time instant when the central zero of the wave transits at point 13. Waves on the time domain in Table 1 are like (a); and waves on the space domain are like (b).

Table 2. Number of waves whose dimensionless height exceeds a given threshold α. N_1, waves on the time domain at point 13 (centre of the horizontal beam); N_2, the same waves on the space domain. Zero up-crossing and zero down-crossing waves on the time domain.

	Zero up-crossing waves—time domain		Zero down-crossing waves—time domain	
α	N_1	N_2	N_1	N_2
0	54039	54039	54032	54032
0.5	52941	49751	53006	49536
1	49076	46173	49153	46031
1.5	42198	40212	42084	40207
2	33335	32419	33248	32464
2.5	24331	23718	24214	23961
3	16347	15610	16519	15941
3.5	10373	9381	10302	9528
4	6022	5021	6030	5261
4.5	3305	2406	3307	2555
5	1673	1075	1710	1157
5.5	788	424	827	467
6	367	146	384	178
6.5	168	49	172	70
7	69	18	72	22
7.5	32	8	28	7
8	11	4	13	1
8.5	4	0	3	0
9	0	0	3	0

zero down-crossing waves. It emerges very clearly that the waves on the space domain are smaller than the waves on the time domain, with both two the wave definitions. The largest of the 54,000 zero up-crossing waves on the time domain had a dimensionless height $\alpha = 8.972$, whereas the largest of the corresponding 54,000 waves on the space domain had a dimensionless height $\alpha = 8.137$. The largest of the 54,000 zero down-crossing waves on the time domain had a dimensionless height $\alpha = 9.300$, whereas the largest of these 54,000 waves on the space domain had a dimensionless height $\alpha = 8.347$. That is the largest wave on the time domain was a 10%–11% greater than the corresponding wave on the space domain! What confirms the prediction of the QD theory.

8 CONCLUSIONS

Concerning the waves on the time domain we have found an impressive agreement between data of $P(\alpha)$ and asymptotic function (10), and function (11). The convergence of data points onto function (10) occurs for α greater than 2.5. This so fast convergence is due to the fact that we have dealt with waves with a very narrow spectrum. These are the pressure head waves yielded by pure wind waves. Equation (8) is close to the data points; then the perfect rendezvous with data is obtained thanks to coefficient K given by equation (9a). Re the waves on the space domain along the y axis orthogonal to the shoreline, we have found the same strict agreement between the actual $P(\alpha)$ and the asymptotic function (10). However the convergence of the data points onto function (10) occurs for $\alpha > 6$. This is because the spectrum of the waves on the space domain is broader than the spectrum of the waves on the time domain: the average narrow bandedness parameter ψ^* proved to be respectively of 0.77 for the waves on the time domain, and 0.50 for the waves on the space domain. Coefficient K of the waves on the space domain plays a role more important than coefficient K of the waves on the time domain, indeed K on the space domain has exceeded the very threshold of 2 in a few sea states, whereas K of the waves on the time domain has remained below 1.2 for all the sea states.

Re the prediction of the QD theory that very large waves must be smaller on the space domain than on the time domain, the experiment definitely proved that the waves on the space domain are smaller than the waves on the time domain. Specifically, we compared each wave on the time domain at the mid of the horizontal beam (like wave (a) in Fig. 7), with the relevant wave on the space domain (like wave (b) in Fig. 7). The comparison yielded a very clear confirmation of the prediction of the QD theory.

REFERENCES

Arena, F. (2005), On non-linear very large sea wave groups, *Ocean Engineering*, 32, 1311–1331.

Arena, F. and Guedes Soares, C. (2009), Nonlinear crest, trough and wave height distributions in sea states with double-peaked spectra, *ASME Journal of Offshore Mechanics and Ocean Engineering*. 131, 1–8. paper 041105.

Arena, F. and Ascanelli A. (2010), Nonlinear Crest Height Distribution in Three-Dimensional Waves, *ASME Journal of Offshore Mechanics and Ocean Engineering*. 132, 1–6. paper 021604-5.

Boccotti, P. (1982a), On ocean waves with high crests, *Meccanica*, 17, 16–19.

Boccotti, P. (1982b), *On the highest sea waves*. Monograph Institute of Hydraulics, University of Genoa, 1–161.

Boccotti, P. (1984), Sea waves and quasi-determinism of rare events in random processes, *Atti Acc. Naz. Lincei, Rendiconti*, 76, 119–127.

Boccotti, P. (1989), On mechanics of irregular gravity waves. *Atti della Accademia Nazionale dei Lincei, Memorie*, 19, 110–170.

Boccotti, P. (1997), A general theory of three-dimensional wave groups. *Ocean Engineering*, 24, 265–300.

Boccotti, P. (2000), *Wave mechanics for ocean engineering*. Elsevier Science, Amsterdam, 1–496.

Boccotti, P. (2008), Quasi-determinism theory of sea waves, *ASME Journal Offshore Mechanics and Arctic Engineering*, 130, 1–9. paper 005802JOM.

Cartwright, D.E. and Longuet-Higgins, M.S. (1956). The statistical distribution of the maxima of a random function, *Proceedings of the Royal Society of London. Series A, Mathematical and Physical Sciences*, 237, No. 1209, 212–232.

Casas-Prat, M. and Holthuijsen, L.H. (2010), Short-term statistics of waves observed in deep water. *J. Geophys. Res.*, 115, 5742–5761.

Cherneva, Z., Tayfun, M.A. and Guedes Soares, C. (2009), Statistics of nonlinear waves generated in an offshore wave basin. *J. Geophys. Res.* 114, 5332–5339.

Fedele, F. and Arena, F. (2005), Weakly nonlinear statistics of high random waves, *Physics of Fluids*, 17, paper 026601, 1–10.

Fedele, F. and Tayfun, M.A. (2009), On nonlinear wave groups and crest statistics, *J. Fluid Mech.*, 620, 221–239.

Forristall, G.Z. (1978), On the statistical distribution of wave heights in a storm. *J. Geophys. Res.*, 83, 2353–2358.

Forristall, G.Z. (1984), The distribution of measured and simulated wave heights as a function of spectral shape. *J. Geophys. Res.*, 89, 10547–10552.

Forristall, G.Z. (2000), Wave Crest Distributions: Observations and Second-Order Theory, *Journal of Physical Oceanography*, 30, 8, 1931–1943.

Hasselmann, K., Barnett, T.P., Bouws, E., Carlson, H., Cartwright, D.E., Enke, K. et al. "Measurements of wind wave growth and swell decay during the Joint North Sea Wave Project JONSWAP" *Dtsch. Hydrogr. Z.*, A8, 1–95.

Longuet-Higgins, M.S. (1952), On the statistical distribution of the heights of sea waves. *J. Mar. Res.*, 11, 245–266.

Longuet-Higgins M.S. (1980), On the distribution of the heights of sea waves: Some effects of nonlinearity and finite band width. *J. Geophys. Res.*, 85, 1519–1523;

Mori, N. and Janssen, P.A.E.M. (2006), On kurtosis and occur-rence probability of freak waves. *Journal of Physical Oceanography* 36, 1471–1483.

Pierson, W.J., and Moskowitz, L. (1964), A proposed spectral form for fully developed waves based on the similarity theory of S.A. Kitaigorodskii. *J. Geophys. Res.*, 69, 5181–5190.

Rice, S.O. (1944), Mathematical analysis of random noise. *Bell Syst. Tech. J.*, 23, 282–332.

Rice, S.O. (1945), Mathematical analysis of random noise. *Bell Syst. Tech. J.*, 24, 46–156.

Shemer, L. and Sergeeva, A. (2009), An experimental study of spatial evolution of statistical parameters in a unidirectional narrow-banded random wavefield. *J. Geophys. Res.* 114, 5077–5087.

Stansell, P., Wolfram, J. and Linwood, B. (2004), Improved joint probability distribution for ocean wave heights and periods. *J. Fluid Mech.*, 503, 273–297.

Tayfun, M.A. (1990), Distribution of large wave heights. *Journal of Waterway, Port, Coastal and Ocean Engineering*, 116, 686–707.

Tayfun, M.A. (1994), Distributions of envelope and phase in weakly nonlinear random waves. *Journal of Engineering Mechanics* 120: 1009–1025.

Tayfun, M.A. and Fedele, F. (2007), Wave-height distributions and non-linear effects. *Ocean Engineering* 34, 1631–1649.

Sustainable Maritime Transportation and Exploitation of Sea Resources – Rizzuto & Guedes Soares (eds)
 ISBN 978-0-415-62081-9

Large scale experimental study of wave-current interactions in the presence of a 3D bathymetry

V. Rey & J. Touboul
LSEET, UMR- 6017 du CNRS, ISITV, Université du Sud Toulon Var, La Valette du Var Cedex, France

F. Guinot
Department of Civil Engineering, Ghent University, Ghent, Belgium

ABSTRACT: Experiments were carried out in the Ocean Engineering Basin (BGO) FIRST, France, of useful length 24 metres and effective width 16 metres. A tridimensional bathymetry consisting of two symmetrical submerged mounds of maximum extension 8.5 m lengthwise was displayed on the flat bed on both sides of the basin. The maximum water depth was of 3 m, the water depth above the top of the mounds was of 1.5 m. Regular waves of frequency corresponding to either deep water or finite water depth above the bathymetry were generated without current and with either following or opposite currents. For the tests of current only, acceleration is observed above the mound over the whole basin width. A decrease of the near surface current intensity is observed downstream not only due to the increase of water depth but also to a vertical mixing. This vertical mixing is the stronger for opposite current conditions, due to higher turbulence generated through the perforated beach. In the presence of current, the mixing is enhanced, particularly for the longest waves. For opposite current conditions, a strong tridimensional behavior is observed for the amplitude. For the longest waves, transversal modes are excited by the current due to the geometry of the basin. For the shortest waves (deep water conditions), a strong focusing of the energy is observed. Indeed, amplitude of twice the incoming wave amplitude is observed in the central part (channel) of the basin. It cannot be ascribed to an increase of the current intensity since this latter is found to be almost uniform in the basin in spite of the tridimensional geometry of the mound.

1 INTRODUCTION

During the last decades, an increasing effort has been focused on the understanding and on the modeling of wave current interactions. Such an effort can be motivated by the effect on the current on the wave characteristics at the shore, the formation of freak waves, the wave damping due to current, or the effect of the waves on the current. The series of laboratory experiments presented hereafter were motivated by the study of the effect of the wave on the current properties in the context of the study of tidal power energy (Guinot, 2010).

Previous laboratory experiments have been carried out in 2D cases. Kemps and Simons (1982) observed for wave-following current conditions a vertical mixing with a decrease of the mean velocity near the surface and an increase near a smooth bottom. For rough beds, the mean velocity was found to decrease due to a modification of the vortices structures. On the contrary, Kemps and Simons (1983) observed for wave-opposite current conditions an increase of the mean velocity near the surface and a decrease near the bottom for either smooth or rough beds. In addition, they observed a much stronger wave damping along the tank for opposite current conditions and significant lateral boundary effect on the wave behavior. These experiments were carried out in a wave tank of width $L = 0.457$ m, a mean water depth $h = 0.20$ m, a regular wave of period $T = 1.006$ s of wave height and current in the ranges 0.27–0.54 m and 0.75–1.50 m/s. Klopman (1994) observed similar behavior during his experiments carried out in a wave tank of water $h = 0.5$ m, a wave period $T = 1.44$ s of amplitude 0.06 m and a current of about 0.15 m/s.

On the basis of in situ observations, Wolf and Prandle (1999) have discussed both the effect of waves on currents and the effect of currents on the waves. Not only bottom friction but also three dimensional effects such as wave refraction and mean horizontal velocity gradients can be significant. The experimental study of these latter effects necessitates 3D facilities. Such 3D experiments have been carried out in the offshore basin of Marin, The Netherlands (Margaretha, 2005) for deep water wave conditions. Measurements were carried out on the horizontal section basin floor. Both unsteadiness and non uniformity of the current in the transversal direction were found to generate time-space modulations.

In the context of the present work, a first series of experiments carried out in 2D cases had shown that velocity profiles depends strongly on the water depth (depth range 1–3 m, smooth bed) for currents in the range 0.3–0.8 m/s (Guinot, 2010) with a stronger vertical velocity gradient for smaller water depth. For wave following current conditions, experiments carried out for h = 1.0 m have shown that initially sheared current become uniform under the wave action (Rey et al, 2007). This process is all the more rapid for longer waves and higher wave amplitudes.

The present study focuses on 3D experiments carried out in a large scale basin. In section 2, the wave basin, the experimental set-up and the instrumentation are presented. Results are then reported and discussed in sections 3 and 4, respectively.

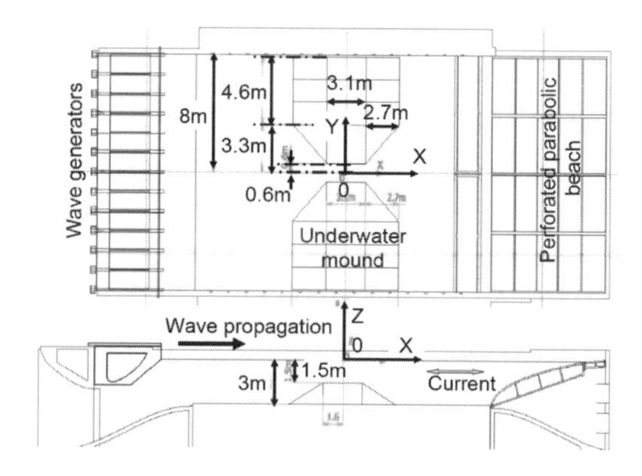

Figure 1. Sketch of the basin and of the underwater mounds.

2 EXPERIMENTAL SET-UP

2.1 *The BGO FIRST wave basin*

The Ocean Engineering Basin (BGO) FIRST has a useful length of 24 m, an effective width of 16 m and a maximum water depth of about 5 m. The BGO is equipped with a current generator, a wavemaker over the entire width and a XY carriage working in a Cartesian coordinate system. The X-axis corresponds to the incoming wave direction, the Y-axis to the cross direction, and the Z-axis is vertical downwards. A permeable wave absorber with parabolic shape is located at the end of the tank. The water depth can be set at the desired value thanks to a mobile bottom, which in turn allows a quick installation of bottom-fixed models and instrumentation. The wavemaker consists of horizontally oscillating cylinders and allows generation of regular and irregular waves with a maximum height—peak to trough—of 0.8 m. The three-dimensional bathymetry consisted of two Underwater Tri-Dimensional Mounds (UTDM) fixed on the mobile bottom on both sides of the central axis of the basin (Fig. 1). The mobile bottom was immersed at 3 m under the water surface.

2.2 *Instrumentation*

Two series of experiments have been carried out. In both series, capacitive wave probes, pressure sensors and current meters were displayed. Velocity profiles were measured by the use of an ADCP Stream-pro from the company RDI during the first series. The first series of measurements was found to be incomplete for the detailed analysis of both wave and current field, additional instrumentation was used in the second series, including a two components electromagnetic current meter (model 802, company Valeport).

Table 1. Location of the wave gauges WG_i and of the current meter CM (X-axis carriage location X = 0).

WG	1, 8	2, 9	3, 10	4, 11	5, 12	6, 13	7, 14
X_{WG} (m)	0,–1	0,–1	0,–1	0,–1	0,–1	0,–1	0,–1
Y_{WG} (m)	6,95	5,80	4,65	3,5	2,2	0,9	0
CM	Pos. 1	Pos. 2	Pos. 3	Pos. 4			
X_{CM} (m)	–0.3	–0.3	–0.3	–0.3			
Y_{CM} (m)	–0.53	–2.23	–3.93	–5.58			

The Wave Gauges (WG) and the Current Meter (CM) were displayed on the moving carriage. The depth of immersion of the current meter was Z = –0.25 m. The positions of the WG along the Y-axis are presented in Table 1. When the carriage positioning is $(X_0, Y_0) = (0, 0)$, WG1 to WG7 are located on the section X = 0, WG8 to WG14 on the section X = –1 m, the CM is located at (X, Z) = (–0.3 m, –0.25 m). Measurements were carried out for the following positions of the carriage along the X-axis, X_n = –8, –6.05, –4.1, –2.8, 0, 2.8, 4.1, 6.05 and 8 m. In addition, the CM could be positioned along Y-axis at the following positions (1 to 4), Y_n = –0.53, –2.23, –3.93 and –5.58 m.

2.3 *Hydrodynamic conditions and measurements*

During the first series of experiments, regular waves of period T = 1.3 s, corresponding to deep water conditions even above the mound, and T = 1.8 s and 2.2 s, corresponding to intermediate water depth conditions, of nominal wave height H = 0.10 and 0.20 m were generated. The current intensity was U = 0.20 m/s in both following and opposite wave conditions. During the second series of experiments, the current intensity was U = 0.15 m/s for the following current and U = 0.25 m/s for the opposite current. Periods were chosen in order to keep the wavelength to depth ratio constant in the presence of current for the incoming wave.

For the deeper wave conditions, T = 1.3 s, λ = 2.64 m for h = 3 m and h/λ = 1.136 without any current. By taking constant h/λ, the wave period was T = 1.2 s for the following current and T = 1.48 s for the opposite current. For the intermediate water depth conditions, T = 2.2 s, λ = 7.46 m for h = 3 m and h/λ = 0.4 without any current. By taking constant h/λ, the wave period was T = 2.15 s for the following current and T = 2.4 s for the opposite current.

The sampling rate of the synchronized instruments (wave gauges and current meters) was 32 Hz. In the presence of current, the wave generation was initiated 120 min after the current. The celerity field of the wave was calculated from the phase lag between the two successive wave gauges on the ramp S_{i+7} and S_i. Calculation of the celerity c = 2π/(kT) from the electromagnetic velocity sensor was also made by use of the relation dispersion

$$(\omega \pm |U|k)^2 = gk\tanh(kh) \qquad (1)$$

where k is the wave wavenumber, g the acceleration due to gravity and ω = 2π/T the angular frequency.

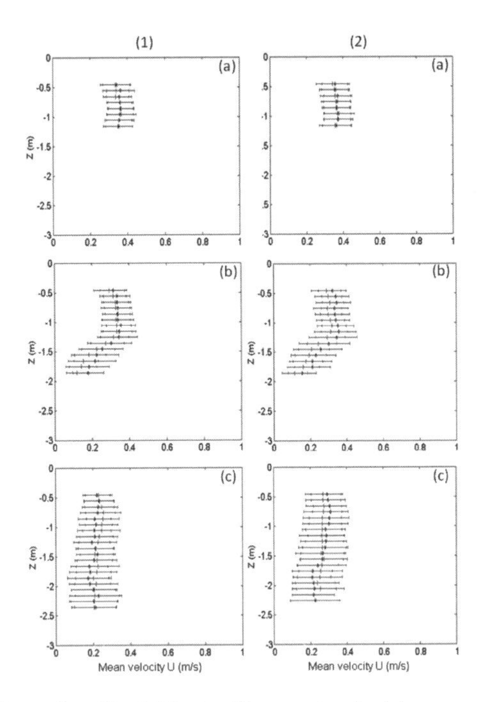

Figure 2. U = 0.20 m/s, (*) current only, (+) wave and current, H = 10 cm, (1) T = 1.3 s, (2) T = 2.2 s, (a) X = 0.0 m, (b) X = 2.8 m, (c) X = 8.0 m.

3 RESULTS

3.1 *Vertical current profiles*

Velocity profiles were measured by the use of an ADCP Stream-pro from the company RDI during the first series of experiments. The sampling rate of the instrument was 1Hz. Results for the current, above, on the sloping descent and down-wave the mound (Y = 5.6 m) are presented in Figure 2 for the following current and in Figure 3 for the opposite current. The error bars are slightly the same without or in the presence of wave, this is mainly due to the noisy measurements due to a very poor signal intensity of the backscattered acoustic beams. However, we can observe at least qualitatively that above the mound (water depth 1.5 m) and down-wave the mound, the vertical profile is quasi-uniform, for both following and opposite current conditions.

The sheared current on the descent is also observed in both cases, certainly due to quite steep slope of the mound. A weak difference is observed between the mean velocities measured either with or without the wave in the descent and down wave, especially for the longer wave. It can be ascribed to a tri-dimensional redistribution of the mean current field as it will be discussed in the following section.

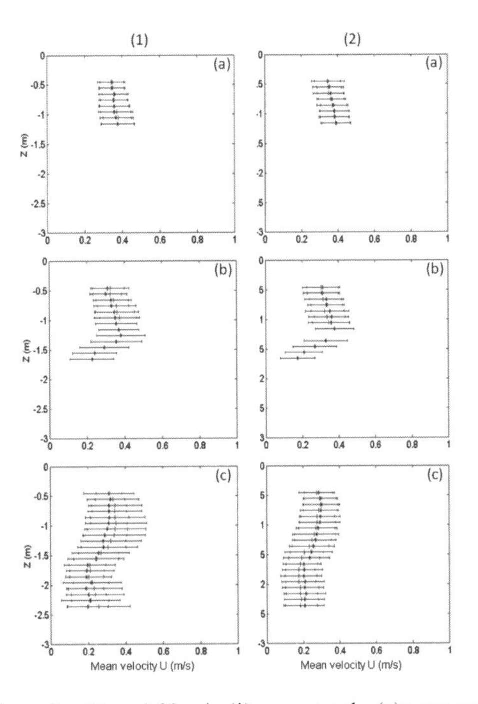

Figure 3. U = −0.20 m/s, (*) current only, (+) wave and current, H = 10 cm, (1) T = 1.3 s, (2) T = 2.2 s, (a) X = 0.0 m, (b) X = −2.8 m, (c) X = −8.0 m.

3.2 *Mean current fields*

During the second series of experiments, the near surface current intensity field at 0.25 m under the mean free surface was measured with an electromagnetic sensor. The sampling rate of the instrument was 32Hz. Analysis of the mean velocities are presented. Results for the following current conditions are presented in Figures 4a and 4b for respectively deep and intermediate water depths, results for opposite conditions are presented in Figures 5a and 5b respectively. Results for the current alone are also given for comparison.

For the following current conditions (current flow toward the right side), we can observe that the near surface current intensity is almost the same in the cross section Y, either without or in the presence of the waves. A general decrease is observed down wave the mounds, but the current intensity remains higher down wave. For both wave conditions, the increase of the current intensity at the location of the mound (around X = 0) is less important in the presence of the waves. This result was observed for the vertical profile measurements as shown in the previous section, especially for the longer waves. Furthermore, the dissymmetry of the current intensity between the up wave and down wave part is less pronounced in the presence of the waves, which are certainly responsible of a part of the vertical mixing.

For the opposite current conditions (current flow toward the left side), we can observe again that the near surface current intensity is almost the same in the cross section Y, either without or in the presence of the waves.

The general decrease observed down wave the mounds is less pronounced. For the longer waves, the current intensity up wave and down wave is almost the same. In any case, the vertical mixing is increased in comparison with the following current due to a more turbulent flow in this latter condition. Indeed, the presence of the perforated parabolic beach generates turbulence. This vertical mixing is enhanced by the presence of long waves as observed in Figure 5b.

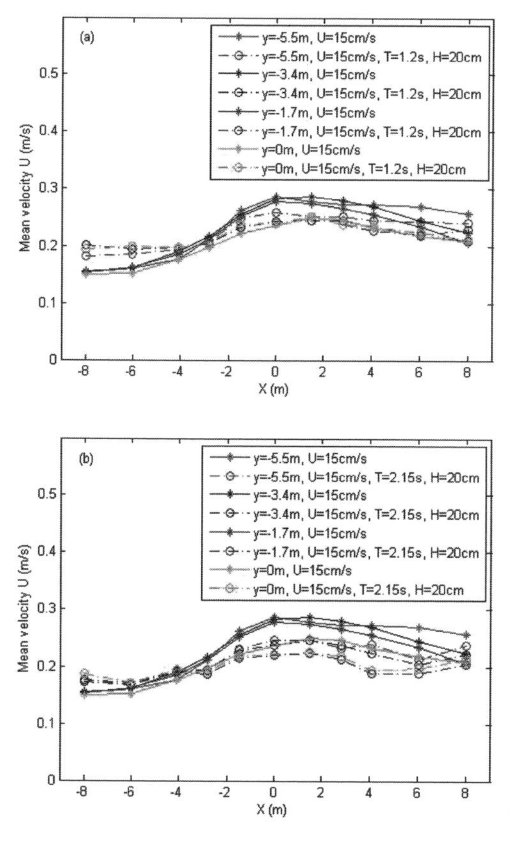

Figure 4. Mean velocity, U = 0.15 m/s, (a) T = 1.2 s, (b) T = 2.15 s.

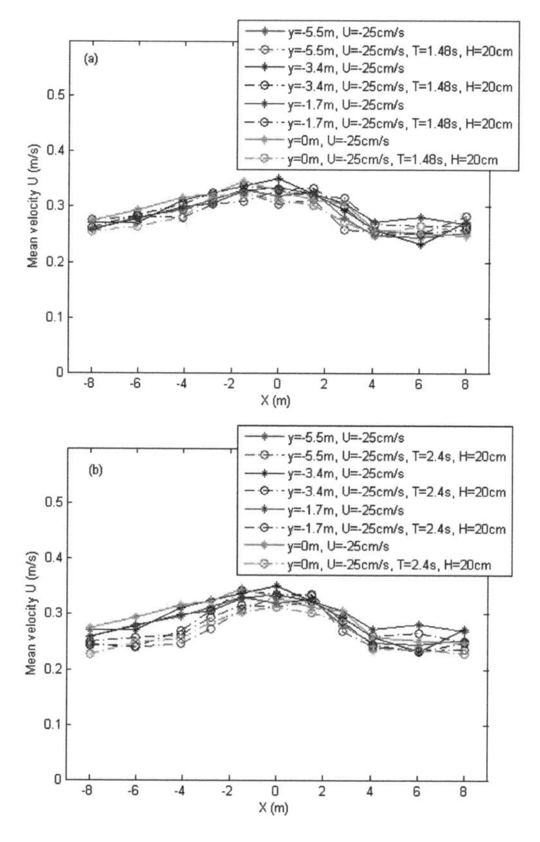

Figure 5. Mean velocity, U = –0.25 m/s, (a) T = 1.48 s, (b) T = 2.4 s.

3.3 *Wave amplitude and celerity*

In this section, the influence of the current on wave amplitude and celerity are studied. Results for the deep water incoming wave conditions are presented in Figures 6–9, results for the intermediate water depth conditions in Figures 10–13.

3.3.1 *Deep water conditions*

In the absence of current and for the following current conditions, the wave amplitude field is almost homogeneous as expected since the wave is not refracted by the mounds and the near surface current was found to be homogeneous.

For the opposite current conditions, a drastically different behavior is observed: a strong wave amplification up to twice the incoming wave amplitude is measured although a strong damping appears on both parts of the wave basin, above the mounds. It is noticeable that this behavior is observed for both H = 10 cm and H = 20 cm (see Figs. 8a and 9a), even if for the latter case, we can observe in a first stage of the wave propagation a

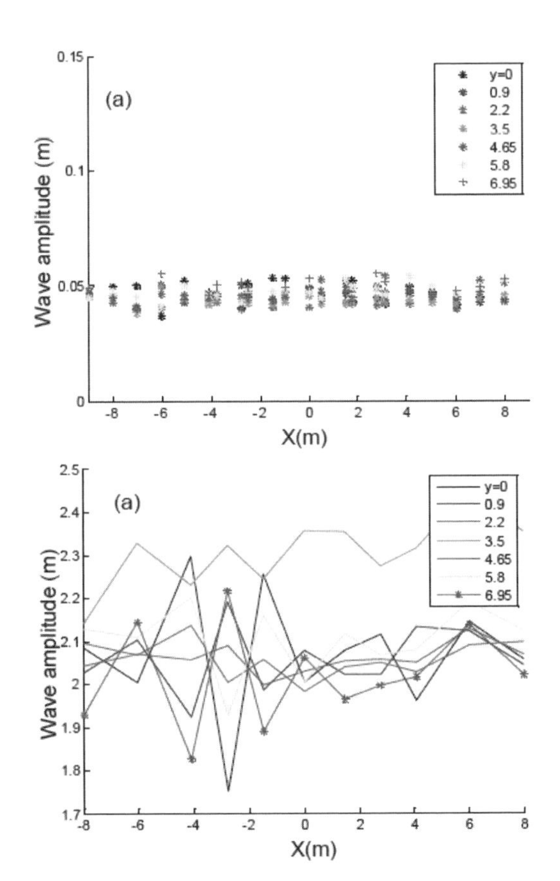

Figure 6. Wave amplitude (a) and celerity (b), U = 0 m/s, T = 1.3 s, H = 10 cm.

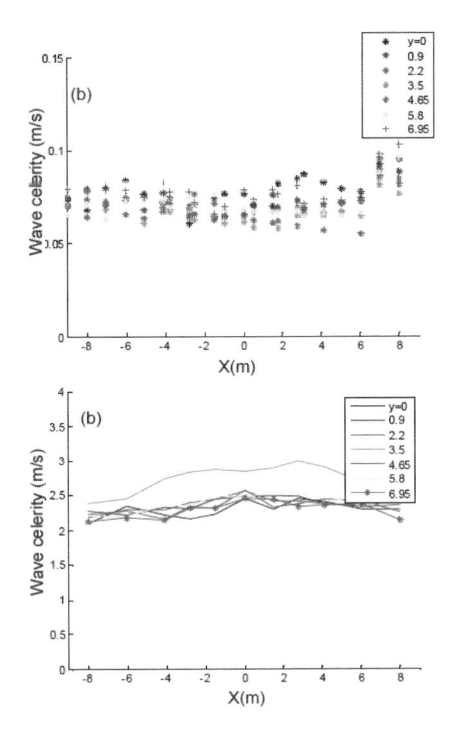

Figure 7. Wave amplitude (a) and celerity (b), U = 0.15 m/s, T = 1.2 s, H = 10 cm.

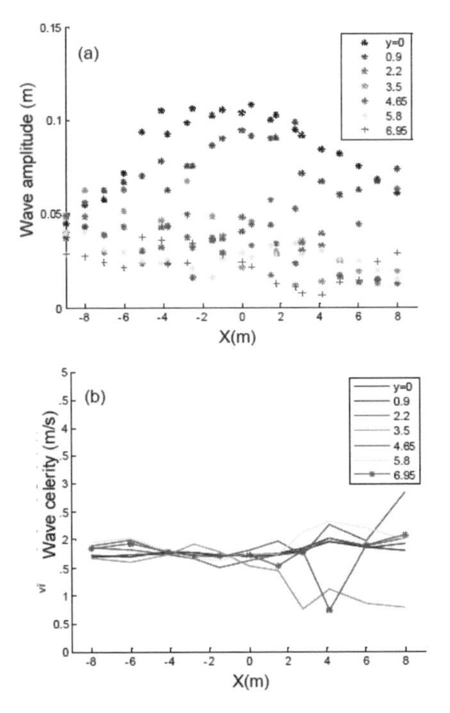

Figure 8. Wave amplitude (a) and celerity (b), U = –0.25 m/s, T = 1.48 s, H = 10 cm.

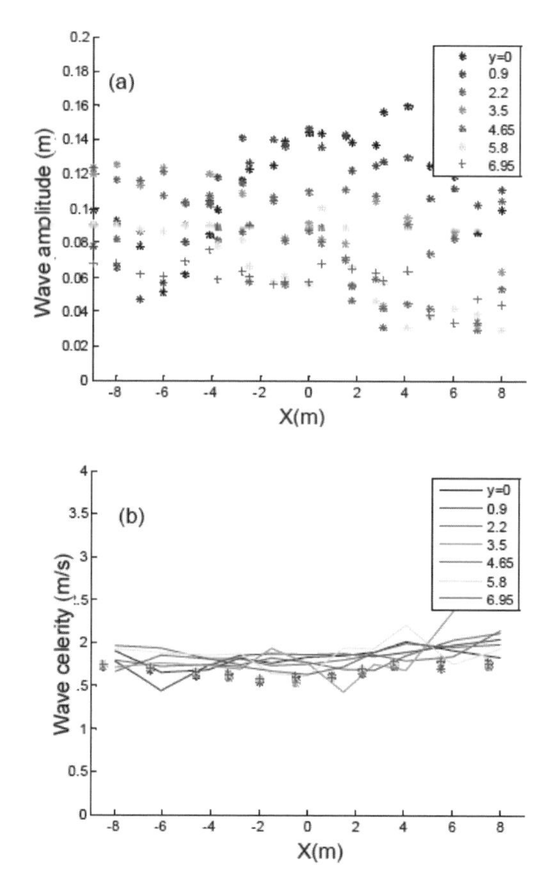

Figure 9. Wave amplitude (a) and celerity (b), U = −0.25 m/s, T = 1.48 s, H = 20 cm. The markers (*) indicate the celerity calculated by use of the dispersion relation.

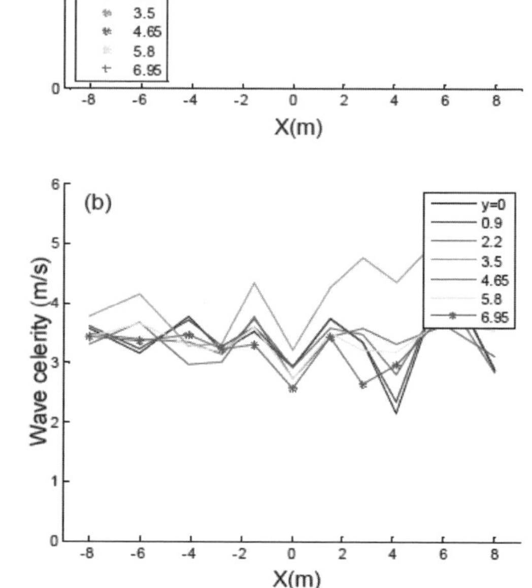

Figure 10. Wave amplitude (a) and celerity (b), U = 0 m/s, T = 2.2 s, H = 10 cm.

decrease of its amplitude in front of the generators due to breaking.

We could expect this amplification to be due to an increase of the opposite current intensity and/or a strong velocity gradient in the cross section. However, the wave celerity measurements presented in Figures 8b and 9b show rather uniform wave celerity in the basin. In addition, such a behavior is observed for both wave height H = 10 cm and H = 20 cm, that suggests a linear process.

3.3.2 *Intermediate water depth conditions*

In the absence of current and for the following current conditions, the wave amplitude field is slightly modulated along the X axis as expected since the wave is partially reflected and refracted by the mounds. Due to relative weakness of the near surface current by comparison with the wave celerity, its effect remains quite limited. For the opposite current conditions, a drastically different behavior is observed: a strong wave amplification up wave and down wave the mound location is observed in the central axis X (Y = 0) and at the top of the transverse slope of the mounds (X = 0, Y = 3.5 – 4.5). Small amplitude oscillations are observed in the middle of the basin (X = Y = 0) and above all at both ends of the mounds (X = 0, Y = 6.95). The measured wave celerity shows the presence of a antinode at this point (see Figs. 12b and 13b). Since relative water depth conditions are kept constant for the experiments in the presence of current, the wavelengths depend slightly on the current intensity. This oscillating behavior may be explained by a resonant condition of wave due to the geometry of the basin and/or the bathymetric shape.

If resonant conditions seems to be almost verified (the wavelength is 7.46 m for h = 3 m), wave amplification is only observed for opposite current conditions. In addition, such a behavior is observed for both wave height H = 10 cm and H = 20 cm, that suggests a linear process.

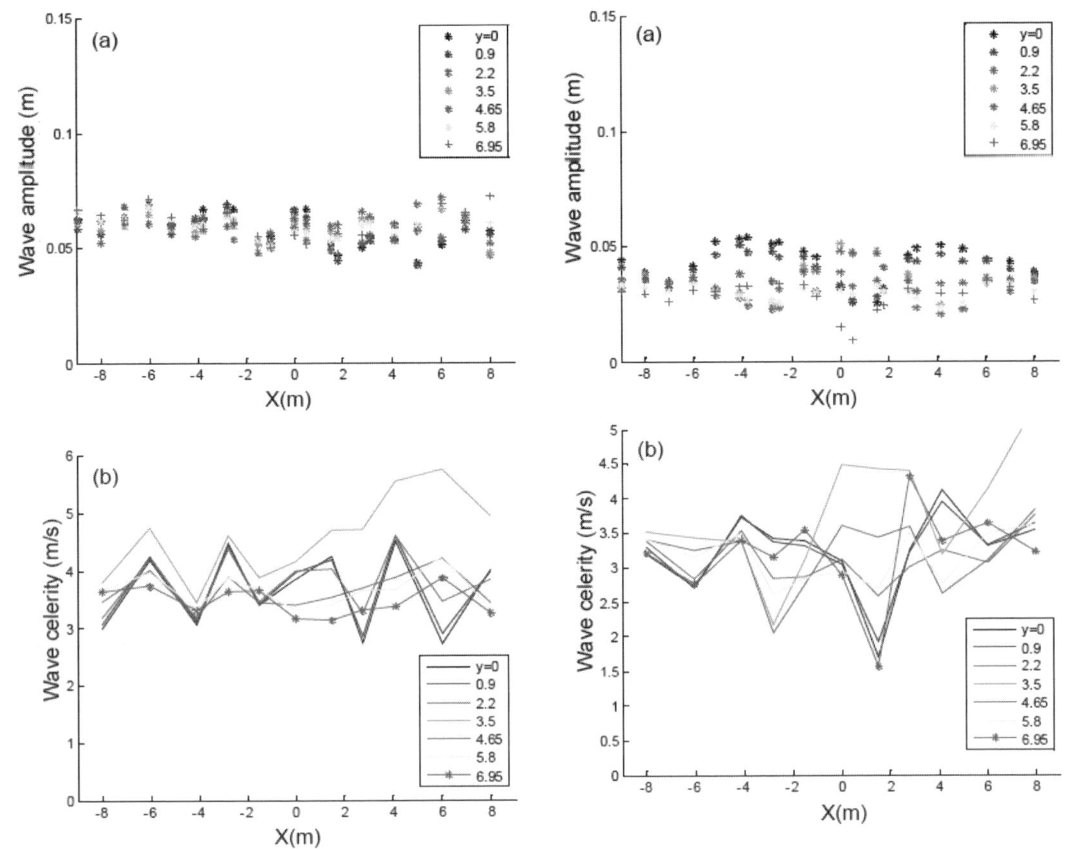

Figure 11. Wave amplitude (a) and celerity (b), U = 0.15 m/s, T = 2.15 s, H = 10 cm.

Figure 12. Wave amplitude (a) and celerity (b), U = –0.25 m/s, T = 2.4 s, H = 10 cm.

4 DISCUSSION AND CONCLUSION

In the present study, laboratory experiment of wave–current interactions was investigated.

Results concerning the mean velocity vertical profile have shown a quasi uniform vertical profile over flat beds. On the contrary, a strong shear is observed on the (steep) descending slopes of the mound. This behavior strengthen the use of the hypothesis of uniform vertical currents in wave propagation models over flat or mild slope bed conditions, validity of use which is much more questionable in the presence of steep slopes. If in the following current conditions, the wave behavior was foreseeable, the results for opposite current conditions have shown noticeable aspects. For the deep water conditions, huge wave amplification up to twice the incident wave was observed. This phenomenon then leads to the formation of a "freak wave".

What is amazing, it is the fact that this amplification is not ascribable to an increase of the wave steepness due to an increase of the opposite current while the wave is progressing (the wave is already generated in the presence of the current) but to an energy focusing due to a tri-dimensional phenomenon. For intermediate water depth conditions, the phenomenon of amplification seems rather due the geometry of the wave basin. However, such tri dimensional configurations may exist in situ and such wave amplification may be observed. Anyway, we can observe that current may enhance resonant oscillations.

At this stage of analysis, the wave current interaction induced hydrodynamics are not fully described. Further data including pressure sensors measurements have to be analyzed. Concerning the electromagnetic sensor, wave analysis can be also carried out through the wave induced velocities. For the case of the opposite current for the deep water wave of height H = 0.20 m, the Standard Deviation (SD) was also calculated for the mean velocity. This SD gives information on the wave induced horizontal velocity.

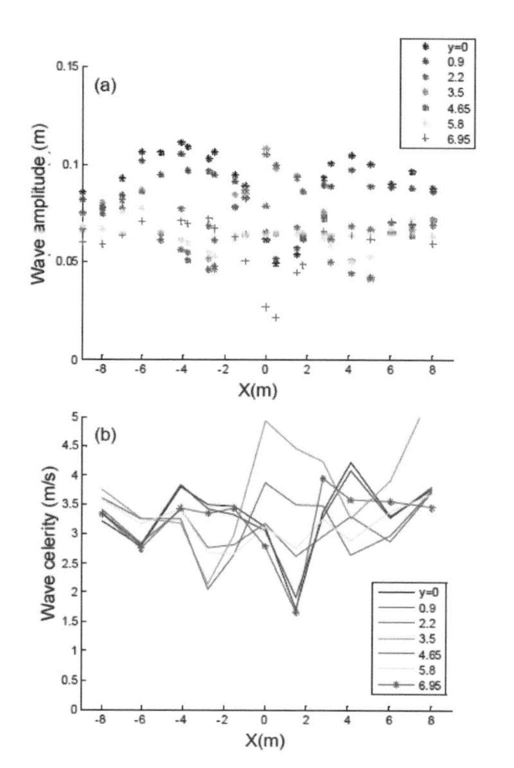

Figure 13. Wave amplitude (a) and celerity (b), U = –0.25 m/s, T = 2.4 s, H = 20 cm.

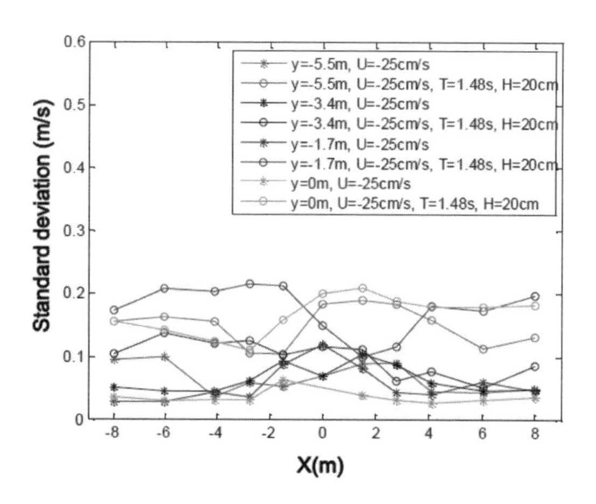

Figure 14. Standard deviation for the mean near surface current. U = –0.25 m/s, T = 1.48 s, H = 20 cm.

For the case of the opposite current for the deep water wave of height H = 0.20 m, the Standard Deviation (SD) for the mean velocity presented in Figure 9a is presented in Figure 14. This SD gives information on the wave induced horizontal velocity. We can observe in Figure 14a strong increase of the SD for Y = 0 corresponding to an increase of the wave amplitude (Fig. 14). On the contrary the increase of the SD for Y = 5.5 does not corresponds to an increase of the wave amplitude. Such a transfer function has then to be analyzed, together with pressure sensors data.

ACKNOWLEDGEMENTS

The authors are grateful to the Conseil Général du Var for its financial support for the experiments carried out in the wave basin BGO FIRST.

REFERENCES

Guinot, F. 2011, Interactions houles courant en bathymétrie variable: approches numériques et expérimentales, Ph-D, University Bretagne Occidentale, France, 14 décembre 2010 (*in french*).

Kemps, P.H. and Simons, R.R. 1982, The interaction between waves and a turbulent current: Waves propagating with the current, *J. Fluid Mech.*, **116**, 227–250.

Kemps, P.H. and Simons, R.R. 1983, The interaction between waves and a turbulent current: Waves propagating against the current, *J. Fluid Mech.*, **130**, 73–89.

Klopman, G. 1994, Vertical structure of the flow due to waves and currents. Progress report H840.30, Part II. Delft Hydraulics.

Margaretha, H. 2005, Mathematical modelling of wave-current interaction in a hydrodynamic laboratory basin, Ph-D, University of Twente, The Netherlands, 23 september 2005.

Rey, V., Guinot, F. et Le Boulluec, M. Interaction houle—courant par profondeur finie: impact sur la cinématique, *XIème Journées de l'hydrodynamique,* Brest, 3-5 avril 2007 (*in french*).

Wolf, J. and Prandle, D. 1999, Some observations of wave-current interaction, *Coastal. Eng.*, **37**, 471–485.

Sustainable Maritime Transportation and Exploitation of Sea Resources – Rizzuto & Guedes Soares (eds)
© 2012 Taylor & Francis Group, London, ISBN 978-0-415-62081-9

Three-dimensional sea wave groups with a superimposed large-scale current

A. Romolo, F. Arena & D. Ciricosta
Department of Mechanics & Materials, 'Mediterranea' University of Reggio Calabria, Reggio Calabria, Italy

ABSTRACT: In the paper, a close-form solution for the free surface displacement and velocity potential of short-crested (three-dimensional) wave groups interacting with an encounter large-scale current, whose direction of advance can be generally different to the dominant wave direction, is achieved through the Quasi-Determinism theory (derived by Boccotti in the eighties). The solution has been applied to investigate the space-time evolution of the wave groups, by taking into account the effects of three-dimensionality, and those associated to the changes of intensity and direction of the current.

1 INTRODUCTION

In this paper the mechanics of Three-Dimensional (3D) sea wave groups propagating in the presence of a large-scale current is analyzed.

It is well-known that the presence of a current is responsible of important modification of water profiles. On one hand an opposite current produces an increment of the wave steepness due both to an enhance of the wave height and a reduction of wave length; on the other hand a following current reduces the wave steepness by increasing the wave length and reducing the wave height.

The effects of strong deformation associated to the interaction among sea waves and adverse currents in deep water could be one of the cause of the occurrence of extreme events as it has been observed by several authors (Mallory, 1974; Lavrenov, 1998; White and Fornberg, 1998). Thus their analysis is of interest for offshore and naval engineering. By considering irregular sea waves, different contributions are given both to describing the change of the spectral shape due to the presence of currents (Tayfun *et al.*, 1976; Hedges *et al.*, 1985; Hedges *et al.*, 1993; Guedes Soares & de Pablo, 2006) and to investigating the mechanics of sea waves propagating onto currents (Peregrine, 1976; Trulsen & Mei, 1993; Jensen, 2002).

Starting from the linear Quasi-Determinism theory of the highest waves (Boccotti 1981, 1982, 1993, 1997, 2000), which is able to fully describe the mechanics of sea wave groups either when a very high exceptional crest amplitude is realized or an exceptional wave height occurs by taking into account the effects of three-dimensionality associated to the finite bandwidth of spectrum of actual sea waves, a closed form solution for wave groups travelling in the presence of a large-scale current, whose direction of propagation could be generally different from the dominant wave direction, has been derived. The deterministic functions of the free surface displacement and velocity potential exact to the first-order (Romolo & Arena, 2003; Arena & Romolo, 2005) and to the second-order in a Stokes' expansion (Nava *et al.* 2006) have been obtained for the case of long-crested (2D) sea waves.

In the paper the solution for sea wave groups with a superimposed large-scale current achieved through the Quasi-Determinism theory is extended to the three-dimensional case by considering the effects associated to the wave energy distribution to both frequencies and directions in order to investigate the space-time evolution of short-crested (3D) sea wave groups.

2 INTERACTION BETWEEN THREE-DIMENSIONAL WIND-GENERATED WAVES AND LARGE-SCALE CURRENTS

2.1 *The theory of the sea states in the presence of a large-scale current*

The theory of wind-generated waves, given by Longuet-Higgins (1963) and Phillips (1967), defines a random sea state as the sum of a very large number, N, of periodic components, with infinitesimal amplitude, a_i, frequencies, ω_i, different from each other, and phase angles, ε_i, uniformly distributed on $(0,2\pi)$ and stochastically independent from each other. In the three-dimensional feature, the single component makes an angle θ_i, with the y-axis (Fig. 1). According to the theory, the linear free surface displacement and the velocity potential are stationary Gaussian random processes.

In this paper we have considered the case of three-dimensional sea waves interacting with a large-scale current, of velocity u, whose direction of advance makes, with respect to the y-axis, an angle θ_C generally different from the angle of the dominant wave direction $\bar{\theta}$ (see Fig. 1).

For the examined configuration, the linear free surface displacement and the velocity potential, through the theory of wind-generated waves, are respectively provided by:

$$\eta(x,y,t) = \sum_{i=1}^{N} a_i \cos(k_{C_i} x \sin\theta_i + k_{C_i} y \cos\theta_i - \omega_i t + \varepsilon_i), \tag{1}$$

$$\begin{aligned}\varphi(x,y,z,t) = {} & u \sin\theta_C x + u \cos\theta_C y \\ & + g \sum_{i=1}^{N} a_i [\omega_i - u k_{C_i} \cos(\theta_i - \theta_C)]^{-1} \\ & \frac{\cosh[k_{C_i}(d+z)]}{\cosh(k_{C_i} d)} \cdot \sin(k_{C_i} x \sin\theta_i \\ & + k_{C_i} y \cos\theta_i - \omega_i t + \varepsilon_i) \\ & + -\frac{1}{2} u^2 t + \frac{1}{\rho} \int_0^t f(t') dt'. \end{aligned} \tag{2}$$

As for the classical treatment of the study of wave-current interaction, two different frames of reference could be assumed: an absolute stationary frame and a relative frame of reference moving in the direction of the current with its velocity.

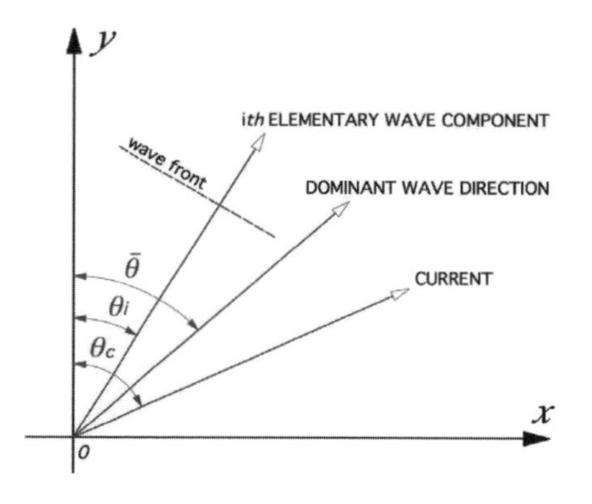

Figure 1. Frame of reference of the present study: current, wave dominant and individual wave component directions relative to the y-axis.

Eqs. (1) and (2) are referred to an absolute stationary frame, and, therefore, for every elementary component of the sea state in the presence of current, for fixed depth, d, the wave number, k_{C_i}, is related to the stationary angular frequency, ω_i, to the velocity of the current, u, and to the angle that the direction of propagation of the elementary component makes with the current, by the following relation (Phillips, 1977).

$$g k_{C_i} \tanh(k_{C_i} d) = [\omega_i - u k_{C_i} \cos(\theta_i - \theta_C)]^2. \tag{3}$$

Moreover amplitudes a_i, frequencies, ω_i, and directions θ_i, of the elementary components define a directional spectrum in the presence of a current $S(\omega,\theta,u)$:

$$S(\omega,\theta,u)\delta\omega\delta\theta \equiv \sum_i \frac{1}{2} a_i^2$$

for i such that $\omega < \omega_i < \omega + \delta\omega$ and $\theta < \theta_i < \theta + \delta\theta$. (4)

The directional wave spectrum of irregular sea waves, which are propagating from a region without current to a region with the presence of a current, otherwise moving from one region to another characterized by different values of the current velocity, was derived by Hedges *et al.* (1993) by considering large-scale currents. They have evaluated the change both of wave amplitude and wave direction (refraction), of the single component of a sea state, induced by an encountering current through the principle of "wave action" conservation valid in the absence of wave generation or dissipation (Bretherton & Garrett, 1968).

From the aforesaid solution, the directional spectrum in the presence of a current $S(\omega, \theta, u)$ may be expressed as a function of the directional spectrum of wind-generated waves $S(\omega,\theta)$ as follows:

$$S(\omega,\theta,u) = \frac{k_C G(k_0 d)}{k_0 \left\{ G(k_C d) + \dfrac{u\cos(\theta - \theta_C)}{d\,[\omega - u k_C \cos(\theta - \theta_C)]} \right\}} S(\omega,\theta) \tag{5}$$

where

$$G(x) = \frac{1}{2x} + \frac{1}{\sinh(2x)}, \tag{6}$$

and k_0 is the wave number in the absence of a current, which is evaluated by the classical linear dispersion rule defined by Eq. (3) with $u = 0$.

By considering the high frequencies of a spectrum for a given direction, it is well known that there is a limit on the grow of the spectral densities due to breaking for sea waves either in quiescent water

($u=0$) or propagating in the presence of a current. Starting from the general approach proposed by Phillips (1958, 1977) for wind-waves and on the definition of an equilibrium range of universal shape, Kitaigordskii *et al.* (1975) examined the problem of an equilibrium spectral range in the case of surface waves on uniform current, where the principle of "wave action" conservation is not longer valid. This latter formulation has been considered by Hedges *et al.* (1993) to evaluate the wave spectral changes for a wave field in the presence of a shearing large-scale current, giving the following solution for the equilibrium range of the directional spectrum

$$S_{ER}(\omega,\theta,u)=\frac{B^{*}k_C^{-3}}{\{d\,[\omega-uk_C\cos(\theta-\theta_C)]G(k_Cd)+u\cos(\theta-\theta_C)\}}D_{ER}(\theta) \tag{7}$$

where B^* is equal to 0.5 times the Phillips' parameter α [its values are $\alpha=0.008 \div 0.012$, usually], the G function is defined by Eq. (6), and

$$D_{ER}(\theta)=\frac{1}{\sigma^2}\int_0^\infty S(\omega,\theta,u)\mathrm{d}\omega, \tag{8}$$

$S(\omega, \theta, u)$ being the directional spectrum in the presence of a current under the validity of the principle of "wave action" conservation [Eq. (5)] and σ^2 the relative total variance.

Eq. (7) is considered for those frequencies and directions such that $S(\omega, \theta, u) > S_{ER}(\omega, \theta, u)$, $S(\omega, \theta, u)$ being given by Eq. (5).

Finally in Eq. (5), as for the directional spectrum $S(\omega,\theta)$, the Pierson Moskowitz-Mitsuyasu spectrum (Pierson & Moskowitz, 1964; Mitsuyasu *et al.*, 1975) is assumed, which is a typical spectrum of wind-generated waves and whose mathematical form on deep water is:

$$S(\omega,\theta)= g^2\omega^{-5}\exp\left[-\frac{5}{4}\left(\frac{\omega_p}{\omega}\right)^4\right] K(n)\left|\cos\left[\frac{1}{2}(\theta-\bar{\theta})\right]\right|^{2n} \tag{9}$$

where ω_p is the peak angular frequency, $\bar{\theta}$ is the angle between the y-axis and the dominant direction of the spectrum, α is the Phillips' parameter, $K(n)$ is the normalizing factor

$$K(n)=\left[\int_0^{2\pi}\cos^{2n}\left(\frac{1}{2}\theta\right)d\theta\right]^{-1} \tag{10}$$

and n is a parameter depending on the frequency

$$n=n_p(\omega/\omega_p)^5 \quad \text{if } \omega\le\omega_p,\qquad n=n_p(\omega_p/\omega)^{2.5} \quad \text{if } \omega>\omega_p \tag{11}$$

$$\text{with } n_p=7.5\cdot10^{-3}\left(\frac{gF_e}{u^2}\right)^{0.825} \tag{12}$$

where F_e is the fetch and u the wind speed.

One of the most important sea state parameters influenced by the spectral changes is the significant wave height, H_S:

$$H_S=4\int_0^\infty S(\omega,\theta,u)\mathrm{d}\omega. \tag{13}$$

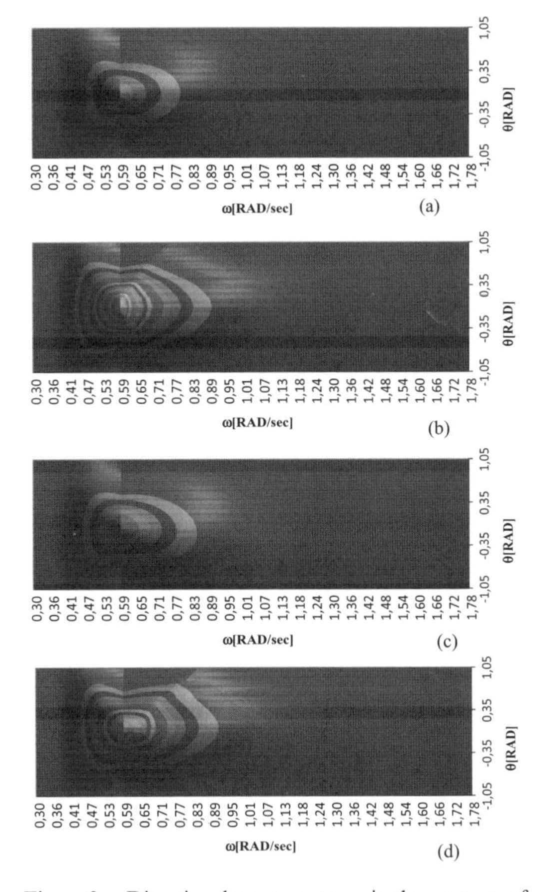

Figure 2. Directional wave spectrum in the presence of a current $S(\omega,\theta,u)$ [Eqs. (5~7)]: (a) $\bar{\theta}=0$, $\theta_c=0$, $u=+1.5$ m/s (concordant to the positive y-axis); (b) $\bar{\theta}=0$, $\theta_c=0$, $u=-1.5$ m/s (concordant to the negative y-axis); (c) $\bar{\theta}=0$, $\theta_c=60°$, $u=+1.5$ m/s (the plus sign means the current velocity vector u is coming out to the origin of the frame of reference of Fig. 1); (d) $\bar{\theta}=0$, $\theta_c=60°$, $u=-1.5$m/s (the minus sign means the vector u is entering to the origin of the frame of reference of Fig. 1).

For example, with respect to a sea state without current with significant wave height of $H_S = 5$m, the presence of a current propagating in a direction concordant to the dominant wave direction (following current) and coincident to the y-axis ($\bar{\theta}=0$, $\theta_C=0$) with a velocity equal to 1.5m/s in absolute value, produces a reduction of H_S of about 23% [$H_{S(u=+1.5m/s)} = 3.8$m]; in the same feature, ($\bar{\theta} = 0$, $\theta_C = 0$) if the current is propagating in the direction discordant to the dominant wave direction (adverse current) an increase by 13% is found [$H_{S(u=-1.5m/s)} = 5.6$m] [Fig. 2 panels (a) and (b)]. These effects grow as greater is the intensity of the current velocity.

In what follows, when the angle ($\bar{\theta} - \theta_C$) is different to zero, the current velocity vector, u, will be identified in direction by the angle θ_c, positive if measured clockwise with respect to the y-axis, and it will be defined adverse if the vector velocity u is entering to the origin of the frame of reference of Fig. 1, otherwise following.

For this last configuration, we observe a reduction of the effects previously highlighted. That is to say, for a current velocity of 1.5m/s, propagating in a direction making an angle of 60° to the dominant wave direction, a reduction of 13% on the significant wave height is found for a following current, and an increase of 8.3% due to the presence of an adverse current [Fig. 2 panels (c) and (d)]. These values are referred to the condition of $H_S = 5$m in the absence of current.

3 THE THEORY OF 'QUASI-DETERMINISM' FOR WAVE GROUPS INTERACTING WITH A LARGE-SCALE CURRENT

The Quasi-Determinism (QD) theory, derived by Boccotti, regards the mechanics of wave groups at sea, conditional to the occurrence either of a very high crest amplitude (first formulation of the QD theory—Boccotti, 1981, 1982) or of an exceptionally high crest-to-trough wave (second formulation of the QD theory—Boccotti, 1997, 2000) in a stationary Gaussian random wave field. The two formulations of the theory match exactly as it was proved by Boccotti (2000), in fact both represent the same wave group, which reaches the apex phase of its development in two different configurations.

In the nineties Boccotti achieved a full validation of the theory by means of some small-scale field experiments at the natural laboratory of Reggio Calabria (Italy)—at present, the Natural Ocean Engineering Laboratory (the N.O.E.L., www.noel.unirc.it), where it is possible to carry out experiments directly in the sea operating through the techniques of the laboratory tanks (Boccotti *et al.* 1993, Boccotti 1997, 2000).

3.1 *The deterministic expressions of the free surface and velocity potential, when a very large crest-to-trough wave occurs*

The linear Quasi-Determinism (QD) theory, formulated by Boccotti (1981, 1982, 1997, 2000), allows for the achievement, under specific assumption, of an expected configuration in space and time of both the free surface displacement and the velocity potential.

If a very high wave height, of elevation H, occurs at a fixed point $\underline{x}_0 \equiv (x_0, y_0)$ at a given time instant t_0, in a stationary Gaussian random process, like a wind-generated sea state, through the QD the expressions of the free surface and of the velocity potential tend, with very high probability, to closed-forms in the limit $H/\sigma \rightarrow \infty$ (σ being the standard deviation of the surface elevation of the wave field where H is realized).

Trough the QD the deterministic surface elevation, in a area surrounding $\underline{x}_0$, before and after t_0 is

$$\bar{\eta}(\underline{x}_0 + \underline{X}, t_0 + T) = \frac{H}{2} \frac{\Psi(\underline{X}, T; \underline{x}_0) - \Psi(\underline{X}, T - T^*; \underline{x}_0)}{\Psi(\underline{0}, 0; \underline{x}_0) - \Psi(\underline{0}, T^*; \underline{x}_0)} \tag{13}$$

and the velocity potential is

$$\bar{\phi}(\underline{x}_0 + \underline{X}, z, t_0 + T) = \frac{H}{2} \frac{\Phi(\underline{X}, z, T; \underline{x}_0) - \Phi(\underline{X}, z, T - T^*; \underline{x}_0)}{\Psi(\underline{0}, 0; \underline{x}_0) - \Psi(\underline{0}, T^*; \underline{x}_0)} \tag{14}$$

with $\Psi(\underline{X}, T)$ and $\Phi(\underline{X}, z, T)$ respectively the auto-covariance of the surface displacement and the cross-covariance of the surface displacement and the velocity potential, defined as

$$\Psi(\underline{X}, T) = < \eta(\underline{x}_0, t)\eta(\underline{x}_0 + \underline{X}, t + T) >, \tag{15}$$

$$\Phi(\underline{X}, z, T) = < \eta(\underline{x}_0, t)\phi(\underline{x}_0 + \underline{X}, z, t + T) >, \tag{16}$$

where η and ϕ are stationary Gaussian random processes, like the processes of free surface displacement and velocity potential achieved through the theory of sea states.

In Eq. (13) and (14), T^* is the abscissa of the absolute minimum of the autocovariance function $\Psi(\underline{0}, T)$ assumed to be also the first local minimum of this function on the positive domain.

That one described is the second formulation of the QD theory (Boccotti 1997, 2000). Expressions (13) and (14) are exact to the first order in a Stokes expansion and, above all, hold for nearly arbitrary bandwidth and solid boundary, as proved by Boccotti (2000). Thus the theory can be applied also for the present case of very high waves interacting with large-scale currents.

3.2 *The 'Quasi-Determinism' theory for wave groups interacting with a large-scale current*

The solutions for the deterministic free surface displacement $\bar{\eta}$ (13) and velocity potential $\bar{\phi}$ (14) when an exceptional wave height, of given elevation H, occurs at a given time instant t_0, at $\underline{x}_0 \equiv (x_0, y_0)$ in a real sea state of wind-waves with an encountering large-scale current, whose direction of advance, with respect to the y-axis, makes an angle θ_C generally different from the angle of the dominant wave direction $\bar{\theta}$, are:

$$\bar{\eta}(\underline{x}_0 + \underline{X}, t_0 + T) = \frac{H}{2}\left\{\int_0^{\infty}\int_0^{2\pi} S(\omega,\theta,u)\{\cos[k_C X \sin\theta + k_C Y \cos\theta - \omega T] - \cos[k_C X \sin\theta + k_C Y \cos\theta - \omega(T - T^*)]\}\,d\theta\,d\omega\right\} \Big/ \left\{\int_0^{\infty}\int_0^{2\pi} S(\omega,\theta,u)[1-\cos(\omega T^*)]\,d\theta\,d\omega\right\} \tag{17}$$

and

$$\bar{\phi}(\underline{x}_0 + \underline{X}, z, t_0 + T) = u \sin\theta_C X + u\cos\theta_C Y + g\frac{H}{2}\left\{\int_0^{\infty}\int_0^{2\pi} S(\omega,\theta,u)[\omega - u k_C \cos(\theta - \theta_C)]^{-1} \frac{\cosh[k_C(d+z)]}{\cosh(k_C d)}\{\sin[k_C X\sin\theta + k_C Y\cos\theta - \omega T] - \sin[k_C X \sin\theta + k_C Y\cos\theta - \omega(T-T^*)]\}\,d\theta\,d\omega\right\} \Big/ \left\{\int_0^{\infty}\int_0^{2\pi} S(\omega,\theta,u)[1-\cos(\omega T^*)]\,d\theta\,d\omega\right\} \tag{18}$$

where $S(\omega, \theta, u)$ is the frequency spectrum in the presence of a current [referring to Eqs. 5~7], u is the absolute value of the current velocity vector and k_C the wave number in the presence of current evaluated by relation (3).

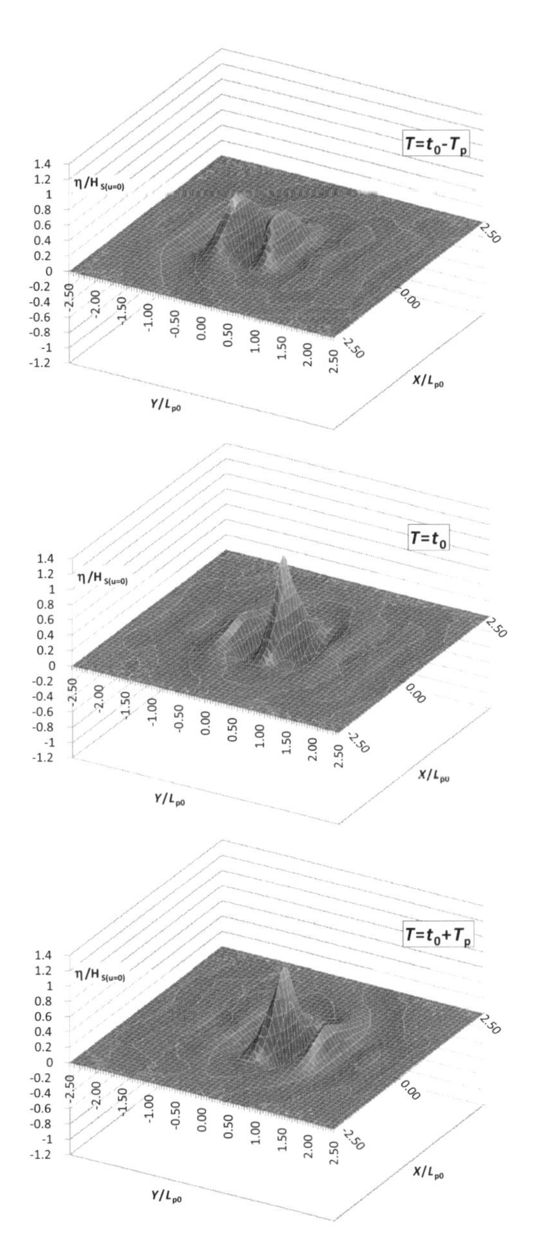

Figure 3. In the pictures is represented the free surface elevation given by Eq. (17), in a area defined by X/L_{p0} [−2.5;2.5] Y/L_{p0} [−2.5;2.5], when an exceptionally high wave height, H, is realized at time instant t_0 (central picture) at point $\underline{x}_0 \equiv (x_0, y_0)$, which is at the centre of the framed area. In the case depicted: $\bar{\theta} = 0$, $\theta_c = 0$, $d = 120$ m, $\alpha_{Phillips} = 0.01$, $n_p = 20$, $H_{S(u=0)} = 5$ m, $u = -1.5$ m/s (that is to say the current are propagating along the negative Y-axis).

4 RESULTS AND DISCUSSION

4.1 *Space-time sea wave groups propagating in the presence of a large-scale current*

In Figure 3 the free surface displacement given by Eq. (17) is represented. It describes sea wave groups propagating in the presence of a large scale current, when an exceptionally high wave height, H, is realized at the time instant t_0 (central picture) at point $\underline{x}_0 \equiv (x_0, y_0)$ located at the centre of the framed area considered in the different pictures. The examined case shows sea waves, whose dominant direction is coincident with the Y-axis, which are propagating in the presence of a large-scale current travelling against the direction of advance of the wave groups (that is to say the current moves along the negative Y-axis—$\bar{\theta} = 0$, $\theta_c = 0$) with a velocity equal to 1.5m/s in absolute value.

In the central picture (time instant $T = t_0$) of Fig. 3 the realization of the highest sea wave is illustrated. According to the QD theory it is determined by twofold effects: *i*) each wave belonging to the wave group, moves inside it with a speed (that is nearly equal to the wave celerity c associated to T_p) greater than that one of the group (that is nearly equal to the group celerity c_G associated to T_p, being $c_G = c/2$ on deep water) and grows in amplitude while approaches to the centre of the group, reaching its maximum value in wave elevation when it is just at the centre of the group; *ii*) the three-dimensional (short-crested) wave group is characterized by a development stage during which the envelop shrinks, and a following decay stage during which the envelop stretches, as clearly highlighted in the sequence of images of Figure 3. Through the theory, the exceptionally high wave height, H, at point $\underline{x}_0$ at time instant t_0 is, with very high probability, the one that occupies the centre of the wave group at the apex of its development. The height H has been assumed equal to two times the significant wave height in the presence of the current, $H_{S(u=-1.5\text{m/s})} = 5.6$m.

The effects on space-time evolution of short-crested (3D) wave groups produced by an encountering current will be quantified in detail in the next section.

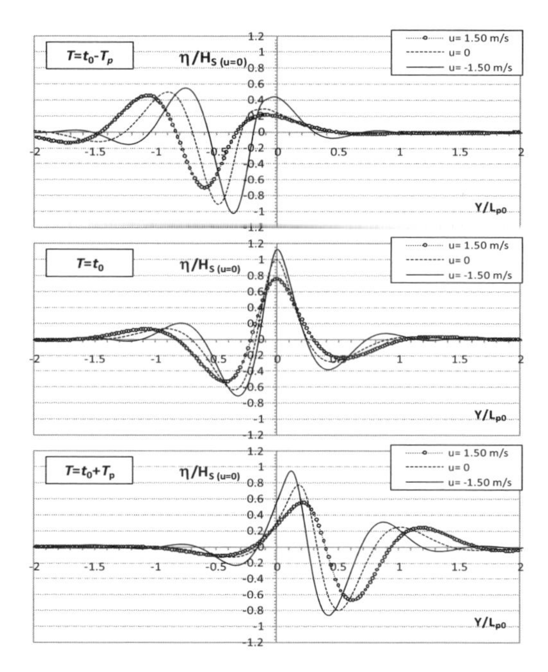

Figure 4. Deterministic water surface profiles along the Y-axis, when an exceptionally high wave height, H, is realized at time instant t_0 (central picture) at point $(X, Y)\equiv(0,0)$. In the case depicted: $d = 120$ m, $\alpha_{\text{Phillips}} = 0.01$, $n_p = 20$, $H_{S(u=0)} = 5$ m, $\bar{\theta} = 0$, $\theta_c = 0$ (that is to say the current and the dominant wave direction are coincident to the Y-axis), three wave field conditions are considered: $u = 0$, $u = +1.5$ m/s, $u = -1.5$ m/s.

4.2 *Analysis of the effects associated to the presence of a large-scale current on the propagation of sea wave groups*

In Figure 4, the three-dimensional wave group of Figure 3 (continuous line) are represented along the direction of the Y-axis, with, for the examined case, is coincident to both the dominant wave direction, $\bar{\theta}$, and the direction of propagation of the current, θ_c.

At time instant t_0 (central panel) the exceptionally high wave height H, equal to two times the significant wave height, H_S, of the wave field where it is realized, occurs at point $(X, Y) = (0,0)$. Three different wave field conditions are considered: the absence of current (that is to say $u = 0$, dotted line), an adverse current with respect to the dominant wave direction ($u = -1.5$m/s, continuous line), a following current to the sea wave group propagation ($u = 1.5$m/s, dots).

As effect of an encounter current, whose direction of advance is coincident to the dominant wave direction, we have that: with respect to the case of absence of current, the highest wave of the short-crested wave group is increased in elevation and reduced in length by an opposite current. On the contrary, a following current determines a decrease on the steepness of the highest wave due to both a reduction of its height and an increase of its length. For an intensity of the current velocity in absolute value equal to 1.5m/s, the height of the highest wave is increased by 13% with respect to the u zero value [$H_{max(u=-1.5\text{m/s})} = 2.27 H_{S(u=0)}$] and its length is reduced about 20% if the current is opposing to the dominant wave direction; on the contrary, if the current is following to the dominant direction of propagation of sea wave group the highest wave is characterized by a reduction on wave height of 24% and an increase on wave length by 17%.

In Figure 5, the evolution of wave front is represented. For this case of $\bar{\theta}$ coincident to the Y-axis, that means to consider the wave elevation profiles along the X-axis for different time instants. The evolution of the sea wave, which at time $T = 0$ will occupy the central position of the group growing in elevation, is considered. We notice that the shape of the wave front is not significantly modified in term of widening by the presence of a current either opposing or following. The increase in the amplitude of the sea waves given by the negative

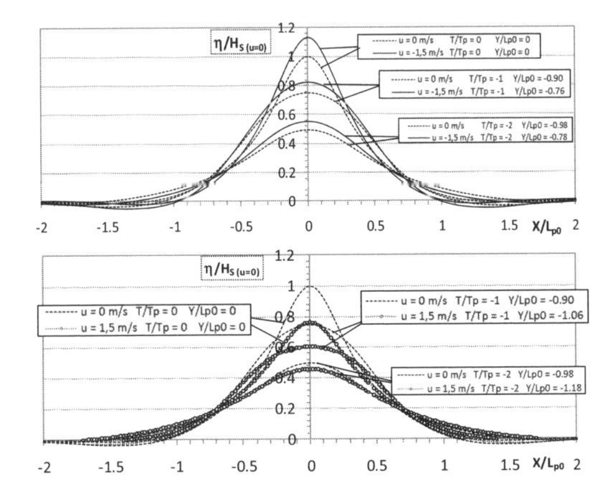

Figure 5. Deterministic water surface profiles along the X-axis, when an exceptionally high wave height, H, is realized at time instant $T = 0$ at point $(X, Y) \equiv (0,0)$. In the case depicted: $d = 120$m, $\alpha_{\text{Phillips}} = 0.01$, $n_p = 20$, $H_{S(u=0)} = 5$m, $\bar{\theta} = 0$, $\theta_c = 0$ (that is to say the current and the dominant wave direction are coincident to the Y-axis), the following wave field conditions are considered: $u = 0$, $u = -1.5$ m/s (upper panel), $u = +1.5$ m/s (lower panel).

encounter current (upper panel of Fig. 5) and the reduction produced by the positive current (lower panel of Fig. 5) is consistent to the results shown in Figure 4. What appears interesting, is the difference on the growing of the wave that leads to the occurrence of the highest wave. According to the QD theory, for $u = 0$, in about one wavelength on deep water referred to the peak period, that is to say L_{p0}, the wave that goes to occupy the central position of the group at the apex of the development stage nearly doubles its height. The same happens in the presence of an adverse current; instead, for sea waves propagating in the presence of a following current the increase in elevation is of about 67%.

Finally, in Figure 6 the wave groups are represented in the time domain. As for the period of the highest wave, T_h, that is reduced by an opposing current and increased by a following one (Fig. 6 lower panel).

The described results on the space-time evolution of sea wave groups propagating in the presence of an encountering current, following or adverse, become, on one hand, more important as the intensity of the current velocity grows in absolute value, and, on the other hand, are not significantly influenced by the water depth conditions.

4.3 *Effects associated to the direction of advance of the current with respect to the dominant wave direction*

In this section the effects associated to variation of the angle $(\bar{\theta} - \theta_C)$ within the dominant wave direction and that of propagation of the current, are analyzed.

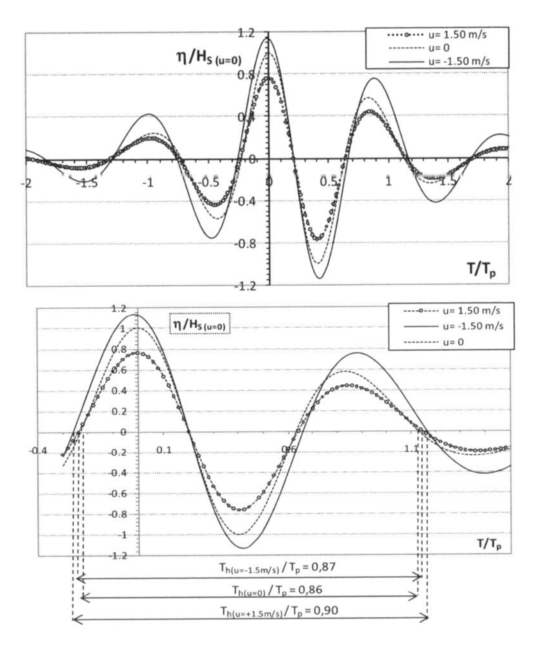

Figure 6. Deterministic water surface profiles in time domain, when an exceptionally high wave height, H, is realized at time instant t_0 at point $(X, Y) \equiv (0,0)$. In the case depicted: $\alpha_{\text{Phillips}} = 0.01$, $n_p = 20$, $d = 120$m, $H_{S(u=0)} = 5$m, $\bar{\theta} = 0$, $\theta_c = 0$ (that is to say the current and the dominant wave direction are coincident to the Y-axis), three wave field conditions are considered: $u = 0$, $u = +1.5$m/s, $u = -1.5$m/s.

In Figure 7, a three-dimensional wave group with $\bar{\theta} = 0$, propagating in the presence of an adverse encountering current, of intensity 1.5m/s and whose direction of advance makes an angle of 60° with the Y-axis, is represented. In the Figure the occurrence of the highest wave height, H, at point $\underline{x}_0 \equiv (x_0, y_0)$ is shown. H has been assumed equal to times the significant wave height of the examined wave field.

As effect of the presence of the current propagating along a direction different from $\bar{\theta}$, the direction of propagation of three-dimensional wave group moves from the dominant wave direction towards the current.

In Figure 8, the three-dimensional wave group of Figure 7 is represented in the time domain at point $\underline{x}_0 \equiv (x_0, y_0)$. The case of the presence of an adverse current is compared to the conditions of absence of current and following current. Also in this case, the presence of an adverse (following) current determines an increment (reduction) in the wave steepness with respect to the case of wave groups propagating without current. The fact

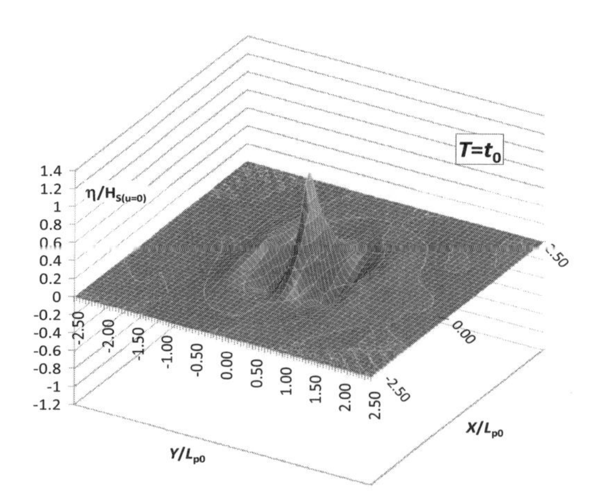

Figure 7. Short-crested wave groups, in a area defined by X/L_{p0} [−2.5;2.5] Y/L_{p0} [−2.5;2.5], when an exceptionally high wave height, H, is realized at time instant t_0 at point $\underline{x}_0 \equiv (x_0, y_0)$, which is at the centre of the framed area. In the case depicted: $\bar{\theta} = 0$, $\theta_c = 60°$, $d = 120$ m, $\alpha_{\text{Phillips}} = 0.01$, $n_p = 20$, $H_{S(u=0)} = 5$ m, $u = -1.5$ m/s (that is to say the current velocity vector $\underline{u}$ is entering to the origin of the frame of reference of Fig. 1).

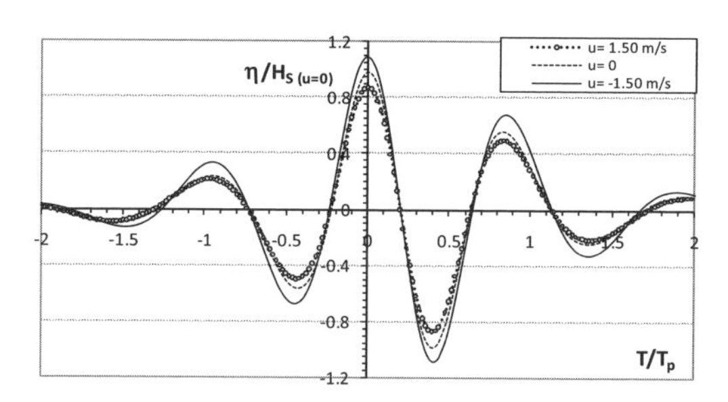

Figure 8. Deterministic water surface profiles in time domain, when an exceptionally high wave height, H, is realized at t_0, at $(X,Y) \equiv (0,0)$. In the case depicted: $\alpha_{\text{Phillips}} = 0.01$, $n_p = 20$, $d = 120$ m, $H_{S(u=0)} = 5$ m, $\bar{\theta} = 0$, $\theta_c = 60°$, three wave field conditions are considered: $u = 0$, $u = +1.5$ m/s, $u = -1.5$ m/s.

that the current is not propagating along the dominant direction reduces the two effects. The highest wave height is increased by 10% for $u = -1.5$ m/s [$H_{max(u=-1.5\text{m/s})} = 2.16 H_{S(u=0)}$] and decreased by 11% for $u = +1.5$m/s.

5 CONCLUSIONS

The space-time evolution of three-dimensional sea wave groups propagating in the presence of an encountering large-scale current has been examined. The closed-form solution of the deterministic free surface displacement has been applied by considering the directional spectrum of irregular three-dimensional sea wave in the presence of a large-scale current achieved by Hedges *et al.* (1993). It is confirmed that, an opposite current produces an increment of the wave steepness due both to an enhance of wave height and a reduction of wave length; on the contrary, a following current reduces the wave steepness. In particular the results obtained for the case of adverse current show a good agreement with the wave field observation on the occurrence of very high waves travelling against a strong current (Mallory,1974; Lavrenov, 1998).

All the effects associated to an encountering current (positive or negative) are, on one hand, greater as the current intensity increases in absolute value, and are not significantly influenced by the water depth conditions; on another hand, they are reduced as greater is the angle within the dominant wave direction and that one of advance of the current.

REFERENCES

Arena F., Romolo A., 2005. Random Forces on a Slender Vertical Cylinder Given by Random High Sea Wave Interacting with a Current. *Int J of Offshore and Polar Engineering*, vol. 15; 21–27.

Boccotti, P., 1981. On the Highest Waves in a Stationary Gaussian Process. *Atti Acc Ligure di Scienze e Lettere*, 38, 271–302.

Boccotti, P., 1982. On ocean waves with high crests. *Meccanica*, 17.

Boccotti P., Barbaro G. and Mannino L., 1993. A Field Experiment on the Mechanics of Irregular Gravity Waves. *Jour. Fluid Mechanics*, 252, 173–186.

Boccotti, P., 1997. A general theory of three-dimensional wave groups. Part I: The Formal Derivation. Part II: Interaction with a Breakwater. *Ocean Eng.*, 24, 265–300.

Boccotti, P., 2000. *Wave mechanics for ocean engineering*, Elsevier Science, Oxford.

Bretherton, F.P. and Garrett, C.J.R., 1968. Wavetrains in inhomogeneous moving media. Proc. *R. Soc. London*, A302:529.

Guedes Soares C., de Pablo H., 2006. Experimental study of the transformation of wave spectra by a uniform current. *Ocean Engineering*, 33, 293–310.

Hedges, S.T., Anastasiou, K., Gabriel, D., 1985. Interaction of random waves and currents. *J. Waterway, Port, Coastal Ocean Eng.*, Am. Soc. Cir. Eng. (ASCE), 111 (2), 275–288.

Hedges T.S., Tickell R.G. and Akrigg J., 1993. Interaction of short-crested random waves and large-scale currents. *Coastal Engineering,* 19, 207.

Jensen J.J., 2002. Conditional short-crested waves in shallow water and with superimposed current, *Proc. of the XXI Int. Conf. on Offshore Mech. and Arctic Eng. (OMAE 2002—28399 paper), by ASME,* Oslo, Norway.

Kitaigordskii S.A., Krasitskii V.P. and Zaslavskii M.M., 1975. On Phillips' theory of equilibrium range in spectra of wind-generated gravity waves. *J. Phys. Oceanogr.*, 5, 410–420.
Lavrenov, I.V., 1998. The wave energy concentration at the Agulhas current of South Africa. *Natural Hazards*, 17, 117–127.
Longuet-Higgins M.S., 1963. The effects of non-linearities on statistical distributions in the theory of sea waves, *Journal of Fluid Mechanics*, Vol. 17, 459–480.
Mallory, J.K., 1974. Abnormal waves on the south-east of South Africa. *Inst. Hydrog. Review*, N 51, 89–129.
Mitsuyasu H., Tasai F. and Suara T., 1975. Observation of Directional Spectrum of Ocean Waves Using a Clover-Leaf Buoy. *J. Phys. Oceanogr.*, 5, 750–760.
Nava V., Arena F., Romolo A., 2006. Non-linear Random Wave Groups with a Superimposed Current. *Proc. ASME 25th Int Conf on Off Mech and Artic Eng (OMAE 2006)*—Hamburg, Germany.
Peregrine, D.H., 1976. Interaction of water waves and currents. *Advanced Applied Mech.*, 16, 9–117.
Pierson W.J., Moskowitz L., 1964. A proposed spectral form for fully developed waves based on the similarity theory of S.A. Kitaigorodskii, *J. Geophys. Res.*, Vol. 69, pp. 5181–5190.
Phillips O.M., 1958. The equilibrium range in the spectrum of wind-generated waves. *J. Fluid Mech.*, Vol. 4, 426–434.
Phillips O.M., 1967. The theory of wind-generated waves, *Advances in Hydroscience*, Vol. 4, 119–149.
Phillips O.M., 1977. The Dynamics of the Upper Ocean, Cambridge University Press, Cambridge, England, 336pp.
Romolo A., Arena F., 2003. Interazione di Gruppi di Onde Alte con Correnti. Proc. *XVI Congresso di Meccanica Teorica e Applicata (AIMETA)*. Ferrara, Italy, September 9–12, 2003, 1–8 (*in Italian*).
Tayfun, M.A., Dalrymple, R.A. and Yang, C.Y., 1976. Random wave-current interactions in water of varying depth. *Ocean Eng.*, 3: 403.
Trulsen, K. and Mei, C.C., 1993. Double reflection of capillary/gravity waves by a non-uniform current: a boundary layer theory. *J. Fluid Mech. 251*, 239–271.
White, B.S., and Fornberg, B., 1998. On the chance of freak waves at the sea. *J. Fluid Mech.*, 255, 113–138.

Sustainable Maritime Transportation and Exploitation of Sea Resources – Rizzuto & Guedes Soares (eds)
© 2012 Taylor & Francis Group, London, ISBN 978-0-415-62081-9

Regional extreme frequency analysis in the North Atlantic ocean during the summer season

C. Lucas, G. Muraleedharan & C. Guedes Soares
Centre for Marine Technology and Engineering (CENTEC), Instituto Superior Tecnico, Technical University of Lisbon, Lisboa, Portugal

ABSTRACT: Accurate extreme wave height (quantile) estimations are vital for the design wave condition for marine structures. Since extreme events are rare, frequency of occurrences are difficult to estimate. *Regional Frequency Analysis (RFA)*: an approach based on *L-moments* resolves this problem by pooling data from identical sites, sites with similar site statistics (*L-moment ratios*). *RFA* is applied for identification of sites that are grossly different from the group of sites, formation of regions (sites with similar site statistics), selection of appropriate regional frequency distribution and estimation of regional extreme quantiles of designated return periods. Incorrect data values, outliers, trends and shifts in the mean of a sample can all be reflected in the *L-moments* of the sample. The deep water daily maximum significant wave heights, HIPOCAS hindcast data of an offshore region in the North Atlantic Ocean in summer season (June to August) and spread over a period from 1958 to 1978 are subjected to *Regional Frequency Analysis*. 25 equally spaced (5°) sites (40°–60°N, $^-$33°–$^-$13°W) of dimensions 0.5° × 0.5° are considered. This study emphasizes the need of *RFA* for estimation of wave height statistics with required precision.

1 INTRODUCTION

Extreme waves, the design condition for marine engineering activities are extremely difficult to estimate with required precision as they are rare events and the data record are often not abundant. Regional frequency analysis resolves this problem by pooling data from several sites in estimating extreme event frequencies at a particular site.

The process of using data from several sites to estimate the frequency distribution of any one site is known as regional frequency distribution. *L-moments* are a recent development within statistics and can be used to facilitate the estimation process in *Regional Frequency Analysis. L-moment methods* are obviously superior to those that have been previously used. *L-moments* have the theoretical advantages over conventional moments of being more robust to the presence of outliers in the data.

Identification of discordant sites, those sites that are grossly incompatible with the group as a whole is the preliminary step in *Regional Frequency Analysis.* Those sites are to be discarded.

Discordance is measured in terms of the *L-moments* of the sites' data. The following stage in a regional frequency analysis involving many sites, the identification of homogeneous regions that is to form groups of sites that possess frequency distributions which are identical apart from a site specific scale factor is the most difficult part. This is achieved by defining for each site of interest a region containing those sites whose data can be advantageously used in the estimation of the frequency distribution at the site of interest. The following step is associated with the task of selection of a single frequency distribution for the data pooled from several sites. The region will be slightly heterogeneous generally and the aim is not to find a "true" distribution but to find a distribution that will give accurate extreme quantile estimates for each site. Since more information is used than in an at-site analysis using only a single sites data, there is potential for greater accuracy in the final quantile estimates. Even though a region may be moderately heterogeneous, regional analysis will still yield much more accurate quantile estimates than at-site analysis (Lettenmaier and Potter, 1985; Lettenmaier et al., 1987; Hosking and Wallis, 1988; Potter and Lettennaier, 1990).

An attempt is made to estimate the regional extreme wave height quantiles of designated return periods for a sample daily maximum significant wave height data of summer season (June to August) in the North Atlantic Ocean extracted from the HIPOCAS data base (Pilar et.al., 2008). The mean of the at-site quantiles is its regional quantile. Hence at-site analysis is also performed along with regional analysis to identify the appropriateness of the regional quantile to stand as a true representative of its at site quantiles. This will also enable us to make corrective measures for regional frequency analysis to be more effective by diminishing site dimensions and space between sites, restructuring sub-regions or if necessary by totally considering another set of sites.

2 EXTREME QUANTILE FUNCTION

Let Q be a random variable that represents the magnitude of an event that can occur at a given site. The frequency distribution $F(z)$ is the probability that the value Q is lower or equal to z:

$$F(z) = \Pr[Q \le z] \quad (1)$$

The quantile function or the inverse function of cumulative distribution function is z (F) which expresses the magnitude of an event in terms of its non-exceedance probability F.

The extreme wave height in the upper tail of the frequency distribution Q_T, is the quantile of the return period T, is represented by equation:

$$Q_T = z\left(1 - \frac{1}{T}\right) \quad (2)$$

Obtaining a good estimate of the quantile Q_T for a practical return period is the aim of the frequency analysis.

An alternative expression is suggested by Muraleedharan et al. (2007), a similar approach is adopted for the analysis of extreme wave heights and for return periods which is explained hereafter.

If Z is a daily maximum significant wave height random variable with a distribution function of $F(z)$, $Z_1, Z_2, \ldots, Z_n$ is a sample of n daily maximum significant wave heights. The probability for the largest maximum significant wave height that does not exceed z_L is given by:

$$G(z_L) = [F(z_L)]^n \quad (3)$$

This implies that the probability for the specified value z_L exceeded is:

$$1 - G(Z_L) \quad (4)$$

A series of observations on the daily maximum significant wave height the probability that the kth observation is the first value that exceeds z_L is given by the geometric law:

$$(1-G)G^{k-1}, k = 1,2,3,\ldots \quad (5)$$

The mean value of the observation k is:

$$E(k) = (1-G)^{-1} \quad (6)$$

where E is the usual expectation operator.

The average number of observations included between two adjacent wave heights that exceed z_L is E (k). As we are observing the largest wave height that occurs every day, the number of observations E (k) equals the number that lies between the appearances of maximum significant wave heights that exceeds z_L. Therefore E (k) represents a period (T) in which an extreme quantile is observed. Therefore:

$$Q_T = z\left(1 - \frac{1}{T}\right)^{1/n} \quad (7)$$

3 *L-MOMENTS* OF PROBABILITY DISTRIBUTIONS

Probability weighted moments of a random variable Z with cumulative distribution function F (.) are defined by:

$$M_{p,r,s} = \mathrm{E}\left[Z^p \{F(Z)\}^r \{1 - F(Z)\}^s\right] \quad (8)$$

For a distribution that has a quantile function $z(u)$, eq. (8) gives

$$\alpha_r = M_{1,0,r} = \int_0^1 z(u)(1-u)^r du, \quad (9a)$$

$$\beta_r = M_{1,r,0} = \int_0^1 z(u)u^r du \quad (9b)$$

In terms of probability weighted moments, *L-moments* are given by

$$\lambda_{r+1} = (-1)^r \sum_{k=0}^{r} p^*_{r,k}\alpha_k = \sum_{k=0}^{r} p^*_{r,k}\beta_k \quad (10)$$

$$p^*_{r,k} = (-1)^{r-k} \binom{r}{k}\binom{r+k}{k} \quad (11)$$

The *L-moment ratios* are defined as

$$\tau_r = \frac{\lambda_r}{\lambda_2}, \qquad r = 3,4,\ldots \quad (12)$$

where τ_3 is the *L-skewness* and τ_4 is the *L-kurtosis*

$$\tau = L - CV = \frac{\lambda_2}{\lambda_1} \tag{13}$$

is the *L-coefficient of variation*. It is analogous to the ordinary *coefficient of variation*. *L-moment ratios* assess the shape of a distribution independently of its scale of measurement.

4 SAMPLE *L-MOMENTS*

Let $z_{1,n} \leq z_{2,n} \leq ..... z_{n,n}$ be the ordered sample of size n arranged in ascending order of magnitude. Then the unbiased estimates of α_r and β_r are respectively a_r and b_r.

$$a_r = n^{-1} \sum_{j=1}^{n} \frac{(n-j)\ (n-j-1).......(n-j-r+1)}{(n-1)\ (n-2)...(n-r)} x_{j,n} \tag{14}$$

$$b_r = n^{-1} \sum_{j=1}^{n} \frac{(j-1)\ (j-2)...(j-r)}{(n-1)\ (n-2)...(n-r)} x_{j,n} \tag{15}$$

$$l_{r+1} = (-1)^r \sum_{k=0}^{r} p^*_{r,k} a_r = \sum_{k=0}^{r} p^*_{r,k} b_r; \qquad r = 0,1,...,n-1 \tag{16}$$

The sample *L-moment* l_r is an unbiased estimator of λ_r. Then the sample *L-moment ratios* are defined as

$$t_r = \frac{l_r}{l_2} \tag{17}$$

and the sample *L-CV* is:

$$t = \frac{l_2}{l_1} \tag{18}$$

5 STEPS IN REGIONAL FREQUENCY ANALYSIS

The Regional Frequency Analysis, an approach based on *L-moments*, introduced by (Hosking and Wallis, 1997) is briefly described here. The steps in regional frequency analyses are:

a. *Screening of the data:* In this case study, the deep water daily maximum significant wave heights off Portugal (41° to 60°N, −33°–−13°W, 25 sites of dimensions 0.5° × 0.5° and 5° spaced, Figs. 4–6) spread over a period from 1958–1978 (HIPOCAS database) in summer months (June–August) are initially subjected to discordance test criteria (Lucas et al., 2011a, 2011b). I.e. to identify discordant sites, those sites that are grossly discordant from the group

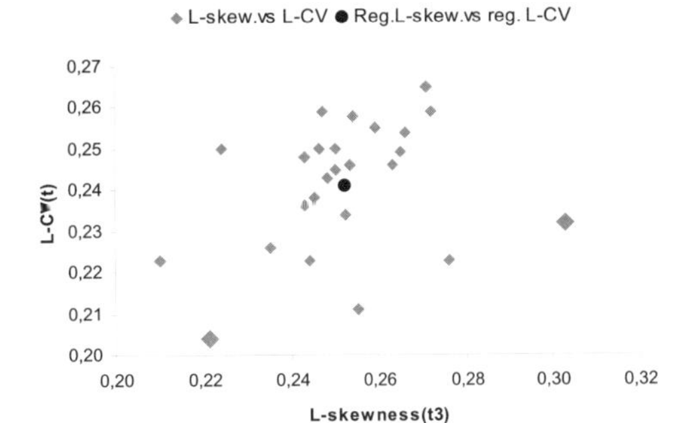

Figure 1a. *L-skewness* (t_3) vs *L-CV* (t) for 25 sites in August. Regional average t_3 vs t. ♦–Discordant sites.

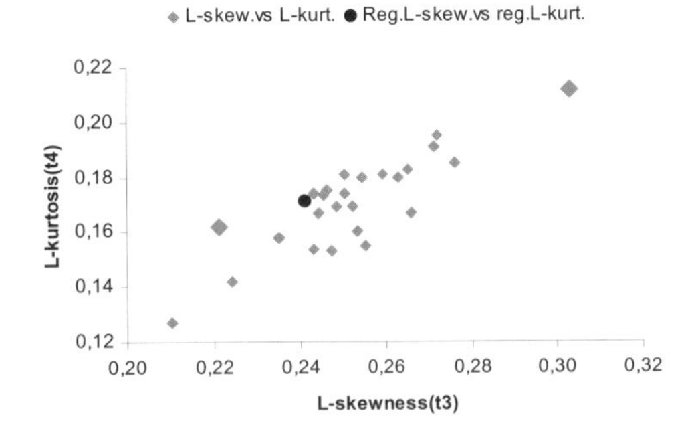

Figure 1b. *L-skewness* vs *L-kurtosis* for 25 sites in August. Regional average t_3 vs t_4. ♦–Discordant sites.

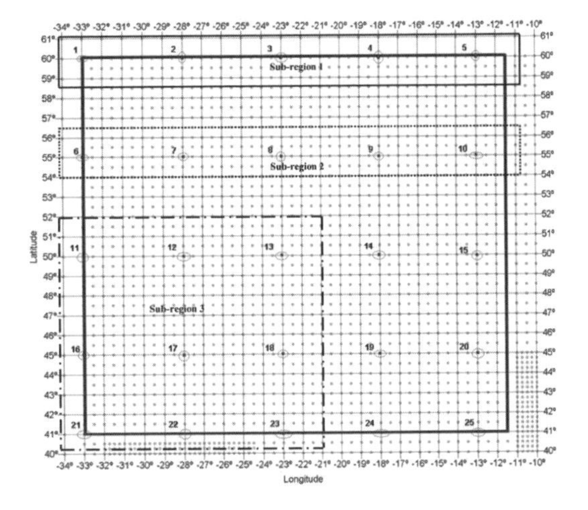

Figure 2. Sub-regions 1, 2 and 3 in June.

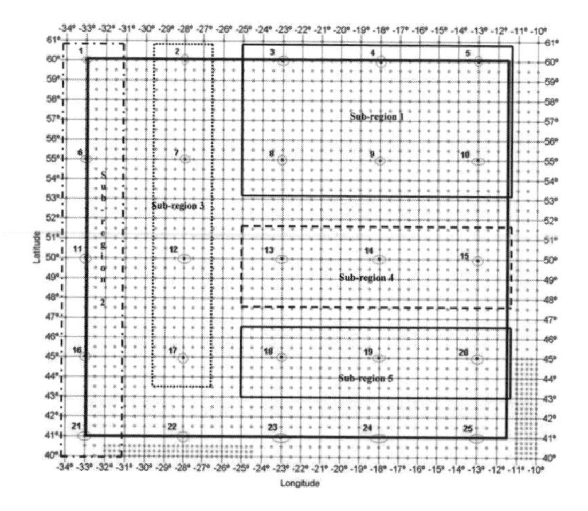

Figure 3. Sub-regions 1, 2, 3, 4 and 5 in July.

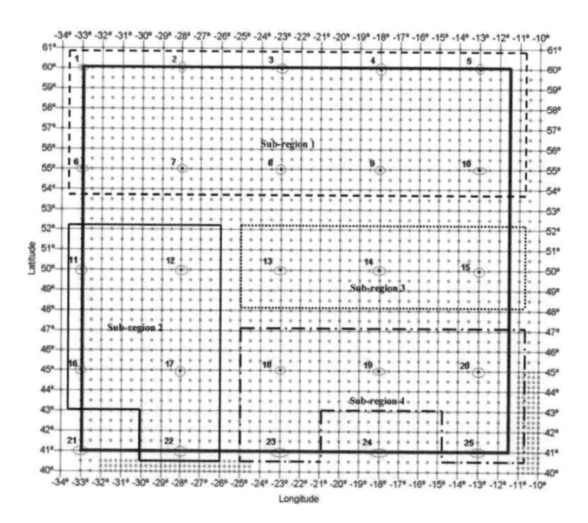

Figure 4. Sub-regions 1, 2, 3 and 4 in August.

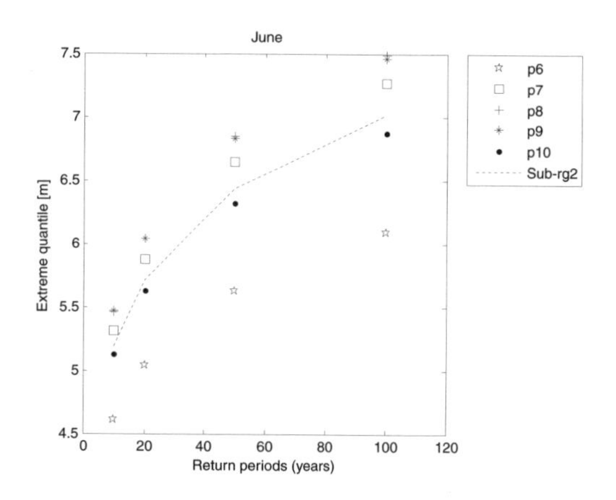

Figure 5. Extreme quantiles vs return periods for the sub-region 1 in June.

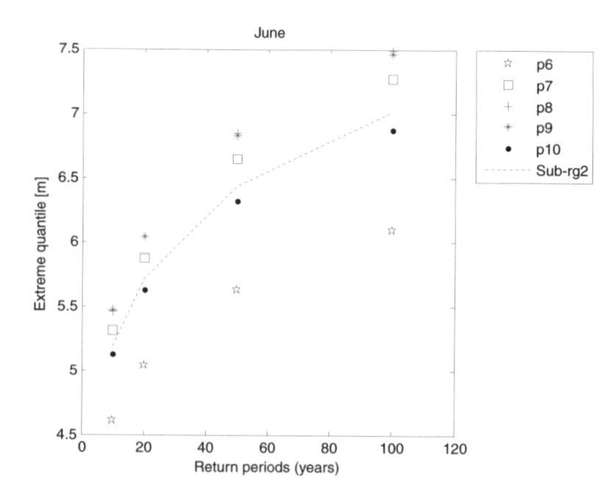

Figure 6. Extreme quantiles vs return periods for the sub-region 2 in June.

as a whole. Discordance is measured based on the *L-moment ratios* of the sites.

b. *Identification of homogeneous regions:* A region consists of sites whose frequency distributions are considered to be approximately the same. Sites with similar site statistics (*L-moment ratios*) constitute a region. The heterogeneity measure compares the between-site variations in sample *L-moments* for the group of sites with what would be expected for a homogeneous region. The between-site variation in *L-CV* has a much larger effect than variation in *L-skewness* or *L-kurtosis*.

Let the proposed region have N sites, with site i having sample size n_i and sample *L-moment ratios* $t^{(i)}$,$t_3^{(i)}$ and $t_4^{(i)}$.

Let regional average *L-CV, L-skewness and L-kurtosis* be denoted by t^{R},t_3^{R} and t_4^{R} weighted proportionally to the sites sample size. I.e.,

$$t^{R} = \frac{\sum_{i=1}^{N} n_i t^{(i)}}{\sum_{i=1}^{N} n_i} \tag{19}$$

Calculate the weighted standard deviation of the at-site sample *L-CV*s,

$$V = \left\{ \frac{\sum_{i=1}^{N} n_i \left(t^{(i)} - t^{R} \right)^2}{\sum_{i=1}^{N} n_i} \right\}^{\frac{1}{2}} \tag{20}$$

A kappa distribution is fitted to the regional average *L-moment ratios*, it is a four parameter distribution which includes the distributions generalized logistic, generalized extreme-value and generalized Pareto as special cases. A large number of Monte Carlo simulations, N_{sim} of realizations of a region with N sites, each having this kappa distribution as its frequency distribution are made. The simulated regions are homogeneous and sites have the same sample size as their original data. For each simulated region V is calculated. The mean (μ_v) and standard deviation (σ_v) of the N_{sim} values of V are computed.

The heterogeneity H measure is calculated as:

$$H = \frac{(V - \mu_v)}{\sigma_v} \tag{21}$$

The region is "acceptably homogeneous" if $-2 < H < 1$, "possibly heterogeneous" if $1 \leq H < 2$, and "definitely heterogeneous" if $H \geq 2$ (Hosking and Wallis, 1997).

If the region is not acceptably homogeneous, the region could be divided into two or more sub-regions, some sites could be removed from the region or a completely different assignment of sites to regions could be tried.

c. *Choice of a frequency distribution*: The selection of an appropriate frequency distribution by goodness-of-fit test, which involves computing summary statistics from the data and testing with the values from postulated distribution is the final stage in regional frequency analysis.

Quality of fit is judged by the difference between the *L-kurtosis.*

A goodness-of-fit measure based on *L-kurtosis* is given as:

$$Z^{fd} = \frac{\left(t_4^R - \tau_4^{fd}\right)}{\sigma_4} \tag{22}$$

Z^{fd}—Goodness-of-fit measure for the frequency distribution;

t_4^R—Regional average *L-kurtosis*;

τ_4^{fd}—*L-kurtosis* of the fitted frequency distribution;

σ_4 – The standard deviation of t_4^R; which is obtained by repeated simulation (Monte Carlo) of a homogeneous region whose sites have the fitted frequency distribution and sample size the same as those of the observed data. *Z statistics* were calculated from 500 simulations. The *Z statistic* has approximately a standard normal distribution. The criterion $|Z| \leq 1.64$ then corresponds to acceptance of the postulated distribution at a confidence level of 90%. The three-parameter (location, scale, and shape) distributions subjected to the goodness of fit test statistic are: generalized logistic, generalized extreme value, generalized normal, Pearson type-III and generalized Pareto.

d. *Estimation of the frequency distribution*: Estimation of the regional frequency distribution can be achieved by estimating the distribution separately at each site and combining the at-site estimates to give a regional average. The distribution is fitted by the method of *L-moments,* its parameters are estimated by equating the population *L-moments* of the distribution to the sample *L-moments* calculated from the data.

6 RESULTS AND DISCUSSION

The *L-moment ratios* and the discordance measures are given in Tables 1–3 for the summer months. According to the discordance measure $D(I) > 3$, the critical value for a region with more than 15 sites, there are two discordant sites in August. Site 21 ($D(I) = 3.29$) and site 24 ($D(I) = 3.37$). The scatter diagrams (Figs. 1 (a,b)), *L-skewness* vs

Table 1. Sample *L-moment ratios* and discordance measure of 25 sites in June.

Site	*L-cv*	*L-skew*	*L-kurt*	*D(I)*
1	0.253	0.246	0.149	0.65
2	0.266	0.281	0.162	1.10
3	0.263	0.260	0.127	2.36
4	0.254	0.241	0.159	1.41
5	0.255	0.260	0.168	0.61
6	0.229	0.243	0.185	1.39
7	0.242	0.272	0.171	0.74
8	0.249	0.275	0.177	0.71
9	0.255	0.268	0.169	0.51
10	0.250	0.254	0.167	0.53
11	0.237	0.265	0.188	1.15
12	0.236	0.237	0.149	0.07
13	0.229	0.216	0.129	0.73
14	0.225	0.216	0.138	0.44
15	0.215	0.193	0.132	1.44
16	0.232	0.243	0.169	0.31
17	0.225	0.222	0.121	1.16
18	0.220	0.252	0.156	1.19
19	0.204	0.239	0.148	2.33
20	0.194	0.196	0.154	1.77
21	0.237	0.258	0.160	0.30
22	0.228	0.242	0.144	0.41
23	0.216	0.215	0.125	0.94
24	0.205	0.230	0.170	1.29
25	0.201	0.189	0.136	1.48
Reg. Avg.	0.233	0.240	0.154	

Table 2. Sample *L-moment ratios* and discordance measure of 25 sites in July.

Site	*L-cv*	*L-skew*	*L-kurt*	*D(I)*
1	0.233	0.233	0.160	0.05
2	0.237	0.229	0.170	0.67
3	0.251	0.225	0.145	0.89
4	0.253	0.248	0.169	0.61
5	0.257	0.234	0.156	0.91
6	0.215	0.192	0.134	1.04
7	0.231	0.210	0.145	0.51
8	0.245	0.236	0.158	0.32
9	0.251	0.248	0.163	0.49
10	0.253	0.253	0.165	0.59
11	0.226	0.226	0.161	0.10
12	0.226	0.239	0.190	1.61
13	0.232	0.236	0.174	0.39
14	0.234	0.245	0.163	0.13
15	0.242	0.267	0.182	0.54
16	0.229	0.277	0.174	1.91
17	0.229	0.269	0.170	1.45
18	0.214	0.250	0.168	0.62
19	0.205	0.246	0.178	0.63
20	0.203	0.245	0.206	2.94
21	0.220	0.285	0.204	1.81
22	0.199	0.224	0.161	0.58
23	0.185	0.204	0.138	2.05
24	0.176	0.192	0.153	2.23
25	0.211	0.185	0.118	1.93
Reg. Avg.	0.226	0.236	0.164	

Table 3. Sample *L-moment ratios* and discordance measure of 25 sites in August.

Site	*L-cv*	*L-skew*	*L-kurt*	*D(I)*
1	0.248	0.243	0.154	0.62
2	0.250	0.246	0.175	0.49
3	0.258	0.254	0.180	0.67
4	0.259	0.247	0.153	1.40
5	0.250	0.224	0.142	1.34
6	0.238	0.245	0.173	0.29
7	0.243	0.248	0.169	0.03
8	0.250	0.250	0.174	0.20
9	0.259	0.272	0.195	0.87
10	0.265	0.271	0.191	1.03
11	0.236	0.243	0.174	0.55
12	0.245	0.250	0.181	0.63
13	0.254	0.266	0.167	1.28
14	0.255	0.259	0.181	0.36
15	0.249	0.265	0.183	0.21
16	0.246	0.263	0.180	0.13
17	0.246	0.253	0.160	0.64
18	0.234	0.252	0.169	0.09
19	0.226	0.235	0.158	0.44
20	0.223	0.244	0.167	0.51
21	0.232	0.303	0.212	3.29*
22	0.223	0.276	0.185	1.54
23	0.211	0.255	0.155	2.81
24	0.204	0.221	0.162	3.37*
25	0.223	0.210	0.127	2.20
Reg. Avg.	0.241	0.252	0.171	

L-CV and *L-skewness* vs *L-kurtosis* also reveal that the discordant sites in August lie away from the cloud of clustered points. Site 21 differs from other sites with higher *L- skewness* and *L- kurtosis* while site 24 has a lower *L-CV*.

Hence site 21 is away from the cluster of points in both figures 1(a) and 1(b) and site 24 is away from the cloud of points in figure 1(a).

Thus sites 21 and 24 in August are identified as discordant sites, sites that are entirely different from the group of sites, discordance measured by *L-moment ratios* and hence discarded from further analysis.

After initial screening of the data, the region containing 25 sites is analyzed for homogeneity. I.e., to identify regions with similar site statistics. This is the most difficult task in regional frequency analysis. There are 3 sub-regions in June, 5 in July and 4 in August by the heterogeneity measure H (acceptably homogeneous criteria: $-2 < H < 1$) (Figs. 2–4).

Appropriate regional frequency distributions are selected by the goodness of fit measure based on *L- kurtosis*. 500 homogeneous regions are generated with the fitted frequency distribution and sample size the same as those of the observed data. The regional growth curve for sub-regions 1 and 3 is Pearson type-III and generalized normal for sub-region 2 in June. In July generalized normal is the appropriate regional distribution for the sub-regions 1–4 and generalized extreme value distribution for the sub-region 5. Of the four sub-regions in August, generalized normal distribution is found to be the regional distribution for the sub-regions 1–3.

None of the five; three parameter distributions is found to be appropriate for the remaining sub-region 4 by the goodness of fit test statistic. Hence the five parameter Wakeby distribution is considered for the purpose. The Wakeby distribution can mimic the shapes of many commonly used skew distributions such as extreme value, log-normal, Pearson type-III etc. It has five parameters and hence can attain a wider range of distributional shapes.

Finally regional extreme quantiles are estimated for the designated periods: 10, 20, 50 and 100 years for the various sub-regions in the summer months (Tables 4–6). The wide range of extreme quantiles of the sub-regions in summer months envisage the significance of regional frequency analysis prior to any maritime activity.

Table 4. Extreme significant wave height for given return periods estimated by Pearson type III and generalized normal distribution for the sub-regions in June.

T (years)	Sub-region 1 height (m)	Sub-region 2 height (m)	Sub-region 3 height (m)
10	4.36	5.19	3.85
20	4.65	5.71	4.09
50	5.02	6.44	4.40
100	5.30	7.02	4.63

Table 5. Extreme significant wave heights for given return periods estimated by generalized normal and extreme value distribution for the sub-regions in July.

T (years)	Sub-reg 1 height (m)	Sub-reg 2 height (m)	Sub-reg 3 height (m)	Sub-reg 4 height (m)	Sub-reg 5 height (m)
10	4.93	4.54	4.56	4.82	4.84
20	5.40	4.96	4.97	5.28	5.39
50	6.03	5.54	5.54	5.92	6.17
100	6.53	5.99	5.98	6.42	6.82

Table 6. Extreme significant wave height for given return periods estimated by generalized normal and Wakeby distributions for the sub-regions in August.

T (years)	Sub-reg 1 height (m)	Sub-reg 2 height (m)	Sub-reg 3 height (m)	Sub-reg 4 height (m)
10	5.09	4.99	5.33	3.80
20	5.59	5.49	5.87	4.02
50	6.27	6.17	6.63	4.30
100	6.81	6.71	7.22	4.50

Table 7. At site parameter estimates (by *L-moments*) for the distribution for sub-regions 1, 2 and 3 in June.

	Sub-region 1	Sub-region 2	Sub-region 3
Sites	Pearson type III (Lo Sc Sh)	Gen. normal (Lo Sc Sh)	Pearson type III (Lo Sc Sh)
1	1.00 0.48 1.48		
2	1.00 0.51 1.69		
3	1.00 0.50 1.56		
4	1.00 0.48 1.45		
5	1.00 0.48 1.57		
6		0.90 0.37 –0.50	
7		0.88 0.38 –0.57	
8		0.88 0.38 –0.57	
9		0.88 0.40 –0.56	
10		0.89 0.39 –0.53	
11			1.00 0.45 1.60
12			1.00 0.45 1.43
13			1.00 0.43 1.31
16			1.00 0.44 1.46
17			1.00 0.42 1.34
18			1.00 0.42 1.52
21			1.00 0.45 1.55
22			1.00 0.43 1.46
23			1.00 0.40 1.30
Reg.Est	1.00 0.49 1.55	0.89 0.38 –0.55	1.00 0.43 1.44

Table 8. At site parameter estimates (by *L-moments*) for the distribution for sub-regions in July.

	Sub-region 1 to sub-region 4	Sub-region 5
Sites	Generalized normal (Lo Sc Sh)	Extreme value (Lo Sc Sh)
1	(2) 0.90 0.37 –0.48	
2	(3) 0.90 0.38 –0.48	
3	(1) 0.90 0.41 –0.47	
4	(1) 0.89 0.40 –0.51	
5	(1) 0.89 0.41 –0.48	
6	(2) 0.93 0.36 –0.40	
7	(3) 0.91 0.38 –0.44	
8	(1) 0.90 0.39 –0.49	
9	(1) 0.89 0.40 –0.51	
10	(1) 0.89 0.40 –0.53	
11	(2) 0.91 0.37 –0.47	
12	(3) 0.90 0.36 –0.50	
13	(4) 0.90 0.37 –0.49	
14	(4) 0.90 0.37 –0.51	
15	(4) 0.89 0.38 –0.56	
16	(2) 0.89 0.35 –0.58	
17	(3) 0.89 0.36 –0.56	
18		0.81 0.27 –0.12
19		0.81 0.26 –0.12
20		0.82 0.26 –0.11
21	(2) 0.89 0.34 –0.60	
	Regional estimations	
	(Lo Sc Sh)	(Lo Sc Sh)
Sub-reg 1	0.89 0.40 –0.50	
Sub-reg 2	0.90 0.36 –0.50	
Sub-reg 3	0.90 0.37 –0.49	
Sub-reg 4	0.90 0.37 –0.52	
Sub-reg 5		0.81 0.27 –0.12

(1) Sub-region 1; (2) Sub-region 2; (3) Sub-region 3; (4) Sub-region 4.

The mean of the at-site quantiles is its regional quantile. In order to emphasize reliability on regional extreme quantiles as the true representatives of it's at-site quantiles, at-site analysis are also performed along with regional frequency analysis. The basic concept in regional frequency analysis is that the at-site distributions of a region are identical apart from a scaling factor, the index flood which is taken to be the mean of the at-site frequency distribution. The at-site model parameter estimations are given along with its regional estimations in Tables 7–9.

The mean of at-site estimates is its regional estimates.

The maximum discrepancy between the at-site and its regional extreme quantile is 1.17 m for site 18 (5.67 m) in sub-region 4 (4.50 m) in August for a 100 years return period. This ambiguity can be reduced by decreasing the distance between the grids (currently it is 5°) to achieve more accurate regional values. The distribution of the extreme quantiles against return periods are shown in Figures 5–16. It shows a logarithmic behaviour. The regional extreme quantile curves determine the locations of the mean of its site extreme quantile curves.

7 CONCLUSION

This case study emphasize the requirement of regional frequency analysis, prior to any marine structure designs, for identification of discordant sites, formation of homogeneous regions, selection of appropriate regional growth curves and estimation of extreme quantiles. The deep water daily maximum wave heights of the 25 equally spaced (5°) sites analyzed reveal that the sites form different homogeneous sub-regions of the oceanic region under investigation and it varies with the summer months. There are 3 sub-regions in June, July has 5 sub-regions and there are 4 sub-regions in August. The mean of the at-site quantiles is its regional quantile. At-site analysis performed along with regional analysis enables to understand the extent of spread of at-site quantiles from its regional quantiles. The maximum discrepancy between the at-site quantiles and the respective regional quantiles is 1.17 m for a 100 years return period. Accuracy of regional extreme quantiles are attained by appropriate estimation and selection of models by *Z-statistics* as a goodness of fit measure by generating 500 Monte-Carlo simulations of the homogeneous region whose sites have the fitted frequency distribution and sample size the same as those of the observed data. Extreme events are rare and hence estimation of their frequencies are difficult. Regional frequency analysis resolves this problem by pooling data from identical sites, sites with similar site statistics measured by *L-moment ratios*.

ACKNOWLEDGEMENTS

The work has been performed in the scope of the project EXTREME SEAS, Design for Ship Safety in Extreme Seas, which has been partially financed by the European Union thought its 7th Framework program under contract SCP8-GA-2009-24175. (http://www.mar.ist.utl.pt/extremeseas/).

Table 9. At site parameter estimates (by *L-moments*) for the distribution for sub-regions in August.

Sites	Sub-region 1 Gen. normal (Lo Sc Sh)	Sub-region 2 Gen. normal (Lo Sc Sh)	Sub-region 3 Gen. normal (Lo Sc Sh)
1	0.89 0.39 –0.50		
2	0.89 0.40 –0.51		
3	0.88 0.41 –0.53		
4	0.89 0.41 –0.51		
5	0.90 0.40 –0.46		
6	0.90 0.38 –0.51		
7	0.89 0.39 –0.52		
8	0.89 0.40 –0.52		
9	0.88 0.40 –0.57		
10	0.87 0.41 –0.56		
11		0.90 0.38 –0.50	
12		0.89 0.39 –0.52	
13	(3)		0.88 0.40 –0.55
14	(3)		0.88 0.40 –0.54
15	(3)		0.88 0.39 –0.55
16		0.89 0.38 –0.55	
17		0.89 0.39 –0.53	
22		0.89 0.34 –0.58	
Reg.Est	0.89 0.40 –0.52	0.89 0.38 –0.53	0.88 0.40 –0.55

	Wakeby distribution				
	Location	Scale		Shape	
Sub reg 4	(ξ)	(α)	(γ)	(β)	(δ)
18	0.421	0.907	0.399	4.321	0.023
19	0.430	0.883	0.403	4.155	–0.009
20	0.439	0.935	0.384	4.435	0.015
23	0.474	1.055	0.440	8.291	–0.066
25	0.428	0.875	0.509	5.789	–0.149
Reg.Est	0.440	0.887	0.426	4.938	–0.036

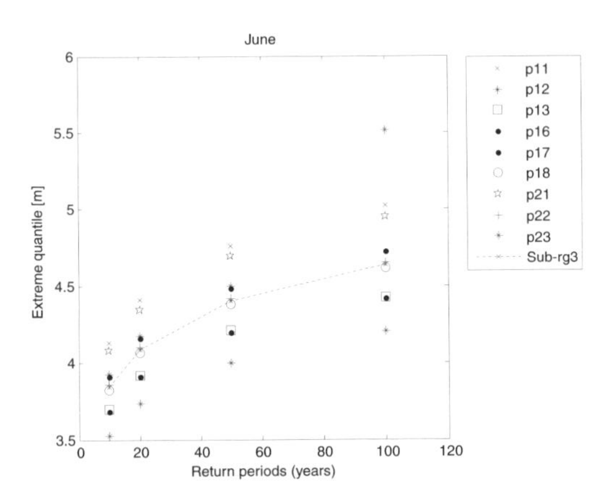

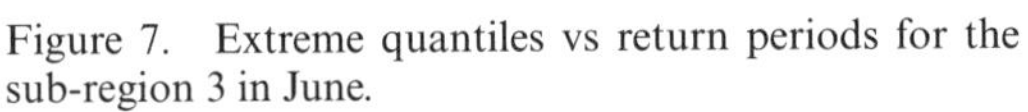

Figure 7. Extreme quantiles vs return periods for the sub-region 3 in June.

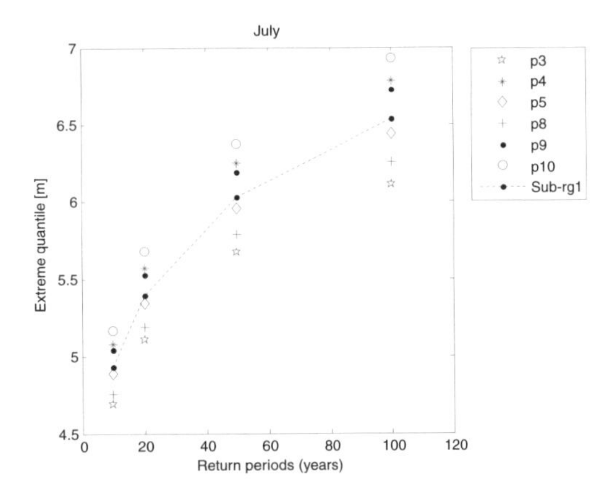

Figure 8. Extreme quantiles vs return periods for the sub-region 1 in July.

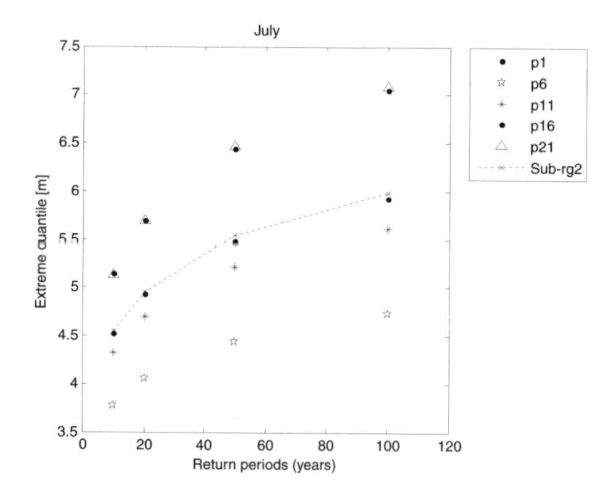

Figure 9. Extreme quantiles vs return periods for the sub-region 2 in July.

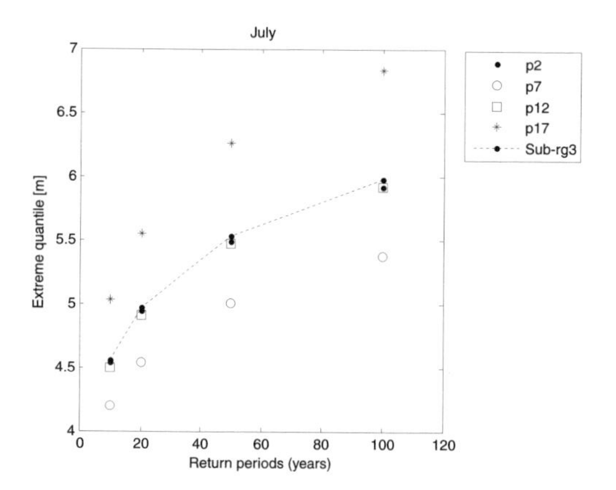

Figure 10. Extreme quantiles vs return periods for the sub-region 3 in July.

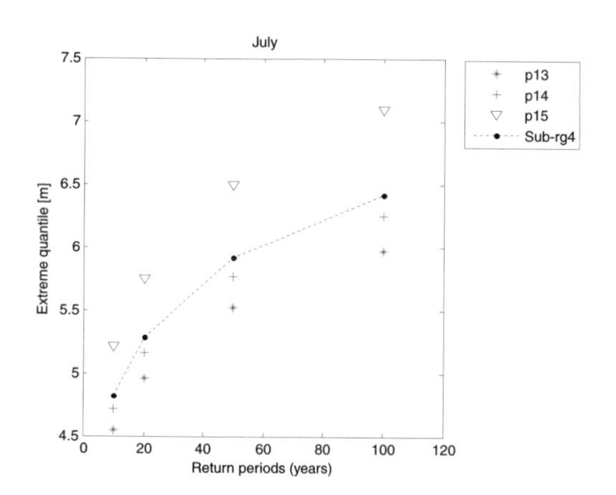

Figure 11. Extreme quantiles vs return periods for the sub-region 4 in July.

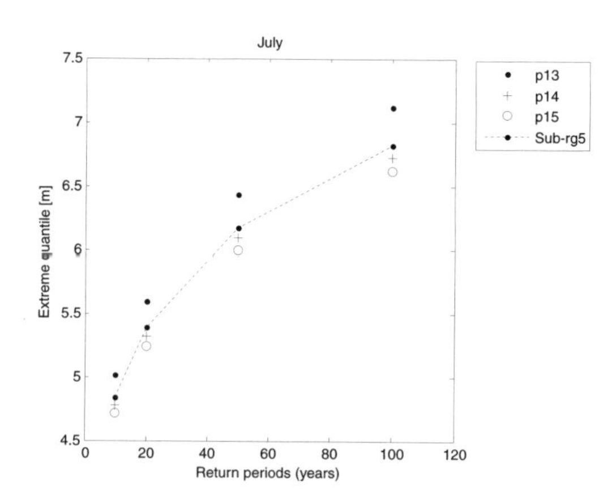

Figure 12. Extreme quantiles vs return periods for the sub-region 5 in July.

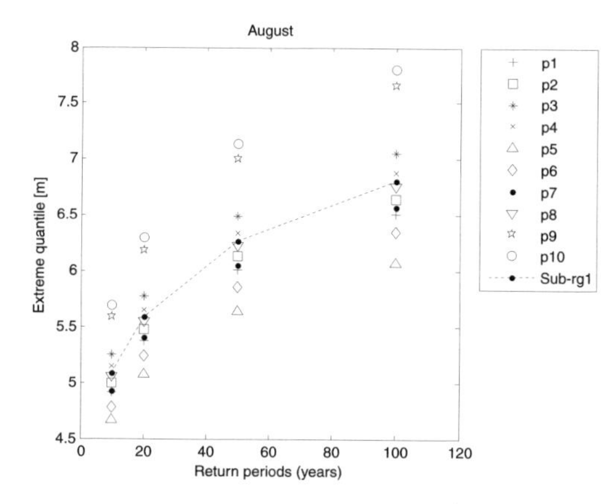

Figure 13. Extreme quantiles vs return periods for the sub-region 1 in August.

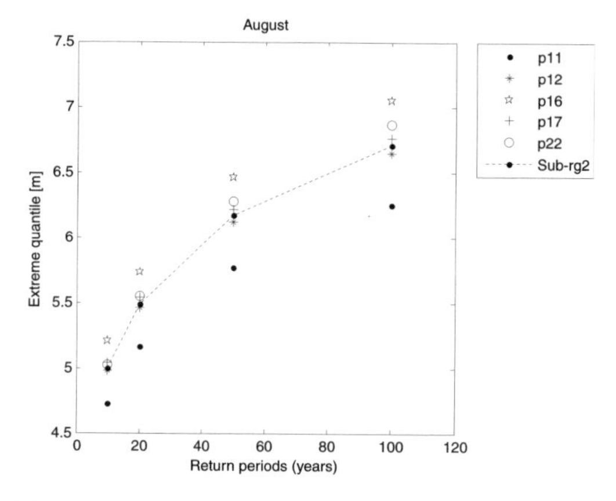

Figure 14. Extreme quantiles vs return periods for the sub-region 2 in August.

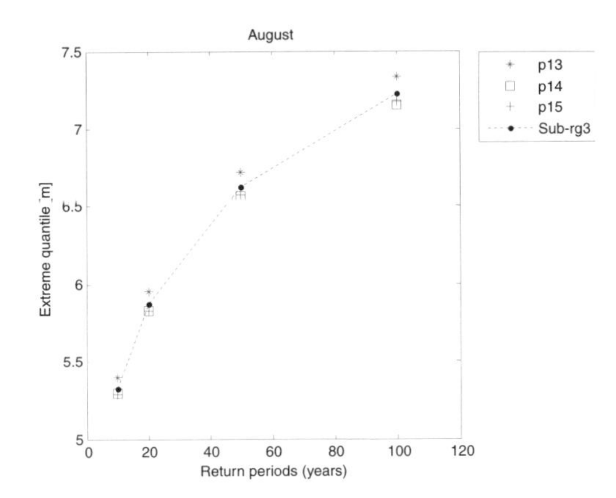

Figure 15. Extreme quantiles vs return periods for the sub region 3 in August.

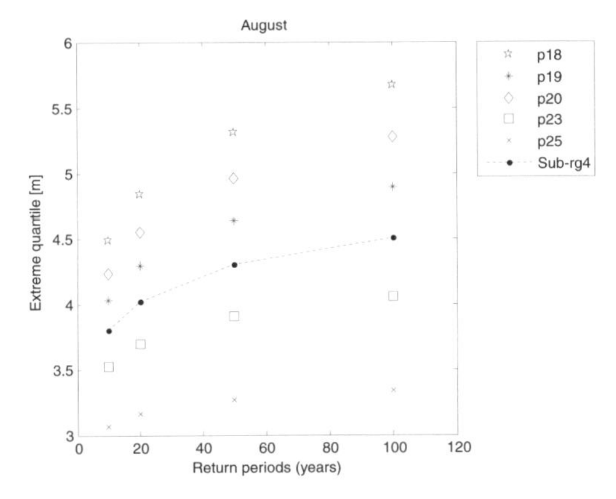

Figure 16. Extreme quantiles vs return periods for the sub region 4 in August.

REFERENCES

Cunnane, C., 1988. Methods and merits of regional flood frequency analysis. *J. Hyd.*, 100, 269–290.

Guedes Soares, C., 2008. Hindcast of Dynamic Processes of the Ocean and Coastal Areas of Europe. *Coastal Engineering,* Vol.55, pp. 825–826.

Hosking J.R.M. & Wallis, J.R., 1988. The effect of inter dependence on regional flood frequency analysis. *Water Resources Res*earch, 24, 588–600.

Hosking, J.R.M. & Wallis, J.R., 1997. Regional Frequency Analysis: An Approach Based on L-moments. *Cambridge University Press, Cambridge*.

Lettenmaier, D.P. & Potter, K.W., 1985. Testing flood frequency estimation methods using a regional flood generation model. *Water Resources Res*earch, 21, 1903–1914.

Lettenmaier, D.P., Wallis, J.R., & Wood, E.F., 1987. Effect of regional heterogeneity on flood frequency estimation. *Water Resources Res*earch, 23, 313–323.

Lucas, C., Muraleedharan, G., & Guedes Soares, C., 2011. Application of regional frequency analysis for identification of homogeneous regions of design wave conditions offshore Portugal. *30th International Conference on Ocean, Offshore and Arctic Engineering (OMAE2011)*, June 19–24, 2011, Rotterdam, The Netherlands.

Lucas., C., Muraleedharan, G., & Guedes Soares, C., 2011. Assessment of Wave Height Extreme Quantiles in North Atlantic using Regional Frequency Analysis. *International Conference on Maritime Technology and Engineering,* May 10–12, Lisbon, Portugal.

Muraleedharan, G., Rao, A.D., Kurup, P.G., Unnikrishnan N., & Sinha, M., 2007. Modified Weibull distribution for Maximum and Significant Wave Height Simulation and Prediction. *Coastal Engineering,* Vol.54, pp. 630–638.

Pilar, P., Guedes Soares, C. & Carretero, J.C., 2008. 44 Year Wave Hindcast For The North East Atlantic European Coast. *Coastal Engineering*, Vol. 55(11):861–871.

Potter, K.W. & Lettenmaier, D.P., 1990. A comparison of regional flood frequency estimation methods using a resampling method. *Water Resources Res*earch, 26, 415–424.

Sustainable Maritime Transportation and Exploitation of Sea Resources – Rizzuto & Guedes Soares (eds)

Wake wave properties generated by HSC in the bay of Naples

G. Benassai
University of Naples Parthenope, Naples, Italy

ABSTRACT: The analysis of the main parameters of high speed vessel wakes based on a field wave measurement campaign in the bay of Naples during the spring of 2007 was performed. The measured wake wash were operated by a pressure gauge in three critical points where the distance from the coastline was less than 700 m. These measurements were taken in shallow water (depth ranging from 4 to 5 meters) in calm weather conditions.

The surface elevation records were processed in order to detect the maximum wave heights and wave energy, together with the mean and s.d.. The main wave parameters were related with the variability of the depth Froude number along the ship tracks, measured with GPS. The analysis of these field data shows the importance of the evaluation of the main wake wave properties for the analysis of the impact waves on the coastline and the subsequent vulnerability assessment of the coastal areas.

1 INTRODUCTION

The introduction of High-Speed Crafts (HSC) in the 1980s, able to carry passengers and vehicles with service speeds up to 40 knots, introduced some new and adverse impacts on the minor vessels and on the recreational activities on the closer beaches. These impacts, in terms of breaking waves, run-up heights, etc. are related to the wave characteristics such as wave height, period, steepness, energy, energy flux (wave power) and wave shape. Most existing studies, however, are only focussed on the maximum height and energy values, which are limited by International Rules.

In the recent past, a study on the wake wash generated by HSC operating in the Bay of Naples was performed by University Parthenope, in order to assess the wake wash risk and to minimize the undesirable effects of wave production. The final objective of the study was to establish guidelines for increasing the safety of the coastal navigation and of the beach recreational activities.

The contribution of the present paper consists on the analysis of the main parameters (maximum height, energy, energy flux) of the waves produced by high speed vessel and their correlation with the route characteristics (speed, depth, Froude number, coastal proximity). This analysis was based on a field measurement campaign in the bay of Naples during the spring of 2007.

The ratio of the vessel speed divided by the maximum wave celerity in shallow water (depth-based Froude number) or to the square root of the gravity by the vessel length (length-based Froude number) is often used to classify the wash. In fact the wash waves produced by vessels that travel at sub-critical Froude numbers are different in patterns (and hence applicable theory) from that produced by vessels which operate at the critical Froude number of 1 or at supercritical Froude numbers.

The study started with the analysis of the vessel fleet operating in the Bay of Naples, and was developed with the route monitoring and determination of critical points along the routes, in order to make a preliminary assessment of the wake wash, to be confirmed by wave measurements in the critical points.

A pressure transducer was used for continuous water surface elevation recording at 2 Hz in three different locations at water depths of approx. 5.0 m at distances from the route between 500 m and 750 m.

The output of the tests were wave-elevation time histories upon which the maximum wave height H_{max} from the wave record was extracted. The wave height reported was therefore the highest wave, peak to through, which occurred in a wave train. The highest waves so obtained were related with the displacement, speed and distance of the crossing vessel.

For each wake wash measurement the vessel route was monitored aboard the crossing HSC and exact speed, distance and water depth was determined. The obtained values of the wake wash were related to the depth Froude number and to the vessel displacements, which govern the wakes in analogous speed and depth conditions.

2 WAKE WASH DESCRIPTION

2.1 *Wake waves and speed regime*

In deep water the ship wave system consists of diverging and transverse waves in restricted wedge-shaped region, known as Kelvin wake, where the cusp angle is about ±19.50° and almost independent of the ship speed. The diverging waves spread on each side of ship at an acute angle of about $\pm 35^0$ relative to the course, whereas the transverse waves move in the same direction as the ship.

In water of medium to low depth, the ship generated wave pattern will be different from "deep-water" pattern. An increase of speed will lead to a wave propagation direction close to 0 degree from the vessel course. In this situation, the diverging and transversal waves are identical, and the wave pattern is characterized by the wave fronts being almost perpendicular to the ship direction as it can be seen from Figure 1.

Increasing of vessel speed over the critical speed will lead to disappearance of the transverse waves while the diverging waves will remain. The wave propagation direction increases for increasing Froude numbers, i.e. the angle between wave fronts and the navigation track decreases, as it can be observed in Figure 2.

In general, it is considered that ship operates with a sub-critical speed when water depth based Froude number F_{nh} is less than 0.6, a trans-critical speed when F_{nh} is about 1, and a super-critical speed when F_{nh} is greater than 1. Typically the largest wake wash will be generated for speeds in trans-critical speed regime.

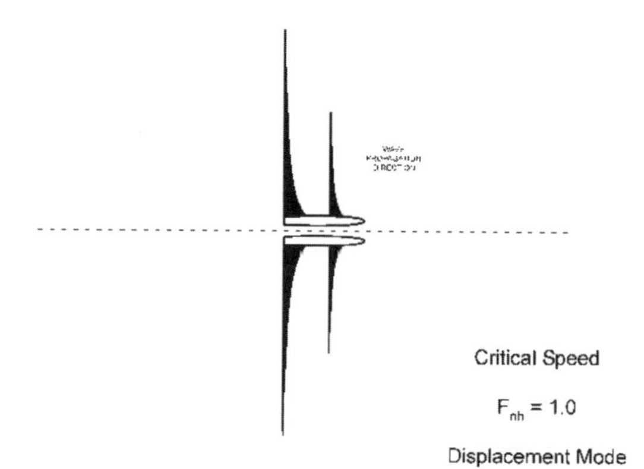

Figure 1. Critical wash pattern for displacement vessels (from Stumbo et al. 2000).

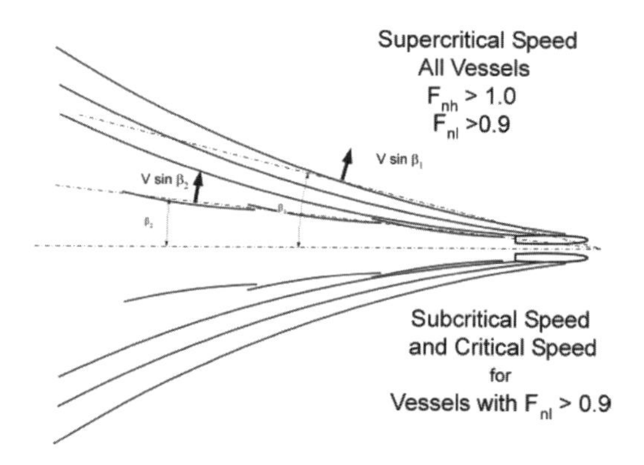

Figure 2. Super-critical wash pattern for displacement vessels (from Stumbo et al. 2000).

2.2 *Propagation from the navigation track*

It is experienced that the wave height decreases with the distance from the navigation track due to the refraction and diffraction process. It can be seen that the wave decay is exponential as expected and predicted by linear theory (Havelock, Kelvin).

Figure 3 shows the maximum wave height versus the perpendicular distance from the navigation track. The measured data were based on several campaigns involving catamarans only and are not affected by depth change effects or shoaling as the water depth was in range of 10 to 30 meters.

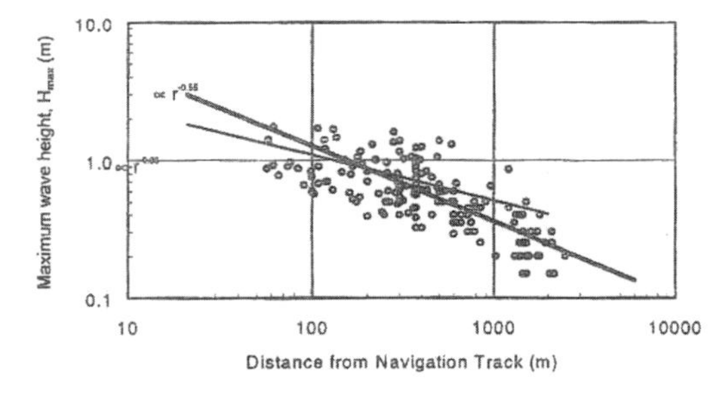

Figure 3. Maximum wave height versus distance from the navigation track (from Kofoed-Hansen et al. 1999).

In case of deep water, theoretical decay rate is $r^{-1/2}$, where r is the distance from the navigation track valid in the interior of the disturbance region. This means that the transverse waves decrease with rate inversely proportional to the square root of the distance from the vessel. At the boundary of the wake, the theoretical decay rate is $r^{-1/3}$ (Kofoed-Hansen et al. 1999) which means that the diverging waves decay slower than the transverse waves in case of a wave pattern generated at sub-critical speeds.

Stumbo et al. (1998) are using the decay rate of $r^{-1/3}$ defining the formula (1):

$$H_2/H_1 = (d_1/\ d_2)^{-1/3} \quad (1)$$

where:
H_1, H_2—wave heights
d_1, d_2—distances from the navigation track.
which has been used for analysis of field data performed by authors.

On the basis of formula (1) the wake wash waves, measured at different distances from the navigation track, were reported to a normalized distance of 300 m, in order to be compared with the limits of the International Rules. In this paper we are concerned with the influence of the route navigation parameters with the wake wash, which have been related to the depth Froude number and to the vessel displacements.

3 WAKE WASH MEASUREMENT CAMPAIGN

The analysis of the main parameters (maximum height, energy, energy flux, mean and standard deviation) of high speed vessel wakes was based on a field wave measurement campaign in the bay of Naples during the spring of 2007.

The wave recorder was a submerged pressure wave gauge, installed in a steel frame which ensured stability and keep the gauge oriented to vertical. The instrument and steel frame weighted about 15 kg and 20 kg, respectively, so the weight of all system ensured self-anchoring.

Figures 4 and 5 report the wave measurement locations with a partial sketch of the navigation routes in proximity to the measurement points. The depth at the measurement site offshore Posillipo (CP 1) was 4.70 m with up to 0.50 m excursion in tidal variation, while in Marina di Puolo (CP 2) was 4.20 m, with the same tidal excursion. The pressure gauge was configured to collect pressure data at a rate of 2 Hz. Sampling time was 15 minutes every hour thus creating 1800 data for each record.

The measurements were made on a three-day period for critical point 1, and on a four-day period for critical point 2. The weather during the field measurements was calm, with essentially no wind.

The HSC fleet which operated in the Gulf of Naples in summer period 2007 consisted in a wide range of vessel types, the investigated ones are reported in Table 1.

There are two sister catamarans (Acapulco Jet and Giove Jet) six sister monohulls (Celestina, Rosaria Lauro, Città di Sorrento, Sorrento Jet, Città di Forio, Europa Jet) and a slightly bigger monohull (Superflyte).

3.1 *Route monitoring campaign*

The monitored tracks of the routes investigated are reported in Figures 4 and 5, the distance from CP1 and CP2, the recorded speed at the measurement points and the mean water depth along the routes are reported in Table 2.

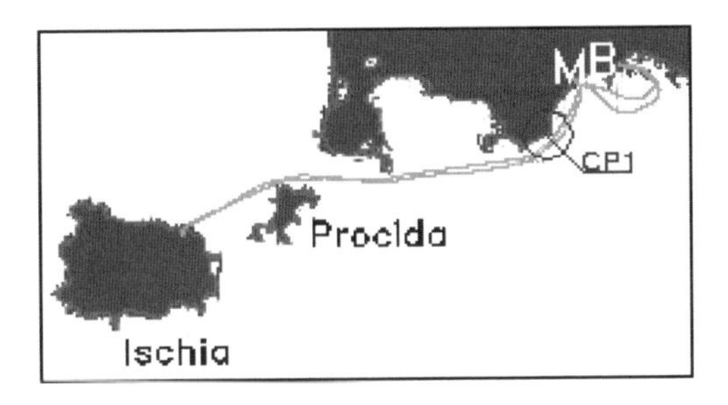

Figure 4. Route Napoli-Ischia.

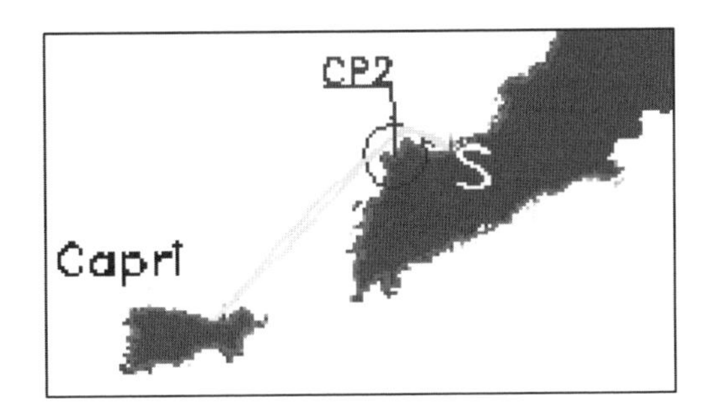

Figure 5. Route Napoli-(Sorrento)-Capri.

Table 1. Investigated HSC on service in the Bay of Naples.

Type	Name	Length (m)	Route	Width (m)	Draft (m)	Speed (kn)
CAT	Acapulco Jet	27,0	Napoli Ischia	9,40	3,46	30,0
CAT	Giove Jet	27,0	Napoli Ischia	9,30	3,52	30,0
MH	Celestina	36,7	Napoli Ischia	7,67	2,80	29,2
MH	Rosaria Lauro	36,7	Napoli Ischia	7,67	2,80	20,0
MH	Superflyte	42,4	Napoli Capri	n.a.	3,25	28,0
MH	Città di Sorrento	34,4	Napoli Capri	7,36	2,65	25,0
MH	Sorrento Jet	36,6	Napoli Capri	7,65	2,84	30,0
MH	Citta' di Forio	34,0	Napoli Capri	7,62	2,70	20,0
MH	Europa Jet	36,7	Napoli Capri	7,67	2,80	26,0

The tracks of the HSC of the route Capri-Sorrento and Sorrento-Napoli were recorded on may 2007, using a GPS with a recording frequency of 2 Hz (0,5 s interval). The corresponding diagrams of the depth along the track, the speed and the corresponding depth Froude number for are given in Figures 6a,b,c and 7a,b,c, for the Capri-Sorrento and Sorrento-Napoli route, respectively.

It is well known that the length and depth Froude numbers, namely:

$$F_{nl} = U/(gL_w)^{0,5} \quad F_{nh} = U/(gh)^{0,5} \tag{2}$$

have a great influence on the wave generating properties of a ship in shallow water, because the wave

Table 2. Parameters measured aboard the HSC.

	Napoli-Ischia	Napoli-Capri
Distance (m)	396.34	688.2
Speed (kn)	25.27	25.9
Mean water depth (m)	20	75

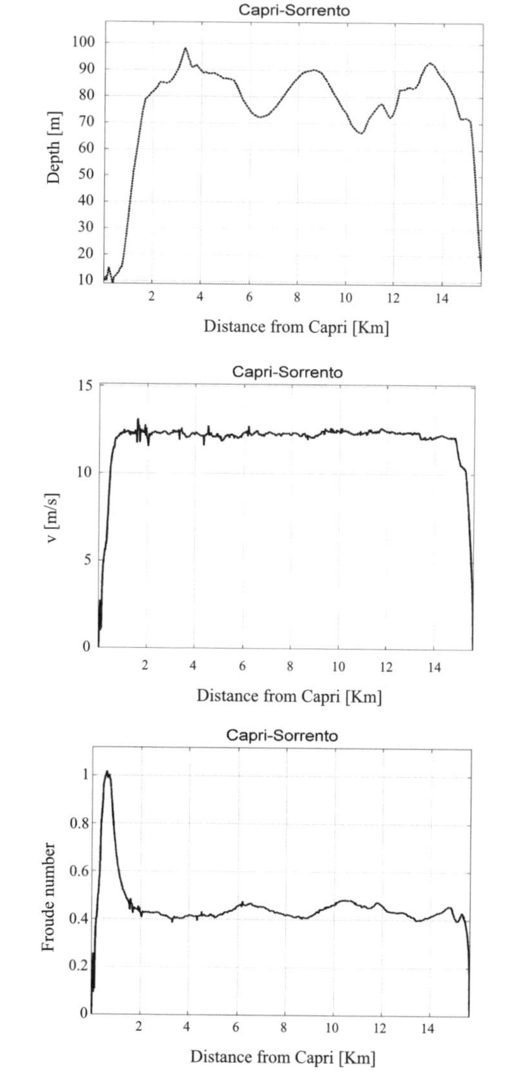

Figuere 6. Depth (a), speed (b) and depth Froude number (c) along the track for Capri-Sorrento route.

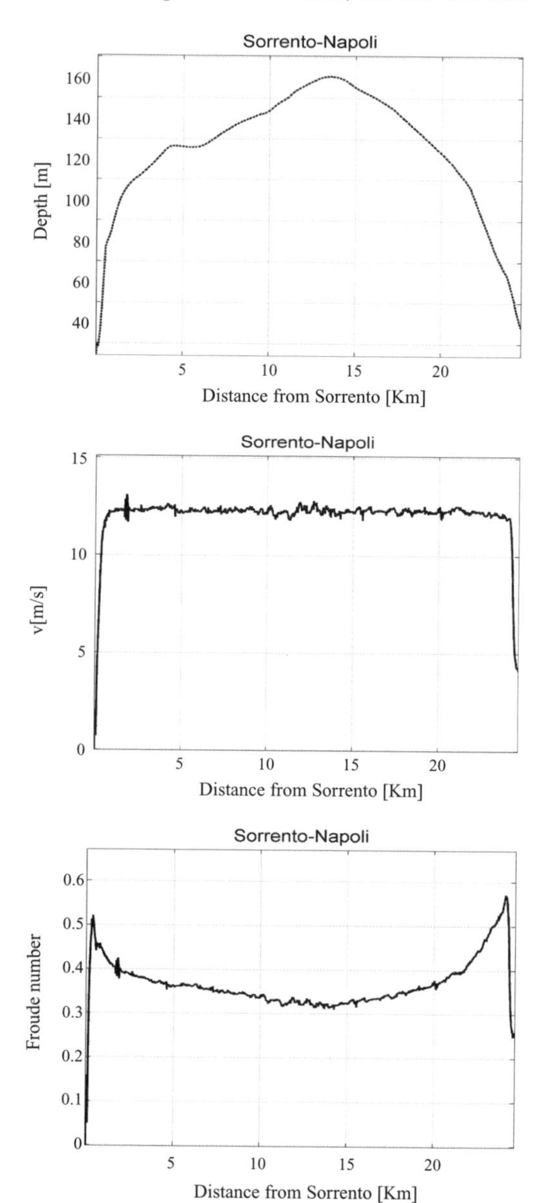

Figure 7. Depth (a), speed (b) and depth Froude number (c) along the track for Sorrento-Napoli route.

making resistance has a maximum in the so-called hump speed region $0.4 < F_{nl} < 0.6$.

In shallow water, the water depth also influences the wave making resistance, the maximum of which occurs near the critical Froude number $F_{nh} = 1$.

Figures 6 and 7 a,b and c shows plots of the ship speed, depth and resulting depth Froude number for the recorded backward ship track from Capri to Napoli, stopping through Sorrento

Figure 6 is relative to the trip from Capri to Sorrento, while Figure 7 is relative to the trip from Sorrento to Naples.

These figures show two distinct features. The first part of the track, between Napoli and Sorrento, cuts the bathymetric curves of the Gulf of Naples, first increasing the depth, assuming a maximum depth of 140 m approximately in the middle of the Gulf, and then decreasing the depth before arriving in Sorrento. The second part of the track, between Sorrento and Capri, first increases the depth and then follows a bathymetry between 70 and 90 m.

The ship maintains a speed of 13 m/s almost throughout the entire length of the bay. This is reflected in the depth Froude number which increases significantly after the start, while in the rest of the track the depth Froude number is between 0.4 and 0.6, that is below the critical speed.

4 DATA ANALYSIS AND RESULTS

4.1 *Form of the wake wash waves*

The wash generated by a high-speed vessel operating above the critical depth Froude number in shallow water can be divided into three zones according to wave period and grouping. The first group comprises the long period super-critical wash waves and is peculiar to high-speed vessels operating above the critical depth Froude number. The second group comprises waves similar in size and height to the wash waves produced by conventional vessels of comparable displacement and length. Finally there is a group of tail waves.

These three groups of wake wash waves are peculiar to high speed vessels in supercritical regime, while in most cases, as Figures 6 and 7 show, the regime is subcritical, sometimes critical, so there is no evidence of the three groups of waves, but, as we can see, we have some conventional waves. Neverthless, some records have evidenced a structure of the waves more grouped, that is the waves recorded at the passage of Superflyte, which is the fastes (and also the biggest) of all HSC considered.

In the vast majority of cases, however, the waves are not grouped in packets of two or three, like in Figure 8, and the most common shape is the one shown in Figure 9.

4.2 *Maximum values of the wake wash waves*

The maximum values of the wake wash recorded for each vessel have been reported in Table 3, together with the corresponding periods, the corresponding value of the wake wash transferred at a distance of 300 m from the navigation track and the corresponding value of wave energy. These two values have been calculated in order to verify the complaints with the International Rules, which have given in previous papers.

The most significant features highlighted in this paper is the comparison between the maximum

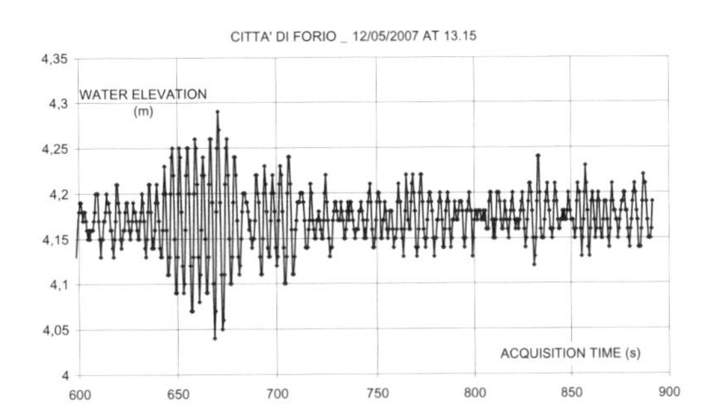

Figure 8. Grouped structure of waves for a MH.

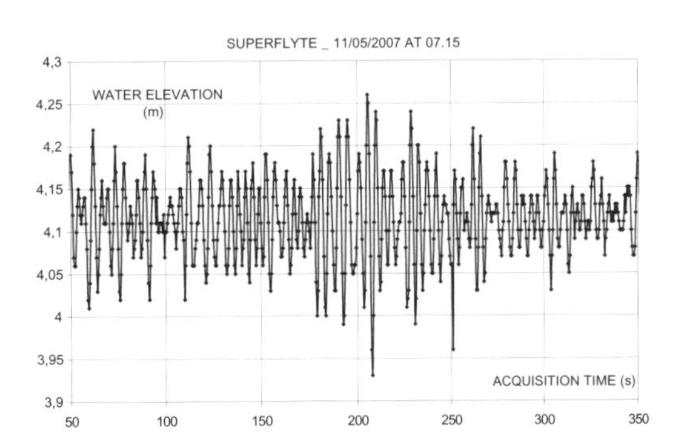

Figure 9. Single group of wake wash waves for a MH.

Table 3. Maximum values of WW for each vessel.

Type	Name	H(m)	T(s)	H_{300} (m)	E (J/m)
CAT	Acapulco Jet	0,33	7,5	0,36321	14551,90
CAT	Giove Jet	0,28	7,5	0,30818	10476,30
MH	Celestina	0,25	4,0	0,33159	3449,79
MH	Rosaria Lauro	0,35	3,5	0,46422	5176,85
MH	Superflyte	0,17	6,0	0,18711	2471,55
MH	Città di Sorrento	0,34	8,0	0,37422	17575,47
MH	Sorrento Jet	0,19	5,0	0,25201	3113,44
MH	Citta' di Forio	0,20	5,0	0,26527	3449,79
MH	Europa Jet	0,33	4,5	0,43770	7607,57

wake wash values of sister vessels in similar conditions of speed and displacement, but in different depth (and so Froude depth number) conditions.

In order to do such comparisons, the two catamarans Acapulco Jet and Giove Jet have been excluded, because they take service only in the route Napoli-Ischia. Besides, the waves produced by Celestina and Rosaria Lauro (recorded in CP1) have been compared with the waves produced by the sister ships Città di Sorrento, Sorrento Jet, Città di Forio and Europa Jet. The waves produced by Superflyte have been excluded because of the significant differences in dimensions.

As we can see from the values of the mean and standard deviation of the records, the circumstance of similar displacement and higher depth Froude number for the route Napoli-Ischia (i.e. closer to the hump speed) leads to higher values of wake wash for the records in CP1. This circumstance is also due to the lower distance of the measurement point CP1 from the navigation track.

5 CONCLUSIONS

The present field study of the wake wash assessment in the Bay of Naples was concentrated on both wake wash measurements in shallow water and vessel velocity and depth measurement through GPS route monitoring.

The study was focused on the comparison between the wake wash values of sister vessels along the two different routes, which have been measured in very similar depth conditions. This is the reason why the analysis excluded the catamarans, which wake wash was measured only in CP1.

The instantaneous calculation of the depth Froude number and of the distance from the wave measurement point gave the opportunity to establish that the vessel speed along the two routes were very similar. The vessels are quite similar in length, beam and draft, so they should have similar displacements.

The speed and depth conditions of the vessels gave always subcritical conditions, particularly for the site CP_2, in which the depth of wake generation is higher than 50 m, that is two times the depth of wake generation for the site CP_1.

The distance from the navigation track is lower for CP_1 (approx. 500 m) while for CP_2 is approx 700 m. This means a lower wake wash decay for CP_1. This circumstance, added to the higher depth Froude number, should lead to a mean value of H_{max} and E_{max} higher for CP_1 than for CP_2, which is confirmed by the calculations.

REFERENCES

Begovic, E., Benassai, G., Nocerino, E., and Scamardella, A., 2006. Field investigation of wake wash generated by HSC in the Bay of Naples. 2nd International Conference on Marine Research and Transportation (Ischia, Naples, 28–30 June 2007), University of Naples "Federico II", 243–252.

Begovic, E. and Benassai, G., 2007. Wake Wash Analysis of HSC Catamarans, *Proc. Int. Conf. ISOPE, Lisboa, 1–6 July 2007;*

Bruno, M., Fullerton, B., Datla, R. 2002. "Ferry Wake Wash in NY Harbor", Technical report SIT-DL-02-9-2812.

Havelock, T.H., 1908. "The Propagation of Groups of Waves in Dispersive Media, with Application to Waves on Water produced by a Travelling Disturbance". Proceedings of the Royal Society of London, Series A, Vol. LXXXI, pp. 398–430.

High-Speed Craft Code, 2000. International Code of Safety for High-Speed Craft. International Maritime Organization, London.

Kofoed-Hansen, H., Jensen, T, Kirkegaard, J,. Fuchs, J. 1999. "Prediction of Wake Wash from HSC in Costal Areas", RINA Int. Conference Hydrodynamics of High Speed Craft, London, UK, paper No.12.

Internal Navigation Association, 2003 Guidelines for managing wake wash from high speed vessels. Report of working group 41, 32 pp.

Marine Accident Investigation Branch, 2000. Report on the Investigation of the Man Overboard Fatality from the Angling Boat PURDY at Shipwash Bank, off Harwich on 17 July 1999. MAIB Report No 17/2000.

Maritime and Coastguard Agency 2001. A physical study of fast ferry wash characteristics in shallow water. Res. Proj. 457.

Maritime and Coastguard Agency 1998. Investigation of high speed craft on routes near to land or enclosed estuaries. Res Proj. 420.

Stumbo, S., Fox, K., Dvorak, F. Elliot, L. 1998. "The Prediction, Measurement and Analysis of Wake Wash from Marine Vessels", SNAME Pacific Northwest Section.

Stumbo, S., Fox, K., and Elliot, L., 2000. "An Assessment of Wake Wash Reduction of Fast Ferries at supercritical Froude Numbers and at Optimised Trim." RINA, The Royal Institution of Naval Architects, London.

Torsvik, T., Soomere T., 2009. Modelling of long waves from high speed ferries in coastal waters. J. of Coastal Research, pp. 1075–1079.

Whittaker, T., Bell, A.K., Shaw, M.R., Patterson K., 1999. An investigation of fast ferry wash in confined waters. Proc. Hydrodynamics of High Speed Craft, RINA, London, Nov. 1999.

Whittaker, T., Doyle, R., Elsaesser B., 2000. A study of leading long period waves in fast ferry wash, Proc. Hydrodynamics of High Speed Craft–Wake Wash and Motions Control, RINA, London, Nov. 2000, paper 7.

Whittaker, T. 2002. A Physical Study of Fast Ferry Wash Characteristics in Shallow Water, MCA Research Project 457, UK.

Sustainable Maritime Transportation and Exploitation of Sea Resources – Rizzuto & Guedes Soares (eds)
© 2012 Taylor & Francis Group, London, ISBN 978-0-415-62081-9

The wind forecast for operating management and risk assessment of port areas

M.P. Repetto & G. Solari
University of Genoa, Genoa, Italy

ABSTRACT: The seaport areas are generally very exposed to high wind velocities. The definition of operating strategies to identify the real risk conditions is an essential tool for planning the work in safety conditions. The present paper illustrates the research activities carried out in the framework of the European Project "Wind and Ports". The project handles the problem of the wind forecast in port areas and proposes an integrated system including the wide network of in situ monitoring, the numerical simulation of wind fields, the statistical analysis of large wind velocity databases and the implementation of algorithms for middle- and short-term wind forecast. The final results are made directly available to the port operators, and can be integrated in a global system for safety management.

1 INTRODUCTION

The seaport areas are generally very exposed to high wind velocities. This can give rise to great risks for structures in the sites, ships and ferries approaching or docking in port, empty containers piled (Fig. 1) and, most of all, health hazard for workers, who need to work in safety conditions also in windy days, or to stop working in extreme windy days. On the other hand, frequent stops of working of port areas can produce large loss of moneys. Thus, the definition of operating strategies to identify the real risk conditions is an essential tools for planning the work in safety conditions.

The present paper illustrates the research activities linked with the European Project "Wind and Ports", financed by the European Territorial Cooperation Objective, Cross-border program "Italia-France Maritime 2007–2013" (www.maritimeit-fr.net, www.ventoeporti.net), which involves the University of Genoa, Department of Civil, Environmental and Architectural Engineering (DICAT), with the role of scientific actuator, and the Port Authorities of five of the main Ports of Tirreno Sea, namely Genoa, La Spezia, Livorno, Savona (Italy) and Bastia (France) (Fig. 2). The project handles the problem of the wind forecast in the port areas and proposes an integrated system including a series of activities. The preliminary steps are related to the realization of a wide network of in situ monitoring, adopting a series of ultrasonic anemometers suitably distributed in the most exposed zones, and the numerical simulation of the wind fields, adopting the mass-consistent model WINDS (Burlando et al., 2007a, 2007b).

a) 

b) 

Figure 1. Example of risk and collapse in seaport areas: a) ferry boats approaching Genoa Harbour during thunderstorm; b) empty containers pile collapsed.

Moreover, a series of existing anemometric stations have been selected in the region of each port and the statistical analysis of the corresponding large wind velocity databases have been computed, in order to obtain information about the general climatology of the port areas.

The measures and simulated wind velocities constitutes the input for the implementation of algorithms for middle-term (1–3 days) and short period (0.5–1 hour) wind forecast, which represents the core of the project and the key issue for the risk assessment of the areas. The final results are made directly available to the port operators, and can be integrated in a global system for safety and operational management.

The paper describes in details the research activity involved, putting in evidence, from one hand, the high innovating level of the project, from the other hand, the critical aspects, linked to the large extension of the considered areas and the complexity and interdependence of the actions proposed.

The conclusions highlight the expected impact on the research activity and the perspectives of the work.

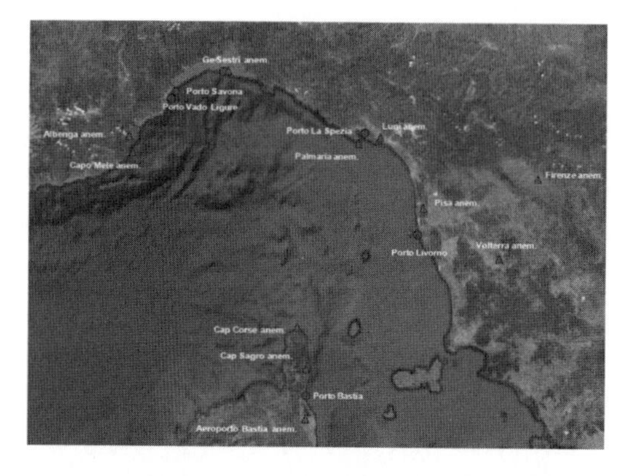

Figure 2. Example of risk and collapse in seaport areas: a) ferry boats approaching Genoa Harbour during thunderstorm; b) empty containers pile collapsed.

2 INTER-PORT MONITORING NETWORK

The first project activity concerns the achievement of a wide monitoring network of wind velocity. Each port area is provided with a network of anemometers appropriately distributed on the area of interest.

Suitable sites for instrument installation were identified in collaboration with DICAT after an appropriate survey. The general criteria for site selection concerns measure optimization and port area covering. In particular, instruments are installed at least at 10 meters above ground level. At this level, measure reliability is not altered by terrain effects.

Instrument network is composed by 22 either biaxial or tri-axial ultrasonic anemometers, that are financed by the project (Fig. 3, circles). Moreover, 9 anemometers co-financed by Port Authority of Genoa are added (Fig. 3, squares).

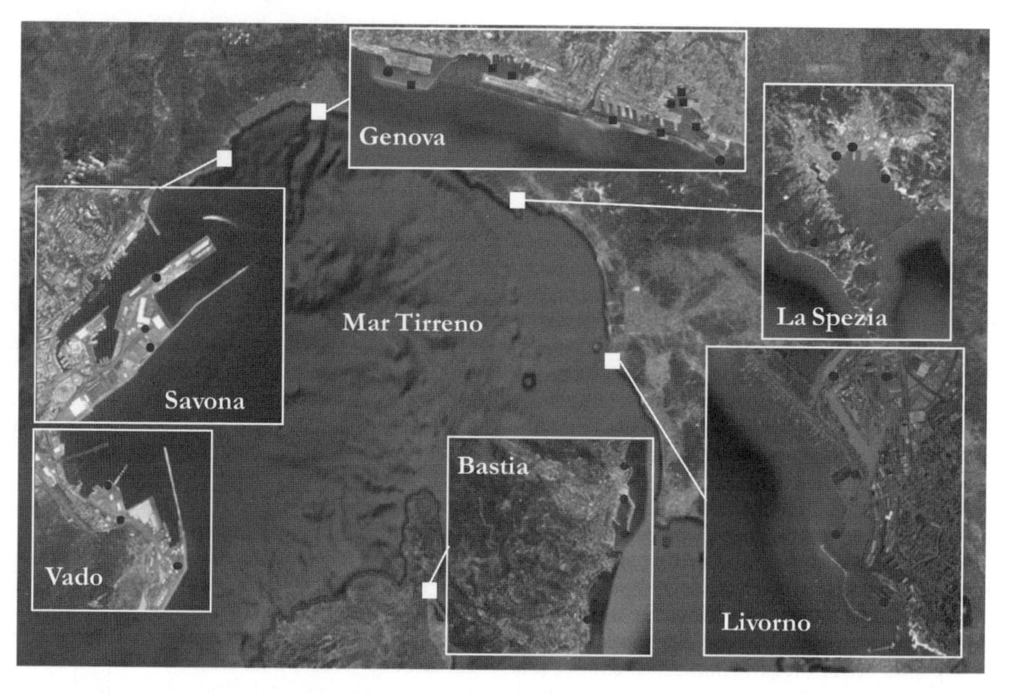

Figure 3. Anemometer stations installed on port areas involved in the project. Stations financed by the project (circles) and by Port Authority (squares).

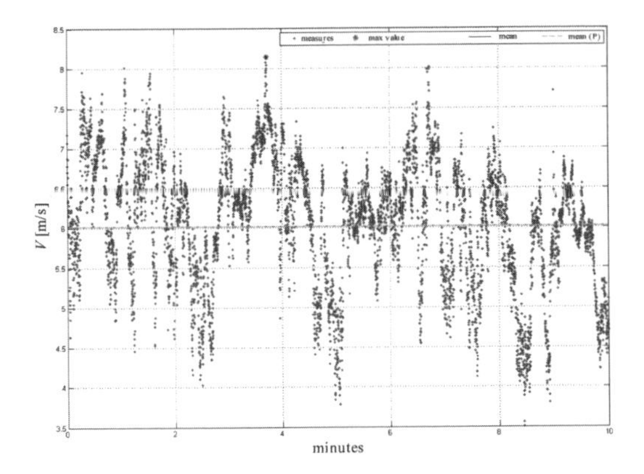

Figure 4. Central acquisition system of anemological measures on port areas: example of 10-minutes 10 Hz registration and their statistics.

Ultrasonic anemometers measure wind velocity components with a sampling frequency of 10 Hz and with a resolution of 0.01 m/s and 1° for intensity and direction, respectively.

Each port is provided with a data-acquisition system allowing to visualize on remote terminals the current values of wind velocity and direction. The main statistics of measures over 10 minutes are computed. Moreover, the system provides a first-control level on the completeness of measures over 10 minutes by assigning a reliability code from 0 to 100 on data package. Measured data are organized in package of 10 minutes. These packages, their statistics and reliability codes are sent in real time from the acquisition system of each port to a central server located in DICAT (Fig. 4) according to a common transmission protocol.

DICAT acquisition center collects all anemometer measurements, organizes them on databases, systematically analyses and verifies their correctness and representativeness.

When the system will be full operating, the real-time joint monitoring on all ports will be included in a final system reachable on the Web by port operators community.

After one year of survey, the databases are statistically analyzed in order to provide an anemological characterization of the sites. Moreover, the complete database will be an asset, constantly updated, available to the research activity related to the modeling of atmospheric turbulence and thunderstorms.

3 NUMERICAL WIND-FIELD MODELING

In parallel with the implementation of the monitoring network, the project involves the numerical-model development in order to simulate wind in each port area.

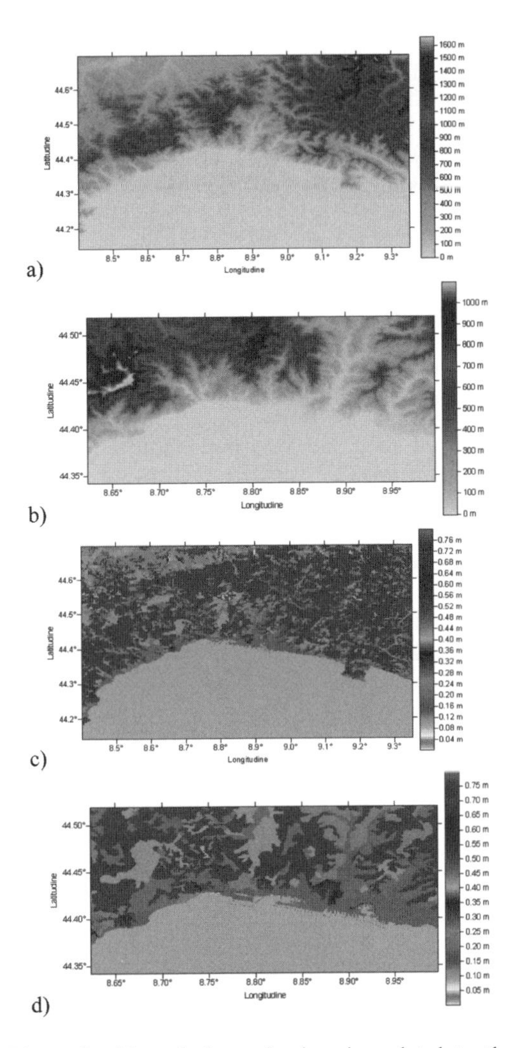

Figure 5. Numerical terrain domains related to the Genoa port: orography domain of a) macro-area and b) micro-area; roughness domain of c) macro-area and d) micro-area.

The simulation of the intensity field and average velocity direction was carried out by using the mass-consistent numerical model WINDS (Burlando et al., 2007a, 2007b). The WINDS initialization uses the Internal Boundary Layer (IBL) concept. When air flow encounters steep change of surface roughness, an IBL develops, within which the wind adapts to the new surface. The growth of the IBL, in general, depends not only on the surface roughness parameter downwind but also on the atmospheric stability.

Numerical simulations are based on orography and roughness domains of the territory adjacent to the area (Fig. 5).

The Digital Terrain Model (DTM) is supplied by the Military Geographic Institut (IGM), while the roughness is supplied by the land cover maps, obtained from the CORINE project (Bossard et al., 2000) by associating to each coverage type an appropriate terrain roughness value. Each model includes the docks, the main anemometer stations with long period time series and the areas which are close to the port and whose properties help to define the climatology.

Each model is implemented through a macro-area and several micro-areas, depending on the characteristics of the considered area. These micro-areas allow a detailed wind simulation in the most significant regions. In particular, five macro-areas are first simulated, each one corresponding to a port area, with a grid step of 270 meters; these macro-areas are intended to simulate the orographic effects. Then, a total number of 9 micro-areas are nested within the macro-areas in order to obtain a higher wind fields resolution in the areas. The micro-area grid resolution is 80 meters. The simulations were carried out under the hypothesis of neutral atmosphere (typical of intense wind conditions) for different scenarios corresponding to 3 values of wind speed and 36 directions at the top of the atmospheric boundary layer. For each scenario examined, the average wind speed and direction are evaluated at grid points, representative of the port area and extracted from the corresponding micro-area (Fig. 6). 11 levels of different heights varying from 5 to 5000 meters above ground level are considered.

The turbulence field in the port area is modeled using the procedures proposed by Engineering Science Data Units (ESDU 1993, 2002), with an accurate ground-roughness domain and a simplified orography model near the port area. The algorithm provides an intensity map of the longitudinal atmospheric turbulence for 12 wind directions for the representative grid point in the areas (Fig.7).

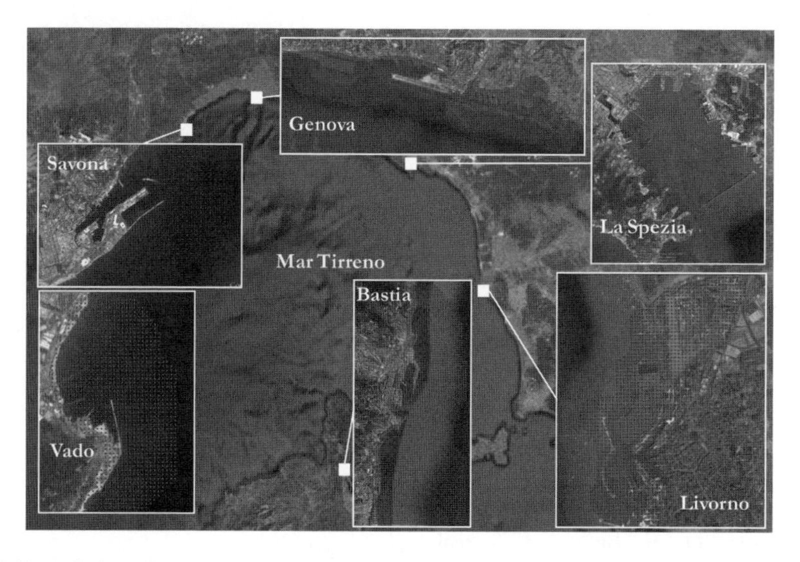

Figure 6. Grid resolution of 80 meters. These points are representative for the areas in the project.

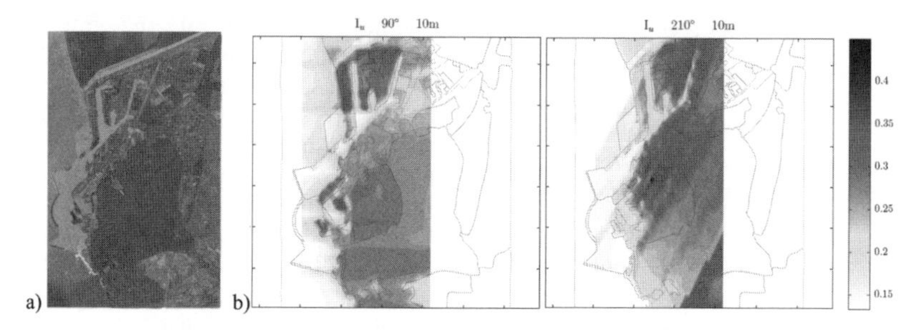

Figure 7. Atmospheric-turbulence field numerical simulation by means of ESDU model in Livorno port: a) terrain roughness pattern; b) turbulence intensity maps at 10 meters a.g.l. for wind directions of 90° and 210°. The colorbar reports the I_u values.

4 WIND FIELD RECONSTRUCTION FROM ANEMOMETRIC MEASURES

The wind field reconstruction consists on a numerical algorithm which transfers anemometric measures to the grid points in the port areas (Fig. 6).

First, the transmission-coefficient matrixes, *kua* and *kva,* are obtained by the numerical wind field on the micro-areas (Burlando et al., 2007a, 2007b), representing, for each direction, the ratios *u*/*ua* and *v*/*va,* where *u*, *v* are the wind velocity components in the grid point corresponding to the measured *ua*, *va* wind velocity components at the anemometer. The transmission matrices. Then, a weight function φ is defined, based on the distance between the anemometer and the grid point $d_{p\text{-}a}$, and the anemometer measure reliability *aff*.

In particular for each anemometer *a*, this function φ is composed by two parts: the principal function γ and the "hole" function β:

$$\varphi_a = \gamma_a \prod_{j \neq a} \beta_j \tag{1}$$

Where:

$$\gamma = \frac{1}{1 + \left(p \cdot d_{p\_a}\right)^q} \tag{2}$$

$$\beta = 1 - \left(\frac{1}{\frac{1}{aff} + \left(r \cdot d_{p\_a}\right)^k} \right) \tag{3}$$

The parameters *p*, *r*, *q* and *k* are suitable choose on the base of port extension.

The transferred wind velocity components (u_{trasf} and v_{trasf}) in each grid point is computed by the following formula:

$$u_{trasf} = \frac{\sum_{a=1}^{N} \varphi_a \cdot aff_a \cdot u_a \cdot ku_a}{\sum_{a=1}^{N} \varphi_a \cdot aff_a}$$
$$v_{trasf} = \frac{\sum_{a=1}^{N} \varphi_a \cdot aff_a \cdot v_a \cdot kv_a}{\sum_{a=1}^{N} \varphi_a \cdot aff_a} \tag{4}$$

The reliability of the transferred wind depends on the number of working anemometers at that time and on the anemometer measure reliability *aff*.

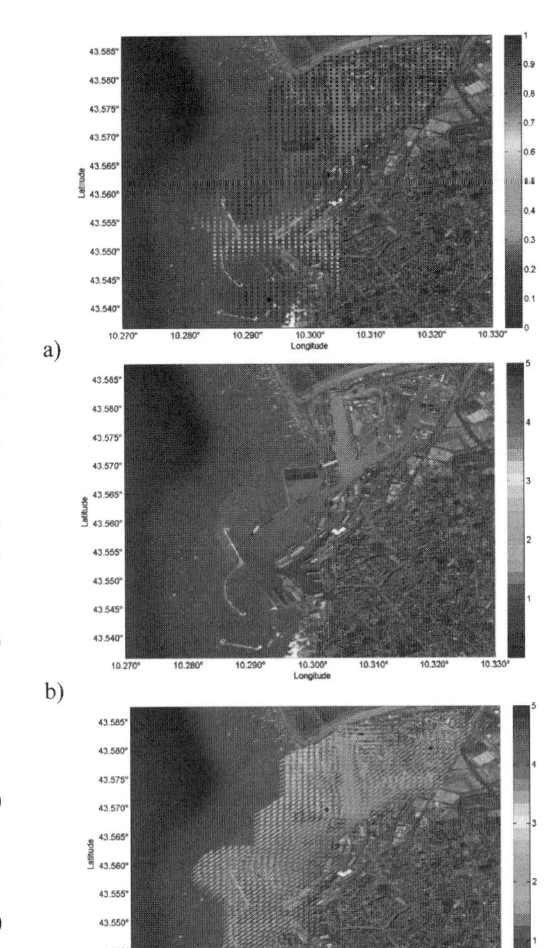

Figure 8. Example of the wind field reconstruction in Livorno port: a) weight function φ; b) instantaneous anemometric measures; c) wind field reconstructed.

Figure 8 shows an example of the wind field reconstruction in Livorno areas, starting from 5 anemometric measures. Colorbars indicate the weigth function values (a) and wind velocities [m/s] (b,c) at each grid points.

5 STATISTICAL ANALYSIS OF HISTORICAL TIME SERIES

The data provided by the anemometers installed in the port areas will be statistically significant only after some years of working. In order to characterize the areas of interest from a climatological point of view, a set of anemometric stations neighbouring port areas and provided with long-period databases of velocity measure was selected. A total of 13 historical databases (Fig. 2, triangles) was acquired; each database includes measurements of velocity intensity and direction and, where possible, of atmospheric pressure and temperature.

Initially, each database is checked and corrected, following the criteria reported in Solari (1996).

Then, probabilistic analyses of both data population and annual maxima are carried out.

The data concerning measures population were analysed and regressed by means of a so-called hybrid Weibull model (Lagomarsino et al., 1992, Solari, 1996), both in a directional and in a non-directional form. In particular, in the present study the current-values directional distributions are obtained for 12 sectors, corresponding to a range of 30° with 0° set for winds from North. Figure 8 a) reports an example of this distribution for the database of Pisa San Giusto. The density function and the distribution function can be expressed in non-directional form by the following equations:

$$f_V(v) = P_0\delta(v) + A\frac{K}{C}\left(\frac{v}{C}\right)^{K-1} e^{-\left(\frac{v}{C}\right)^K} \qquad v \geq 0$$

$$F_V(v) = 1 - Ae^{-\left(\frac{v}{C}\right)^K} \qquad v \geq 0 \tag{5}$$

where K and C are the Weibull-distribution parameters of the whole database and A is the probability of non-zero values in the same database. The first part of Table 1 summarizes the coefficients of the current-values distributions obtained for the Italian databases considered in this project.

The statistical analysis of extreme winds play a crucial role for the analyses of safety and risk in the considered areas. However, in the scientific community there is no unanimous agreement on the probability distribution with the best regression of experimental data. In-progress analyses are addressed to an accurate study of different models of extreme statistics on long-period databases (Torielli et al., 2010). In the present study the probability distribution of V annual maximum is computed via both the asymptotic analysis of type I and the process analysis (Lagomarsino et al., 1992).

The procedure for the asymptotic analysis of type I assigns the Gumbel distribution to the annual maximum v, as expressed in the equation:

$$F_M(v) = e^{-e^{-A(v-U)}} \qquad v \geq 0 \tag{6}$$

where A and U are the model parameters.

The process analysis describes the distribution function of maxima by the equation:

$$F_M(v) = e^{-\lambda f_V(v)} \qquad v \geq 0 \tag{7}$$

where the product $\lambda f_V(v)$ represents the mean number of up-crossings of the v threshold; $f_V(v)$ is the density function of data population given by equation (5); λ is the model parameter evaluated by enumerating threshold up-crossings.

Figure 9 b) reports an example of both I-type and process extreme distributions, referred to Pisa San Giusto database.

Table 1. Summary of the statistical analysis of the Italian historical anemometric stations considered here.

	Population analysis			Asymptotic analysis		Process analysis		V (R) (m/s)						
Anemometric station	A	K	C (m/s)	A	U (m/s)	λ	$\tilde{\lambda}$	5	10	20	50	100	200	500
Capo Mele	0.86	1.27	5.350	0.297	24.50	4584	4918	33.51	35.44	37.27	39.59	41.30	42.98	45.18
Albenga/Villanova	0.54	1.33	4.288	0.510	17.64	3666	5327	24.34	25.73	27.03	28.69	29.90	31.10	32.65
Capo Vado	0.99	1.39	6.232	0.450	27.91	6458	7218	31.63	33.46	35.18	37.36	38.96	40.52	42.54
Genova/Sestri Ponente	0.84	1.67	4.727	0.635	16.44	3334	3498	19.92	20.75	21.53	22.50	23.21	23.90	24.79
Genova/Sestri (M)	0.99	1.65	4.436	0.565	16.47	10549	15669	19.90	20.67	21.40	22.32	22.99	23.65	24.50
Isola di Palmaria	0.93	1.31	5.607	0.233	24.86	4443	5954	34.08	35.92	37.66	39.88	41.51	43.11	45.20
Monte Rocchetta	0.98	1.22	3.179	0.531	19.39	9682	10099	21.21	22.50	23.73	25.29	26.45	27.59	29.07
Sarzana/Luni	0.79	1.04	2.047	0.472	12.47	4348	7285	20.67	21.99	23.26	24.90	26.12	27.34	28.95
Pisa/San Giusto	0.65	1.30	3.323	0.507	15.20	3449	3585	19.66	20.80	21.87	23.23	24.24	25.22	26.50
Volterra	0.81	1.48	5.192	0.424	19.45	4034	5936	25.84	27.08	28.23	29.69	30.76	31.81	33.16
Volterra (M)	0.90	1.42	4.391	0.488	17.73	7354	22728	24.87	25.99	27.05	28.39	29.38	30.35	31.61
Firenze/Peretola	0.49	1.20	2.952	0.472	14.70	4158	4965	19.90	21.18	22.38	23.92	25.05	26.17	27.64

In general, the analysed databases suffer from different kinds of data incompleteness, e.g. due to the presence of missing data and to measure sampling frequency. It can be shown that data incompleteness leads to a systematic underestimate of the extreme distribution.

Thus, it is possible to determine a corrective criterion by substituting the parameter of process-analysis distribution for a value conveniently modified by a corrective factor which takes into accounts data incompleteness (Repetto et al., 2010).

Then, the final extreme distributions are evaluated by applying appropriate corrections to the estimates obtained through the process analysis. The second part of Table 1 summarizes the coefficients of the extreme-values distributions obtained for the Italian databases considered in this project, by applying both the asymptotic analysis of type I and the process analysis. Furthermore, the same table provides the values of parameter $\tilde{\lambda}$, corrected in order to take into account missing data, and the final velocity values related to a given return period, corrected in order to take into account non-continuous samplings.

The statistical analyses described here provide a climatologic characterization of the sites where anemometric stations with long-period time series are present. By making use of the appropriate transformation matrices, it is possible to perform the statistical analysis of the historical time series transferred to the grid points representative of the whole port area, obtaining a statistical wind representation inside the port areas.

Finally, long-period databases are exploited in order to obtain a first validation of the numerical models of wind fields described in section 3. For this purpose, pairs of neighboring anemometric stations are considered and a statistical analysis is carried out on concurrent data. The joint occurrence frequencies of wind intensity and direction at both stations are compared with the corresponding results obtained by numerical simulations of wind fields (Fig. 10).

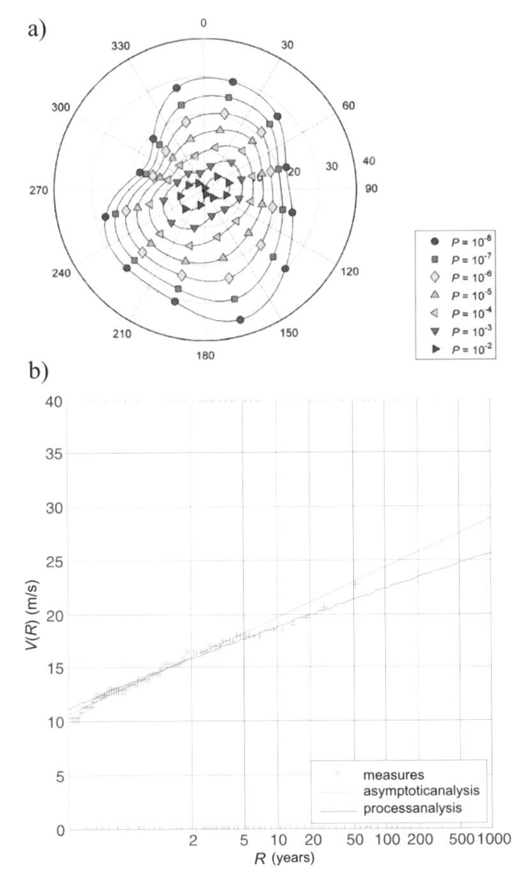

Figure 9. Statistical analysis for the anemometric station of Pisa San Giusto: a) current-values directional distribution; b) extreme distributions.

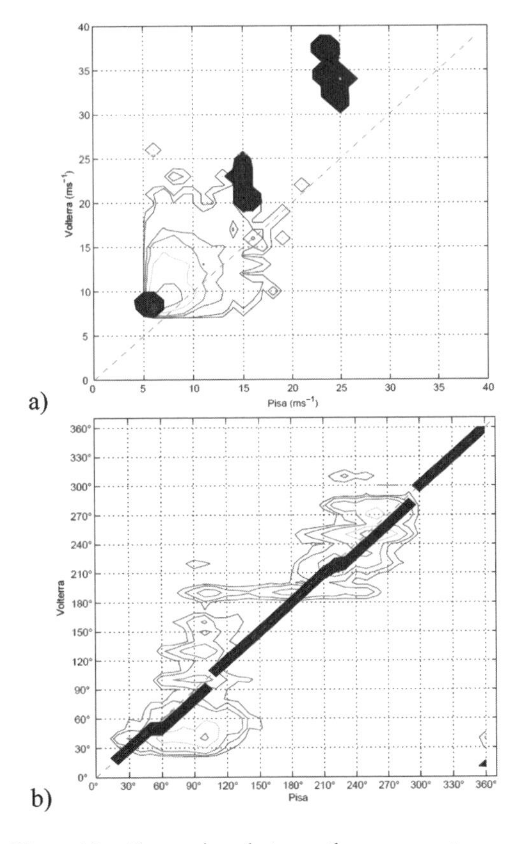

Figure 10. Comparison between the concurrent measures of the anemometric stations of Pisa and Volterra (contour diagrams) and the numerical simulations (black points): a) intensity of the mean velocity; b) direction of the mean velocity.

6 WIND FORECAST IN PORT AREAS

Wind forecast in port area are developed using two algorithms with different approaches: medium-term forecast with a horizon of 1–3 days, short-term forecast with horizon of 0.5–1 hours.

Medium-term forecast is based on a new generation forecast mesoscale system: Weather Research and Forecasting Model (WRF). The model is initialized with variables provided by National Oceanic and Atmospheric Administration (NOAA) Institute of USA. The model starts from the NOAA global circulation model and manages to simulate port area using subsequent nesting with higher resolution. This model chain is operative and provides forecast up to 72 hours, many times a day.

Short-term forecast is based on a probabilistic approach: the conditional model developed in previous researches (Freda et al., 2009) *is* applied to data series measured by port anemometers. The conditional approach provides an accurate evaluation of the expected value of wind velocity for intense events. Since during these events the direction is persistent, records can be considered as a one-dimensional stochastic process, neglecting direction variations in a time period of 2 hours.

The expected value is estimated as a function of its occurrence probability, suitably modelling expected-value probability distribution at time $t_0 + \Delta t$ conditioned to the occurrence of a value at time t_0 (Fig. 11). It is worth noticing that this algorithm does not need assumptions on probability distribution of current values recorded by anemometers. In the original formulation, conditional probability distribution of the future value is modelled by a cut-off Gaussian model. Based on long-period series, it will be possible to revise this hypothesis, e.g. by adopting a Gaussian model for short-term variations. Conditional probability distribution is regressed from the measures of each port anemometric station. After one year of measurements, a preliminary forecast model is improved.

In order to determine the forecast in each grid point, the transmission-coefficient matrixes described on section 3 and based on numerical modelling of port areas are applied to short-term forecast carried out at anemometer site. The obtained values are visualized on maps, highlighting wind expected values at the hour after the last anemometric registration as a function of port plan and altimetry. The results of forecast algorithm constitute the core of the project "Vento e Porti". However currently they are not complete and they will be subject of futher scientific publications.

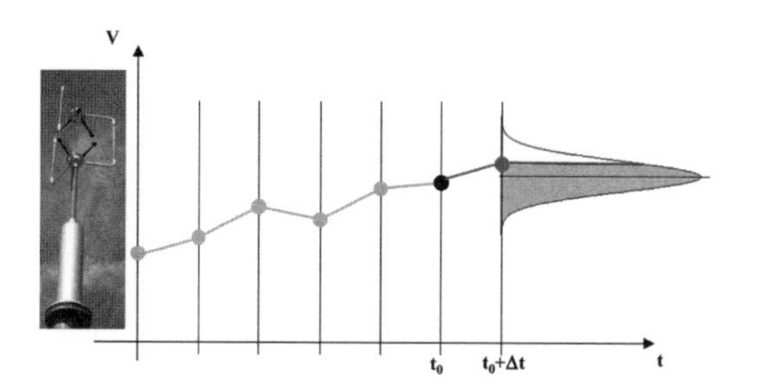

Figure 11. Scheme of short-term forecast procedure.

7 CONCLUSIONS AND PERSPECTIVES

The paper presents the research activities made under the project "Vento e Porti", still in course, providing an inter-port method for the wind forecast in the main port areas of the north Tirreno Sea, through an integrated system including numerical simulation, in-situ monitoring, statistical analysis and algorithm implementation for wind velocity forecast. Once implemented in the port network, the proposed models will bring a significant technologic surplus value for safety management and operation in the port areas. In addition to the direct project impact on port operators, it provides a large number of tools and research ideas. Because of the large spatial extent and the significant acquisition level, the wind velocity monitoring network will be a constantly updated database, available to the research concerning the modeling of atmospheric turbulence and thunderstorms. Moreover, crossed comparisons between monitoring, numerical wind field models and forecast models provide useful insights for research in these areas. The medium-term project horizons include researches concerning wind-potential study for energy exploitation in the port areas, pollutants spread, analysis and forecasting of current and wave.

ACKNOWLEDGEMENTS

The presented study is financed by EU funds, Territorial Cooperation Objective, Cross-border program "Italia-France Maritime 2007–2013.

REFERENCES

Burlando M., Freda A., Ratto C.F., Solari G. 2010. A pilot study of the wind speed along the Rome-Naples HS/HC railway line. Part 1—Numerical modeling and wind simulations, *J. of Wind Engrg. Ind. Aerod.*, 98, (8–9): 392–403.

Burlando M., Georgieva E., Ratto C.F. 2007a. Parameterization of the planetary boundary layer for diagnostic wind models, *Boundary Layer Meteorology*, 25, (3): 389–397.

Burlando M., Carassale L., Georgieva E., Ratto C.F., Solari G. 2007b. A simple and efficient procedure for the numerical simulation of wind fields in complex terrain, *Boundary Layer Meteorology*, 125 (3): 417–439.

Bossard M., Feranec J., Otahel J. 2000. Corine land cover technical guide—Addendum 2000. EEA Technical Report No 40. Copenhagen (EEA).

Engineering Sciences Data Unit, 2002. Strong winds in the atmospheric boundary layer. Part I: Hourly-mean wind speeds, ESDU 82026, London, UK.

Engineering Sciences Data Unit, 1993. Computer program for wind speeds and turbulence properties: flat or hill sites in terrain with roughness changes, ESDU 92032, London, UK.

Freda A., Carassale L., Solari G. 2009. A conditional model for the short-term probabilistic assessment of severe wind phenomena, *CD Proc of EACWE5, Florence, Italy.*

Lagomarsino S., Piccardo G., Solari G. 1992. Statistical analysis of high return period wind speeds, *J. Wind Engng. Ind. Aerod.*, 41.

Repetto M.P., Solari G., Tizzi M. 2010. Il ruolo dei dati mancanti nella statistica dei venti estremi, Proc. *IN VENTO 2010, in press.*

Solari G. 1996. Wind speed statistics, In Lalas D.P., Ratto C.F. (Ed), *Modelling of atmospheric flow fields*. World Scientific, Singapore.

Torielli A., Repetto M.P., Solari G. 2010. Simulation of long-period samples of the mean wind velocity, *J. Wind Engng. Ind. Aerod.*, under revision.

8 *Environmental protection*

8.1 *Sea pollution*

Sustainable Maritime Transportation and Exploitation of Sea Resources – Rizzuto & Guedes Soares (eds)
© 2012 Taylor & Francis Group, London, ISBN 978-0-415-62081-9

An undated analysis of IOPCF oil spill data: Estimation of the disutility cost of tanker oil spills

C.A. Kontovas, N.P. Ventikos & H.N. Psaraftis
Laboratory for Maritime Transport, National Technical University of Athens, Greece

ABSTRACT: Beyond any doubt, the cost of an oil spill is very difficult to be estimated. The approach used in this work is based on the assumption that the cost of an oil spill can be approximated by the compensation eventually paid to claimants. A series of regression analyses of the total costs have been carried out based on the latest available data. These analyses and their results can provide useful insights to the discussion on environmental risk evaluation criteria and in policy evaluation regarding the oil pollution of the marine environment by estimating the damage cost of oil spills and, thus, the benefit from relative regulations that deal with the prevention of oil pollution in cost effectiveness and cost benefit analysis.

1 INTRODUCTION

A downward tendency is apparent in the total annual quantity of oil spilt by crude oil carriers during the last decade as well as in the number of oil spills worldwide, see ITOPF (2010) and Kontovas et al. (2010). However, there is a constant need for designing and operating ships that will lead to minimal consequences to the environment and the society in the case of an accident. A crucial parameter in evaluating designs and policy measures to reduce pollution is the estimation of the cost of an oil spill.

The reduction of oil pollution is one of the stated goals of new regulations, including the implementation of double hulls for tanker vessels. The management of safety at sea is based on a set of accepted rules that are, in general, agreed upon through the International Maritime Organization (IMO) which is a United Nations organization that deals with all aspects of maritime safety and the protection of the marine environment. Many of these regulations aim at reducing environmental risk and, more precisely, the risk that relates with accidental oil spillage. A big chapter that has only recently opened concerns environmental risk evaluation criteria. At the 55th session of Marine Environment Protection Committee (MEPC) that took place in 2006, the IMO decided to act on the subject of environmental criteria. At the 56th session of MEPC (July 2007) a Correspondence Group (CG), coordinated by the third author of this paper on behalf of Greece, was tasked to look into all related matters, with a view to establishing environmental risk evaluation criteria within

Formal Safety Assessment (FSA). FSA is the major risk assessment tool that is being used for policy-making within the IMO. An issue of primary importance was found to be the relationship between spill volume and spill cost and its uses within FSA.

The analysis reported in this paper is an attempt to shed some light into this issue and describes recent regression analyses of oil spill cost data. These analyses have been carried out by the authors and are in the same spirit as those carried out by Yamada (2009) and Psarros et al. (2011) but differ from them on several points. In fact, this work is an updated analysis of a previous work of the same authors that has recently been adopted by the IMO/MEPC as a basis for further discussion on environmental risk evaluation criteria.

These analyses and their results can provide useful insights to the discussion on environmental risk evaluation criteria and in policy evaluation regarding the oil pollution of the marine environment by estimating the damage cost of oil spills and, thus, the benefit from relative regulations that deal with the prevention of oil pollution in cost effectiveness and cost benefit analysis.

The rest of the paper is organized as follows. Section 2 presents a review on the recent literature on oil spill valuation based on compensation data, Section 3 describes the results of the regression analyses and the formulas that can be used to estimate the cost of an oil spills. Section 4 reports on the possible uses of these cost functions and section 5 presents the conclusions.

2 LITERATURE REVIEW

There is a general agreement (Etkin, 1999; Grey, 1999; White and Molloy, 2003) that the key factors that influence the cost of oil spills are the type of oil, location, weather and sea conditions and the amount spilled and rate of spillage. Given that the above parameters are highly variable and cannot be predicted in advance, a usual approach taken in the literature is to connect the cost of an oil spill to its volume. In that sense, a larger oil spill is expected to have a higher cost, all else being equal. Estimates of the cleanup cost and the total oil spill costs which may include the costs for response, third party claims and environmental damages as a function of the oil spill size have been extensively analyzed in the literature and substantial work has been performed over at least the last 30–35 years, mostly in the context of analyzing the economic impact of oil spills and contemplating measures to mitigate their damages. For a short literature review the reader is referred to a previous work of the authors, see Kontovas et al. (2010).

Lately, the subject of estimating the cost of oil spills has been in the center stage of discussions at the International Maritime Organization (IMO) in regards to the establishment of Environmental Risk Evaluation Criteria within the context of Formal Safety Assessment (FSA).

One way to estimate the total cost of oil spills is by using compensation data. For more on the other ways that can be used to derive the total cost the reader is referred among others to Kontovas and Psaraftis (2008). Back on compensation data, the most widely accepted public source that covers compensations paid is provided by the International Oil Pollution Compensation Funds (IOPC Fund).

Among the first recent analyses was the one performed by Grey (1999). A couple of recent cases where the IOPCF data was analyzed were known to the authors prior to their own analysis. It is not our purpose to comment on these in detail here. Friis-Hansen and Ditlevsen (2003) used the 1999 Annual Report (except those accidents that belonged to the categories "loading/unloading", "mishandling of cargo", and "unknown reason" which were removed from their analysis) and converted all amounts into Special Drawing Units (SDR) by an average annual exchange rate taken from the International Financial Yearbook. Then, historic national interest rates for money market rates were applied to capitalize all costs into year 2000 units followed by a conversion into 2000 USD.

Hendrickx (2007) performed an analysis based on data of the 2003 Annual Report and analyzed 91 cases by converting each compensation amount into US Dollars using for each accident the exchange rate on Dec. 31 of the year of occurrence. Exchange rates of the Bank of England were used for the currencies available and for the others an online website (OANDA.com) was used. There is no report that an inflation rate was used to bring these amounts into current Dollars.

The work done in Ventikos et al. (2009) gives a clear picture of oil spill response cost in Greece. In this outline the aforementioned paper takes into account a number of variables to draft a model for the estimation of clean-up cost; namely type of oil, quantity of oil, and impact to shoreline. The results show that oil confrontation in Greece appears to be rather expensive, with a value of about 25,000 euro for the abatement of a spill of one ton of oil.

Yamada (2009) performed a regression analysis of the amount spilled and the total cost by using the exchange rated provided in the IOPCF Annual Report. These rates can be use for conversion of one currency into another as of Dec. 31, 2007 and do not take into account the time of the accident nor is any inflation taken into account. Note that spills less than 1 tonne were excluded by the analysis. His analysis formed the basis of Japan's submissions to the MEPC and, to a large extent, the basis of the MEPC decision to recommend a volume-based approach.

Last but not least, Psarros et al. (2011) used combined data from two datasets, namely the IOPCF report and the accident database developed by EU research project SAFECO II, and thus performed a regression analysis in 183 oil spill incidents. It is not immediately clear from their analysis what the SAFECO II database is and what (if any) biases it introduces to the analysis. The amounts were converted into 2008 US Dollars taking into account the inflation rate.

3 DERIVING COST FORMULAS

3.1 *Original IOPCF regression formulas*

Compensation for oil pollution caused by tankers is governed by four international conventions: the 1969 and the 1992 International Convention on Civil Liability for Oil Pollution Damage ("CLC 1969" and "CLC 1992") and the 1971 and 1992 conventions on the Establishment of an International fund for Compensation for Oil Pollution Damage ("1971 Fund" and "1992 Fund"). These conventions together create and international system where reasonable costs of cleanup and damages are met, first by the individual tanker owner up to the relevant CLC limit through a compulsory insurance and then by the international IOPC Funds, if the amounts claimed exceed the CLC limits. More on compensation for oil pollution damage can be found in Jacobsson (1994 and 2007), ITOPF(2010) and Liu and Wirtz(2009).

The 2008 IOPCF Annual report (IOPCF, 2008) presents the claims that the IOPC Fund dealt with in the past. This report includes 107 accidents that are covered by the 1971 Fund and 33 by the 1992 Fund.

For each accident the time and the place of accident are known and for most of the cases the volume of oil split, as well as, the costs claimed and eventually covered by the Fund are recorded. In order to perform our analysis we followed the steps below:

1. We removed all incomplete entries (cases were no information regarding other the compensation paid or the oil spill size).
2. All claims for the cleanup and the total cost categories (in the case of multiple claims) were added up by converting them to US Dollars at the time of the accident. We note that we are aware of the fact that the year of the accident and the year when the amount agreed was paid are not the same but this was the only available information. Furthermore, the exchange rates used in these conversions were found in various CIA Factbooks and in a list of foreign currency units per dollar that is compiled by Antweiler (2009).
3. The cost of the previous step were capitalized into 2009 US Dollars by using conversion factors based on the Consumer Price Index (CPI).

This way we arrived at two datasets, one having data on the Cleanup Cost (CC) and the Volume (V) and another on the Total Cost (TC) and the Volume (V). These datasets were not disjoint. In fact, the first dataset contained 84 entries, the second had 91 entries, and 68 spills reported both CC and TC.

3.1.1 *Cleanup Cost (CC)*

After removing incomplete entries, a dataset of N = 84 spills for the period 1979–2006 was used for this regression analysis.

The minimum volume was 0.2 tonnes and the maximum was 84,000 tonnes. The average spill was 4,055.82 tonnes with a standard deviation of 14,616.15 tonnes and the median was just 162.5 tonnes. There were only 10 spills above 5,000 tonnes and, thus, one should be very careful when using the regression formulas to extrapolate the cost of large spills. The equation of the fitted model using linear regression was:

$$\text{Cleanup Cost} = 44{,}435\ V^{0.644} \qquad (1)$$

The R-Squared statistic indicates that the model as fitted explains 61.5254% of the variability in LOG10(Cleanup Cost). The correlation coefficient (Pearson's correlation coefficient p) equals 0.7844, indicating a strong relationship between the variables. Furthermore, an average per tonne oil spill cleanup cost using the IOPCF database was calculated by dividing the total amount paid by the Fund for cleanup by the total amount of oil that was spilled. According to our analysis, this value came to 1,639 USD (2009) per tonne.

3.1.2 *Total Cost (TC)*

Following the same methodology as in the previous step, a regression analysis of log (Total Cost) and log (Spill Size) was performed initially for N = 91 spills (for the period 1979–2006). The analysis of the studenized residuals revealed the existence of a total number of 8 possible outliers. These outliers were removed. After three consecutive regressions we arrived at the final dataset of N = 83 spills (see Fig. 4).

The minimum oil spill size was 0.1 tonnes and the maximum was 84,000 tonnes. The average spill was 4,854.29 tonnes, with a standard deviation of 16,064 tonnes and the median is just 140 tonnes. There are only 11 spills above 5,000 tonnes.

The equation of the fitted model using linear regression was:

$$\text{Total Cost} = 51{,}432\ V^{0.728} \qquad (2)$$

The R-Squared statistic indicated that the model as fitted explains 78.26% of the variability in LOG10(Cleanup Cost). The correlation coefficient (Pearson's correlation coefficient p) equals 0.8846, indicating a strong relationship between the variables. As before, an average per tonne oil spill total cost using the IOPCF database was calculated by dividing the total amount paid by the Fund by the total amount of oil that was spilled. According to our analysis, this value comes to 4,118 USD (2009) per tonne.

3.2 *Updated total cost formulas*

Note that in July 2007 a Correspondence Group (CG) was tasked to establish environmental risk evaluation criteria within FSA. At the latest meeting of the MEPC (MEPC 60, March 2010), a Working Group was formed after considerable

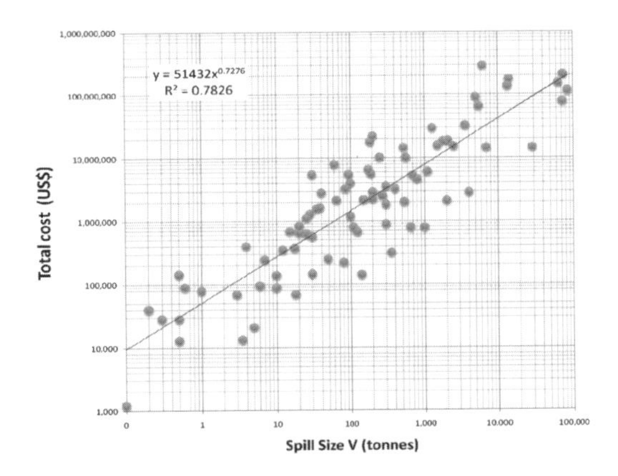

Figure 1. Regression of Log (Spill Size) and Log (Total Cost).

debate, among the three non-linear regressions proposed, the one proposed by the authors of this paper (see Eq. 2) was considered as more conservative and was proposed as a basis.

Following a submission of the United States to the IMO, we reported on recent analysis of oil spill cost data assembled by Oil Spill Liability Trust Fund (OSLTF), see document MEPC 61/18/2. Regression analysis of the cleanup cost is performed arriving at a non-linear function that is compared with the one produced by using data from the IOPC Fund. These analyses are very important since the United States is not a member of the IOPC Fund and, therefore, the IOPCF report contains no information regarding US spills.

The above dataset of US spills includes only the response cost. In order to arrive at the total cost of an oil spill a total cost to cleanup cost ratio can be used. In a submission of the US to the MEPC, see MEPC 61/INF.11, a value of $40,893.64 (in 2005 USD) was given as the 'best estimate' of the avoided volumetric response cost and $102,287.95 for the total avoidance cost. This total cost 'best estimate' is based on literature review. The 4 sources that were used are ICF Kaiser (1997), Brown & Savage (1996), Helton & Penn (1999) and Mercer Management Consulting (1992), which are rather dated papers. Although it is out of the scope of this work to comment on the literature, the 'best estimate' for the per ton response cost ($40,893.64) is based on the median spill size of 0.16 tonnes. The median, as well as, the average of ratios should be used with caution, see Psaraftis (2011). In our opinion, these statistics do not make too much sense. Nor does it make sense to extrapolate to large spills cost statistics derived from very small spills.

In any case, the total cost to response cost ratio for this 'best estimate' is 2.501. Therefore, based on this ratio a 'best estimate' total cost of each oil spill in the IOSLTF dataset can be estimated by multiplying the response cost by this figure. This is in line with Vanem et al. (2007a, 2007b) who taking into account the work of Jean-Hansen (2003), McCay et al. (2004) and Etkin (2004), concluded that a ratio of 1.5 should be assumed for the ratio of socioeconomic and environmental costs divided by cleanup costs. Thus, the total oil spill cost is 2.5 times the cost of cleanup, according to their analysis.

In addition compensation data from 17 spills in Norwegian water is available in a document submitted by Norway to IMO, see doc. MEPC 60/17/1. These are extremely costly cases but it was decided to keep them in our analysis.

A combined dataset of these 3 sources has been jointly developed by a group of researchers from Germany, Greece, Japan and the United States, after a specialized workshop hosted by Germanischer Lloyd in February 2011 and a series of discussions among the group. This paper presents some regression analyses, based on this combined dataset, that was performed by the authors and submitted by Greece to the IMO, see doc. MEPC 62/18/1. Greece's intent was to participate in a joint submission to MEPC 62, but this proved impossible due to lack of time in processing a joint submission.

Note that minor late adjustments in the combined database produced minor differences in some relevant results that were submitted to the IMO by Germany, Japan and the United States, see doc. MEPC 62/INF.24.

Going back to the consolidated dataset, we arrived at a dataset of 357 spill with a minimum oil spill size of 0.0003, average of 12.79 and a maximum of 1,747.55. The median is just 0.16 which means that this dataset consists of extremely small spills. Therefore the combined dataset consists of 476 spills. The min spill size is 0.0003 tonnes and the maximum 144,000. The median is 0.3 and the average 1,251 tonnes. Therefore the dataset consists of very small spills. The equation of the fitted model using linear regression was:

$$\text{Total Cost (2009 USD)} = 68{,}779\ V^{0.593} \qquad (3)$$

Given that the IOPC Fund contains spills greater than 0.1 tonnes, regression analysis for the spills greater that 0.1 tonnes was also performed. The final dataset consists of 343 spills, having a median of just 2.27 tonnes. Obviously there is a great influence of the small spills that dominate the US dataset. The regression formula derived from spills above 0.1 tonnes of the combined dataset is the following:

$$\text{Total Cost (2009 USD)} = 42{,}818\ V^{0.7294} \qquad (4)$$

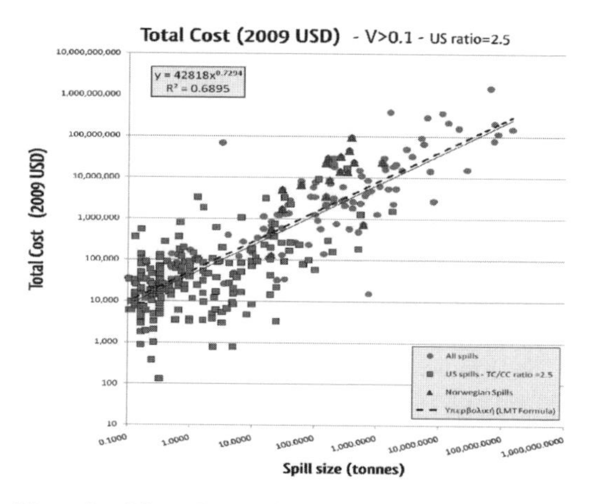

Figure 2. Linear Regression of Log (Spill Size) and Log (Total Cost)—US TC/RC ratio of 2.5.

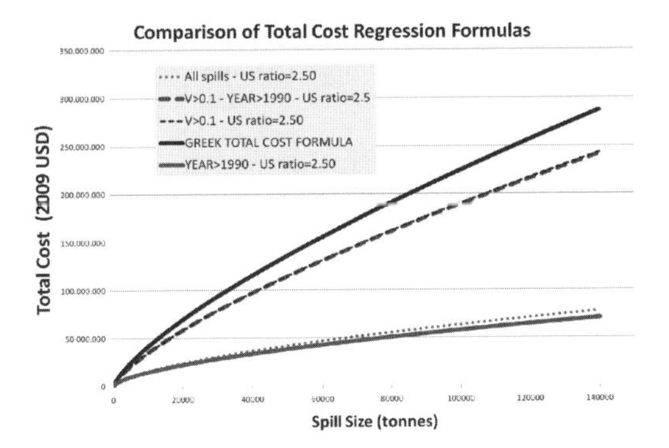

Figure 3. Comparison of various regressions.

Furthermore, various analyses were carried out. Regression formulas were derived for the whole dataset (Fig. 2) and subsets that contain spills greater that 0.1 tonnes (Fig. 3), spills that happened after 1990, and spills that are over 0.1 tonnes and happened after 1990. Although it is not the purpose of this paper to comment on these analysis the main conclusions are

a. based on Figure 2 it seems that the updated formula given by Eq. 3 (based on the combined dataset) underestimates the total cost in comparison to the original regression formula (eq. 2) for spills greater than 10 tonnes and
b. the original regression formula lies above all regression formulas derived from the combined dataset (at least for big spill) which means that it overestimates the total cost, see Figure 3.

4 USES OF THE COST FORMULA

4.1 *Formal safety assessment*

Formal Safety Assessment (FSA) was introduced by the International Maritime Organization (IMO) as *"a rational and systematic process for accessing the risk related to maritime safety and the protection of the marine environment and for evaluating the costs and benefits of IMO's options for reducing these risks"* (IMO, 2007). FSA aims at giving recommendations to relevant decision makers for safety improvements under the condition that the recommended measures reduce risk to the "desired level" and are cost-effective. FSA is, currently, the major risk assessment tool that is being used for policy-making within the International Maritime Organization (IMO), however, until now its main focus was on assessing the safety of human life. No environmental considerations have been incorporated thus far into FSA guidelines. Also note that FSA exhibits some limitations and deficiencies. The reader is referred to Kontovas (2005), Kontovas and Psaraftis (2006a, 2006b and 2009), Kontovas et al (2007a,b), Zachariadis et al (2007) and Giannakopoulos et al (2007) for a discussion on these issues.

The fourth Step of a Formal Safety Assessment is to perform a Cost-Benefit Analysis (CBA) so as to pick which risk control options (RCOs) are most cost effective. According to the FSA guidelines, one stage of this Step is to *"estimate and compare the cost effectiveness of each option, in terms of the cost per unit risk reduction by dividing the net cost by the risk reduction achieved as a result of implementing the option"*.

Up to now in most FSA studies cost effectiveness is assessed by using the so-called Cost Effectiveness Analysis (CEA) and not Cost Benefit Analysis(CBA) both of which will be briefly discussed below. CEA may be considered to be particular form of CBA, where the benefits are usually not monetized, and therefore, net benefits cannot be calculated, see Mishan and Quah(2007) and Krupnick (2004).

Usually, in CEA, one calculates the costs per unit of an effectiveness measure (such as lives saved). Therefore, while CEA cannot help in determining whether a policy increases social welfare, it can help in the choice of policy that achieves the specified goal with the smallest loss in social well-being and can help rank alternative policies according to their cost-effectiveness (Krupnick, 2004).

In theory, the analytical tool of Cost Effectiveness Analysis is the Incremental Cost-Effectiveness Ratio (ICER), also called marginal cost-effectiveness ratio, given by the difference in costs between two actions divided by the difference in outcomes between these two, with the comparison typically being between an action that is proposed to be implemented and the current status.

Currently only one such index is being extensively used in FSA applications. This is the so-called "Cost of Averting a Fatality" (CAF) and is expressed in two forms: Gross and Net. In a similar way, Skjong et al. (2005) and Vanem et al (2007a, 2007b) presented an environmental criterion equivalent to CAF. This is nothing new, but an incremental cost effectiveness ratio to assess the case of accidental releases of oil to the marine environment that measures risk reduction in terms of the number of tonnes of oil averted. This criterion was named CATS (for "Cost to Avert one Tonne of Spilled oil") and its suggested threshold value was 60,000 USD/tonne. According to the CATS criterion, a specific Risk Control Option (RCO) for reducing environmental risk should be recommended for adoption if the value of CATS associated with it (defined as the ratio of the expected cost of implementing this RCO divided by the expected oil spill volume averted by it) is below the specified threshold, otherwise that particular RCO should not be recommended.

Kontovas and Psaraftis (2006) were probably the first to question the SAFEDOR approach, both on the use of any single dollar per tonne figure and on the 60,000 dollar threshold. This paper was the core of a submission by Greece on this issue (doc. MEPC 56/18/1). This submission opened a big chapter concerning environmental risk evaluation criteria and its uses within FSA. At the 55th session of Marine Environment Protection Committee (MEPC) that took place in 2006, the IMO decided to act on the subject of environmental criteria. An issue of primary importance was found to be the relationship between spill volume and spill cost. Given that the cost of oil spill depends on the volume of the spill it is difficult to incorporate the regression formulas within CEA. Besides, most Risk Control Options have multiple effects (for example both in safety and the environment) and CBA should be preferred as it can combine multiple effects. The reader is referred to Kontovas and Psaraftis (2011) for a discussion in this issue.

4.2 *Incorporating a non linear function within FSA*

First of all, assume that the spill cost function is given by Eq. 2; the formula produced after regression analysis of IOPCF data which is as follows (Kontovas et al., 2010):

$$\text{Cost}(V) = 51{,}432V^{0.728} \text{ (in USD, if V is in tonnes)}$$

The use of this particular function causes no loss of generality, as any other function of volume can be tried. The updated formula (Eq. 4) could also be used. RCO evaluation by comparing the benefits (derived by using a function) and the costs is, in theory, presented in Psaraftis (2008) and Kontovas et al. (2010). Yamada (2009), Hammann and Loer (2010) and Yamada and Kaneko (2010) presented a way to incorporate a non-linear cost function within FSA. The latter paper forms the basis of a relevant submission to the IMO, see doc. MEPC 59/17/1 that was submitted by Japan. In reality thing are a little bit more complicated as it will be shown in the following example.

In most FSA studies an event tree is presented. For each sequence of the event tree the expected number of tonnes that will be averted is calculated as the product of the frequency of the event (Pi) and the average consequences (V_i) and is presented as E[V]. This is the so-called Potential Loss of Cargo (PLC) value for each sequence. This value should then be multiplied with the per tonne cost (which is a function of the spill volume) to estimate the risk (denoted as $E[C_i]$) and by summing all the relevant sequences the total risk may be obtained. Another equivalent way to estimate the expected benefit of averting an oil spill by using the cost function (Cost(V) is to multiplying the probability Pi with Cost (V_i). These two ways lead to equivalent results. What is important to stress out is that the expected cost should be estimated before the implementation of the RCO and after it.

According to Yamada and Kaneko (2010), an RCO can be regarded as cost-effective if the following formula is satisfied

$$\Delta B - \Delta S > 0$$

where ΔB is the benefit by implementing the RCO which is the risk reduction (in monetary units) and ΔS is the cost of implementing the RCO. ΔB is the expected cost of an oil spill before the implementation of the RCO ($E[C_{org}]$) and after ($E[C_{new}]$).

Therefore, the criterion becomes

$$\Delta S < \Delta B = E[C_{org}] - E[C_{new}]$$

In practice, the discounted costs and benefits should be compared. Furthermore, in most FSA studies submitted to the IMO the event trees have not been recalculated. In most cases, the RCO was assumed to erase the risk (which means $E[C_{new}]$ equal to zero) or a risk reduction as a percent of the initial risk was estimated by expert judgment. Taking these remarks into account, applying the criterion is straightforward.

4.3 *Probabilistic oil spill risk analysis: Environmental performance of tankers*

In addition the Cost Benefit Analysis presented above as a part of a Formal Safety Assessment (FSA) could also be used to evaluate the environmental performance of alternative tanker designs.

In general, to model the risk of an individual tanker spill, it may be argued that one has to:

1. Determine the probability of an accident;
2. Determine the oil outflow volume given the accident as a probability distribution;
3. Determine the spill consequence given the outflow volume by using the non linear formula.

This methodology has been applied in many studies performing oil spill risk analysis. For example, see Montewka (2009) and Montewka et al. (2010) who present the risk of collision and grounding as a random variable and uses the risk assessment process that is illustrated in the following diagram.

According to their analysis, the risk that tankers colliding or grounding pose to the environment can

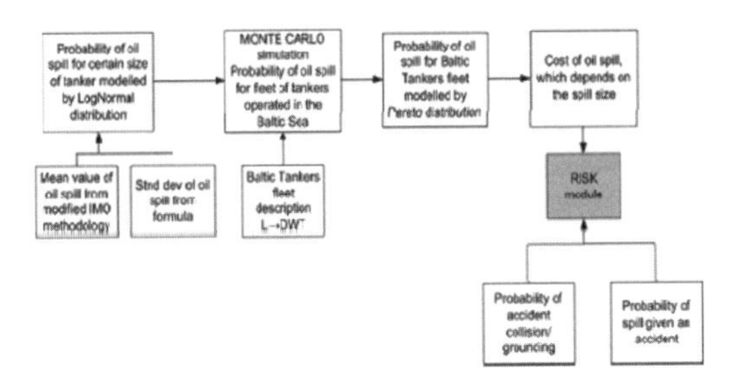

Figure 4. Block diagram of risk assessment process applied in presented study [Montewka, 2010].

be calculated using the general formula, separately for collision and grounding:

$$R = P_A \cdot P_{OS|A} \cdot P_{OS} \cdot C$$

where *PA* means a probability of an accident (collision or grounding), *POS|A* means a probability of an oil spill given an accident, *POS* denotes a probability density function of an oil spill volume, *C* stands for consequences of an accident, which refers to an oil spill clean up costs and used the non linear function derived by Yamada (2009).

Furthermore, probabilistic oil outflow models are being used in risk based optimization of crude oil carriers with respect to loss of cargo. These are in line with the IMO regulations regarding the probabilistic oil outflow for bunker tanks (applied to all spills) and cargo tanks regarding oil carriers. Indeed, MEPC has adopted a revised MARPOL Annex I/23 and 24 applicable to all new oil tankers to provide adequate protection against oil pollution in the event of grounding or collision, see IMO (2006a, 2006b, 2006c).

Regulation 23 applies to new oil tankers, which means all tankers delivered on or after 1 January 2010. The Probability density functions have been determined for the likelihood of damage being encountered at different points in the length of the ship for both side and bottom damage. An assessment is then made of the expected oil outflow from each damaged tank or group of tanks including tidal effects and accounting for any retained oil.

The mean oil outflow parameter is calculated independently for side damage and bottom damage and then combined in non dimensional value as follows:

$$O_M = (0.4 O_{MS} + 0.6 O_{MB})/C,$$

where O_{MS} and O_{MB} are the mean outflows for the side damage and bottom damage respectively and C is the total volume of cargo oil in m^3 for a 98% full tank. Regulation 23 is presented into detail in Annex I.

Thus far, research on risk based ship design has mainly focused on parametric optimization in order to reduce oil-outflow probability and increase cargo carrying capacity (Papanikolaou et al., 2010). However, the economic damage of accidental oil outflow can be estimated by using a non linear function as presented above. That way, alternative designs could be judged for their environmental performance (Sirkar et al., 1997). Ventikos & Swtiralis (2011) present a probabilistic formulation of regulation 23 of MAPPOL to calculate distribution and quantities of oil outflow for all major oil tanker categories and examine numerous cargo tank configurations for tankers by simulating multiple outflow scenarios for the tanker fleet. In addition, they perform an assessment of the cost of these potential oil spills by using some of the formulas discussed above.

5 SUMMARY AND CONCLUSIONS

It is important to comment on the limitations of the compensation funds. First of all, it should point out that costs reported to the public are not 'real' oil spill costs. They refer to the amount of money that was agreed to compensate the claimants. Although the compensation figures are real and cannot be disputed, a question is if compensation figures can be taken to reasonably approximate real spill costs, or, failing that, if they can be used as realistic 'surrogates' of these costs.

Estimates of damages calculated by applying economic valuation methodologies, claims for compensation and the compensation paid to claimants can never be equal (Thébaud et al, 2005). It is further noted that admissible claims cannot be paid in full, especially in the case of large spills, since there exists a limit in total compensation that can be paid. On the other hand, if there are any actual costs that are paid to victims of oil pollution, this is probably as good a source to document such costs as anyone. Plus, this analysis can be amended with additional data, to the extent such data becomes available.

Results of regression analyses that attempted to connect oil spill cleanup cost and oil spill total cost to oil spill size have been reported in various works, most of which were based on data from the International Oil Spill Compensation Fund (IOPCF). A lot of work has been done within the Marine Environment Protection Committee (MEPC) of the International Maritime Organization. In fact, a previous work of the same authors that has recently been adopted by the IMO/MEPC as a basis for further discussion on environmental risk evaluation criteria. The current work has also been submitted by Greece to the IMO and will be discussed in the forthcoming section of MEPC (MEPC 62nd, July 2011).

These kind of analyses and their results can provide useful insights to the discussion on environmental risk evaluation criteria and in policy evaluation regarding the oil pollution of the marine environment by estimating the damage cost of oil spills and, thus, the benefit from relative regulations that deal with the prevention of oil pollution in cost effectiveness and cost benefit analysis. To that extend, these functions can be used within the Formal Safety Assessment (FSA) framework and Probabilistic Oil Spill Risk Analysis as Section 4.

ACKNOWLEDGMENT

The authors would like to thank the experts from Germany, Japan and the United States that were present at the GL meeting for a fruitful discussion and especially Peter Securius and Rainer Hamann of Germanischer Lloyd for making the updated IOPCF dataset available.

REFERENCES

Etkin, D.S. 1999. Estimating Cleanup Costs for Oil Spills, *Proc. International Oil Spill Conference, American Petroleum Institute, Washington, DC.*

Friis-Hansen P. & Ditlevsen, O., 2003. Nature preservation acceptance model applied to tanker oil spill simulations, *Journal of Structural Safety*, 25(1): 1–34.

Giannakopoulos Y., D. Bouros, N.P. Ventikos. 2007. Safety at Risk?, *Proc. of the International Symposium on Maritime Safety, Security and Environmental Protection (SSE07. Athens, Greece, September 2007.*

Grey, C., 1999. The Cost of Oil Spills from Tankers: An Analysis of IOPC Fund Incidents, *Proc. The International Oil Spill Conference 1999, 7–12 March 1999, Seattle, USA.* ITOPF, London.

Hamann, R. & Loer, K. 2010. Risk-based Optimisation of Crude Oil Tanker Cargo Holds, *Proc. The 3rd International Symposium on Ship Operations, Management & Economics, Athens, Greece 7–8 October, 2010.*

Helton, D. & Penn, T.,1999. Putting response and natural resource damage costs in perspective. *Proc. of the 1999 international oil spill conference, 1999.*

Hendricksx, R.2007. Maritime Oil Pollution: Empirical Analysis. In Faure, M. and Verheij, A. (Eds.) *Shifts in Compensation for Environmental Damage*, Springer Verlag.

ICF Kaiser. 1997. *The Economic Impacts of Accidents on the Marine Industry. Washington: U.S. Coast Guard.*

IOPCF. 2009. *Annual report 2008.* International Oil Pollution Compensation Funds, London, UK.

ITOPF. 2010. *Oil Tanker Spill Statistics: 2010*, The International Tanker Owners Pollution Federation. London; UK.

IMO. 2006a. MARPOL 12A—*Res. MEPC 141(5)*,4 Mar 2006.

IMO. 2006b. *Resolution MEPC 122*, October 2006, MARPOL—Explanatory notes on Reg 23.

IMO. 2006c. *Resolution MEPC 117(5)*, 2. October 2006, MARPOL Revised Annex I.

IMO. 2007. *Formal Safety Assessment: Consolidated text of the Guidelines for Formal Safety Assessment (FSA) for Use in the IMO Rule-Making Process.* MSC/Circ.1023–MEPC/Circ.392. London (MSC 83/INF.2.

Jacobsson, M. 2007. The International Oil Pollution Compensation Funds and the International Regime of Compensation for Oil Pollution Damage. In: Basedow J, Magnus U (eds) *Pollution of the sea—prevention and compensation*. Berlin: Springer.

Jean-Hansen, V. 2003. *Skipstrafikken i området Lofoten—Barentshavet*, Kystverket, Transportøkonomisk institutt, 644/2003, 2003 (in Norwegian). ISBN:82-480-0341-8.

Kontovas, C.A. 2005. *Formal Safety Assessment: Critical Review and Future Role.* Diploma Thesis supervised by H.N. Psaraftis. National Technical University of Athens, July 2005.

Kontovas, C.A. &. Psaraftis, H.N. 2006. Assessing Environmental Risk: Is a single figure realistic as an estimate for the cost of averting one tonne of spilled oil?, Working Paper NTUA-MT-06-101, National Technical University of Athens, February (available at www.martrans.org).

Kontovas, C.A., Psaraftis, H.N. & P. Zachariadis 2007a. The Two C's of the Risk Based Approach to Goal-Based Standards: Challenges and Caveats, *International Symposium on Maritime Safety, Security and Environmental Protection (SSE07. Athens, Greece, September.*

Kontovas, C.A., Psaraftis, H.N. & P. Zachariadis 2007b. Improvements in FSA Necessary for Risk-Based GBS, Proc. *PRADS 2007 Conference, Houston, USA, October.*

Kontovas, C.A. & Psaraftis, H.N. 2009. Formal Safety Assessment: A Critical Review, *Marine Technology* 46(1): 45–59.

Kontovas, C.A. &. Psaraftis, H.N. 2008. Marine Environment Risk Assessment: A Survey on the Disutility Cost of Oil Spills, *Proc. 2nd International Symposium on Ship Operations, Management and Economics, Athens, Greece.*

Kontovas, C.A., Psaraftis, H.N. & Ventikos N. 2010. An Empirical Analysis of IOPCF Oil Spill Cost Data, *Marine Pollution Bulletin*, 60(9):1455–1466.

Kontovas & Psaraftis 2011 A primer on the use of non linear disutility cost function within Maritime Risk Assessment, under preparation.

Krupnick. 2004. Valuing Health Outcomes: Policy Choices and Technical Issues, Resources for the Future (RFF) report, Executive Summary.

Liu W., Wirtz, K.W, Kannen, A. & Kraft, A., 2009. Willingness to pay among households to prevent coastal resources from polluting by oil spills: A pilot survey. *Marine Pollution Bulletin*. 58(10):1514–1521.

Mercer Management Consulting. (1992). *OPA-90: Regulatory Impact Analysis Review—Spill Unit Values.* Lexington: Mercer Management Consulting.

McCay, D.F., Rowe, J.J., Whittier, N., Sankaranarayanan, S. & Etkin, D.S., 2004. Estimation of potential impacts and natural resource damages of oil, *Journal of Hazardous Materials.* 107:11–25.

Psaraftis, H.N., 2008. Environmental Risk Evaluation Criteria, *WMU Journal of Maritime Affairs*, 7(2):409–427.

Psarros, G., Skjong, R. & Vanem, E. 2011. Risk acceptance criterion for tanker oil spill risk reduction measures. *Marine Pollution Bulletin* 62(1).

Sirkar, J., Ameer, P., Brown, A., Goss, P., Michel, K., Frank, F. & Wayne, W. 1979. A Framework for Assessing the Environmental Performance of Tankers in Accidental Groundings and Collisions, *SNAME Transactions* 105.

Skjong, R., Vanem, E. & Endresen, Ø. 2005. Risk Evaluation CriteriaSAFEDOR-D-4.5.2-2007-10-24-DNV-RiskEvaluationCriteria-rev-3.0, [Online] www.safedor.org.

Thébaud, O., D. Bailly, J. Hay, & J.A. Pérez Agundez, 2005. The cost of oil pollution at sea: an analysis of the process of damage valuation and compensation following oil spills. in *Economic, Social and Environmental Effects of the Prestige Oil Spill de Compostella*, Santiago. 2005. p. 187–219.

Vanem, E., Endresen, Ø. & Skjong, R., 2007a. Cost effectiveness criteria for marine oil spill preventive measures, *Reliability Engineering and System Safety*, doi:10.1016/j.ress.2007.07.008.

Vanem, E., Endresen, Ø. & Skjong, R., 2007b. CATS—Cost-effectiveness in Designing for Oil Spill Prevention, *Proc. PRADS 2007 Conference, Houston, USA, October.*

Ventikos, N.P., Hatzinikolaou S.D. & Zagoralos, G., 2009. The cost of oil spill response in Greece: analysis and results. *Proc. Int. Maritime Association of Mediterranean. 12–15 October. Istanbul. Tukey.*

Ventikos N.P. & Swtiralis P. 2011. Probabilistic Oil Outflow: The Tanker Fleet in the context of Risk Analysis. *Proc. European Conference on Shipping & Ports 2011, Chios, Greece, June 22–24.*

White, I.C. & Molloy, F. 2003. Factors that Determine the Cost of Oil Spills. *Proc. International Oil Spill Conference 2003, Vancouver, Canada, 6–11 April.*

Yamada, Y. 2009. The Cost of Oil Spills from Tankers in Relation to Weight of Spilled Oil. *Marine Technology* 46(4):219–228.

Yamada, Y. & Kaneko, F. 2010. On the Derivation of $CATS_{thr}$ within the Framework of IMO environmental FSA studies. *Proc. 5th Intern. Conference on Collision and Grounding of Ships, June 14th–16th 2010, Espoo, Finland.*

Sustainable Maritime Transportation and Exploitation of Sea Resources – Rizzuto & Guedes Soares (eds)
© 2012 Taylor & Francis Group, London, ISBN 978-0-415-62081-9

Sustainable maritime transport: An operational definition

S.D. Chatzinikolaou & N.P. Ventikos
Laboratory for Maritime Transport, School of Naval Architecture & Marine Engineering, National Technical University of Athens (NTUA), Athens, Greece

ABSTRACT: The sustainability concept is currently suffering from low credibility as a result of the plethora of definitions and uses which have been launched from a variety of agencies around the globe. The maritime sector is not an exception to this, since there are numerous initiatives within this sector claiming to have a sustainability orientation; however they are diverse in terms of interpretation and implementation of the sustainability principals. Within the maritime industry, there is also a trend of narrow defining sustainability, by focusing for example only on energy consumption and air pollution issues. This narrow definition tends to favour the introduction of particular solutions (mainly technological) that may reach one sustainability goal but simultaneously may contradict with another goal. This paper aims to contribute in the discussion for re-defining in an operational manner the sustainability concept of the maritime transport sector by following the initial notion of this concept.

1 INTRODUCTION

Sustainability has been already adopted either as international or national policy principle (UN, EU, many countries, etc), but also as a key notion for business, industrial, scientific and many other initiatives around the globe. With respect to the maritime transport sector, the EU central policy for the future of this sector is based on values of sustainable development, such as economic development and open markets in fair competition as well as high environmental and social standards (EC, 2009).

Many people consider maritime transport an environmentally sound practice mainly due to the low energy consumption and air emissions per amount of transport work compared to other transport modes. It is evident that within the maritime sector there are several efforts made in policy, technology and research level to reduce the environmental impacts of shipping and to achieve certain sustainability goals.

Yet, in absolute terms, the pollutant emissions from shipping are significant and keep rising, while the emissions from land-based sources are gradually decreasing (T&E Federation, 2010). This is somewhat explained by the massive growth that the maritime transport sector has experienced the previous years, supporting the demand for the international movement of goods and the globalisation of world economy. Only during the last two decades, the international seaborne trade has over-doubled (from 2.253 to 4.742 billion tons) and currently accounts for nearly 90 percent of world trade (UNCTAD, 2010).

Despite its international nature and enormous growth, the maritime transport sector has been extremely slow in achieving global agreements for the reduction of ships emissions and has, so far, "managed" to be left out of the Kyoto Protocol. Hence, a great amount of criticism towards this industry targets its social agenda (i.e. working conditions, safety, ship dismantling practices, etc) as well as certain mechanisms established within the industry that artificially keep the international costs of maritime shipping low at the expense of environmental and labour concerns (McGuire, 2011).

As has been assured for other transportation modes, for sustainability to be successfully implemented it is essential that its concept is adequately understood, quantified and applied (Zietsman, 2000). Thus, the aim of this paper is to contribute to the discussion for defining and assessing the sustainability concept within the maritime sector by following the initial notion of this concept. The structure of the paper is as follows. First a brief report of sustainability issues within shipping is presented. Then, sustainability and transportation sustainability literature is explored in order to provide an operational definition for maritime transport sustainability. Indicators from the broader transport area are reviewed and methods for assessing sustainability in shipping are presented.

2 SUSTAINABILITY IN MARITIME TRANSPORT

In the effort to define the maritime transport sustainability it would be helpful first to identify some of the most significant environmental challenges the sector faces and illustrative unsustainable shipping practises as well.

The main environmental issues of the maritime transport sector currently are the reduction of air emissions from international shipping and energy efficiency solutions form technical and operational perspective.

The contribution of the shipping sector to gases and particles that impact the Earth's climate has only recently begun to be fully understood. In 2007, shipping was responsible for approximately 3.3 percent (over 1 billion tonnes) of global CO2 emissions (Buhaug et al, 2009). In the absence of emission reduction policies, emission scenarios predict a doubling to tripling of 2007 emission levels by 2050 (IMO, 2010).

Emissions from commercial shipping vessels contribute significantly to perturbations in air quality, visibility and climate. The link between Particulate Matter emissions and health effects was recently assessed for global shipping emissions when it was estimated that up to 60,000 premature deaths result annually (Corbett et al, 2007), (Eyring et al. 2007). The primary reason for the negative effect of shipping emissions to health is that 70 percent of shipping occurs within 400 km of land (Corbett et al., 2007), (Wang et al., 2008) and major shipping ports are located in areas surrounded by large populations.

Since the introduction of new fuels in shipping is emerging slowly, the energy efficiency concepts are getting considerable acceptance. From operational point of view energy efficiency refer to speed reductions (slow steaming) an option which also reduces air emissions; nevertheless other issues may arise from speed reductions (i.e. ship out of optimal condition), (Faber et al. 2010). The main driver for slow steaming remains the market mechanisms since currently the global shipping industry is facing an oversupply of ships (Platou, 2010) and the considerable savings in bunker fuels money offered by reduced speeds.

Ocean-going vessels are mainly subject to oversight by the International Maritime Organization (IMO). IMO efforts to mitigate environmental impacts of emissions and wastes from global shipping try to keep pace with the growth of the industry and the evolution of emission and waste control technologies. It is evident that the shipping industry is taking quick steps for introducing technologies for emissions control (e.g. scrubbers, catalysts technologies, quality fuels) energy efficiency, ballast water and waste handling (e.g. ballast water treatment systems, port reception facilities).

Enforcing new international agreements is however complicated by the complex relationships that exist between those nations to which most ships are registered and the large shipping interests (typically headquartered in other nations) that own most of the ships. The above system of registration has been criticised for being fundamentally unsustainable since it enables shipping to avoid internalising its true environmental and social costs in market transactions (McGuire et al, 2011).

An illustrative example of the above inefficiency is the current ship dismantling industry in S. Asia countries, which was settled due to the need of steel and recycled materials and the existence of poverty and availability of cheap (and also childhood) labour. Methods applied in these countries, measured against standards or general norms expected within the industrialised countries generally fail to comply with any environmental, safety or health standards in almost all respects. However, nearly 80 percent of the global volume of end of life ships is still heading to these countries for "recycling" providing this way an extra profit making opportunity to ship-owners (Ventikos et al, 2008).

Sustainability aspects are related also to Ports which tend to increase in size following the expansion of world trade and introduction of larger ships. From economic perspective this makes them attractive as economic growth poles and may provide some social opportunities (e.g. job availability, accessibility). At the same time, it poses environmental and social challenges since it becomes more difficult for port activities to integrate into urban environments. This is particularly due to the increased noise air and optical pollution and annoying security standards of modern ports.

Other environmental issues within shipping are oil pollution and handling of garbage, wastes and antifouling paints. The maritime industry has in general succeeded in reducing the oil pollution over the last years (ITOPF, 2011), especially with respect to the accidental pollution. Still, evidence shows that there is an increasing tendency in operational (illicit) discharges in regional areas such the East Mediterranean (Topouzelis et al. 2007).

Garbage, and solid and non oil liquid wastes produced by the daily operations of ships (some of them are not yet regulated) may be a negligible environmental issue in a global level but there are specific regional sea areas (e.g. Caribbean Sea, Alaska) facing major problems due to the large quantities of wastes produced by certain ship types (Chatzinikolaou et al, 2007).

3 SUSTAINABILITY—DEFINITIONS

3.1 *Sustainability & sustainable development*

Dealing with sustainability has become a fashion in recent years. Today, it would be very difficult for one to find a research project, conference, policy action or other initiative within the broader transport sector that, in one way or another, did not include the term sustainability or sustainable development (Zegras, 2006). While the general idea of this concept is sound, since it emphasizes the integrated nature of the impact of human activities (Litman, 2007), it also establishes links with many issues of concern (i.e. poverty, environmental quality, safety and security, social equity, economic development and so on) which has attracted the interest of many people with diverse backgrounds and objectives.

Currently, there is no globally accepted definition for sustainability or sustainable development (Beatley, 1995), (Jeon, 2005). The most well-known definition for sustainable development is the one introduced by the World Commission on Environment and Development, (WCED, 1987), in the so called Brundtland Report which has set the original notion for this concept: "Sustainable development is development that meets the needs of the present without compromising the ability of future generations to meet their own needs".

In a rigorous interpretation, sustainability or sustainable development can be seen as strictly scientific construct related to, for example, carrying capacities, ecosystem functioning and biological processes (Zegras, 2005). However, the initial Brundtland definition has offered plenty of room for various interpretations of the concept in the years followed. Many of these interpretations have extended the concept to include mainly institutional and political dimensions or various aspects of life and life systems. Inevitably, the concept of sustainability has become to mean different things to different people. These have made the terms of sustainability and sustainable development, arbitrary and user defined (Keiner, 2004). As a result, presently there are many constituencies which perceive the term 'sustainable development' as a vehicle to continue many and varied corporate and institutional interests whilst giving the impression of devotion in environmentally-sound principles (Johnston et al, 2007).

Although the establishment of a standard framework in which sustainable development is considered is still missing, there seems to be a consensus that sustainable development should be made uniformly on at least three fronts or pillars: economy, society, and environment. A fashionable way of expressing these three pillars is known as People, Planet, Prosperity (or PPP or P3), where People represent the social pillar, Planet the environmental pillar, and Prosperity the economic pillar. Prosperity has replaced the term Profit (decision made at the World Summit on Sustainable Development in Johannesburg, 2002), to reflect that the economic dimension covers more than the company profit. Other well-known terms are the Triple Bottom Line (TBL) and the UN's Global Compact.

Many people accept that sustainability and sustainable development have the same meaning. However, a rational distinction between the two definitions states that sustainability is a condition in which economic, social and environmental factors are already optimized, taking into account indirect and long-term impacts, whereas sustainable development is a progress toward this condition of sustainability (Litman, 2010).

A well established approach to define the concept of sustainability is the economists' approach which distinguishes the concept into Weak Sustainability, (WS) and Strong Sustainability (SS), subject to the way that humans chose to utilise the natural capital (i.e. the range of functions the natural environment provides for humans and for itself), (Ekins et al., 2003). Definitions of these two terms are provided below.

- The SS regards natural capital as providing some functions that are not substitutable by man-made (produced) capital and should be maintained.
- The WS regards that man-made (produced) capital of equal value can take the place of natural capital.

A general delineation between the SS and the WS as has been adopted from is presented in the following Table 1.

Table 1. Strong vs. weak sustainability.

Strong Sustainability (SS)	Weak Sustainability (WS)
Natural and man-made capital are complements	Natural capital and man-made capital are substitutes
All forms of capital should be kept intact	Only total capital stock should be kept intact
Not only an economic problem, but also a problem of maintaining non-replaceable recourses	Environment problems may always be treated as economic problems
Accepts precautionary principals & safe minimum standards	Accepts monetary valuation & cost-benefit approach

3.2 *Transportation sustainability*

There is an extended scientific literature available on sustainable transportation (the majority of which refers to the urban and road transportation) and many definitions of this concept may be explored. Most of these definitions answer to the question what essentially is a sustainable transport system (mainly policy oriented definitions) and not the question how to make the system sustainable. To answer the later question for the sustainable transport systems an operational definition is required. Defining a concept in an operational manner is an important prerequisite before trying to measure this concept (Meier, 2002). Therefore, the focus here is to explore the existing operational definitions of sustainability or transportation sustainability in order to come up with an operational definition for the maritime transport sustainability.

The literature review illustrates that many of the available definitions of transportation systems sustainability capture attributes of system effectiveness, and system impacts on the economy, environment, and social quality of life (Jeon, 2005). However, there seems to be a higher focus in addressing the effectiveness of the system as well as some of the resulting environmental impacts (mainly air quality impacts), and less of a focus on economic and social impacts.

The principle of 'eliminating our contribution' has been proposed in the effort to avoid the above weakness and provide sustainability an operational definition (Johnston et al. 2007). According to this approach, operational sustainability principles should develop with the aim to eliminate the human contribution to...:

1. ... systematic increases in concentrations of substances from the Earth's crust.
2. ... systematic increases in concentrations of substances produced by society.
3. ... systematic physical degradation of nature.
4. ... conditions that systematically undermine people's capacity to meet their needs.

This operational approach to sustainability has become known as 'The Natural Step Framework' after the organization promoting it (TNS), (Robèrt et al, 2002).

From the international organisation perspective, for example the World Bank has taken an economic oriented focus by emphasizing the efficient use of resources in the following three dimensions:

a. Economic & financial;
b. Environmental & ecological; and
c. Social.

In contrast, an operational definition of sustainable transport which makes focus more on the environmental dimension of sustainable transportation has been proposed by the Organization for Economic Cooperation and Development, (OECD) (OECD, 1996). This definition states that: An environmentally sustainable transport system is one that does not endanger public health or ecosystems and meets needs for access consistent with (a) Use of renewable resources at below their rates of regeneration, and (b) use of non-renewable resources at below the rates of development of renewable substitutes. The OECD sustainable transport system approach is based on the World Health Organisation, (WHO), guidelines for air pollution, noise levels, acidification and eutrophication as well as climate change and ozone depletion.

A well-known organisational perspective comes from the Centre for Sustainable Transportation of Canada, (CST), which has introduced the so called comprehensive sustainable transportation. The CST definition has been given official status since the EU description for sustainable transportation was taken almost word to word from it. These two similar statements constitute by far the most widely accepted definitions of sustainable transportation (Hall, 2002).

In the EU definition a sustainable transport system is defined as one that:

- allows the basic access and development needs of individuals, companies and societies to be met safely and in a manner consistent with human and ecosystem health, and promotes equity within and between successive generations;
- is affordable, operates fairly and efficiently, offers choice of transport mode, and supports a competitive economy, as well as balanced regional development;
- limits emissions and waste within the planet's ability to absorb them, uses renewable resources at or below their rates of generation, and, uses non renewable resources at or below the rates of development of renewable substitutes while minimising the impact on the use of land and the generation of noise.

All previous definitions of transport sustainability are similar in that they involve the following basic principles:

1. Accessibility;
2. Acceptance of resource constrain (natural, social, economic); and
3. Equity.

Accessibility is essentially the ability to obtain desired goods, services and activities and it will be further discussed in the following sections. The principal of equity essentially reflects the interaction between the other two principals particularly in the sense of inter-generational equity. In addition, equity also refers to a balanced distribution of transport benefits (reflected by access) and costs (reflected by various resources constrains) within the current generation (Zegras, 2005).

Closing this review, it is noted that none of the above efforts provides a comprehensive operational definition of sustainable transport since their focus is more on the description of a sustainable transport condition rather than the course to get to this condition.

4 PROPOSED DEFINITION FOR SUSTAINABLE MARITIME TRANSPORT

Entering into the maritime sector some reasonable questions when trying to define sustainability may emerge. The first of these is whether there is any real value in trying to define and subsequently measuring the maritime transport sustainability. Other questions may refer to the scale of the experiment or time and geographical constraints. For example, whose sustainability do we wish to measure; the whole sector, a country sector, specific ship type, or of just a ship.

The answer to the first question derives from the fact that sustainability initiatives are already a reality in central policy, industry, research and other levels and contributing to the process is at least useful. Hence, the necessity also comes from the fact that the environmental and social standards as well as the economic/financial practices within the shipping industry leave much to be desired even though some positive examples are already in place. The next questions call first for a robust operational definition of maritime transport sustainability which will then guide us to a well functioning framework for studying it.

As has been depicted in the previous sections, when trying to getting closer to an operational definition, one must discriminate between some different approaches.

Considering the economic approach to the problem of transport sustainability, the key choice is whether the society believes that natural capital should be attributed with special protection, or whether it can be substituted by other forms of capital, especially man-made (produced) capital, or in other words whether people are willing to trade natural goods for money. This is the choice between weak sustainability and strong sustainability, (Dietz et al., 2007), (Kosz, 1998). The operational definition presented here is based on the respective definition from urban transport which was first drafted by Zegras (2005).

The operational definition of sustainable maritime transport introduced by the authors of this paper is as follows:

Sustainable Maritime Transport

Maritime transport is sustainable when it has the capability to maintain non-declining and efficient accessibility in time.

The accessibility (or just access) refers to the ease of reaching goods, services, activities and destinations, which together are called opportunities (Litman, 2011b). Accessibility may be seen also in terms of potentials (opportunities that may be reached) or in terms of activity (opportunities that are reached).

By maintaining non-declining accessibility the human capital is increased because more opportunities are provided. The definition therefore, appreciates this way the fundamental role of (maritime) transport in human development.

Simultaneously, while increasing accessibility and human capital other capital stocks such as the natural capital (e.g. fuel consumption) and man-made capital decrease (e.g. land use). Also, the initial notion of sustainable development demands that the welfare provided to the current generation by accessibility does not compromise the welfare of future generations. To overcome these obstacles the concept of the strong sustainability is used. Dally (2002) introduced the concept of throughput as more useful and measurable compared to utility when we are talking about sustainability. He defined throughput as the entropic physical flow from nature's sources through the economy and back to nature's sinks. In his opinion the throughput has to be sustained. This equals to strong sustainability (intact natural capital).

Adopting the definition of Dally (2002) that sustainable development "might more fruitfully defined as more utility per unit of throughput", Zegras (2005), has defined the sustainable transportation as: "more utility, as measured by accessibility, per unit of throughput, as measured by mobility".

It is noted that the above by no means involves the reduction of mobility (i.e. movement of people and goods).

What is proposed is that the target of a sustainable maritime system should be: less mobility consumption per accessibility derived or in other words, efficient accessibility.

The basic features of the definition for sustainable maritime transport that is being proposed here are the following:

1. It integrates the basic principles of sustainable development (intergenerational equity, continuance of development)
2. It is simple
3. It is operational (can be measured)
4. It may be applied in different scales within the maritime transport system (i.e. product scale/ship, fleet, sector, etc)

5 FRAMEWORKS FOR ASSESING MARITIME TRANSPORT SUSTAINABILTIY

5.1 *How to measure maritime transport sustainability*

The existing indicator systems reveal that operationally, transport sustainability is largely being measured by system effectiveness and efficiency as well as the environmental impacts of the system. The differences observed in the mission and policy priorities of various initiatives are accordingly reflected in the selection of indicators (JRC, 2007).

A general observation is that transportation sustainability indicators are typically classified into the following four major categories (Jeon, 2007): transportation system effectiveness-related, economic, environmental, and socio-cultural/equity-related indicators.

Decision makers sometimes focus on easy-to-measure impacts and objectives, while overlooking more-difficult-to-measure impacts and goals (Litman, 2006). For example, accessibility, one of the main goals of most transport activities is difficult to measure, so transport indicators systems tend to include traffic (vessels movement) and mobility indicators (the ability to move people and goods). This reduces the range of impacts and solutions considered in transport planning (VTPI, 2002).

Another illustrative example is the use of Gross Domestic Product (GDP) as an indicator for measuring welfare. Welfare (as used by economists) refers to total human wellbeing and happiness. Economic policies are generally intended to maximize welfare, although this is difficult to measure directly. Instead, monetary income, wealth and productivity such as GDP, are used as economic indicators. These indicators can be criticized on several grounds (Dixon, 2004), (Carvalho, 2011), because they only measure market goods and therefore overlook other factors that contribute to wellbeing such as health, friendship, community, pride, environmental quality, etc.

When selecting indicators for measuring the performance of any system it is important to avoid confusing goals and objectives. Goals are what society ultimately wants. Objectives are things that help achieve goals, but are not ends in themselves and indicators are variables selected and defined to measure progress toward an objective (Litman, 2011a). In Figure 1, the three factors (economy, society, and environment) commonly considered as the essential dimensions of a sustainable transportation system are presented together with the main goals to achieve sustainability in each one of these dimensions.

Zegras (2006) presented the Sustainability Indicator Prism that includes the hierarchy of goals, indexes, indicators, and raw data as well as the structure of multidimensional performance measures (Zegras 2006). In the four-layered pyramid, the top of the pyramid represents the community goals and vision, the second layer represents a number of composite indexes around the selected themes, and the third layer represents indicators or performance measures building from raw data at the bottom of the pyramid.

Figure 1. Essential dimensions of a sustainable transportation system (Jeon, 2007).

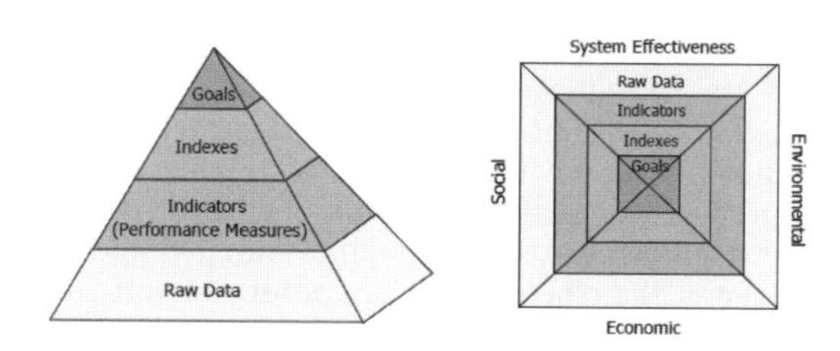

Figure 2. Sustainability Indicator Prism (Zegras, 2006, adapted from Jeon, 2007).

Zietsman et al (Zietsman et al., 2003) introduce a corridor-level index that incorporates travel rates, fuel consumption, local pollutant emissions, travel cost, and safety using multi-attribute utility theory. An international comparative index (Rassafi, 2004) has been developed by using the concordance analysis technique to evaluate transportation system sustainability of selected countries.

The literature on the indicators for sustainability shows that there is a variety of approaches subject to the targets of work. The challenge in the creation of indicators is not only to take into account environmental, economic and social aspects. Although, it could be argued that indicators could serve supplementary to each other, sustainability is more than an aggregation of the important issues, it is also about their interlinkages and the dynamics developed in a system (Singh et al. 2009).

Even if there have been identified many indicators suitable for use in the maritime transport there are no available indicator frameworks explicitly for this sector. Thus, since indicators should be constructed within a coherent framework (i.e. the scale in which the problem should be studied) and this network has not been formulated yet it is decided not to propose any framework of indicators at this paper.

The development of the framework of indicators in accordance with the principals under which the definition of sustainable maritime transport was formulated will be studied in the continuation of this work.

5.2 *Available frameworks—methods*

The literature shows that there is an increasing interest in environmental and sustainability assessment within shipping. A well structured framework for environmental assessing is the Life Cycle Assessment (LCA) a "compilation and evaluation of the inputs, outputs and the potential environmental impacts of a product system throughout its life cycle" (CALCAS, 2009). LCA is standardized through ISO and has some applications in shipping (Fet et al. 1998), (Kameyama, 2000), (Jivén et al. 2004). Problems identified with the use of LCA within shipping are the large amount of data needed and their availability (especially in the life stages of shipbuilding and dismantling).

A method used for environmental reporting is the Corporate Social Responsibility (CSR). Although there are many available definitions (Dahlsrud, 2008), CSR in general is a reporting procedure for companies containing social, environmental or social information of business operations on a voluntary basis.

Other techniques addressing the social and economic life cycle of products such as the Life Cycle Costing (LCC), and Social Impact Assessment (SIA). Cabezas-Basurko et al (2008), proposed a combination of available techniques for the development of a framework for analysis of the ship (as a product) in a life cycle perspective taking into account the three dimensions of sustainability.

An integrated framework for the analysis of the Sustainability has been recently proposed; namely the Life Cycle Sustainability Analysis (LCSA). The LCSA framework broadens the scope of current LCA from mainly environmental impacts only, to covering all three dimensions of sustainability (environmental, social, and economic).

It also broadens the scope from predominantly product-related questions (product level) to questions related to sector (sector level) or even economy-wide levels (economy level).

In addition, it deepens current LCA to also include other than just technological relations, e.g. physical relations (including limitations in available resources and land), economic and behavioural relations, etc (Zamagni et al, 2009).

6 CONCLUSIONS

This paper introduces an operational definition of sustainable maritime transport which is based on the initial principles of sustainable development. In the knowledge of the authors such operational definition is absent from the maritime transport sector. The next step of work in this context will be the development of a framework of indicators for assessing the sustainability of maritime transport.

REFERENCES

Beatley T. 1995. The Many Meanings of Sustainability, *Journal of Planning Literature*, Vol. 9, No. 4, May, 1995, pp. 339–342.

Black, J.A., Paez, A., and Suthanaya, P.A. 2002. Sustainable Urban Transportation: Performance Indicators and Some Analytical Approaches. *Journal of Urban Planning and Development*, Vol. 128, No. 4, 184–209.

Bruntland, G. 1987. Our common future: The World Commission on Environment and Development, Oxford, *Oxford University Press*.

Buhaug, Ø. et al. 2009. Second IMO GHG Study 2009, International Maritime Organization (IMO), London, UK.

Cabezas-Basurko, O., Mesbahi, E., Moloney, S.R. 2008. Methodology for sustainability analysis of ships, *Ships and Offshore Structures*,3:1,1–11.

Carvalho, J.F. 2011. Measuring economic performance, social progress and sustainability using an index. *Renewable and Sustainable Energy Reviews* 15 (2011) 1073–1079.

Chatzinikolaou, S., Ventikos, N.P., Nitsopoulos, S., 2007. Shipboard Wastes: Elements & Critical Review. *Proceedings of the 1st CEMEPE/SECOTOX Conference*, June 24–28, 2007, Skiathos Island, Greece.

Corbett, J.J., J.J. Winebrake, E.H. Green, P. Kasibhatla, V. Eyring, and A. Lauer (2007), Mortality from ship emissions: A global assessment, *Environ. Sci. Technol.*, 41(24), 8512–8518, doi:10.1021/es071686z.

Dahlsrud, A., 2006. Corporate Social Responsibility and Environmental Management Corp. Soc. Responsib. Environ. Mgmt. 15, 1–13 (2008) in *Wiley Inter Science* (www.interscience.wiley.com) DOI: 10.1002/csr.132.

Dally. H.E. 2006. Sustainable Development: Definitions, Principles, Policies. *The Future of Sustainability* 2006, Part 1, 39–53, DOI: 10.1007/1-4020-4908-0_2.

Dietz, S., E. Neumayer 2007. Weak and strong sustainability in the SEEA: Concepts and measurement. *Ecological Economics* Volume 61, Issue 4, 15 March 2007, pp 617–626.

Dixon, F. 2004, Gross National Happiness: Improving Unsustainable Western Economic Systems, *presented at GNH Conference in Thimphu, Centre for Science and Environment* (www.cseindia.org).

Dobranskyte-Niskota, A., A. Perujo, M. Pregl, 2007. Indicators to Assess Sustainability of Transport Activities. Part 1: Review of the Existing Transport Sustainability Indicators Initiatives and Development of an Indicator Set to Assess Transport Sustainability Performance. *Luxembourg: Office for Official Publications of the European Communities.*

EC, 2009. COM (2009) 8 final. Strategic goals and recommendations for the EU's maritime transport policy until 2018.

European Federation for Transport and Environment (T&E), 2010. CO2 emissions from transport in the EU27. An analysis of 2008 data submitted to the UNFCCC.

Eyring, V., Corbett, J.J., S. Lee, D.S., and Winebrake, J.J., 2007. Brief summary of the impact of ship emissions on atmospheric composition, climate, and human health. *Document submitted to the Health and Environment sub-group of IMO.*

Faber, J., M. Freund, M. Köpke, D. Nelissen 2010. Going Slow to Reduce Emissions Can the current surplus of maritime transport capacity be turned into an opportunity to reduce GHG emissions? *Seas At Risk.* at: http://www.seas-at-risk.org/pdfs/speed%20study_Final%20version_SS.pdf.

Fet A.M., Sørgård E., 1998. Life Cycle Evaluation of Ship Transportation—Development of Methodology and Testing, *Research report* HiÅ 10/B101/R-98/008/00.

Hall, R.P. 2002. Introducing the Concept of Sustainable Transportation to the U.S. DOT through the Reauthorization of TEA-21. Master's Thesis, *Massachusetts Institute of Technology.*

Heijungs, R., G. Huppes, J. Guinée. 2009. A scientific framework for LCA Deliverable (D15) of work package 2 (WP2) *CALCAS project* at: http://www.calcas-project.net/

IMO, 2010. Marine Environment Protection Committee, 60th Session. Prevention of Air Pollution from Ships, Assessment of IMO Energy Efficiency Measures for the Control of GHG Emissions form Ships, Note by the Secretariat.

ITOPF, 2011. Oil tankers spills statistics. Retrieved on April 2011 at: http://www.itopf.com/information-services/data-and-statistics/statistics/documents/StasPack2010.pdf

Jeon, C.M. 2007. Incorporating Sustainability into Transportation Planning and Decision Making: Definitions, Performance Measures and Evaluation. Ph.D. Dissertation, *Georgia Institute of Technology*, USA, December 2007.

Jeon, C.M., Amekudzi, A. 2005. Addressing Sustainability in Transportation Systems: Definitions, Indicators, and Metrics. *Journal of Infrastructure Systems*, Vol. 11, No. 1, March 1, 2005.

Jivén K., Sjöbris, A., Nilsson, M., Ellis, J.. Trägårdh, P., Nordström, M., 2004. LCA-ship, Design tool for energy efficient ships - A Life Cycle Analysis Program for Ships, Final report: 2004-08-27.

Johnston P., Everard M,, Santillo D., Robèrt, K.H. 2007. Reclaiming the definition of sustainability. *Environmental science and pollution research international.* Volume: 14, Issue: 1, Publisher: *Springer Berlin/Heidelberg*, Pages: 60–66.

Kameyama, M., 2000. Life cycle inventory analysis of CO2 emission from ship, *Proceedings of the 4th International Conference on EcoBalance*, Nov. 2000, 517–520.

Keiner, M., D. Salmeron, W. Schmid, I. Poduje. 2004. Urban Development in Southern Africa and Latin America. Chapter 1. *Springer*, Dordrecht, The Netherlands.

Kosz, M. 1998. Weak and Strong Sustainability Indicators, and Regional Environmental Resources, Paper presented *at the 38th European Regional Science Association Congress*, Aug. 28 to Sept. 1, 1998, Vienna.

Litman, T. 2007. Developing Indicators for Comprehensive and Sustainable Transport Planning, *Transportation Research Record 2017, TRB* (www.trb.org), 2007, pp. 10–15.

Litman, T. 2010. Sustainability and Livability: Summary of Definitions, Goals, Objectives and Performance Indicators, *Victoria Transport Policy Institute VTPI* (www.vtpi.org).

Litman, T. 2011a. Well Measured. Developing Indicators for Sustainable and Livable Transport Planning. *Victoria Transport Policy Institute VTPI* (www.vtpi.org).

Litman, T. 2011b. Evaluating Accessibility for Transportation Planning. *Victoria Transport Policy Institute VTPI* (www.vtpi.org).

Litman, T., Burwell, D., 2006. Issues in sustainable transportation. *International Journal of Global Environmental Issues*, Vol. 6, No. 4, 2006.

McGuire, C.J., Perivier, H. 2011. The Nonexistence of Sustainability in International Maritime Shipping: Issues for Consideration. *Journal of Sustainable Development* 4.1 (2011): 72-78. At: http://works.bepress.com/chad_mcguire/23

Meier, K.J., J.F. Brudney. 2002. Applied Statistics for Public Administration. Fifth Edition. *Wadsworth/Thomson*, Belmont, CA.

OECD 1996. Towards sustainable transportation. In OECD *Publications (Ed.) Paris, Organisation for Economic Co-operation and Development.*

Platou, 2010. The Platou Report 2010. Oslo: Platou, 2010 http://www.platou.com/dnn_site/LinkClick.aspx?fileticket=75tJnRmn03M=&tabid=119.

Rassafi, A.A., Vaziri, M. 2004. Sustainable transport indicators: Definition and integration *Int. Journal of Environmental Science and Technology.* Vol. 2, No. 1, pp. 83–96, Spring 2005.

Robèrt K-H, Schmidt-Bleek B, Aloisi de Larderel J, Basile, G, Jansen JL, Kuehr P, Price Thomas P, Suzuki M, Hawken P, Wackernagel, M. 2002: Strategic sustainable development—Selection, design and synergies of applied tools. *J Cl Prod* 10, 197–214.

Singh R.K., H.R. Murty, S.K. Gupta, A.K. Dikshit, 2009. An overview of sustainability assessment methodologies. *Ecological Indicators* 9 (2009) 189–212.

Topouzelis, K., Muellenhoff, O., Ferraro, G., Tarchi, D. 2007. Satellite mapping of oil spills in the East Mediterranean. *Proceedings of the 10th International Conf. on Environmental Science and Technology*, Kos Island, Greece, 5–7 Sep. 2007.

UNCTAD, 2010. Review of Maritime Transport, 2010. UNCTAD/RMT/2010 *United Nations Publication,*

New York, N.Y.: United Nations, ISBN: 978-92-1-112810-9.

Ventikos N.P., Chatzinikolaou S.D. 2008. Hazardous Waste Management and Ship Recycling: Friends or FOEs? *Proceedings of the 1st International Conference on Hazardous Wastes Management, Chania*, Greece, Oct. 1–3.

Victoria Transport Policy Institute,VTPI, 2002. Accessi bility and mobility, Online TDM Encyclopedia, www.vtpi.org.

Wang, C., J.J. Corbett, and J. Firestone 2008. Improving spatial representation of global ship emissions inventories, *Environmental Science. Technology,* 42(1), 193–199, doi:10.1021/es0700799.

Zamagni A. - P. Buttol - R. Buonamici - P. Masoni J.B. Guinée – G. Huppes – R. Heijungs – E. van der Voet T. Ekvall – T. Rydberg, 2009. D20 Blue Paper on Life Cycle Sustainability Analysis, Deliverable 20 of Work Package 7 of the *CALCAS project* at: http://www.calcasproject.net/

Zegras, C. 2005. Sustainable Urban Mobility: Exploring the Role of the Built Environment. *Ph.D. Dissertation, Massachusetts Institute of Technology*. September 2005.

Zegras, C. 2006. Sustainable Transport Indicators and Assessment Methodologies. Background paper for Plenary Session 4, *At the Biannual Conf. and Exhibit of the Clean Air Initiative for Latin America Cities: Sustainable Transport: Linkages to Mitigate Climate Change and Improve Air Quality*, Sao Paulo, Brazil.

Zietsman, J. 2000. Incorporating Sustainability Performance Measures into the Transportation Planning Process. *Ph.D. Dissertation, Texas A&M University*, College Station, Texas, December 2000.

Sustainable Maritime Transportation and Exploitation of Sea Resources – Rizzuto & Guedes Soares (eds)
 ISBN 978-0-415-62081-9

Management issues for the safe operation of ships and for pollution prevention

C. Grigorut, C. Anechitoae, A.R. Staiculescu & A.P. Lisievici Brezeanu
University "Ovidius" Constanta, Romania

L.M. Grigorut
National Institute of Economic Research "Costin Kirițescu" Bucarest, Romania

ABSTRACT: The aim of this paper is to summarize the rules of, and to provide comments on the International Management Code (the ISM Code), which was adopted on November 4th 1993 by the International Maritime Organization, and which introduced international legal norms of management for the safe operation of ships and pollution prevention. On this occasion, the Assembly of the International Maritime Organization adopted: 1) The A 443 (XI) resolution, calling upon all governments to take the necessary measures in order to protect the ship captain while adequately fulfilling his responsibilities, related to maritime safety and marine environment protection, and 2) The A 680 (17) resolution, which led to the acknowledgement of the need for the appropriate organization of management, in order to enable seafarers to ensure and maintain a high safety and environmental protection level. By adopting the ISM Code, it was intended to improve the safety management in the maritime activities determined by the commitment, competence, attitude and motivation of the employees at all levels in the field.

1 INTRODUCTION

The International Maritime Organization (IMO), headquartered in London, works as a UN specialized agency, concerned with development of the legal, technical and organizational framework on shipping in international waters. By means of the regulations adopted, the IMO establishes minimum requirements necessary for the safe operation of ships and pollution prevention. Romania became a member of this organization by Decree no. 114/1965 and as such is party to all major international conventions on navigation safety and prevention of pollution from ships.

The IMO has found over time that within the international practice of exploitation activities in the field of ship safety, marine environmental protection and pollution prevention there are discrepancies and deficiencies in ensuring safety at sea, preventing injuries or casualties. It also found insufficient global measures taken to avoid environmental damage, particularly regarding the marine environment.

On 4 November 1993, the IMO adopted the International Management Code for the Safe Operation of Ships and for Pollution Prevention (International Safety Management Code—ISM Code) by Resolution A 741 (18) of 4 November 1993 and it represents the Chapter 9 of SOLAS International Convention (Safety of Life at Sea). The International Maritime Organization has adopted amendments to the ISM Code by Resolution MSC 104 (73) of 5 December 2000 and by Resolution MSC 195 (80). The ISM Code was accepted by Romania through the instrument of ratification by the Romanian Parliament by Law no. 85 of 05.28.1997 (published in Official Gazette no. 107 of 30.5.1997) and, subsequently, the Romanian Parliament, by Law no. 681 of 19.12.2002, accepted the amendments to the Code (published in the Official Gazette no. 964 of 28.12.2002). Also, the amendments to the International Management Code for the Safe Operation of Ships and for Pollution Prevention (ISM Code) adopted by the International Maritime Organization by Resolution MSC 195 (80) of the Maritime Safety Committee of 20 May 2005 were accepted by the Ministry of Transport by Order no. 989 of 11.08.2005, published in the Official Gazette no. 614 of 20.08.2008.

2 THE SHIP FROM A LEGAL POINT OF VIEW

Rarely, there may be doubt as to whether if a building is or is not a ship. The term is understood relatively, it is well established, and the legislature uses it without worrying about the exact definition. However, sometimes there is doubt. The classification should be based on the approach to the case. Is the respective construction a ship, in accordance with the specific law, for example, of the maritime law pledge? The answer does not depend on the fact that the respective building can be regarded as a vessel, in terms of other laws.

Any means of transport, on land, water or air, is defined as a self-propelled mobile.

According to article 23 of Government Ordinance no. 42/1997 (r) on maritime transport and inland waterways, by ships we understand: maritime ships and inland navigation ships of all types, with or without propulsion, sailing on the surface or in immersion, for the transport of goods and/or people, for fishing, towage or pushing, floating devices such as dredgers, floating elevators, floating cranes, floating grips and others, with or without propulsion, and floating installations that normally are not designed for movement or carrying out special works, such as: floating docks, floating jetties, wharves, floating sheds for vessels, drilling rigs and others, floating light houses and barges.

However, if a specific legislation deviates from the "ordinary" laws for ships, we must be careful when classifying a building as a ship

3 MARINE POLLUTION

The Marine pollution has some peculiarities. It is, first, about the pollution of coastal waters and enclosed seas, pollution which has two main causes:

- Pollution from industrial tailings;
- Oil pollution.

Pollution by industrial tailings comes from directly discharging into the sea the wastewater from industries located on the coast, on the (absurd) mentality that the sea is wide enough to take everything.

Seas absorb annually over 6000 tons of hydrocarbons from tanker accidents, waste oil refineries, offshore mining, washing and ballasting of ships in high seas.

"The black blanket" that forms on the surface of the water suffocates the deep sea life. Instead of dispersing, the spilled oil bodies, creating a continuous coating which prevents oxygenation of the water. Towards the shore, the oil film inhibits the activity of photosynthesis of algae and hence the production of oxygen.

The oil is biodegradable under the action of bacteria, but this oxidation process depletes the marine oxygen. In order to break down a liter of oil there are needed 40,000 liters of water.

Thus, in the Indian Ocean (which is bordering countries with a common economic development) there are discharged 20,106 tones of tailings annually, while only about U.S. discharges in its territorial waters about 2105 m^3 of industrial wastes and 11.1010 m^3 of untreated municipal waters.

In the North Sea, the number of microscopic algae has increased four times, following the discharge of water rich in phosphates and nitrates, in addition to heavy metals and the organohalogen (only Rhine discharge 30, respectively 100 tons per day of such pollutants).

The disastrous effects of these practices were revealed for the first time in Mediterranean Sea whose fish production dropped drastically in the last 10 years. Moreover, in the case of a number of species, the content in ions of cadmium and mercury, highly toxic, exceeded the permissible limits; only in the summer of 1989, in the Norwegian fjords, the proliferation of poisoning microscopic algae led to the extinction in 3 weeks of 400 tons of salmon. A tragic example is the Minamata disease (Japan), due to severe mercury poisoning. Present in the sea water as a result of industrial discharges and of its accumulation on the food chain, the mercury affected the fauna in the area (fish), and the man.

The famous ocean specialist J.Y. Cousteau brings another potential issue into discussion: marine creatures also communicate through chemical substances similar, in principle, as function, to insect pheromones. A heavy pollution could "jam" this form of communication, with incalculable serious consequences.

3.1 *The definition of marine pollution*

The newest definition of marine pollution is contained in the UN Convention on the Law of the Sea. Thus, marine pollution means the introduction by man, directly or indirectly, of substances or energy into the sea including estuaries, where it has or may have deleterious effects such as harm to living resources, marine fauna and flora, risks to human health, hindrances to marine activities, including fishing and other legitimate uses of the sea, sea water quality deterioration in terms of its use and the degradation of its recreational values.

3.2 *Forms of marine pollution*

Article 145 of the Convention establishes the obligation of Contracting States to take the necessary steps in order to ensure the effective protection of the marine environment from the harmful effects that may result from these activities. To this end, the

States participating in the Convention shall adopt rules, regulations and procedures, particularly for:

a. the prevention, reduction and control of marine pollution, including the coastline, and in order to face other risks that threaten it, and any disruption of the ecological balance of the marine environment, paying particular attention to the need to protect them from harm activities, such as drilling, dredging, excavations, waste disposal, construction and operation or maintenance of installations, pipelines and other devices used for such activities.
b. the protection and preservation of the area's natural resources and the prevention from causing damage to marine flora and fauna.

The Convention recognizes the sovereign right of Member States to exploit their natural resources, in compliance with their policy in the environmental matter and under the obligation to protect and preserve the marine environment. Article 194 of the Convention sets out the measures to be taken in order to prevent, reduce and control marine pollution, as follows:

1. States are required to take, individually or jointly, as appropriate, all the measures compatible with the Convention, that are necessary to prevent, reduce and control pollution, better adapted measures, depending on their capabilities, and will strive to harmonize their policies in this regard;
2. States shall take all the necessary measures in order to ensure that activities under their jurisdiction or control are managed in such a way as not to cause damage by pollution to other States and their environment, so that pollution due to activities under their jurisdiction or under their control does not extend beyond the areas where they exercise their sovereign rights in accordance with the Convention;
3. The measures taken in order to protect the marine environment cover all sources of pollution of the environment. They include, in particular, the measures which tend to limit as much as possible:
   - The removal of toxic, harmful or noxious substances, especially non-degradable ones, from land-based sources, from or by the atmosphere or by immersion;
   - Pollution from ships, in particular measures to prevent accidents and to cope with emergencies, to ensure the safety of operations at sea in order to prevent spills, whether intentional or not, and to regulate the design, construction, equipment and exploitation of ship and the composition of their staff;
   - Pollution from installations or equipment used for the exploration and exploitation of the natural resources on the seabed and subsoil, in particular the measures to prevent accidents and to cope with emergencies, to ensure the safety of operations at sea in order to regulate the design, construction, equipping and operating of equipment and apparatus and the composition of their staff;
   - Pollution from other installations or devices operating in the marine environment, including measures to prevent accidents, to deal with emergencies to ensure the safety of operations at sea and to regulate the design, construction, equipping and operation of such installations and equipment, and the composition of their staff.

When taking measures to prevent, reduce and control marine pollution, States are obliged to refrain from any undue interference with activities of other States in exercising their rights and obligations under the Convention.

Measures taken to protect the marine environment include the necessary measures to protect and preserve rare or sensitive ecosystems, and the living environment of declining species and marine organisms, threatened or endangered.

According to article 195 of the Convention, when measures are taken to prevent, reduce and control the pollution of the marine environment, States shall act so as not to transfer, directly or indirectly, damage or hazards from one area to another and not to replace one type of pollution with another.

Also, article 96 expressly provides that, where the States will use new techniques or will introduce foreign or new species, they will have to take all the necessary measures to prevent, reduce and control the marine pollution resulting from the use of such techniques in the their jurisdiction or under their control, or resulting from the introduction, either intentionally or accidentally, in an area of the marine environment, of foreign or new species which can cause significant and harmful changes.

The Convention classifies forms of pollution, according to their source, in:

- land-based pollution;
- pollution resulting from activities related to underwater territories subject to national jurisdiction;
- pollution from activities in the area;
- dip pollution;
- pollution from ships;
- atmosphere or trans-atmosphere pollution.

According to article 217 of the Convention, States shall ensure that vessels flying their flag or registered by them to comply with the applicable international rules and standards established through the competent international organization or general diplomatic conferences, as well as the laws and regulations that they adopted under the Convention, in order to prevent, reduce and control the marine pollution from ships, and shall adopt laws and regulations and shall take the necessary measures in order to implement them. The flag State shall ensure that such rules, regulations and laws are effectively applied, regardless of where the infringement was committed.

States are required to take appropriate measures to prohibit vessels flying their flag or which are registered by them to start the race as long as they do not meet the international rules and standards required in order to avoid the pollution of the marine environment, including provisions for the design, construction and equipment of ships. States are required to ensure that vessels flying their flag or which are registered by them have the certificates required and issued pursuant to international rules and standards to avoid marine pollution. States shall ensure that vessels flying their flag are periodically inspected, in order to ensure that the entries made on those certificates conform to the actual condition of the vessel. Other States will accept these certificates as evidence of state of the ship and will recognize them with the same force as the certificates issued by them, unless there are reasonable grounds to believe that the condition of the vessel does not correspond, to a significant extent, with the entries made on those certificates.

4 APPLICATIONS OF THE ISM CODE

The ISM Code was adopted by the International Maritime Organization in order to create an international shipping standard imposing an integrated safety management system for the operation of ships and pollution prevention.

This integrated management must be implemented by the flag State administration or by one of the organizations recognized by it. Thus, the vessel operating company is required to issue certificates for all types of vessels they operate. A certificate, called Safety Management Certificate, should be issued to each ship and the administration must verify whether the company management and the board management are in accordance with the approved safety management system and must periodically check the proper functioning of the management system safety of the vessel, as it was approved. According to paragraph 13.7–13.8 of the Amendment of 05.12.2000, the Safety Management Certificate should be issued to a ship for a period not to exceed five years, by the Administration or an organization recognized by the Administration or, at the request of the Administration, by another Contracting Government. The Safety Management Certificate should be issued after verifying that the company and its shipboard management operate in accordance with the approved safety management system. Such a certificate should be accepted as proof that the vessel meets the requirements of the ISM Code.

The validity of the Safety Management Certificate should be submitted to at least an intermediate verification by the Administration or by an organization recognized by the Administration or, at the request of the Administration, by another Contracting Government. If only one intermediate verification is to be carried out and the validity of the Safety Management Certificate is five years, this verification must take place between the second and the third anniversary date of the Safety Management Certificate.

The Safety Management Certificate means a document issued to a ship which signifies that the company management and the board management are in accordance with the approved safety management system, different from the Document of Compliance which means a document issued to a company that meets the requirements of this code. Under the amendment of 2000, an Interim Document of Compliance may also be issued in order to facilitate the initial implementation of this Code, when: 1) a company is newly founded or 2) new types of ships are going to be added to an existing Document of Compliance.

REFERENCES, SYMBOLS AND UNITS

Anechitoae C. (2011) *Introducere în drept maritim internaţional.* Bren, Bucureşti, pp. 274–276. 4th edition.

Anechitoae C. (2008). *Introducere în drept maritim.* Bren, Bucureşti, p. 103.

Anechitoae C. (2005). *Convenţii internaţionale maritime. Legislaţie Maritimă. Vol. I, Vol. II.* Bren, Bucureşti.

Nicolae F. (2002) *Prevenirea poluării mediului marin.* Academia Navală "Mircea cel Bătrân", Constanţa, p. 1.

Piperea G. (2005). *Dreptul transporturilor.* All Beck, Bucureşti, pp. 144–145, 2nd edition.

Sustainable Maritime Transportation and Exploitation of Sea Resources – Rizzuto & Guedes Soares (eds)
 ISBN 978-0-415-62081-9

Evaluation of the environmental impact of harbour activities: Problem analysis and possible solutions

L. Battistelli & M. Fantauzzi
Department of Electric Engineering, University of Naples "Federico II", Naples, Italy

T. Coppola & F. Quaranta
Department of Naval Engineering, University of Naples "Federico II", Naples, Italy

ABSTRACT: The environmental problem involves all aspects of the human activities and is leading to a severe decisions in order to avoid dramatic consequences on human health. In the marine field, after a very long period of stand by, now the problem is felt mainly in the big water cities where, close to the harbours, a heavy microclimate can be dangerous for those who live nearby. This work aims at analysing the present rules operating in port areas, and at presenting the sustainable actions to cope with the problem of the environmental impact of naval activity.

1 INTRODUCTION

In the marine activity, harbour operations have a dramatic environmental impact because of logistics, energy supply, traffic congestion, etc. The alarm of institutions studying the environmental problem focus on the densely inhabited areas where pollution may cause major damage to the people living close by. Regulations have been adopted in order to reduce the noxious effects of port activities. The marine field has been less touched by this problem, probably because most part of the noxious elements is released by engines far away from inhabited zones and it is not "felt" as dangerous as other fields of human activity (industrial, services, road traffic, etc.).

But, aside from the evident need to save the planet as a whole and release the least possible waste—although in "desert" zones—, there are particular areas where the simultaneous operations of a number of ships may create a heavy impact. Typically, very busy harbours, "water cities" and other contexts where the human life develops close to the sea or the aquatic environment, expose people to major risks of serious damages for their health.

Therefore, it is very important to analyse the complex scenes of the maritime harbour operations and find efficient solutions that should be logistically and economically sustainable, in compliance with the present rules on exhaust emissions from ships.

2 ENVIRONMENT CHARACTERISATION—POLLUTION FACTORS IN PORT AREA

The environmental pollution in the harbours has three main aspects; air, water and acoustic [1,2,3].

The air pollution is due mainly to dusts in the form of suspended particles and gaseous elements that are often very dangerous for the human health.

In the port areas, dusts are mainly due to load/unload phases of bulk carriers; the noxious gases, on the other hand, are emitted by diesel engines onboard ships and by technical vehicles working in the port area or in the surroundings.

Among the gaseous pollutants, the most important are:

a. the nitrogen oxides NOx and sulphur oxides SOx;
b. the Carbon Monoxide (CO) and Carbon Dioxide (CO2);
c. particulate matters PM (in particular PM_{10});
d. the volatile organic compounds (VOCs—benzene, formaldehyde, toluene, etc.).

Apart from the presence of these substances, their combinations may create further noxious elements. For example, the interaction between VOCs, NOx and the sunlight produces ozone; sulphur oxides, combined with water steam, cause the so called "acid rain" which can cause serious damages even in zones far away from the pollution source.

Water pollution is mainly due to oil products and to sewage disposals in harbours. Often accidents involving oil carriers cause severe pollution; this risk is directly proportional to the overall amount of carried oil and to the number of manoeuvres carried out.

Also some maintenance operations normally carried out onboard (for example: painting) can cause pollution; in such cases, antifouling additives may have seriously negative effects on the marine environment.

Finally, there are sources of acoustic pollution both on ships and in port areas; the latter are due to the vehicles moving goods and people and to daily activities developed close to the harbour, especially in inhabited zones.

3 HOW TO COPE WITH THIS PROBLEM: FIRST FUELS AND SETTINGS

As regards the air pollution—the more urgent problem—fuels are at the core of the problem: indeed, emissions certainly depend mainly—but not exclusively—on the quality of fuels burned in the harbours.

First of all, a reduction of the fuel consumption involves the reduction of all emissions from the engine. This means that new generation fuel-saving engines should be preferred to old and "dirty" ones. But, in order to obtain a serious decrease in emissions, also the fuel characteristics must also be changed and adapted to a cleaner use.

Doubtless, problems due to a change in fuels are many and difficult to solve. Ships that have to burn relevant quantities of fuel close to inhabited zones are forced to store different kinds of fuels (dirtier for use in open sea, clearer in ports); and this involves problems of storage, division of thanks for each kind of fuels and so on.

Moreover, the availability of various fuels is not granted everywhere in the world and difficulties could arise in case a ship has not supplies on board of the fuel required for the admission to a port: for example, the low sulphured fuels, required almost in all ports today, are not universally available.

In a next future, probably, the use of non conventional fuels will be more enlarged; the methane added with hydrogen, for example, seems to be able to reduce the NOx emissions from diesel engines; but the arrangement of engines is very invasive and predictably, a long time will pass before such technologies are applied.

However, some emissions are strictly connected with fuels and the most reasonable approach seemed to be limiting the related content of the substance in it. This is the case of sulphur oxides SOx which can be limited by reducing the content of sulphur in fuels.

As for NOx, it can be kept under control either by setting the engine properly, or retrofitting the engine with a system capable of lowering the content of nitrogen oxides in the exhausts.

4 RULES

The awareness of the problem of the environmental pollution dates back to some decades ago: thus, the famous ANNEX VI of the MARPOL 73/78 (and its amendments) created the so called SECA (Sulphur Emission Control Area) where the use of fuels with a controlled level of sulphur was introduced.

The first SECA entered into force in 2006 and involved the English Channel and the Baltic Area.

Sometime later, in order to meet the dramatic conditions of ports located close to big towns, the ECA (Emission Control Area) were introduced as zones where more stringent fuel sulphur and engine NOx limits were imposed. In all European ports, for example, the maximum sulphur contents in the fuels was fixed in the 0.1% (since 2008/1/1).

As regards the air pollution, in particular, the production of NOx, SOx, PM (depending on the sulphur oxides) and CO2 are under observation in order to contain the noxious effects of these substances on the environment.

The ways to handle NOx and SOx in order to clean the exhausts are considerably different: the limitation in SOx is achieved by a reduction of the sulphur in the fuel. The synthesis of present and future limits is given in Table 1.

In addition to these limits, the "EU DIRECTIVE 2005/33/EC", emending "MARPOL Annex VI Air Pollution", limits the sulphur contents of marine fuel in many areas (including all European ports) to 0.1% [4,5].

If low sulphured fuels are not available onboard, the Annex VI permits the use of retrofits that must reduce the maximum value of SOx to 6 g/kWh.

Table 1. Maximum admissible content of sulphur allowed in the fuel (%).

TERMS	NON SECA areas	SECA areas
<1/1/2012	4.5	
>1/1/2012	3.5	
>1/1/2020	0.5	
<1/1/2015		1.0
>1/1/2015		0.1

Table 2. Maximum admissible emissions of NOx (g/kWh).

ENG SPEED	TIER II (<1/1/2016)	TIER III (>1/1/2016)
rpm < 130	14.4	3.4
130 < rpm < 2000	$44/rpm^{0.23}$	$9/rpm^{0.2}$
2000 < rpm	7.7	2.0

Table 3. Sulphur oxides: Exposure with limited injury.

0,06 mg/m^3	Possible bronchitis episodes and chest infections
0,3 mg/m^3	Possible damages to the respiratory system (especially for children and elderly)
0,8–2,6 mg/m^3	Olfactory sensing of the substance (stimulates search for gas mask and refuge)

Table 4. Sulphur oxides: Exposure with serious injury.

3 mg/m^3	Increase of the breathing rhythm and heart pulsation
25 mg/m^3	Irritations to eyes, nose and throat, increase of the heart rate
5 g/m^3	Toxic asphyxia, cardiovascular collapse

Table 5. Nitrogen oxides.

50–150 mg/m^3	(For short periods of time) possible harm to lungs
100 mg/m^3	Serious damages to the breathing apparatus
300–400 mg/m^3	Lethal

The emission of NOx depends on the temperature of combustion and the only way to contain it in the exhaust emissions is to set the engine (pumps, turbochargers, ignition timing etc).

Table 2 presents limits in NOx contents in the exhausts from marine engines; in the field between 130 and 2000 rpm (practically, all the engines destined to the production of electric energy onboard), the limitation are fixed as a function of the engine rpm.

5 AND THE EFFECTS ON THE HEALTH?

All these emissions may have direct effect on the human health and on the vegetal life.

Synthetically, the main effects—identified by WHO—are reported in Tables 3 to 5 [6].

6 MANY SHIPS IN PORT: IS IT A PROBLEM?

Passing through a large European harbour it is possible to count many ships at berth at the same time and "in full activity". Of course cruise ships are only part of the port activity, but they require a large amount of electric energy in order to make the life onboard possible (and pleasant!).

The ships generate electric energy onboard for many applications, from simply feeding lights and air conditioners to more sophisticated systems of control and steering. In order to grant a correct ship management, other systems require electric power such as hull systems (bilge, ballast, balancing, etc.) safety systems, engine room devices, and all tools for the movement of loads and/or the management of passengers.

To produce the electric power needed for services oboard, conventional ships have diesel electric groups that feed only the abovementioned users; the "all electric" systems, more and more widespread nowadays in the cruise ships, feed both the propulsion plants and all electric auxiliary users.

The evaluation of the electric power required onboard must certainly take into account the overall power installed with auxiliary diesel engines; this datum is normally (but not always [7]) available in the case of conventional ships, while it must be derived for all electric ships.

On the other hand, determining of the value of the electric energy required onboard depends on the variability of the electric loads during the berthing period.

The electric power during the berth periods can be evaluated by using the so called "load factor", a coefficient which gives the mean power used onboard with the overall electric power installed onboard as reference. This factor obviously may vary greatly from ship to ship.

Table 6. Load factors for various kinds of ship.

Car carrier	0.24
Bulk carrier	0.22
Container	0.17
Cruise	0.64
General cargo	0.22
Ocean tug/tow	0.22
Reefer	0.30
RoRo	0.34
Tanker	0.67
Others	0.22

The following table reports these coefficients determined in [8].

Tipically, cruise ships and tankers are the vessels that use major rates of electric power when at berth. In order to evaluate the environmental impact of many cruise ships simultaneously at berth in a single harbour, on first approximation, it's enough to evaluate a reasonable value of the electric energy needed for services onboard.

Thus, supposing that 6 large ships are berthed together in a same mooring area (many European ports have a similar situation), and each of them reserves 10–15 MW for the onboard services, according with a given datum (load factor = 0.64, supposed value of power: 11 MW), the overall power delivered by generators will be about 45 MW.

So, while supposing that all of these ships fully respect the present limits for NOx and SOx, an easy calculation permits to determine the amount of emissions released in a small area close to the berthing point:

for NOx (if generators work @ ~ 900 rpm) max 9.20 g/kWh;
for SOx max 6 g/kWh (permitted value as alternative to low sulphur fuels).
If the overall power released is about 45 MW, the correspondent emissions are:
NOx emission: 0.41 t/h
SOx emission: 0.27 t/h

discharged in a relatively limited zone (berthing points are often close to each other) with a possible (probable!) huge concentration of noxious substances in such areas.

7 WHAT CAN (MUST) BE DONE?

The solution to a big problem cannot be easy.

And, most of all, a definitive solution for the limitation of emissions close to the big harbours can be achieved only with a strong cooperation among politicians, technicians and operators in the marine sector.

Thus, intermediate solutions are to be devised and tested, with the risk that partial solutions became definitive in time.

But, in some cases, the environmental situation of areas around large harbours cannot wait for global solutions that could take a long time; improvements in the management of shipping should be achieved right now, the rules introduced—at what cost!—are an example of the action to be taken.

Beside cold ironing—the subject of the following discussion—and the other mentioned systems, some palliative could be considered. Among them, a better distribution of berths separating the largest ships and, when possible, moving some mooring quays offshore; electric energy supply directly from shore with system composed by an electric generators (in a trailer truck size a power of about 7–8 MW is available with current technology) fed by very clean fuels (hydrogen, methane, hydromethane, etc.) provided by a tank self-moved like the generator; restrictions to the access to the port area of old (and polluting) vehicles, reduction in the speed of ships close to shore and of land vehicles in port area, etc.

Table 7. Causes and remedies to the air pollution in ports.

Pollutants
– Particulate matter PM (PM10 in particular);
– VOCs (Volatile Organic Compounds—benzene, formaldehyde, toluene and others);
– Nitrogen Compounds (NOx);
– Sulphur Oxides (SOx);
– Carbon Monoxide (CO);
– Carbon Dioxide (CO2).
Main pollution causes
– Propulsion and auxiliaries diesel engines onboard ships;
– Diesel engines onboard the vehicles destined to the handling of merchandise inside harbours
Possible solutions
– Low sulphured, ecologic fuels;
– Use of new engines with lower SFOC and capable to use cleaner fuels;
– Retrofit (NOx and SOx);
– Cold ironing;
– Restrictions to the access in ports for old pollutant vehicles;
– Restrictions to the speed of ships and road vehicles in port areas;
– Electric energy supply to berthed ships from shore mobile systems fed by ecologic fuels

Table 7 summarises the main causes of the air pollution in port areas together with the possible remedies.

8 ECOSUSTAINABLE PROCESSES: THE COLD IRONING

Cold ironing (or "shore connection") means to supply ships at berth in feeding points close to the mooring sites so that the diesel electric generators can be switched off and no exhausts are delivered during ship stay. In these conditions, the shore electric system can feed the ship in "plug-in" mode.

This technique is not to be considered an absolute innovation since applications of electrified banks have been known for many years and, starting from the 80s, even ferries could use such a technology because, as they have a stable mooring point, the connection with an electric network is very easy.

Table 8. Cold-ironing installations in Europe and USA.

	Connection voltage (kV)	Frequency (Hz)
European ports		
Goteborg (Sweden)	0.4	50
	6.6	
	10	
Stockholm (Sweden)	0.4	50
	0.69	
Helsingborg (Sweden)	0.4	50
	0.44	
Piteå (Sweden)	6	50
Antwerpen (Belgium)	6.6	50/60
Zeebrugge (Belgium)	6.6	50
Lubeck (Germany)	6	50
Kotka (Finland)	6.6	50
Oulu (Finland)	6.6	50
Kemi (Finland)	6.6	50
USA ports		
Los Angeles	0.44	60
	6.6	
Seattle	6.6	60
Washington	11	60
Pittsburg (California)	0.44	60
Juneau (Alaska)	6.6	60
	11	

Anyway, this approach is nowadays particularly appreciated in the case of larger ships—first of all, cruise ship—in order to lower the air pollution in ports.

Indeed, the electrified docks have a considerable success today, especially in North America and Europe; the following tables report European and American installations in use up to 2008 together with the main electric characteristics [9].

Many other applications are in progress; the most interesting ones are in North Europe and North America but remarkable installations regard China and Japan. In Italy the ports of Civitavecchia, La Spezia and Venice have already shown their interest in cold ironing applications.

A real standard solution for cold ironing is not available yet; actually, the kind of installation depends on the specific environmental and commercial vocation of the port as well as on the kind of ship to be fed.

This means that the solution to be adopted is to be evaluated on the basis of the maritime traffic, and of environmental pollution of a particular port area, preferably in a mid long-term scene.

Furthermore, the real plant will depend both on the particular layout of the port (with constraints deriving from the arrangement and the existing electric connections), and on the particular electric power plant onboard the ships [10,11,12,13].

Nevertheless, for practical, economic and dimensional problems, there are some invariants in present installations among which, doubtless, the most important are the MV flexible cables (6÷20 kV) even if in the fed ship the inner electric distribution net is at low voltage.

Other elements almost always present in the terrestrial plant are:

- the availability of an easy connection of the port medium voltage box dedicated to the cold ironing with an adequate power;
- the installation of such a box close to the mooring area.

In Figure 1 [14] a cold ironing system type is reported schematically. The connection is grant by a mid voltage flexible cable while, in order to make them compatible the shore and ship voltages, a medium voltage socket and a transformers are fitted onboard of those ship with a low voltage internal electric distribution system.

In some cases, the shore side connection is made by flexible cables unrolled from the ship and plugged to the berth.

Cold ironing represents only the first step of the road map of eco-sustainable processes in port areas that renewable sources and electric and ICT (Information & Communication) technologies can offer nowadays. The potential of environmental sustainability of these technologies, with the goal of containing the pollution in harbour areas, is certainly very high.

It includes the use of renewable sources as sun, wind and biomasses, production of electric energy with fuel cells, innovative systems of energy storage just to name a few.

These systems should be integrated by microgrids or smart microgrids configured as a local active electrical systems with the abovementioned distributed energy resources able to supply all the electric loads normally in connection with the electric distribution utility [15].

Among the most important technological aspects of modern microgrids, it's worth

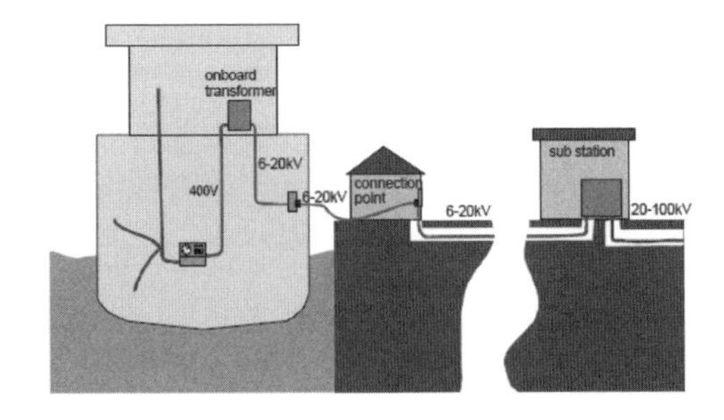

Figure 1. Scheme of cold ironing connection.

mentioning the two-way communication system with the possibility of a direct participation to the electric energy market.

9 CONCLUSIONS

The development of port activities led to evident benefits involving relevant technological resources.

But it also highlighted some important problems that haven't been solved yet.

Among them, the environmental pollution related to the maritime activity in harbours is certainly into close-up; in particular, the simultaneous presence of many ships at berth—with engines working—generates large amounts of toxic substances in limited zones. Air pollution due to the emission of noxious gases in the atmosphere by many diesel engines operating in ports is now under observation by all the Bodies in charge of the safety in such areas.

The situation has become more critical with the spread of cruise ships, actual "giants of the sea" which, although equipped with technologically advanced, high-performance engines, emit large amounts of exhaust emissions due to the high power rates required for the production of electric energy while at berth.

This problem can be faced by adopting temporary solutions, that would allow a normal development of the harbour operations without significant risks for the human health. Among them, the development of the engine technologies that could contain specific emissions, the use of ever clearer fuels—to feed external generators as alternative to those onboard—the use of efficient retrofits to reduce the exhaust emissions ad so forth.

Cold ironing requires a particular mention because it constitutes one of the most efficient (and definitive) solutions to the problem of the environmental pollution in harbours. However problems of standardization and methods of use should be evaluated by means of a dedicated R&D phase, mainly with the goal of a possible integration with electric, automatic and ICT technologies in the specific field of port applications.

In such a complex field, a very robust "system view" will be absolutely necessary in order to stimulate the cooperation of all the protagonists of this scene; apart from the benefits on the human health—that are obviously very important—it is encouraging to consider that focused and efficient operations could turn a negative situation to a condition economically convenient for all the operators involved.

REFERENCES

Amendments to the Annex of the Protocol of 1997 to amend the International Convention for the prevention of pollution from ships, 1973, as modified by the Protocol of 1978 relating thereto (Revised MARPOL Annex VI)—MEPC 58/23.

Battistelli, L. Fantauzzi, M: "Microretielettricheintelligenti in ambito portuale" Convegno Nazionale AEIT (Federazione Italiana di Elettrotecnica, Elettronica, Automazione, Informatica e Telecomunicazioni) Milano 27–29 giugno 2011.

Browning, L. Bailey K:" Current Methodologies and Best Practices for Preparing Port Emission Inventories"—15th International Emission Inventory Conference Session 1—New Orleans, May 15–18, 2006.

California Air Resources Board, "2005 Oceangoing Ship Survey, Summary of Results," September 2005.

Dev Paul, Vahik Haddian; "Shore-to-ship Power Supply System for a cruise ship" IEEE Industry Applications Society Annual Meeting, 2009—Houston 4–8 Oct. 2009.

Directive 2005/33/EC of the European Parliament and of the Counci—Official Journal of the European Union—L 191/59–22.7.2005.

Ericsson, P. Fazlagic I: "Shore side power supply" Master of Science Thesis—Chalmers University of technology—Goteborg Sweden 2008.

http://www.imo.org/conventions/contents.asp?doc_id=678&topic_id=258.

http://www.indoor.apat.gov.it/site/it-it/AGENTI_INQUINANTI/Chimici/Biossido_di_azoto/.

http://www.severnesnow.com/html/emissions.php.

http://www.fi.infn.it/sez/prevenzione/gas/index.html.

http://www.ipta.org.uk/review_of_marpol_annex_vi.htm.

Health effects of transport-related air pollution—WHO 2005—ISBN 92 890 1373 7.

Islam M., Peterson K., Chavdarian P., Cayanan C.; "State of shore power standards for ships"—IEEE PCIC '07 (Petroleum and Chemical Industry Technical Conference)—Calgary 17-19 sept 2007

Islam M., Peterson K., Chavdarian P., Cayanan C.; "Tackling ship pollution from the shore"—IEEE Industry applications magazione—Jan/Feb 2009 pagg. 56-60

Khersonsky, Y. Islam, M. Peterson K: "Challenges of Connecting Shipboard Marine Systems to Medium Voltage Shoreside Electrical Power"—IEEE Transactions On Industry Applications, Vol. 43, No. 3, May/June 2007, pagg. 838–844.

Natural Resources Defense Council (NRDC), 2004. "Harboring Pollution: The dirty true about U.S. Ports." NRDC.

Natural Resources Defense Council (NRDC), 2004. "Harboring Pollution: Strategies to Clean Up U.S. Ports." NRDC.

Presentation by Capt. Pawanexh Kohli—Cold ironing An overview—CrossTree techno-visors available at http://crosstree.info/Documents/ColdIroning.pdf.

Zhuang, N. Zhu K.Z: "Research on prevention and control technologies of harbor pollution" ICEET 2009 (International Conference on Energy and Environment Technology) pagg. 713–716.

Sustainable Maritime Transportation and Exploitation of Sea Resources – Rizzuto & Guedes Soares (eds)
© 2012 Taylor & Francis Group, London, ISBN 978-0-415-62081-9

Status of persistent organic pollutants and heavy metals in perch (*Perca fluviatilis* L.) of the Port of Muuga impact area (Baltic Sea)

L. Järv, M. Simm & T. Raid
Estonian Marine Institute, University of Tartu, Tallinn, Estonia

A. Järvik
Estonian Maritime Academy, Tallinn, Estonia
Estonian Marine Institute, University of Tartu, Tallinn, Estonia

ABSTRACT: The paper provides the insight to a possible impact of operating effects of large harbour on the water quality and fisheries resources. We assessed the DDT, PCB, HCH (in 2007–2008) and heavy metals Hg, Cd, Cu, Zn (in 2006–2009) content in the perch tissues in the impact area of the Port of Muuga (Baltic Sea). Perch have been treated as an indicator species in order to quantify the content of pollutants of coastal waters. The concentration of heavy metals in perch was low and did not exceed the average content in the Baltic Sea. The contamination with DDT (dominant isomer pp'DDE) in perch was slightly and with HCB significantly higher form average content in the Baltic Sea. The PCB (dominant isomers CB 138 and CB 153) content was relatively uniform in studied area. The environmental condition of seawater could be classified according to the quality criteria established for Baltic Sea water as moderate in the impact area of the Port of Muuga.

1 INTRODUCTION

Over 90% of world trade is performed by the international shipping industry. Ships generate residuals during the port operation. The environmental effects of port activities can be seen as direct effects—significant damage to water quality, marine life (fish, shellfish, plankton etc.), and indirect effects—decrease of water transparency, oxygen concentration, bioinvasion, bioaccumulation of toxins in fish etc. (Lodge 1993, Bailey *et al.* 2004). The excretion of toxins is mostly caused by the bilge water combined with other ship wastes and antifouling additives, which can be contaminated with toxic chemicals, including organochlorine compounds: PCBs, DDT, HCH, mercury and other heavy metals, polycyclic aromatic hydrocarbons (PAHs), pesticides etc. (Anon 2008).

Organochlorine compounds and heavy metals are ubiquitous contaminants in various compartments of the environment (Martin *et al.* 2003, Fu and Wu 2005). In spite of numerous countries having withdrawn from registered usage of Persistent Organic Pollutants (POP), these man-made chemicals still persist at considerable levels worldwide (Cleemann *et al.* 2000; Wurl & Obbard 2005; Doong *et al.* 2008).

These compounds are generally generated by anthropogenic processes and can be introduced into the water environment mainly through the industrial sewage and port activities. Due to their toxic, mutagenic, and carcinogenic characteristics, these persistent compounds are considered to be hazardous to the biota and environment (Roots *et al.* 2004, Vuorinen *et al.* 1998, Voigt 2005, Tulonen *et al.* 2007). In addition to the contamination of Persistent Organic Pollutants (POP) the distribution of trace metals can be affected by anthropogenic inputs and are globally distributed in wide range of concentrations in environment and so in biota (Doong *et al.* 2008). The trace metals circulation in aquatic environment results in deterioration of the water quality and long-term implication on ecosystem health (Fatoki & Mathabatha 2001, Ip *et al.* 2007).

The ecosystem of Muuga Bay (Northeasten Baltic Sea) and the adjacent areas have been studied since the 1950s (Järvekülg 1969). Previous surveys were largely focused on water quality issues and covered different aspects of impact assessment such as plankton, benthos, reproduction areas-, reproduction and feeding success of fish and also on fisheries (Järvik *et al.* 2007, Palm, 1985, Järvekülg 1969, Järvik *et al.* 2005, Kotta *et al.* 2007, Järv *et al.* 2009). However, there are only a few studies on the bioaccumulation of persistent organic pollutants and heavy metals.

The present paper provides the insight to the possible impact of operating effects of large harbour on the quality of seawater and on the quality of fisheries resource.

2 MATERIAL AND METHODS

2.1 *Study area*

The Port of Muuga, located in the central part of the southern coast of the Gulf of Finland (59^0 30' N; 24^0 58' E; Fig. 1), is the largest port in Estonia with annual potential turnover of 40 million t of cargo. The port is in active use since 1985. The last major unit—the Coal Terminal was put into operation in 2004 (Järvik *et al.* 2007). During several decades prior to the construction of the Port, the ecosystem of Muuga Bay was under the influence of sewage waters from the Maardu Chemical Plant (was closed in early 1990s) and from the Kehra Paper Mill (the wastewater treatment started in mid of 1990s). During the last two decades the only major source of antrophogenic pollution of bay originates from various activities of Muuga Port: dredging, occasional oil spills, ballast waters discharge etc. Muuga Bay and its adjacent small bays are characterized with low water salinity at 7–8 PSU and have a good connection to the open Gulf of Finland (Soomere *et al.* 2009).

2.2 *Studied species*

Perch (*Perca fluviatilis* L.), is widely distributed in fresh- and brackish coastal waters preferring shallow, sheltered habitats rich in macrophytes. Perch have been historically one of the most important fish species in Estonian coastal sea fishery, so in impact area of the Port of Muuga. Perch muscle tissue is relatively lean which makes perch highly demanded as the diet food on markets. Being a relatively sedentary (Järv 2000), the state of contamination of perch reflects the respective situation in its particular area of living. Therefore, perch has been recommended as an indicator species of biological effects of environmental pollution by HELCOM (2011).

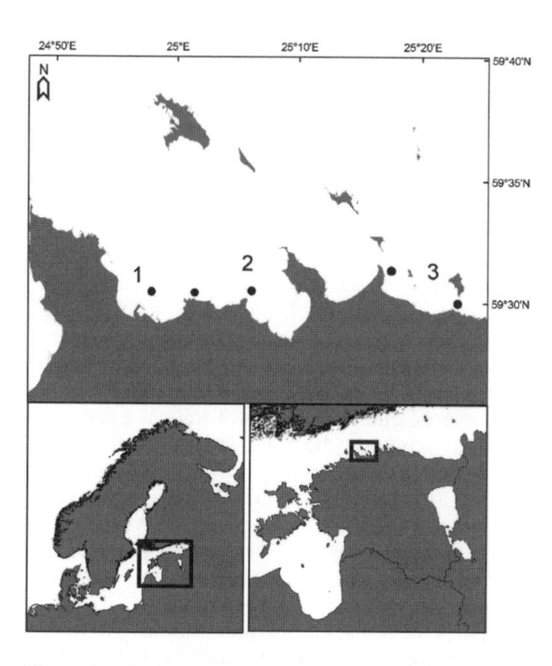

Figure 1. Sudy area: 1. Muuga Bay; 2. Ihasalu Bay; 3. Kolga Bay and sampling sites: •.

2.3 *Sampling and laboratory analyses*

Fish sampling was carried out in Muuga Bay and in its two adjacent areas: the Ihasalu, and Kolga Bays, altogether at five stations in July 2006–2009 (Fig. 1.). In each station the monitoring gill nets (1.8×28.7 m) with mesh sizes of 32, 44, 60, 72, 80, 92, 100 and 120 mm were used. The following parameters were measured: species composition of catches, total length (TL, cm) and weight (TW, g) of all specimens of all species caught by net with given mesh size. In addition age, maturity stage according to the six grade maturity scale (Anon 2007), of perch was assessed. Fallowing the methodology of HELCOM COMBINE programme (HELCOM 2011) only the female perch below 25 cm (TL) were chosen for chemical analysis. Perch was aged using opercular bones according to Tesh (1971).

For determination of zinc (Zn), cadmium (Cd) and copper (Cu) content were determined in liver tissues, while for determination of mercury (Hg) and persistent organic pollutants the pieces of muscle were dissected above the lateral line between the dorsal fin and tail. Samples of 1.5 g of muscle tissue were prepared for mercury analysis and of 10 g for organochlorine compounds. The samples of tissues were frozen immediately after collection and kept frozen until chemical analysis.

2.4 *Chemical analyses*

Prior to chemical analysis of heavy metals: Cu, Zn, Cd, Hg, the samples were melted and the dry weight (d.w.) determined. The tissue samples were homogenized and digested in concentrated HNO_3 and H_2O_2 (6:1). The copper and zinc were analyzed by AAS in flame-, the mercury was analyzed by AAS cold vapour and the cadmium by ETAAS method. The results represent the means of two replicates. Standards for AAS by Merek & Co were used.

For determination of POPs the samples of muscle tissue were homogenized with an IKA T25 homogenizer (Labassco AB, Pertille, Sweden). The samples extracted and cleaned up according Jensen *et al.* (1983), Haraguchi *et al.* (1992), Roots (1996), Roots & Talvari (1999). Internal standard IUPAC 189 was added. The recovery of organochlorines from the extraction and clean up procedures was measured. PCB and chororganic pesticides were analysed by capillary gas-chromatography (Varian 3400/3300) fitted with a Electron Capture Detector (ECD).

The samples were analyzed in the laboratory of Estonian Environmental Research Centre (EERC) which made most of the laboratory tests. The Estonian Accreditation Centre certifies that EERC has competence according to ISO/IEC 17025:2005. All solvents used were of highest quality commercially available; n-hexane, acetone and diethyl ether were obtained from Riedel-de Haen AG (Seelze, Germany) or Fisons Discol Solvent (UK) and methyl-tert-butyl ether (MTBE) from Rathburn Chemical (Walkerburn, Scotland). 2, 3, 4, 5, 3′, 4′, 5′, - heptachlorobiphenyl (IUPAC No 189), were syntesized at the Department of Environmental Chemistry of Stockholm University Wallenberg Laboratory. Clophen A-50 from Bayer AG (Germany) was used for the quantification of the total PCB.

The data were statistically tested with ANOVA analysis package.

3 RESULTS AND DISCUSSION

3.1 *Concentration profile of POPs*

The range and mean (±SE) concentrations of studied POP-s in perch lipids from the three stations: Muuga-, Ihasalu- and Kolga Bay, of the impact area of the Port of Muuga are represented in the Table 1.

The HCH (hexachlorocyclochexan) is an industrial mixture of five isomers: α-HCH, γ-HCH (lindane), β-HCH, δ-HCH and ε-HCH (HELCOM, 1996). The constitution of HCH depends on the environment: sediments contain more α-HCH and living systems more γ-HCH. As a rule the contents of only α-HCH and γ-HCH are analyzed in living organisms. The mean concentrations of α-HCH (0.003 mg*kg lipids^{-1}) and γ-HCH (0.0085 mg*kg lipids^{-1}) in perch from the Muuga system were generally higher compared to the mean values observed in other Baltic Sea regions (Bignert *et al.* 2010; HELCOM 2010), and to the mean concentrations of Estonian coastal sea:

α-HCH 0.0023; γ-HCH 0.007 mg*kg lipids^{-1} and so of the Gulf of Finland: α-HCH 0.002; γ-HCH 0.005 mg*kg lipids^{-1}. The concentrations of both α-HCH and γ-HCH studied in perch were higher in Muuga Bay and were decreasing eastwards ($R^2 = 0.84$, Fig. 2., Table 1). 80% of HCH conceive lindane (Fig. 3.).

The HCH concentration has decreasing trend in the whole Baltic Sea since the 1980s. Lindane concentration has been decreasing approximately 10% per year while the α-HCH has been decreasing even faster (Bignert *et al.* 2010).

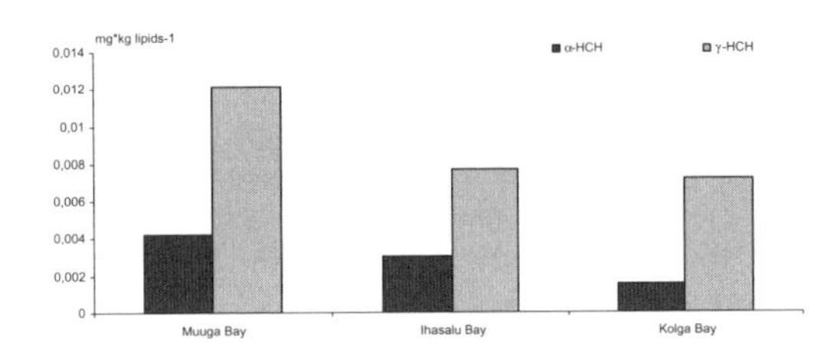

Figure 2. The total concentrations of ά-HCH and γ-HCH in perch from the impact area of the Port of Muuga.

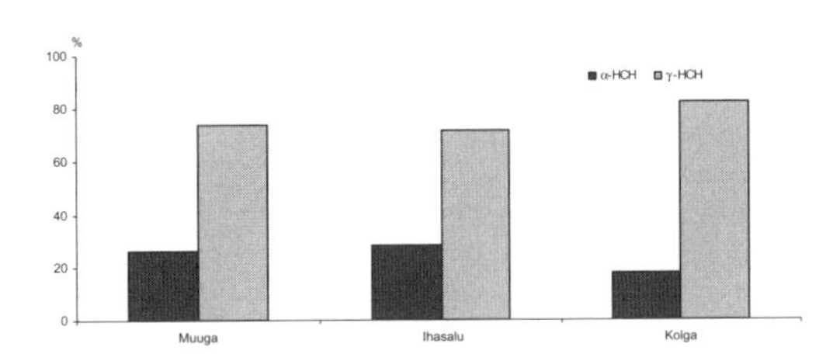

Figure 3. The content of studied HCH isomers: ά-HCH and γ-HCH, in perch from the impact area of the Port of Muuga.

Table 1. The concentration range (mg*kg^{-1}) and mean values (±SE) of POPs in perch lipids in the impact area of Muuga Harbour.

		Muuga Bay	Ihasalu Bay	Kolga Bay
α-HCH	Range	0.0015–0.007	0.0013–0.0048	0.0002–0.0035
	Mean ± SE	0.0042 ± 0.009	0.00304 ± 0.008	0.00152 ± 0.009
γ-HCH	Range	0,0064–0,02	0,005–0,0104	0,0028–0,011
	Mean ± SE	0.0121 ± 0.0024	0.0077 ± 0.0031	0.00717 ± 0.002
Σ PCB	Range	0,0246–0,434	0,0409–0,393	0,061–0,328
	Mean ± SE	0.227 ± 0.31	0.211 ± 0.22	0.194 ± 0.19
Σ DDT	Range	0,034–0,13	0,0226–0,13	0,0126–0,11
	Mean ± SE	0.0815 ± 0.003	0.0794 ± 0.002	0.0548 ± 0.002

DDT (dichlorodiphenylethane) is a hydrophobic persistent organic pollutant strongly absorbed by soil. In the environment DDT starts degraded *via* photolysis, aerobic- and anaerobic biodegradation (Aislabie *et al.* 1997). DDT and its breakdown products: p,p'DDE, p,p'DDT and p,p'DDD, absorbed quickly by organisms and accumulating in the food webs. Because of degradation the summary DDT decrease in environment. Usually the p,p'DDE is the most common isomer of DDT in aquatic organisms (Korhonen *et al.* 2004). The mean concentrations of summary DDT (0.065 mg*kg lipids^{-1}) in perch from the Muuga system (Muuga-, Ihasalu- and Kolga Bays) were higher compared to the mean values observed in the Baltic Sea (Bignert *et al.* 2010; HELCOM 2010) and also higher compared to the mean content found in perch of Estonian coastal sea: 0.026 mg*kg lipids^{-1} and from the Gulf of Finland perch: 0.044 mg*kg lipids^{-1}. The concentrations of summary DDT in perch were higher in Muuga Bay and were decrasing eastwards ($R^2 = 0.79$, Fig. 4., Table 1). Over 50% of summary DDT in perch of Muuda system conceives by DDE (Figure 5.). The concentration of DDE has decreasing trend in the whole Baltic Sea (appromaxely 10% per year) since the end of the seventies (Bignert *et al.* 2010).

PCB (polychlorinated biphenyl) has been in wide use in manufacturing processes especially as plasticizers and as insulators and fire retardants. PCBs distributed in the environment through inappropriate handling of waste, leakage of condensers and hydraulic systems etc. The toxic effect of PCBs is well documented (Alsberg *et al.* 1993, Jensen *et al.* 1977, Bignert *et al.* 2010). Seven CB-congeners: CB-28, CB-52, CB-101, CB-118, CB-138, CB-153 and CB-180, are listed as mandatory contaminants and should be analysed and reported within the OSPARCOM and the HELCOM conventions. The mean contents of summary PCBs (0.21 mg*kg lipids^{-1}) in perch from the Muuga system were generally lower compared to the mean values: 0.6–1.2 mg*kg lipids^{-1} in the Baltic Sea (Bignert *et al.* 2010; HELCOM 2010), but higer from the mean content of Estonian coastal sea perch: 0.1 mg*kg lipids^{-1} and from the Gulf of Finland perch: 0.119 mg*kg lipids^{-1}. The concentrations of summary PCBs in perch were higher in the Muuga Bay and were decreasing eastwards ($R^2 = 0.88$) (Fig. 4., Table 1). The dominant PCB isomers were CB-138 and CB-153 which conseiv 50–60% of summary PCBs in perch of Muuga system (Fig. 6.). In all cases the content of CB-153 was higher as the content of BC-138 thereby the mean concentrations of those two isomers exceeded 0.07 mg*kg lipids^{-1}.

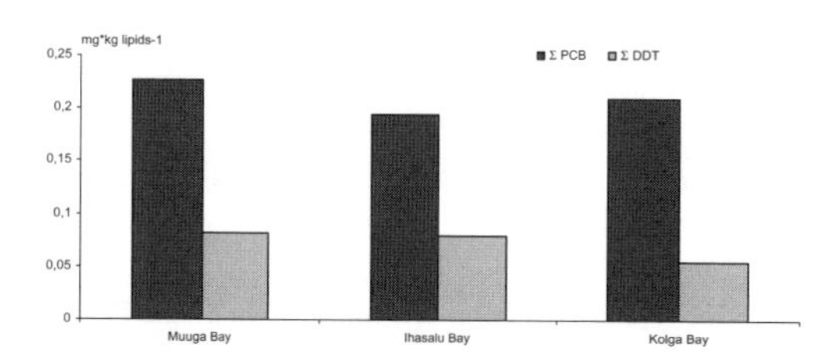

Figure 4. The total concentrations of Σ PCB and Σ DDT in perch from the impact area of the Port of Muuga.

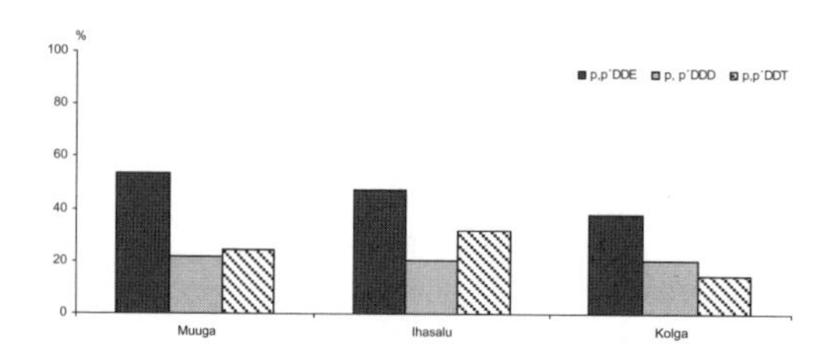

Figure 5. The content of studied DDT isomers in perch from the impact area of the Port of Muuga.

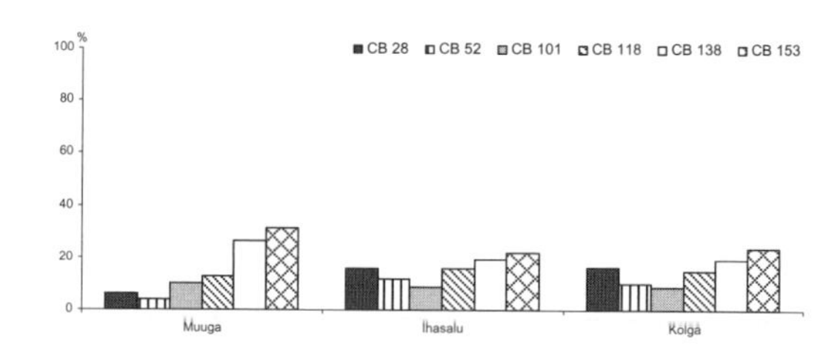

Figure 6. The content of studied PCB isomers in perch from the impact area of the Port of Muuga.

3.2 *Concentration profile of heavy metals*

The concentrations of studied heavy metals in perch from Muuga system (Table 2) were generally higher compared to the respective values reported for the Baltic Sea in other studies (Szefer *et al.* 2003; Voigt 2003 & 2005; Bignert *et al.* 2010; HELCOM 2010).

Only the Cd content was lower compared to the level observed in semi-enclosed Moonsund Archipelago Sea (Voigt 2005) and in the Gulf of Bothnia (Bignert *et al.* 2010). The port activities may substantially increase the heavy metal loading into aquatic ecosystems (Tulonen *et al.*, 2006), what may have pronounced ecological consequences, including the toxic effects (Webb *et al.*, 2006; Sorensen, 1991; Wiener & Spry, 1996). Some heavy metals, like Hg and Cd, and their compounds are also considered as carcinogenic substances (Janssen *et al.*, 2000; Diaconescu, 2008). Accumulation of Hg and Cd

Table 2. The concentration (mg/kg, w.w.) range and mean values (±SE) of trace metals in perch muscle (Hg) and liver (Zn, Cu, Cd) tissue in the impact area of Muuga Harbour.

		Muuga Bay	Ihasalu Bay	Kolga Bay
Zn	Range	13.8–24.5	17.6–22.7	12.8–20.4
	Mean ± SE	20.44 ± 7.33	20.36 ± 6.84	19.93 ± 2.29
Cu	Range	3.11–8.25	2.30–6.06	2.24–4.38
	Mean ± SE	4.65 ± 2.55	4.41 ± 2.59	3.12 ± 0.466
Hg	Range	0.042–0.117	0.058–0.098	0.046–0.091
	Mean ± SE	0.075 ± 0.052	0.068 ± 0.034	0.045 ± 0.025
Cd	Range	0.028–0.144	0.031–0.091	0.032–0.052
	Mean ± SE	0.072 ± 0.083	0.072 ± 0.083	0.037 ± 0.048

in fish tissues affects their consumers. While the fish is very important in the human diet, consumption of contaminated fish could pose a significant threat to human health (Schmitt *et al.* 2006; Kalvins *et al.* 2009).

Perch is one of the most important commercial fresh water fishes in study area. Therefore the state of perch in the area was assessed on the basis of contaminant with heavy metals according to the OSPAR Background Assessment Criteria. The following quality classes were applied (HELCOM 2010): for Cd < 0.026 mg kg^{-1} = "good"; 0.026–0.2 mg kg^{-1} = "moderate"; >0.2 mg kg^{-1} = "poor" and for Hg < 0.035 mg kg^{-1} = "good"; 0.035–2.0 mg kg^{-1} = "moderate"; >2.0 mg kg^{-1} = "poor" (w.w.). The values below the BAC 0.026 mg kg^{-1} and 0.035 mg kg^{-1} respectively, indicate the low concentration of the metal. According to this criteria, the perch found in Port of Muuga impact area fall with respect to Cd and Hg content into the "moderate" respect to Cd and Hg content into the "moderate" quality class (Fig. 7.).

It can be noted that the concentrations of all studied trace metals: Zn, Cu, Hg, Cd, studied in perch were higher in Muuga Bay had a decreasing eastwards (respectively R^2 = 0.83, R^2 = 0.86, R^2 = 0.92, R^2 = 0.89) (Figs. 8. & 9, Table 2), which is in good agreement with the distribution patterns of POPs. The higher concentrations of trace metals detected at Muuga Bay may be attributed to the emissions from port activities. The increasing trend in trace metal concentrations in perch tissues of Muuga system can partly explained by water mixing causing a change in salinity. The heavy metals are usually associated with fine particles in freshwater (Birch & Taylor 2000, Singh *et al.* 2003). When the fine particles move downstream the increasing in salinity enlarges the particle diameters resulting in the deposition of metal concentrations in the sediments (Doong *et al.* 2008). Indissoluble compounds can not enter any more into the food webs.

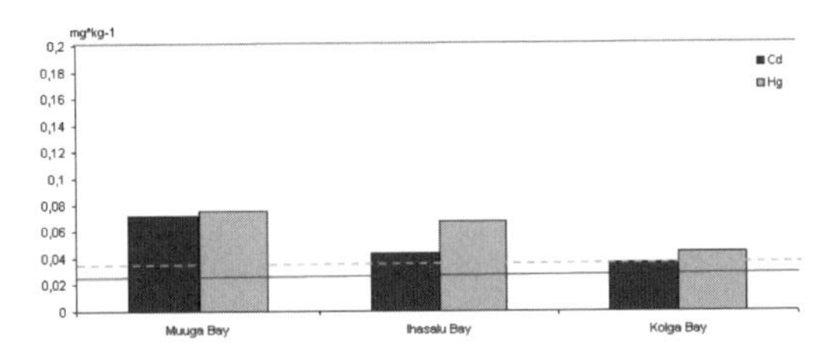

Figure 7. The status of perch in impact area of the Port of Muuga (w.w.) according to the OSPAR Background Assessment Criteria for mercury and cadmium: boder between "good" and "moderate" for -------- Cd and ——— for Hg.

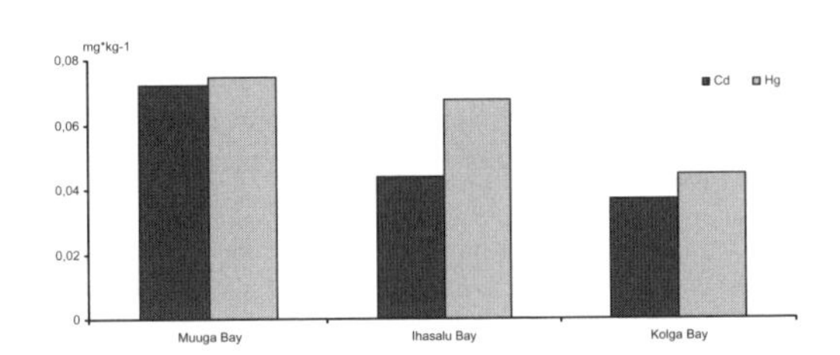

Figure 8. The average content of cadmium in liver- and mercury in muscle tissue of the perch from the impact area of the Port of Muuga (w.w.).

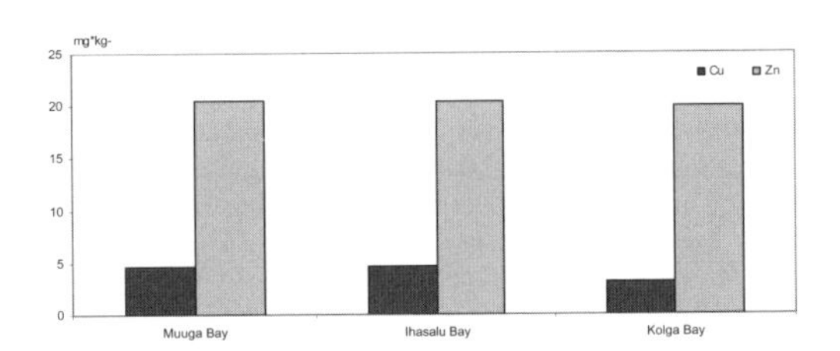

Figure 9. The average content of Zn and Cu in liver tissue of perch from the impact area of the Port of Muuga (w.w.).

4 CONCLUSIONS

1. Elevated content of Hg- and Cd compared the the averages of the Baltic Sea and in the area with no river discharges allow assuming that the higher concentrations result from the operations of the Port of Muuga.
2. Presence of the PCBs in the aquatic environment can usually be attributed to accidents in handling of the chemicals. Lower than average concentration of PCBs in the Muuga Bay indicates at appropriate the technological procedures and presence of adequate environmental monitoring schemes in the port.
3. It may be assumed that mixing of water masses with different salinity as a result of the port operations accelerates sedimentation of pollutants and their removal from the active food webs.

ACKNOWLEDGEMENTS

Funding for this research was provided by target financed project SF0180013s08 of the Estonian Ministry of Education and Research and by the Port of Tallinn.

REFERENCES

Aislabie, J.M., Richards, N.K. & Boul, H.L. 1997. Microbial degradation of DDT and its residues—a review. New Zealand Journal of Agricultural Research, Vol. 40: 269–282.

Alsberg T., Balk L., Nylund K., de Wit C., Bignert A., Olsson M. & Odsjö T. 1993. Persistent Organic Pollutants and the Environment. Swedish Environmental Protection Agency, report 4246.

Anon 2007. Manual for the Baltic International Trawl Surveys. International Council for the Exploitation of the Sea. March 2007, Rostock, Germany: 9–10.

Anon 2008. Certification of Sea Port facilities in the Mediterranean. Project Concept. Waste Environment Cooperation Centre, Draft 2.

Bailey, D.B., Plenys, T., Solomon G.M., Campbell, T.R., Feuer, G.R., Masters, J. & Tonkonogy, B. 2004. Harboring pollution. The Dirty Truth about U.S. Ports. Natural Resources Defense Council, March 2004.

Bignert, A., Danielsson, S., Nyberg, E., Apslund, L., Eriksson, U., Nyland, K.A. & Berger, U. 2010. Comments concerning the National Swedish contaminant Monitoring Programme in Marine Biota, 2010. Swedish Museum of Natural History. Report nr. 1: 2010.

Birch, G.F. & Taylor, S.E., 2000. The use of size-normalized procedures in the analysis of organic contaminants in estuarine sediments. Hydrobiology 431: 129–133.

Cleemann, M., Rigetm, F., Paulsen, G.B., Klungsøyr, J. & Dietz, R. 2000. Organochlorines in Greenland marine fish, mussels, and sediments. Sci. Total Environ. 245: 87–102.

Diaconescu, C., Urdes, L., Marius, H., Ianitchi, D. & Popa, D., 2008. The influence of heavy metal content on superoxide dismutase end glutathione peroxidase activity in the fish meat originated from different areas of Danute river. Roumanian Biotechnological Letters, Vol. 13, No. 4: 3859–3962.

Doong, R.A., Lee, A., Sun, Y.C. and Wu, S.C. 2008. Characterization and composition of heavy metals and persistent organic pollutants in water and estuarine sediments from Gao-ping River, Taiwan. Marine Pollution Bulletin 57: 846–857.

Fatoki, O.S. & Mathabatha, S. 2001. An assessment of heavy metal pollution in the East London and Port Elizabeth harbours. Water SA 27: 233–240.

Fu, C.T. & Wu, S.C. 2005. Bioaccumulation of polychlorinated biphenyls in mullet fish in a former ship dismantling harbor: a PCB-contaminated estuary and nestled coastal fish farms.Mar. Pollut. Bull. 51: 932–939.

Haraguchi, K., Athanasiadou, M., Bergman, A., Hovander, L. & Jensen, S. 1992. PCB and PCB methyl sulfones groups of seals from the Swedish waters. AMBIO, 21: 546–549.

HELCOM 2007. Baltic Sea Action Plan. Adopted on 15 November 2007 in Krakow, Poland by the HELCOM Extraordinary Ministerial Meeting. http://www.helcom.fi/stc/files/BSAP/BSAP_Final.pdf. 2007.

HELCOM 2010. Hazardous substances in the Baltic Sea. An integrated thematic assessment of hazardous substances in the Baltic Sea. *Baltic Sea Environment Proceeding*, No. 120B: 109 pp.

HELCOM 2011. Manual for Marine Monitoring in the COMBINE Programme of HELCOM, Helsinki. http://www.helcom.fi/groups/monas/CombineManual/en_GB/main/ 12.01.2011.

Ip, C.C.M., Xi, X.D., Zhang, G., Wai, O.W.H. & Li, Y.S., 2007. Trace metal distribution in sediments of the Pearl River Estuary and the surrounding coastal area, South China. *Environ. Pollut.* 147 (2): 311–323.

Jensen, S., Reuthergardh, L. & Jansson, B. 1983. Analytical methods for measuring organochlorines and methyl mercury by gas chromatography. Analysis of metals and organochlorines in fish. *FAO Fish Tech. Paper*, Rome, Vol. 212: 21–33.

Janssen, C.R., Schamphelaere, K., Heijerick, D., Muyssen, B., Lock, K., Bossuyt, B., Vanheluwe, M. & Van Sprang, P., 2000. Unceratines in the environmental risk assessment of metals. *Hum. Ecol. Risk. Assess.* 6: 1003–1018.

Järv, L., 2000. Migrations of the perch (*Perca fluviatilis* L.) in the coastal waters of Western Estonia.- Proc. *Estonian Acad. Sci. Biol. Ecol.*, 49 (3): 270–276.

Järv, L., Kotta, J., A.; Raid, T., Järvik, A., Drevs, T., Jaanus, A., 2009. The impact of selected Baltic ports on adjacent fish fauna: revealed effects and uncertainties. *In: Proceedings of the 13th International Maritime association of the Mediteranean (IMAM 2009); Istanbul, Turkey, 12–15 October 2009. Guedes Soares, C. (ed.) Isyanbul Technical University, 2009 (Towards the Sustainable Marine Technology and Transportation)*, 853–862.

Järvik, A., Raid, T., Drevs, T., Järv, L. & Jaanus, A., 2007. Monitoring of the impact of Muuga Port activities on fish communities and fishery in Muuga Bay in 1994–2006. *In: Proceedings of the 12th International*

Congress of the IMAM, Varna, Bulgaria, 2–6 September 2007,(Eds) C.G. Soare &, P.N. Kolev. Taylor & Francis, London/Leiden/New York/ Phyladelphia/ Singapore. Vol.2: 735–740.
Järvekülg, A., 1969. The influence of sewage water to the development of zoobenthos and capacity for self purification in the bays of the Baltic Sea. *In XIV Conference of the water-bodies of inner Baltic Sea, (eds.) Andrusaitis, G.P., Katsalova, O.L., Kumsare, A.J., Laganovskaja, P.J., Leinerte, M.P., Matison &, M.N. & Salna, L.J., Zinatne, Riga*, 57–62. (In Russian)
Kipling, C. & Le Cren, E.D., 1984. Marc-recapture expeiments on fish in Windermere, 1943–1982. *Journal of Fish Biology* 24: 395–414.
Klavins, M., Potapovics, O. & Rodionov, V., 2009. Heavy metals in fish from Lakes in Latvia: concentrations and trends of changes. *Bull. Environ. Contam. Toxicol.* 82: 96–100.
Korhonen, K., Liukkonen, T., Ahrens, W., Astrakianakis, G., Boffetta, P. & Burdorf, A. 2004. Occupational exposure to chemical agents in the paper industry. *International Archives of Occupational and Environmental Health* 77(7): 451–460.
Kotta, J., Lauringson, V. & Kotta, I. 2007. Response of zoobenthic communities to changing eutrophication in the northern Baltic Sea. *Hydrobiologia*, 580: 97–108.
Lodge D.M. 1993. Biological invasions—lessons for ecology. *Trends in Ecology and Evolution* 8: 133–137.
Martin, M., Richardson, B.J. & Lam, P.K.S. 2003. Harmonisation of polychlorinated biphenyl (PCB) analyses for ecotoxicologucal interpretations of southeast Asian environmental media: what's the problem? *Mar. Pollut. Bull.* 46: 159–170.
Palm, T. 1985. The potential impact of heavy metal concentrations in the waters of southern Gulf of Finland on Baltic herring stocks. *Finn. Fish. Research*, 6: 71–76.
Roots, O. & Talvari, A. 1999. Bioaccumulation of toxic organic compounds and their isomers into the organism of seals in West-Estonian Archipelago Biophere Reserve. *Environmental Monitoring and Assessment, Kluwer Academic Publishers*, 54: 301–312.
Roots, O. 2003. Halogenated cotaminants in female perch from the Matsalu Bay (Baltic Sea). *Chemisrty and Ecology*, Vol 19: 1–3.
Roots, O., Järv, L. & Simm, M. 2004. DDT and PCB concentrations dependency on the biology and domicile of fish: an example of perch (*Perca fluviatilis* L.) in Estonian coastal sea. *Fresenius Environmental Bulletin*, 13 (7): 620–625.
Singh, M., Muller, G. & Singh, I.B. 2003. Geogenic distribution and baseline concentration of heavy metals in sediment of the Ganges River, India. *J. Geochem. Explor*. 80: 1–17.
Schmitt C.J., Brumbaugh W.G., Linder G.L. & Hink J.E. 2006. A screening level assessment of lead, cadmium and zinc in fish and crayfish from Northeastern Oklahoma, USA. *Environmantal Geochemical Health*, 28: 445–471. doi: 10.1007/s10653-006-9050-4.
Szefer, P., Domagała-Wieloszewska, M., Warzocha, J., Garbacik-Wesołowska, A. & Ciesielski, T. 2003. Distribution and relationship of mercury, lead, cadmium, copper and zinc in perch (*Perca fluviatilis*) from the Pomeranian Bay and Szczecin Lagoon, southern *Batic. Food chemistry*, 81: 73–83.
Sorensen, E.M. 1991. Metal poisoning in fish. *CRC Press, Boca Raton, FL.*
Soomere, T., Leppäranta, M. & Myrberg, K. 2009. Highlights of physical oceanography of the Gulf of Finland reflecting potential climate changes. *Boreal Environment Research*, 14(1): 152–165.
Tesch, F.W. 1971. Age and growth. In: (ed.) Ricken W.E. Methods for assessment of fish production in fresh waters. Blackwell Scientific Publications, Oxford, UK: 98–130.
Tulonen, T., Pihlström, M., Arvola, L. & Rask, M. 2006. Concentrations of heavy metals in food web compentents of small boreal lekes. *Boreal Environment Research* 11: 185–194.
Voigt, H.-R. 2003. Heavy metal concentrations and condition of some coastal Baltic fishes. *Seminar on Environmental Research, 10 April, Department of Limnology and Environmental Protection, University of Helsinki-Helsingfors.*
Voigt, H.-R. 2005. Condition and cotaminants of Baltic fishes. Bilateral Estonia-Finnland co-operation on the state of the Aquatic Environment of Väike-Väin Strait in Western Estonia. *Estonian Society for Nature Conservation and Finnish Assotiation for Nature Protection. Report.*
Vuorinen, P.J., Haahti, H., Leivuori, M. & Miettinen, V. 1998. Comparison and temporal trends of organochlorines and heavy metals in fish from the Gulf of Bothnia. *Marine Pollution Bulletin*, 36: 236–240.
Webb, M.A.H., Feist, G.W., Fitzpatrick, M.S., Foster, E.P., Schrek, C.B., Plumee, M., Wong, C. & Gundersen, D.T. 2006. Mercury Concentration in Gonad, Liver, and Muscle of White Stugeron Acipenser transmontanus in the Lower Columbia River. Arch. *Environ. Contam. Toxicol*. 50: 443–451.
Wiener, J.G. & Spry, D.J. 1996. Toxicological significance of mercury in freshwater fish. In: (eds) Beyer W.N., Heinz, G.H. & Redmon-Norwood, A.W. Environmental contaminants in wildlife interpreting tissue concentrations, Lewis, Boca Raton, FL, 297–339.
Wurl, O. & Obbard, J.P. 2005. Organochlorine pesticides, polychlorinated biphenyls and polybrominated diphenyl ethers in Singapore's coastal marine sediments. *Chemosphere* 58 (7): 925–933.

8.2 *Noise*

Sustainable Maritime Transportation and Exploitation of Sea Resources – Rizzuto & Guedes Soares (eds)
© 2012 Taylor & Francis Group, London, ISBN 978-0-415-62081-9

Acoustic impact of ships: Noise-related needs, quantification and justification

A. Badino, D. Borelli, T. Gaggero, E. Rizzuto & C. Schenone
University of Genoa, Genoa, Italy

ABSTRACT: The evaluation of the acoustic impact of ships is a complex problem, involving not only different sources but also different kinds of receivers and transmission paths. The problem of health and comfort for crew and passengers on board has been considered since a few decades, leading to quite a structured and detailed framework of Norms and Requirements. On the other hand, only in recent years a growing attention has been devoted to air-borne noise emissions outside the ship, for which requirements are not present. Underwater noise emissions have, even more recently, gained attention for their potential interference with mammals' communications and with the sophisticated use such animals make of acoustic signals for interacting with their living environment. An aim of the present work is to review the state of the art in the three areas above identified outlining the differences in the specific fields as regards: the present knowledge of the phenomena involved; the accuracy of the models available for the description of noise propagation; the accuracy of the models available for the quantification of noise effects.

1 INTRODUCTION

The subject of the environmental impact of anthropogenic activities is nowadays a key issue. The impact is to be assessed in respect to a human environment (workplace, living areas) as well as to a natural environment (flora, fauna, landscape), at a local as well as global level. This has brought to identify and evaluate types of impacts that earlier were not (or only in part) considered. Among these types of impact is noise pollution: for shipping as for other activities in the industrial and transportation sector, this subject has recently gained an increased attention.

The noise generated by the ship has a direct impact on the quality of the workplace and of the living environment for crew and passengers on board, but affects also third parties when ships are sailing along the coast, in bay, channels or are moored at quay.

As regards the impact on the natural environment, it has been pointed out that the diffused source of noise represented by shipping has implied an increase of the background noise of the ocean, with a global modification of the living environment of the marine fauna.

The approach to the control of noise emissions from ships, therefore, results to require a much more holistic approach than earlier applied in this field.

This is the approach followed in the SILENV collaborative project (Ships oriented Innovative soLutions to rEduce Noise & Vibrations), funded by the E.U. within the 7th Framework Programme. The present paper refers to the first phase of the project, related to identification of the needs for noise control and how they can be quantified and justified.

In the following, the subject of the acoustic impact of ships will be addressed with reference to the different affected environments (emissions towards the internal part of the ship, radiation in air, radiation in water).

2 NOISE INSIDE THE SHIP

2.1 *Normative development*

The problem of the impact of noise on workers on board ships has been addressed since a few decades. The first key document establishing requirements in the field is the "Code on noise levels on board ships" issued by the International Maritime Organisation in 1981. This Code has been used by several bodies as a reference for the formulation of norms and requirements. The aim of the Code is to limit noise levels and to reduce the exposure to noise, in order to: prevent communication problems possibly causing danger; protect the seafarer from short or long term physical injuries to the auditory system; provide the seafarer with an acceptable degree of comfort in rest and recreational spaces. As stated above, the focus is on the workers on board and in particular on their health. To reach this goal, in the

Code limit levels are fixed on: the maximum allowable pressure levels in spaces (see Ch. 2.2 below); the maximum exposure time (noise dose: see again Ch. 2.2) and the minimum insulation index (see Ch. 2.2) of vertical divisions between spaces.

In addition to this document, in the 90's, almost all Classification Societies provided additional Class Notations, called Comfort Classes (CC), Following their name, CC focus mainly on the comfort of both crew and passengers. For this reason, the noise limits provided are more severe than those contained in the IMO Code (more health-oriented). Probably for the same reason, in the CC the time of exposure to noise is not considered. Other basic differences are represented by the subjects of vibrations and of impact noise (see Ch. 2.2), covered only in the CC.

In the light of what above it is clear that, as regards the noise on board ships, norms and requirements are quite developed, following a good knowledge in this field, resulting from decades of experience.

Presently an updating of the Code is under discussion within IMO (IMO 2009a), with the aim of making noise limits mandatory through the insertion of an explicit reference to the Code in the text of Chap II-1 Reg.36 of the SOLAS (IMO 1974). Noise level limits, too, are in the process of being lowered, but the general framework is going to be maintained.

Nevertheless, improvements in the formulation of requirements are possible as it will be discussed in the following.

2.2 *Main indicators*

Noise indicators are defined with the goals of simplifying the quantification and the rating of noise effects. As mentioned, methods for avoiding hearing damage and discomfort differ from each other and this reflects in the choice of the indicators adopted and in the limitations expressed in terms of such indicators.

The noise level in dB(A) is a widely used indicator in a major part of norms and requirements. It is based on a filtering of the actual sound spectra. The A-weighting curve follows the hearing sensibility curves (isophonic curves) in a simplified way. For every frequency band, a weighting factor is applied to the dB level of the sound in the same band. The single filtered levels in each band are then summed, in the logarithmic sense and a single dB(A) number is derived.

$$L_{A-weighted}[dB(A)] = 10\log_{10}\left(\sum_i 10^{\frac{dB_i - AF_i}{10}}\right) \quad (1)$$

where dB_i are the levels in dB of the measured sound in the i-th frequency band and AF_i is the value of the A weight in the i-th frequency band.

The main advantage in using this indicator is that the comfort of an ambient is evaluated by means of a single value and its computation is very easily implemented in modern sound level metres. On the other hand, being a sort of global indicator, the dB(A) level is not able to capture and evaluate properly in general the distribution in frequency of the sound energy and in particular possible concentrations (tonal components). In other technical fields, in addition to the $L_{\text{A-weighted}}$ indicator, other methods are used to evaluate the noise annoyance in order to take into account the energy distribution in the whole frequency range. In these methods (e.g the Noise Criteria: NC, the Room Criteria: RC, the Room Noise Criteria: RNC) the sound energy is decomposed in frequency bands and the annoyance is evaluated separately. In addition, some of these methods include an acronym that classifies the sound quality (see e.g. Badino et al. 2011a).

The noise dose is an indicator mainly used to preserve the worker's health in a noisy working ambient (Glorig 1988). It evaluates the total energy (A-Weighted) perceived by an individual during a fixed period (usually 8/24 hours). A maximum allowable level in the 8 or 24 hours is defined; this level, kept constant for the 8/24 hours, corresponds to the maximum allowable energy. Actually, during the working shift, a worker can be exposed to different levels for different time periods. To compare such exposures another parameter is fixed: the exchange rate (e.r.). The e.r. establish how the allowable level increases halving the exposure time; through this parameter is possible for each couple (level + exposure time) to calculate the percentage of the total allowable energy. A limit of 100% is set to the sum of the different contributions: overcoming it, countermeasures must be adopted. The percentage of energy is evaluated by means of the following formula:

$$\%Dose = 100 \times \sum_i \frac{t_i \cdot 2^{\left(\frac{L_i - \text{Maximum}}{\text{exchange rate}}\right)}}{8} \quad (2)$$

where t_i is the ith time of exposure to the ith sound level L_i.

In the norms different maximum allowable levels and exchange rates are present. For example, in the IMO Code the maximum level is fixed at 80 dB(A) for 24 hours and the exchange rate is 3 dB(A); the same e.r. is found in ISO R1999 (ISO 1971) but the level corresponding to 24 hours is there 85 dB(A). In other norms the e.r. is 5 dB(A), see for instance OSHA 1910.95 (OSHA 1974). From an energetic view point the e.r. of 3 dB(A) corresponds to a constant limit on the energy perceived; the e.r. of 5 dB(A) does not have a physical meaning and implies acceptance of higher levels for low exposure times.

The indicators discussed so far refer to the noise that reaches the receiving position. More precisely, limits in dB(A) are referred to a specific ambient, while limits on the noise dose are in principle referred to a specific worker (who could be exposed to noise in different locations during the shift hours). The fulfillment of these limits can be attained at a design stage by a careful acoustic planning on board, acting either on sources (engines, ventilation, etc.) or on transmission paths (or both). In fact sources and propagation paths are, in principle, known because they are internal to the ship and do not vary during the ship life.

On the other hand, there are cases in which the source of noise is represented by people on board. In this case, the only element of the noise chain on which it is possible to act is the transmission path. For this reason, requirements are set, both in the IMO code and in the CC, on specific parts of the transmission path, aiming mainly at ensuring quietness in cabins. These provisions are expressed by means of two indicators: the insulation index and the impact noise levels.

To evaluate the insulation index, the apparent transmission loss has to be calculated in every frequency band (usually 1/3 octave bands). It is defined as follows

$$R' = 10\log_{10}\frac{W_1}{W_2 + W_3}dB \tag{3}$$

where W_1 is the incident acoustic power on the wall, W_2 is acoustic power transmitted to the receiving ambient through the wall and W_3 is the acoustic power transmitted to the receiving ambient through the various transmission paths connecting the two locations.

The meaning of the single terms is clarified in the sketch reported in Figure 1.

This index evaluates the ability, of a subdivision between two spaces, of reducing the sound level transmitted from the source space to the receiver space. On board ships it is referred to subdivisions between cabins and cabins and between public spaces and cabins.

To evaluate the impact noise level, the standardized impact sound pressure level has to be calculated in every frequency band (usually 1/3 octave bands). It is defined as follows

$$L'_{nT} = L_i - 10\log_{10}\frac{T}{T_0}dB \tag{4}$$

where L_i is the impact sound pressure level, T is the reverberation time in the receiving room and $T_0 = 0.5$ s is a reference reverberation time.

In the sketch presented in Figure 2 the meaning of the index is clarified.

This index is referred to the horizontal subdivision of spaces on board. In particular is referred to decks that separate a public space (restaurants, gyms etc.) from a cabin in order to control the noise transmitted to the space below by people walking on deck.

Methods have been developed to convert frequency dependent values of airborne sound insulation, R′ and L'_{nT}, into a single number characterising the acoustical performance. These single-number quantities are intended for rating the airborne noise insulation and for simplifying the acoustical requirements in shipbuilding codes, according to varying needs.

2.3 *Possible improvements*

The indicators described above are widely used in norms and requirements devoted to the control of noise disturbance. Nonetheless, some aspects of the noise annoyance, linked with particular situations that can take place on board, can escape from an evaluation carried out with the classic indica-

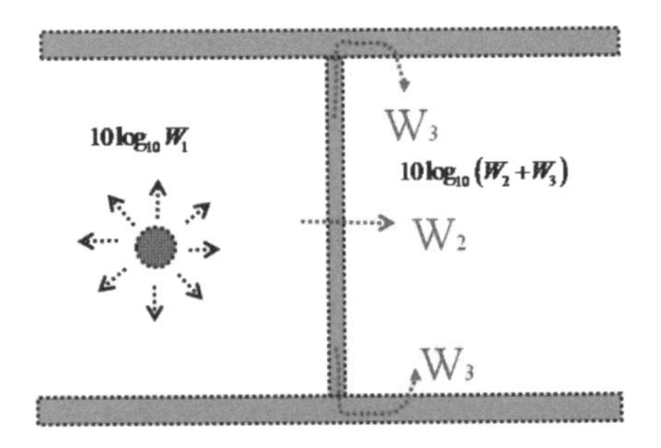

Figure 1. Sketch of the quantities for the definition of the insulation index.

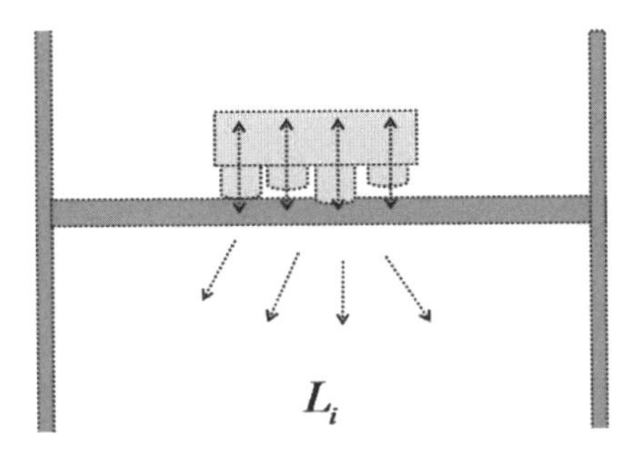

Figure 2. Sketch illustrating the definition of impact noise level.

tors. In the following a description of these cases will be given, together with some possible recommendations to overcome these problems.

The effects that cannot be captured by the dB(A) indicators or the noise dose are those linked with concentrations of acoustic energy in time or frequency. To overcome the problem of the evaluation of the annoyance due to tonal components, several methods are in use in the civil field (Beranek 1988), and it can be extended to ships. Some of this methods are: the Noise Rating (NR), the Noise and Balanced Noise Criteria (NC, NCB), the Room Criteria (RC), the Room Criteria Mark II (RC Mark II), the Room Noise Criteria (RNC). In these methods the noise is analysed decomposing the signal in the single frequency bands, evaluating not only the total energy of the signal, but also how the energy is distributed in frequency.

As regards the concentration of energy in time or transient noises, a possible solution could be to evaluate the 'instantaneous' sound pressure level (by a time record of a few seconds) and to compare the value with the reference limit value of the space considered that is meant to be evaluated on a longer time period (of the order of a few minutes). The difference between the two values should be lower than a fixed value.

The annoyance of these possible concentrations of energy in time and frequency can be stressed by the general trend to lower limit levels set in dB(A) followed by norms and requirements in the last years. In fact a higher broadband background noise can mask these components. A debate is open on this question because if on one side the need of more comfort on board brings to the reduction of limits on broadband noise, on the other side this reduction implies an increased perception of existing noise components that can jeopardise the improvement.

About the noise concentration in time, a specific indicator has been introduced (Finegold 1994), named "% Awakenings". This sleep disturbance criterion is based on the L_{AE} and is obtained from a regression fit of the data from 21 sleep disturbance studies, which gives the following equation:

$$\%\ Awakenings = 7.1 \cdot 10^{-6} \cdot L_{AE}^{3.5} \qquad (5)$$

where L_{AE} is A-weighted sound exposure level

This indicator permits to correlate percentage awakenings and indoor exposure level, so assessing the sleep disturbance of noise.

Another important aspect of the acoustic comfort of an ambient is the speech intelligibility, i.e. the ability of hearing and understanding in a clearly way a person talking, the music etc. This aspect of comfort is quite different from the others treated so far; in fact the goal here it is not to limit the noise but on the contrary is to ensure that a

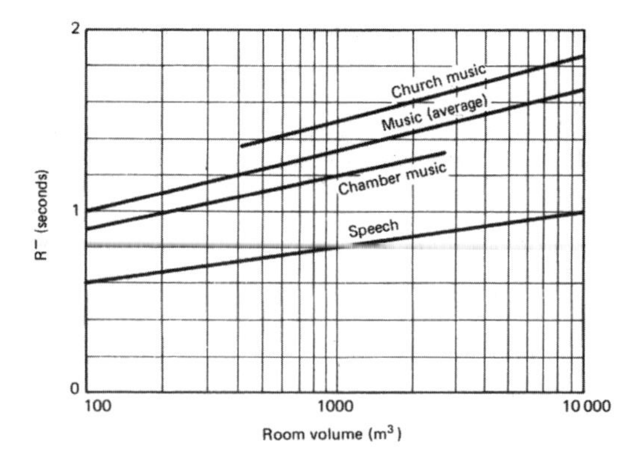

Figure 3. Optimum reverberation time (Pretlove & Turner 1991).

sound reaches the receiver without being altered. To this aim, a possible improvement for requirements on board, could be to evaluate the reverberation time (RT60) of the ambient. It is defined as the time required for reflections of a direct sound to decay by 60 dB below the level of the direct sound. The measured value can be compared with the optimum reverberation time for that kind of room (see Fig. 3) and countermeasures to align the measured reverberation time with the optimum can be adopted: this parameter is a very important issue for public spaces such as auditoriums, restaurants, games rooms and so on.

Concerning the transmission path, it is to be noted that the main aim of the noise insulation index is to ensure the quietness. The index was designed in particular to evaluate the ability of the wall to cut the frequencies of the speech. On the other hand, in the case of specific public space (e.g. a discotheque) adjacent to a cabin the disturbing sound energy could be concentrated in different frequency ranges.

To evaluate properly the isolation of the wall the spectral adaptation term C, introduced in addition to the Insulation Index in the second edition of ISO 1996, could be used in order to account for this effect.

3 NOISE IN AIR OUTSIDE THE SHIP

3.1 *Normative development*

The problem of airborne noise pollution of ships has been addressed only recently. The normative framework is substantially not present. At European level the only directive that deals with the noise emissions of marine vehicles is the Directive 2003/44/EC in which limits for airborne noise are given at source for recreational crafts with stern drive engines in function of the engine power. This Directive is devoted to a very specific typology of vehicles, in the other cases reference can be made to the Directive 2002/49/EC that addresses the problem of noise pollution of industrial plants among which ports are included, but ships are not specifically mentioned.

The need of specific regulations for ship noise is underlined by the increasing number of complaints sent to local Authorities in Europe regarding noise annoyance due to ships at quay, entering ports or in navigation along the coast. Such cases are pushing local administrations to face the problem.

3.2 *Main indicators*

As regards airborne noise emissions from ships, no specific indicator exists. The indicator presently used in norms is the L_{DEN} (Level day-evening-night). This indicator, contained in the Directive 2002/49/EC above, is aimed at representing the impact of noise emitted by major sources (road, rail and aircraft vehicles and infrastructures, industrial plants, equipment and mobile machinery) in the short, medium and long term. The L_{DEN} is defined as follows:

$$L_{DEN} = 10\log_{10}\frac{1}{24}\left(12\cdot 10^{\frac{L_{day}}{10}} + 4\cdot 10^{\frac{L_{evening}+5}{10}} + 8\cdot 10^{\frac{L_{night}+10}{10}}\right) \tag{6}$$

where L_{day}, $L_{evening}$, L_{night} are the A-weighted long-term average sound levels determined over all the day, evening, night periods of a year. The levels regarding the three periods of the 24 hours day are weighted by means of the duration (12 hours for the day, 4 hours for the evening and 8 hours for the night) and by means of a "penalty" (5 dB for the evening and 10 dB for the night) that takes into account that noise in these hours can be more fastidious and can disturb the sleep.

Also the level L_{night} that is the equivalent sound pressure level measured during the night period is considered to be a meaningful criterion to assess the impact of airborne ship noise on population. Sleep can be more easily affected by sound during the night, particularly when isolated high level peaks of noise are present. The EU Directive proposed to utilize L_{night} together with L_{DEN}, in Noise Strategic Mapping, to establish the acceptability of noise from the main noise sources (road, rail, or air-traffic noise, industrial noise, construction activities, etc.), including also ports.

Limits in L_{DEN} and L_{night} are fixed at the receiver position. The sources taken into account by the EU Directive are industrial sources that emit (or can emit) at different levels in the three periods of the day. In the case of railways, roads and airports, for example, the traffic is much different during the 24 hours and the noise emissions are linked to the traffic flow and not to the characteristics of a single vehicle. Moreover these sources are static, in the sense that their position is fixed referring to the receivers, and their annoyance is evaluated by means of levels (L_{day}, $L_{evening}$, L_{night}) averaged on an entire year.

In the case of ships, sources have quite different characteristics:

- The disturbance is linked to a single source moving or stationing at the quay and not to a traffic flow of comparatively similar sources. There are substantial differences in the noise emitted from ship to ship. This suggests that to operate a long term average on the noise emitted could not be the best approach because of the high variability of noise intensity in time.
- The ship is a varying source that emits differently in the navigation phase and during loading/unloading operations.
- The noise produced by the ship is less likely to be dependent from the period of the day (at least if compared with other transportation noises).

For these reasons, the use of the L_{DEN} indicator to assess specifically airborne noise emitted from ships is quite difficult or better it is difficult to correlate the level in L_{DEN} with the emissions of a single ship and with the actual annoyance for the receivers.

3.3 *Problems in fixing limits to airborne noise*

To control the airborne noise emissions from ships the best strategy could be to fix limits at the source, in order to facilitate the comparison between two ships and between a ship and the requirements and in order to decouple the ship-source characteristics from the surrounding ambient. The problem is to bring together limits on the receiver and limits at sources, giving that the transmission path is external to the ship and not controllable in the ship design. In fact a limit fixed at the ship, has the aim to control the sound energy that is radiated by the source but does not provide information on the energy that actually reaches the receiver. For this reason the complex subject of the transmission of sound must be taken into account. It is influenced in particular by: noise spectrum; reciprocal position of source and receiver; reflections on the ground and on the buildings; temperature gradient in air; wind. In the case of ships it is almost impossible to consider all the possible transmission paths and the actual transmission losses. In fact a ship can touch different harbours, can moor at different docks within the same harbour emitting every time through different transmission paths. Moreover, when the ship is sailing, the transmission path varies second by second. It is clear that the solution to the problem is to be reached in the acoustic planning of ports. In fact it should be the port authority that, given that a ship fulfil a certain limit, is authorised to enter in that port or to station at that quay or not. Moreover urbanistic countermeasures such as acoustic barriers must be adopted as in cases when the ship is too near to buildings.

In order to assess and manage the environmental port noise, the Noise Strategic Mapping (NSM) is considered the most suitable instrument for acoustic planning (Badino et al. 2011b). NSM is essential to identify the critical zones and to establish solutions that can allow the development of ship trade without compromising the quality of life in port cities.

In general, Noise Strategic Mapping is obtained by the superimposition of noise fields calculated for every ship-source. That is, the results of outdoor noise propagation analysis are added, taking into consideration all operating noise sources. For this reason, NSM needs the knowledge of all vessels staying into the ports, of their operative conditions and of their acoustical characteristics.

It also implies the definition of methods to assess the noise levels produced by ports activities. A method to identify critical areas may be the comparison of NSM and maximum levels allowed by regulations in terms of L_{den} and L_{night}: critical areas are those zones where the limit values are exceeded.

By using noise indicators in Strategic Noise Mapping it is also possible to find out how much noise can cause harmful effects on the exposed population. Through cross-comparisons between characteristic sound levels of the area, limit values of noise indicators, and the number of people harmfully affected, the actions against noise can be planned and intervention priorities can be set (Badino et al. 2010).

Following the approach to fix limits at the ship, in addition to the problems listed so far, a key aspect is represented by the measurement procedures which could be very complicated. The ship, in fact, is a complex noise source made by the contribution of many single sources distributed all along the ship both in the longitudinal and in the vertical direction generating an acoustic field that propagates in the 3D space. For this reason to fully characterise the source an efficiency grid of measure points should extend also in a vertical plane at height that, sometimes, reaches 50 m; but this is often limited by the lack of free spaces around the ship. Moreover these problems are stressed by the fact that the ship during port approach or coastal navigation is a moving source.

4 NOISE IN WATER OUTSIDE THE SHIP

4.1 *Normative development*

The problem of the impact of UnderWater Noise (UWN) emissions from ships on the marine fauna has been addressed very recently. For this reason no compulsory norms with the aim to limit the emission of waterborne noise from ships exist. However many bodies and treaties deal with the problem of preservation of cetaceans, because of their fundamental role in the ecosystem. Some of the documents include noise within the possible threats for the marine fauna; among them there are the Agreement on the Conservation of Small Cetaceans of the Baltic And North Seas (ASCOBANS 1992), the Agreement on the Conservation of Cetaceans of the Black Sea, Mediterranean Sea and Contiguous Atlantic Area (ACCOBAMS 1996), but they only mention ships as a source without facing directly the problem of noise due to ship traffic. Within the International Maritime Organisation, a working group dealing with the impact of ships' underwater noise on marine mammals started in 2007 (IMO 2007, IMO 2009b,c IMO 2010).

In literature the problem of fixing numeric limits to the underwater noise of ships has been faced by two bodies: the International Council for the Exploration of the Sea (ICES) (Mitson 1995) and by the Det Norske Veritas (DNV 2010).

As regards the ICES limits are set for fishing research vessels with the aim to minimise behavioural changes in fish stocks; for this reason the limit curve is derived starting from a single point of the curve 30 dB above the cod threshold. By this single point a technology based curve is traced.

DNV provide limits for different types of vessels. The limit curves are based on underwater emission performances of existing ships. They represent reasonable limits that are reachable by each ship typology, basing on the fact that other ships of the same typology fulfil that limit. Both these two (ICES & DNV) requirements are then based on the state of the art, while the real impact of noise on the marine fauna should be taken into account.

As regards the standards to measure the underwater signature of the ship, the most recent are those proposed by DNV (DNV 2010) and by ANSI/ASA (2009).

The first one is devoted to verify the fulfilment of the limits provided by the DNV in its Silent Class Notation requirements. The layout for the measures provides a single hydrophone placed on a sloping seabed and the signature is measured with the vessel passing abeam the hydrophone position.

The American Standard is a more general one that applies to all surface vessels with velocity less than 50 kt. It provides three different grades of accuracy depending on the number of hydrophones, the extension of the frequency range of measures etc. In the most precise method three hydrophones at different depths are used in order to capture the source directivity in a vertical plane; the ship under test shall transit several times abeam the hydrophone position and data are recorded in a defined window of time.

4.2 *Improved Criteria to fix limits to UWN*

As stated above the limits to underwater noise existing so far are technological limits not based on a specific goal that must involve the receiver characteristics. In the following three possible methods to set performance-based limits on the waterborne noise emitted from ships are proposed. The aim is to set limits based on the actual perception of sound by the marine mammals. The limits will be set at the source; this because it is impossible to evaluate the noise energy that reaches an animal whose position is extremely variable and whose sensibility varies markedly from species to species.

The proposed methods take into account, increasing in accuracy and complexity, the possible consequences that noise can have on marine mammals; the main effects linked with ship underwater noise are mainly the behavioural change and the communication masking. The behavioural effects are due to relatively higher noise levels and for this reason they affect animals at lower distances from the ship. On the other hand the communication masking takes place when a noise can mask partially or completely a sound and it is mainly due to the general increase of the background noise that reduces the distance at which a sound, emitted by an individual, can be heard by a conspecific.

A first strategy in fixing limits is to impose the noise emitted by the ship to be lower than the background noise at a fixed distance (r) from the ship. With this approach the ship it is not heard at all by any animal at a distance greater than r. This is a quite conservative criterion; in fact in this case the ship is completely undetectable. Theoretically, it can be accepted that the animal detects the ship, provided that the ship does not disturb the animal. The great advantage of this method is that we need to know and model only the characteristics of the ship emission and the background noise. In Figure 4 the spectrum of a ship, shifted to different distances with a spherical spreading law, is reported together with a spectrum of the background noise. As it can be seen, at a certain distance the ship spectrum is completely below the background noise.

A possible improvement is represented by the implementation of the hearing sensibility of the marine mammals in order to compare directly the noise emitted with the actual perception of the receiver. In this case it can be stated that the noise emitted by the ship must not be detected (by a

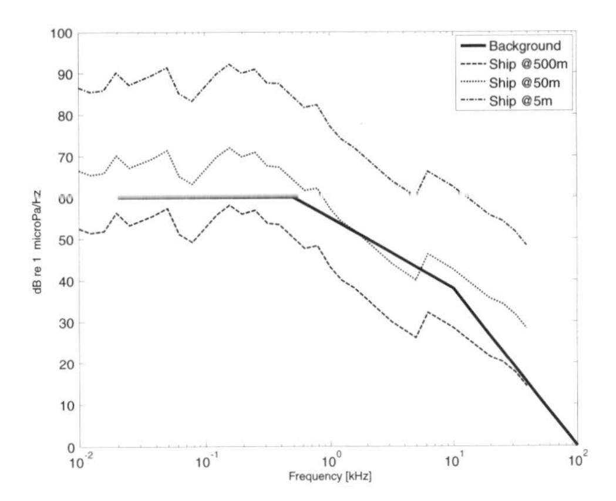

Figure 4. Ship spectrum & background noise.

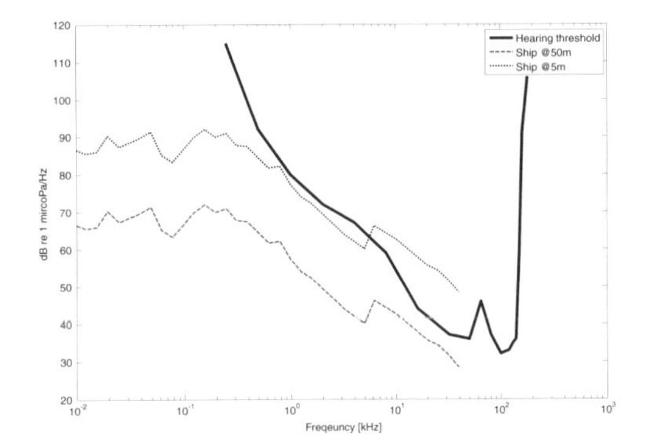

Figure 5. Ship spectrum & hearing threshold of a cetacean.

specific species) at a distance (r) from the ship. This approach is more realistic than the previous one but needs the knowledge of the mammals' audiograms. Such audiograms are derived through laboratory tests, during which the cerebral reaction of the animals to different sounds is monitored in order to find the hearing threshold at the different frequencies (see Fig. 5 solid line). Due to their dimensions, the large mammals (mysticetes) cannot be tested in laboratory. For this reason no data are available on their hearing sensitivity.

This represents the major limitation of the method proposed.

A third approach, particularly devoted to prevent communication disturbances, implies to consider a third element in the problem description: the emitter. By comparing the vocalisation spectrum with the audiograms modified to account for the background noise from the ship the actual perception by the receiver could be evaluated. The difficulty is again the lack of data for mysticetes. In Figure 6 these three elements are represented.

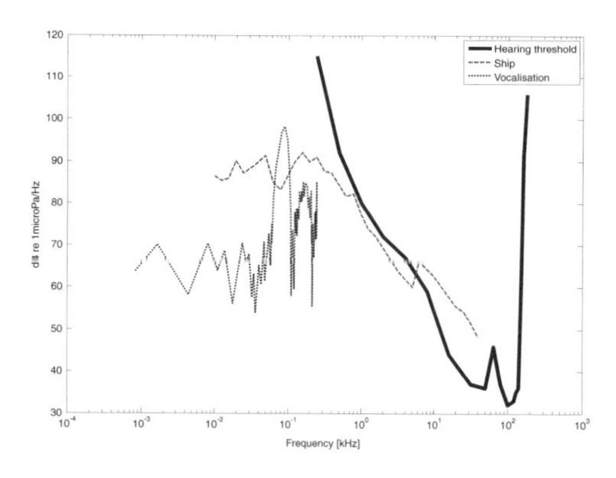

Figure 6. Ship spectrum, hearing threshold and vocalisation spectrum.

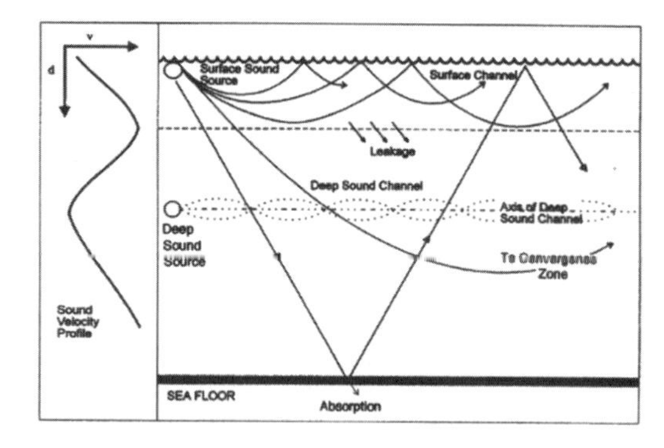

Figure 7. Influence of speed profile on sound propagation in the sea from Urick 1983.

It is important to underline that in this case the spatial distribution of the three elements (receiver, emitter and ship) must be known in order to know propagation path of sound.

What is in common among the three methods analysed above is the problem of noise propagation in the sea, which is influenced by several parameters: frequency of the sound, celerity profile, free surface profile, bottom profile and composition.

As regards the frequency content, in general lower frequencies propagate at greater distances than the higher ones.

The speed of sound in water (celerity) is the variable that most influences the propagation. It depends on: temperature, salinity and pressure (depth). The combination of these parameters results in the final celerity profile that can vary for different seas, seasons and even time of the day. How the profile influence the sound propagation is shown in Figure 7.

As shown in Figure 7, the noise field radiated from the source may include shadow zones (where sound levels are low) relatively close to the source and convergence zones (where levels are higher because several sound rays converge in a restricted zone) far away from the source. In general sound tends to converge in zones where the speed of propagation is lower.

A further phenomenon that influences the sound propagation is represented by reflections from the bottom and the free surface. In particular for ships, that are floating sources, reflections from the free surface play an important role. This effect is influenced by the shape of the free surface, i.e. by the sea state. In fact, in the case of still water, the free surface can be considered as perfectly reflecting, while, when waves are present, a parameter governing reflection is represented by the ratio between the wave length of sound and of waves.

5 CONCLUSIONS

The definition of the needs for the control of noise emissions from ships is definitely at different levels in the various environments affected.

On board, the problem of noise control has been addressed since long time and a well defined normative framework is in place, dealing with the preservation of acceptable working conditions and of suitable comfort levels in the living and recreational spaces on board.

Requirements are evolving towards more restrictive formulations, as a result of technical progress as well as of an increasing attention to the acoustic comfort aspects. Improvements can be achieved by means of more detailed indicators of the human perception, correlating in a more precise way the objective quantification of noise to the subjective feeling of discomfort that it generates. In general, however, the technical problem is clear, and can be tackled at the ship design stage by proper actions on sources and transmission paths (both categories internal to the ship).

As regards the airborne noise emitted from the ship, the situation is complicated by the fact that levels at the receiving positions are correlated to the levels at the source through the characteristics of the transmission path, which is dependent on the geography of the port (or of the channel) and surrounding areas and not on the ship characteristics. This makes the normative problem very complex, due to the number of official bodies involved (Municipalities, Port Authorities, Maritime Authorities….).

Standards for ship airborne emissions to be used at the design stage of the ship are not presently available. On one hand, it should be noted that a proper characterisation of such a complex and large source as a ship is difficult (particularly if the characterisation is carried out in a confined environment like a shipyard). On the other hand, imposing limits to the ship source emission does not in itself guarantee the achievement of the final goal (control of noise at the receiver) as this final outcome depends also on local characteristics of the environment. Due to this possible requirements (limits for the emissions) can only be set with reference to the present technology state-of-the-art and not on the basis of target levels at the receiver.

Finally, the definition of the needs for the control of waterborne noise emissions is difficult because is difficult to define the general objectives of such control: the receiver characteristics are not well defined (different sensitivities to noise of the various species, some of them unknown). Paradoxically, if compared with the situation regarding airborne noise, standards for the characterisation of the ship underwater signature are presently already available, with even formulations of limits to the ship emission. Such limits are based on collection of data and reflect the state of the art of technology more than the way of achieving a definite goal in terms of reduced environmental impact, but nevertheless can represent a starting point from a pre-normative perspective.

In conclusion, the formulation of needs for the control of noise emissions from ships reflects necessarily the definition of the long term goals of such control. When the receiving environment is represented by humans, the target is clear (even though the definition can be improved) when the natural environment is considered, the target is less defined. On the other hand, targets based on the technological state of the art are always possible, but on this issue the situation regarding airborne noise emissions appear to be somehow lacking behind the other ones, probably because of the number of subjects involved in the problem.

ACKNOWLEDGMENTS

This work was developed in the frame of the collaborative project SILENV—Ships oriented Innovative soLutions to rEduce Noise & Vibrations, funded by the E.U. within the Call FP7-SST-2008-RTD-1 Grant Agreement SCP8-GA-2009-234182.

REFERENCES

ACCOBAMS 1996, Final Act of the Negotiation Meeting to adopt the Agreement on the Conservation of Cetaceans of the Black Sea, Mediterranean Sea and Contiguous Atlantic Area, Monaco.

ANSI/ASA 2009. ANSI/ASA S12.64-2009/Part 1, Quantities and Procedures for Description and Measurement of Underwater Sound from Ships- Part1: General Requirements.

ASCOBANS 1992, Agreement on the Conservation of Small Cetaceans of the Baltic And North Seas, United Nations, New York.

Badino, A., Borelli, D., Gaggero, T., Rizzuto, E., Schenone, C. 2011a. Criteria for noise annoyance evaluation on board ships. IMAM 2011.

Badino, A., Borelli, D., Gaggero, T., Rizzuto, E., Schenone, C. 2011b. Analysis of airborne noise emitted from ships. IMAM 2011.

Badino, A., Schenone, C., Tomasoni, L. 2010. Managing the environmental sustainability of ports: noise pollution. ICEMT 2011.

Beranek, L. 1988. Noise and vibration control. Institute of Noise Control Engineering.

DNV 2010, Det Norske Veritas: Rules for Classification of Ships, Silent Class Notation, Part 6, Chapter 24.

EU 2002, Directive 2002/49/EC of The European Parliament and of The Council, 25 June 2002.

EU 2003a, Directive 2003/10/EC of the European Parliament and of the Council, 6 February 2003.

EU 2003b, Directive 2003/44/EC of The European Parliament and of The Council, 16 June 2003.

Finegold, L.S., Harris, C.S., Von Gierke, H.E. 1994. Community annoyance and sleep disturbance: updated criteria for assessment of the impacts of general transportation noise on people. *Noise Control Engineering Journal* 42(1): 25–30.

Glorig, A. 1988. Damage-risk criteria for hearing, in Noise and vibration control, L. Beranek ed. Washington DC: Insitute of Noise Control Engineering.

IMO 1974. SOLAS International Convention for the Safety of Life at Sea.

IMO 1981. Resolution A.468(XII): Code on Noise Levels on Board Ships.

IMO 2007. Document MEPC 57/INF.4 Shipping noise and Marine Mammals.

IMO 2009a. Document DE 53/10, Proposals for the development of amendments to SOLAS regulation II-1/36 and a revision of the Code on noise levels on board ships.

IMO 2009b. Document MEPC 59/19, Noise From Commercial Shipping and its Adverse Impacts on Marine Life.

IMO 2009c. Document MEPC 60/18, Noise From Commercial Shipping and its Adverse Impacts on Marine Life.

IMO 2010. Document MEPC 61/19, Noise From Commercial Shipping and its Adverse Impacts on Marine Life.

ISO 1971. ISO R1999, Acoustics—Assessment of occupational noise exposure for hearing conservation purposes.

ISO 1996. ISO 717-1 1996, Acoustics—Rating of sound insulation in buildings and of building elements -Part 1: Airborne sound insulation.

OSHA 1974. OSHA 1910.95, Occupational noise exposure.

Pretlove, A.J., Turner, J.D. 1991. Acoustics for Engineers. London: Macmillan Education Ltd.

Urick, R.J. 1983. Principles of underwater sound. New York: McGraw-Hill.

Sustainable Maritime Transportation and Exploitation of Sea Resources – Rizzuto & Guedes Soares (eds)
 ISBN 978-0-415-62081-9

Criteria for noise annoyance evaluation on board ships

A. Badino, D. Borelli, T. Gaggero, E. Rizzuto & C. Schenone
University of Genoa, Genoa, Italy

ABSTRACT: Marine and offshore regulations about onboard noise are quite stratified and tangled, being issued by a number of international and national bodies. At the same time, many of the requirements are actually expressed in non quantitative terms and neglect parameters like noise spectra, low frequency pulsations or sound reverberation, which deeply influence the "sound wellness". Aim of this paper is therefore to improve present requirements by introducing more meaningful noise annoyance criteria. These new criteria should be at the same time able to better express the actual reactions to noise from crew and passengers and simple enough to be widely applied on board ships. Such enhanced criteria are derived from the noise rating methods provided to rate indoor background sound in the civil engineering context. The possibility to make use of these indicators in the naval field is analyzed and some recommendations for a proper application on ships are given.

1 INTRODUCTION

In the marine field, rules and requirements about on board noise have addressed so far the problems connected to health and performances of the crew in working areas and, more recently, those related to the comfort of crew and passengers in living, recreation and resting spaces. These regulations are quite stratified and tangled, being issued by a number of international and national bodies, like IMO, ILO, ISO, Class Societies and National Authorities. It may be necessary, therefore, to check a large number of documents to get a complete picture of the normative framework, but at the same time many of the requirements are actually expressed in general terms, without even indications about quantitative measures. Even when they contain limits, most requirements regarding noise only consider simple sound energy levels in dB(A) and not further parameters, like noise spectra, low frequency pulsations or sound reverberation, which deeply influence the "sound wellness".

The challenge is therefore to implement present requirements by considering more meaningful noise annoyance criteria. These new criteria should be at the same time significant in respect of the noise evaluation and simple enough to be widely applied on board ships.

These considerations have been first proposed in the SILENV project (Ships oriented Innovative soLutions to rEduce Noise & Vibrations), funded by the EU within the 7th Framework Programme. Aims of this paper are to analyze standard criteria to evaluate noise annoyance on board ships and to propose improved indicators, able to express better the actual reaction to noise from crew and passengers.

In general it's very difficult to evaluate people's satisfaction achieved with a given level of background sound, since this feeling is determined by many factors. In fact, the elements which have greater impact on the acoustic annoyance are the scale of sound, or sound pressure level and the exposure time, but also other ancillary features, such as the spectral composition of noise or the repetition over time, assume unquestioned importance. In particular, the A-Weighted Sound Level is widely used to state acoustical design goals as a single number, but its usefulness is limited because it gives no information on the spectrum content. The A-weighted sound level is essentially an energetic index and does not necessarily correlate well with the annoyance caused by the noise.

Alternative methods can be derived from the sound rating criteria provided by the ASHRAE (American Society of Heating, Refrigerating and Air-conditioning Engineers) to rate indoor background sound in the civil engineering context. They include the Noise Criteria (NC), the more recent Room Criteria (RC) and Balanced Noise Criteria (NCB), the Room Noise Criteria (RNC) and the new RC Mark II. In the following, the possibility to make use of these indicators in the naval field is analyzed and some recommendations for a proper application on ships are given.

2 RULES ABOUT ON BOARD NOISE ANNOYANCE

One of the first documents available concerning noise on board ships is the International Convention for the Safety of Life at (SOLAS 1974).

The noise problem is treated in chapter II-1 reg. 36.

Anyway, in general, for commercial vessels the rules concerning noise levels onboard are quite simple and few. It is basically a set of dB(A) levels given for a range of different positions onboard. These settings are given in the Resolution A.468(XII) of the International Maritime Organization (IMO) described hereafter.

The resolution, titled "CODE ON NOISE LEVELS ON BOARD SHIPS", has been first published in 1981 and, since then, has been the reference guide for almost all the States members of the Organization.

The IMO Resolution A.468(XII) is designed to provide standards to prevent the occurrence of potentially hazardous noise levels on board ships and to provide standards for an acceptable environment for seafarers.

Recommendations are made for protecting the seafarers from the risk of noise induced hearing loss under conditions where at present it is not feasible to limit the noise to a level which is not potentially harmful; there are also recommendations about how to measure noise levels and exposure, and limits on acceptable maximum noise levels for all spaces to which seafarers normally have access to are given.

The Code applies to ships in service, in particular to new ships of 1600 tons gross tonnage and over, but the provisions relating to potentially hazardous noise levels contained in the Code should also apply to existing ships of 1600 tons gross tonnage and over or to new ships of less than 1600 tons gross tonnage, as far as reasonable and practicable, to the satisfaction of the Administration.

It must also be noted that the Code does not apply to dynamically supported craft, fishing vessels, pipe-laying barges, crane barges, mobile offshore drilling units, pleasure yachts not engaged in trade, ships of war and troopships and ships not propelled by mechanical means.

Another important thing to note is that the Code is not intended to apply to passenger cabins and other passenger spaces except in so far as they are work spaces and are covered by the provisions of the Code. The A-weighted sound pressure level must be measured by a sound level meter in which the frequency response is weighted according to the A-weighting curve (see IEC publication 651), and the noise level limits for various spaces are reported in Table 1.

Table 1. Noise level limits (IMO 1981).

	dB(A)
Work spaces	
Machinery spaces (continuously manned)*	90
Machinery spaces (not continuously manned)*	110
Machinery control rooms	75
Workshops	85
Non-specified work spaces*	90
Navigation spaces	
Navigating bridge and chartrooms	65
Listening post, including navigating bridge wings and windows	70
Radio rooms (with radio equipment operating but not producing audio signals)	60
Radar rooms	65
Accommodation spaces	
Cabins and hospitals	60
Mess rooms	65
Recreation rooms	65
Open recreation areas	75
Offices	65
Service spaces	
Galleys, without food processing equipment operating	75
Serveries and pantries	75
*Normally unoccupied spaces**	
Spaces not specified	90

*Ear protectors should be worn when the noise level is above 85 dB(A).

Personnel entering spaces with noise levels greater than 85 dB(A) are required to wear ear protectors, while the limit of 110 dB(A) given for machinery spaces not continuously manned assumes that ear protectors giving protection meet the requirements for ear muffs as in the following Table 2.

The only information the Code gives about sound spectrum is that in accommodation spaces where the dB(A) limits are exceeded and where there is a subjectively annoying low-frequency sound or obvious tonal components, the ISO Noise Rating (NR, the number found by plotting the octave band spectrum on the NR curves given in ISO Standard R 1996:1967 and selecting the highest noise rating curve to which the spectrum is tangent) number should also be determined; the limits specified may be considered as satisfied if the ISO noise rating number (Fig. 1) does not numerically exceed the specified A weighted value minus 5.

In addition, the Resolution widely describes a noise survey report which should be made for each ship, and should comprise information on

Table 2. Attenuation of ear protectors in dB(A) (IMO 1981).

Type of ear protector	Ear plugs	Ear muffs
125 Hz	0	5
250 Hz	5	12
500 Hz	10	20
1000 Hz	15	30
2000 Hz	22	30
3150 Hz	22	30
4000 Hz	22	30
6300 Hz	22	30

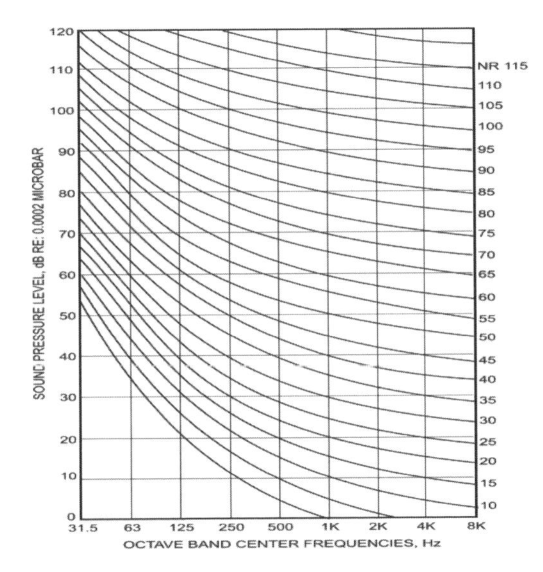

Figure 1. ISO NR curves.

the noise levels in the various spaces on board; noise exposure limits (Fig. 2) are described, too, to be sure that, in spaces with sound pressure levels exceeding 85 dB(A), the requirements concerning the use of suitable ear protection or the applying of time limits for exposure are satisfied, so that an equivalent level of protection is maintained.

The philosophy of the noise exposure limits is to ensure that seafarers will not be exposed to an $L_{eq}(24)$ exceeding 80 dB(A). The $L_{eq}(24)$ is defined in Equation (1).

$$L_{eq}(24\ hours) = 10\log_{10}\frac{1}{24h}\int_{24h}\left(\frac{p_A(\tau)}{p_0}\right)^2 d\tau \qquad (1)$$

where p_A = instantaneous A-weighted sound pressure; p_0 = reference pressure.

In spaces where the level exceeds 85 dB(A) the use of ear protection and/or the limitation

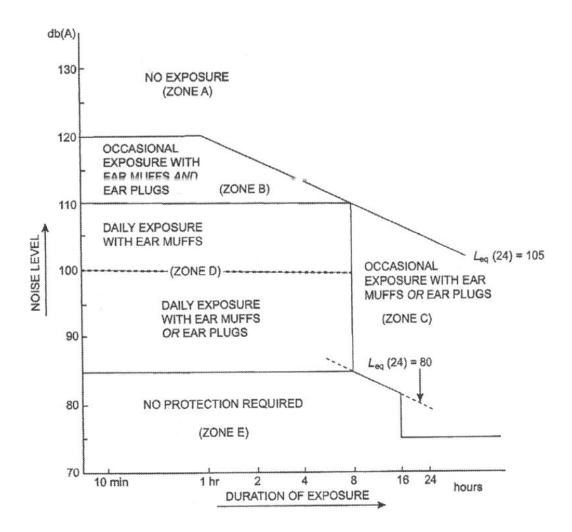

Figure 2. Allowable daily and occasional noise exposure zones (IMO 1981).

of exposure in terms of time are prescribed, as previously described.

In the code there are also about the limits for the airborne insulation index for bulkheads and decks. The noise insulation index is defined in the following Equation (2):

$$I_a = 10\log_{10}\frac{W_1}{W_2 + W_3}[dB] \qquad (2)$$

where W_1 = incident acoustic power; W_2 = acoustic power transmitted to the receiving ambient through the wall; W_3 = acoustic power transmitted to the receiving ambient through boundary elements or other components.

According to ISO Standard R717, insulation index should comply at least with the values given in Table 3.

Overall seen marine and offshore world of rules and regulations is quite complex. It can be necessary to investigate a large range of documents to ensure fulfilment of rules, but at the same time many of the requirements are actually basic regulations.

The basic IMO rules do not specify other relevant issues of sound wellness. Within cruise liners and ferries, more focus is made towards noise, and therefore comfort class rules from all major Class Societies can be seen, and of course the owners can add their own stricter requirements. For offshore modules, additional national rules for this kind of industry are also found, and these are again more related to sound comfort than the IMO Code.

Other rules and guidelines are, for example, the IMO's High-Speed Craft Code (HSC), the Code of

Table 3. Insulation index required by IMO (1981).

Space	Insulation index [dB]
Cabin to cabin	Ia = 30
Mess rooms, recreation rooms to cabins and hospitals	Ia = 40

Practice for Noise Level in Ships of the UK's Maritime and Coastguard Agency, and so on.

The present regulatory framework is mostly still based on the 1981's IMO A.468(XII) which focuses on noise exposure limits in dB(A), with only minor considerations about spectrum evaluations in terms of NR curves (Badino et al. 2011). Moreover time duration of the exposure is considered for crew only. Comfort classes focus mainly on the same aspect of limiting the total perceived acoustic power but lowering limits.

An improvement on regulations should move in the direction of a more specific evaluation of the sound characteristics. This means focusing on the characterisation of the effects of frequency spectra, tonal components and intermittent sound.

3 ENHANCED NOISE ANNOYANCE CRITERIA

As above discussed, most rules within noise only look at simple sound energy levels and not further parameters, i.e. noise spectra, low frequency pulsations or reverberation, which may reduce the "sound wellness". The challenge is therefore to combine these rules and regulations with comfort oriented sound engineering and design. Basic rules are given by IMO, but for many countries the national marine authorities add stronger regulations to ensure the health, safety and work environment for the crews. For the marine vessels all Class Societies additionally have guidelines and comfort classing. For offshore installations rules are mainly given by the national authorities, and these are in most countries the "Energy Departments".

The degree of occupant satisfaction achieved with a given level of background sound is determined by many factors. For example, large conference rooms, auditoriums, and music halls can tolerate only a low level of background sound. On the other hand, higher levels of background sound are acceptable and even desirable in certain situations, such as in banquet rooms where a certain amount of speech and activity masking is essential. Therefore, the system sound control goal varies depending on the required use the space.

The background sound should have a balanced distribution of sound energy over a broad frequency range, and no audible tonal or other characteristics such as whine, whistle, hum, or rumble, but unfortunately no acceptable process easily characterizes the effects of audible tones and level fluctuations.

A-Weighted Sound Level (dB(A)) is the first and most basic and ordinary noise rating method: it is widely used to state acoustical design goals as a single number, but its usefulness is limited because it gives no information on spectrum content. A-weighted sound levels correlate well with human judgments of relative loudness, but give no information on spectral balance: dB(A) is in fact only a merely energetic index, simulating the response of the human ear at the loudness level in the neighborhood of 40 phon, thus they do not necessarily correlate well with the annoyance caused by the noise.

Many different-sounding spectra can have the same numeric rating, but have quite different subjective qualities. A-weighted comparisons are best used with sounds that sound alike but differ in level. They should not be used to compare sounds with distinctly different spectral characteristics; that is, two sounds at the same sound level but with different spectral content are likely to be judged differently by the listener in terms of acceptability as a background sound. One of the sounds might be completely acceptable, while the other could be objectionable because its spectrum shape was rumbly, hissy, or tonal in character.

This is the reason why it could be useful to use some enhanced criterion to evaluate the acoustic comfort of crew and passengers: it is very difficult to evaluate occupant satisfaction achieved with a level of background sound, since this feeling is determined by many factors. There is no doubt, in fact, that elements which have greater impact on the acoustic annoyance are the scale of sound, or sound pressure level, the exposure time but also other ancillary features, such as the spectral composition of noise or the repetition over time, assume unquestioned importance.

In particular, sound rating criteria recommended by the American Society of Heating, Refrigerating and Air-Conditioning Engineers to rate indoor sound in the civil engineering context can be used. They include the Noise Criteria (NC), the more recent Room Criteria (RC) and balanced Noise Criteria (NCB), and the new RC Mark II and Room Noise Criteria (RNC). Everyone of these noise criteria is defined and discussed in the following paragraphs.

3.1 *Noise Criteria (NC) method*

Noise Criteria (NC) were established in U.S. for rating indoor noise, noise from air-conditioning equipment etc. The method (Beranek 1957) consists of a set of criteria curves extending from 63 to 8000 Hz, and a tangency rating procedure. The criteria curves define the limits of octave band spectra that must not be exceeded to meet occupant acceptance in certain spaces. The NC rating can be obtained by plotting the octave band levels for a given noise spectrum on the NC curves (Fig. 3). The noise spectrum is specified as having a NC rating same as the lowest NC curve which is not exceeded by the spectrum.

The NC method is sensitive to level but has the disadvantage that the tangency method used to determine the rating does not require that the sound spectrum approximate the shape of the NC curves. Thus, many different sounds can have the same numeric rating, but rank differently on the basis of subjective sound quality.

"Quality", as referred to here, addresses the judgements of people as to the adequacy of the noise environment in the building space that they occupy, such as that produced by air-conditioning, people's activities, outdoor transportation and the like.

Two problems occur in using the NC procedure: the first one is that when the NC level is determined by a prominent peak in the spectrum, the actual level of resulting background sound may be quieter than that desired for masking unwanted speech and activity sounds, because the spectrum on either side of the tangent peak drops off too rapidly; the second one is about that when the measured spectrum closely matches the shape of the NC curve, the resulting sound is either nimbly or hissy or both.

3.2 *Noise Rating (NR) method*

The Noise Rating (NR) method (Kosten & Van Os 1962) is similar to the NC method. It's mostly used in the EU and the principal difference between the two is that the shape of the curves (Fig. 1) are more permissive in the low-frequency region than the NC curves.

3.3 *Balanced Noise Criteria (NCB) method*

The NCB method (Beranek 1989) is a specification for evaluation of room sound including noise due to occupant activities. The NCB criteria curves (Fig. 4) are intended as replacements for the NC curves, and include both the addition of two low-frequency octave bands (16 and 31.5 Hz) and lower permissible sound levels in the high-frequency octave bands (4000 and 8000 Hz).

The NCB rating procedure is based on the Speech Interference Level (SIL), which is the arithmetic average of the sound pressure levels in the four frequency bands: 500, 1000, 2000, 4000 Hz. Additional tests include rumble and hiss compliance. The rating is expressed as NCB followed by a number, for example NCB 40.

The NCB method is better than the NC method in determining whether a sound spectrum has a shape sufficiently unbalanced to demand corrective action. Also, it addresses the issue of low frequency sound.

The rating procedure is somewhat more complicated than the tangency rating procedure and is not described here for the sake of brevity.

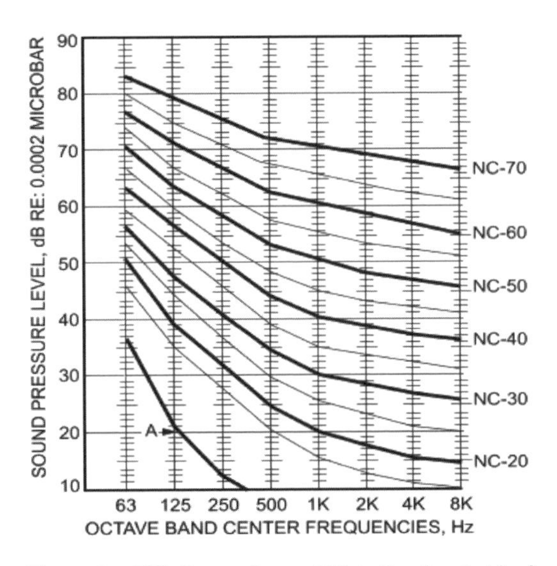

Figure 3. NC Curves (curve "A" is the threshold of hearing).

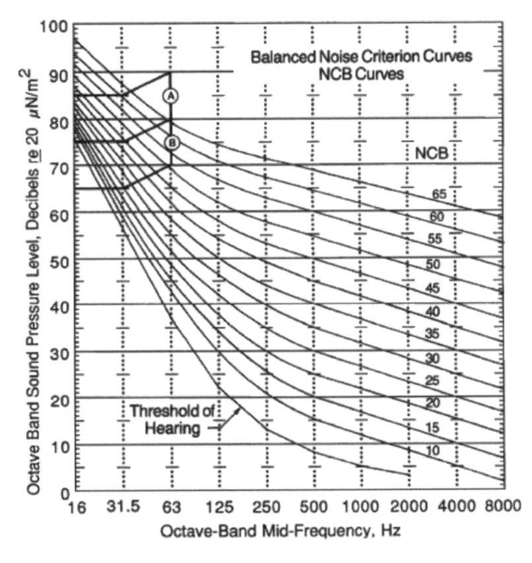

Figure 4. NCB Curves.

3.4 *Room Criteria (RC) method*

For some time, the RC (Fig. 5) method was recommended as the preferred method for rating HVAC-related sound.

The RC curves were intended to establish HVAC system design goals. The revised RC Mark II method (discussed below) is now preferred.

The RC method consists of a family of criteria curves and a rating procedure. The shape of these curves differs from the NC curves to approximate a well-balanced neutral-sounding spectrum, and two additional octave bands (16 and 31.5 Hz) are added to deal with low-frequency sound. This rating procedure assesses background sound in spaces based on the effect of the sound on speech communication, and on estimates of subjective sound quality.

The rating is expressed as RC followed by a number to show the level of the sound and a letter to indicate the quality, for example RC 35 (N), where N denotes neutral.

3.5 *Room Noise Criteria (RNC) method*

The newest method proposed for the evaluation of room noise is the Room Noise Criteria (RNC), which attempts to address cyclic variation of low frequency sound produced by large air ventilation systems, sometimes described as "surging" (Tocci 2000).

The RNC (Schomer 2000a,b) uses a set of curves (Fig. 6) that are similar to NC ones; however, the RNC has an additional rating feature that evaluates fluctuating room noise, such as

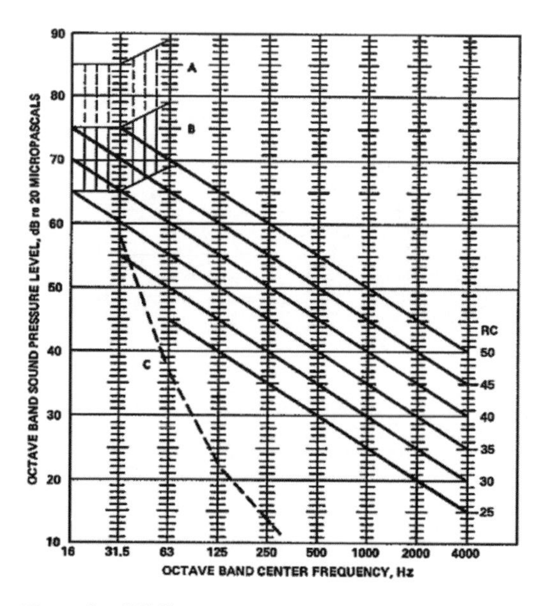

Figure 5. RC Curves.

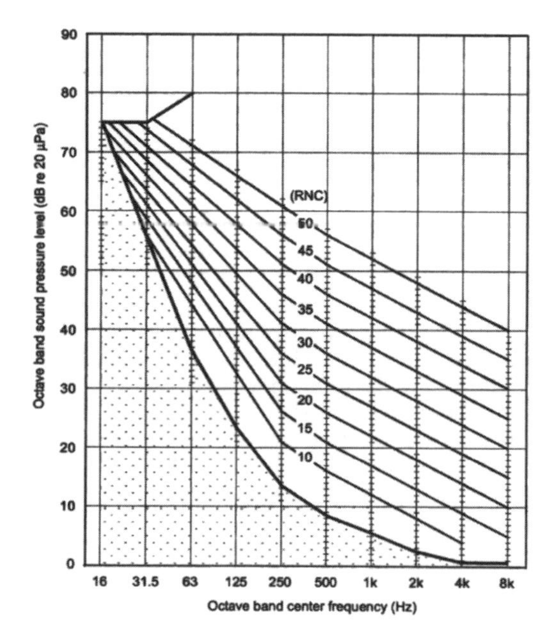

Figure 6. RNC Curves.

that sometimes experienced in spaces near large mechanical systems. The RNC goal is to propose a new set of criteria curves representing an "averaging" of NCB and RC curves that do not "penalize" (meaning to be above the threshold of hearing) a well designed system and, at the same time, prevent a turbulence-producing fan-surging HVAC system from meeting the criteria of acceptability.

The method for determining this adjustment is highly technical (e.g. the use of a tangency method for determining a spectrum RNC rating, and the use of measured or calculated peak-to-peak variation and standard deviation of sound pressure level with time in the 16, 31, 63, and 125 Hz octave band) and it is unlikely that the RNC curves will receive general acceptance because of the complexity of the rating process (Bies & Hansen 2009). Anyway, it must be noted that the method has recently been included in the American National Standard ANSI S12.2-2008, Criteria for Evaluating Room Noise.

3.6 *Room Criteria mark II (RC mark II) method*

In 1997 the RC method was revised to the RC Mark II method (Blazier 1997) which, like its predecessor, is intended for rating the sound performance of an HVAC system as a whole, and it can also be used as a diagnostic tool for analyzing general noise problems. The RC Mark II comprises a procedure for estimating occupant satisfaction when the spectrum does not have the shape of an RC curve (Quality Assessment Index, QAI). The rating is expressed as RC followed by a number and a letter, as in the previous method. The number is the arithmetic average rounded to the nearest integer of the SPL in the 500, 1000 and 2000 Hz octave bands, which are the principal speech frequency region. The letter is a qualitative descriptor that identifies the perceived character of the sound: N for neutral, LF for low frequency rumble, MF for mid frequency roar and HF for high frequency hiss.

The LF is also divided in two different subcategories, LF_A which denotes a noticeable degree of sound induced ceiling/wall vibration, and LF_B which denotes a moderate but perceptible degree of sound induced vibration: these occur respectively if the sound pressure levels exceed 75 dB or 65 dB in one of the first two octave bands considered.

In Figure 7, each reference curve identifies the shape of a neutral bland-sounding spectrum, and the curve number corresponds to the sound level in the 1000 Hz octave band; regions A and B denote levels at which sound can induce vibration in light wall and ceiling constructions that can potentially make light fixtures and furniture rattle, a problem that must be considered seriously without any doubt. The curve T is the octave-band threshold of hearing.

Subjective quality is assigned by calculating the Quality Assessment Index (QAI). This index is a measure of the degree the shape of the spectrum under evaluation deviates from the shape of the RC reference curve.

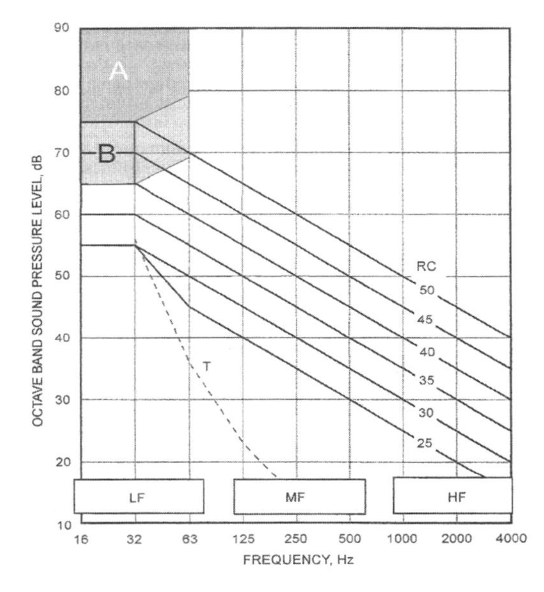

Figure 7. RC Mark II Curves.

The procedure requires calculation of the energy-average spectral deviations from the RC reference curve in each of three frequency groups: low frequency, LF (16–63 Hz), medium frequency, MF (125–500 Hz), and high frequency, HF (1000–4000 Hz); the procedure is given by the following equations:

$$LF = 10\log\left(\frac{10^{0.1\Delta L_{16}} + 10^{0.1\Delta L_{31.5}} + 10^{0.1\Delta L_{63}}}{3}\right) \quad (3)$$

$$MF = 10\log\left(\frac{10^{0.1\Delta L_{125}} + 10^{0.1\Delta L_{250}} + 10^{0.1\Delta L_{500}}}{3}\right) \quad (4)$$

$$HF = 10\log\left(\frac{10^{0.1\Delta L_{1000}} + 10^{0.1\Delta L_{2000}} + 10^{0.1\Delta L_{4000}}}{3}\right) \quad (5)$$

In the above equations the ΔL terms are the differences between the spectrum being evaluated and the RC reference curve in each frequency band. In this way. three spectral deviation factors (LF, MF, HF), expressed in dB with either positive or negative values, are associated with the spectrum being rated. QAI is the range in dB between the highest and lowest values of the spectral deviation factors. If QAI ≤ 5 dB, the spectrum is assigned a neutral (N) rating. If QAI exceeds 5 dB, the sound quality descriptor of the RC rating is the letter designation of the frequency region of the deviation factor having the highest positive value.

The Quality Assessment Index (QAI) is useful in estimating the probable reaction of an occupant when the system does not produce optimum sound quality.

The basis for the procedure for estimating occupant satisfaction can be summarized as follows:

- Changes in sound level of less than 5 dB do not cause subjects to change their ranking of sounds of similar spectral content.
- A QAI of 5 dB or less corresponds to a generally acceptable condition, provided that the perceived level of the sound is in a range consistent with the given type of space occupancy.
- A QAI that exceeds 5 dB but is less than or equal to 10 dB represents a marginal situation in which the acceptance by an occupant is questionable.
- A QAI greater than 10 dB will likely be objectionable to the average occupant.

3.7 *Summary of the described methods*

In Table 4 a comparison of different sound rating methods is reported; it can be noted that the RNC method and the NR method are not represented in the table: that's because RNC is highly technical, very complex and moreover not

Table 4. Sound Rating Methods (ASHRAE 2009).

Method	Overview	Evaluates sound quality	Used for rating of
dBA	• Can be determined using sound level meter • No quality assessment • Frequently used for outdoor noise ordinances	No	Cooling towers water chillers condensing units
NC	• Can rate components • No quality assessment • Does not evaluate low frequency rumble, frequencies <63 Hz	No	Air terminals diffusers
NCB	• Can rate components • Some quality assessment	Yes	
RC	• Used to evaluate systems • Should not be used to evaluate components • Can be used to evaluate sound quality • Provides some diagnostic capability	Yes	
RC Mark II	• Evaluates sound quality • Provides improved diagnostics capability	Yes	

yet largely used, and the NR is quite obsolete, doesn't evaluate sound quality and is originally developed for outdoor environmental noise (Blazier 2003).

4 CRITERIA COMPARISON BASED ON EXPERIMENTAL DATA

To understand better the differences between the new proposed enhanced method and the existing sound evaluation methods, the following experimental data sets have been used. They represent the sound spectra on board for a navigating bridge and a cabin (Table 5), for different operating conditions. For the cabin, the main sound source were the engines, running at 2300 RPM, while for the navigating bridge the main noise source was the HVAC system.

For the data sets below, the application of IMO's criteria and of the Room Criteria Mark II gives very different results, as reported in Table 6.

From the results shown below, it can be easily seen that for the navigating bridge IMO's criteria is satisfied, but a deeper analysis through the RC Mark II criteria shows that there is a prevalence in the medium frequency range, that can result in a marginal roar perceived by the occupants. For the cabin, the analysis shows instead that the IMO's criterion is not satisfied, even in respect of the NR number (that should be smaller than 55). Annoyance assessment by means of the RC Mark II method gives much more information. This criterion suggests that the noise perceived in the cabin is characterized by low frequencies: there is probability that a rumble is perceived, with clearly perceptible acoustically induced vibration and rattle in lightweight wall and ceiling construction, since all the bands in the range from 16 to 63 Hz exceed the 75 dB limit; the QAI larger than 10 provides the information that, assuming the level of specified criterion is not exceeded, the probable average occupant evaluation will be that the noise in the room is objectionable.

Table 5. Experimental sound spectra.

Frequency (Hz)	SPL (dB) navigating bridge	SPL (dB) cabin
16	57.3	79.2
31.5	63.6	84.1
63	59.6	93.5
125	57.5	76.7
250	54.2	68.6
500	49.6	59.8
1000	40.7	53.5
2000	39.7	43.3
4000	33.4	33.8
8000	25.9	28.9

Table 6. Sound evaluation criteria comparison.

Criterion	Navigating bridge	Cabin
IMO's A.468(XII) limit	65 dB(A)	60 dB(A)
Experimental noise level	50.8 dB(A)	71.8 dB(A)
NR	46	73
RC Mark II	43 (MF)	52 (LF)
QAI.	>5 (6.1)	>10 (18.8)

5 DISCUSSION

As noise annoyance primarily depends on the type of activity is being done, noise response on board ships is to be close to that in other indoor environment. The room use determines the wellness in reference to a certain background noise and similar annoyance criteria can be adopted for both civil and on board environments.

Since all spaces on board are more or less exposed to the noise of machineries and equipments, annoyance indexes described in the previous paragraph are to be applied to the whole ship environment. It must be emphasized that instrumental evaluation of these indexes does not involve operative problems, since modern digital devices (e.g. sound level meters, spectra analyzers, data loggers, etc) make easy the measurement of all noise parameters in the whole interesting frequency range and their elaboration. The evaluation criteria previously described are quite simple to apply, so that they can be operated by well trained staff of medium-skills. Some of the methods described before (particularly the ones referring to ISO or ANSI standards) have even been implemented in the recent firmware of sound level meters (Brüel & Kjær 2009). Publications also exist (Bledsoe & Reynolds 1995) which describe algorithms for acoustics indicators and implement some of them in computer programs.

Actually, priority should be given to most recent criteria, as RC Mark II and the related QAI, which also allow evaluating the subjective perception and thus the satisfaction of passengers and crew. RC Mark II appears to be the culmination of a long evolution, and the success obtained in comparison to the other criteria also recently proposed (for example the Balanced Noise Criteria) demonstrates its effectiveness and broad operability. RC Mark II is not included in any standard, but it seems a good compromise between effectiveness of the results given in terms of occupant satisfaction and easiness of implementation.

In general, these criteria should be widely applied to all on board environments regardless of the room use. The results of such analysis should be evaluated on the basis of the design guidelines given by ASHRAE (ASHRAE 2009) by setting appropriate limits, for example by assuming as onboard limit the upper value of the acceptability range for each room type (Table 7).

Enhanced noise criteria could initially work alongside the existing dB(A) noise level limits, as quality values; with time and with the consolidation of the experience of their use, they may later even get to replace dB(A) limits.

At present, the introduction of specific criteria for on board ships is considered neither useful nor justified: the need for comfort at the same intended use of a specific room type is similar to the one of the civil context. Using indexes derived from such use can therefore take advantage of decades of experience in this field.

Table 7. Design Guidelines for onboard limits in different types of rooms.

Room types	RC (N); QAI < 5 dB criterion
Individual rooms or suites	35
Meeting/banquet rooms	35
Corridors, lobbies	45
Service/support areas	45
Executive and private offices	35
Conference rooms	35
Teleconference rooms	25
Open-plan offices	40
Churches, Mosques, Synagogues	35
Wards	40
Natatoriums, indoor sport facilities	45

6 CONCLUSIONS

Acoustic comfort is one of the most important factors that passengers and crew usually consider to assess their on board wellness. At present, rules mostly refer to merely energetic indexes, as the A-weighted sound level, that does not consider elements which have greater impact on the acoustic annoyance, such as the spectral composition of noise or the repetition over time. Some enhanced acoustic criteria to value the noise annoyance on board ships have been then proposed.

These evaluation criteria are derived from civil engineering context and seem to be able to improve present on board ships comfort classes, taking into consideration low frequency sound or relevant tonal components.

The advantages of such criteria over a merely energetic descriptor such as a dB(A) level is, first of all, that the latter would not provide sufficient detail to identify and reduce the annoying frequencies; on the contrary, the examined criteria provide the examiner additional information regarding the spectral characteristics of the noise, giving the possibility to tackle and overcome the problem. Moreover, the Room Criteria Mark II allows the estimating of the occupant satisfaction using the Quality Assessment Index, an indicator whose magnitude is useful to predict the probable reaction of an occupant when the system does not produce the optimum sound quality.

Proposed criteria have been applied as an example to noise spectra measured on board to the aim of better understanding the benefits of new proposed enhanced method. This way, useful information about noise effects on people onboard have been obtained.

These criteria are quite easy to be operated by means of standard instruments and well trained staff of medium-skills. Besides, these single number criteria are also easy to understand, so that they can be shared and communicated without particular problems.

Nowadays, when on board wellness becomes a relevant task in ship design and assessment, more accurate quality indicators are needed to guide shipbuilding. Recommended evaluation criteria for ship noise annoyance seem to be an operative tool to assess sound wellness for both crew and passengers and to obtain a better comfort on board. In fact, a deeper analysis of the effects of noise spectra, low frequency pulsations and sound reverberation on annoyance can provide the improvement of the acoustical design and a better knowledge of the interventions needed for sound damping and noise attenuation. At the present moment, when technological and industrial competition is more and more taut, this know-how can constitute a relevant competitive advantage in the global market fight.

ACKNOWLEDGMENTS

This work was developed in the frame of the collaborative project SILENV—Ships oriented Innovative soLutions to rEduce Noise & Vibrations, funded by the E.U. within the Call FP7-SST-2008-RTD-1. Grant Agreement SCP8-GA-2009-234182.

REFERENCES

ANSI 2008. ANSI S12.2-2008, *Criteria for Evaluating Room Noise*.

ASHRAE 2009. *ASHRAE Handbook. HVAC Applications (SI Edition)*. Atlanta: ASHRAE, Inc.

Badino, A., Borelli, D., Gaggero, T., Rizzuto, E., Schenone, C. (2011). Normative framework for noise emissions from ships: present situation and future trends. In Guedes Soares, C. & Fricke, W. (eds.), *Advances in Marine Structures*. Leiden: CRC Press/Balkema.

Beranek, L.L. 1957. Revised Criteria for Noise in Buildings. *Noise Control* 3(1): 19–27.

Beranek, L.L. 1989. Balanced Noise-Criterion (NCB) curves. *Journal of the Acoustical Society of America* 86: 650–664.

Bies, D.A. & Hansen, C.H. 2009. *Engineering noise control: theory and practice*. London & New York: Spon Press.

Blazier, W.E. Jr. 1997. RC Mark II: A refined procedure for rating the noise of Heating, Ventilating, and Air-Conditioning (HVAC) systems in buildings. *Noise Control Engineering Journal* 45(6): 243–250.

Blazier, W.E. Jr. 2003. Room Noise Criteria—Simple, Level-Sensitive Metrics. In E.H. Berger et al. (eds), *The Noise Manual Revised Fifth Edition*. Fairfax: AIHA Press.

Bledsoe, J.M. & Reynolds, D.D. 1995. *Algorithms for HVAC acoustics*. Atlanta: ASHRAE, Inc. Special Publications.

Brüel & Kjær 2009. Hand-held Analyzer Applications for Types 2250 and 2270—New Indoor Noise Parameters. Brüel & Kjær magazine 1: 28.

IMO 1975. Resolution A.343(IX): *Recommendation on Methods of Measuring Noise Levels at Listening Posts*.

IMO 1981. Resolution A.468(XII): *Code on Noise Levels on Board Ships*.

ISO R 1996:1967. *Assessment of noise with respect to community response*.

Kosten C.W. & Van Os G.J. 1962. Community reaction criteria for external noise. Paper F-5 in *The contol of Noise: National Physical Laboratory Symposium No. 12*. London: HMSO.

Schomer, P.D., 2000. Proposed revisions to room noise criteria. *Noise Control Engineering Journal* 48(3): 85–96.

Schomer, P.D., 2000. A test of proposed revisions to room noise criteria curves. *Noise Control Engineering Journal* 48(4): 124–129.

SOLAS 2009. SOLAS *Consolidated text of the International Convention for the Safety of Life at Sea, 1974, and its Protocol of 1988: articles, annexes and certificates*. London: IMO Publishing.

Tocci, Gregory C. 2000. Room Noise Criteria—The State of the Art in the Year 2000. *Noise/News International* 8(3): 106–119.

Sustainable Maritime Transportation and Exploitation of Sea Resources – Rizzuto & Guedes Soares (eds)
© 2012 Taylor & Francis Group, London, ISBN 978-0-415-62081-9

Innovative pod propulsive and noise performances assessment

B. Saussereau, F. Chevalier & T. Tardif d'Hamonville
DCNS, France

ABSTRACT: The experience from modern submarines has shown that the pump jet propulsion system has many advantages including low radiated noise. In this part, an innovative pump-jet pod developed by DCNS and CONVERTEAM has been assessed in term of noise and vibrations performances and the noise radiated will be computed. In a first stage, an unsteady CFD RANS calculation with Fluent is performed in the case of a ship propelled firstly by conventional pods and secondly by pump-jet pods. This preliminary calculation will highlight how an average propulsive efficiency is far better with the pump jet solution. In a second stage, the results from URANS calculation are coupled to a hydro-acoustic analysis in Sysnoise, based on the acoustic analogies, to evaluate blade rate noise in far field. The beneficial effects of the nozzle and the stator in pump-jet configuration on the acoustic levels will be considered.

1 PROBLEM SET UP

In this 3D study, all the immerged hull of the ship is considered with the complete propeller part. To simplify the flow problem, the ship bow is truncated, the appendages are suppressed and just the portside of the ship is represented. This ship hull is positioned (Fig. 1) without any incidence in a rectilinear uniform flow along the x axes, with a velocity value V. The free surface is represented by a mirror plane.

The half width of the ship is about 14.5 m, its waterline length is 197 m and its draught 6.2 m. The conventional and the pump-jet pods have a number of rotor blades respectively equal to 5 and 4. The rotor diameter, D, is 4.4 m and 3.6 m respectively for the conventional and the pump-jet pods and the rotor speed rotation, N, is 168 rpm and 215 rpm respectively for the conventional and the pump-jet pods.

2 ASSUMPTIONS AND MODELLING TOOLS

2.1 *Flow characteristics*

Simulations are realized in seawater (values for a temperature of 15°C):

- Mass density (ρ_0): 1026 kg/m^3
- Kinematic viscosity (ν_0): $1.18 \cdot 10^{-6}$ m^2/s
- Sound velocity (c_0): 1480 m/s

Figure 1. Problem set up schematization.

- Reference pressure (P_{ref}): $1 \cdot 10^{-6}$ Pa
- Flow velocity (V): 10.29 m/s (20 knots).

2.2 *Flow solving method*

2.2.1 *CFD solving approach*

The velocity choice enforces Reynolds number value in order of $2 \cdot 10^9$. For this range of Reynolds, the simulations must be done in turbulent flow. In this case of a flow around an obstacle, the turbulent model k–ε « Realizable » with a standard wall law is chosen.

Moreover, the flow will be considered in three dimensions, uncompressible and unstationary.

The solver used is Fluent v.12. An unsteady calculation is performed with a time step value of 0.001s, was performed during several rotor turns, and this is done until the convergence is well established.

During the last turn, dipolar source data on blades (which will be used for acoustic calculation) and forces (trust, torque) are recorded.

2.2.2 *Shape and size of the fluid mesh*

The meshes respect the ratio size evolution between elements and the equisize skew criteria. The y_+ values respect also the quality criteria necessary for k–ε turbulence model. They are about 205 in average for the conventional pod and about 200 for the pump-jet pod. The total number of mesh elements is about $3.2 \cdot 10^6$ for the conventional pod case and $2.5 \cdot 10^6$ for the pump-jet one (Fig. 2).

2.2.3 *Fluid boundary conditions*

Fluid boundary conditions associated to:

- the hull and the propeller are walls, no roughness is considered,

- inlet condition: a velocity inlet condition is used,
- outlet condition: pressure outlet condition is used,
- the down (y,z) plane and the portside (x,z) plan: symmetry conditions to model an equivalent infinite domain are used,
- the upper (y,z) plane: symmetry ("mirror") condition to model the sea free surface is used,
- the starboard (x,z) plane: symmetry condition to model the presence of the starboard part of the ship is used.

2.3 *Acoustic solving method*

2.3.1 *Acoustic solving approach*

To extract the acoustic from the unstationary uncompressible turbulent flow modelled with Fluent, Lighthill Acoustic Analogy (LAA, 1952) approach will be considered. This theory enables separating the flow modelling from the acoustic one (Roger 2000). It assumes that acoustic sources can be calculated directly from an uncompressible flow and then, these sources can be put into acoustic equations to solve the noise radiation problem. The first formulation of this theory was used to find the free flow noise radiation:

$$\frac{\partial^2 \rho'}{\partial t^2} - c_0^2 \frac{\partial^2 \rho'}{\partial y_i^2} = \frac{\partial^2 T_{ij}}{\partial y_i \partial y_j} \tag{1}$$

with

$$T_{ij} = \rho u_i u_j + (p - \rho c_0^2)\delta_{ij} - \tau_{ij} \tag{2}$$

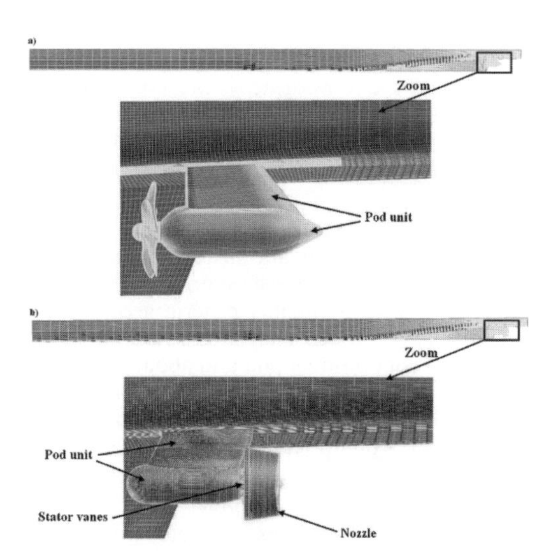

Figure 2. a) Conventional and b) pump-jet pods surface meshes (without fluid volume mesh).

where ρ' is associated to the fluid density little perturbations. The term containing the Lighthill tensor T_{ij} corresponds to a quadrupole source associated with flow velocity fluctuations. This exact formula is just a rewriting of Navier-Stokes equations and can be simplified by assuming:

- viscous effect negligible: $\tau_{ij} \sim 0$,
- any entropic disturbance: $(p - \rho c_0^2)\delta_{ij} \sim 0$
- uncompressible turbulence: $T_{ij} = \rho u_i u_j \sim \rho_0 u_i u_j$

The Lighthill tensor finally becomes:

$$T_{ij} = \rho_0 u_i u_j \tag{3}$$

In the Fourier space, the Lighthill equation becomes the following Helmholtz wave equation:

$$(\Delta + k^2)\rho'(\vec{y},\omega) = -\frac{1}{c_0^2}\frac{\partial^2 T_{ij}}{\partial y_i \partial y_j} \tag{4}$$

with

$$k = \frac{\omega}{c_0} \tag{5}$$

where k is the wave number and ω the pulsation.

With a Boundary Element Method (BEM) approach which considers Green function, $G(\boldsymbol{x},\boldsymbol{y})$, in free space as solution of this equation, the density can be written:

$$\rho'(\vec{x},\omega) = \frac{1}{c_0^2}\int_\Omega T_{ij}\frac{\partial^2 G(\vec{x},\vec{y})}{\partial y_i \partial y_j} d\Omega(\vec{y}) \tag{6}$$

with

$$G(\vec{x},\vec{y}) = \frac{\exp(ik|\vec{x}-\vec{y}|)}{4\pi|\vec{x}-\vec{y}|} \tag{7}$$

For a given space done, the Green function enables explicitly passing from given sources to the radiated acoustic field. The pressure level (for the little perturbations, $p' = \rho' c_0$) in one point of the space (in $\boldsymbol{x}$) results from the integration of velocity fluctuations (in $\boldsymbol{y}$) on the whole source volume area (Ω).

Even if, the Lighthill theory (1952) does not take into account the flow presence effects on the noise propagation, because of the separation between the flow and the acoustic problems, there are other theories which integrate this aspect like the Lilley equation or the Linearized Euler Equation. However, the effects of the flow presence on the propagation can be neglected for low Mach numbers.

This first Lighthill approach has been extended by Curle (1955) to cases of noise generated by a flow on motionless solid surface. The problem is no more in free space and a new term appears in the Fourier formulation (6):

$$\rho'(\vec{x},\omega) = \frac{1}{c_0^2}\int_{\Omega} T_{ij}\frac{\partial^2 G(\vec{x},\vec{y})}{\partial y_i \partial y_j} d\Omega(\vec{y}) + \frac{1}{c_0^2}\int_{\partial\Omega} P_{ij}\frac{\partial G(\vec{x},\vec{y})}{\partial y_i} n_j d\partial\Omega(\vec{y}) \tag{8}$$

The contribution associated with the rigid body presence is repres ented by a new dipolar term corresponding to pressure fluctuations normal to the solid body surface ($\partial\Omega$). This contribution is called the loading noise.

If the sources are compact which means that $D \ll \lambda$, with D the characteristic length of the source and λ the acoustic wave length, the dipolar contribution will be more efficient than the quadrupole one. Generally, this is always true for low Mach numbers.

A last generalization of the Lighthill theory was (1952) done by Ffowcs Williams & Hawkings (1969) to estimate the noise generated by moving solid surface. A new term appears in the formula (8) and also a moving effect more visible in time space (with the Green function as solution in free space):

$$\rho'(\vec{x},t) = \frac{1}{4\pi c_0^2}\int_{\Omega} \frac{\partial^2}{\partial y_i \partial y_j}\left[\frac{T_{ij}}{R|1-M_r|}\right] d\Omega(\vec{y}) - \frac{1}{4\pi c_0^2}\int_{\partial\Omega} \frac{\partial}{\partial y_i}\left[\frac{P_{ij}}{R|1-M_r|}\right] d\partial\Omega(\vec{y}) - \frac{1}{4\pi c_0^2}\int_{\partial\Omega} \frac{\partial}{\partial t}\left[\frac{\rho_0 V_n}{R|1-M_r|}\right] d\partial\Omega(\vec{y}) \tag{9}$$

with $V_n = \vec{V}_{\partial\Omega}.\vec{n}$ the body normal velocity, $R = |\vec{x}-\vec{y}|$ the distance between sources and observer, $M_r = V_{\partial\Omega}\vec{R}/(c_0 R)$ the relative Mach number and $[\ ] = (\vec{y}, t - R/c_0)$ evaluation at the delayed time. This time represents the fact that when the signal is received, the sources position is different from the position they have at the emission time. The noise radiation is amplified by the movement towards the observer and attenuated in reverse. This amplification phenomenon is managed by the Doppler factor $1-M_r$.

The new appeared term is a monopolar one and corresponds to the thickness noise, the noise generated by the solid body when it "pushes" the fluid to progress. In the far field ($R \gg \lambda$), if the object in motion corresponds to a profile thin enough, in rotation and which can be considered as a compact source, this thickness noise can be neglected. This is moreover verified that the Mach number is low.

This rotating source noise problem can be adapted to the rotor one and especially to the blade rate one (Roger 1994). This noise is related to the periodical unstationary part of aero dynamical phenomenon while the broadband noise generated by the rotor corresponds to the random part.

M. Roger (1994) has shown that, in the far field, one element of a compact blade is only able to produce a noise at the rotation frequency harmonics. For the *mth* harmonic, the blade rate noise pressure at the distance R from the segment source is:

$$p'_m(\vec{x},t) = \frac{imNe^{i2\pi mNR/c_0}}{2c_0 R}\sum_{s=-\infty}^{+\infty} e^{i(m-s)(\varphi-\pi/2)} F_s J_{m-s}(mM\sin\theta)\left\{\cos\gamma\cos\theta - \frac{m-s}{m}\frac{\sin\gamma}{M}\right\} \tag{10}$$

with m an integer, N the rotor rotation speed in revolution per second, F_s the Fourier series of the total force on a compact blade segment, J the Bessel function. The other parameters are defined in the blade local frame like shown in Figure 3.

If the propeller is built with Z identical blades, the acoustic field radiated by the rotor is the sum of every blade contributions. For the blade rate noise, each blade contribution corresponds to the one of a reference blade with a dephasing time which creates an interference phenomenon. In the far field, for a free rotor, the complex acoustic pressure amplitude at the mth harmonic is:

$$p'_{mZ}(\vec{x},t) = \frac{imZ^2N}{2c_0 R}\sum_{s=-\infty}^{+\infty} e^{-i(mZ-s)\pi/2} F_s J_{mZ-s}(mZM\sin\theta)\left\{\cos\gamma\cos\theta - \frac{mZ-s}{mZ}\frac{\sin\gamma}{M}\right\} e^{i(mZ-s)[\varphi-\Omega_s(t-R/c_0)]} \tag{11}$$

with

$$\Omega_s = \frac{2\pi mZN}{mZ-s} \tag{12}$$

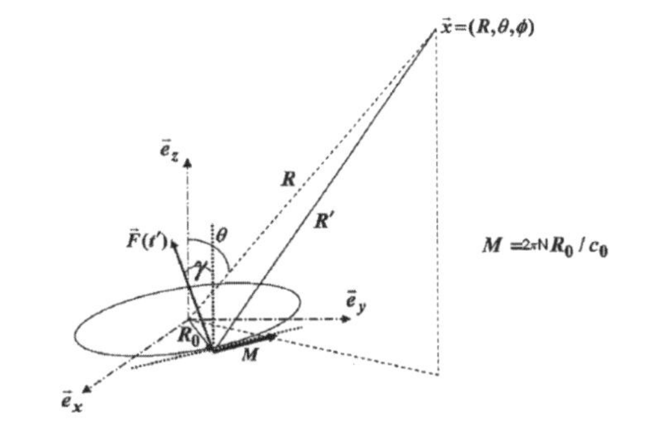

Figure 3. Acoustic source notation schematization.

Other formulations exist to take directly into account the effect of an inlet stator on the noise calculation. Inlet guide vane stator wakes creates a rotor inflow which presents a particular stationary distortion with the fundamental frequency vN, where v is the number of stator vanes.

By assuming that the space between stator and rotor is long enough, the stator presence does not generate any new noise source and the equation (11) remains available. However, the forces on the Z blades, which cross equally spaced guide vane wakes, are periodic with the fundamental frequency vN and this periodicity is translated by replacing s by sv:

$$p'_{mZ}(\vec{x},t) = \frac{imZ^2N}{2c_0R} \sum_{s=-\infty}^{+\infty} e^{-i(mZ-sv)\pi/2} F_{sv} J_{mZ-sv}(mZM\sin\theta) \left\{ \cos\gamma\cos\theta - \frac{mZ-sv}{mZ}\frac{\sin\gamma}{M} \right\} e^{i(mZ-sv)[\varphi-\Omega_{vs}(t-R/c_0)]} \tag{13}$$

with

$$\Omega_{sv} = \frac{2\pi mZN}{mZ-sv} \tag{14}$$

F_s are also called loading harmonics and they are the only unknowns of this problem since the other terms are defined by the geometry and the kinematics.

Finally, this approach is well adapted to solve the acoustic problem for the present study on the blade rate noise especially as the Mach number considered is very low which enables neglecting the flow presence on the noise propagation and also the quadrupole term contribution if the blade considered is compact.

As Sysnoise uses the Roger's approach (1994), to solve this acoustic problem, we just have to extract the F_s values from the CFD results on a reference blade.

With Sysnoise, the BEM approach will be used so only the solid surfaces need to be meshing and there is no fluid mesh. With the Green function, the Helmholtz equation in the whole space has just to be solved on the surfaces taking into account all the boundary conditions of the field. More especially, the BEM direct approach will be considered which supposes that the unknown of the problem are the pressure and the normal velocity on the surfaces.

2.3.2 *Shape and size of the acoustic mesh*

In this study, we are just interested in modelling the first blade rate frequencies which are harmonics of the fundamental frequency:

$$f_{BR,m} = mZN \tag{15}$$

with m a nonzero relative integer. This formula gives a fundamental frequency value of 14 Hz for the conventional pod and 14.3 Hz for the pump-jet pod.

To build the BEM surface mesh, quadrilateral elements are preferred and the mesh is taken the most homogenous as possible. The surface structure description can be simplified and, as the acoustic field observed will be around the propeller and behind the ship, the hull length is shortened in the front part. Moreover, with the fan source approach, the rotor mesh is not necessary.

To have a good description of the acoustic radiation with the BEM method, the minimal element size, $d_{x,min}$, for the acoustic mesh must respect this criteria:

$$d_{x,\min} < \lambda/6 \tag{16}$$

As just the first four blade rate frequencies need to be modelled, this study can be limited to a maximal frequency around 85 Hz which imposes a minimal element size of 2.9 m.

Finally, the conventional pod acoustic mesh gets 3005 elements and the associated maximal frequency is precisely 79.26 Hz. The pump-jet pod acoustic mesh gets 2819 elements and the associated maximal frequency is precisely 87.87 Hz. Even if the hull and the pod meshes were not built precisely with the same resolution for the both pods, the element size remains small enough to solve precisely the acoustic problem for the both cases.

For these frequencies, we can also notice that the propeller length for the two pods considered respects the compactness assumption ($D << \lambda$):

- for the conventional pod, as the maximum frequency is 79.26 Hz, the minimal wave length is about 18 m which is larger than the rotor diameter 4.4 m,
- for the pump-jet pod, as the maximum frequency is 87.87 Hz, the minimal wave length is about 17 m which is larger than the rotor diameter 3.6 m.

2.3.3 *Acoustic boundary conditions*

The acoustic boundary conditions are associated to:

- on the whole structure: motionless condition (surfaces are considered rigid, here there is no vibro-acoustic study),
- on the rotor: pressure loading generating rotating dipolar source which can take into account a stator effect (Generate Fan Source),
- on ship portside: symmetry condition ($v = 0$),
- on the free surface: asymmetry condition ($p = 0$).

Moreover, with these symmetry conditions, the structure mesh is closed and the direct BEM approach becomes suitable to the present study.

2.3.4 *Acoustic measure field points*

The radiated pressure field in the far field hypothesis ($R >> \lambda$) will be calculated in four particular measurement points situated at the distance R from the rotor:

- in rotor x axis, $R = 200$ m rotor downstream,
- in rotor y axis, $R = 200$ m at the rotor portside,
- in rotor z axis, $R = 200$ m below the rotor,
- at 45° in rotor x, y and z axis, at $R = 200$ m from the rotor.

If the rotor center is situated at the coordinates (0,0,0), these four points are placed on a sphere with a radius R = 200 m. Finally, all these measurement points can be schematized in Figure 4.

In following analysis, the acoustic pressure level (L_P, in dB) calculated on the eighth sphere using the acoustic pressure P, corresponds to:

$$L_P = 20\log\left(\frac{P}{P_{ref}}\right) \tag{17}$$

To dispense with the measurement distance, on the four measurement points, the acoustic pressure level calculated at 200 m (L_{P200m}) will be reported at 1 m (L_{P1m}), which means:

$$\begin{aligned} L_{P_{1m}} &= L_{P_{200m}} + 20\log(200) \\ &= 20\log\left(\frac{P_{200m}}{P_{ref}}\right) + 20\log(200) \end{aligned} \tag{18}$$

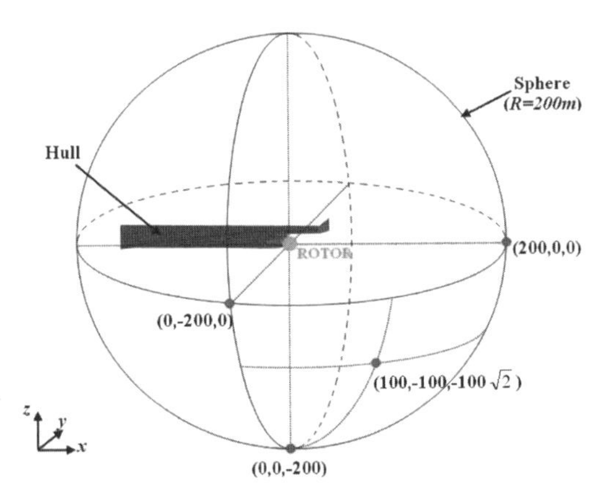

Figure 4. Acoustic measurement field points schematization.

3 CFD RESULTS

3.1 *Nominal wake*

Before calculating the propulsive performances, we have performed a calculation without pod to determine the hull resistance and the wake coefficient.

The wake coefficient is very small and is almost the same for both pod units. The small difference is due to different propeller location: the pump jet pod is closer to the hull.

As the free surface is not modelled, the wave resistance is not computed. This explains why hull resistance is lower than the propellers thrust (2 per ship).

3.2 *Conventional pod results*

3.2.1 *Propulsive performances*

The propeller forces are calculated during the last turn. The propeller propulsive performances are the following:

The thrust deduction factor is about 0.07 and the overall pod efficiency for the complete pod is 0.59.

3.2.2 *Velocity field around the propeller*

The velocity field in the rotor vertical symmetry plane (Fig. 5) highlights a strong recirculation at the ship back. The fluid acceleration through the rotor clearly appears especially below the pod unit.

In the $x = 8$ m plane, the velocity and the vorticity fields (Fig. 6) highlight the presence of a vortex left in the rotor wake and deformed by the pod unit placed behind the rotor. This vortex dissipates itself progressively in the far field but remains visible after 50 m rotor downstream. We can note that the hub vortex is very strong and will probably cavitate.

3.2.3 *Pressure field around the propeller*

The pressure field in the rotor vertical symmetry plane (Fig. 7) highlights the depression on the blade extrados. We can notice the heterogeneous pressure distribution between the top and bottom blades because of the non uniform wake seen by the propeller.

The pressure field evolution on the ship in the $y = 0$ plane during one fifth revolution is visible in Figure 8 at four times to see the propeller rotation and the direct effect on the around pressure field.

3.2.4 *Hull pressure fluctuation*

The maximum hull pressure fluctuations on the hull induced by the conventional pod are about 1000 Pa.

The tip clearance for the conventional pod is 1.3 m.

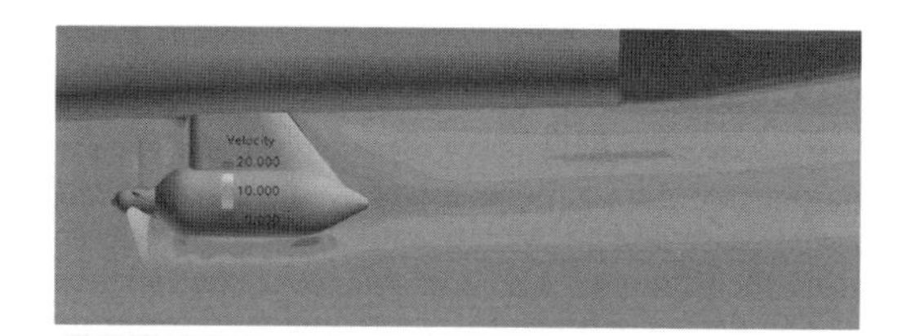

Figure 5. Velocity field around the propeller.

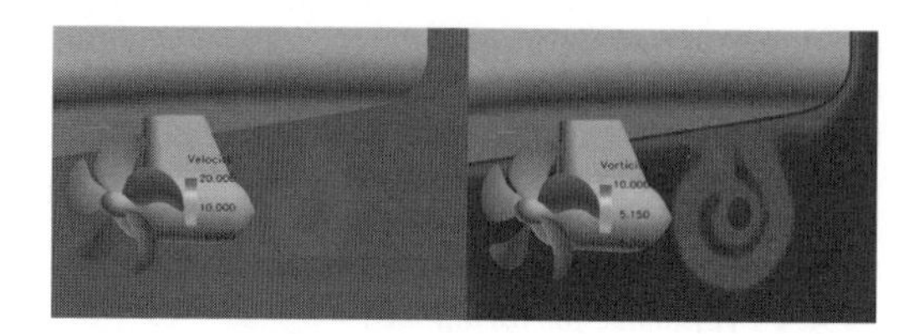

Figure 6. Velocity and vorticity field at the propeller back.

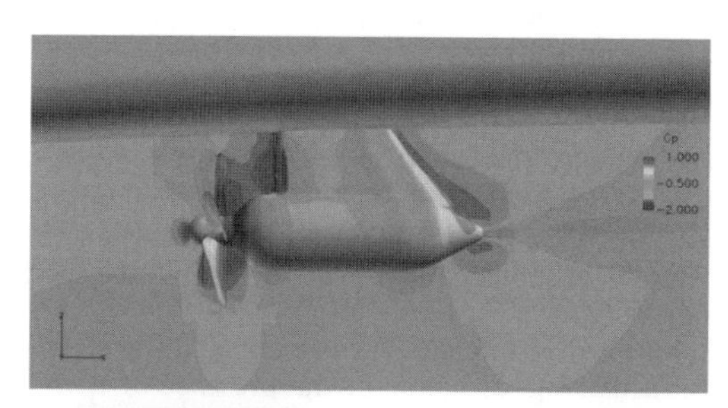

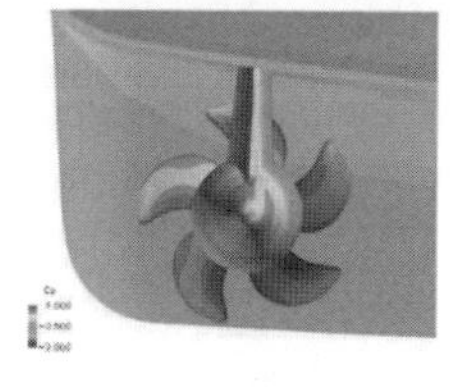

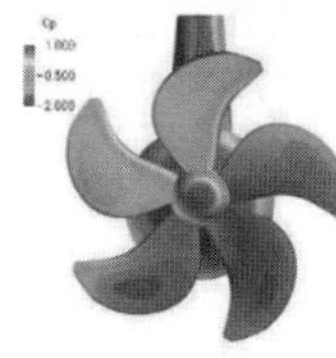

Figure 7. Pressure field around the propeller.

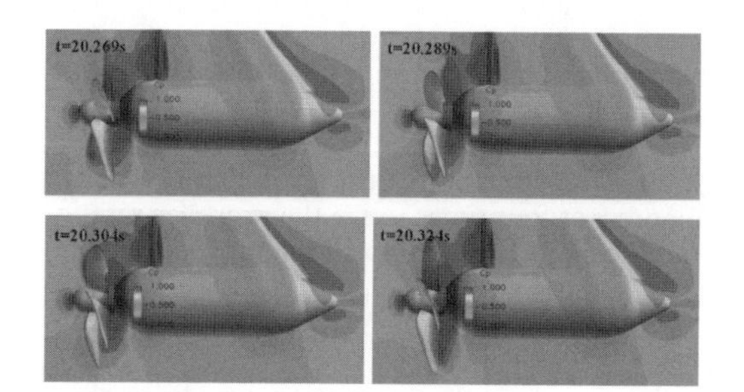

Figure 8. Pressure field around the propeller during one fifth revolution.

Table 1. Nominal wake coefficient.

	Pump jet pod	Conventional pod
Wake coefficient	0.04	0.03

Table 2. Conventional pod propeller thrust performances.

Thrust [kN]	Average value	RMS value*
Pod	–84.9	1.1%
Rotor	586.2	0.1%
Complete pod	501.3	0.5%
Hull drag	–804.9	0.8%
Hull drag without pod	–752.3	–

* % of average value.

Table 3. Conventional pod propeller torque performances.

Torque [kNm]	Average value	RMS value*
Rotor**	455.3	0.1%
1 blade	91.1	15.8%

*% of average value, ** blades + hub + cone.

3.3 *Pump-jet pod results*

3.3.1 *Propulsive performances*

The propeller forces are calculated during the last turn. The propeller propulsive performances are the following:

The thrust deduction factor is about 0.045 and the overall pod efficiency for the complete pod is 0.67.

3.3.2 *Velocity field around the propeller*

The velocity field in rotor vertical symmetry plane axis (Fig. 9) highlights a strong recirculation at the ship back. The fluid acceleration through the rotor clearly appears and is higher than for conventional pod because of a smaller rotor diameter.

In the $x = 8$ m plane, the velocity and the vorticity fields (Fig. 10) highlight the presence of a vortex left in the rotor wake. This vortex dissipates itself progressively in the far field but remains visible after 50 m rotor downstream.

3.3.3 *Pressure field around the propeller*

The pressure field in the rotor vertical symmetry plane (Fig. 11) shows vortex formation leaving the nozzle. The pressure distribution on the blades is very uniform regarding the blade position unlike for the conventional case. This is due to the nozzle and the stator effects which induce generate a more uniform flow on the rotor.

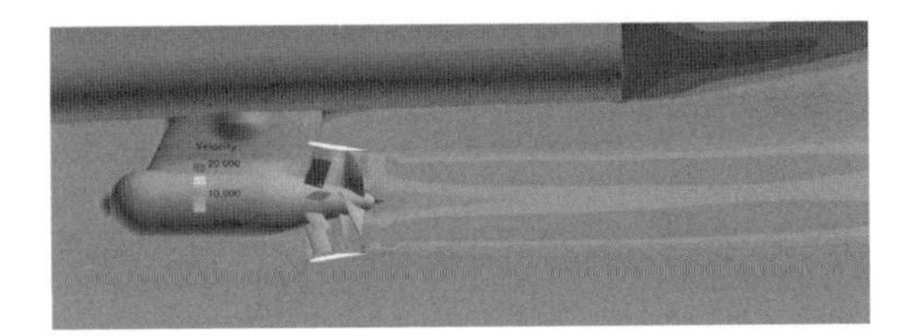

Figure 9. Velocity field around the propeller.

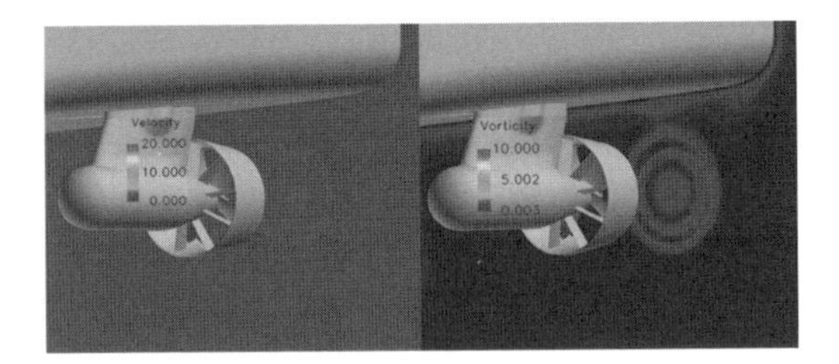

Figure 10. Velocity and vorticity field at the propeller back.

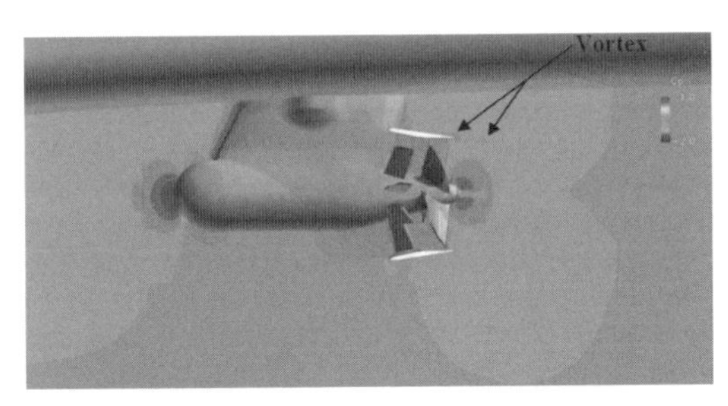

Figure 11. Pressure field around the propeller.

Table 4. Pump-jet pod propeller thrust performances.

Thrust [kN]	Average value	RMS value*
Pod	−69.6	0.6%
Stator	−69.9	0.2%
Nozzle	167.4	0.2%
Rotor	478.6	0.4%
Complete pod	507.1	0.3%
Hull drag	−786.1	0.5%
Hull drag without pod	−752.3	–

* % of average value.

The pressure field evolution on the ship in the $y = 0$ plane during one quarter revolution is visible in Figure 12 at four times to see the propeller rotation and the direct effect on the around pressure field.

3.3.4 *Hull pressure fluctuation*

The maximum hull pressure fluctuations induced by the pump-jet pod are about 300 Pa. This is a low level compared to standards levels on ships at this speed.

The tip clearance for the pump jet pod is smaller (1.2 m) but the nozzle induces lower pressure fluctuations.

3.4 *Results comparison*

3.4.1 *Design considerations*

The design point of the conventional pod and the pump-jet pod are not the same: the rotation rate of the pump jet pod is higher. This choice has been done for architectural purposes.

This higher rotation rate should induce more hull pressure fluctuations, more noise and less efficiency.

But these disadvantages are counterbalanced by the pump-jet and nozzle: the inlet stator induces a smooth and uniform flow on the rotor, which induces lower vibrations. The nozzle allows using a smaller diameter, which is very good for noise.

3.4.2 *Propulsive performances*

The calculations have shown that the pump-jet pod, despite its higher rotation rate, has a better efficiency (+13.6%) than the conventional pod.

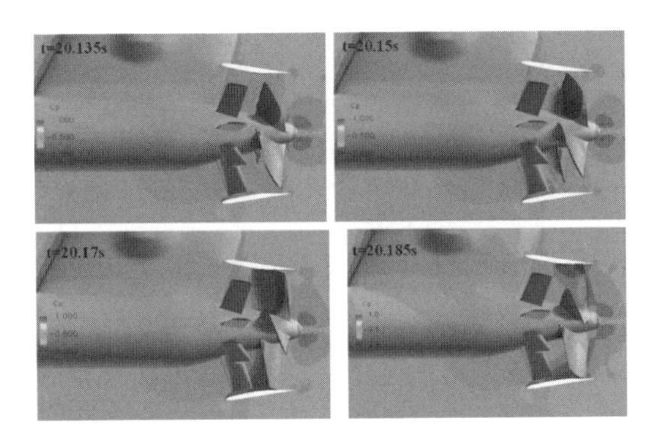

Figure 12. Pressure field around the propeller during one fifth revolution.

Table 5. Pump-jet pod propeller torque performances.

Torque [kNm]	Average value	RMS value*
Rotor**	325.9	0.3%
1 blade	81.5	6.3%

*% of average value, ** blades + hub + cone.

Moreover, the flow seen by the rotor being more uniform, the torque and thrust fluctuations are smaller.

3.4.3 *Hull pressure fluctuation*

The CFD calculations have shown that the hull pressure fluctuations induced by a pump-jet pod are about two times lower than for a conventional pod.

As this hull excitation will induce hull noise (structural response), this is also a very good point for radiated noise and inboard noise and vibrations.

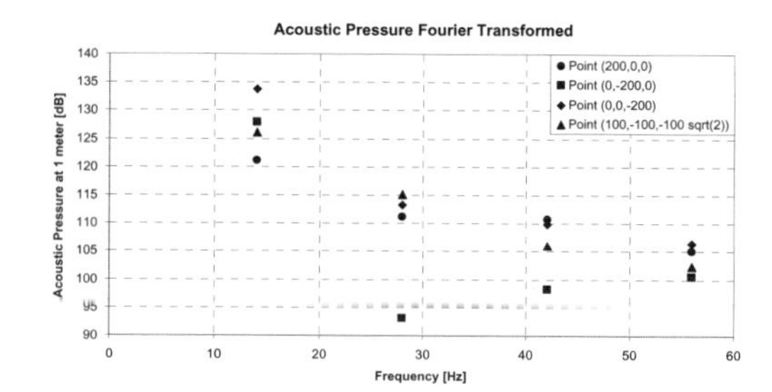

Figure 13. Acoustic pressure Fourier transformed for the conventional pod.

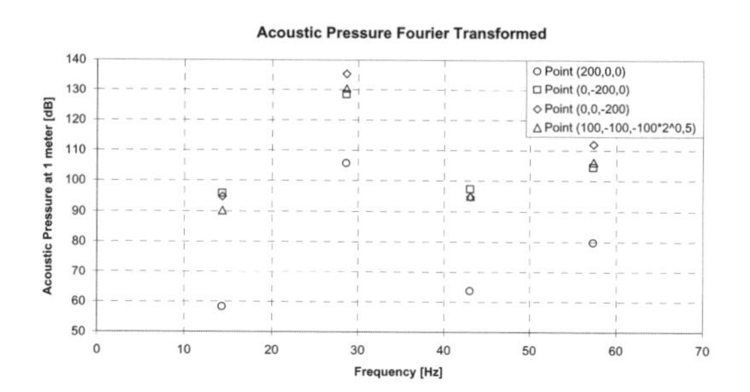

Figure 14. Acoustic pressure Fourier transformed for the pump-jet pod.

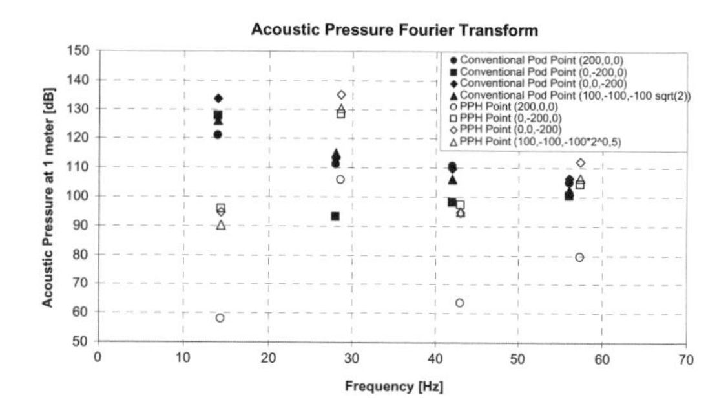

Figure 15. Acoustic pressure Fourier transformed for the conventional pod.

4 ACOUSTIC RADIATION RESULTS

4.1 *Conventional pod results*

The acoustic pressure level is calculated (Fig. 13) for the first four blade rates, on the four reference points.

Except for the point on the rotor y axis (square point), it can be notice that the acoustic pressure of the three other points progressively decreases when the harmonic level increases. The pressure radiation directivity seems being more isotropic for the first blade even if it is found again that there is a slight directivity on the rotor z axis (diamond point).

For the second and the third blade rates, points on the rotor x and z axis (respectively the sphere and the diamond points) have the same evolution. The point on the rotor y axis remains with the lower pressure level even if, on the three last blade rates, its level increases progressively because the directivity rises up slightly on the rotor (x,y) plane.

4.2 *Pump-jet pod results*

The acoustic pressure level reported at one meter is calculated (Fig. 14) for the first four blade rates, on the four reference points.

For the four harmonics, it can be notice that the point situated on the rotor x axis (sphere point) still has the lower pressure level and the three other points have nearly the same behavior: the pressure acoustic level is lower for the first and the third harmonics and the behavior and values of these both harmonics remain close.

4.3 *Results comparison*

On the four reference points, the acoustic pressure level really changes from one pod to an other (Fig. 15).

We can note that the pump-jet pod has a lower radiated noise at blade rates frequencies than the conventional pod despite a higher rotation rate.

The pump-jet pod noise at the first blade rate is 40 dB lower than for the conventional pod. It is interesting to note that for the second harmonic, the level is 20 dB higher. The third and fourth harmonics have nearly the same level.

It can be noticed again that the stator presence really filters blade rates, especially the first and the third ones which become less noisy than the conventional pod.

On the contrary, the nozzle of the pump-jet pod has no effect on the acoustic radiation. But, the nozzle avoids the huge pressure fluctuations on the hull above the rotor visible on the conventional pod which can generate much more noise by fluid-structure interaction.

5 CONCLUSIONS

First, CFD calculations have shown that the efficiency is better for the pump-jet pod than for the conventional one. This gain will also involve significant fuel saving. Moreover, the nozzle presence in the pump-jet pod involves lower pressure fluctuations on the hull. So, these lower hull vibrations are better for the ship acoustic and the onboard comfort.

Then, acoustic calculations have also shown that the inlet stator presence in the pump-jet pod filters and decreases considerably the blade rate pressure levels compared to the conventional pod ones.

Finally, the CFD calculations coupled with acoustic calculations have shown the several advantages of a pump-jet pod compared to a conventional pod.

ACKNOWLEDGEMENT

Authors gratefully acknowledge European Union which has financed the research presented in this paper through the project SILENV (Ships oriented Innovative soLutions to rEduce Noise and Vibrations).

REFERENCES

Curle, N. 1955. The influence of solid boundaries on aerodynamic sound. *Proc. Roy. Soc. London*, 2311(1187), 505–514.

Ffowcs Williams, J.E. & Hawkings, D.L. 1969. Sound generation by turbulence and surfaces in arbitrary motion. *Phil. Trans. Roy. Soc. London*, 264, A1151, 321–342.

Lighthill, M.J. 1952. On sound generated aerodynamically. *I. General theory*, *Proc. Roy. Soc. London*, 211, A1107, 564–587.

Roger, M. 1994. Contrôle du bruit aérodynamique des machines tournantes axiales par modulation de pales. *Acoustica* 80.

Roger, M. 2000. The acoustic analogy some theoretical background. *VonKarman Institute for Fluid Dynamics, Lecture Series* 2000-02, *February* 14–17, 2000.

Sustainable Maritime Transportation and Exploitation of Sea Resources – Rizzuto & Guedes Soares (eds)
© 2012 Taylor & Francis Group, London, ISBN 978-0-415-62081-9

Hydroacoustic characterization of a marine propeller through the acoustic analogy

S. Ianniello, R. Muscari & A. Di Mascio
CNR-INSEAN, Rome, Italy

ABSTRACT: This paper deals with the use of the Acoustic Analogy for the prediction of the underwater noise generated by a marine propeller. Different configurations are treated in order to demonstrate the potentiality of such a numerical approach and to analyze the role played by the different noise sources. Unlike analogous aeronautical propellers, it will be shown that the evaluation of the nonlinear *quadrupole* sources is relevant regardless of the blade rotational speed. On the contrary, the linear sources (*thickness* and *loading* noise) due to the propeller are predominant just in a very limited region, while the only appreciable linear contribution affecting the acoustic pressure far field comes from hull scattering effects.

1 INTRODUCTION

Within the last years the marine scientific community has been paying an increasing attention to many hydroacoustic phenomena concerning the maritime transport and many marine problems. This is due not only to the well-known health and comfort problems onboard or to the operational ability of different ships affected by a high noise level (offshore survey vessels, fishery and ocean research vessels, seismic vessels and others), but also to the environmental pollution of the sea and the negative impact on marine mammals. For this reason, many international organizations are moving towards more and more stringent regulations on underwater noise and some more restrictive certification tests for many types of ships. As a matter of fact, such a general tendency is not related to a satisfactory knowledge of the generating and propagating noise mechanisms in water and looking at the available literature it is easy to recognize a deep lack of both theoretical and, above all, numerical models. At present, the criteria adopted to satisfy the noise emission requirements for a ship are based on empirical basis and the use of some approximated numerical procedure able to provide, to the utmost, a qualitative raw estimation of the acoustic far field.

Among the many sources of sound related to a ship, the propeller plays a primary role. Even at a simple nocavitating condition and at cruising speed, the propulsor acts not only as a direct noise source (being a body moving in the fluid), but also as a sort of *indirect* source, since it excites the stern vault of the hull by an unsteady (periodic) hydrodynamic load. Thus, the hull itself *scatters* the pressure and affects the resulting ship underwater noise field. Unfortunately, many aspects of the acoustic behavior of a marine propeller are still completely unknown, as, for instance, the influence of a *full-unsteady* inflow (as in a manouvering ship), the interactions with hull, rudders and appendages or, above all, the hydroacoustic effects due to the different cavitation phenomena. At present, the selection criteria for a propulsor are strictly limited to the efficiency features (evaluated through the well-known torque and thrust diagrams), while the generated noise is assumed to be a sort of inevitable and unverifiable consequence. This way, the present inability in modeling the underwater noise field could become a critical aspect for many shipyards in the next future, especially in view af the above mentioned and more stringent certification tests.

This situation is rather surprising. Many theoretical and numerical models, originally developed in Aeroacoustics for different rotary-wing propulsion systems, are presently available to analyse the acoustic behaviour of a propeller. For a long time, these models were developed and validated in a lot of national and international research projects and are now used by industry in many applications of practical interest. These models could be successfully adopted to investigate the hydroacoustic behavior of a marine propeller and, in general, to evaluate the noise generated by the *ship-system* and many of its subcomponents.

Early studies for a rotating blade started in the thirties, when it was already known that both the blade loading and the body thickness could generate noise by separate mechanisms. Most of the early works focused on aeronautical propellers and in the fifties the application of the theory of *Acoustic Analogy* (Lighthill, 1952), formulated for the jet noise phenomena, contributed to the enhancement of theoretical approaches. A fundamental step towards a full understanding of noise generated by bodies moving in a fluid was given by Ffowcs Williams and Hawkings (1969), as an extension of Lighthill work. Exploiting the theory of the generalized functions (see Kanwal, 1983), they derived a governing differential equation for the acoustic pressure which has been established as the theoretical basis for a number of modern and sophisticated prediction tools devoted to the prediction of rotating machinery noise. The FWH equation may be considered as an extension of Lighthill equation to take into account the basic mechanisms of noise generation related to the shape of the body and the loads it experiences along its motion through the fluid. In the seventies the research efforts were devoted to the analytical treatment of the FWH equation: the application of the standard Green function technique (Morse, 1953) and some relevant results concerning the generalized derivatives enabled the derivation of useful integral expressions for the sound pressure field. In particular, F. Farassat proposed a number of different forms of solution in time domain (Farassat, 1975 and 1981) which were successfully implemented for the calculation of the FWH linear source terms for an helicopter rotor. Nowadays, these solving formulations represent a standard approach for many computational tools devoted to the aeroacoustic analysis of complex multibodies configurations, and are widely used even in the industrial context.

Unlike the fruitful research in Aeronautics, only few applications of the FWH equation were proposed in the recent years for a marine propeller. A noise prediction was carried out for a noncavitating propeller with and without a duct (Seol, 2001), by coupling the well known Farassat time-domain *formulation 1A* to a hydrodynamic BEM solver based on a potential formulation. Some years later, the same numerical approach was used to account for the presence of sheet cavitation (Seol, 2005), although in that work no particular algorithm was implemented to deal with the occurrence of the bubble or to investigate its own acoustic effects, so that any high frequency content of the noise signals was admittedly removed. The robustness of the acoustic analogy and its advantages with respect to a direct pressure estimation by the Bernouilli equation were discussed later (Testa, 2008), by pointing out the role played by the numerical modeling of the propeller wake. Some preliminary interesting results on a sheet cavitation noise prediction were also published (Salvatore & Ianniello, 2003). There, starting from the knowledge of the bubble shape time evolution, the linear terms of the FWH equation were evaluated on a time-dependent radiating domain (the blade *plus* the bubble) and the expected impulsive waveform of the noise signals was carried out, with an overall hydroacoustic behaviour very similar to a monopole source with a high frequency content. An alternative FWH-based approach to deal with the hydroacoustic effects of a sheet cavitation was introduced (Salvatore, 2006), discussed (Testa, 2008) and used (Salvatore, 2009) to describe the effects of a transient cavitation occurring on a propeller operating in a inhomogeneous flow, even in presence of a scattering plate simulating the aftbody of a ship hull.

A notable limitation, however, of all the aforementioned papers is that the numerical investigations only concern some physical or numerical aspect of the problem and, above all, avoid to perform a comprehensive characterization of the propeller hydroacoustic behaviour. In particular, they always assume that the effects of the FWH nonlinear terms can be neglected because of the low rotational speed of the blade, but such a limiting assumption (rather usual in the aeronautical context) has never been confirmed underwater, either from and experimental and a numerical point of view. The main aim of this paper is to investigate on the actual role played by all the noise sources related to a marine propeller and to prove the potentiality of the Acoustic Analogy in the numerical prediction of the underwater noise.

2 THEORETICAL BACKGROUND

The FWH represents an elegant manipulation of the fundamental conservation laws of mass and momentum which gives rise to the following inhomogeneous wave equation written in terms of generalized functions

$$\bar{\mathbb{D}}^2 p'(x,t) = \frac{\bar{\partial}}{\partial t}[\rho_0 v_n + \rho(u_n - v_n)\delta(f)] - \frac{\bar{\partial}}{\partial x_i}\{[\Delta P_{ij}\hat{n}_j + \rho(u_n - v_n)\delta(f)]\} + \frac{\bar{\partial}^2}{\partial x_i \partial x_j}\{T_{ij}H(f)\} \quad (1)$$

The equation $f = 0$ is an implicit equation which describes an arbitrary surface, whose choice heavily affects the physical meaning of the different terms. The fluid and surface velocity components are indicated by u_i and v_i, respectively, while the subscript n indicates the projection along the outward normal to the surface. The D'Alembert operator is given by

$$\overline{\mathbb{D}}^2 = \frac{1}{c_0^2}\frac{\partial}{\partial t} - \overline{\nabla}^2$$

while $T_{ij} = \rho u_i u_j + P_{ij} - c_0^2\tilde{\rho}\delta ij$ is the Lighthill stress tensor, $\tilde{\rho} = (\rho - \rho_0)$ the density perturbation, c_0 and ρ_0 the speed of sound and the fluid density, respectively, in the undisturbed medium, P_{ij} the compressive stress tensor ($\Delta P_{ij} = P_{ij} - p_0\delta_{ij}$) and δ_{ij} the Kronecker symbol. The presence of the Dirac and Heaviside functions points out the different nature of the source terms: two surface terms directly related to the effects of the discontinuity $f = 0$ in the flow field and a volume term accounting for all noise sources acting outside it. When $f = 0$ coincides with the body surface S, the impermeability condition $u_n = v_n$ simplifies equation (1) and the use of the Green's function approach leads to the following integral equation for the acoustic pressure (at point $\boldsymbol{x}$ and time t)

$$4\pi p'(x,t) = \frac{\partial}{\partial t}\int_S \left[\frac{\rho_o v_n}{r|1-M_r|}\right]_\tau dS + \frac{1}{c_0}\frac{\partial}{\partial t}\int_S \left[\frac{p\hat{n}\cdot\hat{r}}{r|1-M_r|}\right]_\tau + \frac{\partial}{\partial t}\int_S \left[\frac{p\hat{n}\cdot\hat{r}}{r^2|1-M_r|}\right]_\tau dS \; p_Q'(x,t) \qquad (2)$$

This equation is written under the assumption of inviscid flow (thus reducing the compressive stress tensor to the scalar pressure field on the blade surface: $\Delta P_{ij} = (p - p_0)\delta_{ij}$) and isentropic transformations, for which the pressure-density relationship can be approximated by the linear term of its series expansion (i.e. $p' = c_0^2\tilde{p}$, where p' denotes the acoustic pressure disturbance). In equation (2) the subscript r denotes the projection along the source-observer direction, M is the Mach number and r indicates the source-observer distance. The surface integrals represents the linear terms and are usually known as *thickness* and *loading* noise. The last term $p_Q(\boldsymbol{x},t)$ represents the so-called *quadrupole* noise and corresponds to three volume integrals theoretically extended to the whole region flow (V) affected by the body motion

$$p_Q'(x,t) = \frac{1}{c_0^2}\frac{\partial^2}{\partial t^2}\int_V \left[\frac{T_{rr}}{r|1-M_r}\right] dV + \frac{1}{c_0^2}\frac{\partial}{\partial t}\int_v \left[\frac{3T_{rr}-T_{ii}}{r^2|1-M_r|}\right]_\tau dS + \int_v \left[\frac{3T_{rr}-T_{ii}}{r^3|1-M_r|}\right]_\tau \qquad (3)$$

Note that all the integral kernels are determined at the emission (retarded) time τ, which represents, for any observer time t and location x, the instant when the contribution to the noise signature was released. The difference between t and τ is an essential feature of the acoustic integrals and emphasizes that sound propagates at finite speed. By avoiding the computation of the nonlinear quadrupole sources and moving the time derivatives within the integrals, equation (2) gives rise to the *formulation 1A* (Farassat, 1981), the standard retarded time formula for the linear acoustic analysis of rotating blades. This formula is rather straightforward to be implemented and reduces the noise prediction to a simple post-processing of the aero/hydrodynamic data. On the contrary, the computation of the quadrupole noise is not so easy, and requires the knowledge of the three-dimensional velocity, pressure and density fields, besides to a volume integration. It's worth noting that when the projection of the Mach vector along the source-observer direction tends to 1, all the integrals become singular. This is an interesting feature of these integral solution forms of the FWH equation which is related, from a physical point of view, to the occurrence of *multiple emission times* for noise sources moving at supersonic speed (Ianniello, 2007).

Since the last nineties, an alternative integral formulation was proposed to achieve a comprehensive (linear *plus* nonlinear) evaluation of the acoustic pressure and, at the same time, to avoid any direct volume integration. Such an alternative and effective approach is known as *porous formulation*. It consists of integrating equation (1) on a closed surface S_P placed far from the body, which contains all the possible nonlinear sources and where the usual impermeability condition has not to be applied. Although such a method had already been treated by Ffowcs Williams and Hawkings, it was first implemented by Di Francescantonio (1997), by assuming

$$U_i = \left(1 - \frac{\rho}{\rho_0}\right)v_i + \frac{\rho}{\rho_0}u_i$$
$$L_i = P_{ij}\hat{n}_j + \rho u_i(u_n - v_n)$$

In this way, equation (2) is formally not altered and gives rise to the alternative integral solving formula

$$4\pi p'(x,t) = \frac{\partial}{\partial t}\int_S \left[\frac{\rho_o U_n}{r|1-M_r|}\right]_\tau dS + \frac{1}{c_0}\frac{\partial}{\partial t}\int_{S_P} \left[\frac{L_r}{r|1-M_r|}\right]_\tau dS + \frac{\partial}{\partial t}\int_S \left[\frac{L_r}{r^2|1-M_r|}\right]_\tau dS + p_Q'(x,t) \qquad (4)$$

Here, the term p_Q still indicates the noise contribution of the field quadrupole sources, but in the region outside the *porous* surface S_P. Thus, if this surface is suitably placed in order to include all the possible noise sources, the volume integrals tend to zero and an overall noise prediction can be carried out by surface integrals only. Of course, the weak-point of such a numerical approach is the availability of an accurate and reliable set of data in the flow field external to the body-source.

In the following section, the numerical solutions of all integral equations (2), (3) and (4) will be presented, in order to show the role played by the different sources due to a marine propeller and to demonstrate the potentiality and effectiveness of the acoustic analogy in performing an underwater noise prediction.

3 NUMERICAL RESULTS

The hydroacoustic characterization of a marine propeller through the FWH equation is here tested by taking into account two different problems. As a first step, a simple, isolated (*open water*) propulsor in a uniform flow will be treated to show the role played by the FWH source terms and the effectiveness of the porous formulation. Subsequently, a mounted propeller on a complete ship model (hull, rudder and appendages) will be taken into account to achieve a deeper understanding of the predominant generating noise mechanisms taking place in the flow field.

In the following sections we will focus our attention on the hydroacoustic results and avoid any detail on the hydrodynamic numerical simulations used to carry out the requested data. It is sufficient to know that all the simulations are based on full unsteady, incompressible RANSE approaches and provide the pressure and velocity fields around the body-source, as well as the hydrodynamic loads acting on the rigid surfaces. The same (incompressible) pressure signatures computed through the Navier-Stokes simulations in the far field will be used to compare and, somehow, validate the noise predictions. In this context, it is worth noting that from a theoretical point of view, the incompressibility assumption characterizing the RANSE calculations is simply not compatible with any hydroacoustic analysis, since it removes a priori any propagation phenomena (the speed of sound is infinite). In other words, the RANSE pressure does not correspond to what we usually refer to as *noise*. On the other hand, the region concerned is spatially very limited, compared to the underwater speed of sound and the distances covered by the pressure disturbances. Therefore, from a practical point of view, the available RANSE pressure signals can be reasonably identified with the acoustic pressure time histories in the proximity of the propeller and will be used to assess our hydroacoustic analysis. This way, a good agreement with the RANSE pressure signal should represent a roundabout validation of the FWH noise prediction, as well as a sort of consistency proof between the two numerical approaches. Moreover, as usual for aeronautical applications, the water will be considered as a *homogeneous medium*, where the sound propagation speed is constant. Of course, this assumption is quite debatable for the sea water, compared to the air: the presence of not negligible temperature and pressure gradients due to depth, the sea currents, the scattering effects of the free surface, the salinity concentration are all variables which can heavily affect the sound propagation. On the other hand, we are here interested on the sound *generation* phenomena, more than the *propagation* aspects of the problem, so that this assumption should not affect the main results of our numerical investigations.

3.1 *Isolated propeller*

An interesting check on the FWH equation capabilities in the analysis of the acoustic behaviour of a conventional marine propeller may be carried out on a very simple configuration: an isolated (non cavitating) propeller in a uniform flow. As already mentioned, the few papers available in literature on this matter, exploits the rather common conviction of a negligible contribution from the nonlinear source terms, due to the low rotational speed. As well-known, this is a quite usual approximation in Aeroacoustics where the quadrupole terms were proven to be relevant only at a high transonic or supersonic speed. This regime is very far from the usual operating conditions of a marine propeller. In that case, however, this assumption has a well defined physical explanation: the predominant noise generation mechanisms for a blade rotating in air are related to its shape and, above all, the pressure distributions acting on its surface. These mechanisms are represented by the two linear surface terms of the FWH equation. At a low rotational speed the nonlinear, three-dimensional effects are limited to the turbulence and vorticity fields generated by the blade rotary motion, which do not provide a significant contribution to the acoustic pressure field. Thus, they are usually neglected. On the contrary, at high rotational speed the flow field surrounding the blade suffers the occurrence of a shock wave, which represents a highly nonlinear phenomenon and an intense source of sound. For this reason, at transonic or supersonic regime, the quadrupole noise assumes a fundamental role.

For a marine propeller the situation is very different. The usual rotational speed of a marine propulsor is not comparable with the speed of sound in water, and the Mach number itself is never mentioned in a hydroacoustic analysis. This, however, does not mean that the three-dimensional, nonlinear sources can be automatically neglected. For example, it is sufficient to consider a propeller exhibiting bubble or cloud cavitation: when the external pressure around a bubble starts to increase, after a very short time the pressure gradient between the outer and inner pressure decreases and the bubble enters a collapsing stage. This stage creates shock waves and, hence, noise. The phenomenon certainly does not depend on some transonic value of the blade rotational speed but its effects on the hydroacoustic far field may be relevant and depend on both the air bubbles size and the flow field area affected by cavitation. In the FWH equation, the only way to account for such a noise source is through the quadrupole volume terms. Generally speaking, a marine propeller generates a vortical wake, whose structure and breaking phase strongly depend on the operating condition. Due to the different density and viscosity of air and water and unlike the analogous condition in air, the turbulence and vorticity generated by the blade rotational motion persist underwater both in time and space and their effects on the hydroacoustic far field have never been investigated. Moreover, in a real ship configuration the propeller works in the rear part of the hull, with an inflow characterized by an intense turbulence and where the propulsor somehow enforces its own blade passage frequency to the turbulence and vorticity fields, moving a huge mass of water. Within this context and even assuming the absence of any cavitation phenomena, the removal of the only FWH integral terms able to account for the possible field noise sources seem to be a rather rash hypothesis.

In order to clarify the role played by the different FWH source terms, a relatively simple investigation has been performed on a isolated, four-bladed marine propeller, by exploiting the availability of a full set of hydrodynamic data from a RANSE simulation. These data include not only the pressure distribution upon the blade surface, but also the velocity and pressure fields around the propeller, thus allowing a direct estimation of the quadrupole integrals. Figure 1 shows a sketch of the tested configuration. Each blade of the isolated propeller is embedded into a 3D mesh (rigidly attached to the rotating body) used to model the boundary layer and the flow in the proximity of the blade itself. This three-dimensional mesh is here adopted as integration domain for the quadrupole volume integrals. Furthermore, the whole propeller is surrounded by a further, cylindrical grid, whose most exterior layer (reported in the Figure by a grey shaded cylinder) is used as the *porous surface* to carry out the numerical solution provided by equation (4).

From a practical viewpoint, the test refers to the INSEAN scaled model E1630 (with a diameter of 21*cm*), rotating at 10*rps* (rounds per second) and in a uniform flow at a velocity of approximately 1.05*m/s*, so that the corresponding advance coefficient is $J = 0.5$. Figure 2 shows an isocontour map of the pressure field and points out the location of the two hydrophones used for noise predictions. The first is placed very close to the propeller disk plane, where both the FWH thickness and quadrupole source terms usually exhibit a relevant directivity; the second is (x) aligned to the first one, but located downstream to the propulsor.

Next two Figures 3 and 4 report the noise signatures (pressure *vs* time) computed by the FWH solver within a blade revolution period.

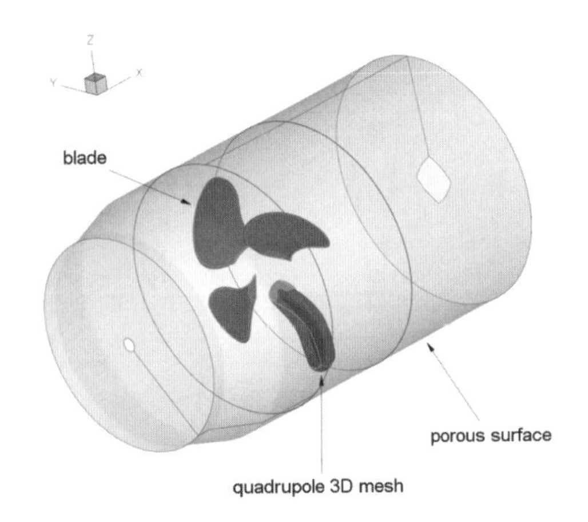

Figure 1. Sketch of the isolated propeller in a uniform flow, used to assess the FWH quadrupole noise contribution to the hydroacoustic pressure far field.

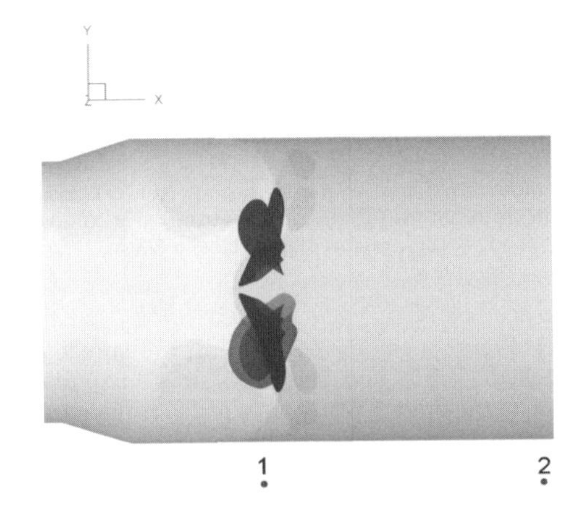

Figure 2. Isocontour pressure map of the tested configuration and the two hydrophones used for noise predictions.

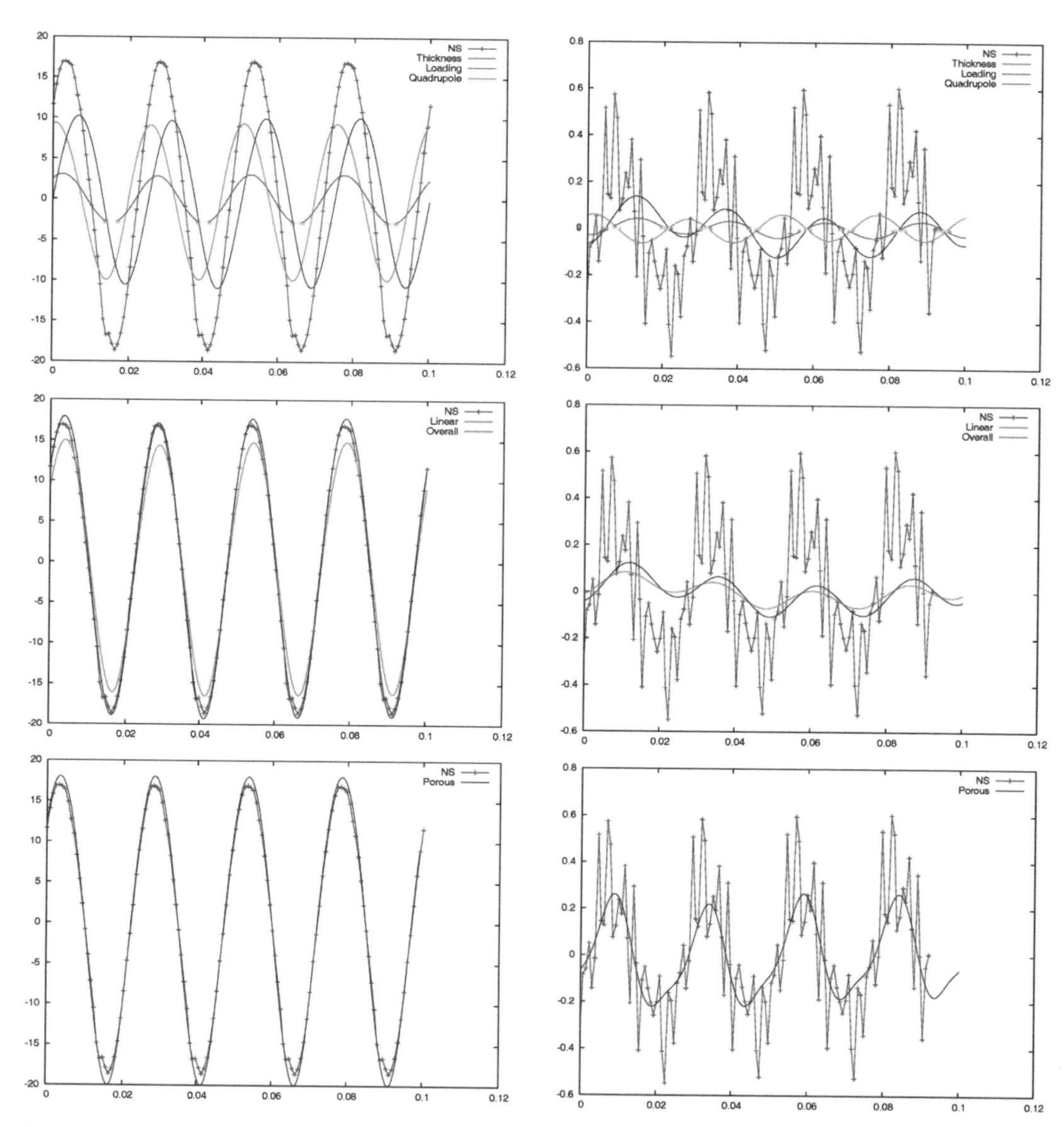

Figure 3. Noise predictions carried out at hydrophone 1. The three contributions from the FWH source terms (top), the pure linear and overall signatures (center) and the result from porous formulation (bottom), are compared with the RANSE pressure signal (label « NS »).

Figure 4. Noise predictions carried out at hydrophone 2. The three contributions from the FWH source terms (top), the pure linear and overall signatures (center) and the result from porous formulation (bottom), are compared with the RANSE pressure signal (label « NS »).

Each Figure includes the RANSE pressure signal (labelled "NS") and the separate contributions from all the three FWH source terms (top picture), a comparison between the Navier-Stokes pressure and both the overall and linear solutions provided by the acoustic analogy (mid picture) and, finally, the signature provided by the *porous formulation* (bottom picture). As expected, at hydrophone 1 the thickness and loading noise represent the predominant sources, but the quadrupole noise contribution does not look like a *negligible* term.

This can be well appreciated in the mid Figure, where the agreement between the FWH signature and the corresponding RANSE pressure becomes excellent just accounting for the nonlinear noise sources too. At the same time, also the porous formulation provides a very good agreement with the RANSE signature, both in amplitude and in the resulting waveform. Moving downstream the propulsor (at hydrophone 2), the pressure signal coming from the hydrodynamic simulation exhibits a very irregular form, probably due to both the effect of the propeller wake and some fluctuation of numerical nature.

All the frames, however, clearly suggest that a pure linear hydroacoustic analysis is not sufficient to determine the actual pressure values. In fact, both the separate contributions from the linear terms (top picture), as well as the overall signature achieved by adding the quadrupole noise (mid picture), give a notable underestimation of the pressure. This result suggests that the selected 3D mesh for the computation of the quadrupole integrals is too much limited from a spatial point of view and a relevant (here missing) contribution from non linear sources is somehow *spread* in the flow surrounding the body. In fact, the comparison between the RANSE pressure and the noise prediction carried out by the porous formulation is really satisfactory: the mean value on the two signatures match very well and the FWH-based solution even seems to *remove* the numerical problems affecting the NS signature. These results give a hint to the following, new and interesting issues:

- in a FWH-based hydroacoustic analysis of a marine propeller, the contribution of the quadrupole noise should be always taken into account, regardless of the (low) rotational speed of the blades. Given the lack of cavitation and the relatively simple configuration tested, this contribution seems to be only related to the vorticity and turbulence fields generated by the propeller. Thus, unlike the corresponding condition in air, these features of the flow field could represent a relevant noise source underwater. This also appears from the quadrupole noise evaluation at hydrophone 2, where the adopted 3D integration domain is strictly limited to the region surrounding the blade and is clearly not sufficient to account for the nonlinear effects taking place in the flow;
- moving far from the propeller, the contributions from the FWH linear terms seems to reduce rapidly. In other words, the body shape and the hydrodynamic loads are still two relevant generating noise mechanisms, but just in a spatially very limited region;
- the porous formulation probably represents the easiest and most suitable way to perform a numerical prediction of noise in the far field. According to the availability of the requested hydrodynamic data, this numerical approach allows to avoid any three-dimensional integration and to account for a comprehensive evaluation of the nonlinear noise sources, represented by the above mentioned vorticity and turbulence fields.

All these issues, of course, should be tested and verified by some direct comparison with experimental data. In any case, it is worth noting that the occurrence of any cavitation phenomena can do nothing but stress the aforementioned assessments on the role played by the nonlinear noise sources.

In the following section, these points will be discussed again (and fully confirmed) by an additional test-case concerning a propeller mounted on a patrol boat scaled model.

3.2 *The role of the nonlinear sources*

The numerical results reported in the last section points out that the rather usual assumption to neglect the nonlinear source terms concerning the propeller (because of the low rotational speed) should be carefully investigated. Such an analysis would require a Navier-Stokes-based simulation of the flow around the whole ship, a simulation very CPU demanding and rather unusual for marine problems. Furthermore, it is useful to remind that the reliability of such simulations is always limited by the available computing resources.

This problem has been faced at INSEAN by accounting for a full unsteady RANSE simulation of a scaled model of a patrol vessel in a steady course. The model includes all the appendages (struts, fins, shaft, bracket) and a propeller working at the propulsion point of the full scale ship. The model length is $L_{pp} = 5.33$ m and moves at a speed $U = 2.52$ m/s. Consequently, the Reynolds and the Froude numbers correspond to 1.18×10^7 and 0.348, respectively. The propeller is a four-bladed, adjustable-pitch, skewed model (INSEAN Model E1630) with the shaft attached to the hull by two brackets arranged in the typical "V" configuration. The diameter of the propeller is $D = 0.21$ m and its turning rate is set to $n = 820$ rpm, so that (in non-dimensional terms) the nominal advance coefficient is $J = 0.878$. Figure 5 shows a 3D sketch of the tested configuration (where the symmetry with respect to the vertical plane is still exploited to limit the computational burden). The hydrophone used to our aims is placed in the vertical symmetry ship plane and rather well aligned with the propellers disk (top Figure 6). As a first step, we limit the calculations to the linear contributions from the propulsors, the rudders and the hull, by still using the *formulation 1A*. The result is reported in the bottom Figure 6 and exhibits a relevant discrepancy with the corresponding pressure signature directly determined by the RANSE solver (and here labeled as 'NS'). It's interesting to note that the rudders contribution is very close to zero and, above all, that the thickness and loading noise contributions generated by the propellers are very limited. Furthermore, the overall linear noise prediction seems

Figure 5. The patrol boat model used to carry out the nonlinear hydroacoustic analysis.

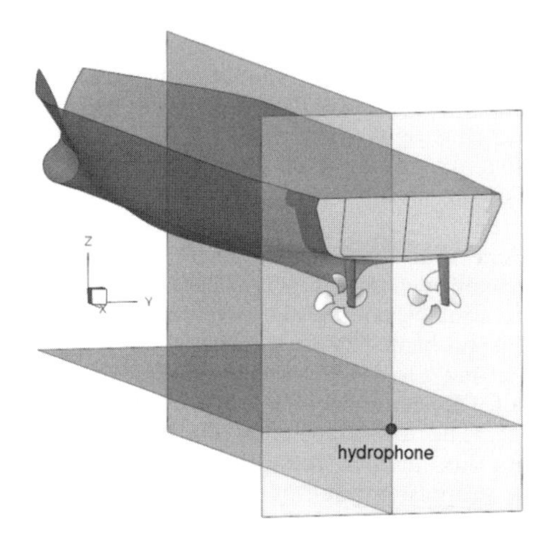

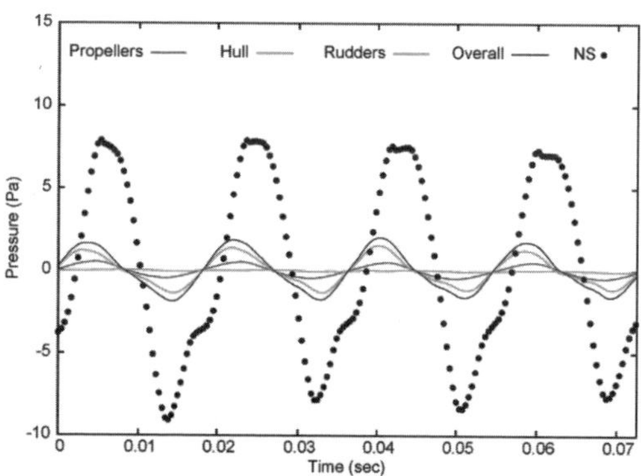

Figure 6. At the selected hydrophone (top frame), the overall linear noise prediction provided by the FWH solver is not comparable with the Navier-Stokes pressure.

to be dominated by the hull *scattering* effects, but is not comparable with the pressure signal determined by the RANSE solver.

Compared to the few results available in literature, this result is really unexpected. The four-peaked waveform of the NS pressure signature somehow confirms that the propulsor represents the dominant noise source in the field, but the very limited contribution coming from the corresponding thickness and loading terms casts serious doubts on the opportunity to neglect the nonlinear sources. As a matter of fact, the (here missing) quadrupole noise can be the only responsible of such a large discrepancy between the two numerical solutions. Furthermore, the hull scattering effects appear to be the only appreciable linear contribution to the underwater pressure far field.

The computation of the FWH quadrupole source term requires the adoption of a suitable integration domain V and the estimation of the Lighthill stress tensor in the corresponding integral kernels. As mentioned above, the four-peaked RANSE pressure signatures indicate the predominant role of the propellers as noise source. Therefore, our analysis will be limited to the ship aftbody, by exploiting the availability of different mesh blocks of the RANSE simulation to model the flow region in the neighbourhood of the propulsors.

Figure 7 shows a sketch of this region. Here, it is possible to identify: i) a grid embedding each blade as a glove (actually made of five separate blocks with a very fine spatial resolution and substantially devoted to the analysis of the blade boundary layer); ii) a toroidal grid around the propeller hub; iii) a cylindrical mesh embedding the whole propulsor; vi) a block surrounding the rudder and made of two adjacent patches; v) three contiguous cylindrical blocks, located downstream to the propeller. It's interesting to note that the first cylindrical block rigidly rotates with the propulsor, while the last three ones are fixed (they traslate with the model) and are obviously used to achieve an accurate description of the propeller wake. Concerning the computation of the Lighthill stress tensor, we first assume a purely linear relationship between the pressure and the density, so that

$$T_{ij} \approx \rho u_i u_j$$

where, by now, u_i and u_j represent the velocity components directly provided by the RANSE code. Note that such an approximation is rather common also for aeronautical problems (at least up to the condition for which the blade rotational velocity gives rise to a weak shock wave). Next Figure 8 shows the contribution from the linear terms (due to the propellers, the hull and the rudders, already depicted in the bottom Figure 6), the quadrupole noise determined in the selected region, the overall noise signature and the corresponding NS-based pressure signature. The result is very surprising. The contribution from the nonlinear sources is very close to zero and the relevant underestimation

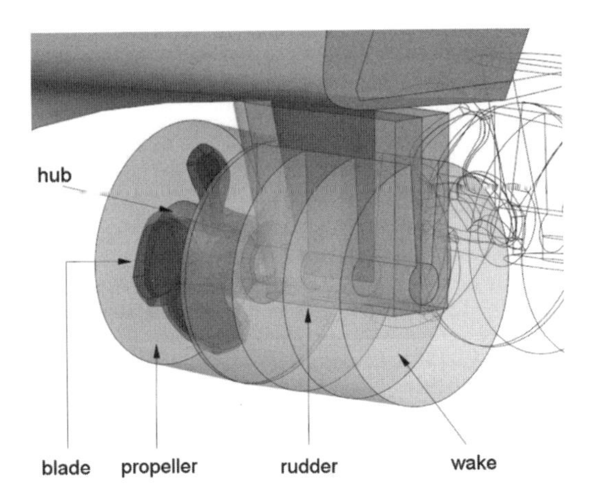

Figure 7. Three-dimensional sketch of the integration domain used to compute the quadrupole noise, constituted by different mesh-blocks in the RANSE simulation.

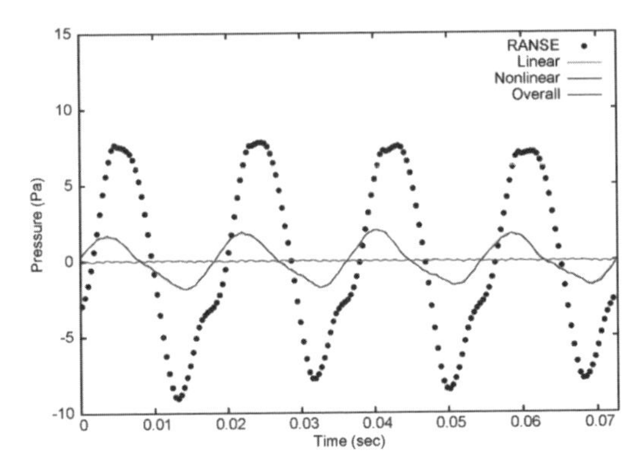

Figure 8. The quadrupole noise contribution, determined by computing the Lighthill stress tensor by the relation $T_{ij} \approx \rho u_i u_j$, is, in practice, equal to zero.

of the FWH noise prediction with respect to the RANSE solution does not change at all!

Here the problem is the adopted estimation of the Lighthill stress tensor, which is completely *unsuitable* in our context. In fact, in a RANSE approach the velocity field is assumed to be the sum of a time-averaged value *plus* a fluctuating term

$$u = \bar{u} + u'$$

so that the second-order tensor $\rho u_i u_j$ should be expressed in the following *Reynolds averaged form*

$$\rho u_i u_j = \rho \overline{u_i' u_j'}$$

From a numerical point of view, the second term on the right-hand side of this equation must be somehow modeled to account for the velocity fluctuations occuring in the region of interest. This term was not included in the signatures of Figure 8, where the Lighthill tensor was directly determined by the output quantities of the RANSE solver, corresponding to the averaged values of the velocity components only. In order to determine the above mentioned term, we follow (in both the RANSE solver and, now, the FWH formulation) the *turbulent-viscosity hypothesis*, introduced by Boussinesq in 1877, which models the averaged contribution of the fluctuating terms in a form similar to the stress tensor for a Newtonian fluid. In general, for an incompressible flow, this hypothesys is expressed in the following form

$$Q\overline{u_i' u_j'} = \rho \nu_T \left(\frac{\partial \bar{u}_i}{\partial x_j} \frac{\partial \bar{u}_j}{\partial x_i} \right) + \frac{2}{3} \rho k \delta ij$$

where the positive scalar coefficient ν_T is the *turbulent* (or *eddy*) viscosity (see, for example, Pope, 2000) and represents a further output variable from the hydrodynamic solver. The RANSE solution already incorporates the term representing the turbulent kinetic energy k in the pressure field. Thus, a suitable representation of the Lighthill stress tensor is given by

$$T_{ij} \approx \rho \bar{u}_i \bar{u}_j - \rho \nu_T \left(\frac{\partial \bar{u}_i}{\partial x_j} + \frac{\partial \bar{u}_j}{\partial x_i} \right)$$

where the velocity gradient can be derived from the knowledge of the velocity field by a simple second-order numerical differentiation. The use of the Lighthill tensor expressed by last equation turns the noise prediction reported in Figure 8 into the one reported in Figure 9. The quadrupole noise signature exhibits the expected waveform, with pressure peak values notably larger than the overall linear terms contribution, and the agreement between the RANSE and the FWH noise prediction becomes excellent.

The last Figure suggests a very important and new result. It demonstrates that, unlike the aeronautical case, the dominant generating noise mechanisms taking place in the flow field are not related to the body shape or the hydrodynamic loads acting on its surface, but rather to the notable fluid velocity gradients (mainly due to the propeller) and then to the vorticity and turbulence fields. For this reason, a reliable ship (or propeller) underwater hydroacoustic analysis cannot leave aside the contribution from the nonlinear sources, regardless of the propeller rotational speed. Such an assertion is

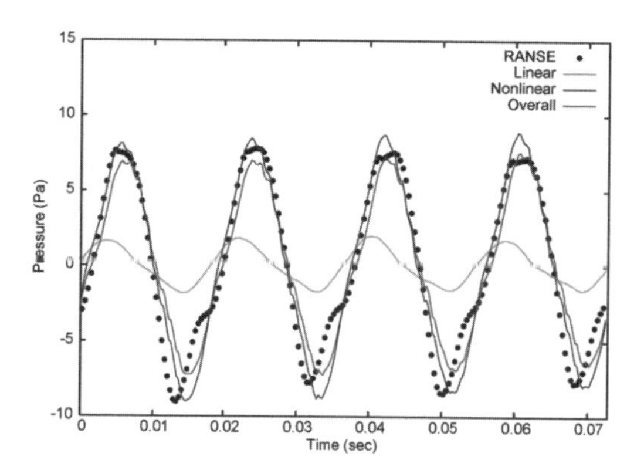

Figure 9. The quadrupole noise contribution, by accounting the velocity fluctuating terms in the Lighthill stress tensor, makes the agreement between the RANSE pressure and the FWH noise prediction excellent.

highly supported by both the comparison between the noise predictions reported in Figures 8 and 9, and the negligible role played by the thickness and loading source terms due to the propeller.

4 CONCLUSIONS

Within this paper, we've tried to show the prediction capabilities of the acoustic analogy in the analysis of a conventional marine propeller and to endorse its own use for hydroacoustic problems. To this aim, different configurations have been analysed: *a*) the hydroacoustic behavior of a propulsor in open water condition, *b*) the overall (linear and nonlinear) analysis of a mounted propeller on a scaled model. The most important and interesting result carried out from these numerical investigations concerns the role of the different noise sources related to the propeller and, in general, the ship. Unlike the analogous aeronautical problem, the pressure far field underwater appears to be heavily affected by the nonlinear noise sources, although the blade rotational speed is very far from a transonic regime. Then, the most important generating noise mechanisms seem not to concern the body shape and the hydrodynamic loads on the blade surface, as the vorticity and turbulence fields generated by the propeller. In other words, a reliable estimation of the underwater pressure field must account for an accurate computation of the velocity gradients taking place in the flow.

ACKNOWLEDGEMENT

This work was partially performed in the framework of the EU-FP7 Research Project SILENV, "*Ships oriented Innovative solutions to rEduce Noise and Vibrations*", grant agreement No. 234182.

REFERENCES

Di Francescantonio P., 1997, A new boundary integral formulation for the prediction of sound radiation, *Journal of Sound & Vibration*, 202.

Farassat F., 1975, Theory of noise generation from moving bodies with an application to helicopter rotors, *NASA Technical Report R-451*.

Farassat F., 1981, Linear acoustic formulas for calculation of rotating blade noise, *AIAA Journal*, 19.

Ffowcs Williams J., E., Hawkings D.L., 1969, Sound generation by turbulence and surfaces in arbitrary motion, *Philosophical Transactions of the Royal Society*, A264.

Ianniello S., 2007, New perspectives in the use of the Ffowcs Williams-Hawkings equation for aeroacoustic analysis of rotating blades, *Journal of Fluid Mechanics*, 570.

Kanwal R.P., 1983, Generalized functions. Theory and technique, *McGraw-Hill*, New York.

Lighthill M.J., 1952, On sound generated aerodynamically. I General theory, *Proceedings of the Royal Society*, A211.

Morse, P.M., Feshbach H., 1953, Methods of theoretical physics, *McGraw-Hill*, New York.

Pope S.B., 2000, Turbulent flows, *Cambridge University Press*.

Salvatore F., Ianniello S., 2003, Preliminary results on acoustic modeling of cavitating propellers, *Computational Mechanics*, 32.

Salvatore F., Testa C., Ianniello S., Pereira F., 2006, Theoretical modeling of un steady cavitation and induced noise, *CAV2005, Wageningen, The Netherlands*.

Salvatore F., Testa C., Greco L., 2009, Coupled hydrodynamic-hydroacoustic BEM modeling of marine propellers operating in a wakefield, *1st International Symposium on Marine Propellers, SMP09, Trondheim, Norway*.

Seol H., Jung B., Suh J.C., Lee S., 2001, Prediction of non-cavitating underwater propeller noise, *Journal of Sound & Vibration*, 257.

Seol H., Suh J.C., Lee S., 2005, Development of hybrid method for the prediction of underwater propeller noise, *Journal of Sound & Vibration*, 288.

Testa C., Ianniello S., Salvatore F., Gennaretti M., 2008, Numerical approaches for hydroacoustic analysis of marine propellers, *Journal of Ship Research*, 52.

Testa C., 2008, Acoustic formulations for aeronautical and naval rotorcraft noise predictions based on the Ffowcs Williams-Hawkings equation, *Ph.D Thesis, Delft University of Technology*, ISBN 978-90-8559-358-4.

Sustainable Maritime Transportation and Exploitation of Sea Resources – Rizzuto & Guedes Soares (eds)
 ISBN 978-0-415-62081-9

Analysis of airborne noise emitted from ships

A. Badino, D. Borelli, T. Gaggero, E. Rizzuto & C. Schenone
University of Genoa, Genoa, Italy

ABSTRACT: To assess and to control the environmental noise due to the main noise sources, including ports, the European Directive 2002/49/CE introduced a tool for the acoustic planning: the Noise Strategic Mapping (NSM). In any harbor, NSM can be obtained by evaluating the sound field for each type of sound sources standing in the port area and by superimposing all noise contributions. The present paper suggests a method to achieve such acoustical analysis of port noise. The method has a general purpose and is implemented in four steps: definition of the vessels category; measurements to characterize ship as a noise source; modeling of outdoor noise propagation; map drawing. All these subjects are discussed in the paper and some specific recommendations to get at a successful noise mapping of ports are reported, too. In the end, limit values to keep under control the problem of noise pollution due to ships are proposed.

1 INTRODUCTION

Ports are nowadays considered as an important source of pollution (water, air, noise, etc.) because of the variety of activities that are carried out and because of the continuous traffic flow from both sea and land. In several cases port areas are situated in close juxtaposition to urban areas and they may even be bounded by, or it may even include, areas of special environmental significance, due to the presence of protected habitats and ecosystems.

In particular, noise pollution is a key issue for a sustainable development of ports (Badino, A. et al. 2010). At the same time, noise analysis of port areas is very complex because of the presence of different types of sound sources and because of the interaction of the various sources. Ports are characterized by a higher degree of complexity and a variety of operations in comparison with other logistic nodes. The various types of cargo, the range of activities, products and services moving within the port area, the multiple use of the areas and the physical impact of the associated infrastructures all identify ports as major logistic nodes with the attendant implications of the generation and impact of noise (NoMEPorts Project 2005–2008).

Due to this complexity, various research projects regarding harbours, such as EcoPorts Project, NoMEPorts Project, SIMPYC Research Project, HADA Project, were realized aiming to define guide lines to manage the environmental noise due to the several sources emitting in ports. These projects followed the recommendations provided by the European Directive 2002/49/EC, which introduced first a tool for the acoustic planning, in order to analyze and to control the environmental noise pollution: the Noise Strategic Mapping (NSM). Day-Evening-Night Level (L_{den}) and Night Level (L_{night}) are the noise indicators recommended for the mapping. In Annex IV, the Directive underlines the importance to make NSMs for areas where the main noise sources are (road, rail, or air-traffic noise, industrial noise, construction activities, etc.), including also ports.

To assess and to manage the environmental port noise, NSM is considered the most suitable instruments of acoustic planning to identify the critical zones and establishing the improvement actions. This is essential to identify solutions that can allow the development of trade, without compromising the quality of life in port cities.

As in the harbour there are various sorts of sources, one NSM for each source category must be drawn, beginning from ships. In general, the analysis of airborne noise emitted from ships is a key-element in defining the noise pollution inside harbour and in controlling the correlated noise annoyance. In such a way, within port the noise contribution of berthed and under way vessels can be analysed by means of this map.

In the present paper, a method to achieve such acoustical analysis of noise emitted from ships is suggested. The method has been proposed in the SILENV collaborative project (Ships oriented Innovative soLutions to rEduce Noise & Vibrations), funded by the E.U. within the 7th Framework Programme. This method has a general purpose and it is implemented in four steps: definition of the vessels category; measurements to characterize ship as a noise source; modeling of outdoor noise propagation; map drawing.

NSM needs a wide set of input data (position and category of sources, ground properties, acoustic characteristics of sources, drawings of port areas and close buildings, time working plan, etc...) that can be get by Port Authority or that can be obtained by measurements or direct surveys.

Moreover, each step in NSM produces an output which feeds the next step, in a knowledge chain having as final result the detailed analysis of airborne noise emitted from ships and other sources inside the ports.

2 NOISE SOURCES IN PORTS

Although several activities are referable to those typical of an industrial area, ports are characterized by different types of sound sources; the numerous noise sources, their interaction and the high sound intensity of a lot of them create a complex acoustic climate, causing harmful effects over exposed population in long periods. For this reason, it is important to start off from the identification and cataloguing of the noise sources normally present in a port, distinguishing them between stationary and mobile. This classification is useful to have a proper approach to legislation and to apply properly past regulations to both mentioned categories.

Since the present paper involves the problem of harbour noise in general and the airborne radiated from ships in particular, the focus is primarily addressed on the noise generated by vessels; as the noise due to other sources, many of which are closely linked to the presence of the ship in port, will be briefly discussed.

Within a port, the noise produced by a ship can be distinguished between direct and indirect noise. Direct ship noise means the noise produced just by the vessel under two conditions:

1. When the vessel is under way in land waterways and harbours and along the coast;
2. When the vessel is alongside a wharf or at anchor.

In the first case the ship can be considered like a mobile source, and in the second case like a fixed (stationary) source.

Indirect ship noise means the noise associated with the various activities that take place in consequence of the presence of a ship. The indirect ship noise involves:

- the activities of handling, loading/unloading goods;
- shipbuilding/ ship repairing processes;
- road and rail traffic to and from the port.

Usually, the noise directly generated from a ship overlaps the indirect noise coming from other sources and all together contribute to the noise port climate. In the assessment of the noise, both direct and indirect, it's very important to consider the ship category (container ship, tanker, ferryboat, cruise ship, sailing ship, harbour vessel, dredger, recreation craft etc ...).

In the present paper, only direct noise emitted from ships will be taken into consideration; the analysis of indirect noise is deferred to a further specific research.

3 DEFINITION OF THE VESSEL CATEGORY

There are different way to classify the ships: by type of cargo, by hull, by engine system, etc.. In order to produce NSM for the vessels, we recommend to classify the ships on the basis of the use category (container ship, tanker, ferryboat, cruise ship, sailing ship, harbour vessel, dredger, recreation craft etc.), because the ships location within port areas and the noise emissions mostly depend on that. Figure 1 shows a vessel classification on this basis.

Some onboard noise sources are common to all ship categories; other ones are specific for a certain category of vessel and strictly depend on that. For this reason, a classification based on vessels category seems to be appropriate to the aim of the present work. For example, in the case of Ro-Ro, Ro-Pax and ferries, forced ventilation of the garage, during the loading and unloading operations, shall be taken into account. For cruise ships the HVAC system shall be considered, because temperature and humidity inside a great number of cabins and rooms must be controlled during the port staying and then AC plants work (and emit) when the vessel is moored.

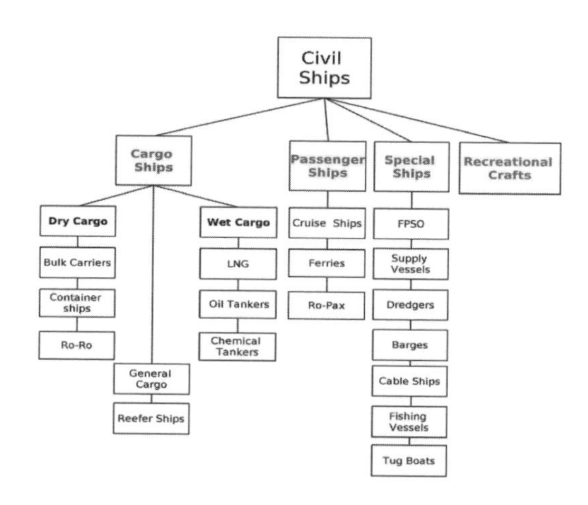

Figure 1. Classification of the vessels.

For special ship, like dredges, the sound emitted by engines for dredging shall be take into consideration.

Common noise sources are:

- intakes and exhausts ducts of main engines
- intakes and exhausts ducts of auxiliary engines;
- air-conditioning and refrigeration systems

In general, although they differ markedly from each other, vessels that belong to different categories present anyway a number of common sources; on the contrary, recreational crafts present particular characteristics and for this reason they are subject to specific standards, as discussed in the next paragraph.

4 RECOMMENDATIONS FOR ACOUSTIC OUTDOOR MEASUREMENTS

Three ISO standards are focused on the assessment of the airborne sound emitted by vessel for acceptance and monitoring tests:

- ISO 2922:2000 for all vessel types with the exception of recreational crafts;
- ISO 14509-1:2008 and ISO 14509-2:2006 for recreational crafts.

As for ships in transit, ISO 2922:2000 proposes to measure the sound level by the noise indicators Sound Event Level, L_{AE}, maximum Sound Pressure Level, L_{pASmax}, for each single passage. These indicators are already used in the assessment of rail and airport noise for the measurement of a single sound event.

In order to better characterize the ship sources and to accurately predict outdoor sound propagation, some complementary measurements, which can be added to those recommended by Organization of the International Standards, should be taken into consideration. These complementary measurements are specifically intended for non recreational craft. For recreational crafts complementary measures seem not to be necessary and ISO 14509-1:2008 or ISO 14509-2:2006 are considered fully significant.

In fact, modeling of outdoor noise propagation (third step) needs input data and benchmarks that standard measurements do not provide: additional measurements near ship, along sound propagation paths and close to the receivers are necessary to a full comprehension and a deep analysis of the sound field.

Contribution by specific sources (vents, funnel, ducts inlet and outlet), noise spectra, sources directivity are obtained by measurements close to the ship. Measurements along the sound propagation paths and close to the receivers are useful to validate the simulation model and to assess its accuracy, by comparing measured and calculated sound pressure levels. Simultaneous measurements near sources and close to the receivers are strongly recommended to get a full cause-effect correspondence.

Measurements procedure varies on the basis of vessel operating conditions: for vessel under way, within the harbour or along the cost, the ship is considered as a mobile source; whereas for vessel at wharf the ship is considered as a stationary source, also during mooring/cast off phases (Moretta et al. 2008).

The possibility to evaluate the sound power levels of noise sources by using sound intensity measurements was taken into consideration (ISO 9614:2009). In general, this measurement technique is rarely used because of its complexity and the need of high-skill staff. To the aim of giving general guidelines, SPL measurements have been considered more suitable for the present analysis.

4.1 *Moored vessel—stationary source*

In general, we recommend adapting the measurement procedure to the characteristics and to the shape of the vessel. A flexible approach can better fit because of the wide variety of vessels and characteristics of noise sources can be run into. Customized measurement procedure can be supported by suitable guidelines, which help the choice of measurements position, duration, and all other parameters for a right on field evaluation, as it is suggested in SILENV Project.

About the position, it is better to distinguish the measurements in two types: local measurements and measurements regarding the sound field.

Local measurements are defined as the measurements to be carried out near specific sources (vents, cranes, heat exchangers, etc..) to the aim of their characterization. Preliminarily, a list of the most significant onboard sources shall be made and relative measurement positions shall be set. Recommendations about the operative procedures can be found in ISO 2922:2000 standard.

Measurements on sound field must be performed both close to the vessel and at a certain distance from that. The target is to characterize the ship as a unique source, in consideration of its global emissions. During the measurements on sound field, the microphone positions shall be lined up in two rows: a first row at the distance of 1 m from the ship hull and a second row at the distance of 25 m, as required by ISO 2922:2000. For each row, various heights must be taken into consideration: the distance between one position and the other shall be chosen as a function of the ship length and the distribution of the most significant noise sources. By the first row measurements the sound power level of the vessel as a sound source (along and above the hull) can be calculated. Second row measurements permit to validate the sound propagation model.

In the ISO 2922:2000 only one height is considered for the measurements at wharf: the microphone shall be located 1.2 m above wharf ground level. Nevertheless, it must be noted that ships present various sizes and shapes; consequently, the sources of airborne noise may be located at a much higher quote: e.g. funnels (which can be located at 50 metres and over for the larger ships), or, ventilation systems whose outlet are high on the deck or on the superstructures (Badino et al. 2011). For this reason microphone positions higher than 1.2 m are proposed to be added, with the aim to evaluate the sound propagation of the tallest sources.

Concerning the heights of the microphone in respect of the wharf floor, four positions are suggested, as described below:

1. H1 = 1.2 m above the wharf floor;
2. $H2 = 2/3\ h_A$;
3. $H3 = h_A + 1/3\ h_B$;
4. $H4 = h_A + h_B$;

where h_A is the distance between the floor of the wharf and the lower deck and h_B is the distance between this last floor and the highest source (usually the funnel).

The lowest microphone position is set at least 1.2 m above the wharf floor, according to ISO 2922:2000. The height of the other positions has been defined taking into consideration EN 12354-4. Similar positions are adopted to evaluate transmission of indoor sound to outside in buildings.

On the basis of the vessel dimensions, if the distance between two heights is less than 5 m, only the lowest height should be taken into consideration. For example: if the distance between H1 and H2 is less than 5 m, only H1 should be considered; if the distance between H1 and H3 is less than 5 m, only the H1 should be considered; if the distance between the H1 and the H4 is greater than 5 m, two heights, H1 and H4, should be considered in this example. In Figure 2 a sketch is showed about microphone positions for height measurements. In this example only three altitude positions are considered because the distance between H1 and H2 is less than 5 m.

In general, the longitudinal distance between the first and the last microphone position along the same row shall be more than the length of the ship, so that sound contribution of sources placed in the stern or/and in the bow can be taken into consideration.

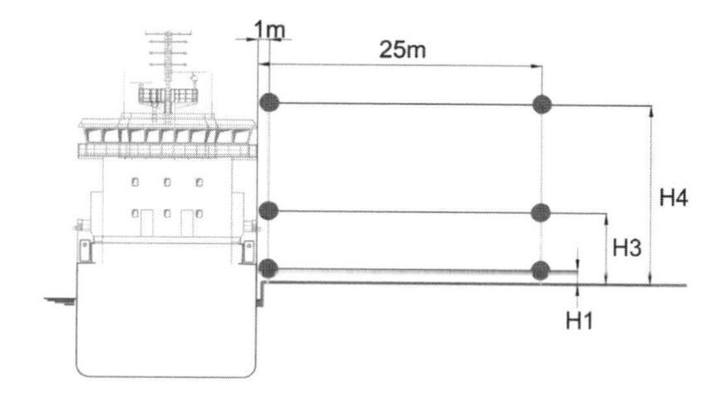

Figure 2. Layout of microphone positions for measurements on a berthed vessel.

The definition of a same time duration for all measurements at wharf is considered not suitable, because different measurement procedures are useful in characterizing the ship. For each measure, it is suggested to watch the time history during the acquisition and to stop the measure when the changes of sound pressure level are small. Anyway the interval shall be more than two minutes.

At the end of the measurements, on the basis of achieved values, some sources shall be neglected if they are not significant compared to other sources.

In this paper the mooring and cast off operations have not been considered, because the manoeuvring time are usually shorter than the time of other phases (Biot, M. et al. 2011). So in the order to make an acoustic map, in terms of L_{den} and L_{night}, these operations can be usually neglected.

4.2 *Moving vessel—mobile source*

For this condition, the aim of the test is to define a significant sound level for a transient phenomenon, as it happens when vessels pass-by. Pass-by tests are commonly utilized for measuring the noise emitted by mobile sources in motion (ISO 7188:1994; ISO 7216:1992; ISO 9645:1990). Concerning pass-by tests for ships, ANSI/ASA S12.64-2009 defines the positions of the hydrophone for pass-by underwater measurements. Following this ANSI/ASA, an example of possible microphone position for pass-by measurements, relating airborne sound propagation, is showed in Figure 3, where d_{CPA} is the distance at the Closest Point of Approach, while DWL is the Data Window Length.

If the surroundings conditions don't allow to use a distance greater than or equal to 100 m or a distance equal to the ship length, it will be possible to consider a shorter distance of d_{CPA}. The positioning of the microphone as showed in Figure 3 allows the knowing of sound directivity. If the sound sources on both sides are located in

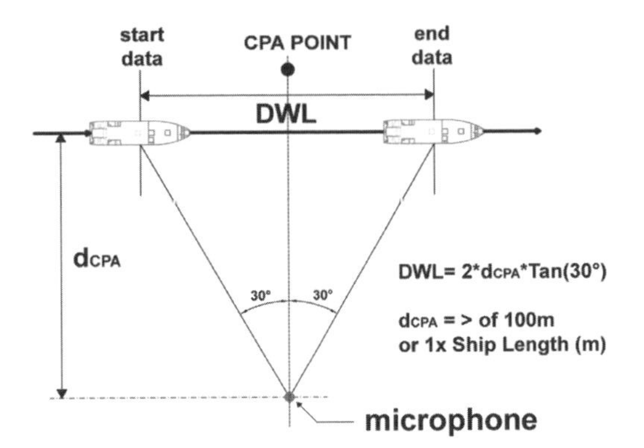

Figure 3. Example of microphone position for pass-by measurements of one ship.

the same way, the measures could be taken on one side only; otherwise, the measures will be taken on both sides.

In particular, for small recreational craft of up to 24 m length of hull ISO 14509-1:2008 specifies the test conditions for obtaining reproducible and comparable measurements of maximum sound pressure level and of sound exposure level for airborne sound generated during the passage.

Thus, for vessels under way an approach close to what standards recommend for other noise sources in motion can be followed. As noise indicator for this specific condition can be set the Single Event Level, L_{AE}, as recommended by ISO 2922:2000 and by other standards referring to sources in motion (ISO 3095:2005; ISO 20906:2009):

$$L_{AE} = 10\log_{10}\left\{\frac{\int_{t_1}^{t_2} p_A^2(t)dt}{p_0^2 T_0}\right\} \quad (1)$$

where $p_A^2(t)$ is the squared, instantaneous, A-weighted sound pressure level as a function of running time t; p_0^2 is the squared, instantaneous, A-weighted sound pressure level as a function of running time T_0 = 1s. The time-interval t_2–t_1 shall be taken as the total time-interval during which the instantaneous value of SPL is within a specified value (not less then 10 dB) of the maximum SPL value.

L_{AE} values measured are used to calculate the sound power level of the ship under way.

For ship under way, at least two heights for the microphone positions must be considered: one 1.2 m above the wharf floor and one at the same height of the funnel. Figure 4 shows the proposed layout of microphone positions for the by-pass measurements of a moving vessel.

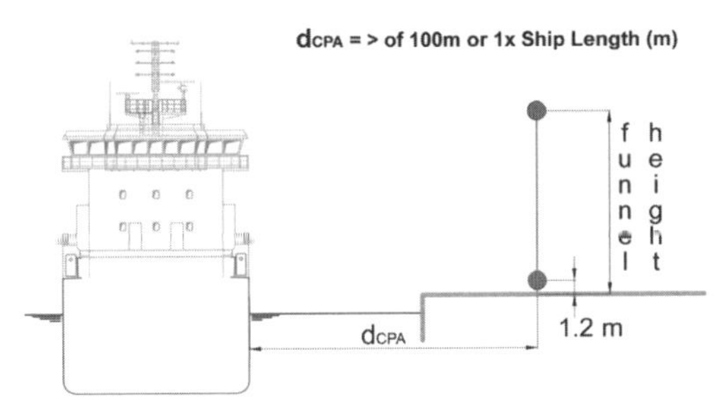

Figure 4. Layout of microphone positions for by-pass measurements of a moving vessel.

Further heights can be added for larger ships or for improving the accuracy.

During pass-by measurements further useful information about noise emitted from ships can be obtained, as spectral Sound Exposure Level and maximum level. The analysis of sound time history under these conditions can also give operative indications about the directivity of on board sources.

5 OUTDOOR NOISE PROPAGATION MODELLLING

Modelling of the outdoor propagation of vessels noise shows some common aspects with modelling of noise from other sources (roads, railways, airports) and some specific aspects. One of the specific key-element is that, as described above, there are for each ship two operative conditions, with different noise patterns: ship on the way and ship at wharf. Therefore, for every condition specific separated modelling shall be carried out.

To apply the propagation model, several parameters are to be known with respect to the geometry of the source and of the environment, the ground surface characteristics, and the spectral sound power levels of sources, as described below:

- drawings of the ship to recreate the vessel shape in the sound propagation software and to place the sound sources, considered during the measuring step. For the operative condition of berth vessel and relating to the sources disposal on the ship, the sources could be considered as point, line and area sources (Moro 2010). For the operative condition of vessel under way, the ship shall be considered as an unique line source.
- drawings of the area surrounding the ship. It is important to put in these drawings the microphone positions where the measurements were taken. The area drawing will be imported in the sound propagation software and receivers shall be inserted in the same place of the microphone positions;

- sound pressure levels measured nearby the sources in order to calculate the sound power levels. The levels in 1/3 octave bands are required;
- description of the ground characteristic in terms of acoustic absorption and acoustic impedance,,
- temperature, humidity, wind speed and direction during the measurements.

Input data must be implemented by means of proper algorithms. At the present moment, specific standards for outdoor noise propagation of ships do not exist. The general purpose standard ISO 9613-2: 1996 can be utilized and its algorithms can be adopted to calculate the following physical effects: geometrical divergence, atmospheric absorption, ground effect; reflection from surfaces; screening by obstacles. The output data are to be expressed in terms of proper noise indicators according to regulations.

Output data will be provided in the form of a set of nodal points in which noise indicators values are known. The results of the simulations shall be validated by the comparison between calculated and measured sound pressure levels. On the basis of this comparison, the accuracy of the model will be stated. A chromatic representation can accompany the numerical definition of the sound field.

6 SHIPS NOISE STRATEGIC MAPPING

The final step of NSM is the drawing of noise maps, which describe values of the noise indicators for all time periods by means of a chromatic representation. This point is very relevant because information about noise pollution must be easily comprehensible to the exposed population and to its representatives. For this reason, colored maps can help a general sharing of NSM results: the choice of the representation criteria and, particularly, of the colors is therefore very important. There are standards that guide the graphic output (European Directive 2002/49/EC) and it is strongly recommended to follow them strictly: a low attention to these aspects and a bad quality in output can vanish a deep work in noise modeling.

In general, Noise Strategic Mapping is obtained by the superimposition of noise fields calculated for every ship-source: the results of outdoor noise propagation analysis for all operating noise sources are added, so obtaining noise indicators values in each nodal point. For this reason, NSM needs the knowledge of position and movements for all vessels staying into the ports, of their operative conditions and of their acoustical characteristics. Thus, the following input data are required:

- The drawings of the port areas where the vessels are usually berthed and the placement of transit lines along which the vessels are usually under way;
- Definition of time period for day, evening and night (beginning hour—end hour)
- Vessels number for every category
  The placement at wharf for moored vessels,
- The number and the time of transit (day-evening-night) for vessel under way
- Acoustic properties for each vessel categories;
- The activity hours of the sources for moored ship;

A numerical simulator can process the input data and can create NSM in terms of L_{den} and L_{night} indicators. L_{den} is defined by the European Directive 2002/49/EC as:

$$L_{den} = 10\lg\frac{1}{24}\left(12\cdot 10^{\frac{L_{day}}{10}} + 4\cdot 10^{\frac{L_{eveningy}+5}{10}} + 8\cdot 10^{\frac{L_{night}+10}{10}}\right) \quad (2)$$

where L_{day}, $L_{evening}$ L_{night} are the A-weighted long-term average sound levels as defined in ISO 1996-2: 1987.

Contour lines are to be defined in reference with noise limit values, as 55 dB(A) for L_{den} and 50 dB(A) for L_{night}, and with significant level ranges. The knowledge of noise indicators values in a net of nodal points permits to spline them and to draw, by means of an automatic or heuristic approach, level curves (WG-AEN 2007). In Figure 5 an example of noise mapping boundaries is showed.

Figure 6 shows an example of NSM for an urban agglomeration nearby a port. The NSM is for 24-hours period and the indicator used is L_{den}.

Figure 7 shows the NSM of the same area for the night period: the used indicator is now L_{night}.

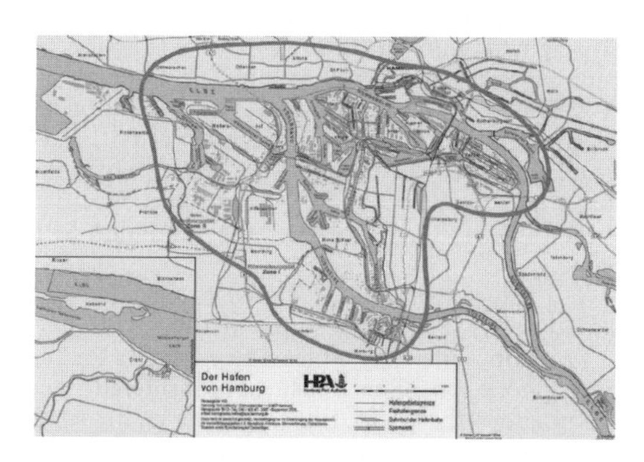

Figure 5. Noise mapping boundaries of the Hamburg Port (NoMEPorts European Project 2008).

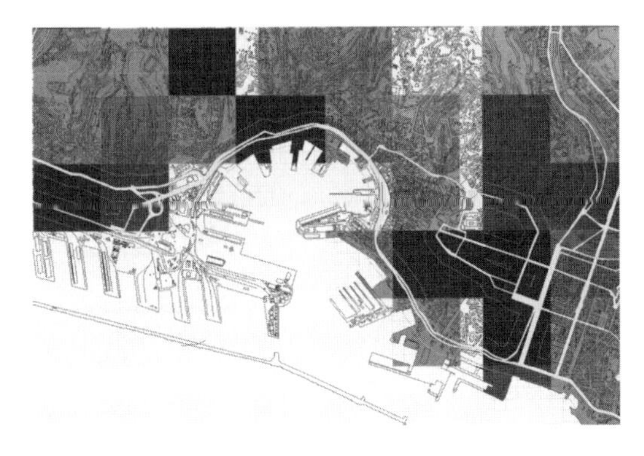

Figure 6. Noise Strategic Map of Genoa Municipality, noise indicator L_{den}.

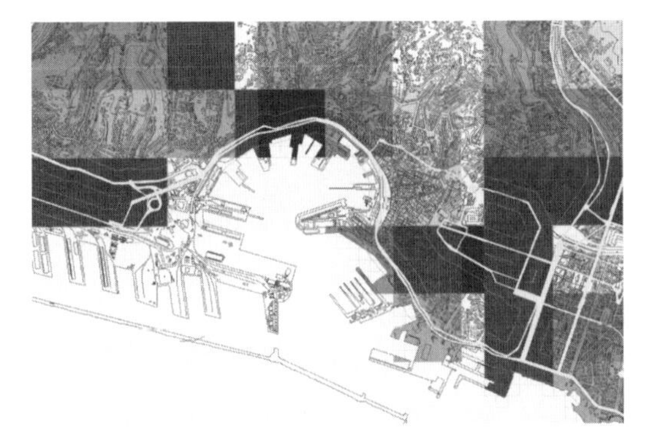

Figure 7. Noise Strategic Map of Genoa Municipality, noise indicator L_{night}.

7 SHIPS NOISE CONTROL

Noise Strategic Mapping is a tool designed to set up a control methodology for noise levels in maritime ports. It also implies the definition of an evaluation methodology of the noise levels produced by ports activities. This led to the introduction of noise assessment criteria and strategies for the analysis and management of noise pollution.

It must be highlighted that different strategies can be followed to this aim. Attention can be focused on sources and their direct emissions can be limited. Else, regulations can fix upper limits to noise emissions at the receivers, i.e. people who are exposed to sound coming from harbour. This second approach emphasizes the preeminent role of population in determining the actual impact produced by ship noise. As public health and wellness should be the main aim of ship noise control, the latter strategy seems to be more suitable: overall noise limits in port and in close urban areas constitute an operative tool for the environmental sustainability of transports (Badino et al., 2010). On the contrary, regulations exclusively restricted to the sources control are bound to fail because of a too narrow perspective. Therefore, limits both on sources and receivers should be adopted.

Anyway, attention must be paid to an excess of regulations. A framework of complex and too numerous constraints is difficult to manage and to apply. A certain interference among different authorities (Municipality, Port Authority, Environmental Agency, Harbour-office, etc.) and between them and the shipping companies can be produced. In general, limits on direct noise emissions from ships are useful, but they should involve mainly shipbuilders and Classification Societies. These direct limits shall be necessarily integrated by rules on environmental noise pollution, which must be governed by public Authorities. When the specific functions of different regulations and authorities are respected, all of them play a positive role in ship noise control.

The impact of ship noise on the environment is to be controlled by means of specific criteria.

A method to identify critical areas with respect to noise pollution may be the comparison between NSM and maximum levels allowed by regulations in terms of L_{den} and L_{night},: critical areas are those zones where the limit values are exceeded.

As described in the previous paragraph, general NSM for a harbour is the result of the sum of noise contributions for each source category staying within the harbour.

The definition of noise limits for the port areas on the basis of source category and operative conditions is in general rather complex and few meaningful. An easier and more effective approach considers limit values depending on land use of the areas nearby the port. In this way, actual annoyance in respect of receivers can be set and properly analyzed. Critical areas can be identified by overlying maps reporting land use categories to NSM for all sources.

Unfortunately, it is very difficult to find specific noise regulations for ports either in an International, European or national scale. Acoustic indicators, currently adopted for assessing environmental noise by European legislation, are substantially focused on noise classified in two great categories: mobile sources (like road-traffic, rail-traffic and aircraft around airport) and stationary sources (like industrial activity sites).

In general, noise nuisance in port areas has only been partly addressed by the European and national legislation, considering often the ports like a single noise source, similar to an Industrial or Easement Area. The European Directive 2002/49/EC introduces two acoustic descriptors to characterize the environmental noise and quantify the harmful effects on human health. They have been formulated in order to correlate the measurement of sound pressure level and the percentage of persons who have negative effects on their health, because of prolonged exposure to examined noise source. To assess the damage caused by noise in the exposed population, the Directive considers annoyance and sleep disturbance, because they are more sensible to the noise levels than the other harmful effects. Annoyance is correlated to L_{den} and sleep disturbance is correlated to L_{night}. The correlation between harmful effects and acoustic indicators is made by means of dose-effect relations.

The Directive establishes that the noise limit levels have to be defined by State Members. Limit levels may vary for different types of noise (road-, rail-or air-traffic noise, industrial noise, construction activities, etc.) or be the same for all source types. At the present time no State Member has established yet noise limit values in accordance with the above mentioned Directive. Previous national regulations are currently in force, which involve different types of sources. In several States the noise limit values are commonly expressed in terms of equivalent sound pressure level weighted A, L_{Aeq}, and they are referred to time periods changing from country to country.

In such a confused picture, in order to define limit values for the noise indicators and particularly for L_{den}, can be useful to refer to General Plan Guidelines of California Governor's Office of Planning and Research (URS Corporation, 2003), which resumes the wide experience of USA about the issue of noise pollution annoyance. In Document 4 a noise classification has been introduced, based on four annoyance levels (normally acceptable—conditionally acceptable—normally unacceptable—clearly unacceptable), that can be a good standpoint in introducing limit levels for port noise. On the basis of the land use and of the above mentioned classification, a specific ranking in terms of Day-Night Level (DNL) and Community Noise Equivalent Level (CNEL), both expressed in dB, has been defined in San Francisco General Master Plan, as shown in Figure 8.

The expression for the Day-Night Level is:

$$DNL = 10\lg\frac{1}{24}\left(15\cdot 10^{\frac{L_{day}}{10}} + 9\cdot 10^{\frac{L_{night}+10}{10}}\right) \quad (3)$$

LAND USE CATEGORY	COMMUNITY NOISE EXPOSOURE 55 60 65 70 75 80
RESIDENTIAL Low density single family	
RESIDENTIAL Multy family	
TRANSIENT LODGING Hotel, Motel	
Schools, Nursing Homes, Hospitals, Libraries, etc.	
Auditoriums, Concert Halls, Amphitheatres	
Sport Arena, Outdoor Spectator Sports	
Playgrounds, Parks	
Golf Courses, Cemeteries, Water Ricreation Areas	
OFFICE BUILDINGS Commercial and Professional	
Industrial, Manifacturing, Utilities, Agriculture	

Figure 8. Guidelines for noise compatible land use (State of California Regulation 1998). Values of L_{dn} and CNEL are expressed in dB.

CNEL is a noise indicator similar to L_{den}:

$$CNEL = 10\lg\frac{1}{24}\left(12\cdot 10^{\frac{L_{day}}{10}} + 3\cdot 10^{\frac{L_{eveningy}+3}{10}} + 9\cdot 10^{\frac{L_{night}+10}{10}}\right). \quad (4)$$

On the basis of these guidelines, noise limit values in terms of L_{den}, valid for all source categories, including vessels, have been proposed, as shown in Table 1. These noise limits can be applied to the external environment for areas nearby ports, so that they can constitute a fundamental tool in noise port control.

For each land use category, a single value limit (and not a range) has been taken into account, with the aim to follow the same approach of the European Directive.

Also L_{night} can help in the evaluation and management of noise nearby the harbour areas. In that case, for L_{night} the same limit values that are adopted for L_{Aeq} during the night period may be introduced: already various countries have defined these limits in their regulations and have related them to the land use.

The use of annoyance indicators in strategic noise maps permit to find out how much noise can cause harmful effects on the exposed population. By means of cross-comparisons between characteristic sound levels of the area, limit values of noise indicators, and the number of people harmfully affected actions against noise can be planned and intervention priorities can be fixed.

Table 1. L_{den} limits proposed for various land use categories.

Land use category	Limits dB
Residential (low density, single-family, duplex)	55
Residential (multi-family)	60
Transient Lodging (motel, hotel)	60
Schools, libraries, hospitals, places of worship, nursing homes	55
Auditoriums, concert halls, amphitheatres	65
Sport arenas, outdoor spectators sports	70
Playgrounds, neighbourhood parks	65
Office buildings, business	60
Mix areas (commercial area + residential area)	60
Industrial areas, utilities, agriculture	75

8 CONCLUSIONS

Noise is a key issue for a sustainable development of ports. This led to the introduction of noise assessment criteria and strategies for the analysis and control of noise pollution. The Strategic Noise Map is considered as the most suitable instrument of acoustic planning to identify the critical zones and establishing the improvement actions. This is essential to identify solutions that can allow the development of trade without compromising the quality of life in port cities.

A method has been defined to produce a successful Noise Strategic Mapping of port areas. The noise mapping is carried out by the means of specific noise prediction models. In order to have the input data for the modeling, the following information must be collected: area topographical features, ground and buildings characterization, noise sources measurements and meteorological data. The monitoring should be set up in four main phases: (1) identification and classification of the sources by type; (2) phonometric measurements near the sources and along sound propagation paths; (3) sound field calculation, (4) noise mapping and assessment of noise annoyance. During the first phase sound sources that generate high levels of noise causing harmful effects to residents on the long term are identified and catalogued, distinguishing them between stationary and mobile. According to this classification, the measurement methodology most appropriate to characterize the sources must be adopted. The second phase concerns measurements to characterize ships noise: measurements procedures must be set, establishing the position (height, distance from the source, distance from one location to the next, etc ...) and the duration. Appropriate sound levels and noise spectra must be recorded and, once acquired, they are used in the third phase to calculate the sound power levels. In the third phase, in addition to phonometric measurements made in the second stage, it is necessary to gather further information for a proper characterization of the analyzed sources. In the case of stationary sources operation time must be considered. For mobile sources available ship traffic data must be used. With this information the sound fluctuations of noise levels in annual averages can be analyzed for the assessment of long-term noise indicators, L_{den} and L_{night}.

As for industrial or urban noise, descriptors have been used to characterize the noise pollution of ports; currently, L_{den} and L_{night} appear the most significant noise indexes. The acoustic indicators have been formulated in order to correlate the measurement of sound pressure level and the percentage of people whose health has been negatively affected by noise pollution. In general, two harmful effects are to be considered: annoyance and sleep disturbance. The problem of correlating these descriptors to the effective disturbance perceived by the population, both within and outside the port area, is still open.

The last step consists in noise mapping, which is a very important planning tool to identify critical noise zones and recognize the causes that have produced them. The maps show the noise climate of the port area in terms of L_{den} and L_{night}. Using these noise indicators in Strategic Noise Maps, it is possible to know how much noise can cause harmful effects on the population exposed. Through cross-comparisons between characteristic sound levels of the area under exam, the limit values imposed by regulations and the number of persons harmfully affected, the map can plan the actions against noise fixing the intervention priorities. To this aim a plan to manage noise issues and effects, including noise reduction if necessary, shall be introduced. Furthermore a monitoring system can allow updating such maps at any time.

Unfortunately, noise limit values for ships in harbours haven't been specifically established yet neither at a International nor at a national level. Due to this fact, it's essential to define some noise indexes values to use as a reference for the comparison with noise levels values measured. In the same way, it's very important to define noise limits in order to protect the population living in urban areas close to harbours and ship's course lines of shipping routes near the coast.

Regarding interventions for noise attenuation in ports, although a lot of different measures have been defined and implemented in ports, there is no evidence of one best solution for noise attenuation as every port has its own characteristics (position and height of the noise sources, direction with respect to the potentially affected zones, operation hours, distribution of buildings and screening elements, volume of activity). Therefore, there is not a universal measure which can be applied to any problem of noise pollution in port surroundings. Nevertheless, in order to achieve the best results for noise abatement, it is essential to develop noise management strategies for each port in a worldwide scale, which will imply a co-ordination and co-operation philosophy at international, national, regional and local levels.

ACKNOWLEDGMENTS

This work was developed in the frame of the collaborative project SILENV—Ships oriented Innovative soLutions to rEduce Noise & Vibrations, funded by the E.U. within the Call FP7-SST-2008-RTD-1 Grant Agreement SCP8-GA-2009-234182.

REFERENCES

ANSI/ASA S12.64-2009, Quantities and Procedures for Description and Measurement of Underwater Sound from Ships—Part 1: General Requirements.

Badino, A., Schenone, C., Tomasoni, L. 2010. Managing the environmental sustainability of ports: noise pollution, *1st International Conference on Environmental Management & Technologies ICEMT2010 proceedings, Amman, Jordan, 1–3 November 2010.*

Badino, A., Borelli, D., Gaggero, T., Rizzuto, E., Schenone, C. (2011) Normative framework for noise emissions from ships: present situation and future trends. In Guedes Soares, C. & Fricke, W. (eds.), *Advances in Marine Structures*. Leiden: CRC Press/Balkema.

Biot, M. & Moro, L. 2011. Methods and criteria to manage airborne outdoor ship noise. In Guedes Soares, C. & Fricke, W. (eds.), *Advances in Marine Structures*. Leiden: CRC Press/Balkema.

EN 12354-4:2000, Estimation of acoustic performance of buildings from the performance of elements—Transmission of indoor sound to the outside.

EU 2002, Directive 2002/49/EC of The European Parliament and of The Council, 25 June 2002.

European Commission Working Group Assessment of Exposure to Noise (WG-AEN), 2006. Good Practice Guide for Strategic Noise Mapping and the Production of Associated Data on Noise Exposure, *Position Paper, Final Draft*.

Imagine European Project. 2007. Guidelines for producing strategic noise maps on industrial sources. *Deliverable [14] of the IMAGINE project.*

ISO 2922:2000, Measurement of airborne sound emitted by vessels on inland waterways and harbour.

ISO 3095:2005, Railway applications—Acoustics—Measurement of noise emitted by railbound vehicles.

ISO 7188:1994, Acoustics—Measurement of noise emitted by passenger cars under conditions representative of urban driving.

ISO 7216:1992, Acoustics—Agricultural and forestry wheeled tractors and self-propelled machines—Measurement of noise emitted when in motion.

ISO 9614:2009, Acoustics—Determination of sound power levels of noise sources using sound intensity.

ISO 9645:1990, Acoustics—Measurement of noise emitted by two-wheeled mopeds in motion—Engineering method.

ISO 14509-2:2006, Small craft—Airborne sound emitted by powered recreational craft – Part 2: Sound assessment using reference craft.

ISO 14509-1:2008, Small craft—Airborne sound emitted by powered recreational craft—Part 1: Pass-by measurement procedures.

ISO 20906:2009, Acoustics—Unattended monitoring of aircraft sound in the vicinity of airports.

NoMEPorts European Project. 2008. Good Practice Guide on Port Area Noise Mapping and Management. *Technical Annex*.

Moretta, M., Iacoponi, A. & Dolinich, F. 2008. The Port of Livorno mapping experience; *Acoustic'08 Paris symp., Paris, France, June 29–July 4, 2008*.

Moro, L. 2010. Setting of on board noise sources in numerical simulation of airborne outdoor ship noise. Proceeding of *9th Youth Symposium on Experimental Solid Mechanics, Trieste, Italy, July 7–10, 2010*.

URS Corporation. 2003. Expansion of Ferry Transit Service in the San Francisco Bay Area. *Final Program—Environmental Impact Report*.

Sustainable Maritime Transportation and Exploitation of Sea Resources – Rizzuto & Guedes Soares (eds)
© 2012 Taylor & Francis Group, London, ISBN 978-0-415-62081-9

Impact of comfort class requirements in a new building ship and possible advantages for owners

A. Cotta & E.P. Tincani
Martinoli & Co., Genoa, Italy

ABSTRACT: This paper would show how the Comfort Class Notation could be feasible for Owners in new ship building. First the paper introduces the main characteristics of a Comfort Rule On Board, presents the general testing conditions and also highlights additional requirements for passenger ships greater than 65 m length (numerical values and reference tables will be shown). An experimental set of testing points is then presented: noise, vibration levels and noise impact values, collected on a new building project without any Comfort Notation, will be the basis for the further critical analysis of these results.

1 INTRODUCTION

1.1 *Comfort on a vessel*

An important area of shipping is today held by cruises, for which comfort is a strategic aim.

In new ship constructions an increasing importance has been given to comfort. This feature is analyzed and surveyed by shipbuilders, owners and also by Class Societies. The idea itself of comfort is changed in the last years and now it is common to consider it as involving both passengers and crew, therefore accommodation spaces but also workshops, mess, galleys, etc.

Aiming to be a concrete support in the construction each Register has developed a special Class Notation.

For instance, it is frequent to find IMO resolution A.486 adopted in 1981 mentioned as reference in ship building noise evaluation and limit and used as target in contractual specification between Owners and Yards.

As comfort also covers vibration issues, noise impact and noise reduction, a large part of preliminary design evaluations are gone through by shipbuilders, numerous tests are usually required by Owner's technical teams and, at the state of art, many different Comfort Notations are suggested by Class Societies.

In order to succeed in the cruise business, an owner finds it essential to ensure his guests the more comfortable living conditions on board, this being traduced on contractual specification by maximum acceptable limit values or making a require having Class Notation for Comfort.

In this paper test results collected on board of M/V Marina are reported, analyzed and commented.

M/V Marina is a cruise vessel of approximately 66.000 tons, with 625 passenger cabins, built by Fincantieri for Oceania Cruises in Sestri Ponente, Italy.

1.2 *Comfort class and tests onboard*

Firs point focused is to have at disposal a real mirror of what is required: limit values, testing procedures and testing conditions evaluated by Registers. A comparison between Germanischer Lloyd, Bureau Veritas and Det Norske Veritas has been developed and is later on presented, starting from the individuation of each regulation main features, moving to the principal steps of Rule certification and ending with the range values accepted for noise, vibration and other significant tests.

The second part of this work is dedicated to the presentation of noise and vibration test results collected onboard during sea trials and to their comparison and analysis.

Finally a parallel between Comfort Class required values and experimental test results will be discussed and conclusions will be reported with authors considerations.

2 COMFORT CLASS

2.1 *What comfort class is*

Comfort Class is an additional class notation provided by Classification Society in order to certify comfort on board referring to Noise, Vibrations and other eventual problems as Climate on board.

In these Rules both passengers and crew spaces are regulated to ensure comfort in all vessel areas, work ones included.

All Rules have similar parameters to be controlled for the obtainment of the certification, but different requirement are used by each Society and various categories of comfort are defined.

When referring to noise, three parameters has to be checked:

- Noise level
- Sound insulation
- Impact sound insulation

When referring to vibration a frequency-weighted rms value or a repetitive peak value limit is defined.

In general design plan approval, material certifications, tests developed during construction and sea trials can be required by the Society to obtain Class Notation.

2.2 *Main class notation characteristics*

As previously mentioned, Comfort Notation can refer to different problems; this work focuses on noise and vibrations starting from a comparison between different Rules.

Classification Societies analyzed are:

- Germanischer Lloyd (GL)
- Det Norske Veritas (DNV)
- Bureau Veritas (BV)

These Rules will also be compared with the Technical Specification Limits (TSL) of Oceania Marina cruise vessel delivered by Fincantieri Group in February 2011.

For a better comprehension of the main characteristics of each Rule two elements will be analyzed:

- Measuring conditions
- Required noise and vibrations limits

2.2.1 *Measuring conditions*

A relevant feature of Comfort Rules is the definition of the data collecting conditions. All Rules refer to ISO 2923, but particular features can be added by each Society.

In Table 1 all measuring conditions required by the three Societies considered in this work are presented.

Table 1. Measuring conditions.

Classification society	Sea mode	Harbour operations	Thruster operations
GL	X	X	X
DNV	X		
BV	X	X	
TSL	X	X	

The sea mode condition is almost uniform in all Rules; it takes into account the propulsion machinery operating in normal conditions to reach the service speed and all other plants and machinery in normal condition during ship life. It also defines water depth as at least five times the draught, a sea state equal to 3 and a wind speed of 4 Beaufort. Relevant importance is given to ventilation and air-conditioning systems that have to work at normal rates because of their relevant influence on noise and vibration.

Concerning harbour condition, a different definition is given by GL and BV; GL checks noise and vibrations during harbour operations whereas BV carries out insulation measurement in quay.

The last measuring condition foreseen only by GL is the thruster one. This condition is also mentioned in IMO Resolution a.468 (XII); measurement shall be carried out with machinery running in normal condition as necessary for thruster operations. This situation results to be very interesting because noise and vibrations due to thruster activities can really be relevant; moreover in many cases manouvering operations are carried out early in the morning or late at night, becoming very uncomfortable for passengers, especially those occupying fore and aft cabins, often suite cabins. As for the testing condition, even the number and the position of testing points are considered by Rules; not all Rules define exactly these items but these must always be approved by the Societies.

2.3 *Noise limits*

The first check concerning noise studies is noise level in all vessel areas. As known each Society defines a number of levels of comfort: the higher the level, the easier for limits to be complied.

The unit of measurement for noise level is sound pressure level expressed in db(A).

All the comparisons presented in this work refer to sea mode measuring condition because this is the only one shared by all Rules and Technical Specifications considered.

2.3.1 *Passenger areas*

For each Rule the maximum, the minimum and a mean level are compared; the idea to define an "Average" is due to the different wide spectrum of values that each Register could accept, in fact GL has 5 different classes whereas DNV and BV only have 3 classes. An example is reported in Table 2.

In Table 3 appears a comparison between noise limits in passenger areas required by Germanisher Lloyd, Det Norske Veritas and Bureau Veritas.

In order to better understand the relationship between these requirements and a real technical specification another comparison is now presented

Table 2. Rules noise limits for open deck.

Open deck					
GL		DNV		BV	
	Limit		Limit		Limit
Class	[db(A)]	Class	[db(A)]	Class	[db(A)]
E	64	1	65	1	65
1	66	2	65	2	70
2	68	3	70	3	75
3	70				
4	75				

Table 3. Rules limit for noise in passenger spaces.

Area	Class society	Max [db(A)]	Min [db(A)]	Mean [db(A)]
First class cabins	GL	52	44	48
	DNV	50	44	47
	BV	50	45	47,3
Standard cabins	GL	54	46	50
	DNV	55	49	52
	BV	56	49	52,7
Public spaces	GL	60	52	56
	DNV	62	55	58,3
	BV*	64,5	58,3	61,3
Open spaces	GL	75	64	68,6
	DNV	70	65	66,7
	BV	75	65	70

Table 4. Technical specification limits for noise in passengers areas.

Area	Technical specification limit [db(A)]
First class cabins	45
Standard cabins	48
Public spaces	55
Open spaces	65

between Oceania Marina Technical Specification Limits (TSL), described in Table 4, and Rules limits.

A diagram can be drawn to resume the limits required by Rules and those required by Owners in technical specification. This will be shown for suites, standard cabins and public spaces for passengers.

In this last diagram BV* means that the achievement of these values implies that many different categories have to be mediate, since the Rule considers many different categories of public spaces, manned or partially manned by passengers, which therefore require or not a very low background noise.

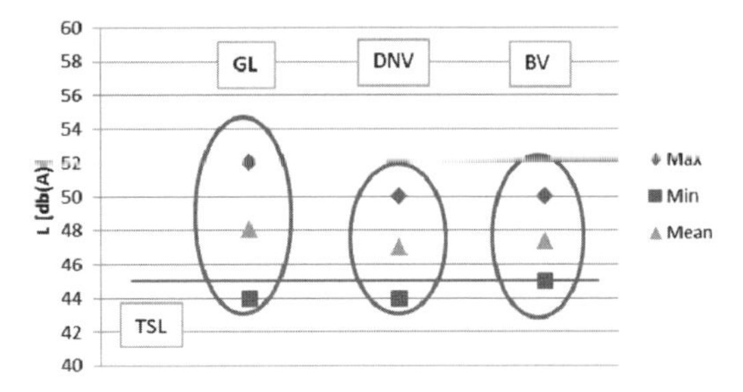

Figure 1. First class cabins noise limits for comfort class notation and TS.

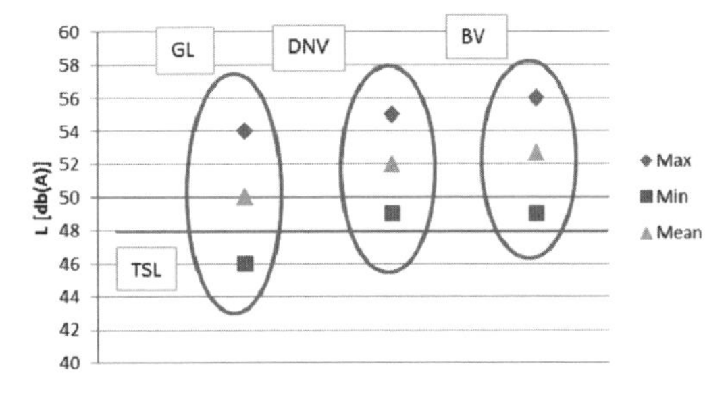

Figure 2. Standard cabins noise limits for comfort class notation and TS.

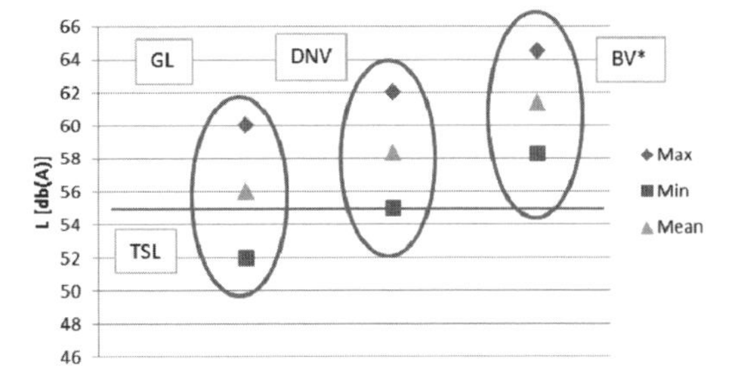

Figure 3. Public spaces noise limits for comfort class notation and TS.

2.3.2 *Crew areas*

Another interesting comparison can be made for noise limit in crew areas; in Table 5 Rules requirements are reported whereas TSL are collected in Table 6. In this case Technical Specification refers to IMO Resolution A.468. (XII) 1981.

Table 5. Noise limits in crew spaces.

Area	Class society	Max [db(A)]	Min [db(A)]	Mean [db(A)]
Cabin	GL	58	50	54
	DNV	60	50	55
	BV	60	52	55,7
Mess rooms	GL	65	57	61
	DNV	65	55	60
	BV	63	57	60
Galleys	GL	75	68	72
	DNV	–	–	–
	BV	76	70	73
Workshop	GL	85	80	83
	DNV	–	–	–
	BV	85	85	85
ECR	GL	75	67	71
	DNV	75	70	71,7
	BV	75	70	72,7

Table 6. Technical specification limits for noise in crew areas.

Area	Technical specification limit [db(A)]
Cabin	60
Mess rooms	65
Galleys	75
Workshops	85
ECR	75

Making a comparison between maximum sound level values accepted by TSL and Classification Societies is possible to underline that for these areas comfort class is stronger than TSL (IMO 468 Res.); this means that also crew comfort becomes a target of these Rules.

As for passenger areas a graphical comparison is presented in the following diagrams where the previous consideration is better comprehensible.

In Figures 4, 5 and 6 where red line is the contractual limit vale (TSL) is important to note that for crew spaces it is accepted to have noise limits equivalent to the lower Comfort Class Notation. This is the opposite of what is required for passenger areas.

2.3.3 *Noise limits in thruster operation*

As said before it is not possible to compare noise limit during thruster operation because only GL defines these limits. However a presentation of these values is interesting because of the great importance of comfort on board also during manoeuvring operations.

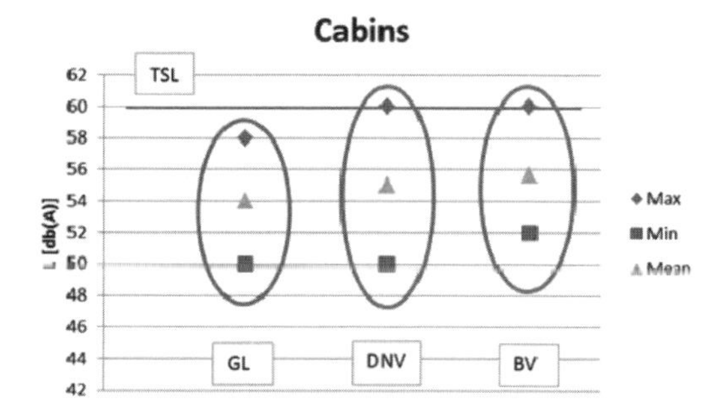

Figure 4. Crew cabins noise limits for comfort class notation and TSL.

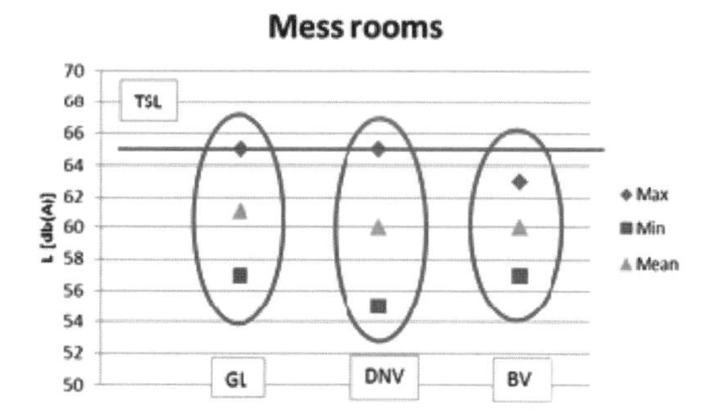

Figure 5. Mess rooms noise limits for comfort class notation and TSL.

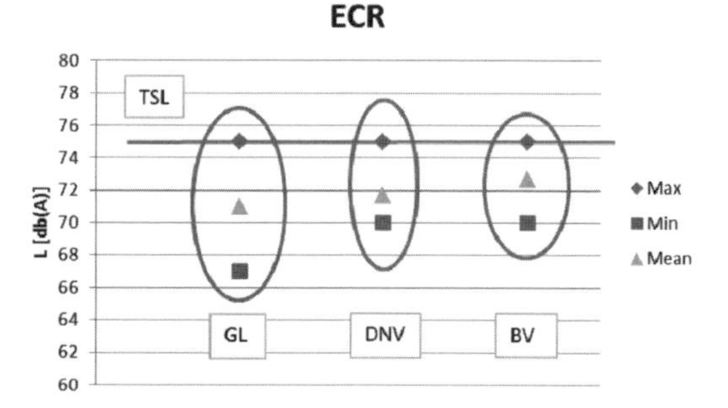

Figure 6. Engine control room noise limits for comfort class notation and TSL.

Analyzing these data is possible to underline as the best class for thruster operation requires the same noise limit than the worst class in sea mode condition. It means that the limitation of thruster noise and vibrations is an important item in the evolution of comfort on board.

2.4 *Noise insulation and impact noise*

Referring to noise, Rules consider also other two relevant items for the definition of comfort on board: noise insulation and impact noise.

Table 7. GL Noise limits in thruster operation.

	Class limits [db(A)]				
Area	E	1	2	3	4
First class cabins	52	54	56	58	60
Std cabins	54	56	58	60	62
Public spaces	58	60	62	64	66
Open spaces	64	66	68	70	72

In this work these problems are not investigated mostly because of the differences in defining the limit for noise insulation, while a homogeneity in limits for impact sound can be found. All Rules and Technical Specification, except GL Rule, define 50 db and 60 db as a limit for impact sound insulation depending on whether a soft or a hard covering is disposed on upper deck. Whereas for noise attenuation in some cases background noise level of the area in analysis is considered and however many different situations are defined in Rules leading to a very hard and substantially non relevant comparison.

2.5 *Vibration limits*

An important part of Comfort Class Rule is the definition of vibration limits in all vessel areas.

In this case it is not possible to carry on a comparison similar to the former one, because of the different definitions of unit of measurement involved.

For example GL and DNV define a root mean squared value, whereas BV and TS consider a repetitive peak value.

Surely vibrations are an important feature speaking of comfort on board but a uniform approach still has not been developed.

3 TESTS ON BOARD AND RESULTS

3.1 *M/V Marina test conditions*

M/V Marina was built during the last two years by Fincantieri in Sestri Ponente Yard for Oceania Cruises, her main dimension and characteristics are expressed in Table 8.

The design of this ship obviously aimed to an optimized comfort on board: the choice of six-blade propellers avoids vibrations in the stern area, hosting restaurants on deck 5, suites from deck 8 to deck 11 and again restaurants on deck 12 and deck 14. In a preliminary design analysis, vibration distribution has been evaluated and on this basis adequate resilient foundations and mechanical solutions have been adopted to ensure insulation in passenger cabins and public areas.

Table 8. M/v Marina main characteristics.

Length	239 m	Passenger decks	11
Breadth	32 m	Passengers	1250
Draft	7 m	Crew	800

As for contractual specification the following test memos have been performed onboard:

- Noise on board
- Vibrations
- Impact Noise
- Noise reduction

The first two series have been carried out in sea mode condition (as said before), during sea trials, with propulsion motors at 85% of power and others auxiliary machineries working in normal operating condition.

For noise and vibrations tests about 100 different points has been chosen in all representative locations throughout the vessel, including accommodation, working areas, machinery spaces, open decks cabins etc. This plan has been submitted to the owner approval before trials.

Noise reduction index and Impact Noise have been instead evaluated in harbour condition only with auxiliary machineries operating.

3.2 *M/V Marina tests results*

Tests performed on M/V Marina represent a fully profile of onboard comfort, a relevant and significant number of point having been checked. Every single point measured is a picture of an area of the ship, obviously major attention has been given to passengers ones: cabins, suites and public spaces. Authors however concentrated also to those places dedicated to crew lifetime and accommodations.

The most relevant sets of measure will be presented:

- Noise in passenger suites, cabins, restaurants, recreation areas and open decks
- Noise in crew cabins, accommodation, mess room, workshops, wheelhouse.
- Noise in suites and crew cabins with thrusters operating at 80% of power.
- Noise insulation between adjacent cabins in passenger area.
- Vibration in passenger suites, cabins ecc.
- Vibration in crew areas
- Vibration in suites and crew cabins with thrusters operating at 80% of power.

Table 9. Passenger accommodations and public areas.

	Min value dB(A)	Max value dB(A)	Mean value dB(A)	Eval points N
Suites	36	45	41	17
Cabins	32	46	36	104
Corridors	37	58	49	27
Public areas	41	58	50	51
Open decks	47	65	59	16

Table 10. Crew accommodations, recreation and workshops or work spaces.

Crew	Min value dB(A)	Max value dB(A)	Mean value dB(A)	Eval points N
Cabins	31	56	40	49
Recreation	50	64	57	3
Galleys	60	73	67	6
Mess room	55	64	60	3
Mach. Spaces	86	108	94	15
Engine CR	62	62	62	1
Wheelhouse	51	51	51	1
Workshops	62	83	71	3

Table 11. Measurement with thrusters operating at 80% of power.

	Min value dB(A)	Max value dB(A)	Mean value dB(A)	Eval points N
Suites	56	61	59	4
Public areas	65	65	65	1
Crew cabins	73	74	74	2

Table 12. Sound insulation between cabins in passengers areas.

	Min value dB(A)	Max value dB(A)	Mean value dB(A)	Eval points N
Cab to cab	35	40	38	14
Cab to corr	41	58	48	42
Cab intercom	41	44	43	2

Table 13. Vibration in passengers areas.

	Min value mm/s	Max value mm/s	Mean value mm/s	Eval points N
Suites	0.1	0.4	0.3	9
Cabins	0.1	0.9	0.3	27
Corridors	0.1	0.2	0.2	2
Public areas	0.1	1.2	0.3	33
Open decks	0.3	1.4	0.9	2

Table 14. Vibrations in crew areas.

CREW	Min value mm/s	Max value mm/s	Mean value mm/s	Eval points N
Cabins	0.1	0.6	0.3	10
Recreation	0.2	0.4	0.3	2
Galleys	0.2	0.2	0.2	1
Mess room	0.3	0.6	0.4	3
Mach. spaces	0.3	0.8	0.5	6
Engine CR	0.4	0.4	0.4	1
Wheelhouse	0.1	0.6	0.4	2
Workshops	0.2	0.7	0.4	3

Table 15. Vibration measurement with thrusters operating at 80% of power.

	Min value mm/s	Max value mm/s	Mean value mm/s	Eval points N
Suites	0.5	3.3	1.5	3
Public areas	0.5	0.5	0.5	1
Crew cabins	4.8	5.9	5.4	2

Table 16. Contractual limits for passenger spaces.

	Limit values dB(A)	Mean measured values dB(A)	Evaluated points N
Suites	45	41	17
Cabins	48	36	104
Corridors	60	49	27
Public areas	55	50	51
Open decks	65	59	16

A table and a brief synthesis of the collected values is given.

In general about 300 different points have been checked throughout the vessel for noise evaluation and about 60 different cases of mutual sound insulation between two adjacent cabins or between cabin and corridor has been spotted measured.

In this table, as well as in the others, for each homogeneous set of spaces the measured minimum and maximum values are presented, while mean value is the average of all the evaluation points of that location typology.

Thrusters are evident sources of noise and vibrations. Not often their influence on comfort is included in testing conditions required by owners. In common cruise however it is quite usual to have thrusters operating also during recreation time or early in the morning during mooring operation: these transitional conditions have to be checked to avoid as much as possible interference with peacefully climate expected onboard by passengers.

Also there is a significant common interference between auxiliary propellers and suites. These are both normally located in the extreme areas of the ships, bow and stern: ones to optimize ship maneuvering capability, the others to allow maximum space for suites and its external spaces, privacy and beautiful landscape.

A noise insulation test has been carried out to evaluate soundproof capacity of the cabins panels. This feature permits to ensure adequate privacy to passengers and to avoid sound propagation between public areas and cabins.

Vibration results hereafter presented have been collected in about 100 different point in the ship, whit the aim to have an exhaustive check of oscillation values onboard.

A vibration test has been carried out in accordance with M/V Marina's technical specification: for each point maximum peak [mm/s] and the relative frequency [Hz] are recorded.

The vibration velocity amplitudes are expressed as time averaged peak values in agreement with ISO 6954 (1984). All the data have a maximum peak in a range of frequencies that are measured in between of 6.88 Hz and 64.38 Hz.

Although not indicated in contractual owner requirements, some tests with thrusters operating have also been evaluated. In this case some spot check in suites and crew cabins located vertically over the bow thrusters and in the main dining restaurant located over the stern thruster has been carried out.

3.3 *Analysis of the results*

Results obtained during noise and vibration tests are in general all in adherence with the expected values, as required in the technical specification by owner.

However, it is important to specify that in contractual specification only the maximum acceptable noise level for passenger areas has been customized: suites, cabins, corridors, public rooms ect., crew accommodation, crew recreation areas and working spaces the contract between owner and shipbuilder referring to IMO Resolution A.468 (XII) adopted on 29 November 1981.

Vibration test highlighted how at the design stage the preliminary analysis could be helpful to evaluate critical points and could permit to avoid excessive vibrations on ship accommodations.

4 CONCLUSIONS

In the paper has been presented the main features that each Register require today to classify a ship with a Class Notation.

From the comparison between Rules Notations and Technical Specification Limits (TSL) it is possible to note how in passenger areas maximum accepted limit values are close to the best comfort class in every conditions analyzed, on the contrary noise and vibration values accepted in crew cabins and accommodations have maximum limits close to the lower Class Notations. Probably this discrepancy is related to the difficulties to have an homogeneous definition of comfort on board.

There is therefore evidence of how in Rule Comfort Class the idea is to define a concern about "safety" of people who live the ship as workers and not simply as guests having a beautiful time.

However it is important to highlight how test values collected on board of M/V Marina are in all condition better of what required in TSL and even in crew spaces very high conditions of comfort have been measured. This is the result of a good preliminary design evaluation of critical points for noise and vibration and probably is also due to the capability and the experience collected by Fincantieri in the last ten years in similar ships construction.

Some few words to note that a particular condition represented by tests of noise and vibrations with thrusters operating is evaluated only by one of the Rules Class. This singularity probably could be in the future involved also in other class notation, since this condition is increasing weight in a cruise ship and could be attractive for owners to have it checked.

Getting to a conclusion, it is probably not strictly necessary for a new ship to have a comfort class but it is true that this notation could be helpful for Owners to define desired target of noise and vibrations on board.

There are at the moment some discrepancies in test conditions between Rules even if they are aligned in the matter of maximum and minimum acceptable values.

ACKNOWLEDGEMENTS

A special thanks to all the New Building Team in Genoa and its chief Salvatore.

Thanks to Oceania Cruises and Robin with his "Good Luck".

REFERENCES, SYMBOLS AND UNITS

Biot, M. & De Lorenzo, F. 2007. Noise and vibrations on board cruise ships: are new standards effective?. *ICMRT'07, Naples*.

Bureau Veritas 2006. Bureau Veritas Rules for the classification of steel ships.

Det Norske Veritas 2011. Rules for classification of ships.

Fincantieri 2007. Ship Technical Specification, Oceania Marina cruise vessel.

Germanisher Lloyd 2003. Rules for Classification and Construction.

IMO Resolution A. 468 1981. Code on noise levels on board ships.

International Standard ISO 2923 1996. Measurement of noise on board vessel.

International Standard ISO 6954 1984. Guidelines for the overall evaluation of vibration.

International Standard ISO 4867 1984. Code for the measurement and reporting of shipboard vibration data.

Sustainable Maritime Transportation and Exploitation of Sea Resources – Rizzuto & Guedes Soares (eds)
 ISBN 978-0-415-62081-9

An international reference vessel, from a noise and vibration point of view, in the framework of the SILENV project

P. Beltrán & R. Salinas
TSI-Técnicas y Servicios de Ingeniería, SL, Madrid, Spain

ABSTRACT: The new and increasingly demanding requirements of reducing the environmental impact of all types of ships from all countries and Regulatory Institutions such as the new "Green Policy" of the EU, leads to the development of a series of Regulations/Directives which will immediately affect both the owner and the builders worldwide. This makes it necessary to introduce modifications in the vessel design in order to fulfill these requirements and, ultimately, improve their exploitation avoiding penalties and/or restrictions.

Awareness of some Spanish owners with these requirements have allowed to the Spanish shipbuilding to be positioned strategically at the forefront of technology to fulfill these requirements. This work, after a review of those Directives that affect the ship design, focuses on the presentation of the results obtained in two Ro-Ro vessels. These vessels, by their design and performances, are a "technological reference" in the new scenario of high environmental performance requirements: Noise & Vibrations on board, Noise Radiated to the Harbour and Noise Radiated to the Water.

1 INTRODUCTION

The growing society interest in the importance of control, monitoring and improvement of the environment, has been the stimulus for different states to respond to this current demand. As many other countries, the European Union (EU), also sensitive to these claims, has started the so-called **"Green Policy"**, which main target is to establish Directives or/and Requirements to improve and maintain the environment.

In this present work, after a detailed revision of the Current Regulatory Framework and the one that is about to come (that will affect design and cost), it is presented the **Case Study** of the obtained results as a clear evidence of the correct training and preparation of the Spanish Marine Sector to answer to these new challenges.

2 NEW REGULATORY FRAMEWORK: CURRENT AND FUTURE COMMUNITY REQUIREMENTS AND DIRECTIVES

This section aims to offer the reader a summarized version of the current and future regulations, with the clear target of raising awareness and approach to the resulting design and production costs.

2.1 *Noise and Vibration requirements justification*

In order to help the reader to recognize in advance the "*technical and economic implications*" in the vessel design and price, in the attached Figure 1 Regulations, Polices and Directives in force and the future ones that are appearing as notations class, have been summarized. To enrich this approach with more details, references to other publications are included along the document.

It is important to remark, that in some R&D European Projects inside the **7th Frame Program** (in which the authors are participating) like **"SILENV"** and **"BESST"**, a future **"Green Label"** for all types of vessels is being debated.

This overview will be separately done describing the different aspects: *Noise and Vibration on board*, *Noise Radiated to Harbour*, and *Underwater*

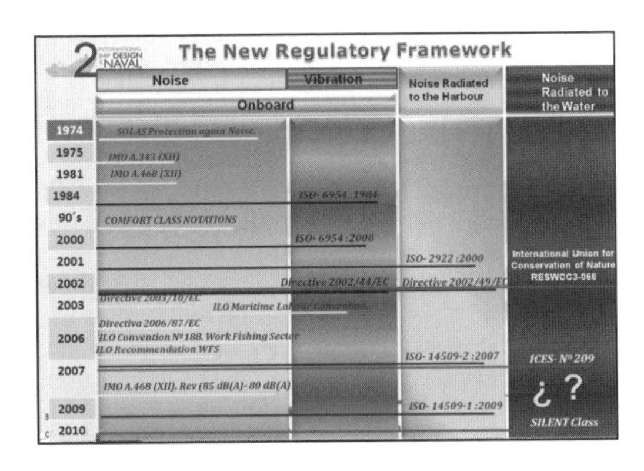

Figure 1. N & V full signature. The regulatory framework.

Radiated Noise, that constitute what the author has called *N&V Full Signature*. For a detailed analysis it is recommended to see Ref (1).

2.2 *Rules and regulations referred to vessel Underwater Radiated Noise*

Taking into account that for the Civilian Shipbuilding the Noise Radiated to Water can be one of the novel aspects, we considered interesting to extend the chronological aspects of the emergence of new Regulations that will affect it.

The recent Framework Directive 2008/56/CE about the strategy of reducing the impact in the marine environment, that represents the first international legal instrument including underwater noise from human source explicitly defined as pollution, along with the most recent publication of an Underwater Radiated Noise Measurement Procedure by the Acoustical Society of America—ASA-URN-(2), are the first pillars upon which rules and regulations will set the preliminary steps: technological and design changes that permit the reduction or control of the environmental impact of vessels in the marine life.

The authors want to point out that both European Projects mentioned before are nowadays working in defining the "technological changes" that will be needed in future vessels so as to be able to accomplish or be near to have the "**Green Label**", that will include the limits for each Acoustic Complete Signature for every kind of vessel: Noise and Vibration on board, Harbour Radiated Noise and Underwater Radiated Noise.

2.3 *How will this affect the New Regulatory Framework in the vessel designing? Advantages and disadvantages. Threats and Opportunities*

In view of this New Regulatory Framework, there are two alternatives, such it was pointed out previously: To **"look the other way"** and be limited to a "sub-standard market" or to "answer efficiently" turning what it looks like a problem into a business opportunity. This last one way will allow us a favorable position in a more selected market niche where minimizing the environmental impact of the future vessels will be a clear indicator of our difference against cheaper competitors.

The tool "Noise and Vibration Integrated Management": As a whole group of studies and experimental tests designed by the authors and which efficiency has been proved in the strict accomplishment of the most demanding requirements of the Oceanographic and Fishing Research vessels, has been revealed as an "efficient tool" that, when is correctly applied, allow us to achieve all the new requirements included in the current and new Regulations and Directives. A complete detail of this tool can be found in other publications (3, 4, 5, 6, 7).

3 CASE STUDY: RO-RO NAVANTIA VESSELS FOR ACCIONA TRANSMEDITERRÁNEA

3.1 *Description and Main Particulars*

At the end of March and in the middle of August 2010, NAVANTIA-Factoría de Puerto Real delivered the RO-RO vessels "José María Entrecanales" (C-509) and "Super-Fast Baleares" (C-510) to the Spanish company ACCIONA TRANSMEDITERRÁNEA. The main characteristics of these vessels and their main machinery are collected in the attached Figure 2.

Finally as important noise sources in these types of vessels, we can highlight the ventilation of the hold and the engine room. Both vessels have fifty-eight (58) ventilation units, some of them with a capacity up to 120.000 m^3/h.

3.2 *Basic objectives of the project. Noise and Vibration Specifications. ISO standard 2992/2000*

The exam of this N&V Contractual Specification allow us to highlight the following comments and observations:

- It is the first time that a fulfilment requirement about noise level **(80 dB (A))** with the vessel in operational conditions in the harbour appears in the Contractual Specification. This is considered a clear example of foresight of the evolution in Environmental Regulations from the Ship-owner's side.
- The ship-owners "show a high sensitivity" to the working conditions of their employees (the crew) and the clients (the passengers), demanding the fulfilment of noise and vibration limits set by the in force rules IMO A.468 (XII) Regulation (8) and ISO Standard 6954/2000 (9).
- Finally, and to be sure that all the appropriate measures will be considered by the shipyard, the ship-owner demands the compulsory "Prediction Studies".

Normally the first position to take is to continue "doing things as they were done for the last years", without taking any preventive measurement that can guarantee the fulfillment of the rules. This position, even when believed to "reduce cost", only leads to the opposite effect: several penalties, last minute corrections with fines for late delivery and excessive cost in the corrections, and, in some cases,

Main Particulars of the Vessel	
Total Length	209,00 m
Length between pp.	190,0 m
Breadth	26,5 m
Depth to Main Deck	9,60 m
Scantling Draft	7,1 m
Design Draft	7,0 m
Death Weight (Design Draft)	9.325 T.
Power	4 × 10.800 kW @ 500 RPM
Load Capacity:	
Crew and Passengers	40 persons
Classification	BV
Main Engines	
Mark	MAN DIESEL & TURBO
Model	9L- 48-60 B.
Nominal Power	4 × 10.800 kW.
Speed	500 RPM
Nª Cylinder	9
Nº Stroke	2T
Gear	
Mark	RENK
Reduction	3,324:1
Nº of units	2
Propeller	
Diameter	5.200 mm
Nº of Blades	4
Pitch	Variable (CPP)
Speed	150 RPM

Figure 2. Main characteristics of the vessels and their main machinery.

vessels that were already finished and it is technical and economically impossible to be modified.

In fact, in the Case Study that will be presented in this work, it will be clear that not taking preventive measures about noise radiated to harbour, could have led to a brand new vessel with limitations in the loading and unloading operations in certain harbours, Spanish and Europeans, and also to possible penalties and complaints.

If the studies required by the shipyard had not been done, the vessels involved in this project would not have accomplished the contractual Specification, mainly in relation to Noise Radiated to the harbour.

The requirement from the ship-owner to fulfill the ISO 2922 Standard "Measurement of airborne sound emitted by vessels on inland waterways and harbors" (10), asking for a contractual limit of **80 dB (A)** at 25 meters from the vessel. The rule has as a main goal avoiding complaints and fines when the vessel ties up near densely—populated areas.

4 SHIPYARD ANSWER: WHAT HAS IT BEEN DONE? HOW HAS IT BEEN DONE?

4.1 *General information*

The shipyard NAVANTIA-Puerto Real with an extended experience in the noise and vibration field related to military and civilian cargo vessels, in an attempt of trying to meet all the strict requirements, showed from the beginning a special attention in covering all the ship-owner's requests, knowing that to fulfill the demanding specifications it was necessary to understand and apply the principle that "dynamic and acoustic design of the vessel" should be mandatory from the earliest stages of this project.

As a consequence, NAVANTIA-Puerto Real required to SENER, as designer, the assistance of a specialized company (TSI) which was in charge of Noise and Vibration Integrated Management and added the previous "design principle" to all its building processes.

4.2 *What has it been done? Basic principles*

Vessels are elastic systems that when under periodic forces, coming from different sources, are "subject to vibrations". The vibration level obtained in the system (vessel) depends, mainly, on three parameters: 1) The intensity or magnitude of the excitation forces. 2) The rigidity of the structure. 3) The dynamic amplification at different frequencies due to resonance phenomena: local and global.

As a consequence, the *possible actions* to keep *vibration levels* under the pre-set limits are the following: A) Minimize the excitation forces of the system. B) Avoid flexible structures from a dynamic point of view. C) Avoid resonance phenomena by coincidence of structural frequencies and excitation frequencies.

Similarly, and from an acoustic point of view, the vessel has built-in Sonorous Focus: Main engine and Auxiliary ones, Propeller, Hydraulic

Systems, HVAC, etc, that are airborne and structure borne noise generators, noise transmitted or spread along the structure of the vessel (Path) that reaches the different locals or spaces (Receptor) as well as it is radiated to water.

In the same way the *possible actions* to maintain *noise levels* under the pre-set limits are the following: 1) Minimize the sound power and vibration power at the noise sources. 2) Reduce or diminish their transmission to the paths. 3) Isolate duly the receivers.

4.3 *How has it been done? Procedures*

The first module, inside the Noise and Vibration Integrated Management and as a control mechanism, should incorporate *"dynamics and acoustics specific requirements"* in the purchase specifications of the different supplies as well as Factory Acceptance Tests Procedures.

In the second module of possible actions, under the direct responsibility of the shipyard, the previously mentioned principle "Dynamics and Acoustics Vessel Design is mandatory" is assumed from the beginning of the project.

With the purpose of minimizing resonance risk and optimizing the acoustic behaviour of the vessel, the Noise and Vibration Integrated Management considers carrying out a Dynamic Acoustic Design of the vessel, that will be accomplished in the following sections: 1) Vibration Prediction. 2) Noise Prediction. 3) Radiated Noise Prediction.

5 PRACTICAL APPLICATION OF THE "NOISE AND VIBRATION INTEGRATED MANAGEMENT" IN THE VESSELS: "JOSÉ MARÍA ENTRECANALES" AND "SUPER-FAST BALEARES"

5.1 *General scope*

The *"first activity level"* it is intended that the shipyard can develop and exert a "control" over those aspects that, as previously described, are in the scope of the suppliers. The intention with this is to achieve what we call "contractual sensitivity" of the suppliers towards the dynamic and acoustic objectives of the project. Its "no-application", leaves the shipyard, in many occasions, "with tight hands" when it comes to finding solutions or countermeasures more economic and technically efficient.

In the *"second level"* of activities, are considered all those aspects that are in direct competence with the shipyard, as it is the supplying of a structure with an appropriate dynamic design: without resonance and enough acoustic isolation to guarantee the minimum requirements.

In the Case Study that we are working on, the development of the Noise and Vibration Integrated Management scope has been, as indicated, "Partial" focusing only and exclusively on the application of "Simulation Techniques" for those aspects required in the Specification: Vibrations, Noise and Noise Radiated to Harbour.

5.2 *Vibration Prediction through Finite Element Method in the building of NAVANTIA-Puerto Real C-509 and C-510*

A mathematical model of calculation through finite element—F.E.M.—was applied, with the intention of avoiding dynamic amplification phenomena due to possible "Resonance Phenomena" in the vessel structure, to identify own frequencies and avoid their coincidence with the main excitation sources of the vessel: Forces and Moments of Main Engines excitation and Forces coming from Pressure Pulses inducted by the propeller on the sternpost.

The application, to this dynamic model of the vessel structure, of the different forces of the excitation source: Main Engines and Propellers, will help us to obtain the "Expected Vibration Levels" in the different areas of all the vessel structure. From the comparison between these "Expected Vibration Levels" with the limits required in the ISO Standard 6954/2000 (9) demanded in the specification, decisions will be taken that will allow us to validate the structure from a dynamic behaviour point of view or introduce those structural modifications that guarantee this fulfillment inside the demanded limits. In Figure 3 we can see a summary of the applied methodology.

5.2.1 *Description of the mathematical calculation model*

To represent the dynamic characteristics (mass and stiffness) of the structure, two-dimensional

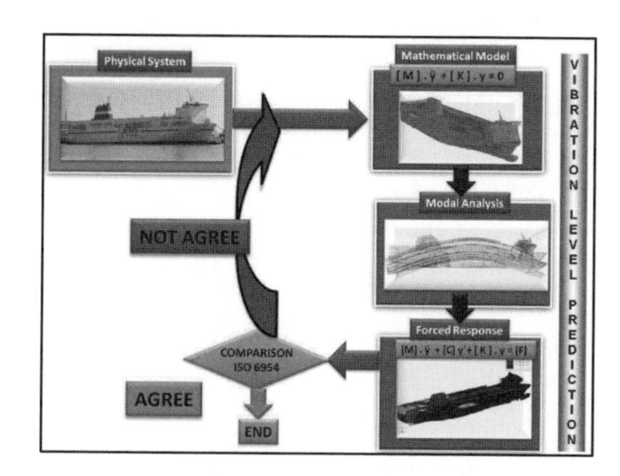

Figure 3. Methodology of the vibration.

elements like plate (shell) were mainly applied, capable of admitting distortions in its plane and perpendicular to this, using one-dimensional elements (beam) for the primary structure and struts. Equipment with a weight over 1000 kg has been considered as point masses distributed in the area where beams are supported, with the exception of the main engine and the reduction gear that has been modeled with two-dimensional elements.

To do the calculation, it had taken in account ballast and consumable tank filling. In Figure 4 the mathematical model used can be seen.

5.2.2 *Fulfilled calculation*

On the mathematical model, two types of calculations were done, as described in the following lines.

Modal Calculation: Through this calculation were obtained the eigenfrequencies or resonance frequencies of the vessels and its vibration mode shapes. The coincidence between eigenfrequencies and excitation frequencies (propellers and engines) produce resonance phenomena and then high level of vibration could be expected. This type of calculation is an essential requisite to be able to make the forced response calculation or to predict vibration levels.

Forced response. Vibration levels prediction: This calculation includes the characteristics of the vessel through the modal calculation and the characteristics of the excitation sources (amplitude and frequency) to obtain the expected vibration levels in the vessel structure.

5.2.3 *Obtained results*

In Figures 5 and 6 vibration modes and shapes associated with the first two modes or eigenfrequencies of the ship-beam obtained, were collected.

In this case, it was preferred to represent the following step of "Forced Response" that will permit, in a more illustrative way, to identify those structural areas with more problems or risks of *"Local resonance"*.

Figure 4. Mathematical model of finite elements.

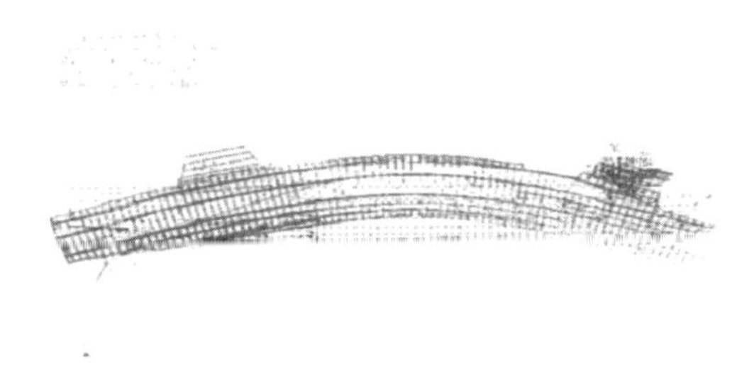

Figure 5. Vertical Bending Mode Shape. 1,78 Hz.

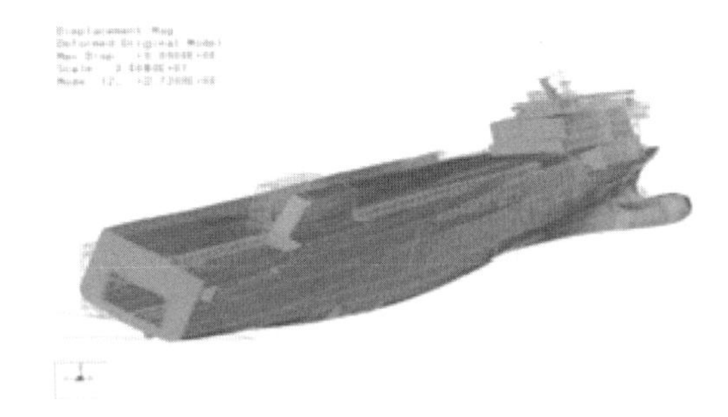

Figure 6. First Torsional Mode Shape. 2,72.

For the *"Forced Response"* calculation, the main excitation forces used were the followings: pressure pulses induced by the propeller in the sternpost at the excitation frequencies 1 × BPF (Blade Pass Frequency) and 2 × BPF, and the forces and free moments coming from the main engines at the excitation frequencies 1 × RPM, 2 × RPM and 4,5 RPM, as harmonic or more significant orders. Auxiliary engines, not considered due to their isolation by elastic suspension and the bow propeller, due to its study can be found inside an analysis of noise presented in this work.

The application of these excitation forces, in the most conservative case that both are acting in phase, the mathematical model has helped us to obtain, through colored maps, the distribution of the "expected vibration levels" in the different locations of the vessel structure.

5.3 *Noise Prediction through SEA method*

5.3.1 *Brief introduction to SEA method*

To make a noise prediction it is necessary to count on the acoustic excitations and their transmission paths, this means working in high frequency. The SEA Method (Statistical Energy Analysis) gives a way to alternative model FEM and BEM—Boundary Element Method-, and to represent the vibratory state of a system. The model represents the means behaviour of a group of similar systems and it also includes an uncertainty factor in the model. The vibratory state is expressed in terms of vibratory energy of individual components. The application of these excitations is expressed in terms of power. And the relation between the excitations and the energy of the elements is expressed in terms of energy flow.

5.3.2 *General description of Noise Prediction*
With the analysis model created we can proceed to solve it and as a result *expected noise levels* in the selected compartments are obtained.

The comparison of theses *expected levels* with the established limits in the technical specification, will take us either to finish the process or, in the contrary, through a repetitive process to simulate special systems of isolation aimed at reducing noise levels in those compartments which do not accomplish the specifications. In Figure 7 a graphic Flow Diagram that corresponds to this methodology is described.

5.3.3 *Mathematical model description*
The noise sources, among others, that have been considered to predict noise in the C-509 and C-510 are the following ones: propellers, main engines, gearboxes, auxiliary engines, thrusters, compressors, harbour generator, purifiers, steering gear, HVAC units, fans of: engine room, garages and technical spaces. In Figure 8 we can see the Acoustic Model used.

5.3.4 *Noise Results obtained with the original Model. Analysis and Acoustic reinforcement*
5.3.4.1 Original Model Results/ Modified Model
Noise level predictions in the spaces of the vessel have been done considering the working conditions in Free Sailing at 90% MCR. Additionally, a prediction of noise level has been done for the Load/Unload condition.

Figure 9 shows the distribution of the expected noise level in C-509/510 with the Original Insulation and the operational conditions tested: *Free Sailing 90% MCR and Load/Unload* and the corresponding Modified Acoustic Model.

5.4 *noise radiated to harbor prediction. Answer to the new requirements*

5.4.1 *Introduction*
The appearance of a "totally new requirement" in the Specification has constituted a "novelty". The reasons for this "new requirement" are two: one is the *Operational* one that answers the need of the Ship-owner to avoid complaints and penalties. The second one is the *Vision and Sensitivity* from ACCIONA TRANSMEDITERRÁNEA in

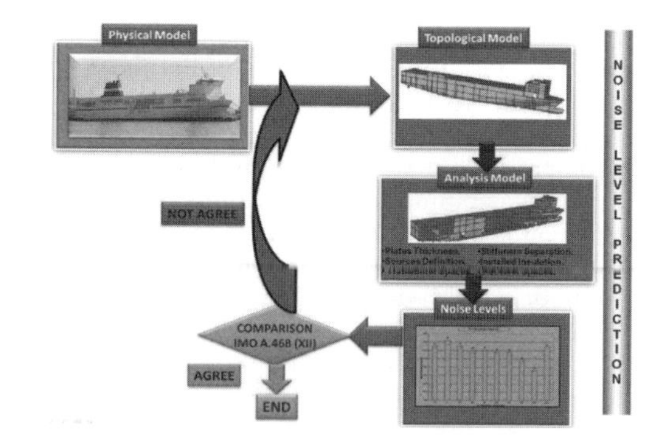

Figure 7. Methodology of the Noise Prediction.

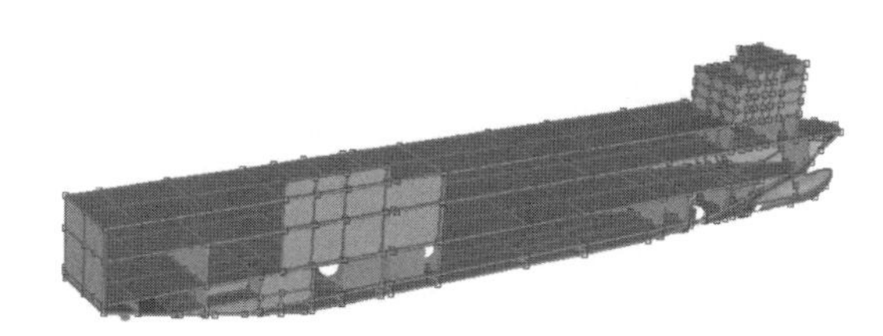

Figure 8. Acoustic Model of the RO-RO's C-509/510.

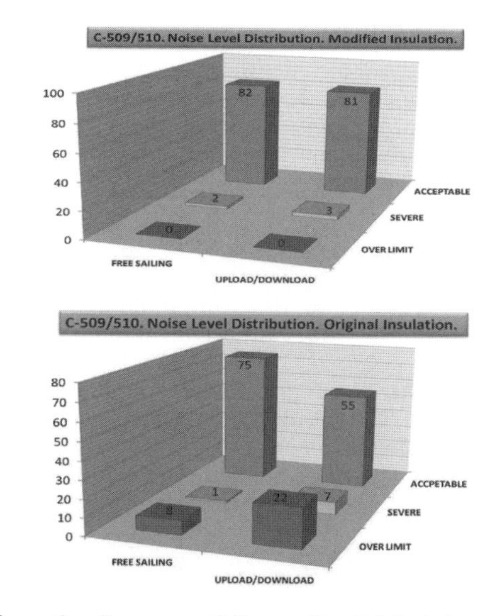

Figure 9. Summary of the results: Original Acoustic Model (A)/Modified Acoustic Model (B).

accomplishing the EU Directives related to Evaluation and Management of Ambient Noise in the Harbours, Directive 2002/49/CE (10) specifically.

5.4.2 *Calculation method applied*
The prediction of the noise generated by the vessel to the harbour has been done counting as a main

noise sources the inlets/outlets of the ventilation system that the vessel can have in use during the stay and operations in the harbour.

The levels of acoustic power, from each inlets/outlets, have been calculated considering the power and noise pressure of each fan, the ducts dimension and the fan chamber, and its acoustic insulation in case of having it.

As a consequence the accuracy of the obtained results has been constrained by the precision of the data given by the supplier. As a result, with the experience of the authors, this data had to be experimentally proved through factory acceptance tests. In this vessel, the main noise sources of Harbour Radiated Noise are integrated by 58 Ventilation fans from the garages and machinery rooms.

With these noise sources characterized, and using the calculation model described in Figure 10 and with the tool developed with TSI, called "NoRaPort", we proceeded to estimate, for the different configurations, noise level at the different distances of the vessel. Not only in the quay but also in the main sections of it, where most of the noise sources to the harbour were found. The comparison of these *"expected noise levels"* at different distances and mainly at 25 m of each side of the vessel, and the comparison of the Specified limit of 80 dB(A), allowed us to optimize the acoustic design of the different ventilation systems.

5.4.3 *Acoustic ORIGINAL design of fans. Obtained results*

The corrections experimentally obtained through FAT tests (Factory Acceptance Tests), as security margins, it was proceeded to do the first prediction of Noise Radiated to Harbour calculation with the Original Acoustic Design of Fans and Silencers.

The analysis of the previously mentioned results shows that the expected pressure level EXCEEDS in +11 dB (A) the contractual required levels at 25 m from the side.

Figure 10. Noise Radiated to the Harbour Model.

For the optional condition analyzed (at 1m from the side) out of the Specification, in the same operational conditions as the previous case vessel, the expected noise pressure levels resultant EXCEEDS in +18 dB(A) the level of 80 dB(A) at 1 m from the side.

5.4.4 *Acoustic MODIFIED design of fans. obtained results*

Based on the deficient results obtained in the prediction of the noise radiated to the harbour with the Original Acoustic Fan Design, the company responsible for the calculation of noise and vibration and exterior noise prediction knowing the needs of the ship-owner and of the Regulations, not mentioned in the Specification but currently in force, recommended the Engineering and the shipyard to do two levels of actions on the silencers: a First Level of actions that we will call Optimization at 25 m and a Second Level of action called Optimization at 1m. In Figure 11 are shown the results obtained in the Optimization at 25 m.

As it can be seen in the analysis of results, with the first action, level noise radiated to harbour by the C-509/510 fulfilled, for both sides of the vessel, with the 80 dB (A) limits required in the Specification.

In the analysis of Figure 12 with the Second Level of action, noise radiated to harbour by

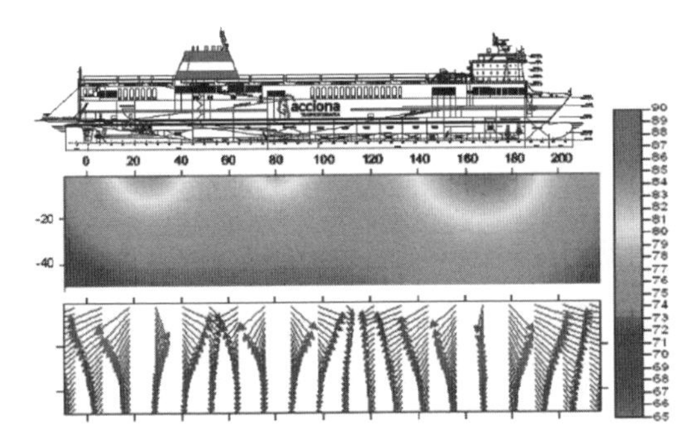

Figure 11. Optimization at 25m. Noise Radiated to Starboard Distribution.

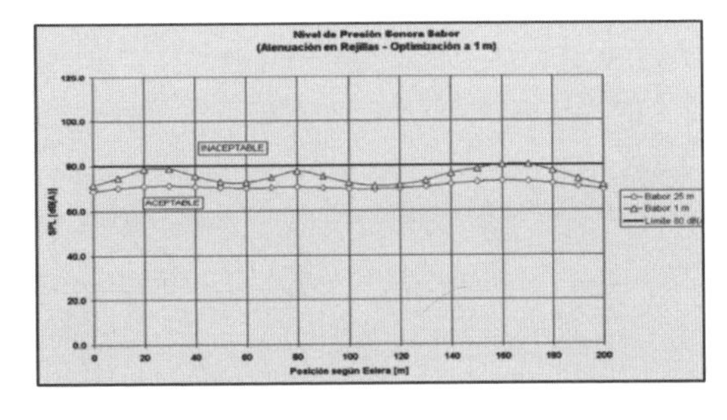

Figure 12. Optimization at 1 m. Distribution of the pressure levels at 1 and 25 m length. Port side.

C-509/510 fulfilled, for both sides of the vessel, with the 80 dB(A) limits required.

As this option: Optimization at 1 m meant an "indisputable improvement" of the vessel features, as it implied not only guaranteeing the fulfillment of the Noise and Vibration Specifications but also with the adaptations of it to the in force more recent Directives, its application was constrained by the decision of NAVANTIA-Puerto Real.

6 SEA TRIAL TESTS C-509/510: CORRELATION MODEL/TESTS

The experimental results obtained in this program of Official Tests, which will be summarized in the following sections, will help us to evaluate the goodness and efficiency of the noise and vibration predictive calculations applied.

6.1 *Vibrations Results*

In Figure 13 there is a summary of the vibration levels obtained during the sea trial tests.

From the experimental data examination of vibration levels obtained in the Official Sea Trial Tests of both vessels we can highlight the following summary:

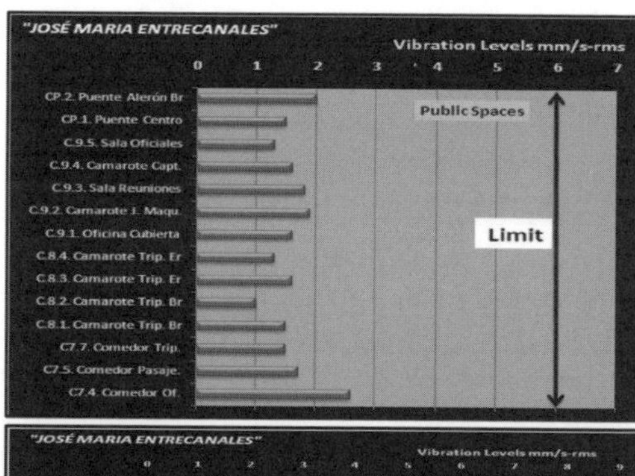

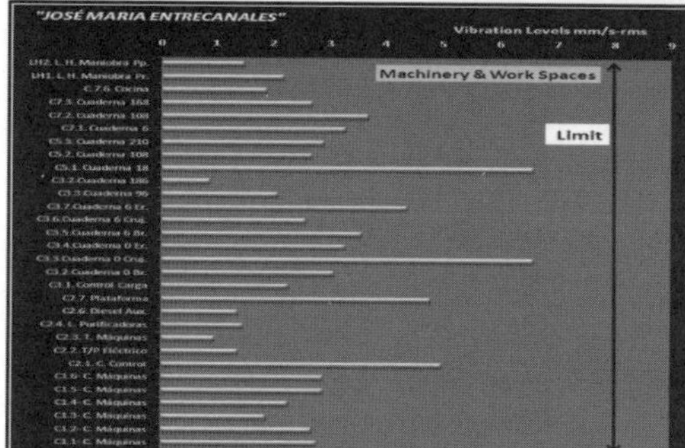

Figure 13. Vibration Levels obtained at different locations.

- In both vessels, vibration levels obtained in the different localization are well below the required limits in the *ISO Standard 6454/2000* (9).
- In both vessels the IMPROVEMENTS (deviation respect the limits) obtained oscillate between 62% and 71% below the contractual limits.
- In a general way, vibration levels obtained in both vessels during the sea trials allow us to confirm that the "Dynamic design" in both constructions delivered by NAVANTIA-Puerto Real, designed by SENER and calculated by the authors, fulfilled satisfactorily the Specification requirements (ISO 6954/2000 (9)) even being able to opt for a Comfort Class 2 according to BV Limits.

6.2 *Noise Results*

In the same way Figure 14 shows a summary of noise levels in different areas: machines and working spaces and crew & passenger cabins, of the Ro-Ro's vessels "José María Entrecanales", measured during the official tests when delivery.

From the experimental data examination, of noise levels obtained in the sea trials of the vessels "José María Entrecanales" and "Super-Fast Baleares", we can also highlight the following summary:

- In both vessels, noise levels obtained in the different localizations are well below the required limits in the IMO A.468 (XII) Regulation (8).
- In both vessels the IMPROVEMENTS (deviation respect the limits) obtained oscillate

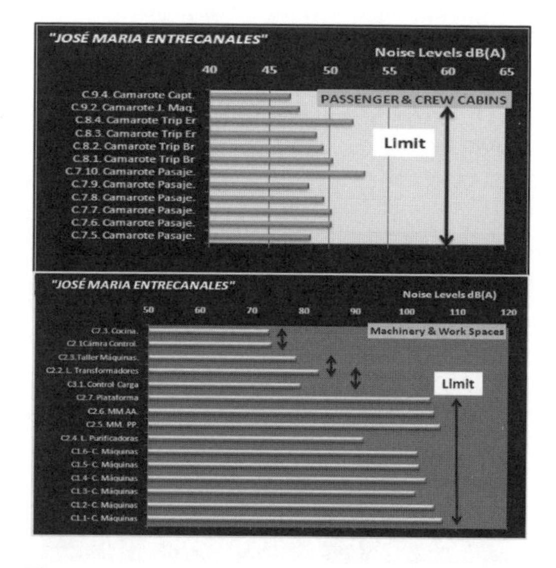

Figure 14. Noise Levels obtained at different locations.

between –6 dB (A) up to –11 dB (A) below contractual limits.

- In a general way, noise levels obtained in both vessels during the official tests allow us to confirm that the "Acoustic design" in both constructions delivered by NAVANTIA-Puerto Real, designed by SENER and calculated by the authors, fulfilled satisfactorily the Specification requirements (IMO A 468(XII) (9)) even being able to opt for a Comfort Class 2 or 3, in the cases that the insulation rate between cabins and public spaces accomplish the minimum requirements of the Classification Societies.

6.3 *Noise Radiated to the Harbour Results*

After the results of the Tests in relation to the measures of Noise Radiated to Harbour by the vessels "José María Entrecanales" and "Super-Fast Baleares" (Fig. 19), the following remarks are highlighted:

- In both vessels the Noise Radiated to Harbour level at 25 m for each side ARE WELL BELOW the 80 dB (A) limits required in the Specification.
- Also the IMPROVEMENTS (deviation respect the limits) obtained oscillate between –8.5 dB (A) and –15.2 dB (A) for starboard side and between –11.5 dB (A) and –15.6 dB (A) for port side, **ARE WELL BELOW THE CONTRACTUAL LIMITS**.

Noise and Vibration Predictions can be considered "solid" tools due to several publications about "Model-Test Correlation" that are endorsed by the experience of the authors in the Silent Vessels and the Oceanographic vessels, in which they have participated (3, 4, 5, 6). With the new requirement of Noise Radiated to Harbour by vessels, it is necessary that the applied "calculation tools" used must be contrasted with the experimental results obtained. Figure 15 shows the results of this "Correlation: Exterior Noise Radiated Model/Tests" for C-509 at 1m from the side.

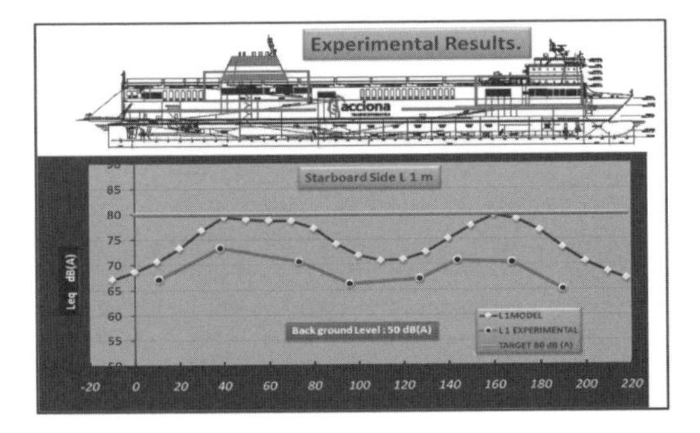

Figure 15. "Correlation: Exterior Noise Radiated Model/Tests" for C-509 at 1m from the side.

Analyzing these Figures it is possible to confirm that this is a "Solid, consistent and conservative tool to predict Noise Radiated to the Harbour by the vessel" as the expected calculation results are between +5 and +10 dB (A) over the values experimentally obtained. Security margins which could be taken as appropriate in noise prediction.

7 CONCLUSIONS: LESSONS LEARNT

Some of the most interesting Conclusions and Lessons learned, according to our judgement, are detailed in the following items:

- The recent delivery of the RO-RO vessels with: vibration levels well below (67%) the ISO 6954/2000 (9) established limits, noise levels -8 dB (A) below the recommended limits in the IMO A.468 (XII) [4], and finally, noise radiated to harbour levels within the realm of 15 dB (A) below the 80 dB (A) limit required not only in the Specification but also in the EU Directives, in force, such as 2002/49/EC (10), achieving an unprecedented fact and becoming an additional technological milestone in the Marine Spanish Sector.
- The quality of the obtained final product allows NAVANTIA-Puerto Real to reaffirm its position in the selected "market niche" of "Silent Vessels", and also, to have a "technically supported reference", in national and international markets where, currently, there is a high demand of these types of vessels.
- This fact that, some could consider "a problem", the author believes that if we are clever enough it could become a "new business opportunity" in our Marine Sector. The authors, from their experience with his participation in the **R&D European projects (SILENV/BESST)** more avant-garde in these aspects, thinks that more would be the advantages than the disadvantages when applying an "efficient technical answer" to the new Regulatory Framework. All of the European shipyards (our neighbours) that participated in these projects understood it in this way and their movements are going in this direction.
- For those who may feel "frightened" in view of this new Regulatory Framework, that is already included in the Contractual Specification of the vessels, or for those who prefer "looking at the other side" facing this new unquestionable reality, we would like to tell them that the answer to the question: Is the National Marine Sector ready to respond to these new and strict requirements? Is definitely **YES**, it is ready.

- In few words, "we know how to do it". We only need to "know how to sell it". It is a hard task but based on our extended experience as a "survival SME" we would like to provide some ideas:
- Ship-owners, nationals and foreigners, must know and understand our skills to help them with guarantees in the fulfilling of the new Regulatory Framework that they are forced too. For this: 1) We must "make noise", this means, our achievements in the sector should appear in foreign publications. 2) Our commercial colleagues must, besides knowing the new Directives and their technical and economic implications in the vessel price know and have a value enhancement strategy.
- We must know that our guarantees on fulfilling depend on the appropriate tools that we use: Noise and Vibration Integrated Management, "contractual commitment" from the suppliers with the shipyard and ship-owner's objectives, data verification provided by reception test in the factory acceptance test (we should not forget that the "absence or inaccuracy" will be paid by the shipyard), we use insulation standardize material (pay now or pay later) and its correct assembly, etc. All this doubtedly "add" and must be considered when making the shipyard understand.
- Finally, these new Directives are seen by some with fear, due to the implications when coming into force. In view of this scene, let's "have a look at" other sectors and see what they have done. In particular, Automobile Sector, in view of the Community Regulations and Directives of emission's reduction, which scared some, what did they do?: Acting as a "lobby", they decided to invest in all the technical improvements to meet the targets, they knew how to make a good "business opportunity" out of what others considered "a problem": "Prever plan" (financed by EU) and the complete renovation, in ten years, of all car fleet in all the European countries. In conclusion **"BUSINESS OPPORTUNITY".**

REFERENCES

ANSI/ASA S12.64-2009/Part 1: Quantities and Procedures for Description and Measurement of Underwater Sound from Ship—Part 1: General Requirements. (2).

Beltrán, P. 2001. Noise vibration on ship prediction: Basic engineering tool to fulfill the current requirements of comfort and quality. *Naval Engineering.* (3).

Beltrán, P. et al. 2006. Silent Fishing Vessels. A milestone for small and big Spanish shipyards. Lessons learnt. *Separata Naval Engineering*. (4).

Beltrán, P. 2007. Oceanographic ship Miguel Oliver. *Naval Engineering.* (5).

Beltrán P. 2008. Oceanographic Ship "Miguel Oliver": The excellence on Noise and Vibration on board fulfilling ICES-Nª209. *First Prize 47th Congress of Naval Engineering and Maritime Industry.* (6).

Beltrán, P. & Tesorero, M.A. Spanish Shipbuilding ready to meet the new and demanding Environmental Requirements: New Challenges and Business Opportunities *Ingeniería Naval* 887. Silver Price 49th Congress of Naval Engineering and Maritime Industry (1).

Directive 2002/49/EC of the European Parliament and of the Council of 25 June 2002 relating to the assessment and management of environmental noise. (10).

IMO Resolution A.468 (XII): Code on Noise Levels on Board Ships. IMO 1.975. (8).

Enciso, C. 2007. About noise and vibration transmission in the MAPA 70 of M.Cies shipyard. *Naval Engineering*. (7).

ISO 2922/2000 Acoustics—Measurement of airborne sound emitted by vessels on inland waterways and harbours. (10).

ISO 6954:2000: Mechanical vibration—Guidelines for the measurement, reporting and evaluation of vibration with regard to habitability on passenger and merchant ship. (9).

9 *Sea resources exploitation*

9.1 *Fishing*

Sustainable Maritime Transportation and Exploitation of Sea Resources – Rizzuto & Guedes Soares (eds)

Carbon footprint and energy use of Norwegian fisheries and seafood products

E.S. Hognes, U. Winther & H. Ellingsen
SINTEF Fisheries and aquaculture, Trondheim, Norway

F. Ziegler, A. Emanuelsson & V. Sund
SIK—Swedish Institute for Food and Biotechnology, Gothenburg, Sweden

ABSTRACT: Fuel consumption and emission of cooling agents are important sources for climate impact from the value chain of wild caught Norwegian seafood products. For products that are exported quickly and/or over long distances transport is also an important source. Pelagic products have low carbon footprints due to energy efficient fishing, modern refrigeration systems and efficient export methods. The fuel consumption per kilo landed products varies a lot in Norwegian fisheries, both within fisheries that use the same gear or that target the same species. This variation shows that there is a high potential to reduce GHG emissions from Norwegian fisheries.

1 INTRODUCTION

The Carbon Footprint (CF) of a product is the sum of Green House Gas (GHG) emissions it causes from a defined part of its life cycle, often from cradle-to-grave.

Carbon footprint of products, including food, is gaining high attention as it is expected to be an efficient tool to promote production and distribution with less climate impact. CF analysis has a system perspective and can provide decision makers and R&D institutions with info on how and where they can reduce the climate impacts caused by their products. If and when CF of products also becomes a competitive parameter, e.g. by using it as an ecolabel, the market economy will push for improvements as those that perform best get an economically benefit by increased prices and/or markets acceptance. Already today important European food retailers request their suppliers to document their environmental- and climate impacts (SINTEF 2011).

In 2009 SINTEF Fisheries and Aquaculture together with SIK, the Swedish Institute for Food and Biotechnology, calculated the carbon footprint and energy use for 22 Norwegian seafood products (Winther 2009). The project was financed by The Fishery and Aquaculture Industry Research Fund (FHF) on request from the Norwegian's Fishermen's Association and The Norwegian Seafood Federation. The goal was to benchmark the average carbon footprint of the most important—in terms of export volume and value—Norwegian seafood products and to identify the central climate aspects in their value chains. In 2008 Norway exported 2.3 million tonne of seafood to more than 100 countries and knowledge on how the climate impact from this activity can be reduced is important.

This project included carbon footprint of both wild caught products and aquaculture products, here only results from wild caught products are represented: The pelagic species mackerel and herring and the demersal species cod, saithe and haddock.

2 METHOD

The carbon footprints was calculated with Life Cycle Assessment (LCA) methodology according to ISO 14040: Environmental management—life cycle assessment—principles and framework (ISO 2006). The scope of the assessment was from catch to delivery to wholesaler in different markets. This includes diesel consumption and emissions of cooling agents from the vessels, fuel and energy use in processing and conservation of the fish and finally fuel use, packaging and cooling agent emissions in transports. Figure 1 illustrates the system boundaries for the assessments.

The functional unit for the assessment was 1 kilo edible product and the different GHG emissions were calculated into CO_2-equivalents according to the IPCC guidelines with a 100 years perspective (IPCC 2007). All results are given as kilos of CO_2-equivalents per kilo edible product.

In the value chain of seafood several processes has multiple outputs, e.g. processing can have output of fillet and byproducts like skin and bones. When processes have multiple outputs the environmental impacts from this process and prior to this processes must be allocated between the different outputs. In this project allocation was done on a mass basis; based on their weight ratio the carbon footprint prior to processing and from processing was allocated to the main product and byproducts. Outputs that are not utilized are waste and no environmental impact are allocated to waste. Figure 2 illustrates such allocation. Data regarding by-products utilization was found in (Bekkevold and Olafsen 2007): Utilization of by products from processing of demersal products was 39% and for pelagic products 95%.

The data for teh life cycle inventory of teh value chains was mainly gathered from official statistics, but sometimes complemented by interviews with commercial actors in the value chain. Processes that underpin the value chain of the product with inputs and infrastructure was modeled with the life cycle inventory database EcoInvent 2.0 (EcoInvent 2009).

2.1 *Calculation of gear and species specific fuel factors*

The species specific fuel factor (liter diesel combusted to land one kilo of product in round weight) was calculated by combining data on fuel consumption from an annual profitability survey on the Norwegian fishing fleet and sales statistics, both statistics was provided by the Norwegian Directorate of Fisheries (personal communication). The profitability survey provided data on the fuel consumption with different fishing gears and the sales statistics how the different species was caught.

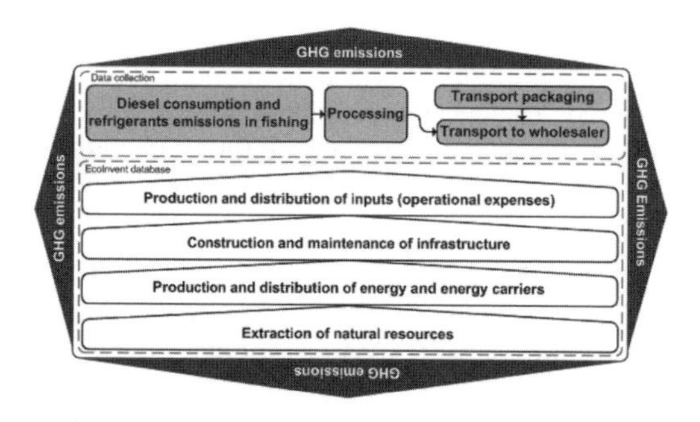

Figure 1. System boundaries for the assessment.

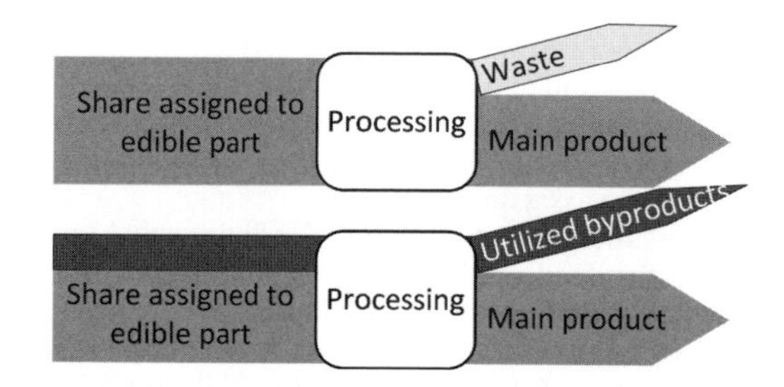

Figure 2. Conceptual figure of allocation effects.

The annual profitability survey covers a statistical representative selection to analyze the economic profitability of Norwegian fisheries. Table 1 presents the proportion of the total landings it covered in 2007. The survey covers almost half of all fisheries. It cover coastal fisheries less than ocean fisheries and that is the main reason that cod is covered less than the other demersal species.

Equation 1 shows how the gear specific fuel factors (FS_j) where calculated. The total fuel used by each boat (D_i) was allocated to the different fishing gears it used (FD_{ij}). The allocation was based on the ratio between the boats landing with each gear type (f_{ij}) and the sum of all its landings (F_i). Finally the gear specific fuel factor was calculated by dividing the sum of fuel allocated to each gear by the sum of landings with that gear.

$$FS_j = \frac{\sum_i^n FD_{ij}}{\sum_i^n f_{ij}} = \frac{\sum_i^n \frac{f_{ij} D_i}{F_i}}{\sum_{i(1)}^n f_{ij}} \qquad (1)$$

Explanation of terms in Equation 1:

- FS_j: Fuel factor for gear j [l/kg]
- FD_{ij}: Fuel allocated to gear j on boat i [l]
- f_{ij}: Landings with gear j on boat i [kg]
- D_i: Total (annual) fuel consumption by boat i [l]
- F_i: Sum of all landings by boat i [kg]
- n: number of boats in profitability survey after data corrections.

Table 1. Proportion of total landings in 2007 covered by the profitability survey.

Species	Proportion %
Herring	71
Mackerel	69
Cod	45
Saithe	66
Haddock	59

Data from 458 vessels was used and the calculations resulted in the gear specific fuel factors in Table 3. The species specific fuel factors presented in Table 4 was calculated by combining statistics on how each species was caught (Table 2) with the gear specific fuel factors (Table 3).

Table 2. Distribution of Norwegian catches in 2007 (from sales statistics provided by Norwegian Directorate of Fisheries).

	Distribution %					
Species	PS	PT	BT	DS	AL	CG
Cod	0	3	29	17	9	42
Haddock	0	0	41	15	22	22
Saithe	16	1	52	5	1	25
Herring	88	12	0	0	0	0
Mackerel	88	3	0	0	0	9

PS = Purse seine, PT = Pelagic trawl, BT = Bottom trawl, DS = Danish seine, AL = Auto-line, CG = Other coastal gears (gillnets, coastal line, jig and other).

Table 3. Gear specific fuel factors from profitability survey in Gear specific fuel factors from the 2007 profitability survey.

	Average	
Fishing gear	Liter fuel/kg catch	Standard deviation
Long-line (autoliners)	0,31	0,12
Bottom trawl	0,43	0,24
Double trawl	0,94	0,31
Pelagic trawl	0,10	0,12
Pelagic pair trawl	0,09	0,02
Hand line/ jig	0,15	0,19
Gillnet	0,15	0,18
Purse seine	0,089	0,03
Danish seine	0,12	0,20

Table 4. Species spesific fuel factors.

	Fuel factors	
	Liter fuel/kg landed round weight	Standard deviation
Cod	0.24	0.10
Haddock	0.29	0.11
Saithe	0.29	0.13
Herring	0.09	0.03
Mackerel	0.09	0.03

2.2 *Cooling agents emission rate*

Cooling agents used in the Norwegian fishing fleet include R22, ammonia and CO_2. The most important in terms of climate impact is R22 (HCFC-22) that has a global warming potential that is 1810 times the effect of CO_2 (IPCC 2007).

R22 is regulated by the Montreal protocol (UNEP 2006) due to its high ozone depletion potential. Since 2002 installation of new R22 systems has been banned and import of R22 was only allowed for refilling of existing systems. From 2010 refilling of R22 was only possible with regenerated R22 from systems that are no longer in use (MD 2004). In 2007 the total import of R22 was 323 tonne, from interviews of the major importers it was concluded that 200 of these tonne went to fishing vessels. It was assumed that this equal the total emissions of R22 from fishing vessels in 2007. This assumption is based on the fact that new R22 systems were permitted, only refilling is allowed and stock piling is not allowed. It was also investigated if R22 from fishing vessels was collected and delivered for secure destruction, but this amount was confirmed to be insignificant by the organization that is responsible for collection and destruction of refrigerants in Norway (Stiftelsen Returgass).

The cooling agent emission rates were calculated in two steps; first a rate was calculated for pelagic fisheries and then what was left was assumed to be emitted by the demersal fleet.

Pelagic vessels mainly use cooling agents in their Refrigerated Sea Water (RSW) systems. In a perfect RSW system refrigerants are not emitted, but in practice emissions occur by leaks and during repairs and services. In the literature cooling agent emission rates, in terms of annual leak of the total volume of cooling agent in the refrigeration system, range from 20–40% (Sandbakk 1991; Klingenberg 2005; SenterNOVEM 2006). Here it was assumed an annual emission rate of 30%. Further it was assumed that 70% of the pelagic vessels still use R22 and that pelagic vessels above 28 m have 1200 kg R22 per RSW system, and that vessels under 28 m have 600 kg R22. These assumptions combined with how much was landed by the pelagic fleet led to an emission rate of 0.023 g R22 per kilo landed fish in round weight.

Emission of R22 from the demersal fleet was calculated by subtracting the amount of R22 emitted by the pelagic fleet from the total mass of R22 emitted by Norwegian fisheries (200 tonne). The remaining amount was divided by the total Norwegian landings in 2007 minus pelagic species (722.148 tonne). This calculation resulted in an emission rate of 0.224 g R22 per kilo round weight for the demersal fleet.

3 RESULTS AND DISCUSSION

Fuel consumption and emission of cooling agents from refrigeration systems are in general the most important climate aspects for Norwegian wild caught demersal- and pelagic products. For products that are exported over long distances and/or fast, transport can be the most important climate aspect.

Pelagic products have the lowest carbon footprints due to energy efficient fishing, refrigeration systems with low emissions and efficient export methods. Figure 3 presents the total carbon footprint for Norwegian wild caught seafood products to different export markets—and on different transport means.

The calculation of fuel consumption in different fisheries resulted in the gear specific fuel factors presented in Table 3 and the species specific fuel factors presented in Table 4.

Among the demersal species cod is caught with the least climate impact. The main reason for that is that a big part of the annual catch of cod is done by coastal fisheries that use passive fishing gears such as gillnet close to shore, while more of the haddock and saithe is caught with trawlers and autoliners further away from land.

The last columns in Table 3 and Table 4 show the high variation in the data behind the average fuel factors. One important source for this is the way they are calculated; some gear will have very low minimum values when they have been used in combination with very energy efficient gear. Even though this methodic source is contributing to the variation the main reason is the fact that there is high variation in energy us in Norwegian fisheries.

Figure 4 shows a plot of the annual fuel use and annual catch by Norwegian factory trawlers from 2001 to 2007, the variation illustrated here is not influenced by calculations since it is only based on the catch and fuel consumption reported by the vessels. These are vessels that use the same type of gear, that target the same species in the same areas and that operate within identical regulatory and environmental frames.

Fuel use in fisheries is a complex function with many variables; the type of fishing gear used and the behavior of individual vessels is only a part of the equation. The fishing sector is a highly regulated and politicized sector (Hersoug 2005) and energy use in fisheries is, among many factors, determined by the framework set by fisheries management systems (Standal 2005; Driscoll and Tyedmers 2009) with components such as:

- Total available quotas and quota allocation policy
- Structural policies to cut down unprofitable overcapacity
- Technical regulations and spatial and temporal limitations of fisheries, these also includes demands on when and where landings can be delivered
- Auctioning systems for pelagic species that influence economically feasible travelling distances and subsidizing of travelling expenses
- Geographical aspects of where and when specific fisheries are open
- Regulations connected to gear adaptations and rules for minimum size to avoid catches of juvenile fish.

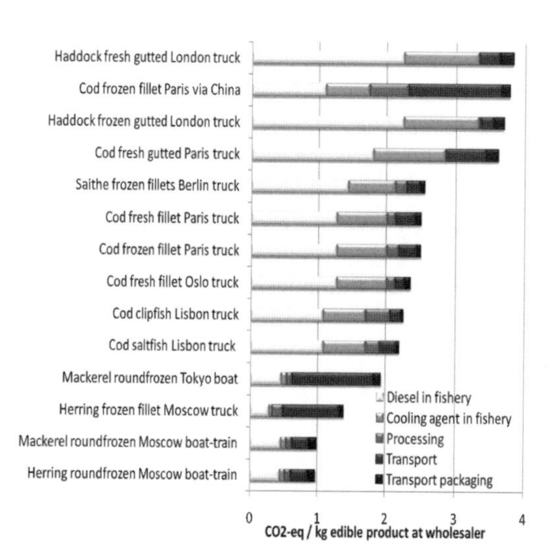

Figure 3. Carbon footprint of Norwegian wild caught seafood.

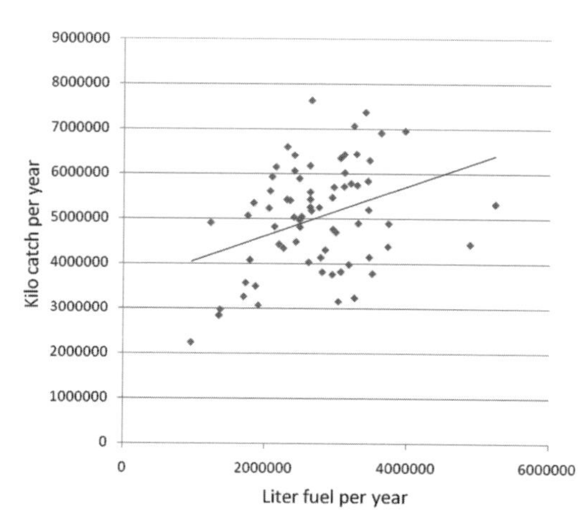

Figure 4. Annual catch and fuel consumption of factory trawlers. Data from profitability surveys, 2001 to 2007.

In addition to these regulatory factors nature also set frames, e.g. stocks are reduced or increased due to dynamics in the food web and stocks change location and time for where they are available for catch.

This list emphasis how important it is to approach energy efficiency improvements in fisheries with a system perspective; technologies needs to be improved, the fishermen need to use the available technology better and the framework this technology is developed and invested in needs to be focused on achieving climate and environmental improvements.

The fuel factors calculated here was compared with data from some of Norway's biggest fishing vessel owners and scientific reports and articles e.g.: (Tyedmers 2001; Eyjólfsdóttir, Yngvadóttir et al. 2003; Thrane 2004; Tyedmers, Watson et al. 2004). Some of these results are represented in Table 5. The calculated fuel factors for the demersal fisheries are lower than the values reported in previous studies. Some reasons are: Higher fuel prices leading to a more fuel-saving behavior, improved stock status with higher catches per effort or higher quotas and regulatory schemes to decrease unprofitable over capacity in the Norwegian fishing fleet. From 2002 to 2007 the number of Norwegian fishing vessels decreased from 2206 to 1709. During the same period the total landings of demersal species increased by 6% and landings of pelagic species decreased by 20%, the decrease in pelagic fisheries was mainly due to lower landings of blue whiting, which is an energy intensive fishery, but the landings of Norwegian spring spawning herring increased. To conclude; fewer boats catch with higher quotas has increased the energy efficiency of Norwegian fisheries.

3.1 *Climate impact from cooling agents emission*

R22 emissions play an important role in the over all carbon footprint for all the products, see Figure 3. The total emissions of R22 from Norwegian fisheries in 2007; 200 tonne, is based on a thorough investigation with interviews of all major importers. Thus the fact that R22 emissions is a major climate aspect is a safe conclusion, but how these emissions are distributed on to the different products are based on somewhat weaker assumptions. It is fair to assume that the emission rate from the pelagic fleet is less than from demersal vessels: RSW systems emit less than freezing and cooling systems used in the demersal fleet, the pelagic fleet is more modern and pelagic vessels have larger catches per vessel compared to the demersal fleet.

Table 5. Gear specific fuel factors from literature and vessel owners.

	Liter fuel/kg round weight	
Fishing gear	Average value	Data range (min–max)
Bottom trawlers	0.63	0.33–1.0
Purse seiners	0.077	0.036–0.11
Long liners	0.31	0.18–0.49

3.2 *Climate impact from processing and exports*

Seafood processing influence on the carbon footprint through the yield (kilo edible product per kilo round fish), how the byproducts are used and how much energy that is used in the processing.

Figure 5 shows the carbon footprint of three different cod products at the French market (Paris).

Because of different levels of yield in filleting and by product utilization the climate impact from fuel consumption and refrigerants emissions in the fishing is different in all three cases. Filleting in China is done by hand and have higher yield than filleting in Norway. Thus it is used less fuel and less refrigerants per kilo edible product produces. When the cod is sent head and gutted to shops in Paris, and filleted by the consumer, it was assumed that byproducts are not used at all and the edible part has to carry all the climate impact prior to and from filleting.

Figure 5 also shows a case where transport is the single most important climate aspect in the value chain of the product. Sending Norwegian cod for filleting in China (frozen on containership) and then returning it to Europe for sale in Paris has more than 50% higher climate impact than filleting the product in Norway and sending directly to Paris by truck.

Processing cod in China for sale in Europe is an example where long transports causes considerable

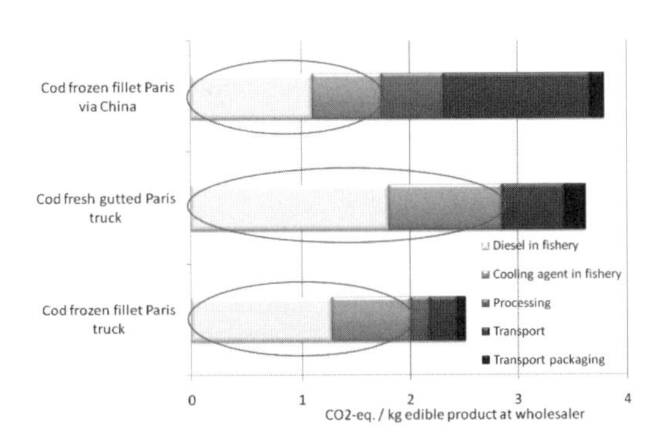

Figure 5. Carbon footprint of cod delivered to Paris.

climate impact, when long transports are also done fast the climate impact is even more considerable.

In exports of fresh salmon from Norway to Tokyo, by airfreight, transport alone causes more than 10 kg CO_2-eq. per kilo edible product. When this fact is held up against the amount of intercontinental export of salmon by airfreight it shows that even small improvements can give substantial reduction in GHG emission caused by Norwegian seafood exports. These improvements can be done by moving transport from road to rail and from air to sea.

4 CONCLUSION

The carbon footprint of wild caught Norwegian seafood consists of many important climate aspects. To reduce the climate impact from Norwegian seafood products more energy efficient fisheries with renewed refrigeration technology is an obvious focus area, but also export method is important;

- Process the products before export; to facilitate the utilization of byproducts and to use as much as possible of the transport capacity to transport high value edible products rather than lower value byproducts and/or waste
- Focus on products that allows for slower transport mean and keep the "food miles" down by direct export too final consumer.

The high level of variation in the fuel consumption of Norwegian fisheries show that there is a potential for massive reductions of GHG emissions if the vessels that today have the highest fuel consumption can lower their expenses to the level of those that use the least. This can be done by improving how the fisheries are performed (fishermen's behavior and their technology), but just as important by making sure that the regulatory frames they work within promotes energy efficiency improvements. By insuring that the carbon footprint of seafood products becomes a competitive parameter in the seafood markets it can promote these improvements; the link between reducing environmental impacts and economic profitability will become strong and direct.

REFERENCES

Bekkevold, S. and T. Olafsen (2007). Marine biprodukter: Råvarer med muligheter http://www.rubin.no/files/documents/rubin_lavoppl._m.omslag.pdf. RUBIN.

Driscoll, J. and P. Tyedmers (2009). "Fuel use and greenhouse gas emissions implications of fisheries management: The case of the New England Atlantic herring fishery." In press Marine Policy.

Ecoinvent (2009). The ecoinvent Centre: Ecoinvent Life Cycle Inventory database, www.ecoinvent.org/home.

Eyjólfsdóttir, H.R., Yngvadóttir, E. et al. (2003). Environmental effects of fish on the consumer´s dish-Life cycle assessment of Icelandic frozen cod products, Research Counsil of Iceland, Ministry of Fisheries in Iceland, Icelandic Fisheries Laboratories and Technological Institute of Iceland.

Hersoug, B. (2005). Closing the commons Norwegian fisheries from open access to private property. Delft, Ebouron academic pubishers.

IPCC (2007). Intergovernmental Panel on Climate Change, Fourth Assessment Report: The Physical Science Basis.

ISO (2006). ISO 14040 Environmental management—life cycle assessment - principles and framework. ISO 14040:2006(E). International Organization for Standardization. Geneva. Switzerland.

Klingenberg, A. (2005). "Enforcement of chlorofluorocarbon regulations on maritime vessels." Proceedings from 7th International Conference on Environmental Compliance and Enforcement, Marrakech, Marocko; www.inece.org/conference/7/vol1/40_Klingenberg.pdf.

MD, M. (2004). "FOR 2004-06-01 nr 922: Forskrift om begrensning i bruk av helse- og miljøfarlige kjemikalier og andre produkter (produktforskriften) http://www.lovdata.no/cgi-wift/ldles?doc=/sf/sf/sf-20040601-0922.html."

Sandbakk, M. (1991). Utslipp av klimagasser fra fiskerinæringen, SINTEF Kuldeteknikk.

Senter NOVEM. (2006). "Environmentally friendly cooling systems for fishing vessels (broschure about pilot projects)" Retrieved November 30, 2009.

SINTEF, F.A. (2011). http://www.tracefood.org/index.php/International:Workshop190110. The future Environmental labelling of seafood and Traceability of Ecolabelled Seafood, Copenhagen, SINTEF Fisheries and aquaculture.

Standal, D. (2005). "Nuts and bolts in fisheries management—a technological approach to sustainable fisheries?" Marine Policy 29(3): 255–263.

Thrane, M. (2004). Environmental impacts from Danish fish products (Ph.D.). Institut for Samfundsudvikling og Planlægning. Aalborg, Danmark, Aalborg Universitet.

Tyedmers, P. (2001). Energy Consumed by North Atlantic Fisheries. School for Resource and Environmental Studies, Dalhousie University, 1312 Robie Street, Halifax, NS, B3H, 3E2, Canada.

Tyedmers, P., R. Watson, et al. (2004). "Fueling global fishing fleets." AMBIO: A Journal of the Human Environment 34:635.

UNEP. (2006). Montreal protocol on substances that deplete the ozone layer.

Winther, U., Ellingsen, H., Ziegler, F., Skontorp Hognes, E., Emanuelsson, A., Sund, V. (2009). Project report: Carbon footprint and energy use of Norwegian seafood products, SINTEF Fisheries and aquaculture.

Sustainable Maritime Transportation and Exploitation of Sea Resources – Rizzuto & Guedes Soares (eds)
 ISBN 978-0-415-62081-9

Fuel efficiency in trawlers under different fishing tactics using a consumption model and VMS data: A case-study for the Portuguese fleet

A. Campos, T. Pilar-Fonseca, J. Parente & P. Fonseca
INRB, IP/IPIMAR, Lisboa, Portugal

M. Afonso-Dias
Centro de Investigação Marinha e Ambiental, Universidade do Algarve, Faro, Portugal

ABSTRACT: Trawling is the most fuel-intensive fishing activity, with trawlers constituting the fleet segment most affected by the high volatility of the fuel prices. This is likely to be a main driver for adopting fuel consumption reduction strategies, which may be related to alternative fishing practices. In this study, individual vessel trajectories obtained from Vessel Monitoring System (VMS) data processed by GIS software (GeoCrust 2.0) are used to characterize a number of operational and economical parameters, for different landing profiles, corresponding to fishing trip types with specific landings composition. This information is combined with fuel consumption estimates derived from a mathematical model for the distinct phases of a round fishing trip, providing an insight into the profitability of fishing operations.

Keywords: trawl fleet, Vessel Monitoring Systems (VMS), fuel consumption, economic efficiency, Portuguese waters

1 INTRODUCTION

The significant amount of time spent during trawling and navigating among fishing grounds, place bottom trawlers among the most fuel-demanding coastal fishing vessels. Energy saving in trawlers has been a subject of research since the 1970's oil crisis, leading to several studies aimed at improving vessel design and power consumption. As an example, special attention has been given to hull resistance (Kasper, 1983) and gains in propulsive efficiency (Basañez, 1975, O'Dogherty *et al.*, 1981, Haimov *et al.*, 2010). Parente *et al.* (2008) addressed the economy of energy in Portuguese coastal trawlers by looking at trawl gear performance and the operational procedures along the different phases of the round trip. Potential increases in the Net Cash Flow (NCF) of up to 27% over the range of operational navigation and trawling speeds were achieved through gear modifications (Parente *et al.*, 2008)

Information on vessel activity (including georeferenced data) on a trip level is of utmost importance for understanding fleet strategies in fishing effort allocation and contributing to the design of efficient energy management solutions. This is now possible due to the existence of high spatial resolution data provided by the satellite-based Vessel Monitoring Systems (VMS), which collect observations on single vessels operation and may provide detailed effort on a highly disaggregated spatial scale. The data can then be related to landings revenues and running costs (i.e., the costs depending on vessel specific effort allocation in time and space, such as fuel costs), providing information on vessel profitability. Taking this information into account, fuel efficiency may be tackled, for instance by adopting the best running point, that is, the vessel's operating speed, both in trawling and in free navigation, that maximizes net income.

The Portuguese VMS (MONICAP), is presently operating in vessels with Length Overall (LOA) higher than 15 m (EC, 2003), including all coastal trawlers, fishing in Western Iberian waters, ICES division IXa. Global Positioning System (GPS) data for each vessel, consisting of a succession of geographical locations (latitude, longitude), dates, times and speed, are received on board, recorded and automatically transmitted via satellite (Loran-C) to the Portuguese Fisheries Directorate (DGPA/IGP), the national monitoring authority responsible for marine surveillance. The frequency of data transmission is at present every two hours, according to EU regulations (EC, 2003). However, until 2004, records were obtained every 10 minutes; the high discrimination of such geo-referenced data explains their usefulness as support to fisheries research. For each vessel, a series of trajectories can be produced by mapping the geo-referenced information, each defining a Fishing Trip (FT). These trajectories tend to form patterns consisting in a succession of points, closer or more distant according to the slower or faster vessel speed. In addition to providing geographical location of the fishing activity, such patterns can also identify the main operational phases of a trip.

At present, a total of around 100 vessels are active in the Portuguese coastal bottom trawling, corresponding to 15000 GT (gross tonnage) and 40000 of engine power (kW). This fleet accounted for approximately 12 and 18% of the total national landings in continental waters, in weight and value, respectively (DGPA, 2010). Fishing pressure is high and the fishing activity taking place all year round and exploiting a large number of fish, cephalopods and crustaceans species.

Previous studies on trawl fleet segmentation (Campos *et al.*, 2007; Pilar-Fonseca *et al.*, 2008, 2009), identified different Landings Profiles (LP), characterized by a species or association of species.

In this paper, VMS processed data, corresponding to a 12-month period of activity (2003), are examined to estimate the duration of the main phases of individual round trips of a coastal trawler, involved in the capture of a diversified amount of species under distinct operational scenarios. This information is combined with data on vessel technical characteristics, used as an input to a mathematical model (Parente, 2009) simulating the fuel consumption associated to each trip phase, to estimate consumptions within trip/landing profile (LP). Several operational and economic indicators (namely fuel efficiency) are calculated in order to characterize each phase of the fishing trip, allowing for comparison among the main LP.

2 METHODS

The sample vessel is a coastal fish trawler 24 m LOA and 600 HP, involved in the capture of horse mackerel, cephalopods and demersal fish species and operating under 3 distinct LP. It presented a regular activity along the entire year, as well as a good coverage in terms of VMS data.

The geo-referenced information was processed using GeoCrust 2.0 (Afonso-Dias *et al.*, 2004; 2006), a dedicated geographical information system (GIS) designed to analyze the spatial dynamics of the trawl fleet by a semi-automatic procedure. High-quality fishing trips, with well individualized tows were selected for this study, representing about 80% of the total number of trips in 2003, identified from 10-minute interval VMS records. The GIS allows for the identification of individual FT and the corresponding main operational phases: navigation from port to the fishing ground; trawling (fishing); navigation between fishing grounds; and navigation to the landing port. For the purpose of this study, only two categories were considered: trawling and non-trawling, for both of which duration (in hours) and average speed (in knots) were estimated. Trawling duration corresponds to the effective fishing effort. This information is used as the input of a mathematical model estimating fuel consumption developed for trawlers. The model, described in Parente, (2009) estimates the engine performance (total fuel consumption, hourly consumption, engine power output and emissions expelled into the atmosphere) for each operational phase of the FT.

The data above were used to calculate a number of statistics and to derive different indices characterizing the two phases of the FT under each LP, in terms of their operational behaviour, energy efficiency and economic results. In other words, average trip duration, number of hauls, trawling speed and duration, percentage of trawling in the overall trip, landings—in weight and value—per trip, target species landings per hour trawled, hourly fuel consumption per trip phase, fuel consumption during trawling over total fuel consumption, landings—in catch and value—per litre consumed, landed value per hour trawled, and landed value over fuel costs were estimated.

3 RESULTS

During 2003, this vessel distributed its activity mainly over three LPs (of the six different LP derived by Pilar-Fonseca et al, 2009 for the finfish bottom trawl fishery), one of them targeting exclusively horse mackerel (LP5), a second one where horse mackerel is targeted along with demersal species (LP3), and another targeting only demersal species, mainly octopuses and pouting (LP1).

This trawler operated over approximately 200 nautical miles along the north-western coast, at a depth interval between 30 and 200 m. Geographical location of fishing activity was not found to differ much between LP (Figure 1). However, most of the trips targeting demersal species started in

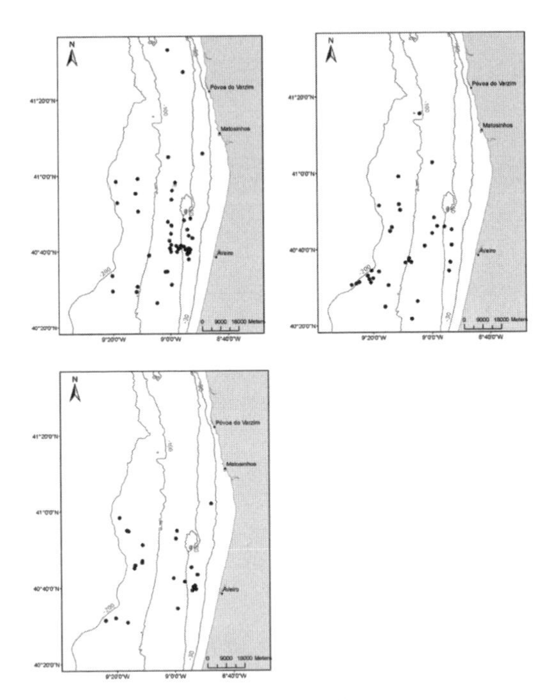

Figure 1. Geographical distribution of vessel activity by LP (LP1, LP3 and LP5) during 2003. Each point corresponds to the first VMS trawling position for each trip. VMS information processed by GeoCrust 2.0. Mapped using ArcGis software.

near-shore fishing grounds, around the 50 m isobath. Landings occurred in the ports of Matosinhos, Figueira da Foz and Aveiro, with the latter accounting for 92% of the landing events.

Figure 2 displays shifts in fishing tactics (sequence in landings for VMS-based trips), showing that the vessel alternated mainly between 3 LP throughout the year. There is a clear dominance of LP1 (targeting octopuses), representing about 41% of the total LP, contrasting with 20 and 27% for LP5 and 3, respectively.

Differences in trip duration between LP are evident, with longer trips (average 28.1 hours at sea) when benthic species are targeted (LP1), reducing to 22.4 hr in horse mackerel directed fishing (LP5), and further still for LP3 (15.7 hr), with a mixed capture of horse mackerel and benthic species (Table 1). The different average fishing trip duration by LP corresponds to differences in average number of hauls carried out (8.1, 4.9 and 6.5, for LP1, LP3 and LP5, respectively) and percentage of time spent trawling (65% for LP1 and about 50% in the remaining ones). The range of average trawling speeds is of small magnitude (less than 0.5 kt), being slightly smaller for the benthic species than for pelagic (3.8 vs. 4.1 kt). Non-trawling speeds provide higher contrasts between LP; however, this parameter cannot be considered as an operational one, since it does not correspond to an individualized trip phase, but rather to a mixture of phases, including navigation (from and to landing port, between fishing grounds), setting and haul-back.

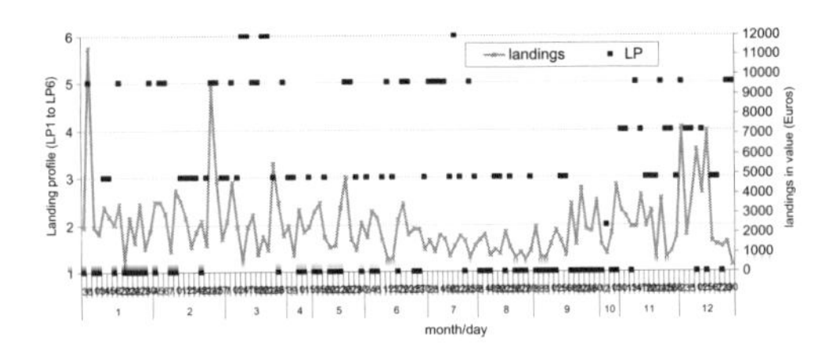

Figure 2. LP (LP1 to LP6) and landings in weight (€) by fishing trip (month/day, x-axis) for 2003.

Fuel consumption was highly correlated with trip duration ($r^2 = 0.97$). Although average estimates of fuel consumption by LP varied between 927 and 1852 litres (for LP3 and LP1, respectively), hourly consumption during trawling was found to be a constant value (85 l/hour trawled). During the remaining trip phases (here illustrated as non-trawling) consumption was, as expected, significantly lower, with values ranging between 28.5 and 32.3 l/hr (Table 1 and Figure 3).

Notwithstanding the unobserved differences among LP in hourly fuel consumption during trawling, fuel efficiency (landings per fuel consumed) varied, both in quantities and in value (Figures 4 and 5). For trips targeting demersal species, values observed for total weight landed for each litre of fuel consumed were about 0.6 kg, increasing up to 1.4 kg/l for LP directed almost exclusively at the horse mackerel. In value, the trend is similar, varying between € 1.5 and € 2.7 per litre. Translating these results into an overall economic index—the average ratio between landed value and fuel costs (assuming an average fuel price of 0.3 €/l in 2003)—figures ranged from between 4.8, for LP1, to 7.7 and 8.8, for LP3 and LP5, respectively. This clearly indicates that FT targeting horse mackerel have a higher profitability when compared to those targeting demersal species. Fuel costs represent 11%–(1/8.8)*100%—of the sales value for LP5 compared to 21%–(1/4.8)*100%—in LP1.

The ratio between landed values and fuel costs is presented in Figure 6. Higher trip profitability was generally observed in the first semester, which is related to trips targeting horse mackerel. However, this increase in profitability is associated to higher variability, which is possibly related to the fact that this period is characterized

Table 1. Average trip estimates for effort, LPUE, fuel consumption and economic indicators by LP for the vessel in study. High-quality data processed by GeoCrust 2.0. LP named according to Pilar-Fonseca *et al.* (2009). Standard deviations in parenthesis.

Estimates	LP		
	LP1	LP3	LP5
Percentage of FT in the LP[1]	41.4	27.3	20.3
Trip duration (h)[2]	28.1 (14.9)	15.7 (9.3)	22.4 (18.9)
Number of hauls	8.1 (4.5)	4.9 (2.6)	6.5 (5.1)
Trawling time (h)[2]	18.9 (10.9)	8.1 (5.5)	12.5 (12.4)
Trawling speed (kt)[2]	3.8 (0.2)	4.1 (0.2)	4.0 (0.2)
Non-trawling speed (kt)[2]	5.9 (1.1)	7.0 (1.0)	6.6 (1.3)
Ratio between trawling time and trip duration	0.65 (0.10)	0.50 (0.12)	0.53 (0.14)
Ratio between fuel consumption in trawling and total fuel consumption	0.8 (0.1)	0.7 (0.1)	0.7 (0.1)
Hourly fuel consumption during trawling (l/hour trawling)	84.5 (0.2)	84.7 (0.2)	84.7 (0.2)
Hourly fuel consumption (l/hour at sea)	65.3 (4.7)	58.9 (5.2)	60.0 (6.7)
Landed weight (kg)	885.8 (459.2)	866.2 (499.0)	1660.0 (1183.5)
Landed value (€)	2375.5 (459.2)	1849.4 (499.0)	3170.3 (1183.3)
Price (€/kg)	2.8 (0.8)	2.3 (0.7)	2.0 (0.9)
Landed weight per fuel consumed (Kg/l)	0.6 (0.3)	1.1 (0.7)	1.4 (0.7)
Landed value per fuel consumed (€/l)	1.5 (0.9)	2.3 (0.7)	2.7 (1.0)
Horse mackerel LPUE (kg/h)	6.2 (5.9)	61.8 (50.4)	130.3 (79.4)
Octopuses LPUE (kg/h)	10.9 (6.3)	5.6 (5.5)	4.7 (5.1)
Landed value per hour trawling (€/h)	148.7 (90.2)	293.0 (204.0)	313.3 (182.3)
Ratio between total revenues and fuel costs	4.8 (2.6)	7.7 (4.7)	8.8 (4.8)

[1]Percentage; [2]Input parameter for the fuel consumption model.

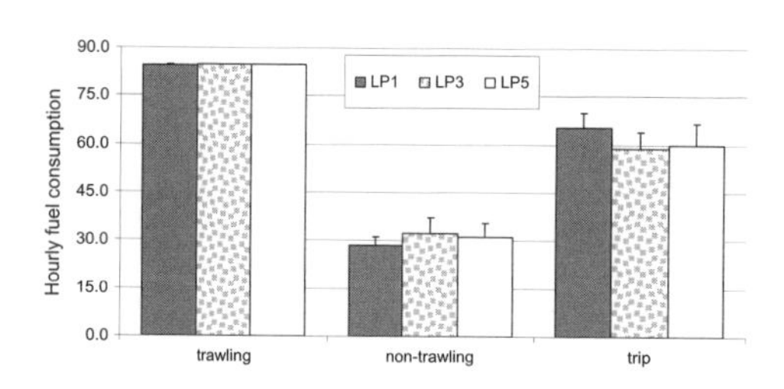

Figure 3. Fuel consumption in l/hr by operational phase (trawling and non-trawling), trip and LP.

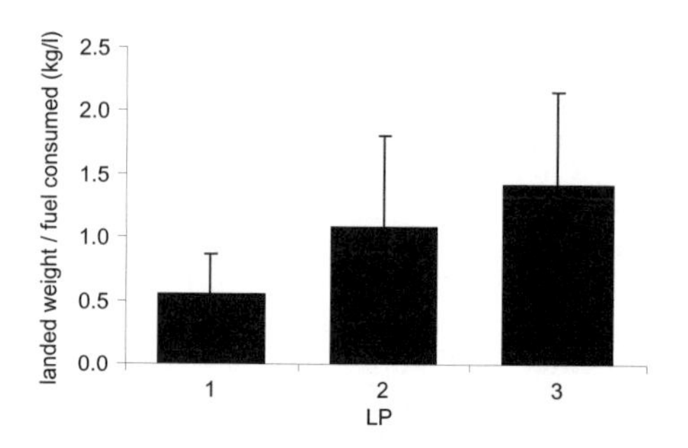

Figure 4. Fuel efficiency ratio (landings in weight per fuel consumed, kg/l) by LP, assuming an average fuel price of 0.3 €/l in 2003, for the sample trawler.

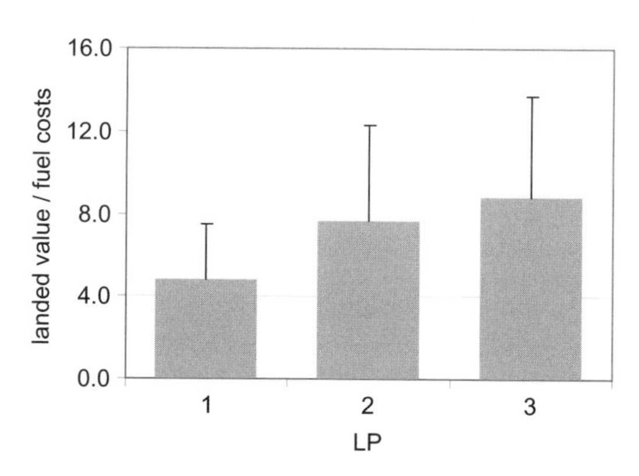

Figure 5. Ratio between landings in value and fuel cost by LP in 2003, for the sample trawler.

by daily switch between LP5, LP3 and LP1 as a result of changes in fishing tactics (Pilar-Fonseca, unpublished). From June onwards and until November, values for this ratio decrease which were related to a reduction in the number of trips targeting horse mackerel, increasing again at the end of the year.

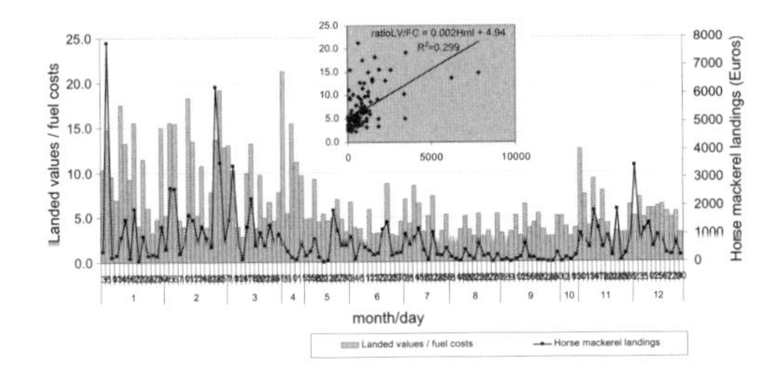

Figure 6. Ratio between Landings in Value and Fuel Cost (ratio LV/FC, y-axis) by date of fishing trip (month and day, x-axis) in 2003. Secondary y-axis contains horse mackerel landings in value (€). Relationship between ratio LV/FC and horse mackerel landings ($r^2 = 0.299$).

4 FINAL CONSIDERATIONS

This case study represents a first attempt to integrate different sources of information obtained at the trip level, with the objective of better characterizing trawling activity, by estimating energy efficiency and economic indicators for the different landing profiles (fishing tactics). Differences in fuel consumption and overall economical efficiency between octopuses and horse mackerel LP may indicate that there is room for changes in operational procedures for LP1 fishing trips, in order to optimize their return.

ACKNOWLEDGEMENTS

We gratefully acknowledge Beatriz Mendes for her valuable assistance using ArcGis software. The fisheries data was supplied to IPIMAR by the Portuguese General-Directorate for Fisheries and Aquaculture (DGPA). The processed VMS data was provided by the University of Algarve (GeoPescas research project). This work is the continuation of previous work under the Programme "MARE: Fishing Technologies" (MARE, FEDER, QCA-III, 22-05-01-FDR-00114: 2000-2007). It was partially funded by the Portuguese Fundação para a Ciência e a Tecnologia, through a PhD grant (SFRH/BD/43409/ 2008) attributed to Tereza Pilar-Fonseca.

REFERENCES

Afonso-Dias, M., Simões, J. & Pinto, C. 2004. A dedicated GIS to estimate and map fishing effort and landings for the Portuguese crustacean trawl fleet. In T. Nishida, P.J. Kailola, & C.E. Hollinworth (eds), *GIS/Spatial Analyses in Fishery and Aquatic Sciences (Vol. 2)*. Fishery-Aquatic GIS Research Group, Saitama, Japan, 323–340.

Afonso-Dias, M., Pinto, C. & Simões, J. 2006. GeoCrust 2.0 – a computer application for monitoring the Portuguese Crustacean Trawl fishery using VMS, landings and logbooks data. ICES C.M. 2006/N:19 (Poster).

Basañez, J. 1975. Resultados obtenidos en arrastreros com helices en tobera. Canal de experiencias hidrodinamicas, El Pardo. Publicacio′ n nu′m. 53. Madrid, p. 21.

Campos, A., Fonseca, P., Fonseca, T. & Parente, J. 2007. Definition of fleet components in the Portuguese bottom trawl fishery. Fisheries Research, 83: 185–191.

DGPA, 2010. Recursos da pesca. Série estatística, Vol. 23A-B, ano 2009. Direcção-Geral das pescas e Aquicultura, Lisboa, Julho 2010, 181p.

EC, 2003. Commission Regulation No. 2244/2003 of 18 December 2003, laying down detailed provisions regarding satellite-based vessel monitoring systems. Official Journal of the European Union 2003. L333:17–27.

Haimov, H., Bobo, M., Vicario, J. & Corral, J. 2010. Ducted Propellers. A Solution for Better Propulsion of Ships. Calculations and Practice. First International Symposium on Fishing Vessel Energy Efficiency. E-Fishing, Vigo, p. 6.

Kasper, E. 1983. Model tests of application of bulbous bows to fishing vessels. In: Paper No. III-2 Presented at the International Symposium on Ship Hydrodynamics and Energy Saving. Canal de experiencias hidrodinamicas, El Pardo, p.13.

O'Dogherty, P., Nunez, J.F., Carlier,M. & O'Dogherty, M., 1981. Nuevas tendencies en el proyeto de buques pesqueros. Canal de experiencias hidrodinamicas, El Pardo. Publicacion nu′m. 67. Madrid, p. 12.

Parente, J., Fonseca, P., Henriques, V. & Campos, A. 2008. Strategies for Improving Fuel Efficiency in the Portuguese Trawl Fishery. Fis. Res. (2008), 10.1016/j.fishres.2008.03.001. (2).

Parente, J. 2009. Simulação do consumo de combustível em arrastões costeiros. Relat. Cient. Téc. IPIMAR, Série digital (http://ipimar-iniap.ipimar.pt), n°50, 26 p.

Pilar-Fonseca, T., Campos, A., Afonso-Dias, M., Fonseca, P., & Pereira, J. 2008. Trawling for cephalopods off the Portuguese coast – Fleet dynamics and landings composition. Fisheries Research, 92: 180–188.

Pilar-Fonseca, T, Campos, A., Afonso-Dias, M., Fonseca, P. & Mendes, B. 2009. Fleet segmentation of the Portuguese coastal trawl fishery: a contribution to fisheries management. ICES CM 2009/O:29.

Sustainable Maritime Transportation and Exploitation of Sea Resources – Rizzuto & Guedes Soares (eds)
 ISBN 978-0-415-62081-9

E-Audit: Energy use in Italian fishing vessels

G. Buglioni, E. Notti & A. Sala
Consiglio Nazionale delle Ricerche—Istituto di Scienze Marine (CNR-ISMAR), Largo Fiera della Pesca, Ancona, Italy

ABSTRACT: Recent oil price increases have brought renewed attention to energy-saving methods in the fishing industry. Due to the European commission restrictions on new constructions, the major opportunities for reducing fuel consumption are chiefly related to improving vessel operation rather than commissioning new energy saving vessels. Large number of fishing vessels is not efficient usually because of outdated technology. Fuel efficiency directly affects emissions causing pollution by affecting the amount of fuel used. In the current experiment some fishing vessels, representing the various fleet sectors of the Italian fisheries, were selected for the fuel efficiency audit. The vessels were divided on the basis of type of fisheries and vessel size. An energy audit template was developed to assess the main vessel and equipment features: engine usage, trip scheduling, propeller, etc. Onsite visual inspections were performed during the audit. during the fishing cruises a data acquisition system allowed to record and diagnose the vessels work parameters, offering a real-time dynamics. Subsequently detailed analysis of energy usage was carried out. In order to assess the energy performance, two energy indicators were proposed: energy consumption indicator (ECI [kJ/(GT · kn)]) and the fuel consumption indicator (FCI [l/(h · GT · kn)]).

1 INTRODUCTION

With the costs of operating a fishing enterprise rising steadily over the past decade, particularly for trawlers, it is crucial for fishermen to find ways to save on fuel in every possible way (Fiorentini et al., 2004; Messina and Notti, 2007). The amount of energy used by a fishing vessel will vary depending on the size (and engineering) of the vessel, weather, fishing gear, location, skill and knowledge. Due to the high fuel consumption in combination with high fuel prices, the trawlers are not profitable and viable anymore (Thomas et al., 2009). Improving the fishing vessel's efficiency at acceptable levels calls for technological interventions, mainly aimed at reducing the fuel consumption. The propulsion systems employed by most of fishing vessels are based on conventional fixed pitch propellers driven, through a reduction gear, by four-stroke medium-high speed diesel engines. The convenience of diesel oil, as energy source, is due to it having a high volumetric energy density and being liquid that is safe to handle, easy distribute and store. Energy efficiency has a direct impact on business operating costs and affects the bottom line of every business, in particular for fishing. Production and use of energy also has a significant impact on the environment. Energy use is not only an economical issue but also environmental conservation issue. The combustion of fossil fuels for these activities produces emissions of various greenhouse gases including carbon dioxide (CO_2), carbon monoxide (CO), oxides of nitrogen (NOx), sulphur dioxide (SO_2), and non-methane volatile organic compounds (NMVOCs). The main goal of the Kyoto Protocol is to achieve sustainable management of natural resources in order to reduce the emissions of greenhouse gases, in particular to reduce the emissions of carbon dioxide (CO_2) from fossil fuel combustion. Energy auditing are currently used by land-based businesses, industries and households to investigate energy use and to identify opportunities for cost-effective in the use of energy (Parente et al., 2008; Thomas et al., 2009). The primary goal of this work is to generate an energy audit system for fishing vessels.

2 MATERIALS AND METHODS

2.1 *Vessels monitored and on-site investigations*

The current study has been mainly conducted for research purpose to investigate energy use and to identify opportunities for improving effectiveness in the use of energy. Four vessels were selected for the energy efficiency audits representing the main fleet sectors throughout the Adriatic coast fishery of the Marche Region. Two vessels were involved in bottom trawling and other two in pelagic pair-boat trawling, with the latter belonging to two different pair trawlers. Table 1 shows the main characteristics of the fishing vessels monitored.

Table 1. Main characteristics of the monitored fishing vessels. L_{OA}: length overall; L_{PP}: length between perpendiculars; B: maximum beam; GT: gross register tonnage; P_B: engine brake power; D: diameter of the propeller. OTB: bottom otter trawler; PTM: midwater pair trawler.

	L_{OA} [m]	L_{PP} [m]	B [m]	GT [-]	P_B [kW]	D [m]
OTB1	21.50	17.02	5.72	82	478	1.78
PTM1	28.60	21.20	6.85	99	940	2.18
OTB2	22.80	19.58	6.21	91	574	1.80
PTM2	28.95	24.32	6.86	138	940	2.20

Following the determination of the vessels to be analyzed, an energy audit template was developed to assess all the main features of the vessels (engine, propeller and gear characteristics, hull type and design, etc.). On-site vessel investigations for a detailed analysis of energy usage were made during typical commercial round trips, which for a trawlers consists of several fishing operations (steaming, trawling, sailing, etc.).

2.2 *Energy performance indicators*

Overall energy consumption is the result of complex web of issue affecting each fishing business. All these aspects interact in term of costs and benefits to the viability and profitability of the business. In order to obtain a broad picture of the operation's outputs and energy input, the audit process can evaluate the energy consumption of vessels. The audit can determine also whether energy use is reasonable or excessive among fishing vessels.

In order to assess the energy performance, two customized energy indicators were defined and calculated. The first is an energy consumption indicator, named ECI ([kJ/(GT · kn)]), which is the overall energy used every 5 seconds, on the basis of the acquisition system setup, standardized for the gross tonnage and the speed of the vessel. The second is a fuel consumption indicator, named FCI ([l/(h · GT · kn)]), which is the fuel consumption standardized for the gross tonnage and the speed of the vessel.

The values of energy performance indicators, ECI and FCI, in steaming and trawling condition are summed. Evaluating and ranking the energy performance of fishing vessels is useful for research purposes.

2.3 *Data collection and metering devices*

The data collection system was conceived at CNR-ISMAR of Ancona (Italy) and consist of a hydraulic and electric power analyzer, a shaft power meter, two load cells for drag resistance, two flow meters for fuel consumption and a GPS data logger. The instruments were linked by RS232/485 serial ports to a personal computer, which automatically control data acquisition and provide the correct functioning of the system in real time through an appropriately developed program.

The hydraulic and electric power analyzer consists of a sensor array that provides flow and pressure signals, and two clamp-on ammeters for measuring the real supply from the alternator connected with the main diesel engine. The shaft power **meter** has a battery powered shaft-mounted strain gauge, which utilizes a short range radio transmission for data (torque and rotational speed) transfer from the rotating shaft to a data recorder. The shaft-mounted strain gauge was powered by a 1600 mAh 4.8 V NiMH battery pack and it is removable for a quick replacement and charging. The battery pack can give to the transducer a run time of approximately four weeks. Shaft rotational speed is measured by the recorder, which has an optical proximity sensor, giving a signal of one pulse per shaft revolution. In order to determine the effective fuel consumption in real-time, two portable ultrasonic flow meters were installed to evaluate the flow inlet and outlet of the main Diesel engine. The ultrasonic flow meter consists of one transmitter and two sensors (Figure 1). The transmitter is used both to control the sensors and to prepare, process and evaluate the measuring signals, and to convert the signals to a desired output variable. The sensors work as sound transmitters and sound receivers and have been arranged for measurement via two traverses as in Figure 1.

The measuring system calculates the volume flow of the fluid from the measured transit time difference and the pipe cross-sectional area. In addition to measuring the transit time difference, the system simultaneously measures the sound velocity of the fluid. This additional measured

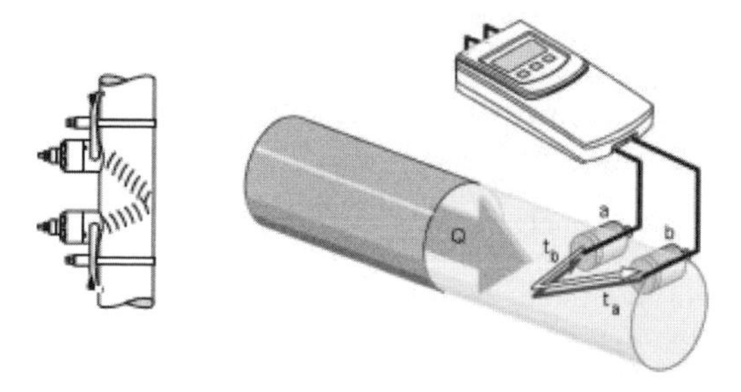

Figure 1. Portable ultrasonic flow meter system (right) with two acoustic sensors (*a, b*) for measuring the volume flow ($Q = v{\cdot}A$) of the fluid from the pipe cross-sectional area (A) and the flow velocity (v) obtained by the transit time difference ($\Delta t = t_a - t_b$; $\Delta t \sim v$). Mounting arrangement (left) for measurement via two traverses.

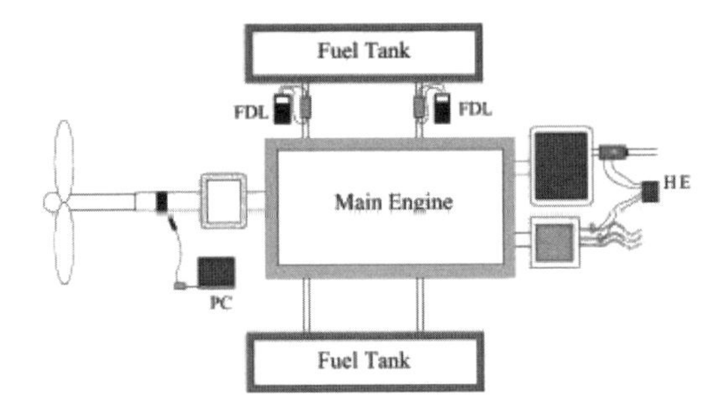

Figure 2. Typical engine room layout and location of the metering devices used during the current energy audit study PC: shaft power meter system with laptop and GPS device; FDL: ultrasonic flow meter systems; HE: hydraulic and electric power analyzer.

variable can be used to distinguish different fluids or as a measure of product quality. Two electronic load cells were used to measure the warp loads during fishing operations. After shooting, load cells were mounted on the warps in order to measure the total gear drag resistance.

2.4 *Machinery room's layout*

The engine room is the heart of the ship providing mechanical, hydraulic and electrical power for the entire vessel (Figure 2). Generally is located aft ship and contains the main diesel engine coupled to a fixed pitch propeller through a reduction gearbox with inverter. Electrical generators, hydraulic pumps and other machineries are connected to fore engine power take-off (Figure 2).

3 RESULTS

The performance of monitored vessels was evaluated during typical daily fishing trip. This allowed for a full characterization of an average trip for each vessel. Figure 3 and Table 2 show the higher power-demanding of a pelagic trawler during towing operations. Vessel PTM1 has the same engine power of PTM2 but it tows a smaller pelagic trawl net.

In particular, the mean total towing force is around 7200 kgf and 5700 kgf respectively for PMT2 and PTM1. While bottom trawlers OTB1 and OTB2 have similar towing forces (around 3800–4000 kgf).

The higher delivered power (PD) might be caused by the efficiency of the propeller system. In general during trawling the two pelagic vessels had higher fuel consumption rates (105–126 l/h) than bottom trawlers (60–65 l/h).

The parameters recorded during the sailing phase are displayed in Figure 4. Afterwards,

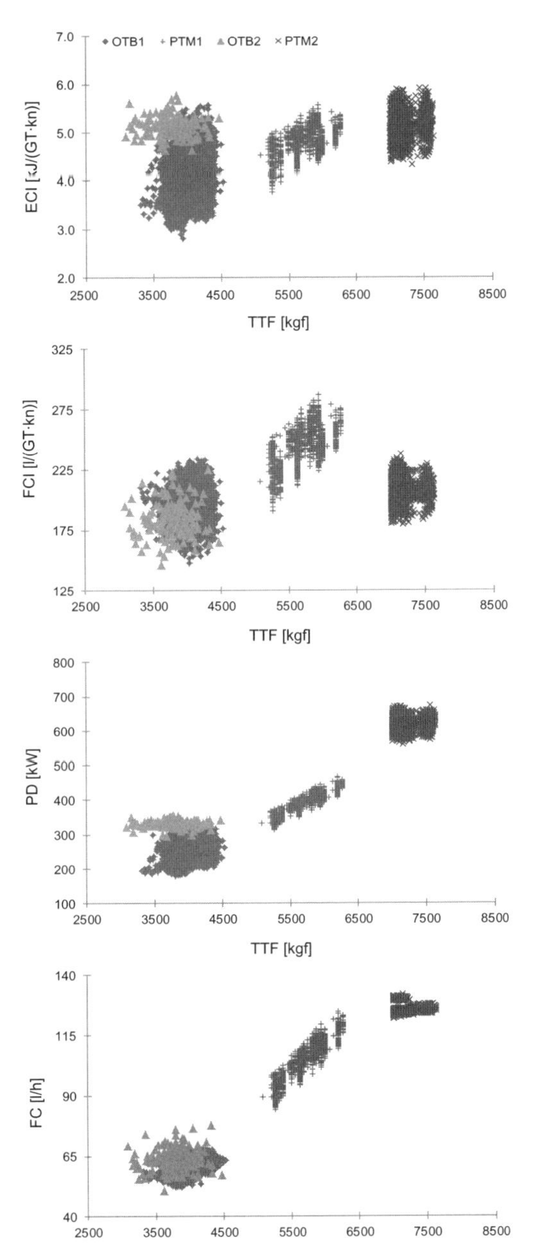

Figure 3. Energy consumption index ECI during the trawling phase. Fuel consumption index FCI, power delivered PD and fuel consumption FC are displayed versus total towing force TTF. OTB: bottom otter trawler; PTM: midwater pair trawler.

regression analysis of sailing data at different speed have been carried out and a comparison at 10 kn is reported in Table 3. Higher energy use of pelagic vessels (85–95 l/h) compared to bottom

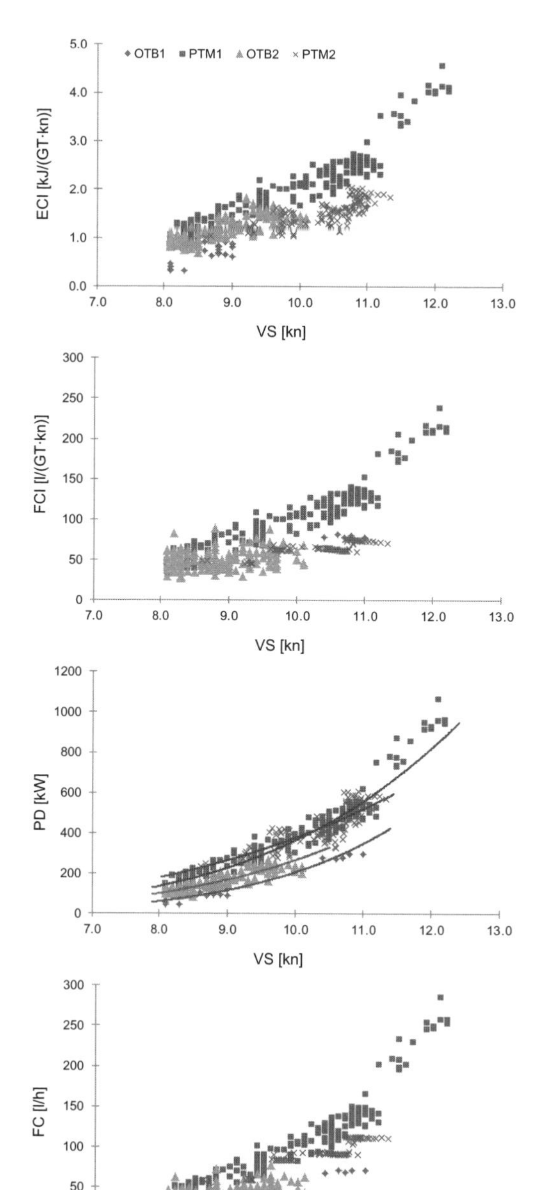

Figure 4. Energy consumption index ECI during the sailing phase, fuel consumption index FCI, power delivered PD and fuel consumption FC are displayed versus vessel speed VS. OTB: bottom otter trawler; PTM: midwater pair trawler.

trawlers (around 55 l/h) is due to higher vessel- and engine-size (Table 3).

Table 4 reports the ranks of the vessels monitored in terms of the energetic indicators ECI and

Table 2. Mean (in bold) and standard deviation (in italics) of vessel speed VS, power delivered PD, fuel consumption FC, total towing force TTF, energy consumption index ECI and fuel consumption index FCI during trawling. OTB: bottom otter trawler; PTM: midwater pair trawler.

	VS [kn]	PD [kW]	FC [l/h]	TTF [kg]	ECI [kJ/(GT·kn)]	FCI [l/(h·GT·kn)]
OTB1	3.81	248	59.8	3994	4.08	191.9
	0.21	26	3.5	186	0.45	12.6
PTM1	4.28	391	104.8	5693	4.82	245.5
	0.18	30	8.1	293	0.30	17.4
OTB2	3.82	332	64.1	3776	5.15	184.4
	0.09	11	4.9	270	0.20	14.7
PTM2	4.42	620	126.1	7225	5.14	206.8
	0.17	21	2.6	184	0.28	10.2

Table 3. Estimated values at 10 kn of vessel speed: power delivered PD, fuel consumption FC, energy consumption index ECI and fuel consumption index FCI during the sailing phase. OTB: bottom otter trawler; PTM: midwater pair trawler.

	PD [kW]	FC [l/h]	ECI [kJ/(GT·kn)]	FCI [l/(h·GT·kn)]
OTB1	217	54.1	1.33	64.5
PTM1	366	94.0	1.93	95.9
OTB2	268	55.4	1.51	59.7
PTM2	378	84.5	1.41	61.3

Table 4. Energy consumption index ECI, fuel consumption index FCI and their respective ranking for each monitored vessel. OTB: bottom otter trawler; PTM: midwater pair trawler.

	ECI [kJ/(GT·kn)]	Rank	FCI [l/(h·GT·kn)]	Rank
OTB1	5.42	1	256.4	2
PTM1	6.75	4	341.4	4
OTB2	6.66	3	244.1	1
PTM2	6.55	2	268.1	3

FCI. OTB1 has the lowest ECI value (5.42 kJ/(GT · kn)) while it has a second rank for the FCI (256.4 l/(h · GT · kn)). OTB2 has the lowest FCI (244.1 l/(h · GT · kn)) while it has the third rank for FCI (6.66 kJ/(GT · kn)).

OTB1 is followed by the pelagic trawler PTM2, which has 5.42 and 256.4 for ECI and FCI, respectively. The high values of both ECI and FCI for the two pelagic trawlers PTM1 and PTM2 (Table 4) allowed a low energy performance characterization of this fisheries, even if the ranks of ECI and FCI are not consistent throughout the two pelagic trawlers.

4 CONCLUSIONS

This paper reports on the development of an energy audit system for fishing vessels. The new energy auditing system was set up for research purpose in order to evaluate the energy performance of fishing vessels under different operating conditions as well as to identify the flow and the amount of energy supplied.

In agreement with Prat et al. (2008) and Sala et al. (2009; 2011), it appears that there is an unexploited potential for fuel savings. For example, the energy performance indicators have shown large differences in fuel consumption within each fishery, in the future it is worth further analyzing the possibilities of changing some of the least efficient vessels, in particular, relating to the propulsion system and the propeller design.

The information provided in the current paper might suggest several ways for achieving fuel-use reductions: such as technical improvements in the efficiency of ship engines, substitution of fishing gear types, and innovation and research into better fishing practices.

REFERENCES

Fiorentini, L., Sala, A., Hansen, K., Cosimi, G. & Palumbo, V., 2004. Comparison between model testing and full-scale trials of new trawl design for Italian bottom fisheries. Fisheries Science 70: 349–359.

Messina, G. & Notti, E. 2007. Energy saving in trawlers: practical and theoretical approaches. Proceedings of the International Conference on Marine Research and Transportation (ICMRT), 28–30 June 2007, Ischia, Naples, Italy: 91–98.

Parente, J., Fonseca, P., Henriques, V. & Campos, A. 2008. Strategies for improving fuel efficiency in the Portuguese trawl fishery. Fisheries Research 93, 117–124.

Prat, J., Antonijuan, J., Folch, A., Sala, A., Lucchetti, A., Sardà, F. & Manuel, A. 2008. A simplified model of the interaction of the trawl warps, the otterboards and netting drag. Fisheries Research 94: 109–117.

Sala, A., De Carlo, F., Buglioni, G. & Lucchetti, A. 2011. Energy performance evaluation of fishing vessels by fuel mass flow measuring system. Ocean Engineering 38: 804–809.

Sala, A., Prat, J., Antonijuan, J. & Lucchetti, A. 2009. Performance and impact on the seabed of an existing- and an experimental-otterboard: comparison between model testing and full-scale sea trials. Fisheries Research, 100: 156–166.

Thomas, G., O'Doherty, D., Sterling, D. & Chin, C. 2009. Energy audit of fishing vessels. Proceedings of the Institution of Mechanical Engineers, Part M: Journal of Engineering for the Maritime Environment 224(2): 87–101.

Sustainable Maritime Transportation and Exploitation of Sea Resources – Rizzuto & Guedes Soares (eds)
 ISBN 978-0-415-62081-9

Assessment of fishing gear impact and performance using Sidescan sonar technology

A. Lucchetti, M. Virgili, F. De Carlo & A. Sala
Consiglio Nazionale delle Ricerche—Istituto di Scienze Marine (CNR-ISMAR), Ancona, Italy

ABSTRACT: The increased sensibility of the International scientific community toward the exploitation of fishery resources, promoted the development of new technologies to study the behaviour and the impact of fishing gears on seabed. In the last ten years the physical disturbances caused by trawling has been widely investigated by using the sidescan sonar technology. In the Mediterranean, changes to marine habitats that are caused by fishing are most pronounced in otter trawls, Rapido and hydraulic dredge fisheries. Sidescan sonar technology permitted to identify these fishing gears impact. Hydraulic dredges and Rapido trawls are basically similar in their seafloor impact by flattening and ploughing seabed features. While the effects of otter trawling varies greatly depending on the amount of gear contact with the bottom, together with the depth, nature of the seabed, and the strength of the currents or tide. Generally otterboards imprint distinct tracks on the seabed, ploughing a groove which can vary from a few cm up to 30 cm deep. Also the present work suggests a further step in using this technology, by analyzing in real time the behaviour, the geometry and the performance of different fishing gears in addition to quali-quantitative evaluation of their impact on the seafloor.

1 INTRODUCTION

The increased sensibility of the International scientific community toward the exploitation of fishery resources, promoted the development of new technologies to study the behaviour and the physical impact of fishing gears on seabed. At the moment the most widespread technology for the evaluation of these physical effects is the "acoustic imaging" mainly through the sidescan sonar technology (Franceschini *et al.*, 2002; Smith *et al.*, 2007). During the past few decades the impact on marine ecosystems due to fishing activities got a growing degree of interest and it became important in crucial considerations on environmental management plans (Hall, 1999; Kaiser, 2002; Hiddink *et al.*, 2007). In the Mediterranean, changes to marine habitats that are caused by fishing are most pronounced in bottom trawl (Demestre *et al.*, 2000), dredges (Morello *et al.*, 2005), rapido and beam trawls fisheries (Pranovi *et al.*, 2000; ICES, 2003).

Bottom trawling impact on the benthic habitat differs among the various bottom trawl fisheries and it also depends on the bottom conditions in the area fished (Valdemarsen *et al.*, 2007). Different damage are observed on bottom community: furrows and scars left by otterboards, seabed scouring and flattening by ground rope and weights, and sediment removing and suspension. In the past, earlier studies of trawl performance were performed by scuba divers direct underwater observations. Despite the clear and immediate overview offered by these kind of methodologies, they did not allowed to investigate bottom areas under 60 m (Caddy J.F., 2000). In the last ten years the physical disturbances caused by trawling has been widely investigated by using the sidescan sonar technology (e.g. Schwinghamer *et al.*, 1998; Thrush *et al.*, 1998; Tuck *et al.*, 1998; Humborstad *et al.*, 2004). Friedlander *et al.* (1999) showed that sidescan sonar can be used to assess trawling impacts over a wide area. Physical changes due to trawling, such as trawl marks, are visible with sidescan sonar imaging systems as tracks (otterboard, dredge and net marks) across the seabed.

In previous studies sidescan sonar has been mainly used for geological surveys or in fishing effort assessment by detection and quantification of the furrows traced by bottom trawls on a georeferenced area of the seabed.

The present work suggests a further step in using this technology, by analyzing in real time the behaviour, the geometry and the performance of different fishing gears in addition to quantitative evaluation of their impact on the seafloor. Traditionally, the trawl geometry is taken under control by using small acoustically linked sensors mounted on the net and on the otterboards, which communicate with the ship via hydrophone/s mounted on the hull (Sala *et al.*, 2009). Nevertheless it is not possible to monitor some important parameters, such as the attack angle of otterboards, that is the angle comprised between the tow direction and the main side of the otter trawl, because no acoustic sensors or other technology are available for this.

The attack angle is essential for the understanding of trawl behaviour, hydrodynamic drag, fuel consumption and bottom impact, nevertheless it could be defined only in flume tank tests or after computing complex formulas (Sala *et al.*, 2009).

2 MATERIALS AND METHODS

2.1 *Fishing gears investigated*

In the current study the physical impact and the performance of the commercial hydraulic dredge, rapido trawl, and two types of Italian bottom trawls have been monitored through a Sidescan sonar technology. The infaunal bivalve *Chamelea gallina* is the target of a large fleet of **hydraulic dredgers** operating in the sandy coastal bottoms (depth of 3–12 m) of the northern and central Adriatic Sea (Italy). The commercial hydraulic dredge comprises a 3 m wide rectangular cage weighing 0.6–0.8 tonnes, mounted on two sledge runners to prevent it from digging into the substratum to a depth of more than 4–6 cm (Morello *et al.*, 2005). The front of the cage is connected by a hose to a centrifugal water pump that ejects water under pressure (1.2–1.8 bar) from the nozzles at the mouth of the dredge and inside the dredge cage (Morello *et al.*, 2005). Once suitable fishing grounds are reached (depth of 3–12 m), the dredge is lowered from the bow and the vessel moves astern, warping on a big anchor for a variable distance depending on how much cable is paid out (Morello *et al.*, 2005), or by using the propeller.

The **Rapido** is a towed gear used only in the Adriatic Sea for fishing pectinids (*Pecten jacobaeus*) in sandy offshore areas and flatfish (*Solea* spp., *Platichthys flesus, Psetta maxima, Scophthalmus rhombus*) in muddy inshore areas, but little is known about the environmental impact of this gear (Hall-Spencer *et al.*, 1999; Pranovi *et al.*, 2000). Modern Rapido gear resembles a toothed beam trawl. The gear consists of a box dredge (3 m wide, 120 kg) rigged with teeth (5–7 cm long) along the lower leading edge and a net bag to collect the catch (Giovanardi et al., 1998). During fishing, the gear is towed at high speed (7 knots), with a spoiler that prevents the gear from rising off the bottom (Pranovi *et al.*, 2000). A commercial vessel typically tows four sets of gear simultaneously.

Bottom trawling (otter trawling) is the most important fishing sector in Italy in terms of fleet dimension, fishing power and incoming. The traditional Italian commercial trawl (*Volantina*, in Italian) is a commonly entirely made up of knotless polyamide netting. The trawl net is a typical asymmetric two-faces net. In the last 5–6 years several Italian bottom trawlers switched their activity from the traditional trawl to a new trawl configuration, so called *Americana*. The Americana net is a four-faces trawl, manufactured with both knotless-PA and knotted-PE netting. Some other changes have been introduced to increase the bosom height and the possibility to tow one or two nets (twin trawls), such as net made with more advanced rules and drawing designs, very short bridles, etc.

2.2 *Sidescan sonar surveys*

Sidescan sonar trials were conducted on fishing grounds of the Central Adriatic normally exploited by local fishermen. The trials took place from 23 to 25 March 2010 off Ancona coast, on board of the Italian Research Vessel RV "G. Dallaporta". The Sidescan sonar system used comprised a Sonar DeepEye 670 kHz towfish (DE670) with a horizontal and vertical beam of 0.9° and 60° respectively, surface unit 12 V (SU1-232), Kevlar coaxial cable 200 m (STC-200), rugged PC touch screen connected to a DGPS system (Garmin GPS).

Following the recommendation of Smith *et al.* (2007), weighted rope (10 m, weighing 10 kg) was attached to the cable in front of the towfish to act as a cable depressor. All sidescan trials were carried out with 50–70 m range on each side, giving a resolution of around 2 cm. The towfish was deployed over the stern of the RV Dallaporta, and counting marks on the cable were used to determine the amount of coaxial wire paid out. Track recording began when the fishing vessel was few hundred meters stern of the towfish. Normal operating height was around 5 m above the seabed, and towing speed was 2–2.5 knots.

During trawling operations, due to higher speeds (4–7 knots), the fishing vessel overcame the RV Dallaporta along the axis of towfish towing. Such parallel pass survey design properly permitted to detect trawl and otterboard marks that ran approximately parallel to the survey track. This made it suitable for quantitative assessment of the dimensions of trawl marks and gear performance, although a simple visual inspection of the sonar images did provide a reasonable qualitative assessment.

2.3 *Sidescan sonograms processing*

Detailed analysis of the sonograms was carried out ashore using the DeepView SE 2.2 software.

In playback in the laboratory, each sidescan track was investigated for measuring the width and height of the different trawl and otterboard marks. Measurement of gear parameters such as otterboard spread and horizontal net opening were also detected. A successful attempt was made to quantify the otterboard attack angles through the video analysis software Image Pro Plus v.4.5.1.

3 RESULTS

The analysis of the sidescan sonograms permitted for the first time to study both the performance of fishing gears and bottom impact. Marks left by fishing gears were very evident as furrows along the sediment surface. The images from Figures 1–5 are screenshots from the sidescan tracks.

3.1 *Hydraulic dredge*

The monitoring of hydraulic dredges took place in an area close to the Ancona coast characterized by shallow waters (3–10 m). Sidescan sonar records showed evidence of considerable physical

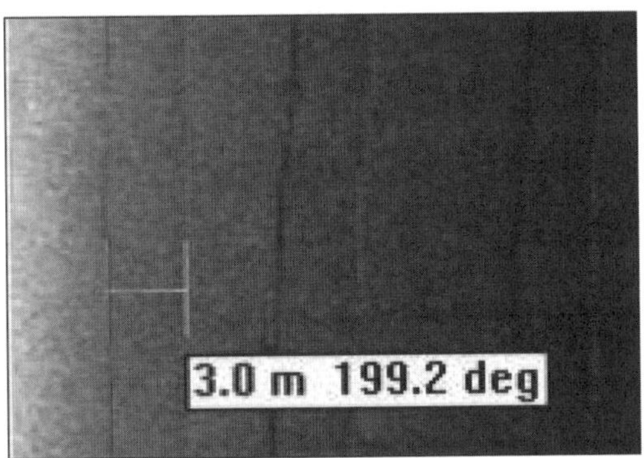

Figure 1. Sidescan sonar sonograms of hydraulic dredging marks (up) and width (down) of furrows left by the dredge on the seabed.

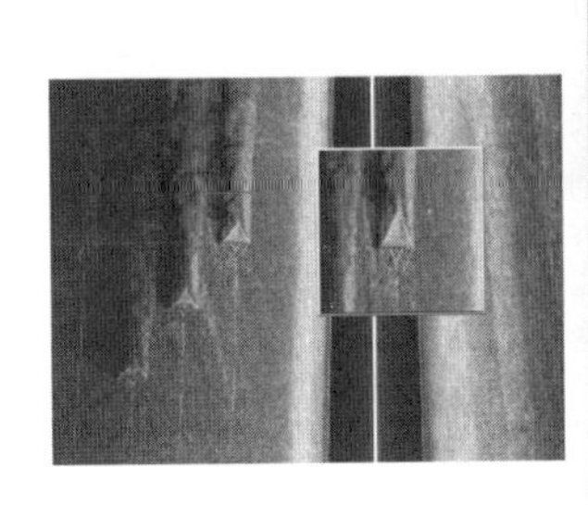

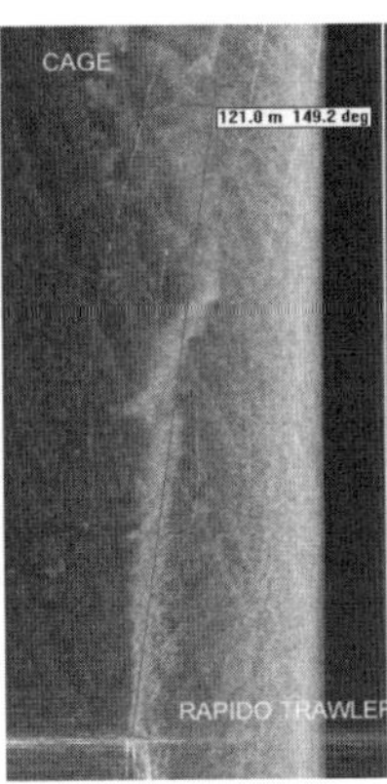

Figure 2. Sidescan sonar sonograms of Rapido trawling marks. Four Rapido trawls towed simultaneously by a commercial vessel (left) and furrows left by Rapido on the seabed (right).

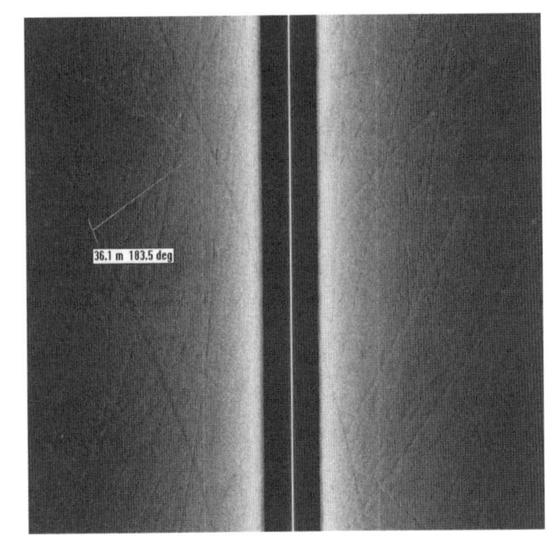

Figure 3. Sidescan sonar sonograms of otter trawling marks and measure of the distance between two doors.

disturbance in the surveyed area, with tracks criss-crossing the area (Figure 1). During the survey, the poor stability of the towfish and the turbulence noise (both from strong tidal flow and the vessel's propeller) was a problem and often obscured the sidescan acquisition completely in areas of low acoustic reflection. Also the resuspended sediments, disturbed the sound transmission and often determined low quality sonograms. Only few sonograms recorded when the resuspension ended permitted an identification of the furrows left by the dredge on the bottom. As showed in Table 1 and Figure 1, the furrows have an average height of 10 cm and a width of 3 m, equivalent to the dredge width.

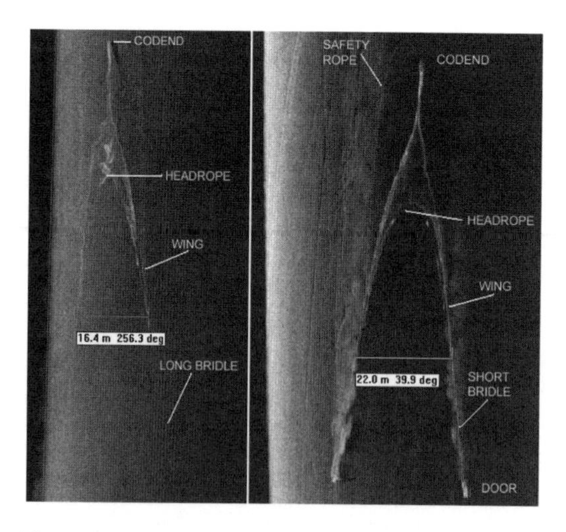

Figure 4. Sidescan-sonar images of a traditional bottom trawl net Volantina (left) and an Americana net (right). The horizontal openings (distance between the wings) are also reported.

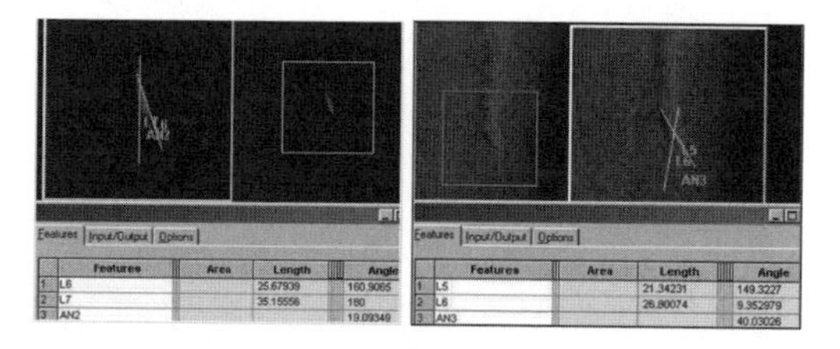

Figure 5. Comparison of otterboard attack angles between the Volantina (left) and the Americana (right) trawl gear, obtained by the video analysis software Image Pro Plus v.4.5.1.

3.2 *Rapido trawl*

Compared to the hydraulic dredge the survey of Rapido trawl was performed in deeper waters (22–24 m). In these conditions the sonograms were relatively clearer and more assessable. Evidence of the resuspension of sediments, i.e. turbidity clouds, can be easily identified from these observations (Figure 2). When towing speeds dropped below 6 kn, the towing warp and bridle periodically came into contact with the seabed, stirring up sediment ahead of the gear. The furrows left by Rapido trawls are 4 m in width, equivalent to the dimension of the frames, and 10 cm in height (Table 1). The teeth of the Rapido projected their full 7 cm length into the sediments and redistributed the surface layer (Figure 2). Sidescan sonar observations taken one hour after trawling revealed extensive sediment redistribution with suspended particles reducing visibility.

Table 1. Measured parameters during the Sidescan sonar surveys of Volantina (VOL) and Americana (AME) otter trawls, Rapido (RAP) trawls and Hydraulic dredge (HYD). HGO: horizontal gear opening; HOS: horizontal otterboard spread; α: attack angle of otterboards; FH: height of the furrow; FW: width of the furrow.

Parameter		VOL	AME	RAP	HYD
HGO	[m]	12-15	15-18	4	3
HOS	[m]	45-50	25-30	-	-
α	[°]	19-20	40-45	-	-
FH	[cm]	20	30	10	10
FW	[cm]	30	40	400	300

3.3 *Bottom trawls*

Sidescan survey of bottom trawls was performed at about 70 m of depth. The most noticeable physical effects of trawling are the furrows produced by the otterboards (up to 20–30 cm high and 30–40 cm wide), whereas other parts of the trawl created only faint marks (Table 1, Figure 3). Less pronounced marks were made by the groundrope, whereas the bridles left no visible marks. Also area between the outer edges of the otterboards was visibly disturbed, i.e. by the two otterboards and the groundgear.

Sediment clouds are produced when trawls were towed over a bottom of muddy sediments. The clouds spread rapidly, reaching a height of 3.0–3.5 m and a width of 4.5–6.0 m in dependence on the type of otterboard used.

In agreement with Krost *et al.* (1990), Jones (1992) and Tuck *et al.* (1998) we noticed that the height of the groove depends on the weight of the otterboard, the angle of attack (higher the attack angles higher the groove's height), and the nature of the substrate, being deepest in soft mud. The two type of trawls investigated in the current study presented different rigging characteristics (Figure 4), in particular the so called Americana trawl was characterized by short bridles and high otterboard attack angle.

As reported in Table 1, the Americana trawl had a higher otterboard impact on the seabed. Parallel tracks could be seen in all the surveyed area. The distance between these tracks varied from 30 to 50 m, which corresponds to the mean measured distance between the otterboards (Table 1).

It can be assumed that all these tracks have been left by otterboards. Horizontal net openings of the Volantina, ranging from 12 m to 15 m, is generally lower than the Americana trawls, which normally are in the range 15–18 m (Table 1, Figure 4).

High resolution and good quality images also permitted to measure the angle of attack of the otterboards by using the video analysis software Image Pro Plus v.4.5.1 (Table 1, Figure 5). In agreement with Sala *et al.* (2009), sidescan sonar sonograms show that the attack angles of the otterboards used in the Americana (40–45°) trawl were higher than that of the Volantina trawl (19–20°) (Table 1). Furthermore, otterboards of the Volantina advanced producing continuous marks due to the sliding movement, while the Americana otterboard advanced jumping on the seabed. The latter produced a series of small but higher furrows into the otterboard path with more evident muddy clouds as showed in Figure 4.

4 CONCLUSIONS

Sidescan sonar technology has become more accurate and more affordable in recent years and now it is a rapid assessment tool for evaluating trawling impacts. From the work performed in the current study, it would appear that hydraulic dredges and Rapido trawls are basically similar in their seafloor impact by flattening and ploughing seabed features. Notably, because the hydraulic dredge operate in restricted coastal areas (i.e. from 3 to 10 m of depth), the density of the dredge tracks is very high.

Gear type and the nature of the seabed are two factors that seem to have a great influence on the level of disturbance caused by fishing activity (Collie *et al.*, 2000). The effects of otter trawling vary greatly depending on the amount of gear contact with the bottom, together with the depth, nature of the seabed, and the strength of the currents or tide. Generally otterboards imprint distinct tracks on the seabed, ploughing grooves which can vary from a few cm up to 30 cm deep.

Chains in the Americana trawl can also leave recognizable tracks and may skim off the surface sediment layers between the two grooves left by the otterboards. Sidescan sonar provided also quali-quantitative information on gear performance during fishing operations and therefore might be useful proof against the information provided by trawl monitoring acoustic systems.

REFERENCES

Caddy J.F., 2000. A short retrospective on personal research carried out in eastern Canada in the period 1966-'75 relevant to impacts of fishing gear on the benthic environment, and some perspectives for further research. *Impact of trawl fishing on benthic communities – Proceedings*, ICRAM: 109–128.

Collie J.S., Escanero G.A., Valentine P.C., 2000. Photographic evaluation of the impacts of bottom fishing on benthic epifauna. *ICES Journal of Marine Science*, 57: 987–1001.

Demestre M., Sánchez P., Kaiser M.J., 2000. The behavioural response of benthic scavengers to otter-trawling disturbance in the Mediterranean. *In*: Kaiser M.J., de Groot S.J. (*eds.*). Effects of Fishing on Non-target Species and Habitats. Blackwell Science, London: 121–129.

Franceschini G., Raicevich S., Giovanardi O., Pranovi F., 2002. The use of side scan sonar as a tool in Coastal Zone Management. Littoral. Porto, 22–26 September 2002: 11–14.

Friedlander A.M., Boehlert G.W., Field M.E., Mason J.E., Gardener J.V., Dartnell P., 1999. Sidescan-sonar mapping of benthic trawl marks on the shelf and slope off Eureka, California. *Fishery Bulletin US*, 97: 786–801.

Giovanardi O., Pranovi F., Franceschini G., 1998. "Rapido" trawl-fishing in the Northern Adriatic: preliminary observations on effects on macrobenthic communities. *Acta Adriatica*, 39: 37–52.

Hall-Spencer J.M., Froglia C., Atkinson R.J.A., Moore P.G. (1999). The impact of Rapido trawling for scallops, *Pecten jacobaeus* (L.), on the benthos of the Gulf of Venice. *ICES Journal of Marine Science*, 56: 111–124.

Hall S.J., 1999. The Effects of Fishing on Marine Ecosystems and Communities. Blackwell Science, Oxford, 274 pp.

Hiddink J.K., Jennings S., Kaiser M.J. 2007. Assessing and predicting the relative ecological impacts of disturbance on habitats with different sensitivities. *Journal of Applied Ecology*, 44: 405–413.

Humborstad O-B, Nottestad L., Løkkeborg S., Rapp H.T., 2004. RoxAnn bottom classification system, sidescan sonar and video-sledge: spatial resolution and their use in assessing trawling impacts. *ICES Journal of Marine Science*, 61: 53–63.

ICES, 2003. Report of the Working Group on Ecosystem Effects of Fishing Activities. ICES CM 2003/ACE:05.

Image Pro Plus, 2005. Image Pro Plus v.4.5.1. Media Cybernetics, Inc. USA.

Jones J.B., 1992. Environmental impact of trawling on the seabed: a review. *New Zealand Journal of Marine and Freshwater Research*, 26: 59–67.

Kaiser M.J., Collie J.S., Hall S.J., Jennings S., Poiner I.R., 2002. Modification of marine habitats by trawling activities: prognosis and solutions. *Fish and Fisheries*, 3: 114–136.

Krost P., Bernhard M., Werner F., Hukriede W., 1990. Otter trawl tracks in Kiel Bay (Western Baltic) mapped by side-scan sonar. *Meeresforschung*, 32: 344–353.

Morello E.B., Froglia C, Atkinson R.J.A., Moore P.G., 2005. Impacts of hydraulic dredging on a macrobenthic community of the Adriatic Sea, Italy. *Canadian Journal of Fisheries and Aquatic Sciences*, 62: 2076–2087.

Pranovi F., Raicevich S., Franceschini G., Farrace M.G., Gionavardi O., 2000. Rapido trawling in the northern Adriatic Sea: effects on benthic communities in an experimental area. *ICES Journal of Marine Science*, 57: 517–524.

Sala A., D'Arc Prat Farran J., Antonijuan J., Lucchetti A., 2009. Performance and impact on the seabed of an existing- and an experimental-otterboard: Comparison between model testing and full-scale sea trials. *Fisheries Research*, 100: 156–166.

Schwinghamer P., Gordon D.C., Rowell T.W., Prena J., McKeown D.L., Sonnichsen G., et al., 1998. Effects of experimental otter trawling on surficial sediment properties of a sandybottom ecosystem on the Grand Banks of Newfoundland. *Conservation Biology*, 12: 1215–1222.

Smith C.J., Banks A.C., Papadopoulou K-N., 2007. Improving the quantitative estimation of trawling impacts from side scan sonar and underwater video imagery. *ICES Journal of Marine Science*, 64: 1692–1701.

Thrush S.F., Hewitt J.E., Cummings V.J., Dayton P.K., Cryer M., Turner S.J., et al., 1998. Disturbance of the marine benthic habitat by commercial fishing: Impacts at the scale of the fishery. *Ecological Applications*, 8: 866–879.

Tuck I.D., Hall S.J., Robertson M.R., Armstrong E., Basford D.J., 1998. Effects of physical trawling disturbance in a previously unfished Scottish sea loch. *Marine Ecology Progress Series*, 162: 227–242.

Valdemarsen J.W., Jørgensen T., Engås A., 2007. Options to mitigate bottom habitat impact of dragged gears. *FAO Fisheries Technical Paper*, 506: 29 pp.

Sustainable Maritime Transportation and Exploitation of Sea Resources – Rizzuto & Guedes Soares (eds)
© 2012 Taylor & Francis Group, London, ISBN 978-0-415-62081-9

Towards sustainable Baltic herring fishery: Trawls vs. pound nets

T. Raid, H. Shpilev & L. Järv
Estonian Marine Institute, University of Tartu, Tallinn, Estonia

A. Järvik
Estonian Maritime Academy, Tallinn, Estonia
Estonian Marine Institute, University of Tartu, Tallinn, Estonia

ABSTRACT: The Baltic herring is traditionally one of the main targets of Estonian fishery. The annual catches of Baltic herring have ranged 22-58,000 t or 15–25% of the total herring catch in the Main Basin and the Gulf of Riga since 1991. The international management of the Baltic herring stocks includes Total Allowable Catch and some technical measures (gear restrictions, closed areas and periods) as management tools. However, the selectivity studies of herring trawl fishery have shown the no efficiency of mesh size regulation. Additionally, the implementation of closed seasons/areas may result in undesired accumulation of excess effort.

The paper is focusing on another management tool—an optimizing the balance between herring trawl and pound net fisheries. The pound net catches include substantially higher share of older age groups with higher mean weights and no by-catch of juvenile fish, often taken by trawl fishery. Due to the different catch structure, the average number of fish, taken in order to achieve the similar catch in tons (population loss) is considerable smaller in pound net fishery than in trawl fishery, indicating that higher share of pound-net fishery might allow to decrease the fishing mortality.

1 INTRODUCTION

The Baltic herring (*Clupea harengus* L.) is traditionally one of the main targets of Estonian fishery. Since 1991, the annual catches of Estonian fishery have been in the range of 22–58,000 t, i.e. 15–25% of the total landings of herring in the Main Basin of the Baltic Sea and the Gulf of Riga. The Estonian fishery exploits two Baltic herring stocks: the Central Baltic Herring (CBH) located in the ICES sub-divisions 25–28, 29 and 32) and the Gulf of Riga herring (Sub-division 28.1; Fig. 1.).

In spite of its centuries-long history, the character of herring fishery in Estonia remained virtually unchanged: the coastal fishery with fixed gears (gillnets and various traps) dominated overwhelmingly until early 1950s.

The implementation of big trap-nets (pound-nets) of Japan origin since 1939 has been the first revolutionary change in the coastal herring fishery in Estonia. The following substantial changes in the structure of fishery started in 1950s with fast development of pelagic trawl fishery in 1960–1962 (Ojaveer 1967). Simultaneously to the development of trawl fishery, the role of fixed started to decline until the 1990s. The role of trawl fishery has increased also in other areas of northern Baltic, e.g. in Finland in early 1980s (Parmanne & Sjöblom 1985). The alteration in the structure of herring fishery towards the use of trawls in Estonia has had effect on:

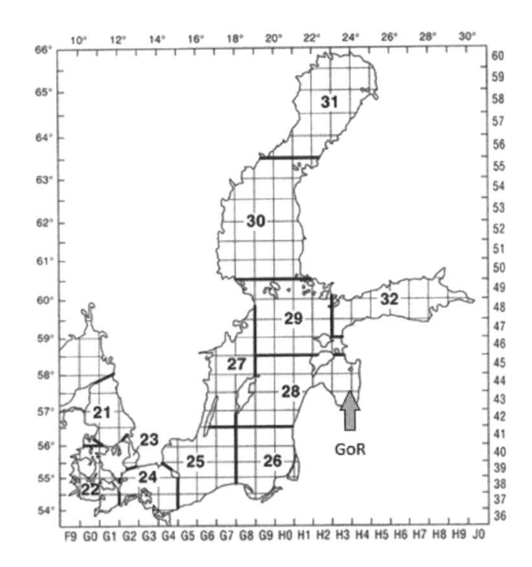

Figure 1. ICES Sub-divisions in the Baltic Sea. GoR: Gulf of Riga.

* seasonality of fishery
* catch composition
* exploitation rate
* social situation.

The both stocks are internationally assessed by the International Council of the Exploration of the Sea (ICES) and managed using the Total Allowable Catch (TAC) and effort restrictions as the main management tools. Some countries involved in the Baltic herring fishery have also applied a number of additional technical measures like gear restrictions, closed areas and periods.

However, the long- term stock dynamics of the Central Baltic herring have shown a clear declining trend throughout of most of the management history despite of all management measures applied. So the Spawning Stock Biomass (SSB) has declined since the beginning of the period of analytical assessments in 1974 from almost 1.8 million t in 1974 to below 400 000 t in early 2000s. The most recent period has revealed the slight increase in SSB, which still remains 2–3 times lower, compared to the situation in 1980s (Fig. 2.). The dynamics of the fishing mortality coefficient F (the rate of removal of fish from the population as catch) indicates that the exploitation of the stock has not been sustainable during the most of the period of observations. The fishing mortality has exceeded the sustainable level $F_{PA} = 0.19$ (ICES 2009) since 1983, showing the increasing trend also in the most recent years (average F = 0.27, ICES, 2010).

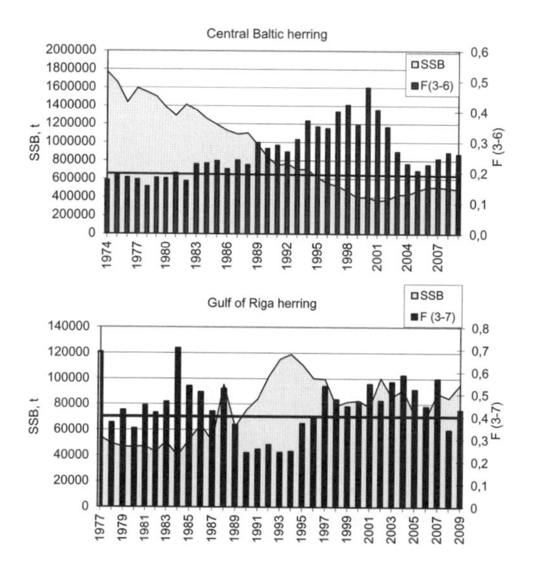

Figure 2. Trends in Spawning Stock Biomass (SSB) and fishing mortality (F) in the Central Baltic herring and the Gulf of Riga herring. The horizontal lines indicate the level of F_{PA} (data source: ICES, 2010).

The general trend of the SSB of the Gulf of Riga herring has shown quite a different pattern: after a period of stability on a relative low level of 50 000 t in the late 1970s- early 1980s, the SSB started to increase as a result of abundant recruitments, peaking at around 119 000 t in 1994. Later, the SSB decreased again and stabilized at 70–90 000 t in the most recent period. On the background of that relatively positive stock trajectory, the fishing mortality (F) in the main age groups of the Gulf of Riga herring has been generally high, exceeding the sustainable levels ($F_{PA} = 0.4$; $F_{MSY} = 0.35$, ICES 2009), both in the period of low stock in the 1970–80s and also during the most recent decades of high stock biomass.

Based on the most recent estimates of fishing mortality, ICES classifies the both herring stocks as being harvested unsustainably (the Gulf of Riga herring) or being at risk to be harvested unsustainably (Central Baltic herring, ICES 2010a). The above indicate that the management measures applied have not been effective in decreasing the fishing mortality and bringing the stocks into the level allowing sustainable exploitation in long run.

The present paper is focusing on another possible management tool—an optimising the balance between herring trawl and pound net fisheries in order to control the fishing mortality.

2 MATERIAL AND METHODS

The biological material used as a basis of present study was collected mostly from Estonian Baltic herring fishery as a part of routine biological sampling for stock assessment purposes. The random sampling method was used on monthly basis to cover all fishing seasons, areas and main gear types (trawls and pound-nets). The material was collected in the ICES Sub-divisions 28.1 (the Gulf of Riga), Sub-division 28.2, 29 and 32 (the Gulf of Finland). The following parameters were measured during the sampling: Total Length (TL), Total Weight (TW), age, sex, and maturity stage in six-degree scale (ICES 2010b). The data extends back to cover most of the period of analytical assessment of the Baltic herring, but in our conclusions we are focusing on the most recent decades (1991–2010). On average, 15, 000–20, 000 individuals were analyzed annually. The official landings′ statistics and the estimates of the catch in numbers calculated on the basis of landings` statistics and mean weights at age were used in order to estimate the fishing pressure on stocks.

3 RESULTS AND DISCUSSION

The fishing mortality is the main imminent effect of fishery on fish stocks. In general, the fish found in the fishing zone of gear will have four options: to be landed, discarded, meshed or escaped. In general, only the landed part of catch is really accounted in the estimates of fishing mortality, leaving certain amount of fish, additionally dying as a result of fishery unknown. The fish meshed, discarded and dead after escape from the trawl are the main source of that unaccounted fishing mortality. Minimizing of the latter is one of the main tasks of fisheries regulation. Besides to the TAC regulation several technical regulatory measures of herring fishery in Estonia have been implemented historically in order to allow maximum social stability in coastal communities, where the herring fishing and fish processing have been traditionally the main sources of employment (Järvik & Raid 2001):

A. capacity regulation,
B. effort regulation,
C. fishing capacity regulation.

The studies of the effect of vessel and gear size on herring the Gulf of Riga herring stock showed that the large vessels using big pelagic trawls cause a substantial additional fishing mortality of younger herring compared to the smaller vessels (Järvik *et al*. 2005). The results yielded in capacity regulation measure prohibiting the use of trawls with vertical opening >12 m (Anonymus 1995). The mesh size regulation is widely recognized as rather powerful tool for achieving the desired catch composition through selective performance of different mesh sizes (e.g. Gulland 1964, Burd 1991 and others). The mesh size regulation in Estonian herring trawl fishery (A_{min} = 24 mm in the Gulf of Finland and A_{min} = 28 mm in the Gulf of Riga), was implemented by the authorities in 1976 as a result of selectivity studies (Järvik 1975, Jefanov 1983). However, the estimates the effect of mesh size regulation on unaccounted mortality of the Baltic herring in pelagic trawl fishery (Suuronen 1995), have shown that the share of survived escapees of Baltic herring is generally very low (8.8%) independently of mesh size in the length groups below 12 cm (TL). That leads to the conclusion that the implementation of mesh size restrictions as main regulatory measure in herring trawl fishery in the Baltic is rather problematic and the additional regulatory measures are needed.

Although, the use of fixed or anchored gears (trap-nets, pound-nets, gillnets) have been dominating during the most of the history of herring fishery in Estonia, the proportion of herring, taken by fixed gears declined after the implementation of active gears like pelagic trawls in early 1960s. So, if the herring trap-net fishery was widely developed on almost all main fishing grounds of herring at Estonian coasts in 1950s, the Gulf of Riga (Sub-division 28.1) has remained the main area for coastal herring fishery at present. As a result, the vast majority of annual landings of the Baltic herring fished in Estonia are taken by pelagic trawl fishery offshore and only the minor part by pound-net or trap net fishery in coastal zone during the spawning period in second quarter. However, the share of trawl catches differ remarkably by the area ranging from 23–60% in the Gulf of Riga to 80–98% in the Sub-divisions 29 and 32 during the recent two decades (Fig. 3.).

The pound net catches include substantially higher share of older age groups and no by-catch of juvenile fish, often taken by trawl fishery (Fig. 4.). Additionally, the mean weight at age of pound-net catches is higher compared to the trawl catches. So, in the Gulf of Riga herring the average mean weight at ages 2–7 in trawl catches made up just 89% of the respective values in pound-nets (Fig. 5.).

The international fisheries regulation system of the Baltic herring relies on the presumption that the weight-based TAC-s, allocated by the conventional assessment units, secures the biologically safe exploitation level for stock(s) in the assessment units involved. Leaving fleets free to operate within the limits of national quotas it is also presumed, that an equal catch in tons, taken within

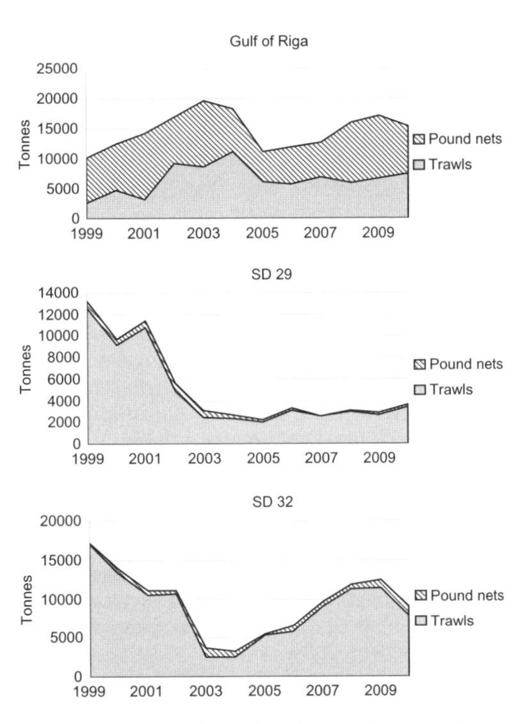

Figure 3. Share of trawl and pound-net catches in Estonian herring fishery in 1999–2010.

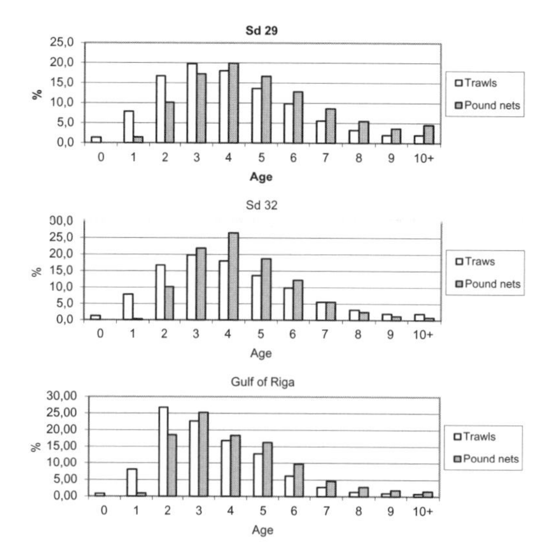

Figure 4. Mean age structure of Estonian herring catches in Sub-divisions 29 and 32 (1992–2010), and in the Gulf of Riga (1996–2010).

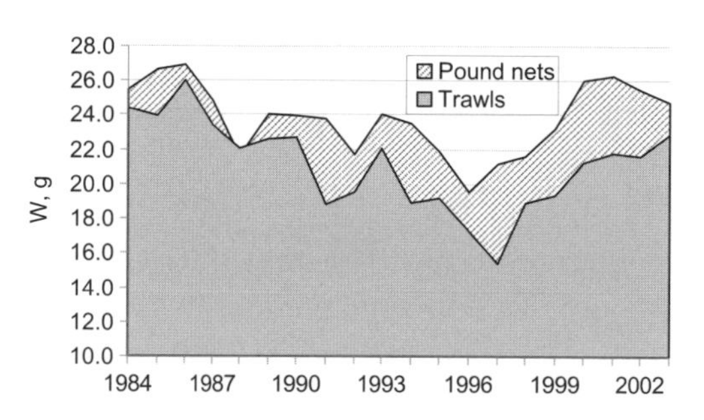

Figure 5. The trends in mean weight of herring at ages 2–7 in the trawl and pound-net catches in Gulf of Riga in 1984–2003.

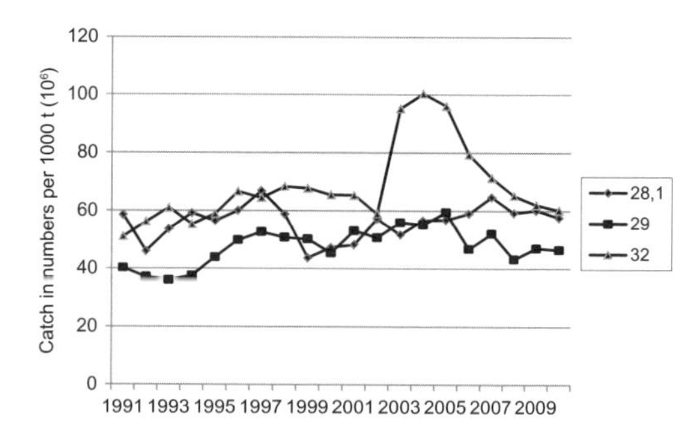

Figure 6. Catch in numbers per 1000 t of landings in Estonian herring fishery in the Baltic Sub-divisions 28.1, 29 and 32 in 1991–2010.

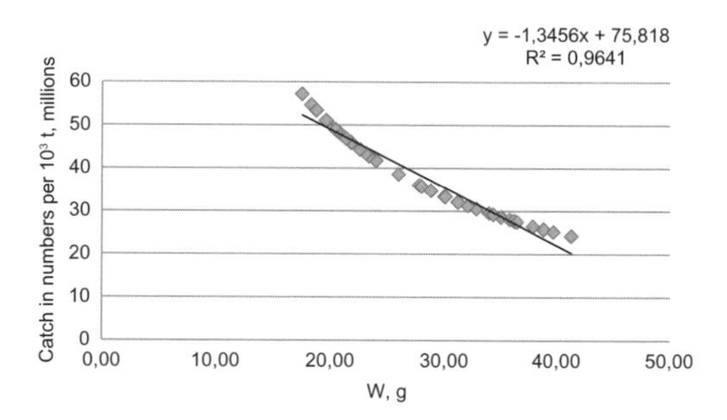

Figure 7. The relationship between the mean weights in age groups 2–7 of the Gulf of Riga herring and the number of catch per 1000 t of landings in 1970–2009.

given assessment unit, has an equal effect on fish stock(s). In fact, as it was shown above, a remarkable variability exists in catch structure of herring even in rather limited area of North-Eastern Baltic (Fig. 6.). Additionally, due to the differences in catch composition, and mean weights at age the effect of trawl and trap-net/pound-net fishery on herring stocks is also rather different. In order to estimate that effect we applied the mean catch in numbers per 1000 t of landings.

Figure 7 presents the observed relationship between the mean weight at age of herring and the average number of fish, taken per every 1000 t of landings in 1970–2009.

So, for example, in the Gulf of Riga herring the increase of mean weight in the age groups 2–7 by 1 g would mean the gain in survivors by 1.35 million individuals. The above also means that due to the differences in mean weights of trawl and pound-net catches, the effect of those on the stock also would be different. Taking into account the differences in mean weight between trawl and pound-net catches we can assume, that if all catch in 2nd quarter in the period of 1984–2003 would have been taken with pound-nets, the resulting gains in the number of survivors would have been on average 3.6 million individuals per 1000 t of landings (Fig. 8.). The latter also indicates that the higher share of pound-net fishery might allow reducing the fishing mortality of the Baltic herring. The extensive development of trawl fishery would lead to unnecessarily big losses in abundance while the quality of catch (condition of fish, mean weight at age) is usually lower than that of in pound-net/trap-net fishery.

Moreover, the trawl fishery is directed, besides to the adult fish, also to the premature fraction of the stock, which is still in the phase of intensive individual growth, and when the increase of biomass, due to growth processes is prevailing above the natural mortality. Therefore, the exploitation of adult fraction of stock that has exhausted its main growth potential looks more reasonable.

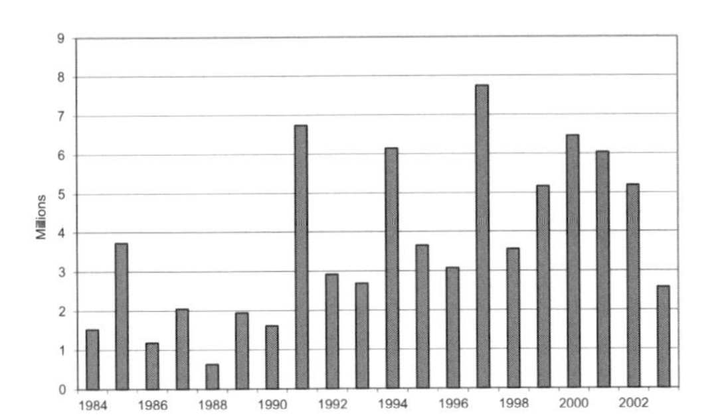

Figure 8. Potential gains in survivors in the 2nd quarter Gulf of Riga herring fishery if all catch would have been taken by pound-nets fishery in 1984–2003.

At the same time, due to its seasonal character and the fact, that it is focused on harvesting the pre-spawning shoals, the increase of the share of pound net fishery must be taken with caution to not allow the decrease of reproduction capacity of herring stocks.

Our results also indicate that, the local distribution of national quotas between different fishing areas and fleets should deserve more attention as a tool for improvement of management quality of Baltic herring stocks. Moreover, further attempts should be made to optimise the pattern of the assessment units for the Baltic herring.

ACKNOWLEDGEMENT

This work was partially financed by the Estonian Ministry of Education and Research GrantNo. F0180005s10.

REFERENCES

Anonymus 1995. Estonian Fisheries Act (RT I 1995, 80,1384) https://www.riigiteataja.ee/ert/act.jsp?id.

Burd, A.C. 1991. The North Sea Herring Fishery: An Abrogation of Management. In: Proc. Int. Herring Symposium Oct. 1990, Anchorage, Alaska. Alaska Sea Grant College Program Report No. 91-01: 1–21.

Gulland, J.A. 1964. Variations in selection factors and mesh differentials. J.Cons. perm. int. Explor. Mer. 29(2): 158–165.

ICES 2009. Report of the Workshop on Multi-annual management of Pelagic Fish Stocks in the Baltic (WKMAMPEL). Copenhagen, ICES CM 2009/ACOM: 38.

ICES 2010. Report of the Baltic Fisheries Assessment Working Group (WGBFAS). ICES CM 2010\ ACOM: 07.

ICES 2010a. Report of the Advisory Committee. Book 8. The Baltic Sea. ICES Advice 2010, Book 8, 118 p.

ICES 2010b. Report of the Baltic International Fish Survey Working Group. ICES CM 2010/SSGESST: 07. Ref. SCICOM, WGISUR, ACOM.

Jefanov, S. 1983. The effect of different cod-ends on the trawl selectivity. Fisherman Handbook. 3: 10–18. (in Estonian).

Järvik, A. 1975. Selectivity of trawls. Fisherman Handbook. 5: 40–47. (in Estonian).

Järvik, A. & T. Raid. 2001. The problem of Mesh Size in Baltic Herring Trawl Fishery. In: Proc. Int. Herring Symposium Oct. 1990, Anchorage, Alaska. Alaska Sea Grant College Program Report No. 91-01: 533–541.

Järvik, A. Raid, T. Shpilev, H. Järv, L. & Lankov, A. 2005. Precautionary approach in Baltic herring trawl fishery: the effect of hauling techniques and engine power on unaccounted mortality estimates. In: Soares, G.C., Garbatov, Y. & Fonseca, N. Proc of the 12th International Congress of the International Maritime Association of the Mediterranean (IMAAM 2005), Lisboa, Portugal, 26–30 September 2005, Taylor & Francis, London/Leiden/New/York/ Phyladelphia/Singapore. Vol. 2: 1223–1330.

Ojaveer, E. 1967. The effect of bottom and demersal trawl fishery on herring stocks.- Proc. BaltNIIRH, 3: 129–145. (in Russian).

Parmanne, R. Sjöblom, V. 1985. Finnish herring fishery in the Gulf of Finland in 1977–1983- Finnish Fish. Res. 6: 15–19.

Suuronen, P. 1995. Conservation of young fish by management of trawl selectivity. Finnish Fisheries Research, 15, 97–116.

9.2 *Off-shore & coastal development*

Sustainable Maritime Transportation and Exploitation of Sea Resources – Rizzuto & Guedes Soares (eds)

An experimental study on brine disposal under wave conditions

B. Bas
Faculty of Naval Architecture and Ocean Engineering, ITU, Istanbul, Turkey
Coastal Sciences and Engineering Division, Graduate School of Science, Engineering and Technology, ITU, Istanbul, Turkey

S.N. Erturk Bozkurtoglu
Faculty of Naval Architecture and Ocean Engineering, ITU, Istanbul, Turkey

S. Kabdasli
Coastal Sciences and Engineering Division, Graduate School of Science, Engineering and Technology, ITU, Istanbul, Turkey

ABSTRACT: Marine outfall is the most common method for disposal of brine which is the concentrated reject from desalination plants. This waste is harmful for the receiving marine environment because of its high salinity and chemical content. In order to develop techniques to reduce these adverse effects, it is important to understand behaviour of this waste in the marine environment. In this study, results of some experiments dealing with vertical mixing of a dense saline layer under wave conditions are presented.

1 INTRODUCTION

Industrial desalination has been used since middle of 20th century in various countries, especially in the Gulf region where the water shortage is a major problem and contributed to development of these countries. Desalination of seawater is achieved by using various treatment technologies which are different from conventional treatment methods. At the end of these processes, brine is produced besides desalinated water which is concentrated wastewater with high salinity concentrations and high density. The most common method for disposal of brine is marine outfalls especially for seawater desalination plants because of cost efficiency. However, brine causes serious harmful effects in the marine environment (Fernández-Torquemeda et al., 2005; Gacia et al., 2007; Dupavillon & Bronwyn, 2009; Malcangio & Petrillo, 2010) despite the general opinion that disposal of seawater with higher salinity than its original environs is not destructive. Disposed brine constitutes a dense layer at the sea bottom because of its higher density compared to the receiving environment eventuating in negative effects especially on benthic organisms. Degree of these effects depends on characteristics of desalination plant and disposed brine, physical features (bathymetry, hydrodynamic properties, morphodynamic characteristics of coastline, etc.) and biological condition of the receiving marine environment. In addition to salinity increase, it gives rise to decrease of Dissolved Oxygen (DO) level, increase of seawater temperature near the discharge point depending on the selected desalination method and increase of turbidity values. There are some experimental and numerical studies concerning with dense jet discharges into stagnant and streamy environment in the literature (Turner, 1966; Zeitoun,vd, 1970; Roberts, 1987; Cipollina, vd, 2005; Shiau et al., 2007). However effect of waves on behaviour of discharged brine is comparatively less investigated (Payo et al., 2010; Marti et al., 2010). Determining the behaviour of disposed brine under wave conditions is important to develop solutions for decreasing the negative effects of it for marine environment. In this context, mixing of a dense layer with salt solutions under regular wave conditions is examined with tests in a wave flume in Hydraulics Laboratory of Istanbul Technical University. During the experiments, salt solutions are used for simulating brine and it is discharged from a horizontal pipe for forming of a dense layer and then mixing of this layer is provided under regular wave conditions. This paper examines the test results as an initial stage of our further studies.

2 BEHAVIOUR OF BRINE IN THE MARINE ENVIRONMENT

Behaviour of brine jet in the marine environment can be examined in two parts as near and far field.

Near-field is the region where momentum flux, buoyancy flux and discharge system geometry is effective on the jet direction and mixing process. Far-field is the region where these parameters leave their places to characteristics of the receiving environment (Vaselali & Vaselali, 2009).

The density difference between the receiving environment and discharged brine is the most important parameter for mixing and transport of that waste. Brine behaves differently then classical jets because of its higher density compared to its receiving environment and causes formation of a dense layer at the sea bottom (Younos, 2005; Malcangio & Petrillo, 2010; Ruiz-Mateo, et al., 2008). This layer comes into far-field region, no matter at what degree the mixing rate in the near-field region is. As the flow develops, width of discharged brine layer increases with decreasing thickness. After this process, an intermediate layer constitutes with salinity values between the receiving environment and this layer. Under the intermediate layer, dense layer continues keeping its own salinity, temperature and density values (Ruiz-Mateo, et al., 2008).

Field studies are also helpful for understanding brine behaviour. In their field study, Payo, et al., (2010) showed that the wave action reduced near bottom salinity, most highly after local wave height maximums. In the same study, it is determined that storm duration is effective on near bottom salinity values. Marti et al., (2010) investigated the near field characteristics of a brine outfall from desalination plant via an offshore diffuser in Australia with field measurements. According to their findings, dilution ratios were compatible with literature sources for densimetric Froude numbers (F) higher than 20.

Potential energy can be used as a measure of stratification due to salinity values in the profile and their vertical location in the water column. By this way it is possible to relate the events causing mixing and stratification. In terms of mixing, external kinetic energy added to the water column gives rise to increasing of potential energy due to the lifting of dense fluid at the bottom (Hodges, et al., 2006).

According to linear wave theory, the total energy per wave per unit width is:

$$E = \frac{1}{8}\rho g H^2$$

and wave energy flux through a unit width is:

$$P = Ec_g$$

where c_g is the group velocity.

In addition, potential energy (E_p) of the water column can be described as:

$$E_p = g\int_0^h \rho z dz$$

3 MATERIAL AND METHOD

3.1 *Experimental system*

Experiments were conducted in a wave flume with dimensions 22.5 m. × 1 m. × 1 m. in Hydraulics Laboratory of Istanbul Technical University. The wave flume has glass walls for easy observation and a wave generator at one end to produce regular waves with a metal slope behind it to prevent the swash behind it. At the other end of the flume, a sandy beach is established for preventing wave reflection. During the experiments water surface elevation was measured with resistant type wave gauges. Wave data was evaluated with zero-crossing method to obtain wave height values. For simulating brine discharges, salt (NaCl) solution with 35 ppt salinity was prepared with freshwater. In this way, same density difference ratio (~0.0256) between the effluent and the receiving environment was provided with the real situation as applied in Ruiz-Mateo, et al.,'s (2008) study. For the flow visualization, the solution was colored with Ponceau 4R red food dye as trace material. Experiments were recorded with a Canon Legria HF M36 digital camera. The wave flume is filled with freshwater and depleted, washed and refilled after every two experiment with the same configuration. Salinity values were measured with salinity/conductivity probes of two WTW MultiLine meters, each one is placed at 0.8 m and 5 m after the discharge point, respectively. Dense jet discharge was implemented from a solution tank at the top of the flume via a plastic tube with 4 mm diameter, 10 cm above the bottom. Schematic diagram and a picture of the experimental set up can be seen in Figures 1 and 2, respectively. At each test, a uniform density layer (~8 cm thick) composed along the flume and then different waves were generated to investigate the effect of wave properties on mixing of this dense layer. Each characteristic regular wave is applied for 5 and 10 minutes for different tests and the vertical distribution of salinity is measured after each test. Experimental procedure is given at Table 1. In each test, water depth is 0.33 m and the waves are transitional with d/L<1/20 ratios. Here d is the water depth and L is the wave length calculated according to the linear wave theory:

$$L = \frac{gT^2}{2\pi}\tanh\left(\frac{2\pi d}{L}\right)$$

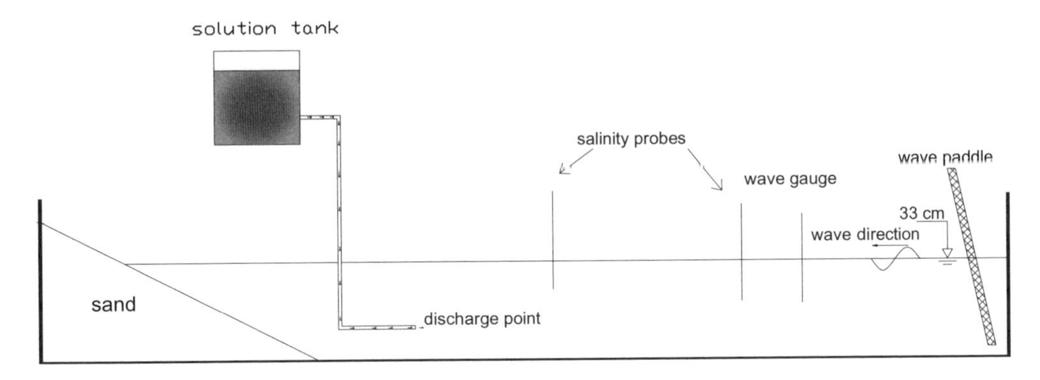

Figure 1. Schematic diagram of the experimental set up.

Figure 2. A picture from the experiments showing formation of dense layer.

Table 1. Test procedure.

Test no	Duration, t (min)	Wave height, H (cm)	Mean wave period, T (s)
1	5	5.81	0.71
2	10	5.81	0.71
3	5	6.36	0.74
4	10	6.36	0.74
5	5	6.93	0.73
6	10	6.93	0.73
7	5	7.86	0.76
8	10	7.86	0.76

4 RESULTS

4.1 *Formation of dense layer*

As mentioned before, a dense layer is constituted in the bottom of the wave flume for each experiment by discharging salt solutions in equal volume and salinity. Salinity values of this layer are measured for at 1 cm distances vertically and are shown in Figure 4 as the stagnant situation. As it can be seen, this dense layer has almost the same salinity and thickness values for each tests.

4.2 *Experiments with regular waves*

During the experiments, it was seen that the dense layer at the bottom of the wave flume was oscillating with water surface with a phase difference (Fig. 3). This situation can be explained by pressure differences at the bottom due to the wave motion. In the experiments, wave motion is the main factor effecting mixing of dense layer besides molecular diffusion.

As it can be seen from the Figure 4, wave motion caused mixing of dense layer with fresh water ending in increased water column height with salinity values for the experiment with 5 minutes duration. In the tests for 10 minute wave

Figure 3. Movement of dense layer with waves.

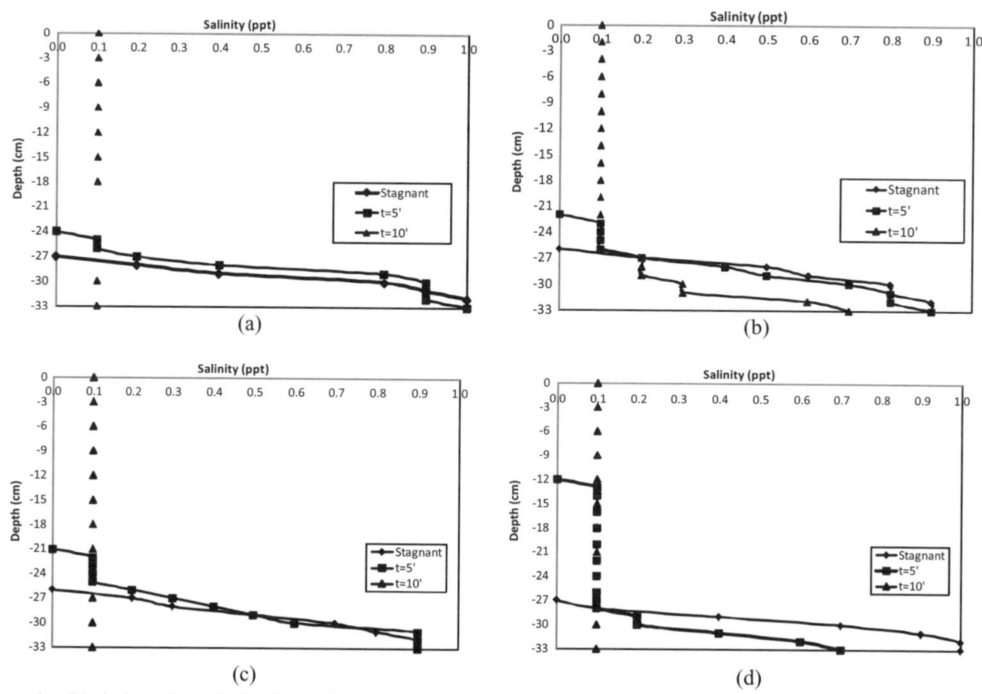

Figure 4. Variation of vertical salinity distribution for the tests a)1–2, b) 3–4, c) 5–6, d) 7–8.

generation, mixing is increased and salinity layer became thicker. As the wave height increased, salinity profiles became uniform along the whole water column with 0.1 ppt value for the two highest wave height values.

Additionally, wave energy and potential energy exchange for each experiment is calculated by using linear wave theory and vertical salinity profile values, respectively.

For the total wave energy calculation per area for test duration (t_e):

$$E_w = \frac{1}{8}\rho g H^2 \frac{t_e}{L}$$

formula is used. For potential energy calculation, density values were estimated from salinity values. Potential energy of the water column was calculated

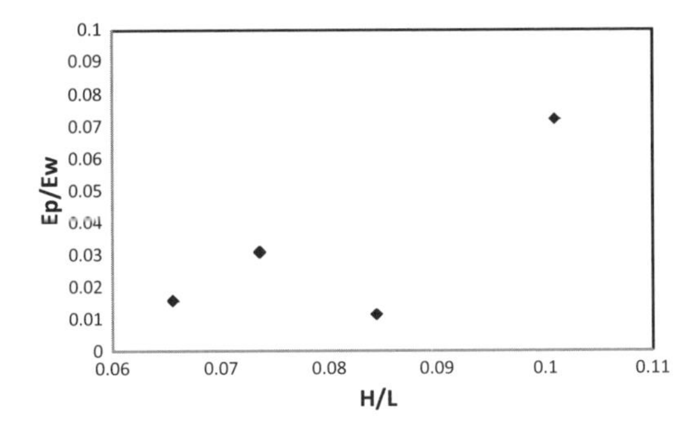

Figure 5. Energy ratio versus wave steepness.

for the tests 1, 3, 5 and 7 by using potential energy formula for each measurement height. These values can be seen in Figure 5 depending on wave steepness (H/L). According to the graphic, ratio of potential energy of the mixed water column to generated wave energy has an increasing trend with wave steepness.

5 CONCLUSION

Wave energy can be effective on the vertical mixing of a dense saline layer in transitional wave conditions for our test conditions. In this process, wave energy in the form of kinetic energy is transferred to the water column and causes mixing. At the end of this process, potential energy of the water column increases due to ascending salinity concentrations in the water column. Wave motion is responsible for this situation.

ACKNOWLEDGEMENT

The authors would like to thank Dr. Ozgur Kirca and Taylan Bagci for their valuable comments and help for the experiments.

REFERENCES

Cipollina, A., Brucato, A., Grisafi, F., Nicosia, S., (2005). Bench-scale investigation of inclined dense jets, Journal of Hydraulic Engineering, 131:11, 1017–1022.

Dupavillon, J.L., Bronwyn, M.G., (2009). Impacts of seawater desalination on the giant Australian cuttlefish Sepia apama in the upper Spencer Gulf, South Australia. Marine Environmental Research, 67, 207–218.

Fernández-Torquemada, Y., Sánchez-Lizaso, J.L., González-Corea, J.M. (2005). Preliminary results of the monitoring of the brine discharge produced by the SWRI desalination plant of Alicante (SE Spain). Desalination, 182, 395–402.

Gacia, E., Invers, O., Manzanera, M., Ballesteros, E., Romero, J. (2007). Impact of the brine from a desalination plant on a shallow seagrass (Posidonia oceanica) meadow. Estuarine Coastal and Shelf Science, 72, 579–590.

Hodges, B.R., Kulis, P.S., David, C.H., (2006). Desalination brine discharge final report, *Submitted to Texas Water Development Board.*

Malgancio, D. & Petrillo, A.F., 2010. Modelling of brine outfall at the planning stage of desalination plants. *Desalination,* 254, 1–3, 114–125.

Marti, C.L., Antenucci, J.P., Luketina, D., Okely, P., Imberger, J., (2010). Near field dilution characteristics of a negatively buoyant hypersaline jet generated by a desalination plant, Journal of Hydraulic Engineering, Posted ahead of print 11 May 2010.

Payo, A., Cortés, J.M., Antoranz, A., Molina, R., 2010. Effect of wind and waves on a nearshore brine discharge dilution in the east coast of Spain, *Desalination and Water Treatment,* 18, 71–79.

Roberts, P.J.W., Toms, G. (1987). Inclined dense jets in flowing current. Journal of Hydraulic Engineering, 113 (3), 323–341.

Ruiz-Mateo, A., Antequera, M., González, M., (2008). Physical modelling of brine discharges to the sea. MWWD 2008—5th International Conference on Marine Waste Water Discharges and Coastal Environment, October 27–31, Cavtat (Dubrovnik, Croatia).

Shiau, B.-S., Yang, C-L., Tsai, B-J., (2007). Experiental observations on the submerged discharge of brine into coastal water in flowing water. ICS 2007 (Proceedings), Journal of Coastal Research, Special Issue 50, 789–793.

Turner, J.S., (1966). Jets and plumes with negative or reversing buoyancy. Journal of Fluid Mechanics, 26, 779–792.

Vaselali, A., Vaselali, M. (2009). Modelling of brine waste discharges spreading under tidal currents. Journal of Applied Sciences, 9, 19, 3454–3468.

Younos, T. (2000). Environmental issues of desalination. *Journal of Contemporary Water Research & Education,* 132: 11–18.

Zeitoun, M.A., Reid, R.O., McHilhenny, W.F., Mitchell, T.M., (1970). Model studies of outfall systems for desalination plants. Part III Numerical simulations and design considerations, Res. and Devel. Progress Rep. No. 804, Office of Saline Water, U.S. Dept. of Interior, Washington, DC.

Sustainable Maritime Transportation and Exploitation of Sea Resources – Rizzuto & Guedes Soares (eds)
© 2012 Taylor & Francis Group, London, ISBN 978-0-415-62081-9

Nozzle-diffuser effects on marine current turbines

V. Díaz Casas, Pablo Fariñas & F. López Peña
Integrated Group for Engineering Research, University of Corunna, Spain

Sara Ferreño
Cetnaga—Centro Tecnológico del Naval Gallego, Spain

ABSTRACT: The work presented in this paper is focused in the improvement of current turbine devices by including a nozzle-diffuser shroud in its design. This technology is being used on the wind turbine technology and is adapted here to the underwater environment. This presents some advantages due to the possibility to install big size ducted turbines in subsea facilities without getting severe structural constrains generated by gravitational forces.

1 INTRODUCTION

A large number of designs and prototypes have been developed during the last years to obtain energy from different sources, among them marine current turbine devices are showed as a promising good alternative in the near future.

Physics of marine current turbines is quite similar to that of wind turbines. In both cases the power available is proportional to the working fluid density and the cube of its velocity. The largest difference between the two resources is that the density of seawater is three orders of magnitude greater than that of air. Therefore the power output from a marine current turbine should be three orders of magnitude higher than a wind energy turbine having similar dimensions and similar working fluid velocities [1][2].

The work presented in this paper is focused in the improvement of current turbine devices by including a nozzle-diffuser shroud in its design [3]. This technology is being used on the wind turbine technology and is adapted here to the underwater environment. However a direct translation from wind technology to subsea systems implies the lost of some advantages that this kind of topology provides.

It is interesting to point out the possibility of installing big size ducted turbines in subsea facilities without getting severe structural constrains generated by gravitational forces, that is a mayor drawback when applying this technology to wind turbines.

The effects of the nozzle-diffuser shroud on the marine current turbine design have been divided into two main aspects:

- Increasing of hydrodynamic performance.
- Positioning and orientation.

Concerning the first aspect, a design methodology has been developed and, in addition, a comparison between a ducted and an equivalent free turbine has been made. Furthermore, the subsequent increment on turbine performance has been evaluated in order to infer the behavior of an actual shrouded turbine.

Positioning and orientation aspects are focused on the study of the nozzle-diffuser shroud as a structural element able to control the orientation and position of the turbine assembly. Within this topic it is also evaluated the structural effects arising by introducing the turbine into a shroud, as well as some geometrical considerations and additional constrains that must be taken into account during the design phase of a shrouded marine current turbine.

2 AVAILABLE ENERGY

A marine current turbine transforms the kinetic energy in the marine flow to mechanical energy that should be transform in electrical energy in a generator. Thus, the maximum available energy is obtained as if the final current velocity could be reduced to zero in the turbine, thus it should be calculated as:

$$P_{\max} = \frac{1}{2}\rho A V^3 \tag{1}$$

where $P_{\max}$ = the maximum available energy, ρ = sea water density, A = area covered by the rotor and V = velocity of the marine current.

However the final velocity cannot be reduced to zero, then the energy that can be transformed results to be:

$$P = \dot{m}\left(\frac{1}{2}V^2 - \frac{1}{2}u^2\right) = \rho A\left(\frac{1}{2}(V+u)\right)\left(\frac{1}{2}V^2 - \frac{1}{2}u^2\right) \tag{?}$$

where u = final velocity.

Introducing an axial induction factor, a, defined as

$$\frac{1}{2}(V+u) = (1-a)V, \tag{3}$$

then the available power can be calculated as

$$P = 2\rho A V^3 a(1-a)^2. \tag{4}$$

From differentiating the previous equation the maximum energy that can be extracted from the marine current, P_{av}, results:

$$\frac{\partial P}{\partial a} = 0 = 4(1-a)(1-3a) \rightarrow a = \frac{1}{3} \tag{5}$$

$$P_{av} = \frac{8}{27}\rho A V^3 = \frac{16}{27}P_{max} \tag{6}$$

This value is known as Betz limit [4] and it becomes an efficiency barrier both for marine and wind turbines.

3 DUCTED ROTORS

In the previous section a limit for the maximum power coefficient of any marine current turbine has been found to be of 59.2%. Therefore design effort have been focused in achieve an efficiency as closer as possible to this limit. Thus a large number of different configurations and blade shapes have been developed in order to improve the aerodynamic performance, however this aerodynamic efficiency rarely achieve a bare 45% [5].

One of the first experiments in the field of shrouded turbines was developed by Kogan in the earlier 60's [6] [7]. In this experiment a convergent-divergent nozzle was applied to a wind turbine. In these case the higher efficiency occurred when the ratio between the duct length and the duct diameter was 7:1 and the drag coefficient of the duct was between 0.18 and 0.22. Research on shrouded wind turbines and applications to pilot plans was continued by Igra [8] who perform experiments using NACA 4412 airfoils and achieving an efficiency of 52%.

However, despite the improvements achieved in designing shroud nozzles for wind turbines, their use is not widespread mainly due to structural problems that become unsolvable for large wind turbines. These structural difficulties have their origin in the heavy weight of the shroud. Thus, one of the advantages of applying this technology to marine current turbines is that this weight can be cancel out by buoyance forces making unnecessary the use of complex support structures.

4 DESIGN

A test case has been developed to serve as illustration to the design process proposed. In this way a basic blade have been designed in order to have a reference value of ducted/not ducted turbine and to compare the forces over the structure.

Thus, the design of the nozzle has been focused on the definition of its requirements for structural strength, stability and hydrodynamic behavior.

4.1 *Blade definition*

In this first step, for the basic definition of the blades a standard hydrodynamic profile has been chosen (i.e. a NACA 63-415) and a constant marine current velocity has been considered (3 knots). The goal is to calculate the optimal distribution of the chord and the angle along the length of the blade that maximize efficiency.

The well-known "Blade Element Method (BEM)" [4] has been used to evaluate blade performance and, ultimately, determining the optimal configuration of the blade. The main parameters of the design are summarize in Table 1.

BEM was originally developed for aircraft propellers and it has been largely improved along time. Today has become the most common design tool for horizontal wind turbines rotors [9]. BEM makes use of two dimensional airfoil data to approximate the full three dimensional rotor. The blade is divided into several sections—elements- and the method integrates the aerodynamic performance of these elements. Modern BEM methods take into account elements mutual interactions, wake influence, tip losses and some other relevant effects. For rotors having a large radius to chord ratio, BEM results may reach the accuracy of those of some 3D computations, which take tens

Figure 1. NACA 63-415 airfoil.

Table 1. Main dimensions of the test case marine current turbine blades.

Airfoil	NACA 63-415
Diameter	D = 8 m
Number of blades	B = 3
Marine current velocity	V_0 = 1.54 m/s
Water density	ρ = 1025 kg/m^3

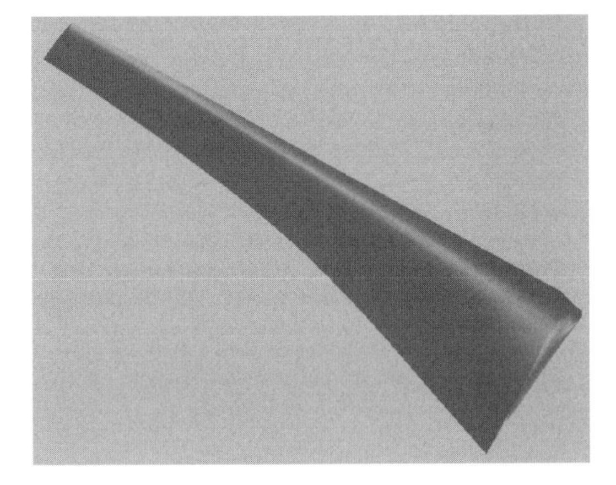

Figure 2. 3D representation of the blade.

of times longer to be computed. The applicability of BEM to marine current turbines has been demonstrated [10].

To define the blade the twist of each section has been selected in order of achieving an apparent angle of attack maximizing the lift/drag ratio. For this airfoil this happens to be at an angle of attack of 4°. Thus, a blade having an apparent wind coming at 4° of angle of attack all along its span will achieve the maximal possible performance for this airfoil.

4.2 *Nozzle design*

A converging/diverging nozzle configuration is used to improve inflow conditions and to increase turbine performance.

The definition of the nozzle is made by three parameters:

- Cross sectional profile
- Angle of attack
- Chord

In addition, in order to obtain an optimal configuration two different parameters have been taken into account; on one hand, its hydrodynamic performance and, on other hand the stabilization platform.

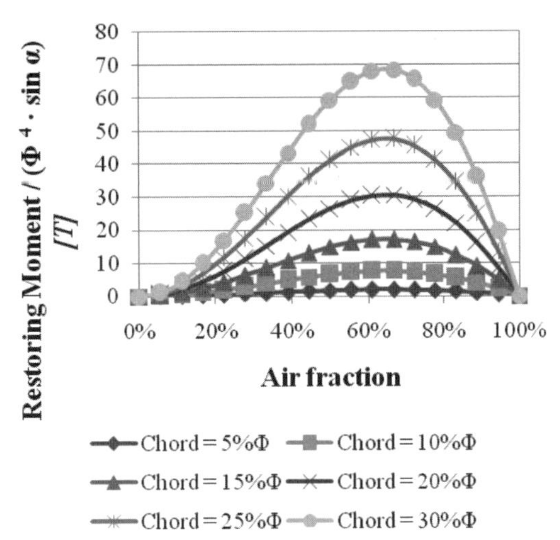

Figure 3. Restoring Moment as a function of the air fraction inside the nozzle.

5 STABILIZATION SYSTEM

It should be considered that the effect of ocean currents as well as the actions of the mooring system could generate some trimming angles on the nozzle. This trimming can produce and adverse effect on the correct operation of the system. Therefore, a stabilization system has been designed in order to achieve a correct positioning of the whole turbine when suffering these actions. This system is based on a ballast/vacuum mechanism placed inside the nozzle.

The nozzle has an inner and an outer hull. The area in between the two hulls is used the ballast tank that can be filled with either air or water. The trimming angle can be adjusted by controlling the air/water ratio inside the nozzle.

Figure 3 shows the relation between the fraction of air inside the nozzle and the Restoring Moment that is obtained (put as function as the diameter of the nozzle, Φ and trimming angle, α).

Six curves have been drawn, corresponding to different values of the nozzle's chord (calculated as a fraction of the nozzle's diameter, and ranging from 5% to 30%). In the x-axis air fraction is represented: 0% denotes that nozzle is filled with water, and 100% denotes that nozzle is filled with air. Values on the y-axis correspond to the ratio between Righting Moment and the product of the diameter of the nozzle and its heeling angle.

6 HYDRODYNAMIC BEHAVIOR

The main effect of the nozzle is to produce an increment of the flow through the rotor. Therefore the presence of the diffuser increases the power for the same thrust coefficient when comparing with a non-ducted turbine.

If the effect of the diffuser is taking into account by the increment of flow it produces, the power coefficient of a shrouded turbine can be defined as:

$$C_{P,d} = \frac{P}{\frac{1}{2}\rho V^2 A} = \frac{TV_d}{\frac{1}{2}\rho V^2 V_d \left(\frac{V}{V_d}\right) A} = C_T \varepsilon \quad (7)$$

where: C_T = Thrust coefficient, V_d = induced velocity produced by the diffusor, $C_{p,d}$ = power coefficient of the ducted turbine and $\varepsilon = V_d / V$.

If an ideal 1D turbine without nozzle is considered, then C_T and C_P can be calculated as:

$$C_P = 4a(1-a)^2 \quad (8)$$

$$C_T = 4a(1-a) \quad (9)$$

Therefore, the increment of the performance due to the nozzle results to be:

$$\frac{C_{P,d}}{C_P} = \frac{\varepsilon}{1-a} \quad (10)$$

7 CONCLUSIONS

In order to evaluate any arbitrary shroud configuration that could be considered, a specific algorithm analysis has been developed. This allows assessing the influence of the shroud on the velocity of the incoming flow in order to optimize the power coefficient of the rotor, and thus maximizing system performance.

In this way, an algorithm for optimal positioning of the nozzle that takes into account the increasing of power generated as well as the losses due to detachment of the flow has been was established (equation 10). The increasing of power generated is calculated from the increment of velocity in the incoming flow.

After the analysis of a test case a NACA 23012 airfoil has been selected for it having a chord of 2 meters and an angle of attack of 6. The value of the angle of attack is limited as a result of the detachment of the boundary layer, since a larger angle can generate an important drop in the performance due to the reduction of the effective area of the rotor generated by an earlier detachment of the boundary layer.

Thus, the proposed nozzle produces an increment on the generated power of 11%. In conclusion the use of a nozzle not only provides a support and protects de turbine, but also enhances the turbine performance.

ACKNOWLEDGEMENT

This work was partially funded by FEDER founds and the Xunta de Galicia through projects 08DPI159E and 10REM007CT.

REFERENCES

J. Twidell, T. Weir, *Renewable Energy Resources*. Second Edition. Taylor & Francis. 2006.

I.G. Bryden, T. Grinsted, G.T. Melville, "Assessing the Potential of a Simple Tidal Channel to Deliver Useful Energy". *Applied Ocean Research*. Vol. 26, no 5. pp. 198–204. 2004.

M.O.L. Hansen, N.N. Sørensen, R.G.J. Flay, "Effect of Placing a Diffuser around a Wind Turbine," *WIND ENERGY*, no. 3, pp. 207–213, 2000.

Martin O.L. Hansen, *Aerodynamics of Wind Turbines*. London: James & James, 2003.

Hans Bernhoff, Mats Leijon Sandra Eriksson, *Renewable and Sustainable Energy Reviews*, vol. 12, no. 5, pp. 1419–1434, June 2008.

A. Nissim, E. Kogan, "Shrouded Aerogenerator Design Study, Two-Dimensional Shroud Performance," *Bulletin of the Research Council of Israel*, vol. 11c, pp. 67–88, 1962.

A. and Seginer, A. Kogan, "Final Report on Shroud Design," Department of Aeronautical Engineering, Technion, Report 32A 1963.

O. Igra, "Research and Development for Shrouded Wind Turbine," in *European Wind Energy Conference*, Hamburg, pp. 236–245, 1984.

C. Crawford, "Re-examining the Precepts of the Blade Element Momentum Theory for Coning Rotors," *Wind Energy*. Vol. 9, pp. 457–478, 2006.

W.M.J. Batten, A.S. Bahaj, A.F. Molland, J.R. Chaplin. "The prediction of the hydrodynamic performance of marine current turbines". *Renewable Energy*. Vol. 33, Issue 5, Pages 1085-1096, 2008.

Sustainable Maritime Transportation and Exploitation of Sea Resources – Rizzuto & Guedes Soares (eds)
© 2012 Taylor & Francis Group, London, ISBN 978-0-415-62081-9

Heuristic approach for solving a pipe layer fleet scheduling problem

M.M. Queiroz & A.B. Mendes
Department of Naval and Ocean Engineering, Escola Politécnica da Universidade de São Paulo, São Paulo, Brazil

ABSTRACT: This research addresses a fleet scheduling problem present in the offshore oil industry. Among the special purpose services one will find the pipe layer activities accomplished by Pipe Layer Vessels (PLV). The jobs are characterized by a release date, which reflects the expected arrival date of the necessary material at the port, or the date the environmental license will be issued. There are compatibility constraints between job and vessel, so that some vessels may not be able to perform a certain job, and the duration of the jobs can be differentiated by vessel. This is a variation of the unrelated parallel machine scheduling problem with release date and weighted tardiness objective function. In this paper it will be addressed a heuristic approach based on GRASP with Path Relinking, which have proved to be an effective solution procedure. Computational experiments were conducted, comparing solutions with bounds provided by linear column generation.

1 INTRODUCTION

Scheduling is present in different production environments and involves a number of activities that require one or more resources for a certain amount of time (Morton & Pentico, 1993). It is a decision process that has an important role in most production and manufacturing systems, besides information networks, transportation systems and other types of industries as well (Pinedo, 2002).

Scheduling resources such as crews, machines and vehicles is a challenging task inasmuch as their problems have a combinatorial structure and several operational constraints must be observed. In the last years this has been the object of study of many researchers such as Graham (1969), Lawler & Labetoulle (1978), Ronen (1983), Lenstra et al. (1990), Wen et al. (2010).

This paper will focus on scheduling a fleet of specialized vessels responsible for launching risers and interconnecting them to subsea equipment in an offshore oil production environment. A mixed integer linear programming model will be proposed to represent the problem. Solutions were obtained by a heuristic algorithm based on GRASP with Path Relinking (Resende & Ribeiro, 2003).

2 PROBLEM DESCRIPTION

The offshore oil exploration and production requires specialized support activities, which are accomplished by appropriate support vessels. An important phase for developing an oil field is the launching of risers (production lines), and its connection to subsea equipment like the submarine manifolds. The vessels responsible for such activities are the PLVs (pipe layer vessels), which are critical resources to be managed.

These lines, also known as submarine pipelines, can be flexible or rigid and are fundamental for transporting oil from oil wells to manifolds, from manifolds to oil platforms and from them to onshore storage tanks. The flexible lines are easier to store when compared with the rigid ones, due to their greater flexibility and smaller tonnage.

PLV usually differs in capacity of storing flexible or rigid lines and in the process of launching these lines. Each process has its own disposal rate, which can be affected by environmental conditions and water depth, among other factors.

Various activities are required to install a new production line. At the beginning of the operation the vessel needs to load the lines and the required equipment at the port. Then the vessel moves to assigned place in order to initiate the launching phase. Depending on the water depth and on the line extension, it may be necessary to return to the port for picking up additional material.

When an oil company holds the operational planning of its PLV fleet, all the necessary trips to the port in order to conclude one installation are aggregated into one single job. This is undertaken by experienced technical staff, who identify which are compatible vessels and estimate the whole job duration, including navigation times, for each of the selected PLV.

The demand is composed by jobs originated from oil fields that are being developed by an oil company. Each job will allow expanding the company's production by an expected daily level. Therefore, if a job is delayed, the return on investment is postponed proportionally to the associated daily production economic value.

Some operational constraints must be observed in order to schedule the fleet such as vessel compatibility and the release date of each job. The former requirement is due to technical compatibility, while the latter depends on when the required material (mostly the lines) will be available at the port and when the environmental license will be issued. The planning horizon should also be taken into account inasmuch as the company works with definite planning periods ranging from 3 to 6 months.

The main objective in schedule generation is to minimize the financial losses due to the late start of each job, while meeting the operational constraints. It is assumed that all the information concerning the jobs and the vessels is known in advance. Therefore a deterministic mathematical programming model must be built and solved.

3 LITERATURE REVIEW

The presented problem can be seen as a variation of the unrelated parallel machine scheduling problem. Cheng & Sin (1990) presented a review of important contributions on parallel machine scheduling problems working on different environments.

Horowitz & Sahni (1976) presented exact and approximate algorithms for parallel machines with the objective of minimizing the completion time of jobs and the weighted mean flow time.

Fuller (1978) made a comparison between optimal solutions and good solutions, analyzing the effectiveness of applying heuristics in the decision making process. The author observed that heuristics simplify the process for the decision maker, allowing decisions to be made faster, by evaluating a reduced number of feasible solutions.

Liaw et al. (2003) addressed the problem of unrelated parallel machines to minimize the total weighted tardiness. They reported which properties are found on optimal schedules and derived a branch and bound algorithm. Shim & Kim (2007) worked with identical parallel machines with the objective of minimizing total tardiness. A branch and bound algorithm was also developed based on properties of optimal schedules.

Panwalkar et al. (1993) heuristically solved the minimum mean tardiness single machine scheduling problem by a method called PSK. The proposed heuristic proved to be effective when the due date was tight. Koulamas (1997) incorporated elements from the Earliest Due Date (EDD) and the Shortest Processing Time (SPT) sequencing rules into the PSK algorithm to solve the problem studied by Panwalkar et al. (1993) with multiple (identical) machines.

Alidaee & Rosa (1997) proposed the Modified Due Date heuristic (MDD) for the identical machines scheduling problem with the minimization of the total weighted tardiness. Cao et al. (2005) studied parallel machines scheduling problems where, besides minimizing each job's tardiness, the machines holding were considered.

Biskup et al. (2008) reported a problem with identical parallel machines with the objective of minimizing the total tardiness. Heuristic algorithms were proposed based on the following sequencing rules: EDD, MDD, TPI and Minimum Slack. These rules were compared with the classic versions of MDD, KPM, TPI and proved to have good overall performance.

Ronen (1993) addressed the issue of ship scheduling, making a review of the most important contributions. The problems found in the literature where usually related to ship scheduling, with the objective of minimizing the routing costs, and the fleet size and composition problem, where fixed and operational costs are included in the objective function.

Brønmo et al. (2007a) presented a multi-start local search heuristic for the ship scheduling problems. The initial solutions were randomly generated by insertions mechanisms. The best initial solutions were improved by a local search heuristic. Brønmo et al. (2007b) proposed a mixed integer linear programming model and a set partitioning model to a routing and scheduling problem of a fleet of ships with loads of flexible sizes.

3.1 *Grasp metaheuristic*

According to Feo & Resende (1989, 1995), GRASP (Greedy Randomized Adaptive Search Procedure) is a meta-heuristic that works with multiple starting solutions having, in each iteration, two main steps: a construction phase and a local search phase. In the first phase a feasible solution is generated by a greedy randomized algorithm, while in the second phase, the neighborhood of the generated solution is explored until a local optimum is reached.

First of all, a list of unscheduled jobs must be ordered according to a specific criterion, for instance, the minimum release date. Then a Restricted Candidate List (RCL) is drawn, containing a subset of the most promising elements to be chosen. One job will be randomly selected from the RCL and added to the partial solution. The RCL is then updated, after having all unscheduled jobs reclassified or simply by following the initial established sequence. After all jobs have been picked a complete solution will have been generated.

Regarding the definition of the RCL size, Prais & Ribeiro (2000) developed an approach known as Reactive GRASP where the best size is iteratively updated. Boudia et al. (2007) confirmed the effectiveness of the reactive approach while solving a combined production and distribution problem.

3.2 *Path relinking*

The Path Relinking (PR) mechanism was originally described by Glover (1996) and also in Glover et al. (2000). It consists in exploring the path between two given solutions named "initial" and "guiding" solutions. Usually the guiding solution belongs to a so called "pool of elite solutions", referring to a set of high quality solutions generated along the iterations. In the PR algorithm a modification is made in the initial solution by introducing attributes found in the guiding solution, until both solutions are the same. This trajectory can be understood as an intensification process that takes place while the solution space between the two solutions is explored.

It is possible that during this process, a new local optimum is reached, as illustrated in Figure 1. The initial solution is given by x' and the guiding solution by x''. To reach x'' from x', new solutions are iteratively produced: $x' = x(1), x(2), \ldots, x(r) = x''$ indicated by dotted line.

Ho & Gendreau (2006) discuss important issues regarding Path Relinking implementation, such as the criteria for admitting a new solution into the set of elite solutions and for excluding one solution from the pool. Also, which solution (initial or guiding) could in fact be the starting solution.

Path Relinking has different forms of implementation. Resende et al. (2010a) discusses two variants for interacting with the pool of elite solutions. In the static case, all the generated GRASP solutions are submitted to the pool, having the acceptance rules verified. After the iteration limit is reached the path between all pairs of elite solutions is explored. In the dynamic implementation, each GRASP solution obtained from the construction phase is submitted to PR, by randomly selection a solution from the pool of elite solution. The resulting PR solution, either from the static or from the dynamic case, always undergoes a local search phase.

Resende & Werneck (2004) merged these two strategies by firstly applying PR to every GRASP solution. Then the final pool of elite solutions was submitted to a so called Evolutionary Path Relinking (EvPR) scheme where new populations were generated by applying PR within the elite solutions. This process was repeated while improved solutions were generated.

Resende & Ribeiro (2003) highlight other variations. In the Forward Relinking the path is constructed starting from the initial solution x' aiming to reach the guiding solution x''; In the Backward Relinking this strategy is opposite: the initial solution in this case is x'' and the guiding solution x'; In the Mixed Relinking two paths are explored simultaneously, the first beginning in x'' and the second in x'', until they meet.

Resende et al. (2010b) discuss the Randomized Path Relinking (RPR). The usual version of Path Relinking adopts a greedy criterion in choosing the move (or attributes) in each iteration, while generating the path from the initial to the guiding solution. This means that the best movement will be applied to the initial solution along the PR iterations. In the RPR a restricted candidates list is introduced, in which the best moves are stored an one of them is randomly chosen. This ensures that different trajectories are generated, increasing the possibility of achieving new solutions, possibly, of good quality.

Ribeiro & Rossetti (2007), Ribeiro et al. (2009) use the "time-to-target solution" concept as being the probability of obtaining a solution less than or equal to a given (target) solution within a specified time. The authors also discussed implementation strategies with parallel metaheuristics in order to accelerate the search and solve bigger problem instances. Two approaches were presented: a parallel independent form where (threads) do not exchange any information and cooperative parallelization with information being shared and used by other threads.

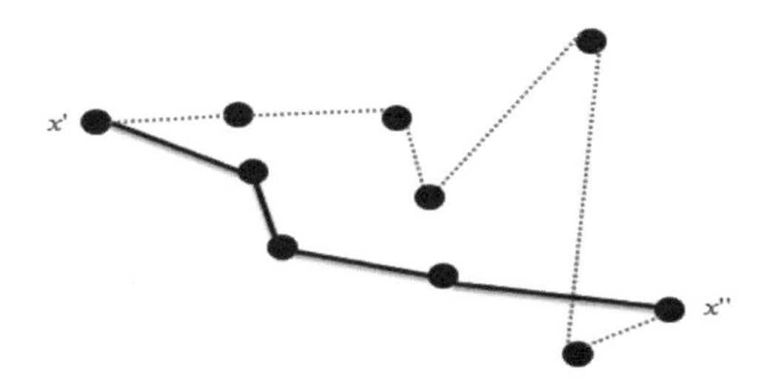

Figure 1. Path relinking example.

4 MATHEMATICAL MODEL

The mathematical model for the problem described is this paper is now introduced as:

Sets

J	Set of jobs	$j = 1, \ldots, n$
M	Set of vessels	$i = 1, \ldots, m$

Parameters

T Planning horizon (index t)
w_{jt} Penalty for starting job j at time t

r_j Release time of job j
p_j^i Processing time for job j by vessel i
a_{ij} Binary parameter which will be equal to 1 if a job j can be done by vessel i.

Decision Variables

$x_{ij}^t = 1$ If job j is assigned to vessel i at time t; 0, otherwise

Mathematical Model

$$min\, C = \sum_{i=1}^{m} \sum_{j=1}^{n} \sum_{t=r_j}^{T-p_j^i+1} w_{jt}\, x_{ij}^t \quad (1)$$

Subject to:

$$\sum_{i=1}^{m} \sum_{t=r_j}^{T-p_j^i+1} aij\, x_{ij}^t = 1 \quad \forall j \quad (2)$$

$$\sum_{j=1}^{n} \sum_{s=t-p_{j+1}^i} x_{ij}^s \leq 1 \quad \forall t, i \quad (3)$$

$$x_{ij}^t \in \{0,1\} \quad \forall j, t, i \quad (4)$$

Equation (1) is the problem objective function and consists of the weighted tardiness function. Constraints (2) ensure that all demand is met. Constraints (3) guarantee that at each time period each vessel is allocated to at most one job. Constraints (4) impose the variables to be binary.

This model was used to generate lower bounds for the test instances, using column generation approach. The framework presented in Van den Akker et al. (2000) was used, in which the problem was reformulated using Dantzig-Wolfe decomposition. The restricted master problem consisted in selecting schedules that would cover all the demand at minimum cost. The subproblem consisted in solving a shortest a path problem in an acyclic network for each vessel, which was to produce a schedule (also called a pseudo-schedule), where jobs could possibly occur more than once on the same route, provided that the vessel capacity would not be violated.

5 PROPOSED HEURISTICS

In this research a GRASP algorithm was implemented according to structure presented in Figure 2. Afterwards, the Path Relinking was add.

Each iteration begins by updating a so called configuration, which consists of three main elements: the size (l) of the Restricted Candidate List (RCL), the ordering rule used to classify jobs, and the seed used in the random generation process. The size of the RCL, is given by: 2, 3, 4, 5, 6 and n.

```
Procedure GRASP(maxIter)
1    readData()
2    for k=1,...,maxIter do
3         update configuration k
4         for j=1,...,n do
5            update restricted candidate list of size l(k)
6            randomly choose one job i in RCL(k)
7            insert job i in the best feasible position
8         end
9         localSearch()
10        update bestSolution
11   end
12   return bestSolution
end GRASP
```

Figure 2. GRASP procedure.

This will allow working with approximately greedy rules (when $l = 2$), or to randomly choose any job from the list (when $l = n$).

As for the ordering rule, it was chosen to use a fixed ordering scheme rather than a dynamic one. This means that after a job is selected from the restricted candidate list the remaining jobs will not be reclassified. Instead, the established sequence given by the ordering rule will be maintained throughout the whole construction phase.

Six rules were proposed to order the jobs, having each rule two criteria. If by the first criterion a tie still holds, then the second criterion will be applied. If the tie persists then the jobs will be ordered by their index. The rules are: (1) Less vessel compatibility and Less release date; (2) Less vessel compatibility and Greater penalty; (3) Less release date and Less vessel compatibility; (4) Less release date and Greater penalty; (5) Greater penalty and Less release date; (6) Greater penalty and Less vessel compatibility. “Less vessel compatibility” rule prioritizes jobs that have few compatible vessels. By applying this rule one is concerned with the generation of feasible solutions inasmuch as those critical jobs will not be left out.

For each ordering rule combined with the each RCL size (6×6) it will be tested 100 different seeds. This will fix the total iteration number to 3600. As for the algorithm described at figure 2, after reading the problem data in line 1, the configuration is updated and the construction phase is applied between lines 4 and 8.

This phase follows the main ideas of updating the restricted candidate list (line 5); randomly choosing one unscheduled job (line 6) and defining the best insertion position for the selected job (line 7). This is done by evaluating all feasible insertion positions in regard to tardiness cost increment. The insertion position with the least cost impact will be chosen.

After scheduling all jobs, a local search routine is called in line 9. Local search is done by applying two types of moves. One is the relocate operator that removes one job form its current position and evaluates every insertion position for all compatible vessels. It will be considered the best move the one that generates the greatest cost reduction in the objective function. After all the jobs are evaluated, the resulting move will be for the job that causes the greater reduction. This process is repeated until no other move can reduce the objective function.

The swap operator exchanges two jobs at the same time provided that they are scheduled in different but compatible vessels. Each job will be designated to its best insertion position of the other vessel. The swap with the best overall reduction will be executed. As in the relocate case, this process is repeated until the objective function cannot be improved.

Finally, if the generated solution is better than the current solution, it will be updated. GRASP will continue until the iteration limit is reached.

The implemented version of Path Relinking (PR) is shown in Figure 3. In regard to interaction with the pool of elite solutions, a dynamic implementation was done in which every new GRASP solution obtained from the construction phase is submitted to PR, by randomly selecting a solution from the pool of elite solution. In the current version of our algorithm, this pool has size 10.

The Path Relinking routine begins by setting the reference solutions (initial and guiding), depending on the chosen strategy (forward, backward or mixed). If the strategy is backward or mixed, the guiding solution will be the starting point. An exception occurs if the informed initial solution is already better than the elite (guiding) solution. In the forward strategy, there is no need of inversion.

It is important to note that our problem involves multiple vessels. Therefore a matching problem as discussed by Ho & Gendreau (2006) has to be solved in order to indicate which route (vessel) in the guiding solution corresponds to each route (vessel) in the initial solution. The main idea is that those routes which are most "similar" in both solutions should be kept for the current vessels. This is done through a greedy procedure.

Path Relinking is an iterative procedure that transforms one solution into another. Each iteration consists of modifying the initial solution hoping that the objective function will possibly improve. Therefore, among the possible modifications it will be chosen the one that will increase the overall objective function performance as in the local search procedure, relocate and swap operators are called and all possible moves are evaluated. Nevertheless, it should be noted that the objective function value may worsen during the PR.

```
Procedure GRASP_PathRelinking(maxIter)
1    readData ()
2    for k=1,...,maxIter do
3         update configuration k
4         for j=1,...,n do
5            update restricted candidate list of size l(k)
6            randomly choose one job i in RCL(k)
7            insert job i in the best feasible position
8         end
9         localSearch()
10        PathRelinking(sol_elite, sol_GRASP)
11        update elitePool
12        update bestSolution
13   end
14   Return bestSolution
end GRASP
```

Figure 3. Basic structure of GRASP with path relinking.

When a job is moved to a new route, it will always be inserted in its best insertion position. It may happen that after all jobs are on the corresponding vessels (taking the guiding solution as reference), that some jobs may not be on their exact positions. As a result, a relocation procedure must be undertaken in order to guarantee that the final solution is exactly the guiding solution.

In the Mixed PR, two paths are simultaneously explored—one from the initial and other from the guiding solution, for which an efficient implementation scheme, proposed by Resende et al. (2010), was implemented.

Finally, in the Randomized Path Relinking instead of making the best move towards the guiding solution, a Restricted Candidate List (RCL) is built with the best moves. Let z_{min} be the cost associated with the best increment in the objective function and let z_{max} be the cost associated with the worst movement. Let α be a random number between 0 and 1, which will be fixed for a complete PR iteration. The movements admitted in the RCL will be those such that its change in objective function z are in the range given by $[z_{min}, z_{min} + \alpha(z_{max} - z_{min})]$.

After executing the PR routine in line 10, the pool of elite solutions is updated. A solution will only be accepted in this set if its objective function value is less than the worst solution belonging to this set. In regard to the solution that will be excluded, the criterion used was to eliminate the one having a worst objective function while being the closest solution, thus contributing to increased diversity among the elite solutions. The measure of distance between two solutions is given by the number of times any given job has a different successor in the other solution.

6 COMPUTATIONAL RESULTS

A scenario generator was developed adapting the mechanism proposed by Crauwels et al. (1998), who generated different instances for the single machine scheduling problem, minimizing the total weighted tardiness.

The test problems were built as follows: for each job j an integer processing time was generated using an uniform distribution between [1, 100]. For the penalty for late conclusion the uniform distribution was applied within the range [1, 10]. The planning horizon was fixed by $\sum_{j=1}^{n} pj / v$, where v is the number of vessels. The release date was calculated in two steps, where RDD is the relative range of due dates, and TF the tardiness factor:

$$dp_j = \left[\sum_{j=1}^{n} p_j (1 - RDD/2), \sum_{j=1}^{n} p_j (1 + RDD/2) \right] \quad (5)$$

$$r_j = max\left(0, dp_j - \left(\min_{k=1}^{n} \{dp_k\}\right)\right) \quad (6)$$

In equation (5) dp_j calculates the preliminary release date while in (6) the actual release for each job is established. The due date of a job is given by $d_j = r_j + p_j$. The values assumed for TF and RDD were {0.2; 0.4; 0.6; 0.8; 1.0}. Each TF factor was combined with all others RDDs, totalizing 25 test instances. These 25 instances were generated for 3 combinations of number of vessels and number of jobs: 4 vessels and 30 jobs (30n × 4 m), 5 vessels and 40 jobs (40n × 5 m) and finally 6 vessels and 50 jobs (50n × 6 m).

For all instances the solution gaps were calculated as being the relative distance of the objective function generated by column generation to the objective function of the solution obtained, calculated as follows:

$Gap = 100(QM - GC)/QM$ where QM = Objective Function of Proposed Method, GC = Objective Function of Column Generation.

The algorithms were processed on a workstation with Intel Core i7 2.80GHz and 16,374 MB of SDRAM implemented in C++ language.

The results are presented in Tables 1, 2 and 3, where the first column shows the instance number, the second column, the value of the linear relaxation of the objective function obtained by column generation, column 3 onward, presents the gap, as it was described of the following algorithms: GRASP, GRASP + Forward Path Relinking (PR Fw); GRASP + Backward Path Relinking (PR Bw); GRASP + Mixed Path Relinking (PR Mxd); GRASP + Randomized Forward Path Relinking (PR Fw Rnd); GRASP + Randomized Backward Path Relinking (PR Bw Rnd); GRASP + Randomized Mixed Path Relinking (PR Mxd Rnd).

Table 1. Gap between Heuristic solutions and lower bounds (30 jobs × 4 vessels).

	Gap % [30n x 4m]							
Instance	OF CG	GRASP	PR Fw	PR Bw	PR Mxd	PR Fw Rnd	PR Bw Rnd	PR Mxd Rnd
1	15417	0.78%	0.73%	0.77%	0.77%	0.73%	0.76%	0.73%
2	9059	1.53%	1.54%	1.52%	1.54%	1.55%	1.54%	1.54%
3	6694	2.22%	2.15%	2.15%	2.15%	2.15%	2.15%	2.15%
4	2107	3.13%	2.59%	2.59%	2.50%	3.61%	2.59%	2.50%
5	1798	9.10%	9.10%	9.15%	9.10%	9.15%	9.15%	9.33%
6	6326	0.28%	0.25%	0.13%	0.13%	0.13%	0.13%	0.14%
7	4366	2.50%	2.50%	2.50%	2.50%	2.50%	2.50%	2.57%
8	2309	3.31%	3.31%	3.31%	3.31%	3.31%	3.31%	3.31%
9	3542	2.93%	2.93%	2.93%	2.93%	2.93%	2.93%	2.73%
10	1711	3.82%	4.25%	3.82%	3.66%	3.82%	3.22%	4.57%
11	9481	0.47%	0.47%	0.47%	0.47%	0.47%	0.47%	0.48%
12	8908	0.69%	0.82%	0.70%	0.76%	0.76%	0.72%	0.60%
13	5634	3.71%	3.84%	3.97%	3.71%	3.71%	3.84%	3.71%
14	4388	4.84%	5.02%	5.04%	4.98%	5.06%	4.98%	5.06%
15	3016	5.34%	5.22%	5.22%	5.10%	5.13%	5.04%	5.04%
16	8652	1.13%	1.06%	1.06%	1.09%	1.07%	1.07%	1.07%
17	7408	1.28%	1.25%	1.25%	1.25%	1.27%	1.25%	1.25%
18	7484	2.36%	2.25%	2.18%	2.21%	2.35%	2.37%	2.25%
19	6461	2.18%	1.99%	1.99%	2.00%	1.99%	1.99%	1.97%
20	1601	4.07%	4.07%	4.07%	4.07%	4.07%	4.07%	4.07%
21	8986	1.48%	1.47%	1.47%	1.47%	1.47%	1.47%	1.47%
22	11008	1.62%	1.62%	1.62%	1.62%	1.62%	1.62%	1.62%
23	14882	1.04%	1.04%	1.04%	1.04%	1.04%	1.04%	1.04%
24	8002	1.47%	1.47%	1.42%	1.42%	1.42%	1.42%	1.42%
25	8466	1.79%	1.78%	1.78%	1.78%	1.79%	1.78%	1.78%
	Mean	**2.52%**	**2.51%**	**2.49%**	**2.46%**	**2.52%**	**2.46%**	**2.50%**

Table 2. Gap between Heuristic solutions and lower bounds (40 jobs × 5 vessels).

	Gap % [40n x 5m]							
Instance	OF CG	GRASP	PR Fw	PR Bw	PR Mxd	PR Fw Rnd	PR Bw Rnd	PR Mxd Rnd
1	12354	0.87%	0.87%	0.87%	0.86%	0.86%	0.86%	0.87%
2	11196	0.80%	0.74%	0.82%	0.75%	0.79%	0.77%	0.75%
3	6130	2.76%	3.16%	2.76%	2.76%	3.08%	2.99%	2.98%
4	2698	5.66%	5.33%	4.56%	4.36%	4.60%	5.43%	5.53%
5	789	6.18%	4.48%	4.48%	4.48%	6.07%	6.07%	3.66%
6	9602	0.40%	0.45%	0.37%	0.37%	0.38%	0.38%	0.38%
7	5918	1.63%	1.53%	1.20%	1.23%	1.30%	1.60%	1.27%
8	4538	0.94%	0.94%	0.59%	0.59%	0.94%	0.59%	0.59%
9	3370	3.69%	3.22%	2.97%	3.16%	3.24%	3.38%	2.99%
10	2169	8.52%	7.62%	7.74%	6.55%	7.23%	9.21%	6.55%
11	16600	0.47%	0.47%	0.47%	0.47%	0.49%	0.50%	0.47%
12	8735	0.95%	1.08%	1.08%	1.01%	1.06%	0.94%	1.14%
13	6746	2.36%	2.47%	1.89%	2.53%	2.37%	2.53%	2.58%
14	4918	4.11%	3.64%	3.66%	3.63%	4.10%	3.93%	4.06%
15	5939	3.67%	3.46%	3.07%	3.51%	3.51%	3.34%	3.46%
16	16139	1.22%	1.19%	1.19%	1.18%	1.18%	1.19%	1.19%
17	10847	1.65%	1.65%	1.61%	1.61%	1.65%	1.67%	1.62%
18	9623	2.30%	2.27%	2.18%	2.18%	2.26%	2.15%	2.15%
19	6392	2.01%	2.01%	2.01%	2.01%	2.01%	2.01%	2.01%
20	4379	4.12%	4.16%	4.05%	4.37%	4.35%	4.45%	4.03%
21	10728	1.69%	1.69%	1.69%	1.69%	1.69%	1.70%	1.69%
22	13275	1.27%	1.28%	1.27%	1.28%	1.27%	1.27%	1.27%
23	13592	1.19%	1.19%	1.18%	1.18%	1.19%	1.19%	1.20%
24	10941	1.60%	1.61%	1.48%	1.65%	1.60%	1.62%	1.59%
25	10350	2.55%	2.46%	2.46%	2.63%	2.50%	2.50%	2.50%
	Mean	**2.50%**	**2.36%**	**2.23%**	**2.24%**	**2.39%**	**2.49%**	**2.26%**

Table 3. Gap between Heuristic solutions and lower bounds (50 jobs × 6 vessels).

	Gap % [50n x 6m]							
Instance	OF CG	GRASP	PR Fw	PR Bw	PR Mxd	PR Fw Rnd	PR Bw Rnd	PR Mxd Rnd
1	19685	0.88%	0.87%	0.86%	0.87%	0.89%	0.92%	0.89%
2	11768	1.43%	1.43%	1.37%	1.42%	1.47%	1.39%	1.40%
3	12450	1.40%	1.34%	1.14%	1.11%	1.25%	1.22%	1.20%
4	7817	2.80%	2.89%	2.49%	2.58%	2.64%	2.73%	2,53%
5	2498	7.24%	6.83%	5.49%	6.23%	6.69%	6.83%	7.03%
6	22599	0.69%	0.68%	0.69%	0.68%	0.71%	0.71%	0.71%
7	15720	0.58%	0.63%	0.59%	0.59%	0.48%	0.48%	0.63%
8	11165	2.16%	1.70%	1.98%	1.53%	1.68%	1.82%	1.66%
9	9334	2.54%	2.62%	1.76%	2.16%	2.15%	2.98%	2.15%
10	2802	3.74%	3.08%	3.58%	2.57%	3.91%	2.88%	2.84%
11	18094	0.33%	0.32%	0.29%	0.34%	0.33%	0.29%	0.32%
12	15734	0.87%	0.73%	0.64%	0.64%	0.84%	0.71%	0.76%
13	9318	2.10%	2.05%	1.93%	2.10%	1.99%	1.90%	2.23%
14	14299	2.37%	2.42%	2.52%	2.62%	2.62%	2.41%	2.10%
15	7972	3.05%	2.96%	2.02%	3.28%	3.45%	2.72%	2.83%
16	24191	0.83%	0.84%	0.83%	0.84%	0.82%	0.84%	0.84%
17	18571	1.46%	1.52%	1.45%	1.46%	1.52%	1.48%	1.51%
18	17025	1.29%	1.24%	1.24%	1.23%	1.26%	1.28%	1.32%
19	11286	2.25%	2.32%	2.08%	2.34%	2.17%	2.44%	2.38%
20	11832	2.20%	2.26%	2.02%	2.22%	2.29%	2.26%	2.27%
21	21524	1.13%	1.13%	1.13%	1.13%	1.13%	1.15%	1.15%
22	18114	1.33%	1.34%	1.32%	1.33%	1.36%	1.35%	1.32%
23	20472	1.42%	1.41%	1.39%	1.41%	1.40%	1.41%	1.41%
24	20639	1.14%	1.16%	1.13%	1.14%	1.16%	1.16%	1.15%
25	14864	1.35%	1.41%	1.43%	1.28%	1.41%	1.43%	1.24%
	Mean	**1.86%**	**1.81%**	**1.65%**	**1.72%**	**1.82%**	**1.79%**	**1.75%**

7 CONCLUSIONS AND FURTHER WORK

This research studied the scheduling of pipe layer vessels responsible for launching risers and connecting them to subsea equipment. This is a variation of the unrelated parallel machine scheduling problem and was solved with the support of GRASP with Path Relinking. To evaluate the quality of our results, lower bounds were derived by linear column generation. It was seen that the heuristics generated solutions with average distance from the optimal value inferior to 2%.

The application of Path Relinking enhanced the solution quality. In all instances, GRASP combined with Path Relinking in all its modes (Forward, Backward, Mixed and Randomized), on average, were more effective. The most difficult instances to solve were those with tardiness factor TF equal to 0.2 and 0.4. Regarding the best sorting rule, which obtained the best solutions were respectively (1), (2) and (4). This shows that it is advantageous to use various ordering rules in the construction phase of GRASP instead of a fixed rule.

It is intended to further improve these results by performing simultaneous processing of several variations of Path Relinking, with features multi-threads and increase the seed of random numbers.

REFERENCES

Alidaee, B. & Rosa, D.1997. Scheduling parallel machines to minimize total weighted and unweighted tardiness. *Computers & Operations Research* 24: 775–788.

Biskup, D.; Herrmann, J. & Gupta, J.N.D. 2008. Scheduling identical parallel machines to minimize total tardiness. *International Journal of Production Economics* 115: 134–142.

Boudia, M.; Louly, M.A.O. & Prins, C. 2007. A reactive GRASP and path relinking for a combined production-distribution problem. *Computers & Operations Research* 34:3402–3419.

Brønmo, G.; Christiansen, M.; Fagerholt, K. & Nygreen, B. 2007a. A multi-start local search heuristic for ship scheduling – a computational study. *Computers & Operations Research* 34: 900–917.

Brønmo, G.; Christiansen, M. & Nygreen, B. 2007b. Ship routing and scheduling with flexible cargo sizes. Journal of the Operational Research Society 58: 1167–1177.

Cao, D.; Chen, M. & Wan, G. 2005. Parallel machine selection and job scheduling to minimize machine cost and job tardiness. *Computers & Operations Research* 32: 1995–2012.

Cheng,T.C.E. & Sin,C.C.S. 1990. A state-of-the-art review of parallel-machine scheduling research. *European Journal of Operational Research* 47: 271–292.

Crauwels, H.A.J.; Potts, C.N. & Van Wassenhove, L.N. 1998. Local search heuristics for the single total weighted tardiness scheduling problem. *Informs Journal on Computing* 10: 341–350.

Feo, T.A. & Resende, M.G.C. 1989. A probabilistic heuristic for a computationally difficult set covering problem. *Operations Research Letter* 8:67–71.

Feo, T.A. & Resende, M.G.C. 1995. Greedy randomized adaptive search procedures. *Journal of Global Optimization* 6: 109–133.

Fuller, J.A. 1978. Optimal solutions versus 'good' solutions: an analysis of heuristic decision making. *Omega The International Journal of Management Science* 6(6): 479–484.

Glover, F., Tabu search and adaptive memory programming—advances, applications and challenges. 1996. In Barr, R.S.; Helgason, R.V. & Kennington, J.L. (Eds.), *Interfaces in Computer Science and Operations Research:* 1–75. Boston: Kluwer Academic Publishers.

Glover, F.; Laguna, M. & Martí, R. 2000. Fundamentals of scatter search and path relinking. *Control and Cybernetics* 29(3): 653–684.

Graham, R.L. 1969. Bounds on multiprocessing timing anomalies. *Siam Journal on Applied Mathematics* 17(2): 416–429.

Ho, S.C. & Gendreau, M. 2006. Path relinking for the vehicle routing problem. *Journal of Heuristics* 12:55–72.

Horowitz, E. & Sahni, S. 1976. Exact and approximations algorithms for scheduling non identical processors. *Journal of the Association for Computing Machinery* 23(2): 317–327.

Koulamas, C. 1997. Decomposition and hybrid simulated annealing heuristics for the parallel-machine total tardiness problem. *Naval Research Logistics* 44: 109–125.

Lawler, E.L. & Labetoulle, J. 1978. On preemptive scheduling of unrelated parallel processors by linear programming. *Journal of the Association for Computing Machinery* 25(4): 612–619.

Lenstra, H.L.; Shmoys, D.B. & Tardos, E. 1990. Approximation algorithms for scheduling unrelated parallel machines. *Mathematical Programming* 46: 259–271.

Liaw, C-F.; Lin, Y-K.; Cheng, C-Y. & Chen, M. 2003. Scheduling unrelated parallel machines to minimize total weighted tardiness. *Computers & Operations Research* 30: 1777–1789.

Morton, T.E. & Pentico, D.W. 1993. Heuristic scheduling systems: with applications to production systems and project management. New York: John Wiley & Sons.

Panwalkar, S.S.; Smith, M.L. & Koulamas, C. 1993. A heuristic for the single machine tardiness problem, *European Journal of Operational Research* 70: 304–310.

Pinedo, M. 2002. *Scheduling: theory, algorithm, and systems*. New Jersey: Prentice-Hall.

Prais, M. & Ribeiro, C.C. 2000. Reactive GRASP: an application to a matrix decomposition problem in TDMA traffic assignment. *Informs Journal on Computing* 12:164–176.

Resende, M.G.C. & Ribeiro, C.C. 2003. Grasp with path relinling: recent advances and applications. *AT & T Labs Technical Report TD-5TU726*: 1–24.

Resende, M.G.C. & Werneck, R.F. 2004. A hybrid heuristic for the p-median problem. *Journal of Heuristics* 10:59–88.

Resende, M.G.C.; Martí, R.; Gallego, M. & Duarte, A. 2010a. Grasp and path relinking for the max-min diversity problem. *Computers & Operations Research* 37, 498–508.

Resende, M.G.C.; Ribeiro, C.C.; Glover, F. & R. Martí, Scatter search and path-relinking: Fundamentals, advances, and applications. 2010b. In Gendreau, M. & Potvin, J.-Y. (Eds.), *Handbook of Metaheuristics:* 87–107. New York: Springer.

Ribeiro, C.C. & Rosseti, I. 2007. Efficient parallel cooperative implementations of GRASP heuristics. *Parallel Computing* 33: 21–35.

Ribeiro, C.C.; Rosseti, I. & Vallejos, R. On the use of run time distributions to evaluate and compare stochastic local search algorithms. 2009. In Stützle, T.; Birattari, M. & Hoos, H.H. (Eds.), *Lecture Notes in Computer Science:* 16-30. Berlin Heidelberg: Springer-Verlag.

Ronen, D. 1983. Cargo ships routing and scheduling: a survey of models and problems. *European Journal of Operational Research* 12: 119–126.

Ronen, D. 1993. Ship scheduling: the last decade. *European Journal of Operational Research* 71: 325–333.

Shim, S-O. & Kim, Y-D. 2007. Scheduling on parallel identical machines to minimize total tardiness. *European Journal of Operational Research* 177: 135–146.

Van den Akker, J.M., Hurkens, C.A.J. & Savelsbergh, M.W.P. 2000. Time-indexed formulations for machine scheduling problem: column generation, *INFORMS Journal on Computing* 12: 111–124.

Wen, C.; Ekşioğlu, S.D.; Greenwood, A. & Zhang, S. 2010. Crane scheduling in a shipbuilding environment. *International Journal of Production Economics* 124: 40–50.

Sustainable Maritime Transportation and Exploitation of Sea Resources – Rizzuto & Guedes Soares (eds)
© 2012 Taylor & Francis Group, London, ISBN 978-0-415-62081-9

On the linear stiffness of tension leg platforms

I. Senjanović, N. Hadžić & M. Tomić
University of Zagreb, Zagreb, Croatia

ABSTRACT: The Tension Leg Platform (TLP) is a type of compliant offshore structures generally used for deep water oil exploration. The platforms are moored by very flexible tendons so that surge amplitude can achieve a large value. The platform hull is considered as a rigid body with six DOF. The total restoring stiffness plays a very important role in the TLP dynamic behavior. In present literature inadequate formulations of the stiffness have been used. The problem is to define realistic centre of rotations. The purpose of this paper is to present a consistent formulation of the stiffness matrix, derived from the general solution established for hydroelastic analysis of ship structures, as a specific case. In reality the total TLP stiffness in dynamic analysis consists of hydrostatic hull stiffness (pressure and gravity), and conventional and geometric tendon stiffness. The new stiffness is compared to the known ones, and discrepancies are analyzed and discussed.

1 INTRODUCTION

A Tension Leg Platform (TLP) is a semi-submersible platform, moored by vertical pretensioned tendons or tethers (Baltrop 1998). The platform constitutive parts are pontoon, columns and deck with drilling equipment (Det norske Veritas 2005). Heave, roll and pitch have high natural frequencies due to high tendon axial stiffness. Surge, sway and yaw are compliant modes due to quite low tendon geometric stiffness. Vertical motions are excited by the first order wave forces, while horizontal motions appear due to the second order wave forces with very low frequency (Natvig & Teigen 1993).

Stiffness plays very important role in dynamic analysis of TLPs (Adrezin et al. 1996). Platform can be considered as a rigid body with tendons as massless quasi-static springs. The hydrodynamic coefficients can be determined by Morison's equation or the radiation-diffraction theory, depending on the ratio of diameters of platform cylindrical segments and the wave length.

Since even linear stiffness is not formulated in the relevant literature in a consistent way, the new formulation is presented in Section 2, and its comparison with the known formulations is elaborated in Section 3. Additional comparison is done via numerical example, Section 4.

2 LINEAR STIFFNESS

2.1 *Definition of total stiffness*

Let us consider a double symmetric TLP with N tendons, Figure 1. The origin of the coordinate system is located in the middle of the waterplane, and the motion components are shown in Figure 2. The platform is treated as a rigid body due to very high stiffness compared to that of tendons. The total stiffness consists of three parts:

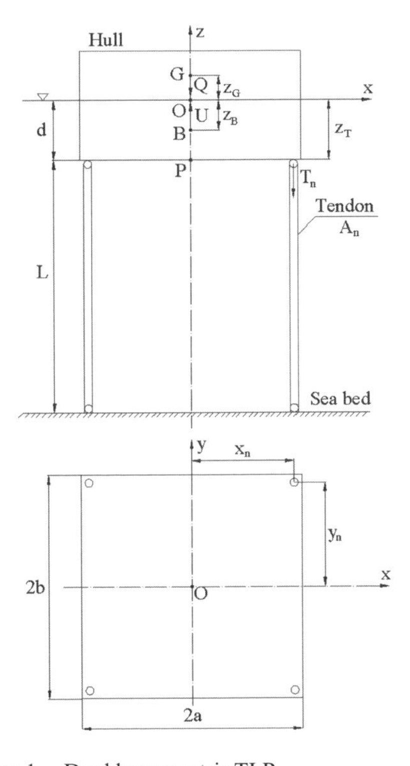

Figure 1. Double symmetric TLP.

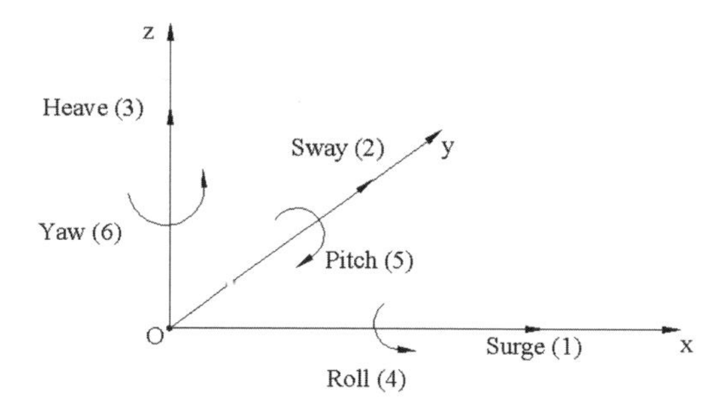

Figure 2. Motion components.

$$[K]_P = [K]^C + [C] + [K]^G, \quad (1)$$

where $[K]^C$ is the conventional tendon stiffness, $[C]$ is the platform restoring stiffness with influence of tendons included, and $[K]^G$ is the tendon geometric stiffness.

2.2 *Conventional stiffness*

The tendons are steel pipes with negligibly small bending stiffness. Heave changes the tension of tendons so that the stiffness is the relation between vertical force and displacement:

$$k_{33}^C = \frac{EA}{L}, \quad A = \sum_{n=1}^{N} A_n. \quad (2)$$

The tension of tendons is also changed by roll and the corresponding stiffness is found from the moment as function of the roll angle. Since

$$M_x = \sum_{n=1}^{N} F_z^n y_n, \quad F_z^n = \frac{EA_n}{L}\delta_z^n, \text{ and } \delta_z^n = y_n \varphi_x, \quad (3)$$

one gets:

$$k_{44}^C = \frac{EI_x}{L}, \quad I_X = \sum_{n=1}^{N} A_n y_n^2. \quad (4)$$

In the similar way, the pitch stiffness is:

$$k_{55}^C = \frac{EI_y}{L}, \quad I_y = \sum_{n=1}^{N} A_n x_n^2, \quad (5)$$

where A, I_x, and I_y are the total cross-section area and moments of inertia about x and y axis of all tendons, respectively. Hence, set of tendons is considered as a beam with distinct fibers.

2.3 *Restoring stiffness*

Hydroelastic analysis of a deformable floating body is usually performed by the modal superposition method. The ordinary restoring stiffness consists of variation of hydrostatic pressure, variation of normal vector and natural mode and gravity part respectively, (Senjanović et al. 2009):

$$C_{ij}^o = C_{ij}^p + C_{ij}^{nh} + C_{ij}^m, \quad (6)$$

$$C_{ij}^p = \rho g \iint_S h_k^i h_3^j n_k \, dS, \quad (7)$$

$$C_{ij}^{nh} = \rho g \iint_S Z h_k^i h_{l,l}^j n_k \, dS, \quad (8)$$

$$C_{ij}^m = g \iiint_V \rho_S h_k^i h_{3,k}^j \, dS, \quad (9)$$

where, according to the index notation, $h_{k,l}^i$ is the l-th derivative of the k-th component of the natural mode h^i, S is the wetted surface, n_k are components of its normal vector, V is the structure volume, while ρ and ρ_s are the water and structure density, respectively.

For the TLP platform only rigid body natural modes are of interest, and the general restoring stiffness is reduced to the formulas of ship hydrostatics for the floating and stability conditions, (SNAME 1988). In the case of double symmetric platform, the centroid of waterplane is located at the vertical line of the buoyancy and gravity centre and the restoring stiffness matrix is diagonal with the following heave, roll and pitch coefficients:

$$C_{33}^0 = \rho g A_{WL}^0, \quad (10)$$

$$C_{44}^0 = \rho g \left[I_{WLX}^0 + V^0 \left(z_B^0 - z_G^0 \right) \right], \quad (11)$$

$$C_{55}^0 = \rho g \left[I_{WLY}^0 + V^0 \left(z_B^0 - z_G^0 \right) \right], \quad (12)$$

where, $A_{WL}{}^0$, $I_{WLX}{}^0$ and $I_{WLY}{}^0$ are the waterplane area and its moments of inertia about x and y axis, respectively, V^0 is the buoyancy volume while $z_B{}^0$ and $z_G{}^0$ are coordinates of buoyancy and gravity centre, respectively.

The buoyancy, U, is larger than the platform weight, Q, due to tendon pretension forces, T_n, which, only for this purpose, can be treated as virtual lumped weights, Figure 3:

$$\sum_{n=1}^{N} T_n = U - Q. \quad (13)$$

The stiffness coefficients, Eqs. (10), (11) and (12) take the following form:

$$C_{33} = \rho g A_{WL}, \quad (14)$$

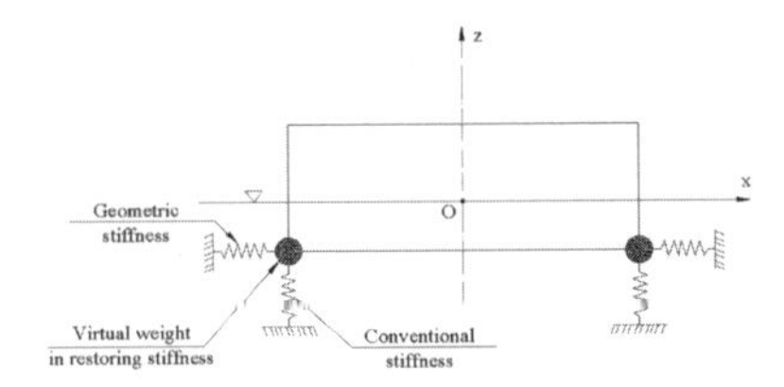

Figure 3. Stiffness model of TLP.

$$C_{44} = \rho g I_{WLX} + U z_B - Q z_G - \sum_{n=1}^{N} T_n z_T, \tag{15}$$

$$C_{55} = \rho g I_{WLY} + U z_B - Q z_G - \sum_{n=1}^{N} T_n z_T, \tag{16}$$

where all quantities are related to the increased platform immersion due to the tendon forces.

2.4 *Geometric stiffness*

TLP can be translated horizontally in x and y direction and rotated about vertical z axis. External forces are equilibrated with internal forces which depend on the tendon geometric stiffness. The stiffness can be determined by the general formulation of geometric stiffness written in the index notation (Huang & Riggs 2000; Senjanović et al. 2010):

$$k_{ij}^G = \iiint_V \sigma_{kl} h_{m,k}^i \, h_{m,l}^j \, \mathrm{d}V, \tag{17}$$

where σ_{kl} is the stress tensor. In the considered case

$$\sigma_{zz} = \sum_{n=1}^{N} \frac{T_n}{A_n}, \quad \mathrm{d}V = \sum_{n=1}^{N} A_n \mathrm{d}z \tag{18}$$

and the surge mode

$$h_x^1 = \frac{(L+d+z)}{L} \tag{19}$$

is defined in domain $-(L + d) \leq z \leq -d$, Figure 4a. Thus, one finds $h_{x,z}{}^{1} = 1/L$ and further

$$K_{11}^G = \sum_{n=1}^{N} \frac{T_n}{L} = K_{22}^G. \tag{20}$$

The yaw mode of the n-th tendon, according to Figure 4b, is:

$$h_r^6(n) = \frac{(L+d+z) r_n}{L} \tag{21}$$

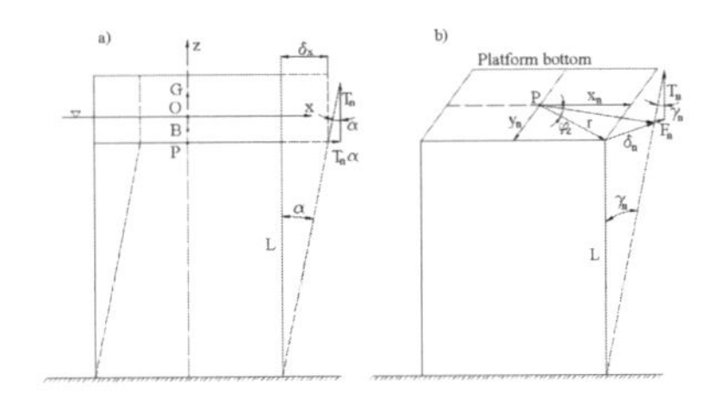

Figure 4. Surge and sway of TLP.

with components

$$h_x^6(n) = \frac{(L+d+z) y_n}{L} \tag{22}$$

$$h_y^6(n) = \frac{(L+d+z) x_n}{L}. \tag{23}$$

In this case, Eq. (16) gives

$$K_{66}^G = \sum_{n=1}^{N} \frac{T_n \left(x_n^2 + y_n^2\right)}{L}. \tag{24}$$

2.5 *Total stiffness and mass matrix*

By summing up the terms of conventional, restoring and geometric stiffness, determined in previous sections, elements of the total stiffness matrix $[K]_P$, Eq. (4), are obtained:

$$K_{11} = K_{22} = \sum_{n=1}^{N} \frac{T_n}{L}, \quad K_{33} = \rho g A_{WL} + \frac{EA}{L}, \tag{25 a, b}$$

$$K_{44} = \rho g I_{WLX} + U z_B - Q z_G - \sum_{n=1}^{N} T_n z_T + \frac{EI_x}{L}, \tag{26}$$

$$K_{55} = \rho g I_{WLY} + U z_B - Q z_G - \sum_{n=1}^{N} T_n z_T + \frac{EI_y}{L}, \tag{27}$$

$$K_{66} = \sum_{n=1}^{N} \frac{T_n}{L} \left(x_n^2 + y_n^2\right). \tag{28}$$

The total stiffness matrix is a diagonal one because the middle point of hull bottom is used as pole P, Figure 4, for the platform rotations. As a result, there is no coupling between degrees of freedom through the stiffness.

On the other side, the mass matrix has some off-diagonal elements since the following inertia forces (designated with i) depend on both displacements and rotations:

$$F_x^i = m\ddot{\delta}_x + m(z_G - z_T)\ddot{\varphi}_y \tag{29}$$

$$F_y^i = m\ddot{\delta}_y + m(z_G - z_T)\ddot{\varphi}_x \tag{30}$$

$$M_x^i = m(z_G - z_T)\ddot{\delta}_y + J_x^P\ddot{\varphi}_x \tag{31}$$

$$M_y^i = m(z_G - z_T)\ddot{\delta}_x + J_y^P\ddot{\varphi}_y \tag{32}$$

where

$$J_x^P = J_x^G + m(z_G - z_T)^2 \tag{33}$$

$$J_y^P = J_y^G + m(z_G - z_T)^2 \tag{34}$$

are the mass moments of inertia. The mass matrix with respect to the pole P, reads:

$$[M]_P = \begin{bmatrix} m & & & & mz_{GT} & \\ & m & & mz_{GT} & & \\ & & m & & & \\ & mz_{GT} & & J_x^P & & \\ mz_{GT} & & & & J_y^P & \\ & & & & & J_z \end{bmatrix}, \tag{35}$$

where $z_{GT} = z_G - z_T$. Due to the off-diagonal terms of matrix $[M]_P$, the vibrations are coupled through the mass matrix.

If the center of gravity G is used as the reference point, the following relations between displacements exist:

$$\delta_x^P = \delta_x^G - (z_G - z_T)\varphi_y, \quad \delta_y^P = \delta_y^G - (z_G - z_T)\varphi_x. \tag{36 a, b}$$

Hence, the displacement transformation matrix reads:

$$[T] = \begin{bmatrix} 1 & & & & -z_{GT} & \\ & 1 & & -z_{GT} & & \\ & & 1 & & & \\ & & & 1 & & \\ & & & & 1 & \\ & & & & & 1 \end{bmatrix}. \tag{37}$$

Now, the stiffness and mass matrix can be adapted to the new coordinate system in the way well known in the finite element method, which is based on the fact that the total energy of a structure does not depend on the chosen coordinate system, (Zienkiewicz 1971):

$$[K]_G = [T]^T[K]_P[T], \quad [M]_G = [T]^T[M]_P[T]. \tag{38 a, b}$$

Thus, one finds:

$$[K]_G = \begin{bmatrix} K_{11} & & & & -z_{GT}K_{11} & \\ & K_{22} & & -z_{GT}K_{22} & & \\ & & K_{33} & & & \\ & -z_{GT}K_{22} & & K_{44} + z_{GT}^2K_{22} & & \\ -z_{GT}K_{11} & & & & K_{55} + z_{GT}^2K_{11} & \\ & & & & & K_{66} \end{bmatrix} \tag{39}$$

while the mass matrix becomes diagonal:

$$[M]_G = \begin{bmatrix} m & & & & & \\ & m & & & & \\ & & m & & & \\ & & & J_x^G & & \\ & & & & J_y^G & \\ & & & & & J_z \end{bmatrix}. \tag{40}$$

In direct analysis of ship motion in seaway, it is assumed that vessel rotates about the centroid of the waterplane. If the same assumption is accepted for TLP, the stiffness and mass matrix can be transformed in the same manner and both have off-diagonal elements.

3 COMPARISON OF THE KNOWN STIFFNESS MATRICES WITH THE NEW ONE

3.1 *Stiffness from Malenica, 2003*

The linear stiffness matrix is derived for the tendon top points and arbitrary origin of the coordinate system.

$$[C] = [C_{WL}] + \sum_{n=1}^{N}[C^{P_0^n}] + \sum_{n=1}^{N}[C^{P_1^n}], \tag{41}$$

where

$$[C_{WL}] = \rho g \begin{bmatrix} 0 & & & & & \\ & 0 & & & & \\ & & A_{WL}^* & & & \\ & & & I_{WLX}^* + V^*z_{BG} & & \\ & & & & I_{WLY}^* + V^*z_{BG} & \\ & & & & & 0 \end{bmatrix},$$

and

$$\left[C^{P_0^n}\right]=\begin{bmatrix}0&&&&&\\&0&&&&\\&&0&&&\\&&&-T_n z_G^n&&\\&&&&-T_n z_G^n&\\&&&&&0\end{bmatrix}, \tag{43}$$

$$\left[C^{P_1^n}\right]=\begin{bmatrix}\left[k^n\right] & -\left[k^n\right]\left[V^n\right]\\ \left[V^n\right]\left[k^n\right] & -\left[V^n\right]\left[k^n\right]\left[V^n\right]\end{bmatrix}, \tag{44}$$

$$\left[k^n\right]=\begin{bmatrix}k_{11}^n&&\\&k_{22}^n&\\&&k_{33}^n\end{bmatrix}, \tag{45}$$

$$\left[V^n\right]=\begin{bmatrix}0&-z_G^n&y_G^n\\z_G^n&0&-x_G^n\\-y_G^n&x_G^n&0\end{bmatrix}, \tag{46}$$

$$\left[k^n\right]\left[V^n\right]=\begin{bmatrix}0&-z_G^n k_{11}^n&y_G^n k_{11}^n\\z_G^n k_{22}^n&0&-x_G^n k_{22}^n\\-y_G^n k_{33}^n&x_G^n k_{33}^n&0\end{bmatrix}, \tag{47}$$

$$\left[V^n\right]\left[k^n\right]\left[V^n\right]=$$

$$\begin{bmatrix}-\left(z_G^n\right)^2 k_{22}^n-\left(y_G^n\right)^2 k_{33}^n & x_G^n y_G^n k_{33}^n & -x_G^n z_G^n k_{22}^n\\ -x_G^n y_G^n k_{33}^n & -\left(z_G^n\right)^2 k_{11}^n-\left(x_G^n\right)^2 k_{33}^n & y_G^n z_G^n k_{11}^n\\ x_G^n z_G^n k_{22}^n & y_G^n z_G^n k_{11}^n & -\left(y_G^n\right)^2 k_{11}^n-\left(x_G^n\right)^2 k_{22}^n\end{bmatrix}, \tag{48}$$

$$k_{11}^n=k_{22}^n=\frac{T_n}{L},\ k_{33}^n=\frac{EA_n}{L}, \tag{49}$$

$$x_G^n=x^n-x_G,\ y_G^n=y^n-y_G,\ z_G^n=z^n-z_G. \tag{50}$$

In the case of a double symmetric platform $x_G^n = \pm a,\ y_G^n = \pm b,\ z_G^n = z_{TG}$, Figure 1, and all off-diagonal elements in the resulting matrix, Eq. (41), vanish. By taking into account the relations

$$\begin{aligned}&\rho g V^0=Q,\ \ U=\sum_{n=1}^{N}T_n+Q,\\&z_{BG}=z_B-z_G,\ \ z_{TG}=z_T-z_G,\\&Ab^2=I_X,\ \ Aa^2=I_Y\end{aligned} \tag{51}$$

one can write for diagonal elements

$$C_{11}=C_{22}=\sum_{n=1}^{N}\frac{T_n}{L},\ C_{33}=\rho g A_{WL}^o+\frac{EA}{L}, \tag{52}$$

$$C_{44}=\rho g I_{WLX}^o+Uz_B-Qz_G-\sum_{n=1}^{N}T_n z_T-\sum_{n=1}^{N}T_n z_{BG}+\sum_{n=1}^{N}\frac{T_n}{L}z_{TG}^2+\frac{EI_X}{L}, \tag{53}$$

$$C_{55}=\rho g I_{WLY}^o+Uz_B-Qz_G-\sum_{n=1}^{N}T_n z_T-\sum_{n=1}^{N}T_n z_{BG}+\sum_{n=1}^{N}\frac{T_n}{L}z_{TG}^2+\frac{EI_Y}{L}, \tag{54}$$

$$C_{66}^P=\sum_{n=1}^{N}\frac{T_n}{L}\left(a^2+b^2\right). \tag{55}$$

By comparing Eqs. (52)–(55) with Eqs. (25)–(28) some differences can be noticed. The 5th and 6th terms in Eqs. (53) and (54) are additional.

3.2 *Stiffness from Jain, 1997*

Nonlinear stiffness matrix presented in (Jain 1997) is ordinary used for dynamic analysis of TLP. It is specified with respect to the centre of gravity:

$$[K]_G^E = \begin{bmatrix} K_{11} & & & & & \\ & K_{22} & & & & \\ K_{31} & K_{32} & K_{33} & K_{34} & K_{35} & K_{36} \\ & K_{42} & & K_{44} & & \\ K_{51} & & & & K_{55} & \\ & & & & & K_{66} \end{bmatrix}. \quad (56)$$

The linear part of Eq. (56) reads

$$[K]_G^L = \begin{bmatrix} K_{11}^L & & & & & \\ & K_{22}^L & & & & \\ & & K_{33}^L & & & \\ & K_{42}^L & & K_{44}^L & & \\ K_{51}^L & & & & K_{55}^L & \\ & & & & & K_{66}^L \end{bmatrix}, \quad (57)$$

where elements $K_{11}{}^L$, $K_{22}{}^L$, $K_{33}{}^L$ and $K_{66}{}^L$ are equal to Eqs. (25a, b) and (28), while the remaining elements are

$$K_{44}^L = \rho g A_{WL} + U z_G - Q z_G - \sum_{n=1}^{N} T_n z_T + \frac{EI_X}{L}, \quad (58)$$

$$K_{55}^L = \rho g A_{WL} + U z_G - Q z_G - \sum_{n=1}^{N} T_n z_T + \frac{EI_Y}{L}. \quad (59)$$

By comparing Eqs. (58) and (59) with Eqs. (26) and (27) it is noticed that the coordinate of the gravity center z_G is accompanied to the buoyancy U instead of its own z_B. Since Eq. (57) is derived with respect to the gravity center it should have additional terms $K_{15}{}^L$ and $K_{24}{}^L$ like Eq. (39).

3.3 *Stiffness from Low, 2009*

Recently, a new formulation of nonlinear stiffness matrix, based on energy approach, is presented in (Low 2009). Its linear part reads

$$[K]^L = \begin{bmatrix} k_1 & & & & & \\ & k_1 & & & & \\ & & k_0 & & & \\ & & & k_0 b^2 & & \\ & & & & k_0 a^2 & \\ & & & & & k_1\left(a^2+b^2\right) \end{bmatrix}, \quad (60)$$

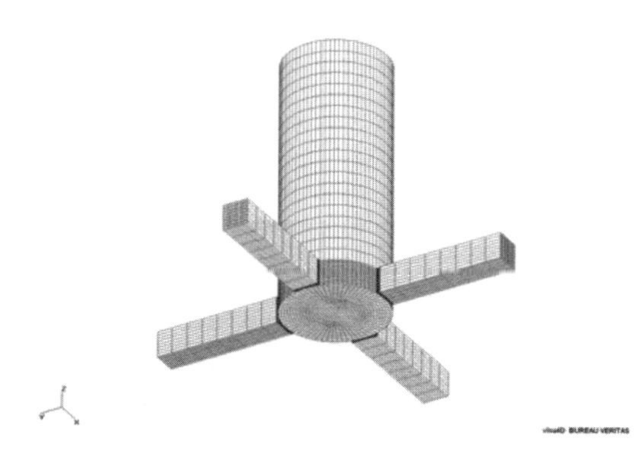

Figure 5. Panel model of TLP.

where

$$k_0 = \frac{EA}{L}, \; k_1 = \sum_{n=1}^{N} \frac{T_n}{L}. \quad (61)$$

It is obvious that only the tendon contribution is taken into account, while the restoring stiffness is ignored.

4 NUMERICAL EXAMPLE

Outlined theory is illustrated by analyzing a TLP spar floater without the installed wind turbine (Withee 2004). Hydrodynamic part of fluid loading (added mass, radiation damping, Froude-Krylov and diffraction loads) and the equations of motion (solved for the centre of gravity of the floater) are calculated using the Bureau Veritas *HYDROSTAR* software (Chen 2004).

The mean wetted surface of a TLP spar floater is discretized into 2272 panels. Two distinctive cases were analyzed, one using high modulus polyester tendons (HMPE, E = 2.5 10^{10} N/m^2) and the other using the usual steel tendons (STEEL, E = 2.06 10^{11} N/m^2). HMPE tendons are used in order to lower the stiffness and thus allow larger oscillation amplitudes. Frequency domain responses (1st order motion transfer functions) were calculated for a range of frequencies in 1m head waves (180 deg wave heading). Please note that the ratio between the response amplitude and wave amplitude is called transfer function in this paper.

In Figure 6 one can see that there are almost no differences in the surge due to different stiffness formulations and obviously these is the most important mode for TLP's. Low's stiffness formulation results with slightly larger values for the heave response, Figure 7, than the other ones that are again very close to each other. The largest dif-

Table 1. Particulars of a TLP spar floater.

Spar diameter	4 m	Water depth	200 m
Spar draft	10 m	Mass	49002.3 kg
Thickness	25.3 mm	Buoyant mass	149887.8 kg
Spoke length	5 m	COG	-6.698 m
Spoke width	1 m	COB	-5.687 m
Line diameter	50 mm	Jxx	1140080.9 kg m^2
Number of lines	1 per spoke	Jyy	1140080.9 kg m^2
Line length	100 m	Jzz	[illegible] kg m^2

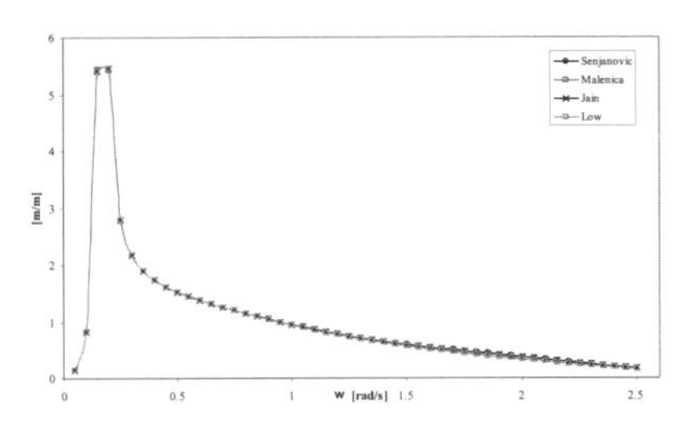

Figure 6. TLP surge transfer function (HMPE tendons).

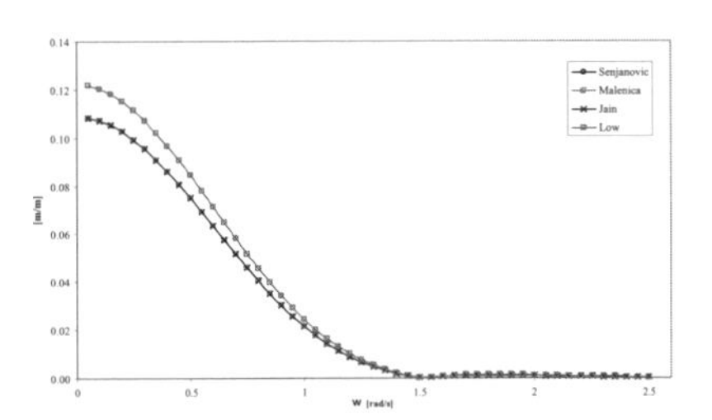

Figure 7. TLP heave transfer function (HMPE tendons).

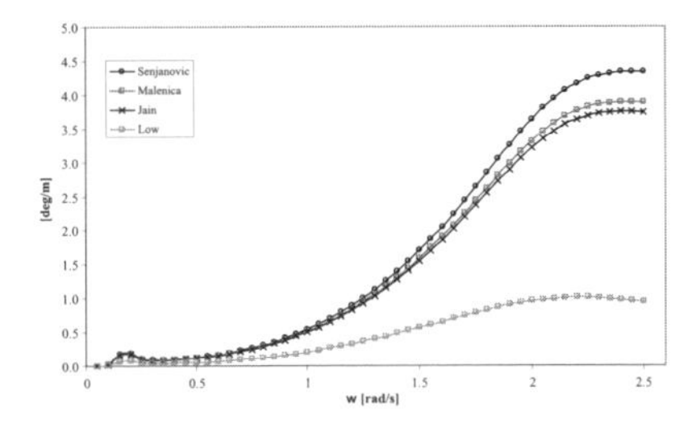

Figure 8. TLP pitch transfer function (HMPE tendons).

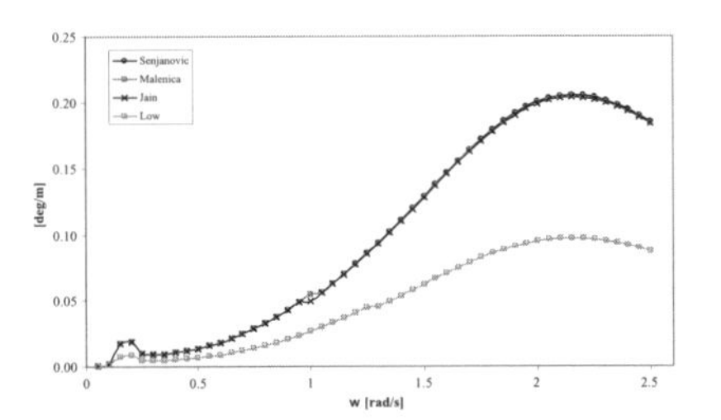

Figure 9. TLP pitch transfer function (STEEL tendons).

ferences are obtained for the rotational modes of motion as can be noted from the Figure 8. That particular case corresponds to HMPE tendons and the lower elasticity obviously will pronounce differences (as compared to steel tendons in Figure 9 were the different formulations give practically identical results). One should also note that the coupling between the modes of motion is influenced by the distance between the center of gravity and the center of buoyancy, so one would expect slightly larger differences between the stiffness formulations by Senjanovic and Malenica for the rotational modes, if the wind turbine itself was included in the model.

5 CONCLUSION

In this paper, the consistent linear stiffness for dynamic analysis of TLP's is derived in a systematic and physically transparent way. It comprises platform restoring stiffness, tendon conventional stiffness and tendon geometric stiffness. The first two components are important for the vertical motions, while the third one is related to the horizontal motions. Comparison of the known stiffness matrices with the new one shows some differences of the former. Their influence on response will be analyzed in further investigation by numerical examples. In general there is a good agreement between different formulations, especially if steel tendons are used. Therefore, in analyzing TLP's one should use formulation which appears to be physically consistent.

REFERENCES

Adrezin, R. et al. 1996. Dynamic response of compliant offshore structures – review, J. Aerosp. Eng., 9(4), pp. 114–131.

Baltrop, N.D.P. 1998. Floating structures: a guide for design and analysis, Vols. 1 and 2, Oilfield Publications Ltd.

Chen, X. 2004. Hydrodynamics in offshore and naval applications – part I., Keynote Lecture, 6th International Conference on Hydrodynamics, Perth, Australia.

Det norske Veritas, 2005. DNV – OS – C105 Structural design of TLP (LRFD method).

Huang, L.L. & Riggs, H.R. 2000. The hydrostatic stiffness of flexible floating structures for linear hydroelasticity, Marine Structures,13, pp. 91–106.

Jain, A.K. 1997. Nonlinear coupled response of offshore tension leg platform to regular wave forces, Ocean Engineering, 24(7), pp. 577–593.

Low, M.Y. 2009. Frequency domain analysis of a tension leg platform with statistical linearization of the tendon restoring forces, Marine Structures, 22, pp. 480–503.

Malenica, Š. 2003. Some aspects of hydrostatic calculations in linear seakeeping, International Conference on Ship and Shipping Research, NAV, Palermo, Italy.
Natvig, B.J. & Teigen, P. 1993. Review of hydrodynamic challenges in TLP Design, Proceedings of the 3rd International Offshore and Polar Engineering Conference, Singapure, pp. 294–302.
Senjanović, I. et al. 2009. Ship hydroelastic analysis with sophisticated beam model and consistent restoring stiffness, Proceedings of Hydroelasticity in Marine Technology, Southampton.
Senjanović, I. et al. 2010. Some aspects of geometric stiffness modeling in the hydroelastic analysis of ship structures, Transactions of FAMENA, Vol. 34, No. 4.
SNAME, 1988. Principles of Naval Architecture, Jersey City, NJ, The Society of Naval Architects and Marine Engineers.
Zienkiewicz, O.C. 1971. The Finite Element Method in Engineering Science, London: McGraw – Hill.
Withee, J.E. 2004. Fully coupled dynamic analysis of a floating wind turbine system, PhD thesis, MIT, USA.

Sustainable Maritime Transportation and Exploitation of Sea Resources – Rizzuto & Guedes Soares (eds)
 ISBN 978-0-415-62081-9

Buckling behavior of catenary risers conveying fluids

I.K. Chatjigeorgiou & S.A. Mavrakos
National Technical University of Athens, Greece

ABSTRACT: The present study investigates the out-of-plane buckling behavior of catenaries conveying fluids and subjected to end imposed axial excitations. The mathematical formulation is generic with no restrictions in the amount of sag in the plane of reference. The theoretical model is treated using Galerkin's expansion in which the actual mode shapes are employed. It is shown that under a parametric excitation the physical model behaves as a multidegree of freedom system. The novelty of the work is that it finds the equivalent multidegree of freedom system that governs the dynamics of the catenary, while in addition it accounts for the internal flow effects which arise from "compressive loading" and Coriolis forces.

1 INTRODUCTION

Catenary-shaped slender structures conveying fluids have many important industrial applications, such as risers in offshore installations, free span pipelines etc. The term "slender" is used to distinguish relatively large diameter line structures from cables for which the flexural rigidity is negligible. In many occasions structures of that type are subjected to motions imposed at the ends, e.g. risers experiencing the action of top imposed excitations originating from the dynamic response of the host floater. Under these conditions the structure experiences extreme bending moments at its lower portion where the curvature obtains a maximum (Passano & Larsen, 2006; Chatjigeorgiou et al., 2007). Bending moment amplifications occur both to the plane of reference and transversely to the out-of-plane direction, i.e. normal to the former (Chatjigeorgiou, 2010a). Apparently, the dynamic amplification of internal loading originates from large motions for which the primarily responsible cause is the axial component of the excursion imposed at the top (Passano & Larsen, 2006).

In fact, under axial motions the catenary performs cycling motions which are more profound at the lower portion due to the smaller available effective tension which acts as a mechanical constrain. This remark implies that the study of the dynamic behavior on the 2D plane of reference of the catenary is definitely a short approximation as the out-of-plane vibrations may be equally important. It is evident that out-of-plane motions occur primarily due to the excitations with the same orientation. This is the obvious condition. Nevertheless, they may also occur due to other causes which are much more interested from the academic point of view such as the vortex shedding (LeCunff et al., 2005) widely referred as Vortex-Induced-Vibrations, or buckling phenomena. The content of the present study appertains to the last category.

If it is assumed that a catenary experiences buckling due to axially imposed motions which apparently act as parametric excitations, then, to a certain extent, the structural model could be approximated by an equivalent straight beam subjected to a follower load. Indicative examples of works on the buckling behavior of pipes, formulated as beams, which additionally convey a fluid are those due to Plaut (2006), Langthjem and Sugiyama (2000) and Sorokin and Terentiev (2003). In relevant cases it can be shown that the mathematical analysis results in a multidegree of freedom system. For a catenary-shaped slender structure the equivalent multidegree of freedom system should be found. That system must contain the terms arising from the complexity of the geometry and if a flow is included, the "compressive loading" terms as well as the Coriolis effects (Paidoussis, 1998).

To study only the out-of-plane buckling behavior of a catenary, the associated motions must be considered decoupled from the motions in the 2D plane of reference. Chatjigeorgiou (2010b) showed that decoupling is indeed achieved under employing certain and valid assumptions. The dynamic model is still influenced by the axial motions imposed at the top of the structure which eventually act as parametric-type excitations. It is widely known that parametric inhomogeneous terms lead to parametric instabilities which for the structural model at hand could be one of the reasons for the extreme amplification of dynamic components especially at the lower portion of the catenary.

The scope of the present study is to show that indeed the out-of-plane vibrations of a catenary-shaped slender structure might be influenced by potential instabilities due to axially imposed motions. To this end the governing multidegree of freedom system is derived. The novelty herein is the inclusion of terms arising from a hypothetical internal flow with steady velocity, namely "compressive loading" and Coriolis effects. It should be also mentioned that the governing multidegree of freedom system is derived using the actual mode shapes appertaining to the real catenary. To this end, the associated eigenvalue problem is formulated and solved in its complete form. Occurrences of potential instabilities are investigated by employing Floquet theory.

2 MATHEMATICAL FORMULATION

The study refers to a catenary-shaped slender structure with the following properties: effective mass m, effective weight w_0, inner diameter d_I, outer diameter d_O, suspended length L, elastic stiffness EA and flexural rigidity EI. All quantities refer to the unstretched length and they are considered uniform along the structure. It is also considered that the structure conveys a fluid with mass per unit length M_f and steady velocity V. The flow inside the pipe corresponds to the so called "plug-flow" model (Paidoussis, 1998). Under these conditions, the out-of-plane vibrations, uncoupled from the in-plane motions, will be governed by the following system of Partial Differential Equations (PDEs) (Chatjigeorgiou 2010a)

$$(m+M_f)\frac{\partial^2 r(s,t)}{\partial t^2}+M_f V\frac{\partial^2 r(s,t)}{\partial s\partial t} - M_f V^2\Omega_{21}(s,t)=\frac{\partial S_{b1}(s,t)}{\partial s}-S_{n0}(s)\Omega_{30}(s)\theta_1(s,t) - T_0(s)\Omega_{21}(s,t) - (w_0+M_f g)\sin\phi_0(s)\theta_1(s,t) - c\omega|r(s,t)|\frac{\partial r(s,t)}{\partial t} \tag{1}$$

$$\theta_1(s,t)=-\frac{\partial r(s,t)}{\partial s} \tag{2}$$

$$EI\frac{\partial\Omega_{21}(s,t)}{\partial s}-EI\Omega_{30}(s)\theta_1(s,t)=S_{b1}(s,t) \tag{3}$$

$$\frac{\partial\theta_1(s,t)}{\partial s}=\Omega_{21}(s,t) \tag{4}$$

In the following the total mass $m + M_f$ and the total weight $w_0 + M_f g$ per unit unstretched length will be denoted by M and w respectively where g is the acceleration due to gravity.

The indexes 0 and 1 in the above system denote static and dynamic components respectively while s denotes the Lagrangian coordinate that takes values along the unstretched suspended length of the catenary. The motion and loading parameters which are involved into the dynamic system are the out-of-plane displacements r, the in-plane and the out-of-plane shear forces S_n and S_b, the in-plane and the out-of-plane curvatures Ω_3 and Ω_2, the in-plane and out-of-plane rotations φ and θ and finally the effective tension T. Apparently, the same system can be applied to non-ordinary applications, e.g. when the structure is completely submerged. The latter is the case where the last term in Equation (1) has a vital contribution as it denotes the distributed drag forces. In Equation (1) these forces are given in their linearized form and ω denotes the excitation frequency provided that the structure is excited in the out-of-plane direction, while the constant c is equal to $(4/3\pi)\rho C_d d_O$ with ρ, C_{db} being respectively the density of the surrounding fluid and the drag coefficient.

Clearly, the dynamic system at hand is linear. According to Equations (1)–(4) the out-of-plane vibrations are completely uncoupled from the associated in-plane motions of the structure in its 2D plane of reference. The conditions that lead to decoupling have been extensively discussed by Chatjigeorgiou (2010b).

Further, Equations (2)–(4) can be integrated into the governing Equation (1) to yield the following compact formulation for the out-of-plane dynamic motions

$$M\frac{\partial^2 r(s,t)}{\partial t^2}+M_f V\frac{\partial^2 r(s,t)}{\partial s\partial t}=-EI\frac{\partial^4 r(s,t)}{\partial s^4} + \left(S_{n0}(s)\Omega_{30}(s)+w\sin\phi_0(s)\right)\frac{\partial r(s,t)}{\partial s} + \left(T_0(s)-M_f V^2\right)\frac{\partial^2 r(s,t)}{\partial s^2}-c\omega|r(s,t)|\frac{\partial r(s,t)}{\partial t} \tag{5}$$

For simplification, we omitted from Equation (3) the virtually insignificant term $EI\Omega_{30}(s)\theta_1(s,t)$. Typically dynamic systems in the form of Equation (5) are treated in the literature using the mode superposition principle by employing Galerkin's expansion. It is also true that the most popular mode shapes which are commonly used for formulating Galerkin's expansion are harmonic functions of the spatial variable s, such as $\sin(n\pi s/L)$ where n is the mode consecutive number. Admittedly, that simplified approximation satisfies the boundary conditions (i.e. zero motions and bending moments at the ends) while in addition facilitates further elaboration steps. Nevertheless it contains a discouraging drawback as it is valid only for weightless structures, free of initial geometric distortions. In other words, is a short approximation for complex geometries like catenaries which experience the effect of large bending moments in the lower end zone. In concluding, when employing Galerkin's expansion it is obviously advisable to use the actual mode shapes.

3 THE EIGENVALUE PROBLEM

The elaboration of any dynamic system using a semi-analytical formulation requires by principle the treatment of the associated eigenvalue problem that provides the natural frequencies and the mode shapes of the dynamic components.

The associated matrix system was treated as a two point boundary value problem taking into account the compressive loading term and it was solved using MATLAB's *bvp4c* function. Special attention is given to the differentiations caused by the inclusion of the inner flow. The calculations refer to a completely submerged catenary structure that was taken from the study of Passano & Larsen (2006) and its properties are listed in Table 1. The same model is used throughout this paper.

Figures 1–2 show respectively the first three mode shapes for the out-of-plane displacement

Table 1. Properties and installation characteristics of the catenary-shaped slender structure.

Suspended length	2024 m
Outer diameter	0.429 m
Inner diameter	0.385 m
Effective mass per unit length	390.85 kg/m
Weight per unit length	927.36 N/m
Elastic stiffness	0.5829×10^{10} N
Bending stiffness	0.1209×10^{9} Nm^2
Drag coefficient	1.0
Installation depth	1800 m
Pretension at the top	1860 kN
Fluid mass per unit length	23.28 kg/m
Fluid velocity	5 m/s

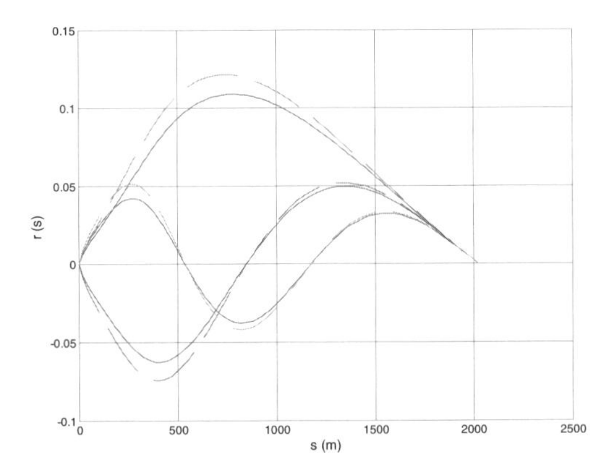

Figure 1. Mode shapes $n = 1 \div 3$ of the out-of-plane displacement $r(s)$. Solid lines: no fluid flow; dashed lines: inner flow is included.

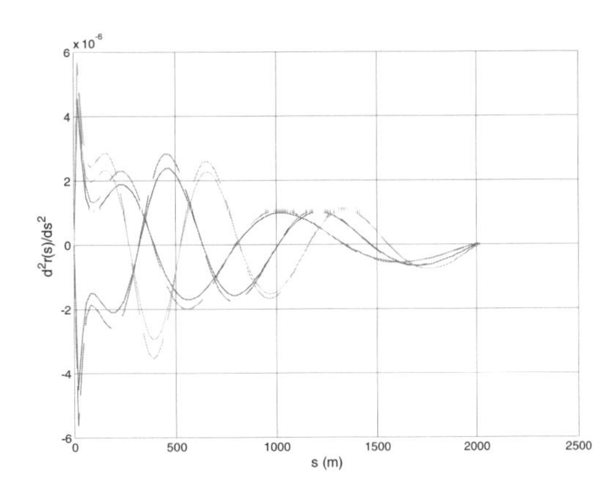

Figure 2. Mode shapes $n = 1 \div 3$ of the out-of-plane rotation $dr(s)/ds = -\theta_1(s)$. Solid lines: no fluid flow; dashed lines: inner flow is included.

Table 2. Out-of-plane natural frequencies (rad/s).

Mode no	With flow	w/o flow
1	0.0778	0.0761
2	0.1521	0.1482
3	0.2252	0.2191
4	0.2991	0.2908
5	0.3740	0.3635
6	0.4500	0.4372
7	0.5268	0.5118
8	0.6044	0.5871
9	0.6829	0.6633
10	0.7622	0.7403

and its second derivative. It is reminded that the second derivatives is a measure of the bending moment. It is immediately apparent that the actual mode shapes of the catenary behave differently than the commonly used sinusoidal eigenmodes. Firstly, the mode shapes are not symmetrical as can be easily deduced. Secondly, the maximum transverse excursion in the out-of-plane direction should be expected to occur at the lower part of the structure. The final and most important remark is associated with the bending moments (or equivalently the curvatures) which according to Figure 2 will exhibit a sharp increase very close to the lower end. Also, the effect of the internal flow is immediately reflected in the results of the eigenvalue problem by changing both the natural frequencies and the mode shapes. The former are directed to lower values whereas the modes shapes, although they retain the same trend are different in magnitude. In particular, the inclusion of the internal flow allows the structure to perform larger displacements by stimulating the "compressive loading" mechanism which reduces the available effective tension along the structure. It should be mentioned however that the eigenvalue problem provides only the qualitative characteristics of the system's response while the actual dynamic behavior could be different. In fact, the eigenvalue problem is set by omitting the all-important Coriolis forces $M_f V \partial^2 r(s,t)/\partial r \partial s$ which may introduce an effective damping due to the energy flow through the boundaries of the structure.

4 SEMI-ANALYTICAL EXPANSION

The calculation of modes shapes enables the employment of the mode superposition principle. Thus, the out-of-plane motions of the structure $r(s,t)$ are assumed to be of the form $r(s,t) = \Sigma\xi_n(t)$ $\varphi_n(s)$, where $\varphi_n(s)$ denotes the nth order mode shape and $\xi_n(t)$ is the associated time dependent generalized variable. Next, it is assumed that the structure is excited at the top by combined axial and transverse motions in the out-of-plane direction. Both motions are considered harmonic with the same circular frequency ω and no phase difference between them. A fundamental assumption is also made according to which the stretching along the structure due to the axially imposed motion varies linearly with the spatial coordinate s. This is definitely the case for cables (Papazoglou et al., 1990) and it holds also for slender structures with considerable flexural rigidity provided that they are excited with relatively small amplitudes. Thus, the out-of-plane transverse motions $r(s,t)$ and the in-plane axial deformations $p(s,t)$ will be given respectively by $r(s,t) = (s/L)r_a(t) + \Sigma\xi_n(t)$ $\varphi_n(s)$, $p(s,t) = (s/L)p_a(t)$ where $r_a(t)$ and $p_a(t)$ are the harmonic excitations imposed at the top of the structure, whereas the corresponding amplitudes are denoted by r_{a0} and p_{a0}. In addition the dynamic tension is approximated by its linear component $T_1 = EA\partial p(s,t)/\partial s$.

Using the above mentioned the governing Equation (5) gives Equation (6). The primes in Equation (6) denote differentiation with respect to s while the index iv denotes fourth-order differentiation. The dots denote differentiation with respect to time t. The use of the correct mode shapes has a collateral effect to the solution procedure. In particular, although the mode shapes of the out-of-plane motions are indeed orthogonal, we cannot speculate that a similar relation is valid by default for the associated mode shapes of curvature. The validity of orthogonality for the mode shapes of motion has been verified numerically and accordingly this is the only relevant property that is employed in the present.

$$
\begin{aligned}
& M\left[\frac{s}{L}\ddot{r}_a(t)+\sum_{n=1}^{N}\ddot{\xi}_n(t)\varphi_n(s)\right] \\
& + M_f V\left[\frac{1}{L}\dot{r}_a(t)+\sum_{n=1}^{N}\dot{\xi}_n(t)\varphi'_n(s)\right] = -EI\sum_{n=1}^{N}\xi_n(t)\varphi_n^{iv}(s) \\
& -\left(S_{n0}(s)\Omega_{30}(s)+w\sin\phi_0(s)\right)\left[\frac{1}{L}r_a(t)+\sum_{n=1}^{N}\xi_n(t)\varphi'_n(s)\right] \\
& +\left(T_0(s)-M_f V^2\right)\sum_{n=1}^{N}\xi_n(t)\varphi''_n(s)+\frac{EA}{L}p_a(t)\sum_{n=1}^{N}\xi_n(t)\varphi''_n(s) \\
& -c\omega\left|\frac{s}{L}r_a(t)+\sum_{n=1}^{N}\xi_n(t)\varphi_n(s)\right|\left(\frac{s}{L}\dot{r}_a(t)+\sum_{n=1}^{N}\dot{\xi}_n(t)\varphi_n(s)\right)
\end{aligned}
\tag{6}
$$

5 DYNAMIC BUCKLING UNDER AXIAL LOADING

The term *buckling* is usually correlated with straight beams subjected to axial loading. Further elaboration of the mathematical model of the catenary structure under investigation shows that even non typical geometries such the one examined in the present could experience the so-called buckling phenomenon due to axial imposed loading which eventually acts as a parametric excitation. The final results can manifest phenomena analogous to straight beams, e.g. occurrences of instabilities governed by the properties of the excitation.

It should be mentioned that the dynamic system which is considered in the present can be also used to different applications. The principal approach was to formulate a mathematical model that applies mainly to submerged catenaries. Nevertheless the formulation is generic arising from a Newtonian derivation procedure (Chatjigeorgiou, 2010a; Chatjigeorgiou, 2010c) and can be used for non immersed structures as well provided that the nonlinear drag forces will be omitted. When the structure is surrounded by water the effect of drag forces is determinative and they explicitly cancel the effects of potential instabilities. In fact, possible instability regions can be detected only in the absence of drag. On the other hand, instabilities are related to the physical and mechanical properties of the structure, its geometry and the excitation characteristics. Therefore they are not correlated with the distributed drag forces which must be considered as an externally imposed energy absorbing mechanism. Here instabilities must be regarded as the cause of unbounded responses in the absence of drag and the cause of the amplification of motion when drag is presented.

Coming back to the mathematical formulation and in particular to Equation (12), we omit the last term of the right hand side and we apply the relation of orthogonality of the mode shapes of the out-of-plane motions. The harmonic excitations are taken in their conventional form with equal frequency, namely $r_a(t) = r_{a0}\cos\omega t$ and $p_a(t) = p_{a0}\cos\omega t$ and eventually Equation (12) is transformed into

$$\ddot{\xi}_k(t) = \beta_k \cos(\omega t + \vartheta_k) - \sum_{n=1}^{N} \gamma_{kn}\dot{\xi}_n(t) + \sum_{n=1}^{N}(-\Lambda_{kn} + \cos\omega t \mathrm{K}_{kn})\xi_n(t) = 0 \quad (7)$$

The tensors β_k, ϑ_k, γ_{kn}, K_{kn} and Λ_{kn} are products of the physical and the mechanical properties of the pipe, the excitation properties and the mode shapes. Due to their extended form their detailed writing is omitted.

It is easily seen that the inner flow, and in particular Coriolis forces, introduce a phase shift ϑ_k between the axial and transverse out-of-plane excitations which varies with the mode of response k.

According to Equation (7) the time dependent generalized variables $\xi_k(t)$ are the unknown components of a multidegree of freedom parametrically excited system. The system of Equation (7) has apparent similarities with the system defined and solved by Nayfeh & Mook (1979) using Floquet theory. Nevertheless the present multidegree of freedom system involves two more terms, namely the forcing term $\beta_k \cos(\omega t + \vartheta_k)$ and the truncated superposition of damping components $\sum_{\nu=1}^{N} \gamma_{kn}\dot{\xi}_n(t)$ which originate from Coriolis forces.

In the following we consider only axial excitations and we base our analysis in Floquet theory (Nayfeh & Mook, 1979; Nayfeh & Balachandran, 1995). Firstly, after omitting the out-of-plane forcing term, Equation (7) is transformed into a more suitable form according to

$$\ddot{\xi}_k(t) = -\sum_{n=1}^{N} \gamma_{kn}\dot{\xi}_n(t) + \sum_{n=1}^{N} a_{kn}(t)\xi_n(t) = 0 \quad (8)$$

where $a_{kn}(t) = -\Lambda_{kn} + \cos\omega t \mathrm{K}_{kn}$.

The basics of the solution procedure outlined in the following were taken from Nayfeh & Mook (1979) and Nayfeh & Balachandran (1995). For convenience Equation (8) is expressed as a system of $2N$ first-order differential equations by defining

$$u_k = \xi_k, k = 1, 2, \ldots, N \quad (9)$$

$$u_k = \dot{\xi}_k, k = N+1, N+2, \ldots, 2N, \quad (10)$$

Thus, Equation (8) becomes

$$\dot{u}_k = u_{N+k}, k = 1, 2, \ldots, N \quad (11)$$

$$\dot{u}_{N+k} = -\sum_{n=1}^{N} \gamma_{kn} u_{N+n} + \sum_{n=1}^{N} a_{kn}(t) u_n, k = 1, 2, \ldots, N \quad (12)$$

Next, Equations (11) and (12) can be written in the compact form

$$\dot{\mathbf{u}} = [F(t)]\mathbf{u} \quad (13)$$

where $\mathbf{u}$ is a column vector with the elements $u_1, u_2, \ldots u_{2N}$ while $[F(t)]$ is an $2N \times 2N$ matrix. For the system of Equation (13) one can define a fundamental set of solutions $u_{1k}, u_{2k}, \ldots, u_{Jk}$, $k = 1, 2, \ldots, J$ where $J = 2N$. This fundamental set can be expressed in the form of an $J \times J$ matrix $[U]$ called a fundamental matrix solution as

$$[U] = \begin{bmatrix} u_{11} & u_{12} & \cdot & \cdot & \cdot & u_{J1} \\ u_{12} & u_{22} & \cdot & \cdot & \cdot & u_{J2} \\ \cdot & \cdot & \cdot & \cdot & \cdot & \cdot \\ \cdot & \cdot & \cdot & \cdot & \cdot & \cdot \\ \cdot & \cdot & \cdot & \cdot & \cdot & \cdot \\ u_{1J} & u_{2J} & \cdot & \cdot & \cdot & u_{JJ} \end{bmatrix} \quad (14)$$

The elements of matrix $[F(t)]$ are $F_{i,N+i} = 1$, $F_{N+i,j} = f_{ij}(t)$, $F_{N+i,N+j} = -\gamma_{ij}$ and zero elsewhere. Clearly the fundamental matrix solution satisfies the matrix equation $[\dot{U}] = [F(t)][U]$. It is also evident that $[F(t)]$ is periodic meaning that $[F(t+T)] = [F(t)]$. According to the previous remarks $[U(t+T)]$ is also a fundamental matrix solution given by $[U(t+T)] = [A][U(t)]$. Here $[A]$ denotes a nonsingular constant $J \times J$ matrix. Let's further assume that $[U(t)] = [P][V(t)]$ where $[P]$ is also a nonsingular $J \times J$ matrix. Thus, $[V(t+T)] = [P]^{-1}[A][P][V(t)] = [B][V(t)]$. The matrix $[P]$ is chosen in such a way, so $[B] = [P]^{-1}[A][P]$ is a Jordan canonical form. Thus, if λ denotes the eigenvalues of matrix $[A]$, and these eignevalues are distinct, the matrix $[B]$ has a diagonal form while the diagonal elements coincide with the eigenvalues of matrix $[A]$. This the only case that is considered herein as in all numerical simulations performed for the purposes of the present contribution, the eigenvalues of $[A]$ were distinct. Nevertheless, it should be mentioned that in case where the eigenvalues of $[A]$ are not distinct, the matrix $[B]$ is not diagonal and obtains a more complicated form. Assuming that the matrix $[B]$ is diagonal then the elements of matrix $[V]$ can be written as $\mathbf{v}_j(t+T) = \lambda_j \mathbf{v}_j(t)$, $j = 1, 2, \ldots$ and $\mathbf{v}_j(t+nT) = \lambda_j^n \mathbf{v}_j(t)$ where

n is integer and denotes the periods of the simulation. Thus it follows that the elements $\mathbf{v}_j(t) \to \infty$ for $|\lambda_j| > 1$. For the same reasons the response of the dynamic system will be unbounded. Here, for the identification of instabilities there is no need to calculate the transformation matrix $[P]$. The latter remark is extremely assisting due to the difficulties involved in the numerical calculation of a Jordan canonical form. So only the matrix $[A]$ is required. This can be determined numerically using the initial conditions $[U(0)] = [I]$ as after one period of numerical simulation $[A] = [U(t)]$.

Figures 3–11 show the results that were obtained by applying Floquet theory for the structure defined in Table 1 for three different excitation

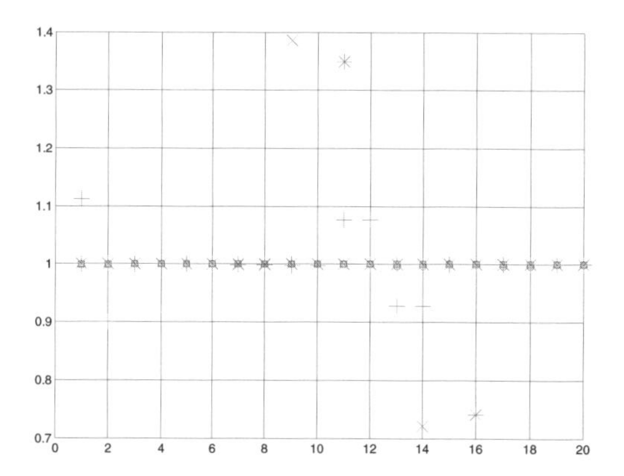

Figure 3. Eigenvalues $|\lambda_j|$ of matrix $[A]$ against mode number for the structure conveying fluid and excited axially with amplitude $p_{a0} = 0.1$m; "o" $\omega = 0.1$rad/s; "+" $\omega = 0.5$ rad/s; "*" $\omega = 1.0$rad/s; "×" $\omega = 1.5$rad/s; "□" $\omega = 2.0$rad/s; "Δ" $\omega = 2.5$ rad/s.

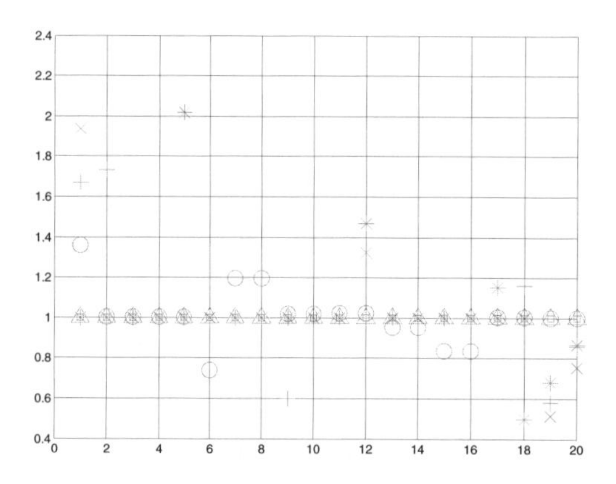

Figure 4. Eigenvalues $|\lambda_j|$ of matrix $[A]$ against mode number for the structure conveying fluid and excited axially with amplitude $p_{a0} = 0.2$m; "o" $\omega = 0.1$ rad/s; "+" $\omega = 0.5$ rad/s; "*" $\omega = 1.0$ rad/s; "×" $\omega = 1.5$ rad/s; "□" $\omega = 2.0$ rad/s; "Δ" $\omega = 2.5$ rad/s.

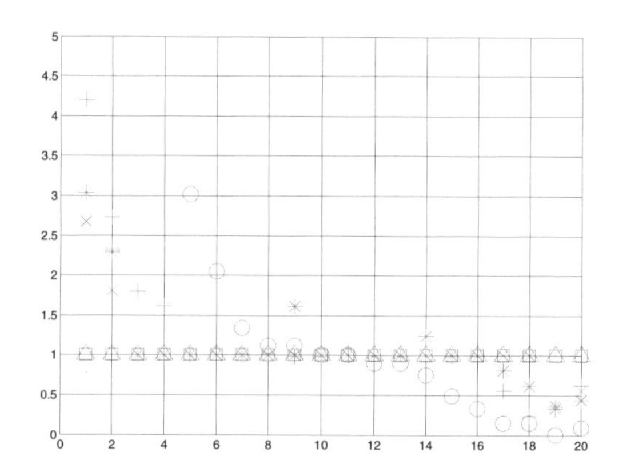

Figure 5. Eigenvalues $|\lambda_j|$ of matrix $[A]$ against mode number for the structure conveying fluid and excited axially with amplitude $p_{a0} = 0.3$ m; "o" $\omega = 0.1$ rad/s; "+" $\omega = 0.5$ rad/s; "*" $\omega = 1.0$ rad/s; "×" $\omega = 1.5$ rad/s; "□" $\omega = 2.0$ rad/s; "Δ" $\omega = 2.5$ rad/s.

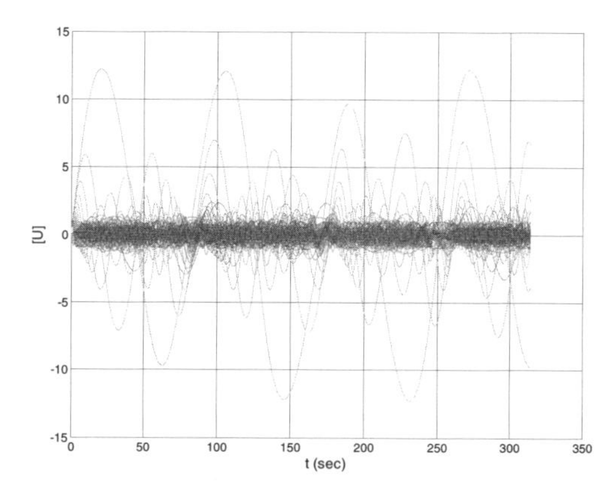

Figure 6. Time histories of the $J \times J = 400$ elements of the fundamental matrix solution $[U]$ for the structure conveying fluid under axial excitation; excitation properties $p_{a0} = 0.1$ m, $\omega = 0.1$ rad/s; simulation time: 5 periods.

amplitudes ($p_{a0} = 0.1$, 0.2 and 0.3 m) and various frequencies ranging between 0.1 and 2.5 rad/s. The calculations were performed using $N = 10$ eigenmodes and the time histories of the 400 elements of matrix $[U]$ were obtained by applying Runge-Kutta method to equation $[\dot{U}] = [F(t)][U]$. The absolute values of λ_j that define whether the response will be bounded or not, are given in Figures 3–5. Each figure corresponds to different parametric excitation amplitude. In all figures a number of eigenvalues fall exactly on the axis $|\lambda_j| = 1$. In these cases the eigenvalues are complex and the associated response is periodic as shown in the following, for selected excitation properties (Figs. 6, 9 and 11). Absolute values above unity are found in Figure 3 ($p_{a0} = 0.1$m) for $\omega = 0.5 \div 1.5$ rad/s, in

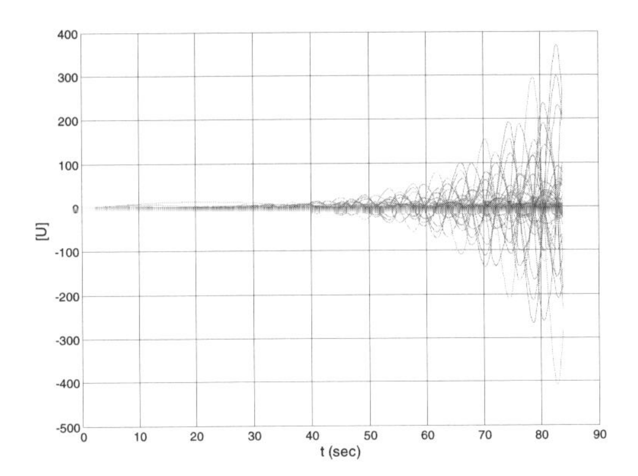

Figure 7. Time histories of the $J \times J = 400$ elements of the fundamental matrix solution $[U]$ for the structure conveying fluid under axial excitation; excitation properties $p_{a0} = 0.1$ m, $\omega = 1.5$ rad/s; simulation time: 20 periods.

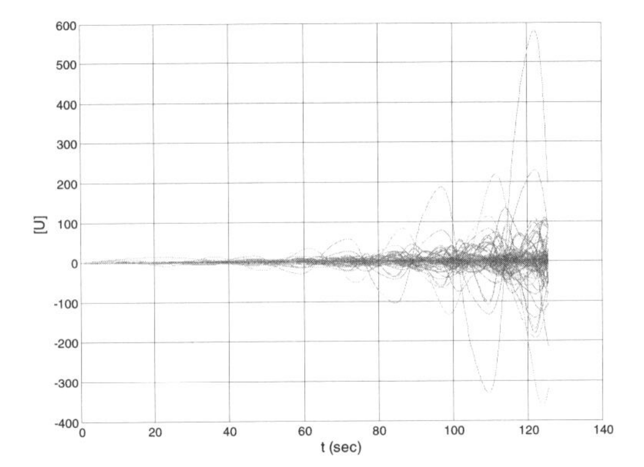

Figure 8. Time histories of the $J \times J = 400$ elements of the fundamental matrix solution $[U]$ for the structure conveying fluid under axial excitation; excitation properties $p_{a0} = 0.2$m, $\omega = 0.5$ rad/s; simulation time: 10 periods.

Figure 4 ($p_{a0} = 0.2$m) for $\omega = 0.1 \div 1.5$ rad/s and in Figure 5 ($p_{a0} = 0.3$m) for $\omega = 0.1 \div 1.5$ rad/s. As the excitation amplitude is increased, more eigenvalues are detached from axis $|\lambda_j| = 1$ whereas they obtain higher values. The common characteristic of the three investigated excitation amplitudes is that for high excitation frequencies (2.0 rad/s and above) the response is stable. In general, lower frequencies are more susceptible to stimulate parametric instabilities.

To show descriptively how the dynamic system responds, Figures 6–11 are given which depict the time histories of the elements of the fundamental matrix solution for sufficient simulation periods. In fact these figures contain the output signals of all 400 elements of $[U]$. To identify whether the response will be bounded or not the

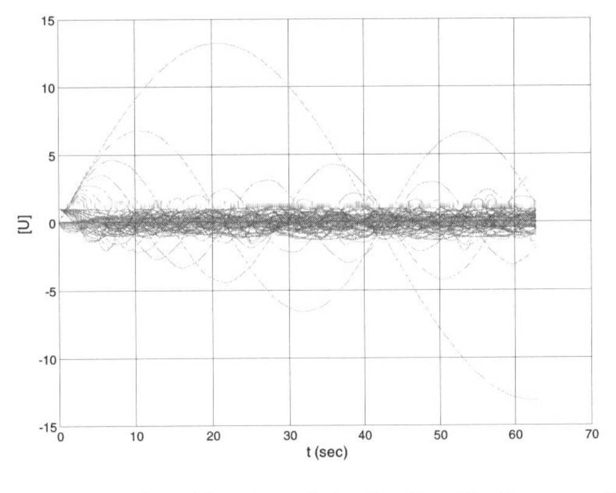

Figure 9. Time histories of the $J \times J = 400$ elements of the fundamental matrix solution $[U]$ for the structure conveying fluid under axial excitation; excitation properties $p_{a0} = 0.2$ m, $\omega = 2.0$ rad/s; simulation time: 20 periods.

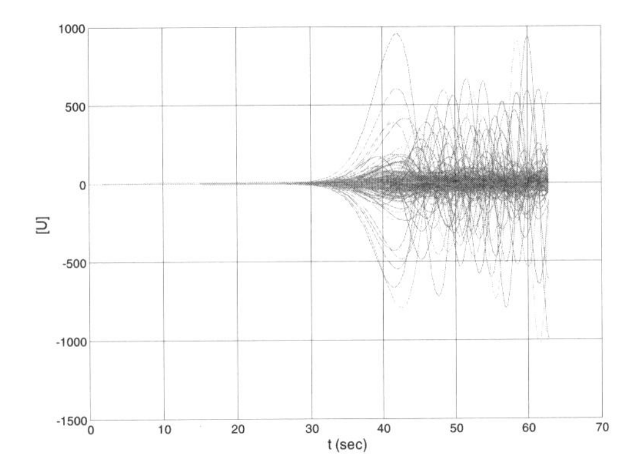

Figure 10. Time histories of the $J \times J = 400$ elements of the fundamental matrix solution $[U]$ for the structure conveying fluid under axial excitation; excitation properties $p_{a0} = 0.3$ m, $\omega = 0.1$ rad/s; simulation time: 1 period.

solution of equation $[\dot{U}] = [F(t)][U]$ is required. If only one simulation period is considered then the solution provides the eigenvalues λ_j, i.e., the values which govern the qualitative characteristics of the dynamic system under a specified excitation condition. Admittedly, only one simulation period is inadequate to provide visual indications regarding the stability of the system. Therefore, the question which easily arises is whether $|\lambda_j| \le 1$ and $|\lambda_j| > 1$ indeed reflect stable and unstable solutions respectively. Apparently, the time histories of Figures 6–11 support the affirmative answer.

According to Figure 3, excitations driven by $\omega = 0.1$ rad/s and $\omega = 1.5$ rad/s should be respectively stable and unstable. This is demonstrated graphically in the time histories of Figures 6 and 7. Also, for $p_{a0} = 0.2$ m, the results depicted in

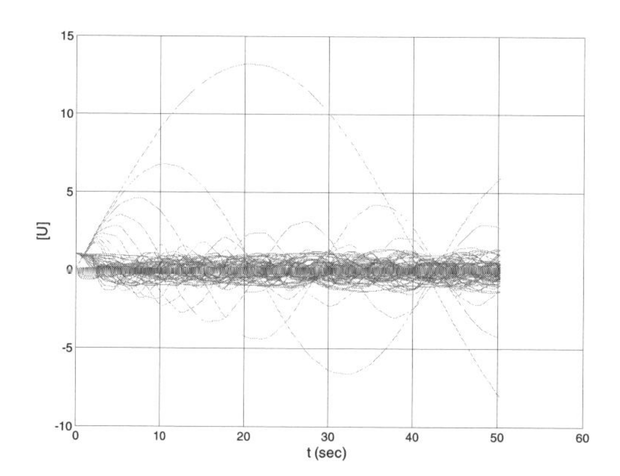

Figure 11. Time histories of the $J \times J = 400$ elements of the fundamental matrix solution $[U]$ for the structure conveying fluid under axial excitation; excitation properties $p_{a0} = 0.3$ m, $\omega = 2.5$ rad/s; simulation time: 20 periods.

Figure 4 show that at least one of $|\lambda_j|$ for $\omega = 0.5$ rad/s is greater than unity whereas for $\omega = 2.0$ rad/s all eigenvalues $|\lambda_j|$ fall on the axis $|\lambda_j| = 1$. This implies that the response should be respectively unstable and stable. This is obviously verified by the time histories of Figures 8 and 9. Analogous conclusions are drawn by examining Figures 10 and 11 in relation to the eigenvalues of Figure 5. It is evident that a bounded response implies a periodic motion.

6 CONCLUSIONS

The out-of-plane buckling behavior of catenary-shaped slender structures transporting fluid was investigated. The theoretical model was treated using the mode superposition principle that results in a convenient semi-analytical formulation for the solution of the dynamic parameters. To this end, the actual mode shapes which appertain to the geometry of the catenary were employed.

These were derived by solving the complete out-of-plane eigenvalue problem. The mathematical analysis showed that out-of-plane dynamics can be described by a multidegree of freedom system. This system contained all terms that govern the specific non ordinary geometry, including the properties of the inner flow. The equivalent multidegree of freedom system was treated by employing Floquet theory. It was shown that indeed buckling due to axial imposed loading could lead to potential instabilities.

REFERENCES

Chatjigeorgiou, I.K., Passano, E. & Larsen C.M. 2007. Extreme bending moments on long catenary risers due to heave excitation, *Proceedings 26th International Conference on Offshore Mechanics and Arctic Engineering,* (OMAE 2007), San Diego, California, USA, Paper No 29384.

Chatjigeorgiou, I.K. 2010a. On the effect of internal flow on vibrating catenary risers in three dimensions, *Engineering Structures*, 32, 3313–3329.

Chatjigeorgiou, I.K. 2010b. Linear out-of-plane dynamics of catenary risers. *Journal of Engineering for the Maritime Environment*, 224, 13–27.

Chatjigeorgiou, I.K. 2010c. Three dimensional nonlinear dynamics of submerged, extensible catenary pipes conveying fluid and subjected to end-imposed excitations, *International Journal of Non-Linear Mechanics*, 45, 667–680.

Langthjem, M.A. & Sugiyama, Y. 2000. Dynamic stability of columns subjected to follower loads: a survey, *Journal of Sound and Vibration*, 238, 809–851.

LeCunff, C., Biolley, F. & Damy, G. 2005. Experimental and numerical study of heave induced lateral motion (HILM), *Proceedings 24th International Conference on Offshore Mechanics and Arctic Engineering* (OMAE 2005), Halkidiki, Greece, Paper No 67019.

Matlab 2010. Version 7.10.0.499 (R2010a), The MathWorks™.

Nayfeh, A.H. & Mook, D.T. 1979. *Nonlinear Oscillations*, Wiley Interscience, John Wiley and Sons Inc, New York.

Nayfeh, A.H. & Balachandran, B. 1995. *Applied Nonlinear Dynamics*, Wiley Series of Nonlinear Science, John Wiley and Sons Inc, New York.

Paidoussis, M.P. 1998. *Fluid-structure interactions: Slender structures and axial flow*, Vol. 1, London Academic Press.

Papazoglou, V.J., Mavrakos, S.A. & Triantafyllou, M.S. 1990. Nonlinear cable response and model testing in water, *Journal of Sound and Vibration*, 140, 103–115.

Passano, E. & Larsen, C.M. 2006. Efficient analysis of a catenary riser. *Proceedings 25th International Conference on Offshore Mechanics and Arctic Engineering* (OMAE 2006), Hamburg, Germany, Paper No 92308.

Plaut, R.H. 2006. Postbuckling and vibration of end-supported elastic pipes conveying fluid and columns under follower loads, *Journal of Sound and Vibration*, 289, 264–277.

Sorokin, S.V. & Terentiev, A.V. 2003. Nonlinear statics and dynamics of a simply supported nonuniform tube conveying an incompressible inviscid fluid, *Journal of Fluids and Structures*, 17, 415–431.

9.3 *Wind*

Sustainable Maritime Transportation and Exploitation of Sea Resources – Rizzuto & Guedes Soares (eds)
© 2012 Taylor & Francis Group, London, ISBN 978-0-415-62081-9

Fatigue analysis of tripods and jackets for offshore wind turbines

N. Alati, F. Arena, G. Failla & V. Nava
Department of Mechanics and Materials, Mediterranea University of Reggio Calabria, Reggio Calabria, Italy

ABSTRACT: It is common understanding that large offshore wind farms may supply an increasing portion of the energy demand in the next years. To minimize the visual impact, future offshore wind energy projects are planned in deeper water than in most of the current sites. For these purposes, tripod or jacket structures must be generally resorted to. At present a considerable theoretical and experimental research effort is devoted to assess whether a tripod or a jacket structure may be the preferable option in deeper water. In general, the choice should take into account technical and economical aspects but up to now no consensus has been reached on the preferable option. In an attempt to contribute to the research effort in this field, this paper carries out a comparative study of the structural performances under fatigue. A Mediterranean site at a water depth of 45 m is considered as a reference site. The tripod and jacket structures are conceived according to typical current design. The fatigue behaviour is assessed in the time domain under combined stochastic wind and wave loading and results are compared in terms of lifetime damage equivalent load.

1 INTRODUCTION

A fatigue damage assessment is critical in the design of support structures for Offshore Wind Energy converters (OWECs). In principle the driving dynamic excitations for fatigue, i.e. wind and waves, should be accounted for in an integrated, non-linear time domain simulation approach. However, this may involve a high computational effort as a major drawback, which is hardly compatible with the iterative nature of the design process, especially in its early stages. On the other hand, there exist well-established methods in the wind energy industry, as well as in the offshore industry, which allow wind and wave responses to be individually computed in a very efficient manner. For these reasons, the development of simplified methods that may combine separate fatigue responses to wind and wave loading has attracted the interest of several researchers (Kuhn, 2001, Van Tempel, 2006).

As a result of the research effort of the last decade, a certain consensus has now been attained on the following concepts (Kuhn, 2001):

a. wind and wind-induced wave loadings can be regarded as independent and stationary processes on a short time scale, within a period of ten minutes to three hours; this is the typical time length of time-domain simulation for fatigue analysis;
b. nonlinear constitutive relations and geometric nonlinearities do not play a significant role for fatigue conditions;
c. the magnitude of the stress ranges in the soil due to either aerodynamic or hydrodynamic fatigue loading is small compared to the ultimate strength; soil behaviour can be then taken as a linear behaviour and separate analysis of the wind and wave responses with the same soil model are suitable.

This theoretical basis has led to simplified methods, either in time or frequency domain, where wind and wave responses are separately computed and corresponding stress ranges are then superposed for fatigue damage assessment. In these methods, a first issue to deal with concerns how to compute the separate wind and wave responses while still accounting for the inherent, mutual interaction between aerodynamic loading and hydrodynamic response. In this regard, numerical tests have shown that the aerodynamic loading on the tower top can computed as the tower top was held fixed on a rigid support, while the hydrodynamic loading can be computed by neglecting the structural velocities of the submerged part (Van Tempel, 2006). Both loadings can be then simultaneously applied on a full structural model where additional modal damping is added in the fundamental fore-aft mode. This additional damping is known in literature as aerodynamic damping and accounts for the fact that the tower top vibration motion induced by aerodynamic and hydrodynamic loading results in a reduction of the wind thrust that would be encountered if the tower was rigid. Several approaches to estimate aerodynamic damping are now available in literature (Kuhn, 2001). Another issue involved in the simplified methods concerns how to superpose, for fatigue damage assessment, the corresponding stress ranges given by the separate wind and wave responses. In this regard, numerical simulations have shown that even a straightforward linear superposition does yield accurate results (Kuhn, 2001).

In view of the accuracy of the results and the relatively easy implementation, the above described simplified methods have become popular in the design process of OWECs. In the design process, a key driver is a fatigue damage assessment. Therefore, based on a time-domain simplified method this paper will pursue a comparative fatigue analysis of typical tripod and jacket support structures. These are generally resorted to in current design for deep waters (>20 m), where monopile support structures would require too large diameters and masses. No general consensus, however, has been attained on whether the tripod or the jacket is preferable (Schaumann & Böker, 2005) and extensive experimental tests are still ongoing (see the homepage http://www.beatricewind.co.uk for the "Beatrice" project and the homepage: http://www.alpha-ventus.de for the "Alpha Ventus" project). This paper aims to contribute to the research effort in this field.

2 STRUCTURAL MODELS

A site offshore the Mediterranean island of Pantelleria has been considered as a reference site. The design water's depth is 45 m. Statistical data on the correlation between sea states (wave period, wave height) and mean wind velocity are available at the website www.waveclimate.org.

2.1 *Tripod and jacket support structures*

Two test structures have been designed according to current design practice. Details on the structural members are given in Figures 1–2 and Tables 1–2. Both the models have been conceived to have almost the same first natural period. The latter has been computed by considering a total lumped mass of $3.4 \cdot 10^6$ kg at the top of the tower, that includes rotor, nacelle and blades masses. The following parameters have been set for constructional steel: Young's modulus $E = 2.1 \cdot 10^6$ kgfcm^{-2}, Poisson's coefficient $\nu = 0.3$, mass density 7800 kgm^{-3}.

2.2 *Turbine*

The turbine is a standard 5MW turbine whose full details can be found in report NREL/TP-500-38060 by Jonkman et al. (2009). Tables 3a and 3b summarize the turbine and the aerofoils parameters.

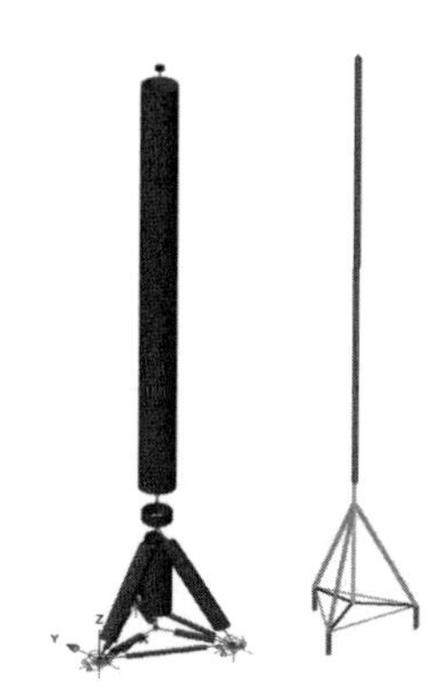

Figure 1. Tripod support structure.

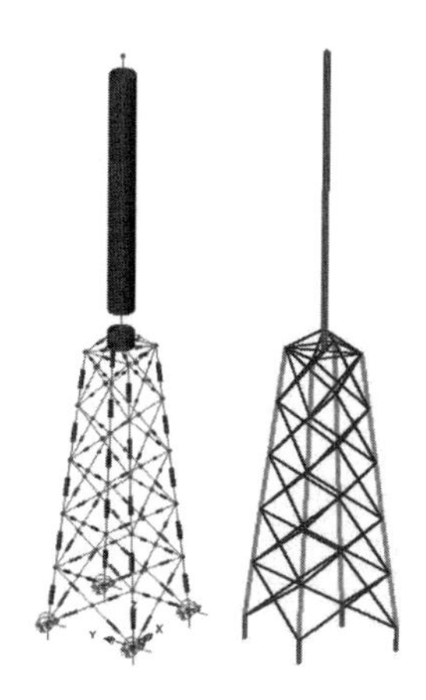

Figure 2. Jacket support structure.

Table 1. Tripod support structure.

Component (colour)	Diameter m	Thickness m	Length m
Orange	0.90	0.025	20.00
Blue	1.20	0.035	11.644
Green	2.60	0.05	27.54
Cyan	4.40	0.12	28.5
Red	5.50	0.25	100

Table 2. Jacket support structure—details.

Component (colour)	Diameter m	Thickness m	Length m
Green	0.87	0.05	12.48
Blue	0.58	0.016	–
Red	5.50	0.25	68.00
Magenta	1.82	0.05	0.50

*Length varies.

Table 3a. Turbine parameters.

5MW turbine	
Power	5MW
Orientation	Upwind
Rotor diameter	126 m
Number of blades	3
Hub diameter	3 m
Hub height (water level)	87.4 m
Cut-in speed	3 ms^{-1}
Cut-out speed	25 ms^{-1}
Nacelle length	19 m
Nacelle height	7 m
Nacelle depth	6 m
Drag coefficient	0.8
Tower height	8.5 m
Hub vertical offset	2.4 m
Hub total height	87.4 m
Cone angle (C)	2.5°
Tilt angle (T)	5°
Overhang (O)	5 m
Mass	56780 kg
Nacelle mass	$2.4 \cdot 10^5$ kg
Centre of mass height	2.5 m
Centre of mass position	1.9 m
Yaw inertia	$2.61 \cdot 10^6$ kgm^2

Table 3b. Aerofoils parameters.

Aerofoils	Distance of root m	N° Reynolds
Cylinder 1	0.0	$6.0 \cdot 10^6$
Cylinder 2	2.0	$6.0 \cdot 10^6$
DU40-A17	11.75	$4.5 \cdot 10^6$
DU35-A17	15.85	$6.0 \cdot 10^6$
DU30-A17	24.05	$7.0 \cdot 10^6$
DU25-A17	28.15	$8.0 \cdot 10^6$
DU20-A17	36.35	$9.0 \cdot 10^6$
NACA64-A17	44.55	$9.2 \cdot 10^6$

3 FATIGUE ANALYSIS

Fatigue damage analysis is cast within the framework of the Miner-Palmgren theory (Sutherland, 1999). Denoting by N the maximum number of stress cycles that a material can withstand under a stress range S, be

$$N = k \cdot S^{-m} \tag{1}$$

where m is the Wöhler integer exponent depending on the material and k depends on the structure. For constructional steel, $m = 4$ is generally set (Sutherland, 1999). Given the number of cycles n_i to which the material is subjected at a given stress range S_i, a total damage index D can be then evaluated by linear superposition of the damage indexes d_i associated to each stress range S_i, according to the Miner-Palmgren law

$$d_i = \frac{n_i}{N_i}; \quad D = \sum_i d_i = \sum_i \frac{n_i}{N_i}. \tag{2a,b}$$

A damage equivalent stress S_{eq} can be then defined as the mean stress that would produce, in an equivalent number of cycles n_{eq}, the same amount of damage produced by the actual, individual stress ranges, according to the equivalence formula

$$D = \sum_i \frac{n_i}{N_i} = \sum_i \frac{n_i}{k} S_i^m = \frac{n_{eq}}{k} S_{eq}^m \quad \Rightarrow$$
$$S_{eq} = \sqrt[m]{\frac{k}{n_{eq}} \sum_i \frac{n_i}{k} S_i^m} = \sqrt[m]{\sum_i \frac{n_i}{n_{eq}} S_i^m} \tag{3}$$

In Eq.(3), n_{eq} is conventionally chosen as $n_{eq} = f_{eq} \cdot T_{eq}$ where $f_{eq} = 1$ and T_{eq} is equal to the time series length. In a similar manner, a Damage Equivalent Load (DEL) can be defined for stress resultants such as normal and shear forces, bending and twisting moments.

The comparative analysis of the tripod and jacket fatigue responses is carried out in terms of a lifetime DEL. In this context, it shall be taken into account that each load case is a stochastic load case that corresponds to a wind velocity class featuring a certain probability of occurrence. The wind velocity class and the related probability of occurrence will be set as explained in section 4 (see Table 4). Based on these data, the lifetime DEL can be introduced as follows (Thomsen, 1998).

$$DEL_{lifetime} = \sum_{j=1}^{M} \sqrt[m]{\frac{1}{n_{eq,lifetime}} \sum_{i=1}^{N_s} \frac{DEL_i^m}{N_s} n_{eq} \cdot P_j \cdot n_t} \tag{4}$$

Table 4a. Lumped load cases.

V_w [ms^{-1}]	3	5	7	9	11
H_s [m]	2.94	3.36	3.54	3.83	4.26
T_z [s]	6.92	6.68	6.45	6.50	6.87
P_j(%)	6.4	12.2	16.9	17.9	16.1

Table 4b. Lumped load cases.

V_w [ms^{-1}]	13	15	17	19	21
H_s [m]	4.93	5.84	6.34	7.53	9.27
T_z [s]	7.20	7.56	8.30	9.05	10.8
P_j(%)	12.3	7.8	4.6	2.1	1.0

where M is the number of wind velocity classes and P_j is the probability associated to each wind velocity class; N_s is the number of samples generated for each load case and DEL_i is the damage equivalent load associated to each sample; n_t is the ratio of the total lifetime to the fatigue simulation time length and $n_{eq,lifetime}$ is the number of equivalent stress cycles in the lifetime, conventionally chosen as $n_{eq,lifetime} = f_{eq,lifetime} \cdot T_{eq,lifetime}$ where $f_{eq,lifetime} = 1$ and $T_{eq,lifetime}$ is the design lifetime. For an OWEC the latter is typically 20 years.

4 WIND AND WAVE LOADING

To reduce the computational effort while preserving accuracy, the load cases are lumped according to the procedure outlined by Andersen (2008) where, for a given wind velocity V_w, the following equivalent wave height $H_{s,eq}$ and equivalent wave period $T_{z,eq}$ are defined:

$$H^m_{s,eq} \sum_{i=1}^{N_H} p_i = \sum_{i=1}^{N_H} p_i H^m_{s,i} \quad \Rightarrow$$

$$H_{s,eq} = \sqrt[m]{\sum_{i=1}^{N_H} p_i H^m_{s,i} \Big/ \sum_{i=1}^{N_H} p_i} \tag{5}$$

$$\frac{1}{T_{z,eq}} \sum_{i=1}^{M_T} p_i = \sum_{i=1}^{M_T} p_i \frac{1}{T_{z,i}} \quad \Rightarrow$$

$$\frac{1}{T_{z,eq}} = \sum_{i=1}^{M_T} p_i \frac{1}{T_{z,i}} \Big/ \sum_{i=1}^{M_T} p_i \tag{6}$$

In Eq.(5) and Eq.(6) p_i are the probabilities ($0 < p_i < 1$) associated to each environmental state ($H_{s,i}$; $T_{z,i}$), N_H and M_T denote the number of wave height and wave period bins. All these data are site depending and, for the reference site of this study, they are available at the website www.waveclimate.org. As a result, the following lumped load cases, along with the related probabilities P_j to be used in Eq.(4) have been considered:

As already mentioned in the introduction, a time-domain simplified approach will be pursued. Specifically, as suggested by Van der Tempel (2006) aerodynamic and hydrodynamic loadings are computed separately and then applied on an finite element model of the tower and the support structure, where the aerodynamic damping effect is also accounted for.

The aerodynamic loading on the tower top is computed as the tower and the support structure were rigid, using the BLADED standard onshore wind turbine design tool (Garrad Hassan and Partners, 2009). As customary, it is assumed that the turbulent part of the wind process is a zero mean Gaussian stationary process, a sample of which can be cast in the form

$$v(t) = \sum_{k=1}^{N_{fv}} \sqrt{2S_v(f_k)\Delta f_k} \cos(2\pi f_k t + \varphi_k) \tag{7}$$

where Δf_k is a constant step on the frequency axis, $f_k = k\Delta f_k$ for $k = 1,2, \ldots N_{fv}$, N_{fv} is the number of frequency steps, φ_k are independent random phases uniformly distributed on $[0,2\pi]$ and $S_v(f)$ is the von Karman power spectral density

$$S_v(f) = \frac{4\sigma_V^2 L_V / V_w}{\left[1 + 70.8\left(f L_V / V_w\right)^2\right]^{5/6}} \tag{8}$$

In Eq.(8) f is the frequency, V_w is the mean wind velocity, to be set according to the load case considered (see Table 4), σ_V is the standard deviation of the wind velocity and L_V is the integral length scale. The latter are site depending and, in this study, have been set as in Table 5:

The hydrodynamic loading is computed by neglecting the structural velocities of the submerged part, as compared to the water particle velocities. Waves are considered collinear and unidirectional, which results in conservative fatigue loads with respect to circumstances where a misalignment between wind and wave direction (Kuhn, 2001).

Specifically, for both the structures under study the wave loading is built based on the slender body theory in an undisturbed wave field, adopting the definition of Chakrabarti (1987). The wave loading is computed based on the Morison equation, according to which the wave forces can be computed as a summation of two contributions, that is the inertia force and the drag force. They are given as

$$f_{Morison}(x,z,t) = f_d(x,z,t) + f_i(x,z,t) \tag{9}$$

Table 5.

Wind velocity ms^{-1}	Intensity turbulence %	L_V m	σ_V ms^{-1}
3	2.88	23.36	0.087
5	4.94	54.20	0.247
7	5.80	78.12	0.406
9	6.34	99.99	0.570
11	6.75	121.13	0.742
13	7.08	141.69	0.920
15	7.36	161.56	1.104
17	7.60	180.65	1.292
19	7.81	189.90	1.484
21	8.00	216.38	1.680

$$f_d(x,z,t) = C_d \cdot \frac{1}{2}\rho_{water} D \cdot |u(x,z,t)| u(x,z,t) \quad (10)$$

$$f_i(x,z,t) = C_m \frac{\rho_{water} \pi D^2}{4} a(x,z,t) \quad (11)$$

where D is the diameter of the generic cylinder section, C_d is the hydrodynamic drag coefficient, C_m is the hydrodynamic inertia coefficient and ρ_{water} is the water density. Generally, the hydrodynamic coefficients C_d and C_m can be determined as a function of the adimensional Reynolds and Keulegan-Carpenter numbers (see Sarpkaya and Isaacson, 1981). In this work $C_d = 0.62$ and $C_m = 1.85$ have been assumed, i.e. the asymptotic values for these coefficients which can be found for value of the ratio between Reynolds number and Keulegan number greater than 10000. The water density has been set as $\rho_{water} = 1030$ kgm^{-3}. Further, in Eq.(10) and Eq.(11) symbols u and a denote the velocity and the acceleration respectively; they take into account both the horizontal and the vertical components.

According to the sea states theory (for a complete state of art, see Longuet-Higgins, 1973 and Phillips, 1967) for an irrotational wave motion, the free surface displacement can be considered as the summation of elementary components

$$\eta(x,t) = \sum_{k=1}^{N_{f\eta}} \sqrt{2S_{JS}(f_k)\Delta f_k} \cos(k_k x - 2\pi f_k t + \phi_k) \quad (12)$$

where symbols Δf_k, f_k and φ_k retain the same meaning as in Eq.(7), $N_{f\eta}$ is the number of frequency steps,

$$k_k \tanh(k_k d) = \frac{(2\pi f_k)^2}{g} \quad (13)$$

being d the water depth, and $S_{JS}(f)$ is the JONSWAP (Hasselmann et al., 1973) power spectral density

$$S_{JS}(f) = \alpha g^2 (2\pi f)^{-5} \exp\left[-\frac{5}{4}\left(\frac{f_p}{f}\right)^4\right] \cdot \exp\left\{\ln\gamma \exp\left[-\frac{(f-f_p)^2}{2\sigma^2 f_p^2}\right]\right\} \quad (14)$$

where f_p is the peak spectral frequency, Phillips' parameter is α = 0.01, and the parameters characterizing the JONSWAP spectrum are

$$\gamma = 3.3; \quad \sigma \begin{cases} = 0.07 & \text{if } f \le f_p \\ = 0.09 & \text{if } f > f_p \end{cases}. \quad (15)$$

If the wave motion is supposed to be irrotational, the potential of velocity function can be determined as:

$$\varphi(x,z,t) = \sum_{k=1}^{N_{f\eta}} \sqrt{2S_{JS,potential}(f_k,z)\Delta f_k} \cdot \sin(k_k x - 2\pi f_k t + \phi_k) \quad (16)$$

where

$$S_{JS,potential}(f_k,z) = g\left[k_k \tanh(k_k d)\right]^{-1} S_{JS}(f_k)\left\{\frac{\cosh\left[k_k(d+z)\right]}{\cosh(k_k d)}\right\}^2 . \quad (17)$$

Under the assumption of irrotational motion, the horizontal and vertical components of velocity and acceleration can be computed as follows:

$$u_x = \partial\phi/\partial x \quad (18a)$$

$$u_z = \partial\phi/\partial z \quad (18b)$$

$$a_x = \partial^2\phi/\partial x\,\partial t \quad (18c)$$

$$a_z = \partial^2\phi/\partial z\,\partial t \quad (18d)$$

Monte-Carlo simulations of the wave loading have been carried out. The algorithm adopted for the numerical simulations is based on Fast Fourier Transform (FFT) method. The main advantage is that this method is not time consuming; however, in order to get an appropriate sampling rate, it is convenient to generate shorter time samples. Also in this case, free surface displacement and kinematic components have been generated in order to determine the forces on the structures.

Therefore, based on Eq.(7) and Eq.(12) two independent samples of wind and wave loading can be computed and applied on a structural model of the tower and the support structure. The effect of aerodynamic damping is accounted for by introducing in the fundamental fore-aft mode an additional modal damping. Still according to Van der Tempel (2006), a satisfactory engineering estimate is 4%.

5 NUMERICAL RESULTS

Only fatigue caused by the production load cases is considered herein (note that parked or transitional behaviour would require, in general, a different modelling of the aerodynamic damping, see for this Kuhn, 2001). For each lumped load cases in Tables 4, 10 samples of wind and wave processes are generated. It is assumed that both mean wind and wave loading act along the x-direction (see Figures 1 and 2). Separate wind and wave responses are then computed and linearly superposed. The rainflow counting algorithm is applied to determine stress ranges and corresponding number of cycles for fatigue damage analysis. The lifetime DEL is then computed based on Eq.(4) for each stress resultant. Note in this regard that Mx and My denote, respectively, the bending moment on the yz plane (moment vector along the x-axis) and on the xz plane (moment vector along the y-axis).

5.1 *Wind loading*

For wind loading only in a calm sea the lifetime DELs are illustrated in Figures 3 and Table 6.

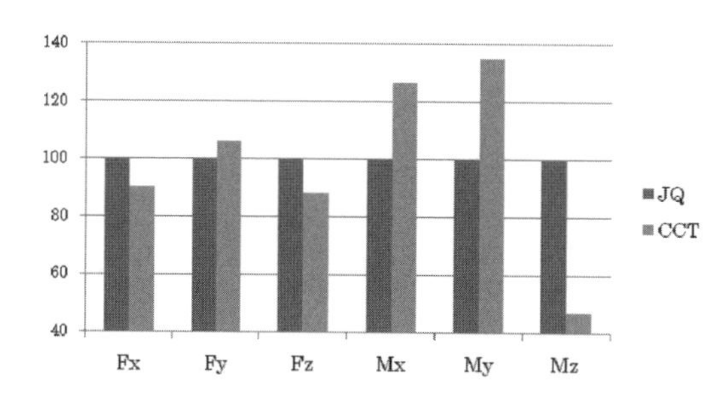

Figure 3a. Lifetime DEL at the tower base for wind loading.

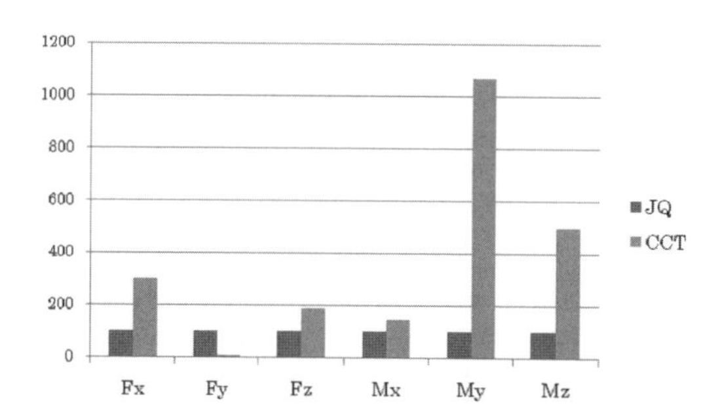

Figure 3b. Lifetime DEL at the foundation level for wind loading.

Table 6a. Lifetime DEL at the tower base for wind loading.

Tower base	Tripod	Jacket
Fx [MN]	0.927	1.026
Fy [MN]	0.096	0.090
Fz [MN]	2.754	3.119
Mx [MNm]	4.270	3.364
My [MNm]	66.995	49.679
Mz [MNm]	0.745	1.584

Table 6b. Lifetime DEL at the foundation level for wind loading.

Tower base	Tripod	Jacket
Fx [MN]	0.825	0.276
Fy [MN]	0.024	0.286
Fz [MN]	5.628	2.975
Mx [MNm]	0.143	0.097
My [MNm]	0.896	0.084
Mz [MNm]	0.058	0.012

At the tower base the Fx and Fy DELs appear quite similar in the JQ and the CCT, the slight differences being due to the different inertia forces on the two structures. The latter result from the different flexibility along the x- and y-direction. Specifically, by applying a unit load at the tower top it can be readily seen that the JQ is quite more flexible than the CCT along the x-direction and that, on the contrary, the CCT is slightly more flexible than the JQ along the y-direction. As a consequence, the JQ experiences quite higher tower accelerations along the x-direction, resulting in higher inertia loads despite the tower mass is larger in the CCT (longer tower); on the contrary, the CCT experiences slightly higher tower accelerations along the y-direction that, taking into the account the larger mass of the CCT, do result in higher inertia loads.

Whereas the differences between the Fx and Fy DELs, however, do not appear significant, certainly more remarkable is the difference in terms of Mx and My DELs. This is expected since the tower is longer in the CCT and, as a result, higher bending moments Mx and My are experienced at the tower base.

As far as the Fz DEL is concerned, it shall be pointed out that the JQ and the CCT exhibit a quite different behaviour along the z-direction. Specifically, by applying a unit load at the top it can be readily seen that the JQ is more flexible along the z-direction. On the other hand, modal analysis reveals that the natural frequency of the first axial vibration mode is lower in the JQ (2.31 rads^{-1}) than in the CCT (5.94 rads^{-1}). Since the Fz loading exhibits a dominant frequency around 0.6 rads^{-1} (Fz samples and a pertinent Fourier analysis are not reported for brevity), the dynamic amplification factor is higher in the JQ. As a result, the tower accelerations are significantly higher in the JQ; the latter determine higher inertia loads in the JQ, despite the tower mass is larger in the CCT (longer tower).

Finally, recognize that the Mz DEL is significantly higher in the JQ than in the CCT. By applying a unit twisting moment it can be seen that the JQ is more flexible than the CCT. On the other hand, modal analysis reveals that the natural frequency of the first twisting vibration mode is lower in the JQ (6.93 rads^{-1}) than in the CCT (20.97 rads^{-1}). Since the Mz loading exhibits a dominant frequency around 1.21 rads^{-1} (again, Mz and a pertinent Fourier analysis are not reported for brevity), the dynamic amplification factor is higher in the JQ. For these reasons, higher accelerations are encountered in the JQ and, consequently, higher twisting inertia loads.

At the foundation level the JQ and the CCT are supported by four and three piles, respectively. For each stress resultant, Figure 4 compares the DEL at the top of one the four JQ piles to the DEL at the top of one of the two CCT piles aligned along the *y*-direction (see Figure 2).

The Fx and My DELs are higher in the CCT than in the JQ. This reveals higher inertia loads on the CCT; therefore, despite the JQ is a more flexible structure along the *x*-direction, it is apparent that higher accelerations do not compensate for the overall larger mass of the CCT. For both the JQ and the CCT, recognize that Fz increases while My decreases significantly, at the foundation level, with respect to the corresponding values at the tower base. Indeed it is mainly F_z that balances the bending moment with respect to the foundation level; the Fz DEL is higher in the CCT due to the higher inertia loads along the *x*-direction.

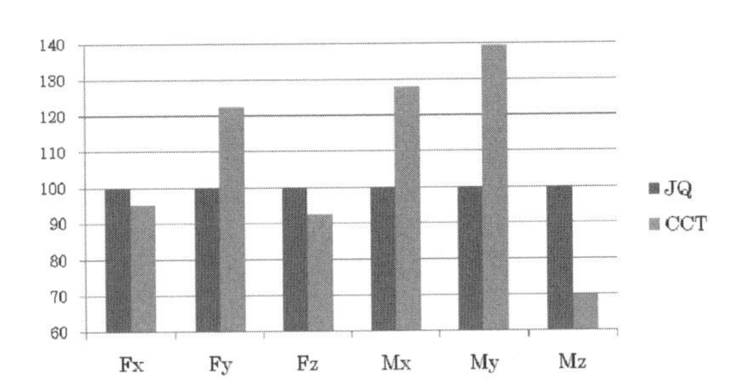

Figure 4a. Lifetime DEL at the tower base for wind + wave loading.

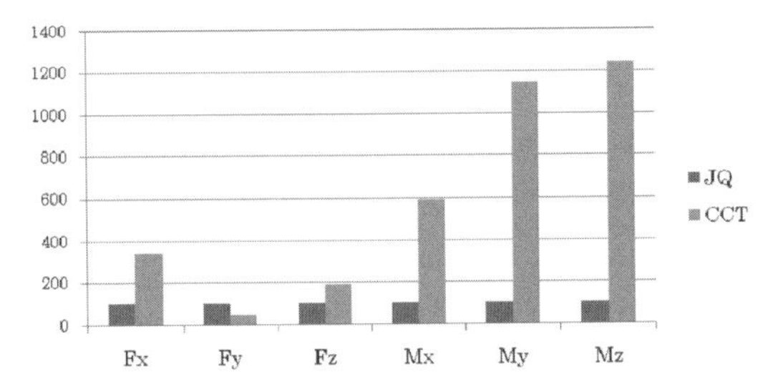

Figure 4b. Lifetime DEL at the foundation level for wind + wave loading.

The Mx DEL is higher in the CCT than in the JQ. This reflects the corresponding DEL ratio at the tower base and it is due to the higher inertia loads on the CCT along the *y*-direction.

Finally, recognize that the Fy and Mz ratios are opposite with respect to the corresponding ratios at the tower bases. This is mainly due to the geometry of the foundation (three piles in the CCT versus four piles in the JQ): in the JQ the force Fy at the tower top + the inertial loads along the whole structure do not induce twisting effects and, as a result, they are balanced by Fy forces at the foundation level, which are equally distributed at the four pile tops; on the contrary, in the CCT the force Fy at the tower top + the inertia loads along the structure do induce twist at the foundation level (see CCT geometry in Figure 2) and, as a result, Fy decreases while the twisting moment Mz increases.

5.2 *Wind + wave loading*

For wind + wave loading the lifetime DELs are illustrated in Figures 4 and Table 7.

The DEL ratios appear consistent with the corresponding DEL ratios for wind loading only. For the six stress resultants, prevalence of the DEL in the JQ or in the CCT can be explained based on the same reasoning followed for the wind loading only. On the other hand it can be stated that, due to the wave loading effects, the stress resultants are generally larger in the CCT than in the JQ. This is expected since the diameters of the structural members are generally larger in the CCT.

Table 7a. Lifetime DEL at the tower base level for wind + wave loading.

Tower base	Tripod	Jacket
Fx [MN]	0.977	1.024
Fy [MN]	0.109	0.089
Fz [MN]	2.883	3.119
Mx [MNm]	4.302	3.363
My [MNm]	67.520	48.474
Mz [MNm]	0.748	1.660

Table 7b. Lifetime DEL at the foundation level for wind + wave loading.

Tower base	Tripod	Jacket
Fx [MN]	0.972	0.284
Fy [MN]	0.142	0.287
Fz [MN]	5.719	2.997
Mx [MNm]	0.564	0.095
My [MNm]	1.043	0.091
Mz [MNm]	0.247	0.020

6 CONCLUDING REMARKS

A comparative fatigue analysis has been pursued on typical tripod and jacket support structures for OWECs. In general, at the tower base the DELs have been found more significant in the CCT than in the JQ, except for the Mz DEL that results higher in the JQ due to its more twisting flexibility. Even more significant appears the prevalence of the DELs in the CCT at the foundation level as a result, also, of the twisting effects induced by the foundation geometry (three piles in the CCT vs. four piles in the JQ). Finally, it has been seen that hydrodynamic loads are higher in the CCT, with a consequent increase of the DEL for all the six stress resultants.

REFERENCES

Andersen, U.V. 2008. *Load reduction of support structures of offshore wind turbines*. Master thesis, Danish Technical University (DTU), Copenhagen.

Chakrabarti, S.K. 1987. *Hydrodynamics of offshore structures*. Berlin: Springer-Verlag.

Garrad H. & Partners, 2009. *Bladed for Windows*. Bristol, England.

Hasselmann, K., Barnett, T.P., Bouws, E. et al. 1973. Measurements of wind wave growth and swell decay during the Joint North Sea Wave Project (JONSWAP), *Deut. Hydrogr. Zeit.* A8, 1–95.

Homepage "Alpha Ventus": http://www.alpha-ventus.de.

Homepage "Beatrice Windfarm Demonstrator Project": http://www.beatricewind.co.uk.

Jonkman, J., Butterfield, S., Musial, W. & Scott., G. 2009. *Definition of a 5-MW reference wind turbine for offshore system development,* National Reneweable Energy Laboratory (NREL), Technical Report NREL/TP-500-38060-2009.

Kuhn, M., 2001, *Dynamics and design optimisation of offshore wind energy conversion systems*. Delft University Wind Energy Research Institute (DUWIND), Report no. 2001.002.

Sarpkaya, T. & Isaacson, M., 1981, *Mechanics of Wave Forces on Offshore Structures*, Van Nostrand Reinhold Company.

Schaumann, P. & Böker, C. 2005. Can jacket and tripods compete with monopiles? *Proceedings of the Copenaghen Offshore Wind Conference, October 2005*. Copenhagen.

Sutherland, H.J. 1999. On *the fatigue analysis of wind turbines*, Report no. SAND99-0089, Sandia National Laboratories.

Thomsen, K. 1998. *The statistical variation of wind turbine fatigue loads*, Risø National Laboratory, Denmark, Report Risø-R-1063(EN).

Van der Tempel, J. 2006. *Design of support structures for offshore wind turbines*. Delft University Wind Energy Research Institute (DUWIND), Report no. 2006.029.

Sustainable Maritime Transportation and Exploitation of Sea Resources – Rizzuto & Guedes Soares (eds)

Hiiumaa offshore wind Park EIA: Main EIA results and description of the required supplementary environmental investigations

A. Järvik
Estonian Maritime Academy, Tallinn, Estonia
Estonian Marine Institute, University of Tartu, Tallinn, Estonia

H. Agabus
Nelja Energia OÜ, Tallinn, Estonia

R. Aps & T. Raid
Estonian Marine Institute, University of Tartu, Tallinn, Estonia

ABSTRACT: The Nelja Energia OÜ (LLC) initiated the project of an offshore wind park in coastal waters of Hiiumaa Island within Estonian EEZ (Baltic Sea) in 2006. Altogether 200 wind turbines with total output about 1000 MW are planned to build in five sections, located in the shallow water at the depth of 15–20 meters. Environmental Impact Assessment (EIA) was initiated in June 2006. However during the scoping of the EIA and the public hearing of the EIA Programme, the lack of essential environmental information was revealed. Because of that, several special studies were performed in 2006–2008. In parallel, a number of technical amendments to the project were made. The paper is focusing on the results of EIA and of related environmental studies. The possible environmental and socio-economical impacts and necessary mitigation measures are discussed. As a result, the two of the initially proposed five sections were excluded from the wind park, whereas the rest three of them were suggested to be moved into deeper and more distant locations. These implemented amendments would decrease the total output of Hiiumaa offshore wind park up to 594–730 MW. As probably the only important environmental factor the wind turbines impact on migrant birds should be highlighted. The visual pollution effects of wind turbines were found to be more problematic from the socio-economic point of view.

Keywords: offshore wind park, EIA, environmental investigations

1 INTRODUCTION

1.1 *Estonian electric industry*

Historically, above 90% of electricity has been produced by the large thermal power plants using oil shale in Estonia. Recently, about 15 million-tons of oil shale are used annually to produce a total of about 9–10 TWh of electricity. Despite the substantial technical development of oil shale power plants during last decades, the problem of emissions of greenhouse gases and another environmental pollution is remained considerable. In 2007 the produced 1 MWh of electricity resulted in emission of 1.1 tons of CO_2. After the full renovation of power plants currently in progress, this figure will be still 0.8 tons in 2013 (Anon., 2010).

Estonian energy management strategy foresees the 10–15% growth of the consumption of electricity in 2015 compared to 2010 (Fig. 1). Simultaneously, the share of oil shale based electricity production should be reduced below 90% and the using of renewable energy resources should be 8% as minimum.

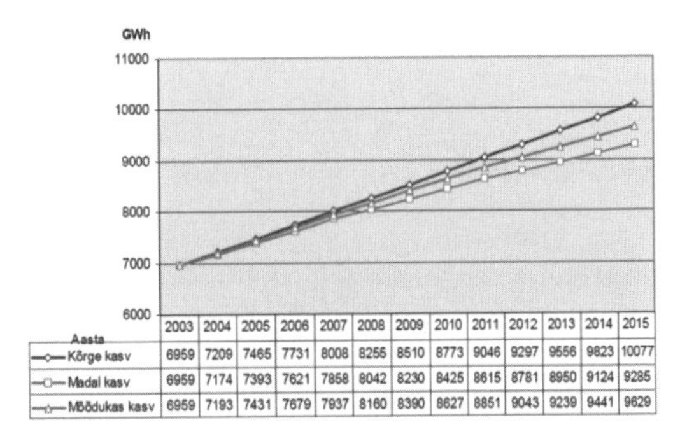

Aasta	2003	2004	2005	2006	2007	2008	2009	2010	2011	2012	2013	2014	2015
Kõrge kasv	6959	7209	7465	7731	8008	8255	8510	8773	9046	9297	9556	9823	10077
Madal kasv	6959	7174	7393	7621	7858	8042	8230	8425	8615	8781	8950	9124	9285
Mõõdukas kasv	6959	7193	7431	7679	7937	8160	8390	8627	8851	9043	9239	9441	9629

Figure 1. The electricity consumption forecast for Estonia (Estonian Energy Management Strategy 2005–2015).

Since the biomass, sun and hydro energy resources are quite limited in Estonia then the wind is accounted to be main resource for the renewable energy production. Since the beginning of 2000 a total of 148.5 MW of wind parks were installed in Estonian mainland. However, because Estonian villages have usually widely dissipated houses and the distances between neighbors are some hundred meters, finding transparent large wind—rich areas without any near housing is not an easy task.

On the other hand, the building of wind turbines close to dwellings may cause the negative impact on inhabitants. Also, many people do not agree with building of wind parks in their backyard. Therefore most of the mainland wind parks are small, consisting of 3–10 turbines only.

In order to overcome the problem of building offshore wind parks in Estonian coastal sea, geologically acceptable grounds, where are many shallows with depths of 10–30 meters, would be the best solution for future offshore development. Furthermore, the wind energy resources in coastal sea are substantially higher (Fig. 2).

The biggest wind parks′ owners in Estonia have initiated recently several offshore wind projects for construction. The experience with offshore wind parks around the world during the last decade has indicated that the offshore wind parks can be the potential sources of large-scale environmental impacts. Therefore, the responsible Environmental Impact Assessment (EIA) is crucial prior to the construction of any offshore wind park.

The present paper describes the EIA of one of the offshore wind park projects in Estonia developed by LLC Nelja Energia.

1.2 *Hiiumaa offshore wind park project*

Integration of offshore wind power into Estonian power system is becoming more realistic and planned the Hiiumaa offshore wind park is one potential large-scale wind park construction in with annual output of 2 ... 2.8 TWh/year. The final total capacity and annual output will depend on the selection and positioning of wind turbines. The planned wind turbine unit capacity is between 3 ... 5 MW and project turbine will be selected as a result of actual procurement.

According to the plans, construction of the offshore wind park will take place in several stages, starting in 2014. The period of construction of an open sea wind park of the specified size could be up to three years. The designed useful life period of the wind turbines is 20 years. According to the current grid connection proposal from Estonian TSO (*Elering*), the grid connection is planned to establish via 330 kV AC network.

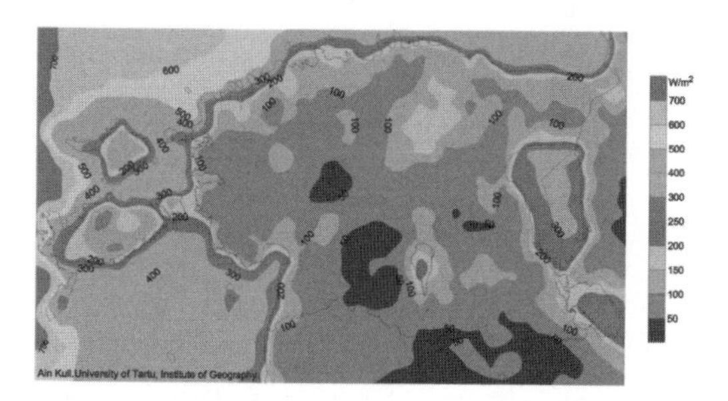

Figure 2. Estonian wind energy atlas. The figures show the year average wind energy density W/m² at altitude of 30 m (Anon., 2008).

1.3 *EIA Description*

According to Estonian legislation the EIA procedures must include:

1. Application of special water using permission, submitted by the Developer;
2. Decision of Ministry of Environment regarding the necessity of EIA;
3. Preparing the EIA Programme by licensed EIA expert, selected by Developer and accepted by Ministry of Environment;
4. Public hearing of EIA Programme;
5. Updating the EIA Programme due regard the results of public hearing;
6. Approving the EIA Programme by Ministry of Environment;
7. Supplementary investigations, if needed;
8. EIA Report compilation;
9. EIA Report public hearing;
10. EIA Report updating following to the results of public hearing;
11. EIA Report approving by Ministry of Environment.

The first five steps of EIA were crossed in 2006–2007. The participation in public hearing was very active and many locals as well as the NGOs provided useful remarks and suggestions, which were taken into account by the Expert group of EIA.

However, due to the legislation problems, the further progress of EIA Programme management was delayed. Only in February 2010 the special section in Estonian Water law was adopted by the Parliament and the EIA Programme was approved in June 2010.

During the scoping process, the Expert group realized that additional environmental data are needed to perform EIA. As result, the Expert group in cooperation with the Developer activated several additional special investigations.

Short description of special investigations performed and the results were drawn up in the following reports:

1. Socio-economical aspects of the construction of Hiiumaa offshore wind park, including modeling of visualization of wind park across the Hiiumaa Island sea shore (Kartau et al., 2008);
2. Investigation of the bottom sediments, including texture and pollution determination, within the area affected by the Hiiumaa offshore wind park (Kask & Kask, 2008);
3. Hydro-meteorological accounting of the sea area potentially affected by the Hiiumaa offshore wind park and the environmental risks, including oil-spill and navigation accidents (Elken et. al, 2008);
4. Inventory of the habitats of sea bottom fauna and flora in the area of the Hiiumaa Island coastal sea, planned to be localities of the Hiiumaa offshore wind park (Martin, 2008 et. al, 2008);
5. The ichthyological and fisheries inventory of potential Hiiumaa offshore wind park localities (Vetemaa et. al, 2008);
6. Overview of the bird, bat and sea mammal communities within the area potentially affected by the Hiiumaa offshore wind park (Leito, 2008);
7. Estimation of the impacts of the Hiiumaa offshore wind park on Estonian state security (Sisask, 2007).

Additionally, the Developer continued the observing of the worldwide progress in offshore wind park construction and commenced a number of special technical investigations, including wind park production assessment study, wind turbulence study, pre-construction (i.e. foundation) study and grid integration inquiry.

2 RESULTS AND DISCUSSION

As the outcome of the investigations, the Expert group found, that the main negative environmental impacts potentially caused by construction of the Hiiumaa offshore wind park would be:

a. The mechanical damages caused by the turbines to the migrating birds;
b. The interferences of bird migrations birds;
c. The expulsion of birds from their usual habitations;
d. The violation (worsening) of sea sight, especially in the Ristna area during sunset.

Planned sites of Hiiumaa offshore wind park are not the important spawning areas of thus commercial fish species, as Baltic herring, sprat, cod and flounder. The nearest notable herring spawning grounds located in more shallow coastal sea around the Hiiumaa Island, and flounders ones on the Hiiumadal shallow between sections Neupokojev and Vinkov. The distances from both sections to spawning grounds of herring and/or flounder are above 3 km, which is enough to be sure, that the suspended sediments during wind park construction should not spread to the spawning grounds, mentioned. The spawning areas of sprat and cod are located in the more distant regions of the Baltic Proper (spawn at the open sea far sufficiently (Rannak 1971, Raid 1989).

The possible negative impacts on fishery were estimated to be not considerable because the shallows planned for the sites of sections of Hiiumaa Wind Park, are not important fishing areas.

Also, the probability to impact the Baltic seals was found to be marginal, if the construction of Hiiumaa offshore wind park in Vinkov shallow will be not done during the gray seal breeding period in winter (Tougaard & Teilmann, 2007).

The special restrictions on leisure navigation and sport fishing inside the wind park should be implemented. Additionally, the potential impacts of the wind park on state security has been evaluated. It emerged, that two smaller preliminary sections of the wind park would disturb the coast guard radars and therefore those sections were discarded from the project. As result, only the three others sections are accounted to be acceptable, in the final version of EIA (Fig. 4).

In January 2011 the Baltic Environmental Forum raised an issue, that one of the shallows planned for the wind park—Apollo shallow, should be taken under the protection as the Natura 2000 bird protection area, since it is mostly

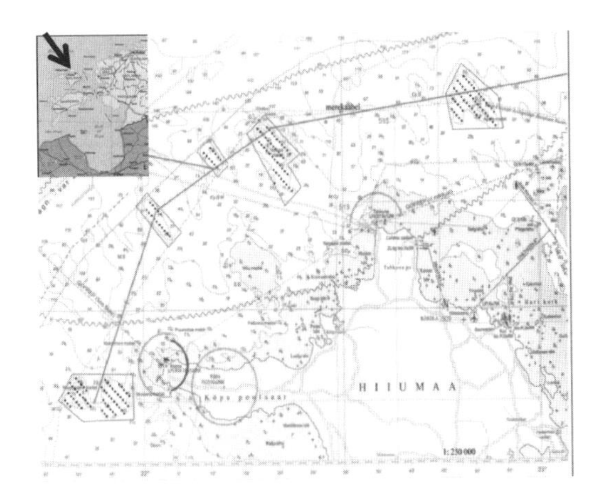

Figure 3. The preliminary plan of Hiiumaa offshore wind park sections disposition submitted by LLC Nelja Energia in 2006.

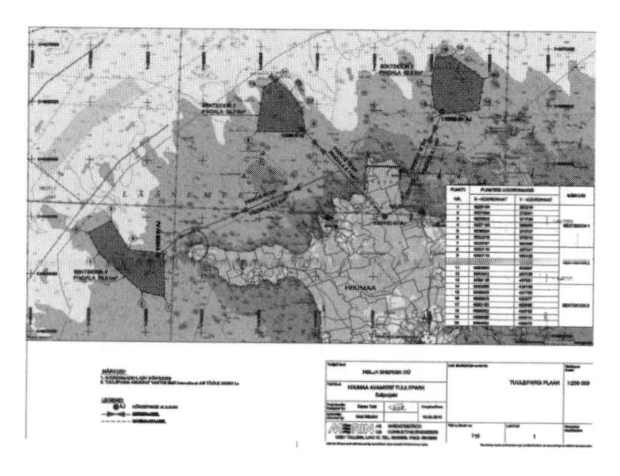

Figure 4. Final location of the Hiiumaa offshore wind park sections. Shallow's (wind park's sections) names from W to E: Neupokojev, Vinkov and Apollo.

as the wintering area of long-tailed duck (Fig. 4). However, the impacts of offshore wind parks on long-tailed ducks are not fully understood. Some authors have indicated that long-tailed ducks avoid the offshore wind parks at the long distance (above 3 km) while others claim, that current distance is some hundreds meters only. Furthermore, the Hiiumaa wind park turbines are planned to be located at the depths between 15–30 meters, while the feeding depth of long-tail ducks do not exceed 10–12 meters (Anon., 2007, Clausen, 2008; Fox et al., 2006; Nilsson, L. 2005; Petersen and Fox, 2007; Peterson & Kalamees, 2006; Wilhelmsson, 2010.)

3 CONCLUSIONS

The Expert group recommended the following mitigation measures, in order to bring the environmental impacts of the building of the Hiiumaa offshore wind park on the acceptable level:

1. The disposition of turbines within the each section should be like aggregated whole, where turbines should be located at the burls of the jalousies and they should be like limbs to avoid the overshadowing effect toward coast guard radars. The distance between turbines should be at minimum 625 meters and the rate of shaded and clear areas should be at minimum 80:20.
2. To minimize the probability of damages to the migrating as well as, the wintering and breeding birds, it is reccommended to position the turbines at the all shallows in series in direction of SW-NE.
3. The construction of offshore windpark was not allowed in February–March in sections of Vinkov and Apollo to avoid the disturbing the seals during their breeding time (Fig. 4).
4. To minimize the negative impacts on fish communities, it is not reccomended to perform the construction works during the main freproduction period—in May.
5. The permanent monitoring of birds should be established during the building and exploitation of the wind park with the option of immediate shutdown of the turbines during the active bird migrations.
6. During the construction phase, to minimaize the navigation risks, the adequate navigation management should be implemented inside each section and the good seafaring traditions should be followed.
7. The next potential additional measures of bird protection should be implemented:
    7.1. Option for operative shutdown of turbines during the mass migrating of birds, especially in the period of low visibility (rainfall, fog etc.);
    7.2. In order to make the turbines visible for the birds, the special technical measures may be implemented, (UV marking, the illumination of turbines, etc.), using the further achievements in this direction.
8. The marine electric cables should be dredged in sea bottom at minimum of 1 meter to avoid the negative impacts of magnetic field on marine fauna and flora.
9. The navigation of wind park service vessels should be regulated taking into account the location of fish spawning grounds and seasonal characteristics (i.e. birds nesting and moulting areas and breeding areas of sea mammals etc.);
10. To minimize the probability of oil-spills during the construction of wind park, it is recommended to avoid the construction works at wind speed above 10 m/s.

ACKNOWLEDGEMENTS

Authors' would like to thank all the coordinators and participators who participated in these EIA special investigations. Our gratitude goes out to all the collaborated institutions, organizations and people.

REFERENCES

Ahlén, I., Bash, L. Baagøe. H. & Pettersson, J. 2008. Bats and offshore wind turbines studied in southern Scandinavia—Oral presentation in Bats & Wind Energy Cooperative Workshop, 8–10 January 2008. BCL, Austin, Texas, U.S.A.

Anonymous, 2008. Eesti Tuuleatlas. In Estonian.

Anonymous. 2006. Eesti elektrimajanduse arengukava 2005–2015 –arengugava aastani 2018. In Estonian.

A. Kask, J. Kask. 2007. Investigation of the bottom sediments, including texture and pollution determination, within the area affected by the Hiiumaa offshore wind park. In Estonian, mimeo, 20 pp.

A. Leito. 2008. Overview of the bird, bat and sea mammal communities within the area potentially affected by the Hiiumaa offshore wind park. In Estonian, mimeo, 48 pp.

A. Leito. 2008. Overview of the bird, bat and sea mammal communities within the area potentially affected by the Hiiumaa offshore wind park. In Estonian, mimeo, 48 pp.

Fox, A., Desholm, M., Kahlert, J., Christensen, T.K. and Krag petersen, I. (2006), Information needs to support environmental impact assessment of the effects of European marine offshore wind farms on birds. Ibis, 148: 129–144. doi: 10.1111/j.1474-919X.2006.00510.x.

G. Martin (eds), L. Rostin, T. Püss, T. Möller, K. Kaljurand, M. Pärnoja, et al. 2008. Inventory of the habitats of sea bottom fauna and flora in the area of the Hiiumaa Island coastal sea, planned to be localities of the Hiiumaa offshore wind park. In Estonian, mimeo, 147 pp.

http://www.avibirds.com/euhtml/Long-Tailed_Duck.html.

http://www.seaduckjv.org/infoseries/ltdu_sppfactsheet.pdf.

Jakob Tougaard, Jonas Teilmann. 2007. Rødsand 2 Offshore Wind Farm Environmental Impact Assessment—Marine mammals. National Environmental Research Institute, Roskilde, Denmark. 77pp.

K. Petersen, A.D. Fox. 2007. Changes in bird habitat utilization around the Horns Rev 1 offshore wind farm, with particular emphasis on Common Scoter. Report request Commissioned by Vattenfall A/S 2007 National Environmental Research Institute University of Aarhus . Denmark. 40 pp.

K.Kartau (eds.), T. Oidjärv, V. Kärbla, K. Peet, M. Öövel, Ü. Reimets ja E. Zirk. 2008. Socio-economical aspects of the construction of Hiiumaa offshore wind park, including modeling of visualization of wind park across the Hiiumaa Island seashore. In Estonian, mimeo, 61 pp.

L. Rannak. 1971. On the recruitment to the stock of the spring herring in the North-eastern Baltic. Rapp. P.-v. Réun. Cons. int. Explor. Mer, 160: 76–82.

M. Vetemaa. 2008. The ichthyological and fisheries inventory of potential Hiiumaa offshore wind park localities). In Estonian, mimeo, 67 pp.

Niels-Erik Clausen. 2008. Environmental Impact of Offshore Wind Farms. IEA Task 23—Offshore Wind Energy Ecology and Regulation Workshop Petten, Holland 28–29 February 2008.

Nielsen, S. 2006. Offshore wind farms and the environment—Danish experience from Horns Rev and Nysted. The Danish Energy Authority, Copenhagen.

Nilsson, L. 2005. Offshore-windmills and sea ducks in Sweden.

T. Raid. 1989. The influence of hydrodynamic conditions on the spatial distribution of young fish and their prey organisms. Rapp. P.-v. Réun. Cons. int. Explor. Mer, 190:166–172.

U. Lips (eds), J. Elken, T. Liblik, V. Alari, G. Väli 2007. Hydro-meteorological accounting of the sea area potentially affected by the Hiiumaa offshore wind park and the environmental risks, including oil-spill and navigation accidents. In Estonian, mimeo, 72 pp.

Wilhelmsson, D., Malm, T., Thompson, R., Tchou, J., Sarantakos, G., McCormick, N., et al. (eds.) (2010). Greening Blue Energy: *Identi fying and managing the biodiversity risks and opportunites of off shore renewable energy.* Gland, Switzerland: IUCN. 102pp.

Sustainable Maritime Transportation and Exploitation of Sea Resources – Rizzuto & Guedes Soares (eds)
© 2012 Taylor & Francis Group, London, ISBN 978-0-415-62081-9

Kite towed ships and kite generated power—intuitive methods for reducing marine pollution

E. Barsan & N.V. Grosan
Constantza Maritime University, Constantza, Romania

ABSTRACT: The non-conventional propulsion methods such as electrical propulsion, wind assisted ship, hybrid propulsion system or solar power were already tested and part of them denoted that, in conjunction with the main engine of a ship, it is possible to reduce the pollutant emissions and to keep nearly the same speed of the vessel. Kite towed ships or kite generator power are two "non-classical" methods with real application possibilities. The studies of these methods can be performed separately: kite propulsion of ships or kite generator energy. The best situation is to use the kite for both goals but even if the methods appear the same in reality there are few contrasts regarding the operation mode, a certain elevated altitude and other technical arrangements.

1 INTRODUCTION

Reducing air pollution due to naval propulsion engines began to gain more ground in the field of shipping. Over the past 25 years carbon gases generated from marine engines were doubled, about 3.6% of total emissions being generated by maritime engines.

It is estimated that this percentage will grow with 6–7% in the next years, considering that the most commercial vessels are equipped with internal combustion engines which develop the power required for relatively high speed march.

But higher speed means higher fuel consumption and thus emissions as high, despite the fact that it uses different methods of recovery and reuse of these gases. The Green Ship concept, still unfortunately misunderstood by those involved in shipping starts to gain more ground, especially due to international regulations and limitations regarding the quality of fuel used.

The appearance of SECA zones in Europe and the U.S. was a first step leading to the revision and rethinking of the way how the propulsion of ships can be done.

During that early period no one would have thought it possible over a hundred years to return to those methods that have been ignored in favour of classical propulsion.

Thus, more and more researchers have reproduced what had been abandoned in the early twentieth century, propulsion by other methods, that don't use or partially use fuel for the propulsion engine operation.

Improperly called unconventional methods such as wind power, electric propulsion, photovoltaic cells mounted on rigid or movable sails, rotating sails (Flettner system, Magnus effect), high sail at a certain altitude, have been tested over the time and despite the fact that they gave good results, the owners have not shown interest in this matter, whether they are not confident in these ways whether they prefer to ignore the growing threat of air pollution due to increased marine engines.

Using the kite for hauling the ship does not mean a complete renunciation of conventional propulsion, it is a combined method in which the main propulsion engine is used to a lower capacity (so the degree of pollution is lower) the loss of speed being partly covered by traction exerted by the kite. Usually for the mixed propulsion motor-kite it is used a single kite, built at an altitude where the winds are almost constant in direction and velocity.

Kites can be used for energy production, in this case being arranged in the shape of the ladder, the system "Laddermill" being eloquent.

In this paper we intend to show in general the present state of international research on the kite's use for power generation and in particular the influence exerted by the kite propulsion on the stability of the ship's route in different wind conditions at the sea surface.

2 PRODUCING ENERGY USING THE KITE (KITE ENERGY GENERATOR)

Recent studies regarding energy production by using the kite's power are based on the same principle but opinions are divided regarding the number of kites used. Some researchers consider that the use of a single kite is more convenient while others support the "Ladermill" system which provides a number of kites placed in line.

The motivations for which some experts consider the use of a single kite to be beneficial are: the power developed by a kite increases by approximately 3 times upon the power of the wind, the kite can be raised to higher heights; it does not require special arrangement and can cover a wide range of actions (Houska B. & Diehl M. 2006).

The concept underlying the energy production is based on the kite's ability to pull a cable that is connected to a generator located on the deck of the ship. Under the influence of airflow and the wind, the kite describes a series of upward and downward movements that pull or loosen the cable drum located on a power generator (Fig. 1).

The power generated by the kite's greatest high altitude is about eight times higher than the power generated by wind turbines from the ground. Wind speed is directly proportional to the altitude up to a certain limit. A kite, found at the altitude of 800–1200 m is in an approximately constant current of air in direction and velocity, without being influenced by friction surfaces found at smaller altitudes near the ground, where the airflow can be diverted by various obstacles.

Research by experts at Delft University of Technology (Lansdorp B. et al. 2008.) and Politecnico di Torino (Fagiano L., Milanese M., Piga D. 2010) on the energy production with kite, whether there is a single kite or the "Laddermill' system, practically shows how energy can be produced (Fig. 2).

The kite, under the action of air currents, moves upward and downward, the movements being called "in eight" (Fig. 1) and a towing cable whose end is connected to a drum reel out or wrap on the drum, depending on the kite's position. The movement is cyclical, after reeling out the cable from the drum, wrapping movement begins.

The speed when the kite executes its movements depends on the ratio between the lift force and the

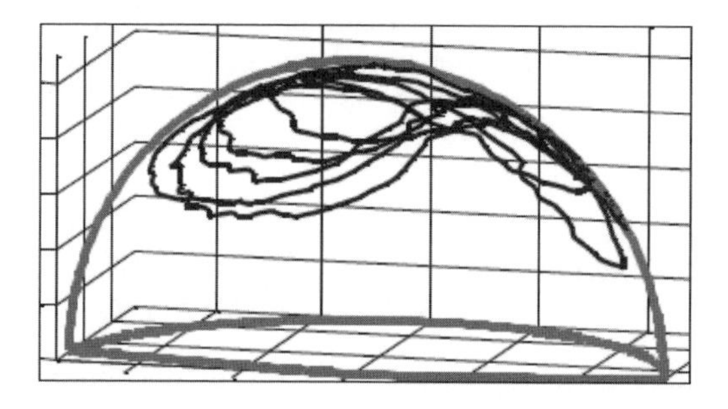

Figure 1. Three-dimensional path plot for experimentally recorded position angles (Dadd G.M., et al. 2010).

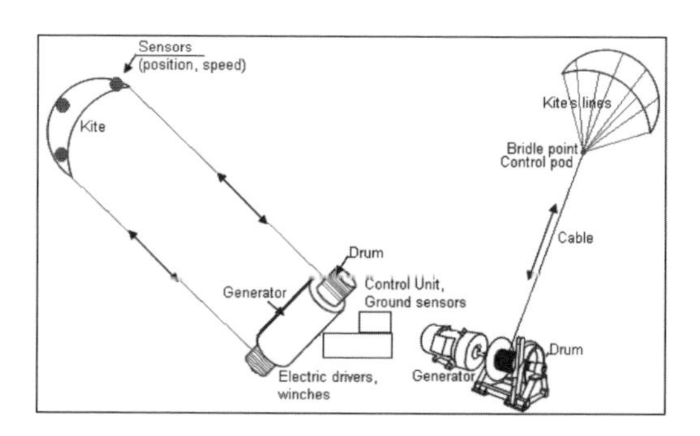

Figure 2. The simple scheme of kite energy generator.

drag force. In the section marked in blue (solid line), better said the optimal pathway, the kite's speed movement is about six times greater than the wind velocity (Lansdorp B. et al. 2008).

This is one of the advantages of using the kite for energy production: by flying across the wind, the resulted power is obtained more easily than in the case of aeolian turbines from the ground.

During the retraction phase, the unstable pathway of the cable, when the kite is not anymore in the phase of maximum traction, the incidence angle between wind direction and kite plane's flight is reduced (Houska B. & Diehl M. 2006).

It is very important that the angle of incidence remains as constant as possible at the values where the resultant force from the composition between buoyancy force and resistance force is maximal. If the angle of incidence increases with more than 20°, the kite loses stability and may reverse, actually it begins falling to the ground.

The study effectuated by Boris Houska & Moritz Diehl (Fig. 3), shows clearly the possibility of producing energy by using one or more kites (Houska B. & Diehl M. 2006), arranged in line, the registered power generator having approximate 5 MW for a wind speed of 10 m/s at an altitude of about 1200 meters. To reduce the drag effect they made studies on two kites and besides the greater power gained, the kite's inefficiency on the total aerodynamic force during the retraction phase is improved by using the two kites that move in opposite directions

In their paper, scientists at the University of Torino (Fagiano L., Milanese M., Piga D. 2009), in the KiteGen project, they use a 500 m^2 kite at an altitude of 800 meters, where the power generated by wind is about 200 kW/m^2. The kite's dynamic shaping is done by using a fast model predictive control method.

The energy generated by the kite (during the two phases, the traction when the kite develops maximum traction power and by kite steering unit producing energy in generator and recovery when

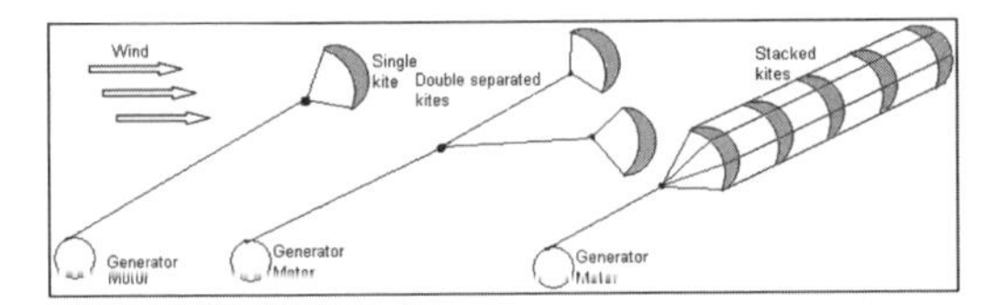

Figure 3. Different kite systems.

the cable's recovery consumes about 20% of energy generated in the first phase) is 10 MW for a wind speed of 15 m/s and a total area covered by the kite's movements of 15000 m^2 at an altitude ranging between 850 m–1150 m. Their research found its applicability to the ground systems, where the energy production system (generator-motor) is positioned on the fixed platform, which is not moving.

3 USING THE KITE FOR SHIP'S PROPULSION

In this part we intend to analyze the behaviour of the vessel's stability in terms of course stability under the combined action of wind from sea surface and the traction exerted by the kite. The vessel chosen for our study has the following characteristics: chemical tanker, displacement 8600 MT, length 105 m, breadth 16 m, draft fore/aft 6.85 m, aerial draft 17.65 m, free board 3.75 m, partly loaded, fixed-pitch propeller, without bow thruster.

3.1 *Consideration about forces on a kite*

A kite can be considered as a wing surface which enables the application of three principle forces (Dinu D. 2010) acting on the kite: the weight, the tension in the towing cable, and the aerodynamic force (Fig. 4).

The aerodynamic force (in some publications named total aerodynamic force Taf) is a result of the composition of two forces, lift force and drag force.

Lift occurs when the wind is turned on a kite. The flow is turned in one direction and the generated lift in the opposite side. This is the Newton's third law of motion: when one object exerts a force on another, the second object exerts on the first a force equal in magnitude but opposite in direction (Grosan V.N., Dinu D., Scupi A. 2011).

$$L = C_y \rho \frac{v_\infty^2}{2} S \tag{1}$$

Drag is the force that is acting on the opposite side of the relative motion of the kite.

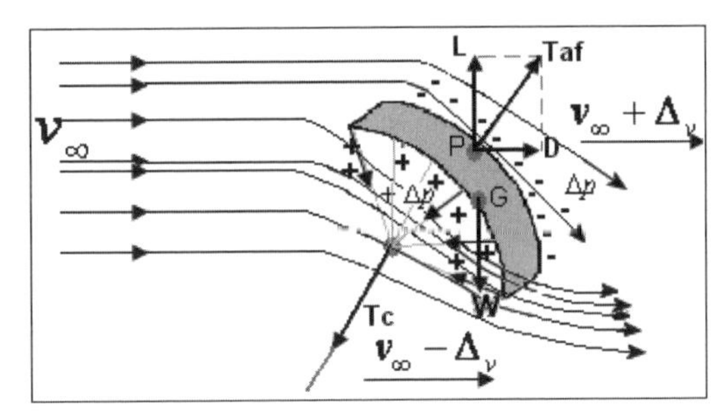

Figure 4. Air flow and forces on a kite.

$$D = C_x \rho \frac{v_\infty^2}{2} S \tag{2}$$

As a result of (1) and (2), the T_{af} is:

$$T_{af} = \sqrt{L^2 + D^2} \tag{3}$$

where C_x is the coefficient of resistance at advancement and C_y is the lift coefficient, ρ is air density, S total area of kite, v is the wind velocity.

Weight is the force acting downwards exerted by the gravitational force and also the opposite force of lift force.

Cable tension is the pulling force exerted by a towing cable attached to the kite. This tension is very important in order to get a stationery flight, to balance the tension force with the drag and lift force. The value of cable tension force, noted Tc, is the same with aerodynamic force value.

3.2 *Calculation of the total aerodynamic force T_{af}*

Considering the incidence angle of the wind being 15° and the total area of the kite being 200 square meters, we have the possibility to calculate the values of the lift, drag and total aerodynamic forces. The following values were considered: $C_y = 0.9250$ (lift coefficient), $C_x = 0.2421$ (drag coefficient), $\rho = 1.2047$ Kg/sqm, $v = 15$ m/s, $S = 200$ m^2.

Both coefficients are on the basis of the incidence angle of 15° on the aerodynamic profile G 417-a, which was selected.

As a result of (1), (2), (3), the total value of the aerodynamic force Taf, for the kite's profile described above, is 22029.11 N and the value of resistance force D = 6562.30 N.

Having calculated all the force values we have introduced them in the simulation for different scenarios with different wind direction while keeping constant the wind force and angle of incidence.

The results are the following: the aerodynamic force of a specific value has the possibility of moving the ship with a slow speed while sheering her off course. The effect of the rudder when the speed is slow is negligible and because of this it is required the use of the main engine. The sheer itself is not of great value on short distances but on long distances is considerable. (Grosan N., Dinu D., 2010)

3.3 *The combined influence of the kite and wind surface on ship's capacity of manoeuvring*

For the beginning we considered only the influence that the kite has on the course stability for a ship whose speed is zero. (Fig. 5) We also considered the wind from the west, 240°–260°. The value of traction force Fx is lower than in cases when the wind acts in a direction from S to SW (180°–210°) but the value of lateral deviation force Fy is greater in this case. The ship's speed is lower, moving forward but the lateral deviation is becoming greater as the vessel progresses.

Using the kite for ship's propulsion is effective when the wind comes from a sector between 310°–180°–050°.

In our research we followed the ship's reaction for different wind directions and selected three of the situations where the wind comes from 180°-270° and that we believe being indicative of what we intend to follow.

For situations where the wind blows from 180°–090°, the results are nearly similar except that the vessel is drifted to port side.

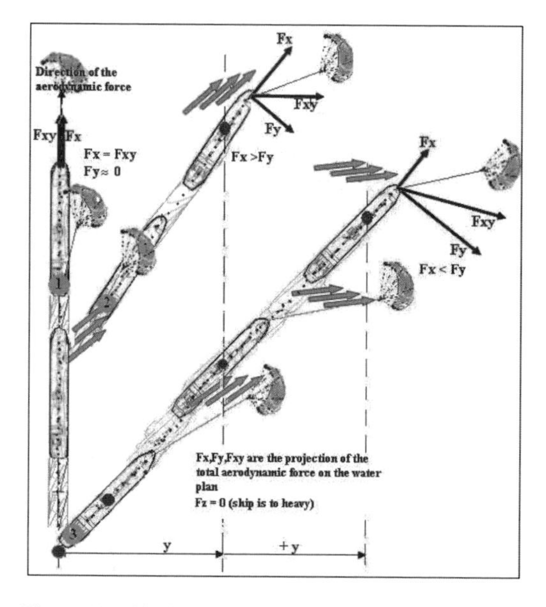

Figure 5. Air flow and forces on a Kite influence on ship's course stability for wind between 180°–260°.

When the wind blows from the sector between 50° port/starboard the propulsion action of the total aerodynamic force Taf is void.

In the first case, for the ship's position noted with 1, the wind direction is from 180° and the kite's forces projection on the water plan are represented by Fx = Fxy. The value of the force Fy that led the vessel out off normal course is zero or nearly zero. The vessel remains on the initial course and the speed is between 0.5 and 0.9 knots.

Forures the second ship's position, noted with 2, the wind direction is considered 30° from ship's axis, from 210°. The traction force Fx is greater than Fy. The value of Fy is not so great but the vessel is taken out from the initial course, the speed of the ship is between 0.6–0.7 knots and the value of yawing (Y) is about 1 cable (185 meters).

For the last position, noted with 3, the wind is considered 60° from the ship's axis, from 240°. The value of traction force Fx is smaller than in the previous case but the value of Fy is greater. The ship's speed is about 0.7 knots and the yawing (Y) is between 2 and 3 cables.

In all presented cases, the initial ship's speed is considered to be zero. One of the effects of kite traction is the yawing of the ship, which is reduced by the use of the rudder. But at slow speed, according to our results obtained regarding traction force, the rudder's effect is ineffective and it is necessary to use the main engine to bring the ship on the ordered course.

Figure 6 presents the effect of the wind from the sea surface combined with the forces exerted by the kite on the ship's route stability. For a good exemplification the ship's speed is considered to be zero and the wind direction at the sea surface in the western sector, between 240–270 degrees.

The line marked with red represents the trajectory of the vessel under the combined influence of all forces and the blue line represents the kite's path under the altitude wind's influence. Performed simulations show that the ship has a forward movement on an irregular path on the traction's force direction exerted by the kite, but also under the influence of the wind surface.

Vessel speed is between 0.7–1.1 knots and the yawing is observed to be about 1.1–1.2 cables (position 1 of the ship in fig. 6).

When the wind from the surface maintains its velocity and direction and the altitude wind changes its velocity too, there is a notable decrease of the force Fx, a growth of Fy force that moves the ship to starboard, but also a growth of Fxy force, the advancing force. There are no rates for the vessel because in this situation the yawing increases, amounting to approximately two cables (position 2 of the ship in Figure 6) and requiring the use of

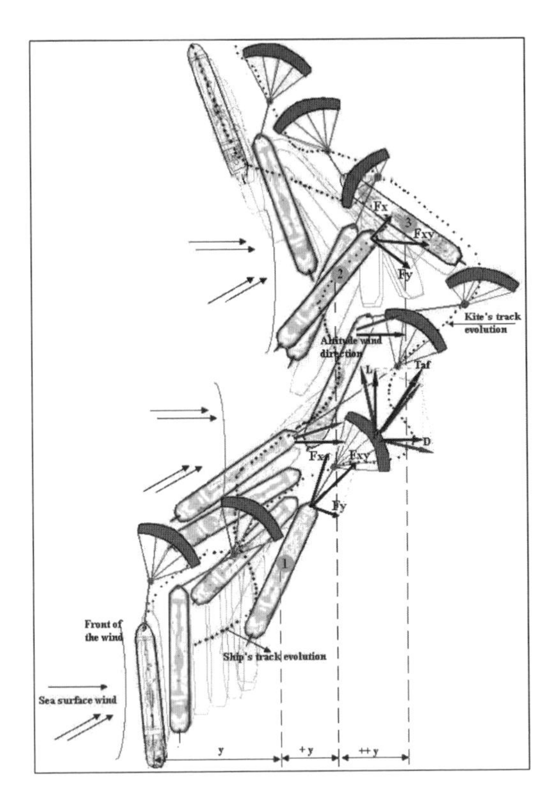

Figure 6. Ship's and kite track evolution under combined effect of sea surface wind and kite forces.

the main engine and rudder to bring the ship on the desired route.

It is noted that the vessel tends to be oriented in the direction commanded by the kite's force (Grosan N., Dinu D. 2010), whose movement is in "eight" shape.

Vessel's speed remains at about the same value and the use of the rudder only is not sufficient because the speed is too low.

During simulation, the vessel doesn't lean to a board or another, nor rises, the force Fz being zero. (Fz force as lift force projection on the water plan). To correct the ship's irregular movements and especially for kite's control, it is necessary to maintain it under the same angle of incidence and this can be achieved only through a control device on the strings that connect the kite and bridle point or otherwise said a "control pod" is necessary.

4 CONCLUSIONS

One of the issues that remain to be solved is the control of the kites or more correctly said the control that has to be exercised over the kites for maintaining the optimum angle of incidence and the effect that the connecting cable between the generator/motor and the kite, has upon the kite's movement.

Also, the question arises whether the use of kite systems in line or scale, "stacked kites", "laddermill", "single kite" or "double separate kites" (which we find in specialized publications under different names like "Ladder Mill", "Reel", "Fly Gen", "Buggy", "Sail", etc.) can be tested and mainly applied in the field of shipping, on board vessels.

However, it is already proven that the uses of a kite as a combined propulsion method lowers the fuel consumption and successfully participate in reducing emissions of SOx, NOx, CO. We hope that through research and our attempts, those who care about the reduction of pollutants' emission, by attending conferences and symposiums will be able to draw attention and more over, to persuade the ship owners to apply on board the combined propulsion system for ships that reduce pollution.

Understanding the concept of Green Ship from the people involved in shipping is a great accomplishment and even greater theoretical achievement if solutions are put into practice when resulted from researches and studies.

This shows that this research and simulations were not in vain.

REFERENCES

Houska B. & Diehl M. 2006. Optimal Control for Power Generating Kites. *Conference on Control and Decision*, San Diego, US.

Lansdorp B., Ruiterkamp R., Williams P., Ockels W. 2008. Long-Term Laddermill Modeling for Site Selection. *AIAA Modeling and Simulation Technologies Conference and Exhibition*. Honolulu, Hawaii, US.

Fagiano L., Milanese M., Piga D. 2010. Optimization of high-altitude wind energy generators, *Airborne Wind energy Conference*. Stanford University, California, US.

Fagiano L., Milanese M., Piga D. 2009. High Altitude Wind Generation: renewable energy cheaper than oil, *EU Conference "Sustainable development: a chalenge for European research"*. Brussels, Belgium.

Dadd G.M., Hudson D.A. & Shenoi-Kitve R.A. 2010. Comparison of Two Kite Force Models with Experiment, *Journal of aircraft*, vol.47, No.1, January-February 2010.

Dinu D. 2010. *Fluid mechanics for seafarers*. Ed. Nautica, Constanta.

Grosan V.N., Dinu D., Scupi A. 2011. Mixed propeller ship-influence of sea surface wind and wave about kite towed ship's manoeuvering capacity, *ECONSHIP Conference*, Chios, Greece.

Grosan N., Dinu D. 2010. Consideration regarding kite towed ship's manoeuvring, *WSEAS 3rd International Conference on Maritime and Naval Science and Engineering MN'10*. Constanta, Romania.

Sustainable Maritime Transportation and Exploitation of Sea Resources – Rizzuto & Guedes Soares (eds)
© 2012 Taylor & Francis Group, London, ISBN 978-0-415-62081-9

Author index